D1061969

GEOCHIMICA ET COSMOCHIMICA ACTA

SUPPLEMENT 2

PROCEEDINGS

OF THE

SECOND LUNAR SCIENCE CONFERENCE

Houston, Texas, January 11–14, 1971

GEOCHIMICA ET COSMOCHIMICA ACTA

Journal of The Geochemical Society and The Meteoritical Society

SUPPLEMENT 2

PROCEEDINGS

OF THE

SECOND LUNAR SCIENCE CONFERENCE

Houston, Texas, January 11–14, 1971

Sponsored by
The Lunar Science Institute

Edited by

A. A. LEVINSON

University of Calgary, Calgary, Alberta, Canada

VOLUME 3

PHYSICAL PROPERTIES

SURVEYOR III

THE MIT PRESS

Cambridge, Massachusetts, and London, England

Supplement 2
GEOCHIMICA ET COSMOCHIMICA ACTA
Journal of The Geochemical Society and The Meteoritical Society

BOARD OF ASSOCIATE EDITORS

Volume 3 — Contents

W. N. Houston and J. K. Mitchell: Lunar core tube sampling . . 1953

W. D. Carrier, III, S. W. Johnson, R. A. Werner, and R. Schmidt: Disturbance in samples recovered with the Apollo core tubes 1959

N. C. Costes, G. T. Cohron, and D. C. Moss: Cone penetration resistance test—an approach to evaluating in-place strength and packing characteristics of lunar soils 1973

H. Heywood: Particle size and shape distribution for lunar fines sample 12057,72 1989

J. O. Isard: The formation of spherical glass particles on the lunar surface . 2003

E. L. Fuller, Jr., H. F. Holmes, R. B. Gammage, and K. Becker: Interaction of gases with lunar materials: Preliminary results 2009

H. Görz, E. W. White, R. Roy, and G. G. Johnson, Jr.: Particle size and shape distributions of lunar fines by CESEMI 2021

J. Borg, M. Maurette, L. Durrieu, and C. Jouret: Ultramicroscopic features in micron-sized lunar dust grains and cosmophysics . . . 2027

G. Mueller: Morphology and petrostatistics of regular particles in Apollo 11 and Apollo 12 fines 2041

C. H. Greene, L. D. Pye, H. J. Stevens, D. E. Rase, and H. F. Kay: Compositions, homogeneity, densities and thermal history of lunar glass particles 2049

R. B. Gammage and K. Becker: Exoelectron emission and surface characteristics of lunar materials 2057

R. Roy, D. M. Roy, S. Kurtossy, and S. P. Faile: Lunar Glass I: Densification and relaxation studies 2069

S. J. Pickart and H. Alperin: Neutron diffraction study of lunar materials 2079

G. L. Connell, R. F. Schneidmiller, P. Kraatz, and Y. P. Gupta: Auger electron spectroscopy of lunar materials 2083

J. W. Freeman, Jr., H. K. Hills, and M. A. Fenner: Some results from the Apollo 12 Suprathermal Ion Detector 2093

C. L. Herzenberg, R. B. Moler, and D. L. Riley: Mössbauer instrumental analysis of Apollo 12 lunar rock and soil samples 2103

R. M. Housley, R. W. Grant, A. H. Muir, Jr., M. Blander, and M. Abdel-Gawad: Mössbauer studies of Apollo 12 samples . . . 2125

P. A. Estep, J. J. Kovach, and C. Karr, Jr.: Infrared vibrational spectroscopic studies of minerals from Apollo 11 and Apollo 12 lunar samples . 2137

J. J. Grossman, J. A. Ryan, N. R. Mukherjee, and M. W. Wegner: Microchemical, microphysical, and adhesive properties of lunar material, II . 2153

D. R. Stephens and E. M. Lilley: Pressure-volume properties of two Apollo 12 basalts 2165

T. Gold, B. T. O'Leary, and M. Campbell: Some physical properties of Apollo 12 lunar samples 2173

J. B. Adams and T. B. McCord: Optical properties of mineral separates, glass, and anorthositic fragments from Apollo mare samples . . . 2183

Contents

R. C. Birkebak, C. J. Cremers, and J. P. Dawson: Spectral directional reflectance of lunar fines as a function of bulky density 2197

P. A. Ade, J. A. Bastin, A. C. Marston, S. J. Pandya, and E. Puplett: Far infrared properties of lunar rock 2203

W. B. White, E. W. White, H. Görz, H. K. Henisch, G. W. Fabel, R. Roy, and J. N. Weber: Physical characterization of lunar glasses and fines . 2213

N. N. Greenman and H. G. Gross: Luminescence of Apollo 11 and Apollo 12 lunar samples 2223

D. B. Nash and J. E. Conel: Luminescence and reflectance of Apollo 12 samples 2235

H. P. Hoyt, Jr., M. Miyajima, R. M. Walker, D. W. Zimmerman, J. Zimmerman, D. Britton, and J. L. Kardos: Radiation dose rates and thermal gradients in the lunar regolith: Thermoluminescence and DTA of Apollo 12 samples 2245

J. E. Geake, G. Walker, A. A. Mills, and G. F. J. Garlick: Luminescence of Apollo lunar samples 2265

G. F. J. Garlick, W. E. Lamb, G. A. Steigmann, and J. E. Geake: Thermoluminescence of lunar samples and terrestrial plagioclases . . . 2277

A. Dollfus, J. E. Geake, and C. Titulaer: Polarimetric properties of the lunar surface and its interpretation. Part 3: Apollo 11 and Apollo 12 lunar samples 2285

A. F. H. Goetz, F. C. Billingsley, J. W. Head, T. B. McCord, and E. Yost: Apollo 12 multispectral photography experiment 2301

C. J. Cremers and R. C. Birkebak: Thermal conductivity of fines from Apollo 12 2311

W. S. Baldridge and G. Simmons: Thermal expansion of lunar rocks . . 2317

H. Kanamori, H. Mizutani, and Y. Hamano: Elastic wave velocities of Apollo 12 rocks at high pressures 2323

H. Wang, T. Todd, D. Weidner, and G. Simmons: Elastic properties of Apollo 12 rocks 2327

B. R. Tittmann and R. M. Housley: Surface elastic wave propagation studies in lunar rocks 2337

N. Warren, E. Schreiber, C. Scholz, J. A. Morrison, P. R. Norton, M. Kumazawa, and O. L. Anderson: Elastic and thermal properties of Apollo 11 and Apollo 12 rocks 2345

R. A. Robie and B. S. Hemingway: Specific heats of the lunar breccia (10021) and olivine dolerite (12018) between 90° and 350° Kelvin . . . 2361

T. J. Katsube and L. S. Collett: Electrical properties of Apollo 11 and Apollo 12 lunar samples 2367

D. H. Chung, W. B. Westphal, and G. Simmons: Dielectric behavior of lunar samples: Electromagnetic probing of the lunar interior . . . 2381

P. Dyal and C. W. Parkin: The Apollo 12 magnetometer experiment: Internal lunar properties from transient and steady magnetic field measurements . 2391

C. P. Sonett, G. Schubert, B. F. Smith, K. Schwartz, and D. S. Colburn: Lunar electrical conductivity from Apollo 12 magnetometer measurements: Compositional and thermal inferences 2415

S. SULLIVAN, A. N. THORPE, C. C. ALEXANDER, F. E. SENFTLE, and E. DWORNIK: Magnetic properties of individual glass spherules, Apollo 11 and Apollo 12 lunar samples 2433

G. W. PEARCE, D. W. STRANGWAY, and E. E. LARSON: Magnetism of two Apollo 12 igneous rocks 2451

T. NAGATA, R. M. FISHER, F. C. SCHWERER, M. D. FULLER, and J. R. DUNN: Magnetic properties and remanent magnetization of Apollo 12 lunar materials and Apollo 11 lunar microbreccia 2461

R. B. HARGRAVES and N. DORETY: Magnetic properties of some lunar crystalline rocks returned by Apollo 11 and Apollo 12 2477

C. E. HELSLEY: Evidence for an ancient lunar magnetic field . . . 2485

C. S. GROMMÉ and R. R. DOELL: Magnetic properties of Apollo 12 lunar samples 12052 and 12065 2491

J. L. KOLOPUS, D. KLINE, A. CHATELAIN, and R. A. WEEKS: Magnetic resonance properties of lunar samples: Mostly Apollo 12 2501

F.-D. TSAY, S. I. CHAN, and S. L. MANATT: Magnetic resonance studies of Apollo 11 and Apollo 12 samples 2515

D. HANEMAN and D. J. MILLER: Clean lunar rock surfaces; unpaired electron density and adsorptive capacity for oxygen 2529

G. CROZAZ, R. WALKER, and D. WOOLUM: Nuclear track studies of dynamic surface processes on the moon and the constancy of solar activity . . 2543

R. L. FLEISCHER, H. R. HART, JR., G. M. COMSTOCK, and A. O. EVWARAYE: The particle track record of the Ocean of Storms 2559

G. M. COMSTOCK, A. O. EVWARAYE, R. L. FLEISCHER, and H. R. HART, JR.: The particle track record of lunar soil 2569

G. ARRHENIUS, S. LIANG, D. MACDOUGALL, L. WILKENING, N. BHANDARI, S. BHAT, D. LAL, G. RAJAGOPALAN, A. S. TAMHANE, and V. S. VENKATAVARADAN: The exposure history of the Apollo 12 regolith . . 2583

N. BHANDARI, S. BHAT, D. LAL, G. RAJAGOPALAN, A. S. TAMHANE, and V. S. VENKATAVARADAN: Spontaneous fission record of uranium and extinct transuranic elements in Apollo samples 2599

N. BHANDARI, S. BHAT, D. LAL, G. RAJAGOPALAN, A. S. TAMHANE, and V. S. VENKATAVARADAN: High resolution time averaged (millions of years) energy spectrum and chemical composition of iron-group cosmic ray nuclei at 1 A.U. based on fossil tracks in Apollo samples 2611

P. B. PRICE, R. S. RAJAN, and E. K. SHIRK: Ultra-heavy cosmic rays in the moon 2621

F. HÖRZ and J. B. HARTUNG: The lunar-surface orientation of some Apollo 12 rocks 2629

M. R. BLOCH, H. FECHTIG, W. GENTNER, G. NEUKUM, and E. SCHNEIDER: Meteorite impact craters, crater simulations, and the meteoroid flux in the early solar system 2639

J. L. CARTER and D. S. MCKAY: Influence of target temperature on crater morphology and implications on the origin of craters on lunar glass spheres 2653

C. M. STEVENS, J. P. SCHIFFER, and W. A. CHUPKA: Search for stable, fractionally charged particles (quarks) in lunar material 2671

T. Gold: Evolution of mare surface   2675

Surveyor III
N. L. Nickle: Surveyor III material analysis program   2683
T. E. Economou and A. L. Turkevich: Examination of returned Surveyor III
    camera visor for alpha radioactivity   2699
D. J. Barber, R. Cowsik, I. D. Hutcheon, P. B. Price, and R. S. Rajan:
    Solar flares, the lunar surface, and gas-rich meteorites   2705
M. D. Knittel, M. S. Favero, and R. H. Green: Microbiological sampling
    of returned Surveyor III electrical cabling   2715
F. J. Mitchell and W. L. Ellis: Surveyor III: Bacterium isolated from
    lunar-retrieved TV camera   2721
W. F. Carroll and P. M. Blair: Discoloration and lunar dust contamination
    of Surveyor III surfaces   2735
R. F. Scott and K. A. Zuckerman: Examination of returned Surveyor III
    surface sampler   2743
D. L. Anderson, B. E. Cunningham, R. G. Dahms, and R. G. Morgan:
    X-ray probe, SEM, and optical property analysis of the surface features of
    Surveyor III materials   2753
B. G. Cour-Palais, R. E. Flaherty, R. W. High, D. J. Kessler, D. S.
    McKay, and H. A. Zook: Results of the Surveyor III sample impact
    examination conducted at the Manned Spacecraft Center   .   .   .   2767
D. Brownlee, W. Bucher, and P. Hodge: Micrometeoroid flux from
    Surveyor glass surfaces   2781
E. A. Buvinger: Replication electron microscopy on Surveyor III unpainted
    aluminum tubing   2791
Apollo 12 Lunar Sample Inventory   2797
Author Index to Vol. 3   2799

GEOCHIMICA ET COSMOCHIMICA ACTA

Supplement 2

PROCEEDINGS

OF THE

SECOND LUNAR SCIENCE CONFERENCE

Houston, Texas, January 11–14, 1971

Proceedings of the Second Lunar Science Conference, Vol. 3, pp. 1953–1958
The M.I.T. Press, 1971.

Lunar core tube sampling

W. N. HOUSTON and J. K. MITCHELL

Department of Civil Engineering, University of California, Berkeley,
California 94720

(*Received* 15 *February* 1971; *accepted in revised form* 30 *March* 1971)

Abstract—A core-tube sampling study has been conducted to determine the influence of core-tube diameter and wall thickness on the quality of samples obtained. A basaltic silty sand was used as a lunar soil simulant for this study. The sampling tests were performed using several sizes and types of tubes, including tubes similar to those used on Apollo 12 mission and to new core tubes approved for Apollo 15. Emphasis in the testing program was placed on the new, larger tubes. Parameters measured during and after sampling included sample recovery ratio (length of sample/length of drive), percent densification during sampling, friction between the sample and the tube's inner walls, percent densification during extrusion, and force required to advance the tube into the soil. Similar measurements were made for tubes driven by hammering. Estimates of lunar soil in-situ density based on Apollo 12 core tube data and the results presented herein give a range of 1.55 to 1.90 g/cm^3.

Assuming the soil gradation is similar to that of Apollo 12, the new core tubes which have been proposed for Apollo 15 and subsequent missions are likely to retrieve considerably more sample, due both to larger core tube area and greater sample recovery. It appears that driving a 2.54–5.08-cm (1–2-inch) core tube deeper than about 45 cm (17.7 inches) may be a relatively unproductive means for retrieving samples of lunar soil at depth, if the original density is relatively low. The relationships developed in this study may be applied to the Apollo 12 and 14 core-tube samples as well as to Apollo 15 and later samples to be obtained using the new larger tubes.

INTRODUCTION

CORE-TUBE SAMPLING of the lunar surface has comprised an important part of lunar surface exploration and study during past Apollo missions and is expected to do so in future missions. Data which may be obtained from core tube samples include in-place density, stratification, and fabric or particle arrangement of the near-surface lunar soil. The quality of these data depends on the degree of disturbance caused by sampling and earth return. Additional properties which are less sensitive to sample disturbance may also be determined if returned core-tube samples are available for testing.

A core tube-sampling study has been conducted to determine the influence of core-tube diameter and wall thickness on the degree of sample disturbance. While the absence of change in density as a result of sampling does not preclude the possibility of disturbance, a change in density provides positive evidence that there has been some disturbance; thus density change was used as a primary indicator of disturbance for this study. Variation of sample recovery ratio (length of sample/length of drive) with depth was also used as an indicator of disturbance.

A basaltic silty sand was used as a lunar soil simulant for this study. This simulant has been tested extensively and is considered to represent the lunar soil well, (HOUSTON and NAMIQ, 1970; HOUSTON *et al.*, 1971).

Table 1. Sampling tube dimensions.

Tube	Inside Diameter, D_i		Outside Diameter, D_0		Area ratio, A_r $= \dfrac{D_0^2 - D_i^2}{D_i^2}$ %	Length of drive	
	cm	(inches)	cm	(inches)		cm	(inches)
1	4.242	(1.670)	4.445	(1.750)	10	40.64	(16)
2	4.155	(1.636)	4.445	(1.750)	14	40.64	(16)
3	3.828	(1.507)	4.445	(1.750)	35	40.64	(16)
4	3.810	(1.500)	4.445	(1.750)	36	40.64	(16)
5	1.974	(0.777)	2.540	(1.000)	142*	30.48	(12)

* Sample disturbance is controlled mainly by the area ratio at the core tube bit. The bit used with this tube has an average outside diameter of about 3.073 cm (1.21 inches).

Description of Tests

The sampling tests were performed using several sizes and types of tubes shown in Table 1.

Tube 5 is a split tube which is held together by a threaded bit on one end and a cap on the other end. It has essentially the same dimensions as the tube used on the Apollo 12 mission, but without a removable liner. Tubes 2 and 4 are one-piece and made of aluminum with sharpened cutting edges. Tubes 1 and 3 are stainless steel and tube 3 has a removable liner whose inside diameter is given in Table 1. Tubes 1–4 were studied as a basis for design of new core tubes for Apollo 15 and later missions.

Parameters measured during and after sampling included the following: (1) force required to advance the tube into the soil by pushing; (2) sample recovery ratio; (3) % densification during sampling and extrusion; (4) friction between the sample and the tube's inner walls. Only the data in connection with (2) and (3) are discussed herein. Sets of data were obtained for tube advancement by both hammering and pushing because preliminary tests showed that the mode of tube advancement affected the results appreciably. Combinations of pushing and hammering were not used.

Differences between the test results for tubes 1 and 2 ($A_r = 10$ and 14%) were smaller than ordinary experimental errors and could not be distinguished. Therefore, data from these two tubes were considered together. The same was true for tubes 3 and 4 ($A_r = 35$ and 36%).

Sample Recovery Ratio

Values of sample recovery ratio for all tubes are plotted vs. the void ratio, e, in Fig. 1. The average void ratio, e, over the core-tube length was used to relate the lunar soil simulant to the actual lunar soil because the void ratio is independent of the specific gravity of solids ($e = n/(1 - n)$ where n = porosity). The corresponding density based on the specific gravity of actual lunar soil is shown at the top of the figure.

The data in Fig. 1 show that the sample recovery ratio is consistently higher for the larger tubes ($A_r \leqslant 36\%$) than for the smaller tubes ($A_r = 142\%$). Although the difference is small, it appears that the sample recovery ratio is slightly higher for

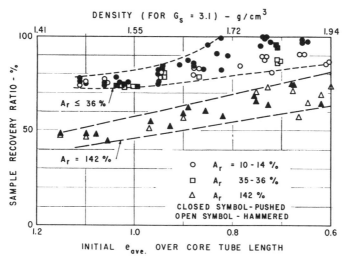

Fig. 1. Sample recovery ratio vs. initial density.

$A_r \leqslant 14\%$ than for $A_r = 35\text{–}36\%$. It also appears that pushed tubes have slightly higher sample recovery ratios than hammered tubes.

For most of the tubes the sample recovery ratio was measured at intervals of 4 cm during tube advancement, thus allowing the computation of an incremental sample recovery ratio. Some typical results for hammered tubes are shown in Fig. 2, which shows the variation (with depth) of the amount of soil collected for each 4 cm of advancement, whereas Fig. 1 shows the cumulative or total sample recovery ratio. The tubes used to obtain the data in Fig. 2 have the same inside and outside diameters as tubes 2 and 4, but are 81.3 cm (32 inches) long.

The curves show that sample recovery was little affected by the difference between the two area ratios, but initial soil density has an important effect. For the loose soil no additional soil moved into the tube after a depth of 45 cm (about 10 tube diameters) had been reached. At even shallower depths some soil was flowing around the tube bit. For example, an incremental sample recovery ratio of 60% means that 40% of the soil is apparently not going into the tube for this increment of advancement, since densification due to driving (discussed in the next section) can account for only a 10–15% change in length.

These results for the loose soil are consistent with data obtained with a cone penetrometer for the same soil (HOUSTON and NAMIQ, 1970). Penetration resistance was found to increase linearly with depth until a maximum was reached at a depth of 6 to 10 diameters for loose soil and 10 to 20 diameters for dense soil beyond which no further increase was observed. This effect is ascribed to arching in the soil. If the wall friction between the core and the tube becomes nearly equal to the final residual penetration resistence of the soil, it is reasonable to expect that little or no additional core can be obtained.

The limiting penetration resistance of the dense soil was apparently not reached although the sample recovery ratio decreased to about 65% (Fig. 2).

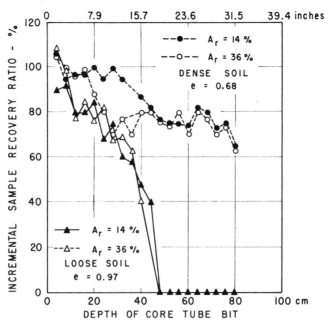

Fig. 2. Incremental sample recovery ratio vs. depth of core tube bit—hammered tubes.

Sample Densification

By measuring the weight and volume of the soil collected in each tube it was possible to compare the density after sampling with the initial density. In a few cases where the soil was initially very dense, sampling caused a loosening of the soil, indicated as a negative percent densification. (This loosening of the soil also occurred at very shallow depths as indicated by the sample recovery ratio values greater than 100% in Fig. 2.) Some example results for hammered tubes are shown in Fig. 3. Similar data were obtained for all tubes, and the results are summarized in Fig. 4.

Fig. 4 can be used to obtain an estimate of the in-place density corresponding to the core-tube samples returned from the Apollo 12 mission. The measured densities in the returned tubes were found (Scott, Carrier, Costes, and Mitchell, 1970) to range from 1.77 to 2.02 g/cm³ for a tube inside diameter of 2.00 cm. By trial and error solution using the curve for hammered tubes with $A_r = 142\%$ in Fig. 4, it can be estimated that the corresponding range in % densification due to sampling is 5.5 to 14%. The corresponding in situ or "undisturbed" density range is 1.55 to 1.90 g/cm³. This estimate is based on hammered tube data because hammering was the predominant mode of advancement for the Apollo 12 core tubes.

Densification during extrusion was measured for a total of 19 samples obtained with tubes 1, 2, and 4. Extrusion was accomplished with a close-fitting piston driven by a hydraulic jack. The results appeared to be insensitive to the initial soil density and to the method of obtaining the samples. The average value of % densification for all samples was 0.4%. All values were less than 1.7% and 90% of the values were less than 1%.

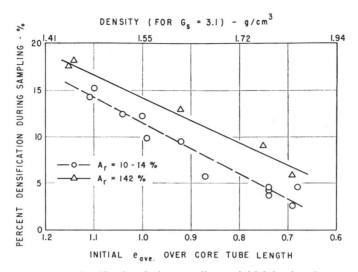

Fig. 3. Percent densification during sampling vs. initial density—hammered tubes.

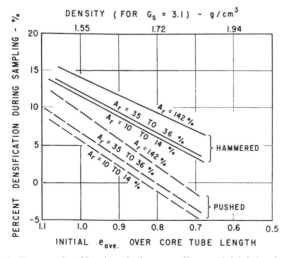

Fig. 4. Percent densification during sampling vs. initial density—summary.

CONCLUSIONS

(1) Estimates of in situ density based on Apollo 12 core tube data and the results presented herein give a range of 1.55 to 1.90 g/cm³. It is not yet known whether the actual in situ density of soil on the moon may range so widely or whether the apparent range was due to uncertainties associated with core sample density measurements, method of tube advancement, or disturbance during earth return or processing in the LRL. Additional studies of core tube data are needed.

(2) If the 4.445 cm (1.75 inch) outside diameter core tubes (with an area ratio of 10–14%) which have been proposed for Apollo 15 and subsequent missions are used, considerably more sample per tube is likely to be returned, due to both the larger area and the greater sample recovery ratio (Fig. 1).

(3) The sample recovery ratio decreases very significantly with depth for all tubes tested, especially for loose soil. It appears that driving a 2.54–5.08 cm (1–2 inch) diameter core tube deeper than about 45 cm (17.7 inches) may be a relatively unproductive means for retrieving samples of lunar soil at depth, if the original density is relatively low. Percent densification during sampling is affected very strongly by the method of tube advancement and by the area ratio of the tube for area ratios greater than about 35%. Available data indicate that the best way for estimating the in situ density from core-tube samples obtained during future Apollo missions is to (a) make careful note of the procedure followed by the astronaut during sampling, including the depth of drive, number of blows, etc; (b) repeat the procedure in detail in the laboratory using a lunar soil simulant prepared to a range of initial densities; (c) adjust measured densities of returned samples in accordance with results of (b).

Acknowledgments—The studies described in this paper were conducted under NASA Contract NAS 9-11266, Principal Investigator Support for Soil Mechanics Investigation (S-200). This support is acknowledged with appreciation.

References

Carrier W. D., III, Johnson S. W., Werner R. A., and Schmidt R. (1971) Distortion in samples recovered with the Apollo core tubes. Second Lunar Science Conference (unpublished proceedings).

Houston W. N. and Namiq L. I. (1970) Penetration resistance of lunar soils, *Proceedings*, ISTVS-TRW Off-Road Mobility Symposium, November 3–4, 1970, Los Angeles, California.

Houston W. N., Namiq L. I., Mitchell J. K., and Treadwell D. (1971) Lunar soil simulant studies, lunar surface engineering properties experiment definition (Chap. 1, Vol. I). Final Report, NASA Contract NAS 8-21432, University of California, Berkeley, Space Sciences Laboratory.

Scott R. F., Carrier W. D., Costes N. C., and Mitchell J. K. (1970) Apollo 12 Soil Mechanics Investigations. *Geotechnique* (submitted).

Proceedings of the Second Lunar Science Conference, Vol. 3, pp. 1959–1972
The M.I.T. Press, 1971.

Disturbance in samples recovered with the Apollo core tubes

W. David Carrier, III and Stewart W. Johnson*

NASA Manned Spacecraft Center, Houston, Texas 77058

and

Richard A. Werner and Ralf Schmidt

Lockheed Electronics Company, Houston, Texas 77058

(*Received 9 February* 1971; *accepted in revised form* 17 *March* 1971)

Abstract—The Apollo 11, 12, 14, and 15 core tubes were tested in a layered lunar soil simulant to assess the influence of tube geometry on the quantity and quality of lunar sample recovery. It was found that the Apollo 11 core tube samples were most distorted and that the Apollo 15 core tube samples were least disturbed. The depth in the core tube samples has been related to the original depth in the lunar surface for all the core tubes. For example, it was found that an element of soil located at 10 cm in the double core tube sample from Apollo 12 is actually representative of lunar soil from a depth of approximately 13 cm; similarly, soil at 40 cm came from a depth of 61 cm. These depth relationships can be used by the various core tube sample investigators to interpret their test results. These data are of particular relevance to those investigators seeking to relate property change to depth in the lunar surface.

INTRODUCTION

Two DIFFERENT TECHNIQUES have been used to gather soil samples on the lunar surface. Bulk samples of a highly disturbed nature have been scooped from the surface, and core tubes have been driven to recover samples with relatively well-preserved stratigraphy. The internal volume of the driven core tubes actually filled with lunar soil has ranged from 61% to less than 47%; so there is a problem in determining the original depth of lunar soil at various points in the core tube soil column. The less than 100% core recovery results from several factors including core tube wall thickness and bit characteristics. The wide range of recovery, from 61% to 47%, results chiefly from differences in bit design for the Apollo 11 and Apollo 12 tubes.

Some have argued that the core tubes plugged at depths corresponding to the length of samples and therefore depths in the soil column correspond on a one to one basis to depths below the lunar surface. Others have expressed belief that the core tube samples represent continuous columns of lunar soil including some material from the complete depth to which the tube was driven and that the material in the tube has been compressed by the action of driving the tube. Obviously, the view taken will have an important impact on how core sample investigators interpret their results such as ages, solar wind effects, mixing rates resulting from meteoroid impact, chemistry and mineralogy, and particle size and shape.

In our investigation, the various core tubes were driven into the layered soil which has properties matching the important properties of undisturbed lunar soil. Our test results from this lunar soil simulant permit an assessment to be made of the influence

---

* National Academy of Sciences-National Research Council Postdoctoral Fellow.

of bit and core tube geometry on the quantity and quality of core tube samples recovered for several lengths of core. From our tests there is a sound basis for inferring that the Apollo 11 core tube samples were severely distorted and that the Apollo 12 samples were somewhat disturbed but were superior to Apollo 11 samples because of improved bit design. The Apollo 14 core tube samples should be similar to the Apollo 12 samples in terms of quality, as the core tube design is the same. The Apollo 15 core tubes have significant design improvements, and our simulations indicate a higher percentage of core recovery and less sample disturbance.

Data gathered in this investigation have been used to estimate the original depth in the lunar surface as a function of location in core tube for each of the Apollo 11 and Apollo 12 core samples. Results vary significantly from a one to one correspondence, and should be very useful to the various core tube sample investigators in interpreting their test results. The technique and test data presented can also be applied to core tube samples to be recovered on future missions.

The Apollo Core Tubes

Core tube geometry

The design of the Apollo core tubes has thus far evolved through three main stages. Shown in Fig. 1 are the dimensions at the bit-end of the three designs, where the most significant changes have occurred (the overall length has remained approximately constant at 35 cm). The Apollo 11 core tubes have a bit with an inward flare, opposite in shape to standard terrestrial samplers. The internal area of the bit decreases from the leading edge to the neck by a factor of greater than 2; it is this constriction which produces the severe sample distortion mentioned above.

The Apollo 12 and Apollo 14 core tubes are all alike and utilize a more conventional bit shape. The wall thickness, however, is relatively large compared to the internal diameter of the tube. One way of rating samplers is by means of the *area ratio* A_r (Terzaghi and Peck, 1967), defined as

$$A_r(\%) = 100 \frac{D_e{}^2 - D_i{}^2}{D_i{}^2}$$

where D_e = external diameter and D_i = internal diameter. In terrestrial applications, it has been demonstrated that in order to minimize sample disturbance, A_r should be kept as small as possible. The area ratio for the Apollo 12–14 tubes is approximately 140%, as compared with the value of 14% for a 2-inch diameter tube commonly used on Earth. The Apollo 12–14 tubes thus produce significant sample disturbance. On the other hand, the nature of this disturbance is primarily one of a change in soil density, rather than distortion of the stratigraphy, as in the Apollo 11 bit, and therefore, the Apollo 12–14 core tube is a considerable improvement. (Additional discussion on the relative merits of these two core tubes may be found in Scott *et al.*, 1970.)

Whereas the Apollo 12–14 core tube was achieved merely by replacing the screw-on Apollo 11 bit with a new and better bit, the Apollo 15 core tube represents a complete new design. The sample diameter is more than double that of the previous core tubes, and the wall thickness is reduced, resulting in an area ratio of only 7.4%. On previous core tubes, the bit was unscrewed on the lunar surface by the astronaut after the

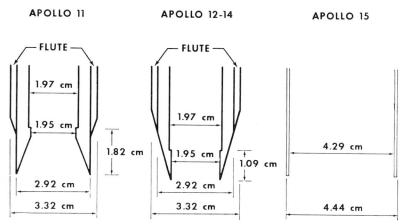

APOLLO 11   APOLLO 12-14   APOLLO 15

Fig. 1. Comparison of Apollo core tube bits. Shown are cut-away views giving diameters at various sections. The Apollo 11 bit was made of aluminum alloy; the other two are of stainless steel. All have a circular cross-section.

sampling operations and a cap screwed on in its place to retain the sample. Normally, the bit, containing the deepest sample, was discarded. On the other hand, the bit on the Apollo 15 core tubes is an integral part of the tube and will be capped by a Teflon retainer, thereby automatically returning the deepest sample from every tube.

An additional improvement in the Apollo 15 core tube is the elimination of the Teflon follower present in the Apollo 11 and Apollo 12–14 core tubes. This device is basically a plug which is pushed up by the column of soil inside the core tube during the sampling process and which prevents the soil from moving during its return to Earth. Unfortunately, the follower is just a simple friction device which applies a force to the soil during sampling. If this force is too high, it can seriously affect the quantity of sample recovered. The Apollo 15 core tube will not have a follower; instead the astronaut will insert a retainer into the core tube and lock it in place in contact with the soil after the sampling is completed.

The force required to move the follower in the Apollo 11 and Apollo 12–14 core tubes varies between 2 and 3 pounds, depending on the position of the follower in the tube; furthermore, each tube-follower combination is different. Thus, the presence of the followers in the core tubes contributes the most uncertainty to the results of this study. In all of the simulation tests discussed below, except for the Apollo 15 cores, which have no followers, the follower drag was set at 2 pounds.

Returned core tube samples

Table 1 summarizes the basic data concerning the core tube samples returned from Apollos 11 and 12: Apollo 11 has two single core tubes; and Apollo 12 had four core tubes, used to return two singles and a double (two tubes connected end to end). The locations of these core tubes at the two landing sites are shown in SHOEMAKER *et al.* (1969, 1970, respectively). Table 2 presents a listing of NASA photographs of the core tubes and their samples.

Table 1. Returned core tube sample data.

Core tube serial S/N	LRL sample	Returned sample weight, g	Returned sample length, cm	Bulk[a] density, g/cm^3	Total[b] sample length, cm	Core[c] tube depth, cm	%[d] core recovery
Apollo 11							
2007	10005	52.0	10.0	1.71	11.8	>25	<47%
2008	10004	65.1	13.5	1.59	15.3	<32	>48%
Apollo 12							
2011	12027	—	17.4[e]	—	18.5	~37	~50%
Double ⎡ 2010	12025	56.1	9.3	1.98 ⎫	42.2	69	61%
⎣ 2012	12028	189.6	31.8	1.96 ⎭			
2013	12026	102.9	19.4	1.74	20.5	37	56%

[a] Based on a sample diameter of 1.97 cm.
[b] Adjusted to include the length of the discarded bit sample; for Apollo 11, an additional 1.82 cm; for Apollo 12, 1.09 cm (see Fig. 1).
[c] Determined from mission photography.
[d] Total sample length ÷ core tube depth.
[e] This core tube has not been opened but has been kept in storage in the LRL. HEIKEN (1970) determined the sample length by means of X-radiography of the core tube.

Table 2. NASA core tube photography.

Core tube serial	Lunar surface photographs	Lunar receiving[a] Laboratory photographs
APOLLO 11		
2007	AS11-40-5963 to AS11-40-5964	S-69-45047 to S-69-45049
2008	None	S-69-45535 to S-69-45537 S-69-45929 to S-69-45931
APOLLO 12		
2011	AS12-48-7068 to AS12-48-7070 AS12-49-7279 to AS12-49-7280	S-70-18021 to S-70-18022
2010		S-69-23722 to S-69-23727 S-69-23803 to S-69-23818 S-70-20400 S-70-21302 to S-70-21309
	AS12-48-7077 AS12-59-7285 to AS12-49-7288	
2012		S-69-23396 to S-69-23412 S-69-23728 to S-69-23758 S-69-60570 to S-69-60572 S-69-62763 to S-69-62765 S-69-64424
2013	AS12-47-7008 to AS12-47-7009	S-69-60356 to S-69-60362 S-69-60477 to S-69-60481 S-69-60488 to S-69-60493 S-69-61191 to S-69-61195 S-69-62744 to S-69-62762

[a] The LRL photograph numbers listed here represent a revision of those listed in WARNER, 1970.

It is presently planned that Apollo 14 will carry six core tubes, which will be used to return a triple (100 cm tube length), a double, and a single. Apollo 15 will carry nine core tubes. Although the arrangement of these tubes is not final yet, the most likely distribution will be one triple, two doubles, and two singles.

CORE TUBE SIMULATION STUDY

Soil simulant

The key to a meaningful core tube study is obviously the use of a soil in our 1 g laboratory experiments which closely simulates the behavior of undisturbed lunar soil in $\frac{1}{6}$ g and vacuum. Gravity and atmosphere are known to be significant parameters in determining soil behavior (cf. JOHNSON *et al.*, 1970). The mechanical properties of lunar soils are so far known only within certain bounds, based largely on data obtained by the Soil Mechanics Surface Sampler Experiments of Surveyor III and VII (SCOTT and ROBERSON, 1969) and the soil behavior observed during the Apollo 11 and 12 traverses. Far more data are required, both in situ and on returned lunar samples, before really accurate modeling is possible. In the meantime, a simulated lunar soil has been prepared which appears to behave satisfactorily. It consists of an air-dry (water content = 0.6% by weight) mixture of League City sand (65%) and kaolinite clay (35%), packed to a density of 1.33 g/cc. The grain size distribution of this mixture is:

Sieve no.	Particle size in millimeters	% Finer by weight
20	0.85	99.4
40	0.42	97.7
60	0.25	95.8
100	0.149	72.0
200	0.074	40.2
	0.0036	34.8
	0.0011	30.2
	0.00049	19.9

Comparison with published lunar grain size distributions (cf. DUKE *et al.*, 1970) will reveal that this simulant is considerably finer than the lunar soil, due to the presence of the kaolinite. The kaolinite, however, simulates the cohesive nature of soil on the lunar surface. Several different simulants have been examined and each requires a compromise of some sort. It was decided that matching the cohesive nature and frictional strength of the lunar soil is far more important to successful simulation for this case than duplicating the grain size distribution.

A comparison of the mechanical properties of the lunar soil and the simulant is shown in Table 3. To account for the difference in gravity between the Moon and the Earth, the absolute value of the cohesion of the simulant has been increased proportionately, such that the ratio of the cohesion to the unit weight of the simulant in 1 g is the same as that for the lunar soil in $\frac{1}{6}$ g. It can be seen in Table 3 that this ratio for the simulant falls within the bounds imposed by the current uncertainties in the knowledge of lunar soil properties.

Table 3. Soil mechanics properties.

Property	Lunar soil	Terrestrial simulant
Friction angle	37°–35°[a]	37°
Cohesion	3430–6860 dynes/cm²[a]	19,330 dynes/cm²
Bulk density	1.6–2.0 g/cm³[b]	1.33 g/cm³
Unit weight	261–326 dynes/cm³	1305 dynes/cm³
Cohesion ÷ unit weight	10.5–26.3 cm	14.8 cm

[a] SCOTT and ROBERSON (1968); [b] SCOTT *et al.*, (1970).

Corroborative evidence that the simulant behaves properly is also provided by the data obtained in the core tube study itself. The % of core recoveries in the simulant compare very favorably with those in the returned core tubes of Apollo 11 and 12. As an example, consider the % of core recoveries for the Apollo 12 core tubes (Table 1). The two single core tubes both had smaller recoveries than the double, which is opposite to what one would intuitively expect. And yet, our simulation produced the same results, with a recovery of 50% for the single (vs. $\sim$ 50% and 56% for Apollo 12) and 63% for the double (vs. 61%). The core recovery in our simulation then decreases to a value of 52% for the case of the triple. This phenomenon is evidently the result of a complex interaction among the Teflon follower, the core tube liner, and the confining pressure of the soil. When core tube simulations are performed without the followers present in the tubes, two things happen: the % core recovery for each core tube combination increases; and the recovery decreases with increasing depth, as is normally the case. The values are approximately 80%, 70%, and 60%, for a single, double, and a triple, respectively. Apparently, the anomalous core recovery for both the actual and the simulated Apollo 12 core tubes is caused by variations in the follower drag along the liner.

Additional evidence that the simulant behaves properly is provided by the relative changes in the bulk density during sampling. In the case of the Apollo 11 core tube simulation, the shape of the bit caused a decrease from the in situ density (that is, the soil expanded) of approximately 13%; the Apollo 12 double core caused an increase of 15%. These data correlate well with the lunar soil behavior which apparently occurred during the driving of the core tubes on the lunar surface, considering the error bracket associated with the in situ density of the lunar soil. There is one inconsistency in the data, however. In the Apollo 12 single core tube simulation, the soil densified approximately 15%, i.e., the same as the double core. Whereas for S/N 2013 (see Table 1), the calculated bulk density is much less than the actual Apollo 12 double and is, in fact, very close to S/N 2007 (Apollo 11). The low % core recoveries for the Apollo 12 single cores would normally imply significant densification, as was the case in the simulation; therefore, we suggest that the returned sample weight for S/N 2013 shown in Table 1 (102.9 g) is in error. A 10 g error in reading the scales in the LRL during the preliminary examination could account for the discrepancy in the density. The alternative explanation is that this core tube was driven into a distinctly different soil on the lunar surface. However, we feel there is insufficient evidence to support this hypothesis.

We look forward to an accurate determination of the bulk density of the sample in the other Apollo 12 single core tube, S/N 2011, as well as the single from Apollo 14, as these data should shed some light on this question.

In the meantime, the percentage changes in density during the core tube simulation that were mentioned above, permit us to predict that the lunar surface in situ bulk density falls in the range of 1.7 to 1.9 g/cc. This range falls squarely in the middle of the range predicted by Scott *et al.* (1970) and shown in Table 3.

Simulant preparation

The soil simulant described above, which is light yellow in color, was compacted in layers in a specially built soil bin; dark layers nominally one-inch thick were prepared by adding powdered manganese dioxide to the mixture. These dark layers were established at different depths in the bin; the final location of the various layers was made on a clear plastic panel on one side of the soil bin after the entire soil column was in place. By measuring the location of these layers inside the core tubes after sampling, a "before and after" depth relationship was developed, such that the original depth of an element of lunar soil in a core tube could be estimated.

Apollo 12–14 simulation

A dissected sample from an Apollo 14 triple core tube simulation is shown in Fig. 2 as an example of the test results. The boundaries between the light and dark soils can be seen to be sharp and quite straight, indicating very little sample distortion, except at the core tube edges where smearing has occurred. This smear zone is approximately 1.5 mm thick and produces maximum longitudinal displacements of about 5 mm.

CENTIMETERS

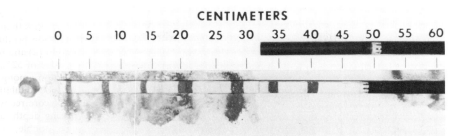

Fig. 2. Dissection of simulated Apollo 14 triple core tube sample. The alternating layers of simulated lunar soil are used to develop depth relationships for the Apollo core tubes. The triple core tube to be driven into the lunar surface on Apollo 14 will return less than two full tubes. Two Teflon followers, shown at the top of the sample, are encountered by the soil column during driving. Before driving, the followers are located at the lower ends of each tube.

The depth to the various layers was measured, and the results are shown in Fig. 3, along with the results from an Apollo 12–14 single and Apollo 12–14 double. It can be seen that at least a portion of every layer the core tube passes through is recovered in the tube, except at the bit. The bit penetrates a few centimeters deeper than is actually recovered; evidently the soil is pushed ahead of the advancing bit like a bow wave. Not shown is the fact that part of each level of soil in the path of the core tube

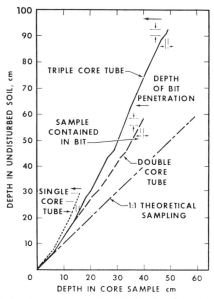

Fig. 3. Depth relationships: Apollo 12–14 simulated core tube samples. Shown are results from driving and dissecting single, double, and triple lengths of core tube with the bit design shown in Fig. 1b. The arrows denote (on the ordinate axis) the depths to which the core tubes were driven in the simulations.

is not recovered, but flows around the outside of the tube. In the case of the triple core tube, density and length measurements of the sample in the bottom core tube indicated that nearly half of the soil mass flowed around the tube and was not recovered. This is an average for the entire bottom tube; the proportion for the sample near the bit is probably much higher, perhaps close to the point at which no additional sample is recovered. This phenomenon must be borne in mind when studying core tube data. The tube recovers some sample nearly to the maximum depth of bit penetration, but the amount of material recovered from various strata decreases with depth.

It should also be noted that the curve for the single core falls to the left of the curve for the triple, rather than to the right of the double, as one might expect. This is a result of the anomalous % core recovery for the Apollo 12 single core tubes (both simulated and actual) discussed in the section on Soil Simulant.

Apollo 12 returned core tube samples

By making minor corrections to the data in Fig. 3 to account for slight differences between the depths of the driven core tubes in the simulation and the actual depths in the lunar surface, the depth relationships for the Apollo 12 double and two single core tubes can be determined. This is shown in Fig. 4. It will be recalled that single tube S/N 2011 was taken in the bottom of a trench on the rim of Sharp Crater. The astronauts reported that the trench was approximately 8 in. deep (20 cm) and that the tube was driven completely into the soil. (p. 142, SHOEMAKER *et al.* 1970). Unfortunately, the bottom of the trench is lost in the blackness of the shadows and cannot be

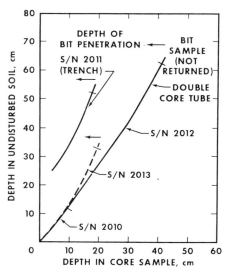

Fig. 4. Depth relationships: Apollo 12 returned core samples. These relationships have been prepared by interpolation from data presented in Fig. 3 and from a knowledge of actual core sample lengths and depths (see Table 1) to which the Apollo 12 core tubes were driven. The arrows denote (on the ordinate axis) the actual depth to which each core tube was driven.

seen in the mission photography, and therefore the precise depth of this core tube cannot be determined. In Fig. 4, we have assumed that the crew was correct in their observations and have offset the curve for S/N 2011 by a depth of 20 cm. Furthermore, the X-radiography of this core tube sample taken by HEIKEN (1970) and presented by WARNER (1970) suggests that the top 4 cm of sample is disturbed, possibly due to fragments falling down from the sides of the trench onto the bottom. Due to this additional uncertainty, that portion of the curve is not shown in Fig. 4.

The absolute accuracy of the depth relationships shown in Fig. 4 is about $\pm 5\%$ of the depth in the core sample for the double and single S/N 2013. That is, at a depth of 10 cm in the core sample, the depth in the undisturbed soil is accurate to ± 0.5 cm; and at 40 cm, ± 2 cm. Because of the uncertainties discussed above, the accuracy for single S/N 2011 is ± 2 cm over its entire length.

Figure 4 will be an aid in the interpretation of the Apollo 12 core tube data. For example, it can be inferred that an element of soil located at 10 cm in the double core tube sample is actually representative of lunar soil from a depth of approximately 13 cm; similarly, soil at 40 cm came from a depth of 61 cm. Furthermore, it should be interesting to compare the test results from different core tube samples originally located at the same depth in the lunar surface. For example, samples located at 7.2 cm in S/N 2011, 18.1 cm in S/N 2013 and 21.8 cm in the double (or 12.5 cm in S/N 2012) all correspond to a depth of approximately 30 cm in the undisturbed soil at the Apollo 12 site. Similar comparisons between different Apollo sites should also yield meaningful results.

Apollo 14 returned core tube samples

When the Apollo 14 core tubes are returned, another figure comparable to Fig. 4 will be prepared on the basis of the data presented in Fig. 3. In the meantime, it is possible to predict, assuming no rock is encountered, that the total sample length from the Apollo 14 triple core tube will be 45 to 55 cm, depending on how deeply the Astronauts drive the tube. A corollary to this conclusion is that the topmost tube of the triple will most likely contain no sample, as the length of just a double core tube is 69 cm (see Table 1), or 16 to 23 cm longer than our predicted triple core sample length. As a result of this study, it was suggested that the Apollo 14 astronauts check the third tube for sample and if none is visible, then use it as an additional single core tube, time permitting.

Apollo 11 returned core tube samples

The Apollo 11 samples are in a special class by themselves.

Two core tubes were driven into the soil near the LM in the closing minutes of the first lunar EVA. In the rush to complete the EVA, it was impossible for the astronauts correctly to document the core tube samples. As a result, no photographs were taken to show maximum penetration of the core tubes, nor was the location of the 2nd core tube documented. Furthermore, Astronaut Aldrin could not recall the order in which he had driven the two tubes. However, he did have difficulty removing the second core tube from the extension handle, and later in the LRL when it was found that one of the tubes (S/N 2008) was missing its top plug (which connects to the extension handle) it was deduced that this tube must have been the second sample.

In the LRL it was also discovered that the liners for S/N 2008 had been installed backwards before the mission, with the teflon follower at the top of the tube, rather than at the bottom. As a result, there was nothing to restrain the sample from moving inside the tube during transit from the Moon to the Earth, and that is precisely what happened: all of the sample was in the top of the core tube when it was opened in the LRL. Our investigation has shown that the soil is significantly disturbed during sampling. The effect of the additional disturbance of S/N 2008 because of lack of restraint can only be the subject for speculation. For that reason, a depth relationship has been developed only for the first core tube sample, S/N 2007, which was restrained in the tube by the follower after sampling.

A dissected sample from an Apollo 11 core tube simulation is shown in Fig. 5. It is clear from this figure that the soil has undergone considerable distortion during the sampling process. This distortion was not noted in tests with the Apollo 12–14 bit and is due solely to the shape of the Apollo 11 bit. Distortion increases with distance measured radially from the axis of the core tube. The bit has produced stretched-out strata, with longitudinal displacements as great as 5 cm near the edges. This is in sharp contrast to the moderate smearing observed in the Apollo 14 triple core tube tests in which the displacements were only 5 mm and were confined to a 1.5 mm zone at the core tube edge.

The zone of cracked soil near the top of the sample in Fig. 5 bears a striking

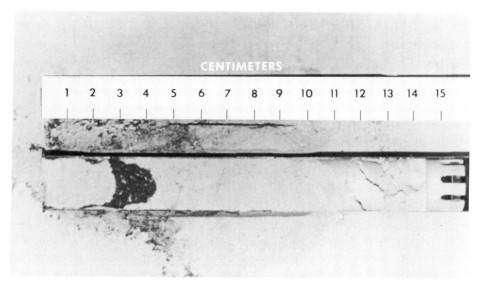

Fig. 5. Dissection of simulated Apollo 11 single core tube sample. The badly distorted, originally horizontal layer is shown near the left or bit end. On the right is the follower. The sample rests in a half-section of the core tube liner.

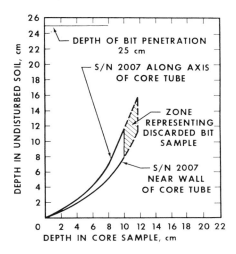

Fig. 6. Depth relationship: Apollo 11 core tube S/N 2007. The severe distortion caused by the Apollo 11 bit (see Fig. 1a and Fig. 5) necessitates the presentation of these data using one curve for material near the core tube wall. The bit penetrated to a depth of at least 25 cm but the core tube recovered material from less than 12 cm depth.

resemblance to the Apollo 11 core tube sample, S/N 2007 (Fig. 4 35 in COSTES *et al.*, 1969 or Fig. 19(b) in COSTES *et al.*, 1970). This feature is apparently also due to the Apollo 11 bit, as it was not observed in any of the core tube simulations with the Apollo 12–14 bit.

Using the same technique as described above with the Apollo 12 core tube samples, the depth relationship for S/N 2007 has been determined and is shown in Fig. 6. Unlike the others, however, the depth is shown as a function of the radial distance from the core tube centerline. By examining Fig. 5, it can be seen that in fact the core tube sample is not axisymmetric; Fig. 6 is therefore an oversimplification. However, the important trends are clearly indicated. The absolute accuracy is about $\pm 10\%$ of the depth in the core sample.

The significance of the sample disturbance in the Apollo 11 core tubes cannot be overemphasized. Obviously, the interpretation of any of the data from these core samples will depend not only on the longitudinal position of the sample, but its radial position as well. If a sample investigator is searching for property gradients in the core tube samples, clearly he must know the original vertical position of his samples. As shown in Fig. 6, a sample taken at the edge of the core tube could be representative of the same material located as much as 4 cm higher in the tube along the centerline. Failure to recognize this fact could lead to erroneous or misleading interpretations of the core tube sample data.

Also illustrated in Fig. 6 is the fact that the bit penetrated strata located much deeper than was actually sampled and returned. Astronaut Aldrin drove this tube more than 25 cm and sample was retrieved in the tube (not counting the bit sample, which was discarded on the lunar surface) from a depth of only about 12 cm.

Apollo 15 simulation

As discussed above, the Apollo 15 core tubes represent a completely new design. They are so new, in fact, that flight equipment had not yet been built when our study was made. Instead, it was necessary to use some early prototypes which although quite satisfactory in terms of diameter and wall thickness, precluded the testing of a triple core tube. We were able, however, to run a single and a double.

The results for the Apollo 15 single core tube indicated a significant improvement over the previous core tube design. The % core recovery was 88%, compared with 50 to 56% for the Apollo 12 singles (Table 1). In addition, the depth relationship was virtually 1:1; i.e., the depth in the core sample corresponded directly to the original depth in the undisturbed soil. The significance of these results should not be overlooked. Besides returning larger and less disturbed samples, a far more accurate estimate of the in situ density of the lunar soil will be possible. This will enhance the models of the lunar surface considerably.

The Apollo 15 double core also represented an improvement, with a core recovery of 74% (vs. 61% for Apollo 12). The depth relationship, shown in Fig. 7, is also much closer to 1:1 than the previous core tubes (compare Fig. 7 with Fig. 3). A sample taken from a depth of 50 cm in the core tube column represents material from a depth of 60 cm.

As the Apollo 15 core tube equipment becomes available, we will conduct additional tests so that depth relationships can be determined for each of the Apollo 15 returned core tube samples.

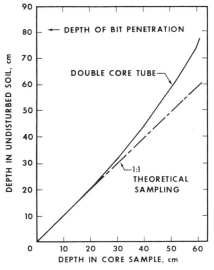

Fig. 7. Depth relationship: Apollo 15 simulated double core tube sample. The bit (see Fig. 1c) penetrated to a depth of 81.3 cm. The length of core recovered was 60 cm; and the core contained material from depths to 77.8 cm. The actual flight equipment will have single core tubes 36 cm in length. Linked together, the Apollo 15 core tubes will form double core tubes 73 cm long and triples 111 cm long.

SUMMARY

(1) The core tube samples recovered on Apollo 11 were severely distorted, due to the shape of the bit. The amount of material recovered and depth of recovery were also adversely affected by the bit. The bit became plugged so that material recovered in the core tube represented soil from less than one-half the maximum depth of bit penetration. Soil recovered was less dense than in situ soil. (2) The core samples recovered on Apollo 12 were a considerable improvement over the Apollo 11 samples but were still significantly disturbed. The Apollo 12 bit and core tube continued to collect sample to within a few centimeters of deepest bit penetration. Part of each level of soil penetrated was recovered. The amount recovered for each layer decreased with depth. Some of the soil is pushed ahead of the advancing bit like a bow wave so that it flows around the outside of the tube. Densification is noted in the Apollo 12 core tube sample, particularly in the case of the double core tube. Our results permit us to predict that the lunar surface in situ bulk density falls in the range of 1.7 to 1.9 g/cm^3. (3) The new Apollo 15 core tubes will return a larger and much less disturbed sample. A far more accurate estimate of the in situ bulk density of the lunar surface will also be possible. (4) Depth relationships for the Apollos 11 and 12 core tube samples have been established which may be used in the interpretation of data from core sample analyses. (5) Simulation data have been gathered which will be used to establish similar depth relationships for the Apollos 14 and 15 returned core tubes. (6) This simulation technique will also be used to investigate the sample disturbance caused by the Apollo lunar surface drill planned for Apollo 15. The data gathered in this additional investigation may also be relevant to the interpretation of the results from the Luna 16 drill.

REFERENCES

COSTES N. C., CARRIER W. D., MITCHELL J. K., and SCOTT R. F. (1969) Apollo 11 soil mechanics investigation. Apollo 11 Preliminary Science Report, NASA SP–214, 85–122.

COSTES N. C., CARRIER W. D., MITCHELL J. K., and SCOTT R. F. (1970) Apollo 11: Soil mechanics results. *ASCE J. Soil Mech. Foundations Div.* **96**, 2045–2080.

DUKE M. B., WOO C. C., BIRD M. L., SELLERS G. A., and FINKELMAN R. B. (1970) Lunar soil: Size distribution and mineralogical constituents. *Science* **167**, 648–650.

HEIKEN G. H. (1970) X-radiography of core 2, Apollo 12 sample number 12027. NASA Manned Spacecraft Center, Houston, Texas, Memorandum to TN12/Acting Curator, Lunar and Earth Sciences Division, July 17.

JOHNSON S. W., PYRZ A. P., LEE D. G., and THOMPSON J. E. (1970) Simulating the effects of gravitational field and atmosphere on behavior of granular media. *Spacecraft and Rockets* **7**, 1311–1317.

SCOTT R. F., CARRIER W. D., COSTES N. C., and MITCHELL J. K. (1970) Mechanical properties of the lunar regolith. Part C of Preliminary geologic investigation of the Apollo 12 landing site. Apollo 12 Preliminary Science Report, NASA SP–235, 161–182.

SCOTT R. F. and ROBERSON F. I. (1968) Soil mechanics surface sampler. In Surveyor project final report: Part 2, Science Results, Jet Propulsion Laboratory Technical Report 32–1265, 195–207.

SHOEMAKER E. M., BAILEY N. G., BATSON R. M., DAHLEM D. H., FOSS T. H., GROLIER M. J., GODDARD E. N., HAIT M. H., HOLT H. E., LARSON K. B., RENNILSON J. J., SCHABER G. G., SCHLEICHER D. L., SCHMITT H. H., SUTTON R. L., SWANN G. A., WATERS A. C., and WEST M. N. (1969) Geologic setting of lunar samples returned by the Apollo 11 mission. Apollo 11 Preliminary Science Report, NASA SP–214, 41–83.

SHOEMAKER E. M., BATSON R. M., BEAN A. L., CONRAD C., JR., DAHLEM D. H., GODDARD E. N., HAIT M. H., LARSON K. B., SCHABER G. G., SCHLEICHER D. L., SUTTON R. L., SWANN G. A., WATERS A. C. (1970) Geology of the Apollo 12 landing site. Part A of Preliminary geologic investigation of the Apollo 12 landing site. Apollo 12 Preliminary Science Report, NASA SP-235, 113–156.

TERZAGHI K. and PECK R. B. (1967) *Soil Mechanics in Engineering Practice* (2nd Edition). John Wiley.

WARNER J. (compiler) (1970) Apollo 12 lunar sample information, NASA Technical Report R-353.

Proceedings of the Second Lunar Science Conference, Vol. 3, pp. 1973–1987
The M.I.T. Press, 1971.

Cone penetration resistance test—an approach to evaluating in-place strength and packing characteristics of lunar soils

N. C. Costes

National Aeronautics and Space Administration,
G. C. Marshall Space Flight Center, Huntsville, Alabama 35812

G. T. Cohron

Cohron Industries, Inc., Huntsville, Alabama

and

D. C. Moss

National Aeronautics and Space Administration,
G. C. Marshall Space Flight Center, Huntsville, Alabama 35812

(Received 3 March 1971; accepted in revised form 31 March 1971)

Abstract—Cone penetration resistance tests on fine-grained, granular lunar soil simulants of varying grain-size distribution and consistency, performed under terrestrial gravity conditions and onboard the U.S. Air Force KC-135 aircraft during $\frac{1}{6}g$, $1g$, and $2g$ flight trajectories, are described. The test results indicate that the average resistance to penetration, $\bar{q}_c$, and its average rate of change with depth, G, of these simulants decrease monotonically with decreasing g level and are sensitive indicators of the soil dry bulk density, void ratio and relative density. It is also shown that $\bar{q}_c$ and G can be used with bearing capacity theory to determine the in-place shear strength characteristics of the soils tested.

Based on the analytical methods developed herein and on crude soil penetration data obtained from the Apollo 11 and Apollo 12 missions, upper bound estimates of the in-place bulk density and the angle of internal friction of the lunar soil are of the order of 1.85 g/cm³ and 45 deg for the soil at the Apollo 11 site, and of the order of 1.82 g/cm³ and 42 deg for the soil at the Apollo 12 site. On the basis of measured soil bulk density values of 1.71 g/cm³ and 1.74 g/cm³, obtained respectively from Apollo 11 and Apollo 12 core tube samples, the in-place angle of internal friction of the lunar soil at the corresponding Apollo 11 and Apollo 12 core tube sites is estimated to be 36.8 deg and 37.7 deg, and its cohesion to be of the order of 9×10^{-2} N/cm² and 7×10^{-2} N/cm². These findings are in agreement with independent evaluations of lunar soil mechanical properties.

The functional characteristics of a portable, self-recording penetrometer design concept, which is currently being brought to flight status for use in future Apollo missions, is described.

Introduction

Because no special soil mechanics testing or sampling equipment were included in the Apollo 11 and Apollo 12 missions and no force or deformation measuring devices were utilized during the surface activities, analysis of the in-place physical characteristics and mechanical properties of the lunar soil from data obtained during the LM landing or the astronaut surface extravehicular activities (EVA) has been primarily descriptive and qualitative (Costes et al., 1970; Scott et al. 1970).

To gain a better insight on the strength and deformation characteristics of the lunar soil at the Apollo landing sites, studies on the mechanical behavior of simulated lunar soils have been carried out at the MSFC Space Sciences Laboratory prior to and since the Apollo 11 and 12 missions (Costes *et al.*, 1969; Olson, 1969; and Smith, 1969). Parallel studies, under contract with the MSFC, have been conducted by the University of California, Berkeley (Mitchell *et al.*, 1969, 1970a, b) and by the USAE Waterways Experiment Station (WES), in connection with studies relating to the Lunar Roving Vehicle program (Freitag *et al.*, 1970; Green and Melzer, 1970).

Parallel with these studies, considerable effort has been spent at the MSFC Space Sciences Laboratory to develop suitable instruments and techniques for measuring directly the mechanical properties of lunar soils. Developments in soil mechanics in the last 30 years have shown that penetrometer tests are proper techniques for quick determination of the sub-surface soil properties that influence the bearing capacity and settlement of foundations. The cone penetration resistance has also been used extensively as a sensitive and convenient measure of soil strength in mobility research (WES, 1945). Dimensional analysis has been introduced to derive "mobility performance numbers" (Freitag, 1965; Powell and Green, 1965) which lend themselves to the establishment of rational criteria for the design and evaluation of lunar roving vehicles based on cone penetration resistance measurements performed on the lunar surface.

The cone penetration resistance has been shown to be not only a sensitive indicator of the in-situ packing characteristics of cohesionless soils, such as their dry bulk density, void ratio and relative density (Terzaghi and Peck, 1948), but also of the type of cohesionless soil being penetrated. Also, the mean grain size, the uniformity and the compactibility of cohesionless soils have been found to have considerable influence on the penetration resistance versus relative density relations for such materials. Extensive reviews of recent research in this field are given by Sanglerat (1967); Schultze and Muhs (1967); Vesic (1967); Melzer (1968); Hvorslev (1970); and Melzer (1971).

In this paper, emphasis is placed upon the cone penetration resistance test, and the design characteristics of a portable, self-recording penetrometer, developed by the MSFC Space Sciences Laboratory for use during the Apollo missions, are discussed. Comparison is made between the physical characteristics and mechanical properties of the soils tested and the actual lunar soil mechanical behavior observed during the Apollo 11 and Apollo 12 EVA's and the preliminary examination of returned lunar samples at the Lunar Receiving Laboratory (LRL). From this comparison, inferences are drawn regarding the in-situ density and shear-strength characteristics of lunar soil encountered at the two Apollo landing sites.

Apollo Self-Recording Penetrometer

A cone penetrometer design concept that meets the operational constraints of an Apollo manned mission is the device shown in Fig. 1, developed at the MSFC Space Sciences Laboratory. The instrument attaches to the Apollo Lunar Hand Tool (ALHT) Extension Handle to conserve weight and space. As shown in Fig. 1, it

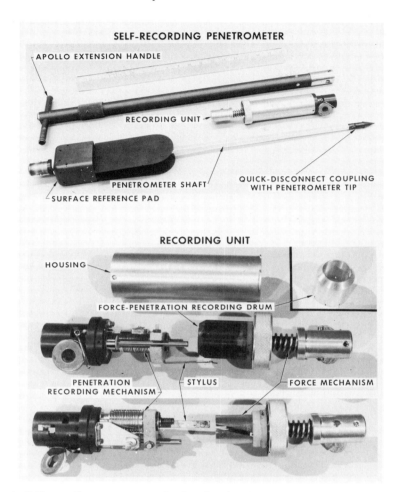

Fig. 1. Self-recording penetrometer developed at the MSFC Space Sciences Laboratory for use on the lunar surface during Apollo missions. Overall length, excluding ALHT extension handle: 114 cm (45 in.). Weight: 8.9 N (2 lbs). Maximum penetration: 76.2 cm (30 in.). Maximum force applied by spring: 111 N (25 lbs), compatible with limiting capabilities of suited astronaut in lunar gravity. Weight of recording drum: 0.28 N (1 oz).

consists of three assemblies: (1) a recording unit; (2) a penetrometer shaft with replaceable penetrating elements (tips); and (3) a surface reference pad.

Through the cable mechanism shown in Fig. 1, the relative motion between the surface reference pad and the penetrometer shaft actuates the stylus to move axially along the recording drum. A sinkage of 76.2 cm (30 in.) produces a 2.5 cm (1 in.) record. Forces are applied through the extension handle and deflect the calibrated coil spring. Axial deflection of the spring is converted to rotation of the recording drum in an amount proportional to the applied force. The maximum force that can be applied through the spring causes a 30° rotation of the drum. The rotation of the drum and the axial deflection of the stylus produce a unique force-penetration diagram inscribed on the recording drum. According to this concept, only the data drums, each containing 12 or more individual force-penetration diagrams, would be returned to Earth.

This basic design concept has been endorsed by the soil mechanics team and presently is being brought to flight status by the Manned Spacecraft Center for inclusion on Apollo 15 and subsequent missions.

Cone Penetration Resistance Tests on Simulated Lunar Soils

Physical properties of lunar soil simulants

In order to assess the type of scientific information that can be derived from penetration resistance data obtained from the lunar surface by a device such as the cone penetrometer described in the previous section, an extensive theoretical and experimental program has been carried out at the MSFC Space Sciences Laboratory using simulated lunar soils at various gradations and consistencies. The basic material selected for these studies has been a crushed basalt from Napa, California, having angular grains and a specific gravity of 2.89. The various gradations at which this lunar soil simulant (LSS) was mixed are shown in Fig. 2 along with grain-size distribution curves from lunar soil samples collected during the Apollo 11 and 12 missions.

In addition to the lunar soil simulation studies with the LSS, lunar roving vehicle mobility performance simulation studies prior to the Apollo 11 landing were conducted on a wind-deposited sand from the desert near Yuma, Arizona (FREITAG *et al.*, 1970).

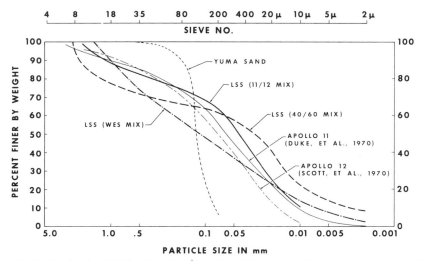

Fig. 2. Grain-size distribution characteristics of lunar soil simulants compared with gradation curves of Apollo 11 and Apollo 12 lunar soil samples.

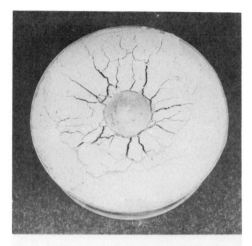

a. APOLLO 11 LUNAR SOIL BULK SAMPLE

b. LSS (40-60 MIX)

Fig. 3. Comparative analysis of penetration resistance tests on Apollo 11 lunar soil bulk sample (COSTES *et al.*, 1970) and LSS, both placed at an average void ratio of 0.7. Note mode of deformation of both samples at failure. Triaxial compression tests performed on LSS at the same void ratio verified the assumption by COSTES *et al.* (1970) that bearing capacity theory can be used with soil bulk density and penetration resistance data to determine probable range of shear strength parameters of soil tested.

The gradation of this soil is shown also in Fig. 2. In this paper, reference to this material is made only for the purpose of testing the validity of methods used in analyzing soil penetration resistance data.

The mechanical behavior of the LSS has been studied extensively under terrestrial gravity conditions with basic test variables: the soil grain-size distribution, dry bulk density, moisture content, confining pressure and testing techniques. For bulk densities ranging between 1.50 g/cm³ and 1.80 g/cm³, the apparent cohesion and angle of internal friction of LSS (40/60 mix), placed at water contents ranging between 0.4% and 2.0%, bracket the values of the shear strength characteristics of the actual lunar soil (cohesion: of the order of 7×10^{-2} N/cm²; friction angle: 35 deg to 39 deg in the normal pressure range of a few N/cm²) as deduced from soil mechanics data obtained during the Surveyor missions (SCOTT and ROBERSON, 1968).

The LSS (40/60 mix) was used in simulation studies prior to the Apollo 11 mission and has since been used extensively in predictions and comparative analyses, such as the one shown in Fig. 3, relating to the mechanical behavior of actual lunar soil encountered during the Apollo 11 and Apollo 12 missions (COSTES *et al.*, 1970; SCOTT *et al.*, 1970). Since the Apollo 12 mission, the gradation of the LSS has been modified as shown by the curve "LSS (11/12 mix)" in Fig. 2 to match more closely the grain-size distribution of returned lunar soil samples. The grain-size distribution of the lunar soil simulant used in lunar roving vehicle mobility performance studies (GREEN and MELZER, 1970) is shown in Fig. 2 by the curve "LSS (WES mix)."

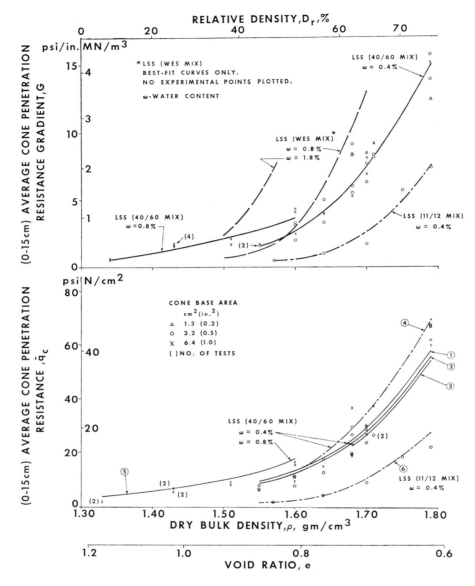

Fig. 4. Cone penetration resistance tests on LSS under terrestrial gravity conditions. (Top) (0–15 cm) average cone penetration resistance gradient, G, vs. relative density, dry bulk density and void ratio. (Bottom) Comparison of (0–15 cm) average cone penetration resistance, $\bar{q}_c$, data from same tests with theoretical curves computed from bearing capacity theory (MEYERHOF, 1961). Curves ①, ②, ③, and ⑤ have been computed, respectively, for cone base areas of 6.4 cm² (1.0 in.²), 3.2 cm²(0.5 in.²), 1.3 cm² (0.2 in.²), and 6.4 cm² (1.0 in.²) according to method II. Curves ④ and ⑥ have been computed for a cone base area of 3.2 cm² (0.5 in.²) according to method I.

Cone penetration resistance tests under 1 g *conditions*

The results of cone penetration tests performed by standard equipment and according to standard procedures (Department of the Army, 1959) on LSS (40/60 mix) (SMITH, 1969), on LSS (WES mix) (GREEN and MELZER, 1970), and on LSS (11/12 mix) have been plotted in Fig. 4. As shown in Fig. 4, both the average cone penetration resistance, $\bar{q}_c$, and the average cone penetration resistance gradient, G, over a depth of 0 to 15 cm, of the LSS depend upon its grain-size distribution and its consistency and apparent cohesion (as indicated by its water content at a given void ratio). They are also strong functions of the packing characteristics of the material as expressed by its dry bulk density, void ratio, and relative density.

Analysis of the same data has also indicated that bearing capacity theory relating to foundations with perfectly rough conical tips (MEYERHOF, 1961) can be used to determine the in-place shear-strength characteristics of the soil, if its dry bulk density is known or can be estimated from G-versus-density or G-versus-void ratio plots.

The ultimate bearing capacity of a conical-shape foundation is given approximately by the Terzaghi equation (TERZAGHI and PECK, 1948), modified by MEYERHOF (1961),

$$q_c = cN_c + \rho g R N_\gamma + \rho g z N_q \tag{1}$$

in which c and ρ are, respectively, the cohesion and the bulk density of the soil, g is the acceleration of gravity, R is the radius of the cone base, z is the depth to the base of the cone, and N_c, N_γ, and N_q are "bearing capacity factors" that depend on the angle of internal friction of the soil, ϕ, the depth-cone base radius ratio, z/R, the roughness of the surface of the cone and on the cone apex angle which was 30 deg in all of the tests reported herein. In this theory, the soil is assumed to behave as an incompressible, rigid-plastic material and to fail in shear along bulb-shaped failure surfaces surrounding the cone. The traces of these surfaces on planes containing the cone axis can be described by linear segments and segments of a logarithmic spiral. Accordingly, the term "ultimate bearing capacity" refers to the applied unit load at which the shear strength of the soil mass is exceeded and the material fails in incompressible shear.

To test the applicability of equation (1) to the analysis of cone penetration resistance data, two methods were used, designated respectively as method I and method II. According to method I, the rate of change of the bearing capacity q_c with respect to depth in equation (1),

$$\frac{dq_c}{dz} = \rho g N_q, \tag{2}$$

was assumed to be equal to the average cone penetration resistance gradient, G. Accordingly, for a given cone penetration test on LSS of known density, N_q was computed from equation (2) and the indicated angle of internal friction, ϕ_{ind}, of the soil was then determined from the N_q versus ϕ plot for shallow, rough cones with a semi-cone angle of 15 deg (MEYERHOF, 1961). Having determined ϕ_{ind}, the bearing capacity factors N_γ and N_c were obtained from the corresponding N_γ and N_c versus

ϕ plots appearing in Meyerhof (1961). Finally, $\bar{q}_c$ was calculated from equation (1) at $z = 7.5$ cm, by entering the value of the cone base radius, R, of the instrument used, and cohesion values, c, obtained from triaxial compression and trenching tests at comparable densities (Costes et al., 1969; Olson, 1969).

According to method II, the bearing capacity factors were determined from the N_c, N_γ, and N_q versus ϕ plots (Meyerhof, 1961) by entering ϕ values obtained from triaxial tests at comparable densities and the rest of the procedure was identical to that according to method I.

Comparison between values of $\bar{q}_c$ calculated from equation (1) according to methods I and II, and experimental data on the average cone penetration resistance of the LSS mixes over a depth of 0 to 15 cm can be made in Fig. 4b. It can be seen that reasonably good agreement exists between computed and experimental data, indicating that equation (1) can satisfactorily describe the dependence of the average cone penetration resistance of the LSS on its grain-size distribution, dry bulk density, void ratio, relative density and water content which in turn govern its shear strength characteristics. Also, reasonably good agreement is indicated regarding the dependence of $\bar{q}_c$ on the penetrometer cone size.

To test the validity of equation (1) in expressing the mechanical behavior of another material of different mineral composition, grain size and shape characteristics, and different gradation, a similar analysis was performed on data from cone penetration resistance tests conducted on Yuma Sand (Figs. 2 and 5) by Melzer (1971) and other investigators (Schultze and Muhs, 1967; Turnage, 1971).

The results of this analysis shown in Fig. 5 again indicate very good agreement between the theoretical and the experimental results. The "theoretical curve" in Fig. 5 was calculated from equation (1) according to method I for a standard cone base area of 3.2 cm² (0.5 in.²) used in these tests.

Cone penetration resistance tests under varying gravity levels

The dependence of both $\bar{q}_c$ and G on the gravity level, as indicated by equations (1) and (2), was checked by performing a series of cone penetration tests on LSS (11/12 mix) and Yuma Sand under varying gravity conditions, simulated onboard the U.S. Air Force KC-135 aircraft during $\frac{1}{6}$-g; 1-g; and 2-g parabolic flight trajectories. During these tests, vertical accelerometers located at the center of gravity of the aircraft were monitored by the pilot who controlled the $\frac{1}{6}$-g lunar gravity level to $\pm 0.01\,g$. These conditions could be produced for about 30 seconds during each "parabola," giving adequate time to perform a cone penetration test at the standard rate of 3 cm/sec (1.2 in./sec). The 2 g level during entry was controlled to within $\pm 0.5\,g$. However, at recovery, the g level varied between parabolas from 1.5 g to 2.5 g.

Both soil samples were placed in containers 35.6 cm by 35.6 cm in cross section and 43.2 cm deep, by sprinkling each soil from a drop height of 5 cm (2 in.). During the tests, the dry bulk density of the soils was determined from settlement measurements recorded in flight. At the top of each box, there were nine access holes (Fig. 6), spaced at 10.2 cm centers. The minimum distance between the center of the holes

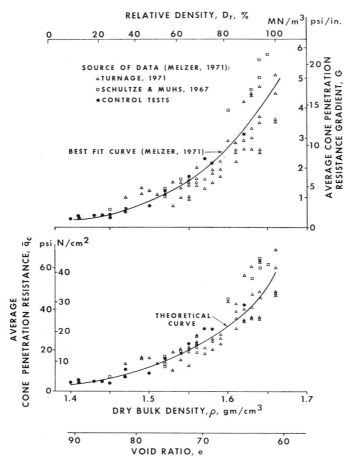

Fig. 5. Cone penetration resistance tests on Yuma Sand under terrestrial gravity conditions. (Top) (0–15 cm) average cone penetration resistance gradient, G, vs. relative density, dry bulk density, and void ratio. (Bottom) Comparison between (0–15 cm) average cone penetration resistance, $\bar{q}_c$, data from same tests with theoretical curve computed from bearing capacity theory (MEYERHOF, 1961) according to method I.

and the container wall was 7.6 cm. To ascertain the influence of the boundary wall and adjacent holes on the cone penetration resistance test results, for each g level there was at least one test performed at an edge hole and one at a corner hole (Fig. 6).

The apparatus for performing the cone penetration resistance tests consisted of a cone penetrometer with a standard 30° cone of 3.2 cm² (0.5 in.²) base area and an associated electronic system for sensing and recording the resistance to penetration of the soil at a given penetration. The overall system from force sensor to digitized force data is accurate to ±4.1%. Records of depth pulses at 2.5-cm (1-in.) increments are accurate to approximately 0.1%. The results from these tests, shown in Fig. 6, indicate that inasmuch as the tests at each g level were performed during successive

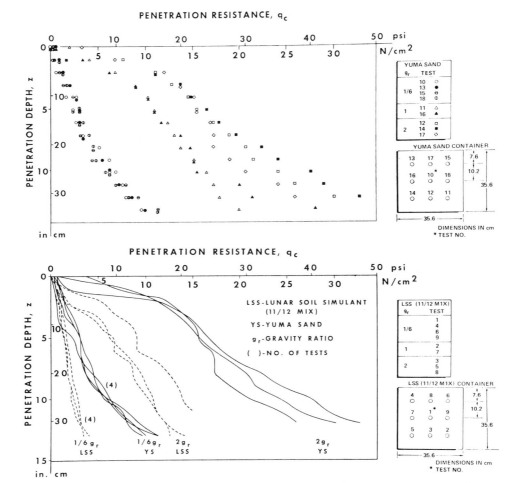

Fig. 6. Cone penetration resistance, q_c, versus penetration from tests on Yuma Sand and LSS (11/12 mix) performed under varying gravity conditions. Note effect of soil type on q_c.

parabolic flights, possible disturbances that could have occurred in either soil due to the in-flight conditions did not affect materially the test results. From the same plot, it appears also that the cone penetration resistance and its average rate of change with depth for both soils decrease with decreasing gravity level.

To test the linear dependence of the bearing capacity equation (1) on the gravity ratio g_r, $\bar{q}_c$ was calculated for both soils as a function of g_r using equation (1) according to method I. The G values employed in these calculations were obtained from the G versus g_r linear diagrams shown in Fig. 7, passing through the origin and the experimental points at $g_r = 1$. The calculated $\bar{q}_c$ values are shown by the solid lines plotted in Fig. 8. These plots indicate close agreement between the theoretical and experimental values of $\bar{q}_c$ at g_r levels of $\frac{1}{6}$ and 1. Furthermore, the values of G and $\bar{q}_c$ for both soils

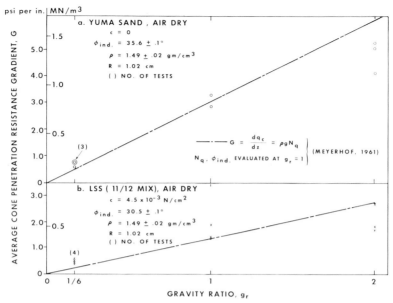

Fig. 7. Cone penetration resistance tests on Yuma Sand and LSS under varying gravity conditions. Average cone penetration resistance gradient, G, over 0–15 cm depth vs. gravity ratio, g_r.

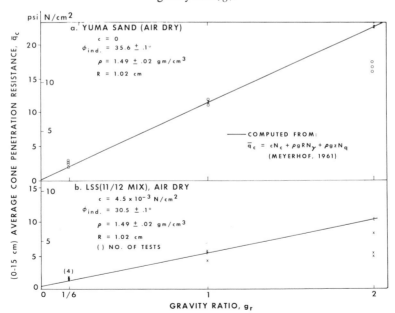

Fig. 8. Cone penetration resistance tests on Yuma Sand and LSS under varying gravity conditions. Comparison between $\bar{q}_c$ experimental data and theoretical values calculated from bearing capacity theory (MEYERHOF, 1961), according to method I at $g_r = 1$.

at $g_r = 1$ are in close agreement with results of cone penetration tests performed on both soils at comparable densities under terrestrial gravity conditions (Figs. 4 and 5). The apparent discrepancies from linearity in the G and $\bar{q}_c$ values for both soils at gravity ratio levels of "2 g" are tentatively attributed to uncertainties regarding the exact g level achieved by the aircraft during the 2 g maneuvers.

EVALUATION OF LUNAR SOIL PENETRATION DATA
FROM APOLLO 11 AND APOLLO 12

The analysis of the cone penetration tests reported herein indicates that penetration resistance profiles in soils similar to those found on the lunar surface provide sensitive indicators of the variation of the in-place dry bulk density of the soil with depth and can be analyzed by bearing capacity theory to estimate the in-situ shear strength characteristics of the material. During the Apollo 11 and Apollo 12 missions, various shafts, staffs, and tubes were inserted in the lunar surface (Costes et al., 1970; Costes and Mitchell, 1970; and Scott et al., 1970). Although the circumstances under which these penetrations were performed, the type of apparatus used and the uncertainties in estimating the forces exerted by the astronauts in pushing these shafts into the lunar surface do not permit a rigorous analysis of these penetrations, the data can be used in conjunction with the penetration resistance tests on the LSS reported herein to make a first-order evaluation of the in-place density and the shear strength characteristics of the lunar soil in the upper few centimeters of the two Apollo sites.

From these considerations, upper bounds of the lunar soil in-place bulk density and angle of internal friction were obtained (Table 1) by analyzing the Apollo 11 and Apollo 12 soil penetration data according to the methods developed in this paper and making the following assumptions: (1) the hollow shafts used as penetrators became plugged at a shallow depth and subsequently behaved as flat-ended penetrators; (2) the maximum downward force exerted by a suited astronaut in pushing these penetrators is of the order of 90 N (Costes and Mitchell, 1970; Werner, 1970); (3) the frictional and adhesive forces mobilized along the penetrator shafts were negligible as compared to the resistance of the soil to penetration at the "flat-ended" penetrator tips; and (4) the term cN_c is very small as compared to the sum of the other two terms on the right-hand side of equation (1) and may be neglected.

As shown in Table 1, assuming that at a given value of G the void ratio of the lunar soil (specific gravity of solids: 3.1) is the same as the corresponding void ratio of the LSS (specific gravity of solids: 2.89), upper bounds of the in-place lunar soil bulk density at the Apollo 11 and Apollo 12 sites should be of the order of 1.85 g/cm³ and 1.81 g/cm³, respectively. Lunar-soil bulk density data (1.74 g/cm³, 1.96 g/cm³, and 1.98 g/cm³) obtained from Apollo 12 core tube samples (Scott et al., 1970) tend to corroborate these indicated density values. On the other hand, possible disturbances in the soil during sampling, caused by the design of the two Apollo 11 core tubes, may have resulted in underestimating the in-place bulk density of the lunar soil at the Apollo 11 site, as determined from the corresponding core tube sample bulk density measurements (1.59 g/cm³ and 1.71 g/cm³) (Costes et al., 1970; Costes and Mitchell, 1970; Scott et al., 1970).

Table 1. Evaluation of in-place soil mechanical properties at Apollo 11 and Apollo 12 sites from in-situ soil penetration data

	(1) R cm	(2) z cm	(3) q_z N/cm²	(4) $G_L{}^*$ N/cm³	(5) $\rho_L{}^*$ g/cm³	(6) $\phi_L{}^*$ degrees	(7) G_L N/cm³	(8) ϕ_L degrees	(9) c_L $\times 10^{-2}$ N/cm²
				Apollo 11					
Contingency Sampler Holder	0.80	20	45	2.2	1.92	48			44†
Flagpole	1.11	18	23	1.3	1.86	45			21†
Core Tube	1.66	7.5–13	10	1.3–0.75	1.86–1.82	45–43	0.18†	36.8†	9.2–8.1†
Solar Wind Composition Experiment Staff	1.70	14	9.9	0.66	1.81	42			7.5†
				Apollo 12					
Core Tube	1.66	10–18	10	0.96–0.55	1.84–1.80	44–42	0.24‡	37.7‡	7.5–5.6‡
Solar Wind Composition Experiment Staff	1.70	15	9.9	0.62	1.80	42			5.9‡

(1) Penetrator radius.

(2) Penetration depth.

(3) Average resistance to penetration of lunar soil at penetrator tip to an applied force of the order of 90 N.

(4) Average penetration resistance gradient of lunar soil under lunar gravity conditions, estimated from equation (1) under the assumption that the term cN_c can be neglected.

(5) Upper-bound estimate of in-place lunar soil bulk density, determined from the G_E vs. ρ plot for LSS, assuming that $G_E{}^* = 6\ G_L{}^*$ and that at a given G value the void ratio of the LSS is the same as that of the lunar soil. G_E = penetration resistance gradient of lunar soil under terrestrial gravity conditions.

(6) Upper-bound estimate of in-place angle of internal friction of lunar soil, based on $G_L{}^*$ and $\rho_L{}^*$ and determined according to method I.

(7) Average in-place penetration resistance gradient of lunar soil, estimated from equation (1) on the basis of measured value of ρ_L.

(8) In-place angle of internal friction of lunar soil, estimated from equation (1) on the basis of measured value of ρ_L.

(9) In-place cohesion of lunar soil, estimated from equation (1) on the basis of measured value of ρ_L.

† Estimated from equation (1) on the basis of a measured average bulk-density value of 1.71 g/cm³, obtained from Apollo 11 core-tube sample 10005 (COSTES and MITCHELL, 1970).

‡ Estimated from equation (1) on the basis of a measured average bulk-density value of 1.74 g/cm³, obtained from Apollo 12 core-tube sample (Core Tube Serial No. 2011) (SCOTT et al., 1970).

On the basis of triaxial compression tests performed on LSS (COSTES et al., 1969; MITCHELL et al., 1970), the indicated upper-bound values for the in-place angle of internal friction of the lunar soil at the two Apollo sites shown in Table 1 are compatible with the corresponding upper-bound estimates of the in-place lunar soil bulk density.

Finally, by solving equation (1) for c on the basis of the measured bulk density values of 1.71 g/cm³ and 1.74 g/cm³, the angle of internal friction of the lunar soil at the corresponding Apollo 11 and Apollo 12 core tube sites is estimated to be 36.8 deg and 37.7 deg, and its cohesion to be of the order of 9×10^{-2} N/cm² and 7×10^{-2} N/cm², respectively. Based on the same density measurements and penetration data from the locations shown in Table 1, the lunar soil cohesion is estimated to vary between 7.5×10^{-2} N/cm² and 4.4×10^{-1} N/cm² at the Apollo 11 site and between 5.6×10^{-2} N/cm² and 7.5×10^{-2} N/cm² at the Apollo 12 sites. These estimates are in agreement with independent observations and measurements associated with the analysis of lunar soil mechanics data obtained from the Surveyor and Apollo 11 missions (Scott and Roberson, 1968; Costes *et al.*, 1970).

Conclusions

(1) Studies on the mechanical behavior of simulated lunar soils have assisted in gaining a better understanding of the actual lunar soil observed during the Apollo 11 and Apollo 12 EVA's. (2) The cone penetration resistance of soils similar to the actual lunar soil is a sensitive indicator of the grain-size distribution and in-place packing characteristics and consistency of the soil. Therefore, it can furnish corroborative information relative to the in-situ physical properties and stratification of lunar soil that can prove to be very useful to other scientific disciplines. (3) Through bearing capacity theory, results of cone penetration resistance tests on lunar soil simulants can be used in conjunction with soil bulk density data to determine the in-place shear strength characteristics of such soils. (4) Estimates of the in-situ bulk density of the lunar soil in the upper few centimeters at the Apollo 11 and Apollo 12 sites, based on cone penetration resistance tests on lunar soil simulants and crude soil penetration resistance data from the two missions, are in agreement with lunar soil bulk density measurements from Apollo 12 core tube samples. Estimates of the in-place cohesion and angle of internal friction of the lunar soil at the same locations, based on the same data, are in agreement with previous estimates based on independent measurements and observations. (5) The mechanical properties of the lunar soil at the Apollo 11 and Apollo 12 sites, as evaluated from this analysis, appear to be of the same order of magnitude as those of the soils encountered at the Surveyor equatorial landing sites.

Acknowledgements—The assistance offered in this research program by: the Simulation Branch, Astronautics Laboratory, NASA/MSFC; Zero-g Test Office, Wright-Patterson Air Force Base; and the Astronaut Office and Flight Crew Support Division, NASA/MSC is greatly appreciated.

The senior author wishes to express his sincere appreciation to all of his former and current colleagues at the MSFC Space Sciences Laboratory for their excellent cooperation and enthusiasm in contributing to various facets of these investigations. Finally, many thanks go to Dr. K. J. Melzer of the U.S. Army Engineer Waterways Experiment Station, Vicksburg, Mississippi for his generous permission to analyze cone penetration data collected by him and to be reported in a forthcoming publication.

References

Costes N. C. *et al.* (1969) Lunar soil simulation studies in support of Apollo 11 mission. NASA Technical Memo, Geotechnical Research Laboratory, Space Sciences Laboratory, Marshall Space Flight Center, Huntsville, Alabama.

Costes N. C., Carrier W. D., Mitchell J. K., and Scott R. F. (1970) Apollo 11: Soil mechanics results. *ASCE J. Soil Mech. Foundations Div.* **7704**, 2045–2080.

Costes N. C. and Mitchell J. K. (1970) Apollo 11 soil mechanics investigation. *Proc. Apollo 11 Lunar Sci. Conf., Geochim. Cosmochim. Acta* Suppl. 1, Vol. 2, pp. 2025–2044. Pergamon.

Freitag D. R. (1965) A dimensional analysis of the performance of pneumatic tires on soft soils. Technical Report 3-688, U.S. Army Engineering Waterways Experiment Station, CE (USAE WES), Vicksburg, Mississippi.

FREITAG D. R., GREEN A. J., and MELZER K. J. (1970) Performance evaluation of wheels for lunar vehicles. Technical Report M-70-2, USAE WES, Vicksburg, Mississippi.

DEPARTMENT OF THE ARMY (1959) Soils trafficability. Technical Bulletin TB ENG 37, Headquarters, Department of the Army, Washington, D.C.

DUKE M. B., WOO C. C., SELLERS G. A., BIRD M. L., and FINKELMAN R. B. (1970) Genesis of lunar soil at Tranquillity Base. *Proc. Apollo 11 Lunar Sci. Conf., Geochim. Cosmochim. Acta* Suppl. 1, Vol. 1, pp. 347–361. Pergamon.

GREEN A. J. and MELZER K. J. (1970) Performance of Boeing-GM wheels in a lunar soil simulant (basalt). Technical Report M-70-15, USAE WES, Vicksburg, Mississippi.

HVORSLEV M. J. (1970) The basic sinkage equations and bearing capacity theories. Technical Report M-70-1, USAE WES, Vicksburg, Mississippi.

MELZER K. J. (1968) Sondenuntersuchungen im Sand (Penetrometer investigations in sand), Dissertation, Technical University of Aachen. Published in Mitteilungen Institut fur Verkehrswasserbau, Grundbau und Bodenmechanik 43, Technical University of Aachen, Germany.

MELZER K. J. (1971) Measuring soil properties in vehicle mobility research. Report 4, Relative density and cone penetration resistance. Technical Report 3-652, U.S. Army Materiel Command.

MEYERHOF G. G. (1961) The ultimate bearing capacity on wedge-shaped foundations. *Proc. Fifth Int. Conf. Soil Mech. Foundation Eng.*, Vol. 2, pp. 105–109, Dunod, Paris.

MITCHELL J. K., CARMICHAEL I. S. E., GOODMAN R. E., FRISCH J., WITHERSPOON P. A., and HEUZE F. E. (1969) Materials studies related to lunar surface explorations. Final Report on NASA Contract NSR 05-003-189 submitted to George C. Marshall Space Flight Center by Space Sciences Laboratory, University of California, Berkeley.

MITCHELL J. K., HOUSTON W. N., GOODMAN R. E., WITHERSPOON P. A., VINSON T. S., DURGUNOGLU T., NAMIQ L. I., THOMPSON J. B., and THREADWELL D. (1970a, b) Lunar surface properties experiment definition. Final Report on NASA Contract NAS 8-21432 submitted to George C. Marshall Space Flight Center by Space Sciences Laboratory, University of California, Berkeley.

POWELL C. J. and GREEN A. J. (1965) Performance of soils under tire loads; analysis of tests in Yuma Sand through August 1965. Technical Report 3-666, Report 2, USAE WES, Vicksburg, Mississippi.

SANGLERAT G. (1967) El penetrometro y el reconocimiento des los suelos (Penetrometers and subsoil exploration). Servico de Publications, Ministerio de Orbas Publicas, Madrid.

SCHULTZE E. and MUHS H. (1967) *Bodenuntersuchungen für Ingenieurbauten (Soil Investigation for Engineering Structures)*. Second ed., Springer-Verlag, Berlin.

SCOTT R. F. and ROBERSON R. I. (1968) Soil mechanics surface sampler. Chapt. V, pp. 135–185. Surveyor Project Final Report, Part II: Science Results, Jet Propulsion Laboratory, California Institute of Technology, TR32-1265.

SCOTT R. F., CARRIER W. D., COSTES N. C., and MITCHELL J. K. (1970) Mechanical properties of the lunar regolith. Apollo 12 Preliminary Science Report, NASA SP-235, pp. 161–188.

SMITH R. E. (1969) Evaluation of the physical properties of lunar soils. ASEE/NASA Faculty Fellowship Program.

TERZAGHI K. and PECK R. B. (1948) *Soil Mechanics in Engineering Practice*. John Wiley.

TURNAGE G. W. (1971) Performance of soils under track loads; model track and test program. Technical Report (in preparation), USAE WES, Vicksburg, Mississippi.

VESIC A. S. (1967) Bearing capacity and settlement of foundations. Proc. of a Symposium held at Duke University, April 5–6, 1965, Durham, North Carolina.

WERNER R. A. (1970) Human vertical force exertion capability at $\frac{1}{6}g$. Lockheed Electronics Company Technical Working Paper LEC/HASD No. 645 D. 21.052, NASA Contract NAS 9-5191, NASA Manned Spacecraft Center, Houston, Texas.

WES (1945) Trafficability of soils. Technical Report (unnumbered), USAE WES, Vicksburg, Mississippi.

Proceedings of the Second Lunar Science Conference, Vol. 3, pp. 1989–2001
The M.I.T. Press, 1971.

Particle size and shape distribution for lunar fines sample 12057,72

H. Heywood

Chemical Engineering Department, University of Technology, Loughborough, England

(Received 10 March 1971; accepted in revised form 12 April 1971)

Abstract—The particle size distribution for the sample 12057,72 was obtained by sieving on woven-wire sieves and electroformed micromesh sieves, combined with sedimentation, Coulter counter, optical, and electron microscope examination. The graph relating percentage by weight undersize with corresponding particle size derived from these measurements had the following characteristics:

Median diameter (50%)	86 μm by microscope	80 μm by square sieve aperture
Mean diameter	153 μm by microscope	123 μm by square sieve aperture
Modal diameter	13 μm by microscope	10 μm by square sieve aperture

Measurements of the distribution of characteristic shapes of particles, such as, scoriaceous glassy agglomerates, relatively smooth opaque particles, and transparent crystalline particles showed that the proportion of glassy agglomerates decreased from about 40% for 800 μm particles to 10% at 50 μm, whilst the proportion of crystalline particles increased from 30% to 70% over the same size range. This change was most rapid at about 100 μm. The proportion of smooth opaque particles varied only slightly. The shape factor relating particle volume and diameter as measured by the microscope was fairly constant over the whole size range, with a mean value of 0.32, but exhibiting a slight maximum of 0.37 at about 120 μm. This effect may be due to the change in composition noted above.

These measurements, when compared with similar determinations on other samples should advance an understanding of the evolution of the lunar fines.

Introduction

This report describes the procedures used to determine the particle size distribution on a sample of lunar fines returned by the Apollo 12 mission. An additional feature of the research was to study the shape of the particles, especially as regards any variation of shape with particle size. The 1 g sample as received had been passed through a 1 mm sieve, though the largest particle present had a size of 870 μm, and it was intended to measure the size distribution to a lower size limit of 0.1 μm. Various procedures for size analysis must be used to cover such a wide range of sizes and since these measure different characteristic sizes according to the particle property which is being sensed, precise definitions of particle size are essential. These definitions are given in the section on notation which follows:

Particle Size Analyses

Notation

Microscopical measurement

d_a = projected area diameter; the diameter of a circle which encloses the same area as the profile of the particle.

d_p = projected perimeter diameter; the diameter of a circle which has the same perimeter as the profile of the particle.

Sieving and microplates

A = size of square aperture through which the particle will just pass.
H = size of circular aperture through which the particle will just pass.

Sedimentation analysis

d_{st} = equivalent Stokes diameter; the diameter of the sphere with the same density as the particle which has the same terminal velocity of fall in a fluid.

Coulter counter measurement

d_v = equivalent volume diameter; the diameter of a sphere which has the same volume as the particle.

The relative magnitude of these dimensions varies with the shape of the particles, but from previous experience of many types of particles, the numerical relationships between them for the lunar dust particles have been determined as follows:

$$d_a/A = 1.24 \qquad H/A = 1.12$$
$$d_a/d_v = 1.28 \qquad d_a/d_{st} = 1.32$$

Procedures for size analysis

The following methods of analysis were used. Woven wire sieves (stainless steel) from 14 mesh (1200 μm) to 150 mesh (103 μm); micromesh sieves with square apertures from 90 μm to 75 μm; microplates with circular apertures from 75 μm to 20 μm; sedimentation in water to obtain fractions from 30 μm to 2 μm; Coulter counter analyses from 75 μm to 20 μm and below 20 μm; electron microscope measurement on particles smaller than 10 μm; optical microscope measurement of the particle size distribution on all graded fractions obtained by the above methods; scanning electron microscope photographs of selected particles.

The data from these measurements is given in Table 1 and the results are plotted in Fig. 1. The curves for the different methods of measurement are not continuous, for the reasons explained in the introduction and are therefore converted to a common basis of projected area diameter by means of the factors given in the section on notation. The projected area diameter has been chosen as a standard of reference since it is the only dimension that can be measured for all sizes of particles ranging from hand-sized lumps of rock to particles that are only visible through the electron microscope. The magnitude of the projected area diameter is greater than other equivalent diameters as it is a function of the two major dimensions of the particle. The shape of the curve in Fig. 1 indicates that the particle size distribution approximates to the log normal type of distribution and this is confirmed by plotting to probability log size scales as shown in Fig. 2.

Since sieving was performed in the dry condition down to 75 μm, the fine material is inevitably underestimated in this analysis, because of some adherence of these particles to the larger ones. Wet sieving was found to be necessary below 75 μm, and in future analyses it is intended to use wet sieving throughout, also to use plates with circular apertures from the upper size limit downwards. As only a single sample has been examined, definite conclusions cannot yet be deduced, but it is noticable that the lunar dust is deficient in fines below 10 μm in comparison with a number of terrestial desert dusts which have been examined. This conclusion was also deduced by DUKE et al. (1970).

Relationship between mean particle weight and size

The number of particles per milligram was determined for each sieved fraction of the sample to a lower limit of 30 μm. Direct counting of up to 3000 particles and

Table 1. Size analysis data

Size fraction limits in μm	Fraction wt. in mg/g sample	Cumulative under-size % smaller than upper size limit	Corresponding mean projected diameter in μm
Woven sieves and square aperture micromesh (A)			
1200–860	1	100.0	1490
860–605	17	99.9	1065
605–430	33	98.2	750
430–305	50	94.9	533
305–211	79	89.9	378
211–155	70	82.0	262
155–103	110	75.0	192
103–92	58	64.0	128
92–75	64	58.2	114
Circular aperture microplates (H)			
75 sq–75	36	51.8	93
75–60	86	48.2	83
60–50	62	39.6	66
50–40	86	33.4	55
40–30	74	24.8	44
30–20	78	17.4	33
20–0	96	9.6	22
Sedimentation analysis (d_{st})			
30–20	112	27.1	40
20–10	88	15.9	26.5
10–5	41	7.1	13
5–2	26	3.0	8
Coulter counter analysis (d_v)			
75–60	47	48.2	83
60–50	46	43.5	77
50–40	55	38.9	64
40–30	73	33.4	51
30–25	40	26.4	38
25–20	37	22.1	32
20–15	25	18.4	26
15–10	65	15.9	19
10–7	47	9.4	13
7–5	25	4.7	9
5–4	10	2.2	6.5
4–3	7	1.2	5
3–2	4	0.5	4
2–0	1	0.1	2.5

weighing on a microbalance was the procedure for sizes greater than 90 μm, below this size a wax dispersion technique described in the Appendix to this paper was adopted. The data is given in Table 2. If all particles had the same shape and density, then the weight would be proportional to the cube of the size or the ratio cube root weight to size would be constant. In fact, the data in Fig. 3 indicates some variation, but a small correction is needed for the varying apparent density of the different sized fractions.

Size distribution on a basis of number of particles

Although the size distribution of industrial powders is usually expressed on a weight basis, a number basis has been used by some other research workers to characterize lunar fines and counts have been determined from photographs obtained by the Surveyor Missions, as described by SHOEMAKER *et al.* (1970). It is therefore necessary

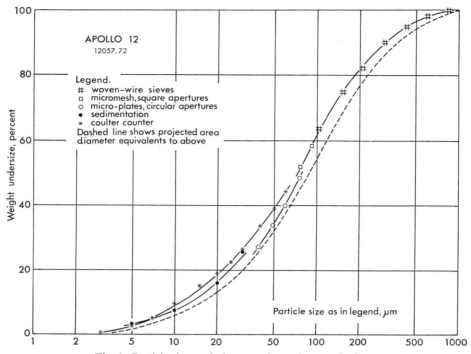

Fig. 1. Particle size-analysis curves by various methods.

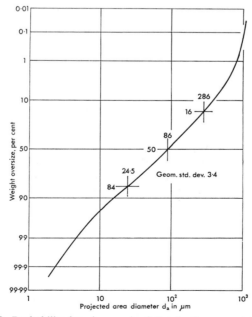

Fig. 2. Probability-log size graph of distribution on weight basis.

Table 2. Mean weight of particles and volume coefficient, $\alpha_{v,a}$; $\rho = 2.9$ (measured value) (see section on "Derivation of Shape Coefficients" for definitions).

Size fraction limits in μm	Mean sieve square aperture in μm	No. of particles per mg	Mean weight in μg	$\alpha_{v,a}$
860–605	733	2	523	0.24
605–430	518	5	193	0.25
430–305	368	13	75.6	0.28
305–211	258	36	27.8	0.34
211–155	183	91	11.0	0.32
155–103	129	228	4.40	0.37
103–92	97	546	1.83	0.36
92–75	83	1142	0.875	0.28
75–67	71	1827	0.548	0.34
67–54	60	2865	0.350	0.30

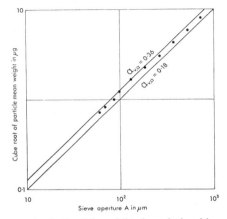

Fig. 3. Particle weight-size relationship.

to determine the mathematical relationship between these two methods of interpretation. Both these properties, namely, weight and number distribution, may be expressed either by a cumulative graph which relates the weight or number of particles greater, or alternatively less, than a specified particle size or by the frequency distribution graph which shows the percentage weight or the number of particles per unit interval of particle size. This latter graph is the differential of the corresponding cumulative graph.

As an illustration, the weight frequency distribution graphs have been plotted to arithmetic and logarithmic scales in Fig. 4. In curve (a), both scales are arithmetic; this curve shows the general characteristics of the particulate system and that the modal size is relatively small compared with the maximum particle size. The scale at the relatively small particle sizes is extended by plotting the particle size to a logarithmic basis as in (b). This enables the shape of the distribution curve in the modal region to be studied more exactly. Finally, by plotting both scales logarithmically curve (c) is obtained, and reference to this will be made subsequently.

It has been observed that the curve for the cumulative number of particles oversize approximates to a straight line over a certain range of particle sizes. An investigation was made to ascertain whether this straight line should continue into the region

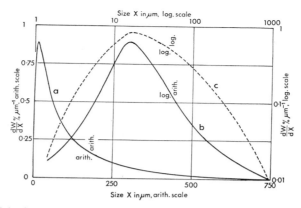

Fig. 4. Weight frequency distribution curves to various scales (see section *Size distribution on a basis of number of paricles*).

of very small sizes. In the first instance, it was proved mathematically that if the cumulative weight curve plotted to probability scale against log size, as in Fig. 2 follows the straight line relationship of a log normal distribution, then the cumulative number against size (log scales) is represented by a curve with decreasing slope as particle size tends to zero. It is incompatible therefore, that both these methods of plotting size distribution should be represented by straight lines.

The number of particles per milligram in each graded fraction of the lunar fines was measured as described in the previous section of this report, and hence knowing the weight of each, the number of particles could be calculated; the cumulative numbers were obtained by summations of the fractions from 30 μm to the maximum size. This graph is shown in Fig. 5 and was extended to a lower size limit of 1 μm by calculation from the weight distribution shown by Fig. 4, on the assumption that the particle shape remained constant. The mathematical relationships are shown as follows:

If W_0^x is the cumulative weight undersize below particle size x, the method of defining size not being specified, then dW/dx is the slope of the cumulative weight curve at size x and is the weight frequency at this size, expressed as percent per micrometre or $\% \, \mu m^{-1}$.

$N_x^{x \, \text{max}}$ is the cumulative number oversize at particle size x, so that dN/dx represents the number frequency at this size. Since the weight of a particle is proportional to x^3, then $dW/dx = \rho \alpha_v x^3 \, dN/dx$, α_v is the volume coefficient, later defined, and ρ is particle density. Therefore

$$N_x^{x \, \text{max}} = \int_x^{x \, \text{max}} \frac{dN}{dx} \cdot dx = \frac{1}{\rho \alpha_v} \int_x^{x \, \text{max}} \frac{dW}{dx} \cdot \frac{1}{x^3} \, dx.$$

As the frequency law is not generally known, this integration is performed by summing successive small finite intervals. If the straight line portion of the cumulative number—size relationship is represented by the equation $N_x^{x \, \text{max}} = cx^{-n}$, where $-n$

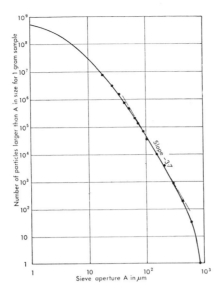

Fig. 5. Graph of cumulative number against particle size, log scales.

is the slope and c is a constant, then

$$dN/dx = -cnx^{-(n+1)}$$

and

$$dW/dx = -\rho\alpha_v cnx^{-(n+1)}x^3 = -\rho\alpha_v cnx^{-(n-2)}.$$

Hence log dW/dx plotted against log x must also result in a straight line with slope 2 units less than the slope for the cumulative number—size relationship. Examination of curve (c) in Fig. 4 shows that the slope of the right hand portion of the curve does not have a constant value, and in no circumstances could such a state continue as x approaches the weight modal size at which the slope is zero. Consequently, there is no doubt that the cumulative number graph cannot be a straight line over more than a relatively limited size range, as indicated by the curve in Fig. 5.

It should be emphasized that it is inadvisable to extrapolate the straight line representing log $N_x^{x\,\max}$ against log x (size), the so-called power law, beyond the region where it has been shown experimentally to be valid, since at particle sizes approaching or smaller than the weight modal size, the relationship between number and size must change and the numerical value of the index in the law will decrease.

Summarizing this section, the parameters of the size distribution curves for the sample 12057, 72 are, on a weight distribution basis:

Median diameter (50%)	86 μm by microscope	80 μm by square sieve aperture
Mean diameter	153 μm by microscope	123 μm by square sieve aperture
Modal diameter	13 μm by microscope	10 μm by square sieve aperture

Geometric standard deviation (from Fig. 2) 3.4

on a number distribution basis:

Slope of cumulative number-size graph −3.7 over the size range 50 to 300 μm, decreasing to −0.35 at 1 μm size.

In comparison with the above results, the following data has been quoted by other research workers: Weight distribution for core tube samples, median diameters 41 to 72 μm, excluding one at 390 μm. SELLERS et al. (1971). Weight distribution for sample 12057,47; median diameter 84 μm, mean diameter 130 μm, geometric standard deviation 2.7. QUAIDE et al. (1971). Weight distribution on several Apollo 12 samples as determined by KING et al. (1971) indicated median diameters of 60 to 70 μm by sieving. Number distribution. Slope of cumulative number graph for sample 12070 equal to −3.6 over the size range 20 to 700 μm. GOLD et al. (1971).

DERIVATION OF SHAPE COEFFICIENTS

The numerical definition of particle shape has been studied extensively by geologists (KRUMBEIN, 1963), but the writer's treatment differs from other researches in that the two components of *shape*, namely, the proportions of the particle (elongation and flatness) and the geometrical form (degree of approximation to the sphere, cube, or tetrahedron) are considered independently.

Notation and theory

L, B, and T = the limiting dimensions of a particle in decreasing order of magnitude.
$\quad$ n = elongation = L/B
$\quad$ m = flatness = B/T
$\quad$ α_a = area ratio = $\pi d_a^2/4LB$
$\quad$ p_r = prismoidal ratio = mean thickness$/T$
$\quad$ $\alpha_{v,a}$ = volume coefficient = volume of particle$/d_a^3$
$\quad$ r = rugosity coefficient = perimeter of particle profile, including minor irregularities and corrugations divided by perimeter of smooth curve circumscribing particle profile.
$\quad$ ϕ = circularity of profile WADELL (1932) = d_a/d_p
$\quad$ ϕ_r = circularity including effect of rugosity = ϕ/r.

If the circularity is corrected for the effect of particle elongation, then the specific circularities are denoted by ϕ^* and ϕ_r^*. The correction for elongation is made by multiplying ϕ by the factor $(n+1)/2n^{1/2}$.

The volume coefficient $\alpha_{v,a}$ is the most convenient general indication of particle shape, but it is a composite factor derived from the first four ratios in the list above. The derivation of $\alpha_{v,a}$ has been described in detail elsewhere (HEYWOOD, 1937, 1963) and results in the following equation:

$$\alpha_{v,a} = \frac{\pi^{3/2}}{8} \cdot \frac{p_r}{\alpha_a^{1/2}} \cdot \frac{1}{mn^{1/2}}$$

Hence the value of $\alpha_{v,a}$ may be determined by direct measurement of the volume of the particle or if this is not possible by calculation from the equation if the factors therein can be estimated. It should be noted that the numerical value of the volume coefficient is dependent upon the characteristic diameter used to represent the particle size. The projected area diameter, d_a, has been chosen because this can be determined for all sizes of particles and in the notation adopted, the subscript a denotes that the diameter d_a has been used for calculation of the volume coefficient.

The circularity coefficient was introduced by Wadell to distinguish between angular and rounded particles; in fact, as defined, it is a function of both the geometrical form and the elongation of the particle. Another critisism is that it changes very little for considerable alteration in angularity, for instance the values for a circle, square and equilateral triangle are respectively 1, 0.886, and 0.777. Elongation can be allowed for by the factor given above, and if rugosity is included then the specific circularity ϕ_r^* does give some indication of the irregularity of the profile. At present these measurements are rather laboriously made on enlarged photographs of the particle profiles, but electronic counters are being developed which will obviate much human effort.

For the purpose of shape assessment, particles have been classified into the groups scoriacious, smooth opaque, and transparent crystalline; typical profiles of these are shown in Fig. 6. These divisions are somewhat arbitrary and do not correspond

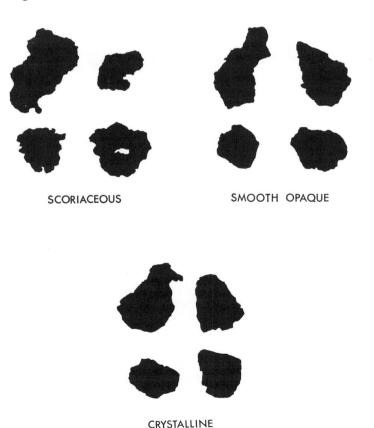

SCORIACEOUS   SMOOTH OPAQUE

CRYSTALLINE

Fig. 6. Typical shapes of 700 μm diameter particles.

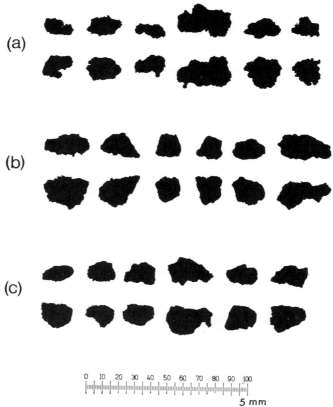

(a)

(b)

(c)

0 10 20 30 40 50 60 70 80 90 100
|ını|ını|ını|ını|ını|ını|ını|ını|ını|ını|

5 mm

Fig. 7. Particle profiles; elevation and plan views (upper and lower rows respectively) (a) Scoriaceous agglomerates (b) Smooth opaque particles (c) Crystalline or transparent particles.

with precise mineralogical components. As shown by the profiles of relatively large particles in Fig. 6, the shapes may be somewhat difficult to differentiate, but as observed through the microscope the differences are much more definite. The distinction between these characteristic shape types is more precise with the finer particles. Examples of the derivation of these shape coefficients for the coarsest size fraction are shown by the profile elevations and plans in Fig. 7. The particles in group (a) are scoriaceous, in group (b) smooth opaque, and in group (c) crystalline. Values of the measured shape coefficients for these particles are given in Table 3 and may be summarized as follows.

The elongation does not vary greatly for the three types of particles: the flatness is greatest for the scoriaceous particles and the area ratio is least. The volume coefficients are not sensibly different for all types of particle and the specific circularities based on the smoothed profiles are identical. The rugosity factor shows the greatest variation and this also influences the true specific circularity ϕ_r^*. Further remarks on this subject are made in the section on conclusions.

Table 3. Shape coefficients measured on 30 particles 700 μm projected
area diameter (see section "Derivation of Shape Coefficients" for definitions)

Shape coefficient	Scoriaceous	Particle type smooth opaque	Crystalline
n	1.33	1.38	1.32
m	1.37	1.17	1.20
α_a	0.70	0.72	0.74
$\alpha_{v,a}$	0.32	0.35	0.34
r	1.10	1.04	1.04
ϕ^*	0.93	0.93	0.93
ϕ_r^*	0.84	0.89	0.90

MODAL DISTRIBUTION OF PARTICLE TYPES AND GLASS SPHERULES

The relative numbers of the three shape groups shown in Fig. 6 in each sized
fraction of the lunar fines were determined whilst making the count for mean particle
weight. The results are shown graphically in Fig. 8, the proportion of crystalline particles increases in the fractions smaller than 100 μm and the scoriacious particles
decrease. These latter consist of rough agglomerations of glass-bonded mineral
particles and the significance of these variations in composition will have to be discussed.

It was not part of this assignment to make a detailed study of the glass spherules,
important and fascinating though these are. A rough estimate is that such particles
are present to the order of 1 in 500 other particles in the coarser sized fractions, though

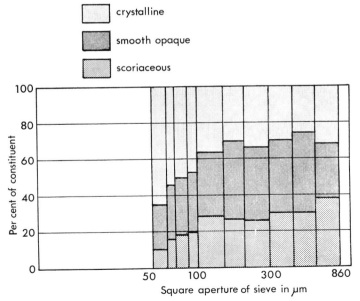

Fig. 8. Variation of the relative proportions of the three shape groups according to
particle size.

becoming more frequent in the finer sizes and spheres of less than 0.5 μm diameter have been observed by means of the electron microscope. It is considered, however, that these spherules are not numerous enough to have a significant effect on the soil mechanics of lunar fines.

CONCLUSIONS

In the main, this report presents a number of facts concerning a single sample of lunar dust and it would be unwise to make any rigorous conclusions therefrom, especially as the investigation of the finest dust fractions is not yet completed. The various procedures for sizing analysis are in concordance and the size distribution curve on a weight basis approximately follows a log normal law. The median size is about 80 μm based on a sieve square aperture, or 86 μm on projected area diameter. The size analysis may be deficient in the estimation of very fine particles because some of these will have adhered to the coarse particles, which were sieved in the dry state. The cumulative number of particles has been plotted against the size, both to logarithmic scales.

The curve approaches a straight-line relationship having a slope of -3.7 over the mid-region of the size range, but it has been shown that a completely straight-line relationship is incompatible with a log-normal weight distribution of particles. The increased slope at the coarse end of the range is an artificial effect due to the separation of the fines by a 1 mm sieve.

The shape of the particles does not seem to vary much throughout the size range investigated up to the present; photo and electron micrographs of particles from all fractions are very similar in visual appearance. Detailed measurements of the volume coefficient $\alpha_{v,a}$ which are quoted in Table 2 show that the intermediate sized fractions are the most nearly equi-dimensional and that the coarsest and the finest particles tend to be slightly more flattened in shape. The mean value of $\alpha_{v,a}$ based on a specific gravity of 2.9 is about 0.32. The proportion of transparent crystalline particles increases in the size fractions smaller than 100 μm and this factor may correspond with changes in other properties of the dust.

In the evolution of a dust there is a conflict between the processes of comminution and agglomeration; in the case of a terrestial dust of desert type, the process of comminution is generally predominant, but in the lunar dust it would appear that the process of agglomeration exerts a considerable, if not predominating, influence on its evolution.

REFERENCES

DUKE M. B., WOO C. C., SELLERS G. A., BIRD M. L., and FINKELMAN R. B. (1970) Genesis of lunar soil at Tranquillity Base. *Proc. Apollo 11 Lunar Sci. Conf., Geochim. Cosmochim Acta* Suppl. 1, Vol. 1, 347–361. Pergamon.

SHOEMAKER E. M., HAIT M. H., SWANN G. A., SCHLEICHER D. L., DAHLEM D. H., SCHABER G. G., and SUTTON R. L. (1970) Lunar regolith at Tranquillity Base. *Science* **167**, 452–455.

SELLERS G. A., WOO C. C., BIRD M. L., and DUKE M. B. (1971) Descriptions of the composition and grain-size characteristics of fines from the Apollo 12 double core-tube. Second Lunar Science Conference (unpublished proceedings).

QUAIDE W., OBERBECK V., BUNCH T., and POLKOWSKI G. (1971) Investigations of the natural history of the regolith at the Apollo 12 site. Second Lunar Science Conference (unpublished proceedings).

KING E. A., BUTLER J. C., and CARMAN M. F. (1971) The lunar regolith as sampled by Apollo 11 and Apollo 12: Grain size analyses, modal analyses, origins of particles. Second Lunar Science Conference (unpublished proceedings).

GOLD T., O'LEARY B. T., and CAMPBELL M. J. (1971) Physical properties of the Apollo 12 lunar fines. Second Lunar Science Conference (unpublished proceedings).

KRUMBEIN W. C. and SLOSS L. L. (1963) *Properties of Sedimentary Rocks*. Freeman.

WADELL H. (1932) Shape of rock particles. *J. Geol.* **40**, 433–451.

HEYWOOD H. (1937) Numerical definitions of particle size and shape. *Chem. Ind.* (*London*) **56**, 149–154.

HEYWOOD H. (1963) Evaluation of powders. *J. Pharm. Pharmac.* **15**, 56T–73T.

APPENDIX. WAX DISPERSION METHOD OF DETERMINING MEAN PARTICLE WEIGHT

The mean particle weight of a closely graded fraction, i.e. with size limits not exceeding the ratio $\sqrt{2}$ to 1, is determined by direct weighing and visual counting of a small sample of particles, provided that the number of particles in a weighable quantity does not exceed about 3000. The lower size limit for this procedure is about 90 μm, and for smaller particles a method of dispersing the weighed sample must be devised so that only a known proportion need be counted. Dispersion in a viscous liquid and the removal of a measured volume for counting is suspect as it is difficult to avoid settlement of particles and drainage errors during sampling. The following method of wax dispersion was therefore devised.

The fraction graded between 75 and 60 μm micromesh sieves is used as an example of the method. Molten paraffin wax, previously filtered, to an amount of about 3 g is poured onto a sheet of clean glass so that it solidifies as a flat disk about 4 cm diameter. This is removed before hardening completely, and when cold weighed accurately on a micro-balance. Small quantities of the particles on the point of a knife blade are then dropped onto the surface of the wax to the amount of about 10 mg. The procedure is then to fold and re-fold the wax many times until the particles are uniformly dispersed. This is accomplished by taking the wax in clean dry hands and holding it within a drying oven until it becomes soft enough to manipulate without fracture. After the initial folding to enclose the particles, the wax is pressed flat, re-folded and pressed again; this process is repeated for about half an hour. The lump of wax is then rolled into a cylinder about 1 cm diameter and after cooling weighed to check for any loss.

Thin cross-sectional slices weighing 40 to 50 mg are cut from various parts of the cylinder by means of a razor blade and each of these is placed on a glass microscope slide which has been weighed and previously ruled into say 1 mm squares with a diamond point. The slide is re-weighed with the wax slice in position to determine the weight of the latter. Finally, a cover glass is placed over the slice and the slide put onto a warm plate to melt the wax. The particles are most easily counted by using a binocular microscope at about 30 magnifications, but in order to render the particles clearly visible, the slide must be placed on an electrically heated copper plate with a central hole, so that the wax remains in the molten state.

The following figures are for the example chosen:

Weight of particles	10.1 mg
Weight of wax and particles	2.7332 g

mg particles per g of wax and particles 3.695

Count no.	1	2	3
Weight of slice, g	0.0448	0.0390	0.0473
Weight of particles, mg	0.166	0.144	0.175
Number counted	454	425	511
Number per mg	2735	2950	2920

Weighted mean number of particles per mg 2865
Mean particle weight μg 0.349

The particles may easily be recovered by the use of a suitable wax solvent.

Proceedings of the Second Lunar Science Conference, Vol. 3, pp. 2003–2008
The M.I.T. Press, 1971.

The formation of spherical glass particles on the lunar surface

J. O. ISARD

Department of Glass Technology, University of Sheffield, Sheffield 10, U.K.

(*Received* 27 *February* 1971; *accepted in revised form* 30 *March* 1971)

Abstract—Lunar glass spheres are assumed to be formed by the break-up of jets of liquid rock resulting from impact events. It is shown that spheres up to about 1 cm diameter could be formed in the cooling time available. Cavities must be nucleated and supported by gas pressure.

INTRODUCTION

SAMPLES 12070,36, 12001,72, and 12057,59 of lunar dust supplied as "less than 1 mm fines" were sieved by hand, and the fractions remaining on the sieve with 100 μ square mesh openings were examined under a hand lens and a low power microscope. Altogether 35 spherical particles were observed; they were separated by hand and their sizes measured. Their size distribution is shown in Table 1. Some were distinctly ellipsoidal and a few had irregular lumps of material adhering, but the majority were nearly spherical, and several, with diameters in the range 200–400 μ, had constant diameters to $\pm 1 \%$. Most appeared quite opaque under the microscope but a few transmitted an appreciable amount of brown light. One, in particular, showed a spherical "seed" or bubble a few μ in diameter.

No spheres greater than 600 μm diameter were observed and no fragments of larger spheres were found. These observations prompted the following investigation of the mechanism by which glass spheres could be formed to ascertain what, if any, physical parameters limit their size. However the proportion of spheres to irregular particles is small and the frequency of occurrence of all particles drops off steeply with increasing size; it is not therefore claimed that a size limitation has been *observed*. On the other hand, spheres of 10 mm radius or greater would be so conspicuous on the lunar surface that a very low rate of occurrence would have been detected.

FORMATION OF SPHERICAL GLASS PARTICLES

Spherical droplets can be formed by condensation from a gas, by the heating of irregular solid particles above their melting temperatures while in flight, or by the break-up of liquid jets. The latter mechanism seems the most likely, since descriptions of glassy deposits on lunar rocks are consistent with melting under impact; the area of impact becomes glazed and some liquid glass is splashed around onto neighbouring surfaces. Lunar glass is evidently close in composition to lunar rocks, suggesting a lunar origin.

The criteria for formation of a lunar glass sphere can therefore be stated as follows: liquid is formed under impact, it is ejected with sufficient vertical velocity to allow time for both the break-up of the liquid into drops and the subsequent cooling of the

Table 1. Distribution of diameters of spherical (and near spherical) lunar glass particles.

Diameters	12070,36 0.3 g	12001,72 1.5 g	12057,59 1.5 g	Total	Cumulative total
1000–900					
900–800					
800–700					
700–600					
600–500		1	3	4	4
500–400		3	5	8	12
400–300	3	1	2	6	18
300–200		7	2	9	27
200–100	1	1	6	8	35
TOTAL	4	13	18	35	

drops to a temperature where they are rigid, before they strike any other surface. The liquid must initially be at a certain minimum temperature to give a low enough viscosity for spherical drops to form in free flight. However if the cooling of the liquid by radiation is too rapid, insufficient time will be available for drop formation and the process will be intrinsically impossible. The times for break-up of liquid jets and for cooling of spheres have been investigated therefore to ascertain if physical limits exist.

PROPERTIES OF LUNAR GLASS

No precise data is available for the material properties of lunar glasses. The density and specific heat can be taken as 2.5 g cm^{-3} and 0.3 cal g^{-1} °C^{-1} within limits of $\pm 50\%$ since these properties vary comparatively little with composition of oxide glasses containing negligible amounts of heavy elements (MOREY, 1960). A value of 300 ergs cm^{-2} can be attributed to the surface tension but with, perhaps, rather less confidence. The measurements of AKHTAR and CABLE (1968) on soda silica and soda lime silica glasses in various atmospheres at high temperatures (viscosities less than 10^3 poise) gave values in the range 300 to 400 ergs cm^{-2} with only slight dependence on temperature. PARIKH (1958), working at lower temperatures (viscosities greater than 10^7 poise), found that measurements made in a vacuum agreed fairly closely with those made in dry air or in dry H$_2$. Most changes in composition do not have effects greater than about 10–20%.

The viscosity curve is likely to be steep and to give very low viscosities at very high temperatures due to the content of FeO, CaO, etc. Curve A in Fig. 1 is the viscosity curve of a typical soda lime glass containing 21% Na$_2$O, 9% CaO given by LILLIE (1931), while the probable shape of a lunar glass viscosity curve is shown at B. It is assumed that the viscosity is 10^3 poise at 1400°K (1127°C) and 10^{12} poise at 1000°K. The viscosity of a glass increases continuously with decreasing temperature, but when it is about 10^{12} poise the material exhibits typical properties of a solid at short times, and, in particular would not distort on impact. It is therefore assumed that lunar glass droplets must cool to 1000°K before returning to the lunar surface. The viscosity must be less than about 10^3 poise for typical liquid properties to be manifest in short times, such as flow under surface tension forces. Hence the lunar glass must be raised to 1400°K to form liquid drops, and the time during which it remains above 1400°K is the time available for the jet to break up into drops.

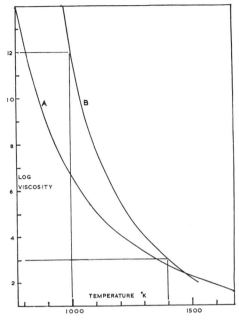

Fig. 1. Curves of log viscosity against temperature for A soda-lime-glass; B estimated curve
for typical lunar glass.

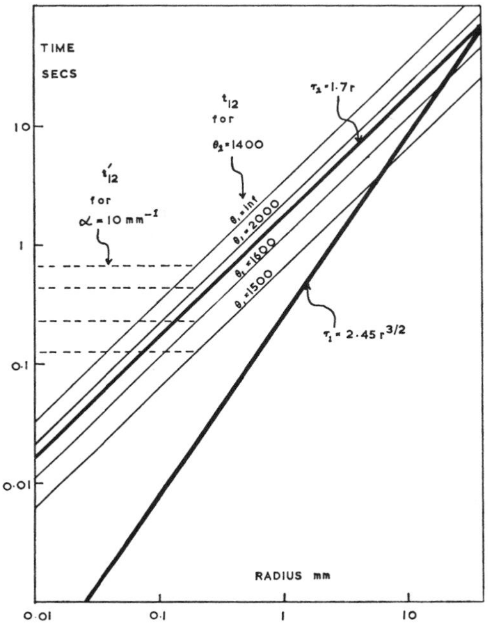

Fig. 2. Cooling times from $\theta_1°$K to $\theta_2 = 1400°$K
against radius for t_{12} = black body cylinder, and
t_{12}' = nearly transparent cylinder; droplet formation times τ_1 = negligible viscosity; τ_2 = viscosity
of 1000 poise.

COOLING TIMES AND DROPLET FORMATION

Cooling by black-body radiation laws is a reasonable approximation for a sphere
with $\alpha > 1/r$ where α is the absorption coefficient and r the radius. Most of the spheres
with $r \sim 0.1$ mm were opaque, or nearly opaque, hence $\alpha > 10$ mm^{-1} in the visible.
Furthermore the glasses contain relatively large concentrations of ferrous iron—up to
25 % FeO according to analyses of Apollo 11 materials (DUKE et al., 1970). Fe^{2+}
in silicate glasses gives rise to a broad absorption band centred on a wavelength of
1.05 μm. STEELE and DOUGLAS (1965) give the extinction coefficient as 28.9 l mole^{-1}
cm^{-1}; for a glass containing 20 % FeO of density 2.5 g cm^{-3} this would give rise to an
absorption coefficient of 46.5 mm^{-1} if the absorption were proportional to concentration over this range. All silicate glasses show strong absorption with $\alpha > 1$ mm^{-1} at
wavelengths of 5 μm and greater with a particularly intense band at about 10 μm
due to the silicate structure itself. The values of the optical constant K given by
CROZIER and DOUGLAS (1965) give values of $\alpha \sim 100$ mm^{-1}. These estimates suggest
that a lunar glass sphere with $r \sim 1$ mm will approximate closely to a black body at
temperatures of 1000–2000°K but that spheres with $r < 0.1$ mm should possibly be
regarded as nearly transparent radiators.

It is assumed that the liquid jet is quickly ejected into surroundings effectively at zero temperature. The heat balance equation for a radiating black cylinder gives

$$t_{12} = (r\rho s/6\sigma)[(1/\theta_2)^3 - (1/\theta_1)^3]$$

where t_{12} is the time taken to cool from θ_1 to θ_2 °K

$$\rho = \text{density (assumed constant)}$$
$$s = \text{specific heat (assumed constant)}$$
$$\sigma = \text{Stefan's constant}$$

Using $\rho = 2.5$ g cm^{-3}, $s = 0.3$ cals g^{-1} °C^{-1},

$$t_{12} = 9.2r[(1000/\theta_2)^3 - (1000/\theta_1)^3] \text{ for } r \text{ in mm}$$

Figure 2 shows the relationship between t_{12} and r on a logarithmic plot calculated for $\theta_2 = 1400$°K and $\theta_1 = 1500, 1600, 2000$°K, and infinity. It is seen that the cooling time is not very dependent on the upper temperature, θ_1, once this is 100 deg greater than θ_2. The calculations show that a liquid jet of radius r mm will stay liquid only for about 1 to 3 times r in seconds if it is cooling freely into cold surroundings.

For a nearly transparent cylinder ($\alpha \ll 1/r$) the cooling time is

$$t_{12}' = (\rho s/3\alpha\sigma)[1/\theta_2^3 - 1/\theta_1^3]$$
$$= 2t_{12}/\alpha r.$$

The value of t_{12}' is the value of t_{12} for $r = 2/\alpha$ but is independent of r. For example, if $\alpha = 10$ mm^{-1}, t_{12}' is the value of t_{12} for $r = 0.2$ mm; the values are shown by the dotted lines on Fig. 2 for the same values of θ_1, θ_2 as used for calculating t_{12}. Clearly t_{12}' only applies in the limit of small r and t_{12} in the limit of large r. A precise calculation based on a detailed knowledge of the spectral distribution of α would give the form of the curve between the limiting lines.

LEVICH (1962) gives the approximate formula $\tau_1 = 8.46\sqrt{\rho r^3/\gamma}$ for the time, τ_1, taken for a liquid cylinder of negligible viscosity to break up into drops, where $\rho = $ density and $\gamma = $ surface tension. Using $\rho = 2.5$ g cm^{-3} and $\gamma = 300$ ergs cm^{-2}, this becomes

$$\tau_1 = 2.45r^{3/2} \quad \text{for } r \text{ in mm.}$$

If the viscosity controls the deformation of the cylinder, LEVICH (1962) gives the formula

$$\tau_2 = 5r\eta/\gamma$$
$$= 1.7r \quad \text{for } \eta = 1000, \quad r \text{ in mm}$$

Both τ_1 and τ_2 are shown plotted on Fig. 2.

The dependence of τ_1 on $r^{3/2}$ and of t_{12} on r means that there must be a limiting radius above which a liquid jet cools too rapidly to form drops. The intersection of the τ_1 line with the t_{12} lines in Fig. 2 occurs at radii greater than 10 mm, and suggests that jets up to 10 mm radius can form spheres—the spheres being of rather greater radius. It is concluded that spheres greater than about 5 cm diameter cannot be formed by this process.

The position of the τ_2 line relative to the t_{12} lines shows that at the viscosity of 10^3 poise a liquid jet cools too rapidly at temperatures less than 2000°K to be able to break up. This means that, as stated above, only if the viscosity stays well below 10^3 poise does the jet break up under surface tension forces.

The position of the t_{12}' lines suggests that very fine threads have plenty of time to break up into drops even at viscosities as high as 10^3 poise.

A large sphere (dia > 5 cm) could be formed from a glass cylinder of large radius if the glass were nearly transparent, but the absorption coefficient would have to be less than 0.1 mm^{-1} over the whole effective black body spectrum and this is unlikely in view of the high absorption of silicate glasses in the infra red beyond 5 μm.

Alternatively, the time available for formation of a sphere would be prolonged if the jet were surrounded by other material at high temperature, but it then seems unlikely that it could travel freely without collisions while still at a low viscosity.

Finally, it can easily be shown that the spheres will cool quickly to 1000°, and that material ejected as a liquid will normally have time to solidify before returning to the lunar surface. However, it is noted that large spheres will be likely to break up on impact especially if they strike rock rather than the lunar dust on landing.

FORMATION OF CAVITIES IN COOLING GLASS

In the commercial production of small glass spheres (ballotini) by passing powdered glass fragments into a high temperature flame a large proportion of the spheres contain seed—small spherical bubbles. These are almost certainly caused by the exsolution of dissolved gases, especially H_2O and CO_2, at the flame temperature. In the formation of large glass spheres (10 cm or more diameter) cavities may form in the core of the sphere during cooling; these are essentially vacuities although they may grow from tiny gas-filled seed as nuclei. Many lunar glass spheres have been observed to contain seed, and it is of interest to estimate if they could have been formed as vacuities during the chilling of the glass or whether they must have been filled with gas at elevated temperatures.

First it is observed that a vacuity can arise only from a temperature difference between the core and outer shell; hence a nearly transparent sphere ($r < 1/\alpha$) cannot produce a vacuity as the whole particle radiates energy uniformly and there is no temperature gradient. Seed in very small spheres ($r < 0.1$ mm) must almost certainly be gas-filled.

The temperature difference between the surface and the centre of a black sphere cooling by radiation is of the order of $\Delta\theta = a\sigma\theta_s^4/K$ where K is the thermal conductivity of the sphere of radius a and θ_s the surface temperature. For a completely opaque material (α very large) the true thermal conductivity applies, a value of 0.003 cal-cm-s units being appropriate for a glass. Under certain circumstances the true conductivity will be enhanced by an effective "radiation conductivity," but this will only be important for large spheres of low absorption. For $\theta s = 1000$°K the formula gives $45a$ for the excess temperature at the centre of a sphere of radius a mm.

Strain may arise in the liquid core of a sphere whose surface has "set" from two effects: (i) the expansion coefficient, α_r, of the liquid above the transformation

J. O. ISARD

temperature, Tg, is larger than that, α_s, of the solid below Tg. This effect will give a strain of order $(\alpha_r - \alpha_s)\,\Delta\theta$. (ii) The temperature gradient becomes less severe during cooling, since the heat loss varies as $\theta_s{}^4$, hence the liquid core cools faster than the solid shell. This effect will give a strain of order $\alpha_r(\Delta\theta_1 - \Delta\theta_2)$, where $\Delta\theta_1$ is the temperature excess when the shell sets and $\Delta\theta_2$ is the temperature excess when the centre sets. The volume expansion coefficient of the solid glass is probably about $3 \times 10^{-5}\,°C^{-1}$ while that of the liquid is probably about $10^{-4}\,°C^{-1}$ by comparison with silicate glasses generally. Hence the maximum strain attainable from both effects together will be about $10^{-4}\Delta\theta$, i.e., about $0.45a\%$ (a in mm). This will be negligible for spheres less than 1 mm diameter and becomes significant only for spheres greater than about 10 mm diameter.

It is concluded that "seed" in small lunar glass spheres must be supported by a gas pressure (against the surface tension effect of the spherical cavity) and are formed while the glass spheres are molten as in the ballotini process.

Acknowledgments—The author wishes to acknowledge the interest of all members of the Department of Glass Technology in this work, and especially the help of Dr. M. Cable in many discussions of the problem.

REFERENCES

AKHTAR S. and CABLE M. (1968) Some effects of atmosphere and minor constituents on the surface tension of glass melts. *Glass Technology* **9**, 145–151.
CROZIER D. and DOUGLAS R. W. (1965) Study of sodium silicate glasses in the infra-red by means of thin films. *Physics and Chemistry of Glasses* **6**, 240–245.
DUKE M. B., WOO C. C., BIRD M. L., SELLERS G. A., and FINKELMAN R. B. (1970) Lunar soil: Size distribution and Mineralogical constituents. *Science* **167**, 648–650.
LEVICH V. G. (1962) *Physicochemical Hydrodynamics*, Prentice-Hall.
LILLIE H. R. (1931) Viscosity of glass between the strain point and melting temperature. *J. Amer. Ceram. Soc.* **14**, 502–511.
MOREY G. W. (1960) *The Properties of Glass*, 2nd ed., Reinhold.
PARIKH N. M. (1958) Effect of atmosphere on surface tension of glass. *J. Amer. Ceram. Soc.* **41**, 18–22.
STEELE F. N. and DOUGLAS R. W. (1965) Some observations on the absorption of iron in silicate and borate glasses. *Physics and Chemistry of Glasses* **6**, 246–252.

Proceedings of the Second Lunar Science Conference, Vol. 3, pp. 2009–2019
The M.I.T. Press, 1971.

Interaction of gases with lunar materials: Preliminary results

E. L. Fuller, Jr., H. F. Holmes, R. B. Gammage, and
K. Becker

Reactor Chemistry and Health Physics Divisions, Oak Ridge National Laboratory,
Oak Ridge, Tennessee 37830

(*Received* 24 *February* 1971; *accepted in revised form* 31 *March* 1971)

Abstract—Adsorption of carbon monoxide, nitrogen, oxygen, and argon at $-196°C$ shows the surfaces of lunar fines to be quite nonpolar. There is no evidence of inherent porosity in the range of 20 to 500 Å and the specific surface area is slightly greater than 1 m²/gm. Water vapor adsorption appears to be specifically oriented in the monolayer region and penetrates the surface at $P/P_s = 0.9$. This penetration is reversed at $P/P_s = 0.8$. Repeated and prolonged treatment with water vapor increases the capacity for water adsorption but the capacity for nitrogen is unaltered. This suggests a rather unique micropore structure. The observations are interpreted in light of the amorphous, radiation damaged, silicious surface region. Water vapor adsorption seems to be an informative means of probing these latent radiation damage tracks.

Introduction

Vapor adsorption studies are useful for determining specific surface areas, surface energy, and porosity of particulate matter. Each of these parameters is intricately controlled by such variables as temperature, pressure, crystal habit and mode of formation, and transpiring changes that occur in the history of said samples. Reviews of these effects are readily available (Gregg and Sing, 1967; Flood, 1967) and preclude any lengthy discussion here. The knowledge of the nature of these surfaces should be quite helpful in elucidating the lunar history. The effect which extended exposure to such an extremely rarified atmosphere has on surface properties has not been studied.

No attempt has been made as yet to evaluate the nature of the interaction of life-supporting vapors with these lunar materials even though elaborate precautions have been taken to avoid such an occurrence. A complete understanding of such an inter-action is of prime importance if man is ever to coexist extraterrestrially with an appreciable amount of lunar soil. Any selective sorption of reactive (toxic or essential) vapors should be investigated in light of any helpful (or detrimental) effects that may be involved. Our atmosphere is preassumed to contaminate these materials but there is little or no available data at present as to the nature or amount of such contamination.

Experimental Techniques

The adsorption apparatus designed around a vacuum microbalance has been described in an earlier report (Fuller *et al.*, 1965) and has been shown to perform excellently when due note is taken of inherent thermomolecular flow effects and inherent beam buoyancy effects (Fuller *et al.*, 1970). The resultant reliability of a fraction of a microgram is of prime importance in the studies of lunar materials in light of the limited amount of material available at this writing.

As a means of evaluating the specific surface area of the samples as well as gaining some knowledge of the surface energy we have employed four probe gases: carbon monoxide, nitrogen, oxygen, and argon. The polarizabilities and multipole factors (HIRSCHFELDER *et al.*, 1954) of these molecules (and hence their anticipated degree of association with a polar inorganic oxide surface) decrease in the order of the above presentation (NAIR and ADAMSON, 1970). We have chosen for convenience, and as the standard practice, to study the adsorption of these vapors at $-196°C$ where liquid nitrogen is used for a thermostat. Simultaneous measurements were made of the respective saturation vapor pressures (FULLER *et al.*, 1965) throughout this investigation.

Water vapor obtained by $100°C$ vacuum dehydration of $BaCl_2 \cdot 2H_2O$ was stored in the reservoir of the precision pressure control device (FULLER *et al.*, 1970). Incremental control of pressure allowed construction of isotherms and equilibrium attainment, with no contributing wall or container effects that are known to plague water sorption experiments (DEITZ, 1970). Equilibrium was assured or disclaimed by extended continuous observations of the sample mass for hours, days, and even weeks at constant (± 0.001 torr) pressure of water vapor. Such continuous observations at constant pressure also gave valuable information as to the kinetics of adsorption with assurance that we were dealing with a phenomena truly characteristic of the sample. Equilibrium was assumed to be attained when the sample weight was constant to ± 0.2 μgm (the instrumental precision) over a 16-hour (overnight) exposure.

The samples studied were lunar fines 12033,46 and 10087,5. Nitrogen adsorption on 12033,46 showed that the sample had a very low (less than 0.05 m²/gm) specific surface area and further verified the reported density of 3.1 gm/cm³ (COSTES *et al.*, 1970). No further studies were made on this material as the probability of achieving valuable data seemed nil.

The sample 10087,5, however, did have a significant specific surface area and is the major subject for this report. The sample was taken as received with no separation as to size, chemical, or physical state so as to have a somewhat representative sampling of the lunar soil. Optical microscopic examination of this material revealed a high predominance of irregularity shaped particles of dimensions ranging from 1 to 5 μm. There were no sharp edges or corners or obvious cracks or crevices. The aliquot of 100.084$_6$ mg was placed on the balance and the system was sealed with no atmospheric exposure during the course of our investigation. We were not able to start with a vacuum sample, however, since the material had been opened to the ORNL atmosphere for a week or so for other studies. At the relevant time (August) our laboratory temperature was $25 \pm 2°C$ and relative humidity was $60 \pm 5\%$. Evacuation to 10^{-5} torr overnight decreased the sample weight to 100.054$_4$ mg. Outgassing at $300°C$ for 24 hours at 10^{-5} torr further decreased the weight to 99.960$_5$ mg. All data taken was corrected for sample buoyancy using the reported (COSTES *et al.*, 1970) specific gravity of 3.1 gm/cm³.

The general attack of this problem was designed to measure the interaction of reactive vapors, i.e., water, with this material and intermittent monitoring of the specific surface area with inert vapors, i.e., nitrogen. Conceptually this should allow one to distinguish surface phenomena (chemical and/or physical adsorption) from bulk reactions (hydration, phase transformation, etc.) and measure any induced surface area or surface energy changes.

RESULTS

Figure 1 shows the isotherms obtained for the four probe gases at $-196°C$ after outgassing the sample for 24 hours at $300°C$ and 10^{-5} torr. The isotherms are completely reversible, that is, the adsorption and desorption data are identical. This is noted by the fact that alternate adsorption and desorption points form the curve above $P/P_s = 0.5$. The amount adsorbed is expressed in molar units to aid in comparison. All are plotted as a function of relative pressure with respect to the measured saturation pressure, P_s. The amount of adsorption does decrease with the polarizability of the gas as anticipated (NAIR, 1970). The CO and N_2 isotherms have the general "type II" (BRUNAUER *et al.*, 1940) sigmoidal shape characteristic of most terrestial materials.

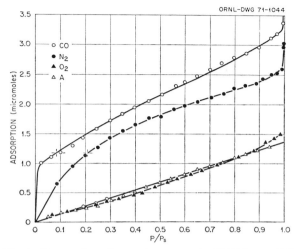

Fig. 1. Vapor adsorption on lunar sample 10087,5 at $-196°C$.

The monolayer capacity, derived by a BET treatment (BRUNAUER *et al.*, 1938) of the data is 1.1_8 micromole for both CO and N_2. The relative pressures of monolayer formation occur at 0.08 and 0.21, respectively, as shown by the indicators on the "knees" of the isotherms. In contrast to the behavior of most inorganic oxides, the argon and oxygen isotherms are quite well characterized by a Henry's Law (GREGG and SING, 1967) equation:

$$X = 1.36(P/P_s) \qquad \text{(micromole)} \qquad (1)$$

The slight undulation $(0.15–0.4P/P_s)$ noted in the adsorption of argon does not appear to be significant since it is absent in the desorption values (open circles on lower curve). In contrast the oxygen data seems to have a very real and persistent deviation from the linearity given by eq. (1).

The contrasting nature of the water sorption data is shown in Fig. 2. The initial adsorption appears to form a monolayer of 0.10 mg/gm at P/P_s of about 0.05 and remains relatively constant until *ca* $0.5P/P_s$ where a second plateau is reached at 0.20 mg/gm. The sudden uptake at $0.90P/P_s$ breaks away to a more gradual approach to the "infinite uptake" (GREGG and SING, 1967; FLOOD, 1967) expected at saturation pressure. The hysteresis on desorption is marked even at high relative pressures and a gradual decrease in slope continues until the sharp inflection at $0.80P/P_s$. There is then a near linear decrease to a BET monolayer value of 0.19 mg/gm. This monolayer is lost on further evacuation with a small vacuum retention that could not be removed at 10^{-5} torr (as witnessed in the constant weight after 16 hours of pumping and verified by 24 to 96 hours of continued evacuation).

Repeated cycling (adsorption-desorption) led to more conventional isotherms with a more and more pronounced sigmoidal shape. Cycle 8 is indicative of the changes wrought by the cycling processes. An adsorption knee is found to give a BET mono-layer of 0.382 mg/gm at $0.06P/P_s$ followed by a very nearly linear portion with a slight upturn at higher pressures breaking to the discontinuity at $0.90P/P_s$ and again an

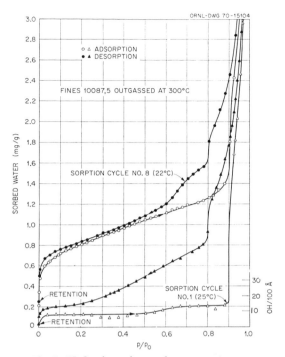

Fig. 2. H_2O adsorption on lunar sample 10087,5.

asymptotic approach to infinity at saturation. The desorption branch persistently falls to a break at $0.80P/P_s$ which is now followed by gradual increase in desorption slope. Below $0.5P/P_s$ only a slight disparity remains with respect to the adsorption branch. This nearly constant disparity is maintained to essentially zero pressure as is noted in the indicated retention.

The pressure changes imposed on the system to evaluate the isotherm points led to virtually instantaneous mass changes with three exceptions. The discontinuities noted at 0.8 and $0.9P/P_s$ were slow processes requiring 3 to 6 hours to attain steady state. The data acquired above $0.95P/P_s$ (not in Fig. 1) required ever more extended periods (up to 24 hours as P/P_s approaches unity) as the pressure was increased. The data above $0.9P/P_s$ was perturbed, however, by a small (1.2 μgm/day) linear mass gain which was not dependent on pressure. The disparity below $0.5P/P_s$ could be accounted for in all cases by the period of time the sample was exposed to $P/P_s > 0.9$.

Nitrogen adsorption on this conditioned material gave a specific surface area of 1.1_3 m²/gm, essentially identical to the initial material. The N_2 monolayer was formed at $P/P_s = 0.19$ (only slightly altered).

Discussion

The virtually identical monolayer capacities for the isoelectronic CO and N_2 gave 1.1_5 m²/gm for the specific surface area, when the coareas are employed (16.2_3 and 16.2_7 Å²/molecule, respectively) (McClellan and Harnsberger, 1967). The "average" interaction energy is deduced from the BET C constant (GREGG and SING, 1967; FLOOD, 1967):

$$E_1 - E_L = RT \ln C \tag{2}$$

where

$$C = (P_s/P_m - 1)^2 \tag{3}$$

and P_m is the pressure where monolayer formation is complete. The energy of adsorption, E_1, is assumed to be always greater than the energy of liquification, E_L, in the BET theory (GREGG and SING, 1967).

Gas	P_m/P_s	$E_1 - E_L$ (cal/mole)	C
CO	0.08	750	130
N_2	0.21	400	14

The low values noted here are indicative of a low surface energy when compared to the nitrogen C values of 100 to 1000 usually noted for inorganic oxides (GREGG and SING, 1967; FLOOD, 1967).

A more striking revelation of the low energy is noted in the linear relationship for argon and oxygen instead of the anticipated sigmoidal isotherm. Henry's Law behavior has been noted (GREGG and SING, 1967; FLOOD, 1967) for extremely low coverages. Many adsorption theories (including the BET theory) predict and describe this behavior at extreme dilution. The BET relationship of eq. (3), which rearranges to

$$P_m/P_s = \frac{\sqrt{C} - 1}{C - 1}, \tag{4}$$

predicts monolayer formation at $P/P_s = 1$ for C approaching zero. However, the BET theory predicts a type III isotherm for this condition and does little to aid in a mechanistic evaluation.

INNES and ROWLEY (1941) have derived the linear isotherm equation

$$n_a = KP \tag{5}$$

by assuming a two-dimensional perfect gas equation of state for the adsorbed molecules, n_a, i.e.,

$$\pi A = n_a RT \tag{6}$$

and using the Gibbs adsorption equation. If we assume a linear relation to a boundary condition of complete monolayer surface coverage at saturation pressure we have

$$n_a = n_s P/P_s \tag{7}$$

suggesting that the area of the sample, A, can be obtained from the Henry's Law constant (eq. 1) and the cross sectional area, σ, of the adsorbed molecule, since at complete coverage $n_s \sigma = A$. Such a calculation for our argon data gives a specific surface area of 1.1_4 m²/gm, based on the accepted (MCCLELLAN and HARNSBERGER, 1967) 13.8_5 Å² for the occupancy area of an argon molecule.

The excellent agreement with the nitrogen and carbon monoxide area gives a high degree of confidence for 1.15 ± 0.01 m²/gm for comparative purposes in our studies. The absolute surface area may differ from this value by as much as $\pm 30\%$ (EMMETT, 1959). Furthermore this value is consistent with a particle (assumed spherical or cubic) dimension of 1.7 μm when the geometric relationship (GREGG and SING, 1967) of specific surface area Σ(m²/gm) to particle size, $l(\mu$m) is employed with the aid of the known density:

$$l = \frac{6}{\rho\Sigma} \tag{8}$$

Such a relationship would not correlate to our optical 1–5 μm observation if the material were grossly porous and leads us to conclude that the roughness factor cannot be much greater than 10 and more probably is near unity.

The specificity of water for this surface is quite intriguing as noted in the complex isotherms of Fig. 2. A monolayer of physically adsorbed water can be predicted to be 0.320 mg/gm based on 10.6 Å²/molecule (hexagonal close-packing of liquid-like water). The initial plateau at 0.10 mg/gm is consistent with a coarea of 31 Å² per molecule and breaks to the second plateau at 0.20 mg/gm (or 16.6 Å²/H_2O). The near coincidence of this latter value to the anticipated 0.209 mg/gm (16.0 Å²) for lattice site areas for oxides leads one to believe that the water molecules are associated with these sites by dipole or hydrogen bond attractions. The initial plateau corresponds quite well to a bridging arrangement where each water molecule has hydrogen bond orientation to two sites. Such a bridging orientation concept is a direct result of the 31 Å²/H_2O calculated for the first plateau coverage. When the pressure becomes great enough (ca $0.5 P/P_s$) the bridging type collapses and continues to fill to a 1:1 site-oriented adsorption. The first desorption monolayer shows the persistence of the 1:1 orientation as does the second adsorption BET value (not shown).

It is informative to compare the "conditioned" adsorption isotherm to the sigmoidal isotherms observed for most nonporous oxides. BRUNAUER *et al.* (BRUNAUER, 1969) have derived the standard isotherms which, when reduced to a unit area (BET) and equal energy (BET C), are superimposable for a number of inorganic oxide substrates. Figure 3 shows the comparison of 10087,5 to such a curve where the coincidence is maintained in the monolayer coverage. The multilayer region of 10087,5 is less than the standard as noted in the $\Delta\theta$ curve. This deficiency is lost at $P/P_s = 0.90$ where an excessive uptake is noted for 10087,5. Extrapolation of the $\Delta\theta$ to $P/P_s = 1$ indicates a maximum deficiency of $1.08\ \theta$ at saturation. The same value is obtained as a point of departure by the extrapolation of the $\Delta P/P_s$ curves to the point of identity for the two isotherms.

The expanded pressure curves for high P/P_s point out the virtual step nature at $P/P_s = 0.9$ followed by an asymptotic increase in uptake at high pressures. The questionable nature of the finite extrapolation for the standard data at $P/P_s = 1$ is also highlighted. A further discrepancy at low pressures is noted in the two first points; where we find $0.45\ \theta$, BRUNAUER tabulates $0.54\ \theta$ (BRUNAUER, 1969).

A more facile comparison of the two curves can be afforded by n, θ, or t plots, depending on your preference. The n and θ are numerically identical, the number of

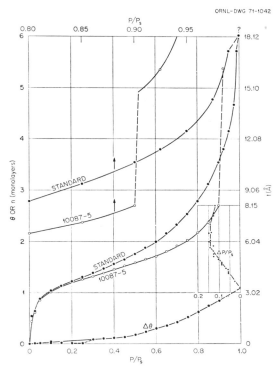

Fig. 3. Comparative plot of H_2O on lunar sample 10087,5.

layers and fractional coverages which can be converted to thickness t by a constant multiplicative factor, 3.0_2 (BRUNAUER, 1969). The latter involves assumptions as to molecular diameters and mode of packing involved in the concept of the statistical thickness of a monolayer normal to the surface. Such plots are naught more than the plot of experimental data vs. anticipated results and circumvent the visual problems of comparing two sigmoidal curves. Our results are given in Fig. 4.

The initial slope of the curve should give the monolayer capacity (μgm/monolayer) and prevail throughout if the sample is nonporous. Any deviations from this ideal behavior are normally attributed to interaction within pores. The slope of the line (38.4 μgm/θ) corresponds excellently (if the explained shift of the first point is effected) to that of the BET monolayer (38.2 μgm/θ). The initial slope is not continued, as the data for 10087,5 falls to a linear relationship of 21.5 μgm/θ. Thus it seems that some microporosity is filled in the monolayer region giving an erroneously high knee or monolayer value. The latter linear relation corresponds to the open surface (0.215 mg/gm as compared to the anticipated 0.209 mg/gm), based on single site-oriented water molecules. The magnitude of micropore volume is related to the difference in the BET monolayer and the open surface capacity ($0.382 - 0.215 = 0.167$ mg/gm) if the described condition indeed prevails. The recommended extrapolation of the second linear portion gives 19.8 μgm (0.198 mg/gm) as the amount of water incorporated in micropores (GREGG and SING, 1967; FLOOD, 1967). The initial linear relationship of

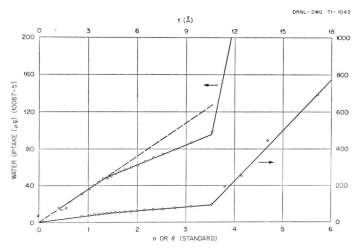

Fig. 4. *n* plot for water vapor adsorption on lunar sample 10087,5.

Fig. 4 would indicate that the micropores are being generated as the water monolayer is being formed and desorption allows them to close. Site orientation of water molecules is normally associated with a high energy surface but the results obtained from these calculations are consistent with the nitrogen data.

The distinct vertical rise and fall of the hysteresis envelope between 0.8 and $0.9 P/P_s$ is identical to the type A hysteresis described by deBoer (1958). Furthermore the relation of adsorption, $(P/P_s)_a$, and desorption, $(P/P_s)_d$, indicates condensation in slit or cylindrical pores:

$$(P/P_s)_d = (P/P_s)_a{}^2. \tag{9}$$

If such a physical phenomena prevails, the pore radii can be calculated by the Kelvin equation (LINSEN, 1970)

$$r - Zt = -\frac{2\gamma V_L}{RT \ln (P_d/P_s)}. \tag{10}$$

r = pore radius
γ = surface tension of liquid adsorbate
V_L = molar volume of liquid adsorbate
t = thickness of adsorbed layer
$Z = 1$, for cylindrical pores
$Z = 2$, for slit-shaped pores

Such a calculation indicates that the pores have a radii > 50 Å and they should be available to nitrogen. Our initial N_2 results and subsequent extensive attempts showed that these pores were not present when the water is removed. Here again the linear excessive uptake noted at high coverage (Fig. 4) would seem to indicate that the penetration phenomena is induced by water adsorption and sealed on water removal. The latent damage tracks may be the avenues for such a penetration. Water vapor adsorption may prove to be a good means of probing these latent damage tracks.

It is difficult to make a direct correlation of this penetration to the intercalation noted for clays (BARRER and MACLEOD, 1954), since these laminar minerals are not found (nor anticipated) in lunar materials. Furthermore the relationship of eq. (9) generally does not prevail and nitrogen porosity is usually present (BARRER and MACLEOD, 1954) in the clays. Comparison shows the clay intercalation occurs at much lower pressures than the lunar soil penetration.

The low energy character of this surface appears to be related to its unique history. Several factors may have contributed to form a surface energy lower than that normally found for terrestrial analogs. The extremely long period of exposure to radiation has surely brought about much more radiation sintering than has been imposed in laboratory experiments. In the review of the subject, TAYLOR (1968) tabulates the results for a number of materials. Specific surface area changes varied from no effect to as much as 68% decrease under the relatively minor integrated doses. One single study showed that TiO_2 underwent a 10% loss in area, and the remaining surface showed a marked (17%) shift of high energy to lower energy sites (ADAMSON, 1961). The grinding action of meteoritic impact and/or roiling of the soil itself would also lead to an attrition and an alteration of surface energy as noted or SiO_2 (MACRIDES and HAKERMAN, 1959). A further source of surface energy diminution lies in the exposure of these materials to the ultra high vacuum for 10^9 years. Stringent dehydroxylation renders silica and alumina hydrophobic (ASHER, et al., 1965). This phenomena is attributed to the closure of the metal-oxygen surface bonds in such a rigid manner that they are not readily attacked by water vapor to form surface hydroxyl groups. Furthermore the surface is so nonpolar that even the polar water molecules are not attracted with any great tenacity. Under the stringent conditions of lunar history this phenomena may have been enhanced. A further consideration must be taken of the fact that this sample was exposed to the atmosphere prior to these analyses. If contamination has rendered the sample so relatively inert, it is quite significant. An outgassing of 300°C will remove most of the contamination from terrestial oxides and generate a high energy surface. The original sample was either very inert or very reactive and rendered inert by adsorption of atmospheric vapors. Our results favor the former but studies on a virgin sample are the final criteria.

The general concept of radiation sintering may be the sole answer. MAURRETTE (BORG et al., 1971) observes a thin (200–1000 Å) layer of "amorphous" material rather uniformly distributed on most if not all lunar fines. This is undoubtedly radiation damage, but it can also be explained by vapor deposition (HAPKE, 1971). The existence of a completely random layer of this sort could account for a low surface energy. A complete breakdown of the crystal lattice would no longer leave alternating layers of oxide ions and metal ions which in turn are the source of a field gradient at the surface. The surface normal dipole is the predominating factor in normal crystalline materials. Thus our data show only the last vestige of order where the surface can interact with the dipole of CO and only very weakly with the quadrapole of N_2.

The interactions of water vapor are consistent with the above picture, where initially the surface dipoles are so weak that two are required to hold a water molecule. The water appears to bring some order into this melange to orient the surface species in such a way that the dipoles become strong enough to each orient a water molecule. The one to one orientation persists and the micropores generated in the water conditioning may be minute molecular interstices between small domains which occur as the "annealing" process proceeds. The hysteresis loop is possibly the result of penetration of water in and around chains or laminae of such domains, expanding this amorphous layer. This expansion then is reversed as the water is removed.

These preliminary results are for a single sample, and full credence as to their representative nature awaits studies on other samples. Analyses of a virgin sample will shed considerable light on the mechanism which produced such surfaces and/or prove if the observed states are inherent in the material or brought about in transit.

Acknowledgments—The authors are extremely appreciative of the encouragement and interpretive discussions given by P. H. EMMETT prior to and during the course of this investigation. This research was sponsored by NASA with the U.S. Atomic Energy Commission under contract with the Union Carbide Corporation.

REFERENCES

ADAMSON A. W. and LING I. (1961) Effect of radiation on the surfaces of solids. *Advan. Chem.* **33**, 51–71.

ASHER R. C., GOODMAN J. F., and GREGG S. J. (1965) The adsorption of water vapor by some inorganic oxides. *Proc. Brit. Ceram. Soc.* **5**, 125–132.

BARRER R. M. and MACLEOD D. M. (1954) Intercalculation and sorption by montmorillonite *Trans. Faraday Soc.* **70**, 980–981.

DEBOER J. H. (1958) *The Structure and Properties of Porous Materials.* Butterworth.

BORG J., DURRIEN L., DRAN J. C., JORET C., and MAURETTE M. (1971) Irradiation, texture and habit histories of the lunar dust. Second Lunar Science Conference (unpublished results).

BRUNAUER S., EMMETT P. H., and TELLER E. (1938) Adsorption of gases in multimolecular layers. *J. Amer. Chem. Soc.* **60**, 309–314.

BRUNAUER S., DEMING L. S., DEMING W. S., and TELLER E. (1940) On a theory of van der Waal's adsorption of gases. *J. Amer. Chem. Soc.* **62**, 1723–1730.

BRUNAUER S., HAGAMASSY J., and MIKHAIL R. SH. (1969) Pore structure analysis by water vapor adsorption, 1. *t*-Curves for water vapor. *J. Colloid Interfac. Sci.* 584–491.

COSTES N. C., CARRIER W. D., MITCHEL J. K., and SCOTT R. F. (1970) Mechanical properties of Apollo 11 soils. *Science* **167**, 739–741.

DEITZ V. R. and TURNER N. H. (1970) Introduction of water vapor into vacuum systems and the adsorption by the walls. *J. Vacuum Science Tech.* **7**, 577–580.

EMMETT P. H. (1959) Adsorption and catalysis. *J. Phys. Chem.* **63**, 449–456.

FLOOD E. A. (1967) *The Solid-Gas Interface.* Marcel Dekker.

FULLER E. L., HOLMES H. F., and SECOY C. H. (1965) Gravimetric adsorption studies of thorium dioxide surfaces. *Vacuum Microbalance Tech.*, **4** Plenum Press.

FULLER E. L., HOLMES H. F., and GAMMAGE R. B. (1970) System evaluation for high temperature studies. *Vacuum Microbalance Tech.* **10**, in press.

GREGG S. J. and SING K. S. W. (1967) *Adsorption, Surface Area and Porosity.* Academic Press.

HAPKE B. W., CASSIDY W. A., and WELLS E. N. (1971) Analysis of optical coatings on Apollo fines. Second Lunar Science Conference (unpublished results).

HIRSCHFELDER J. O., CURTISS C. F., and BIRD R. B. (1954) *Molecular Theory of Gases and Liquid.* John Wiley.

INNES W. B. and ROWLEY H. H. (1941) Relationships between the adsorption isotherm and the spreading force. *J. Phys. Chem.* **28**, 158–165.

LINSEN B. G. (1970) *Physical and Chemical Aspects of Adsorbents and Catalysts.* Academic Press.

MAKRIDES A. C. and HACKERMAN N. (1959) The silica-water system. *J. Phys. Chem.* **63**, 594–598.

MCCLELLAN A. L. and HARNSBERGER H. F. (1967) Crossectional areas of adsorbed molecules. *J. Colloid Interfac. Sci.* **23**, 577–599.

NAIR N. K. and ADAMSON A. W. (1970) Physical adsorption of vapors on ice. III Argon, nitrogen and carbon monoxide. *J. Phys. Chem.* **74**, 2229–2230.

TAYLOR E. H. (1968) Effects of ionizing radiation on catalysts. *Adv. Catalysis* **18**, 111–248.

Proceedings of the Second Lunar Science Conference, Vol. 3, pp. 2021–2025
The M.I.T. Press, 1971.

Particle size and shape distributions of lunar fines by CESEMI

H. Görz, E. W. White, R. Roy, and G. G. Johnson, Jr.

Materials Research Laboratory, The Pennsylvania State University,
University Park, Pennsylvania 16802

(*Received* 23 *February* 1971; *accepted in revised form* 29 *March* 1971)

Abstract—Size analyses have been carried out on six sets of lunar fines from Apollo 12 by computer evaluation of scanning electron microscope images (CESEMI). The grain size of the particles in the fines range from 400–0.2 μm or less. Particle shapes are presented for one sample from ellipses calculated by least square fits to each particle perimeter. Most of the grains are slightly to medium elongated and are quite angular.

Introduction

Size analyses of Apollo 12 fines have been initiated using recently developed procedures for the computer evaluation of scanning electron microscope images (CESEMI). Details of this approach to morphological characterization of particulate materials have been published recently (McMillan *et al.*, 1969; White *et al.*, 1970a). This paper briefly outlines the techniques used and presents results on the size distribution analysis for six samples including 12001,1(77), 12003,43, 12033,27, 12042,32, 12057,63, 12070,42 and the shape distribution analysis of sample 12001,1(77).

Instrumentation

The SEM used in this work is a Japanese Electron Optical, model JSM. The output of the secondary electron detector is an analog signal that generally varies between about -3 to $+2$ V. This signal is fed, in parallel, to the CRT of the SEM and to the signal preprocessor. The preprocessor module was especially designed in our laboratory to manipulate the secondary electron signal. It consists of a bias level adjust, amplified with gain adjustable continuously from 1 to 10 and a time constant variable continuously from 0 to 10 msec. The output is continuously monitored on a Tektronix scope.

The entire system is shown schematically in Fig. 1. The recording instrumentation including the multiplexer, buffer controller, analog to digital converter and tape deck was designed to our specifications by Instrument Technology Corp., Northridge, California. The control module is the ITC Model 8104 B Buffer Controller. The system uses an Astrodata Series 3000 analog to digital converter and a Precision Instrument Corp. Model 1200 digital magnetic tape transport. Various signals from the SEM are fed, in parallel, into a six channel multiplexer, which facilitates the recording of from one to six channels of information simultaneously. The first channel is generally reserved for the secondary electron signal while the other channels are used for electron backscatter, absorbed electron current and X-ray signals. A six-decade thumb-wheel switch is used to enter a header of information at the start of each record. When the first raster line of a picture is initiated the system converts analog information from the SEM (secondary electron, electron backscatter, X-ray signals, etc.) into digital data which are recorded in Binary Coded Decimal (BCD) format on seven level magnetic tape. A record gap is automatically inserted at the end of each raster line. This process continues until the end of the last raster line is reached whereupon a end of file is recorded and the recording stops. The tape transport records data at 556 bits per inch at a speed of $37\frac{1}{2}$ inches per second. A single picture composed of 250 lines and 250 sample points per line is recorded in 15 seconds. The input sensitivity is ± 10 V so that the corresponding "gray" scale ranges from -999 to $+999$. The output is compatible with the University's IBM 360/67 computer.

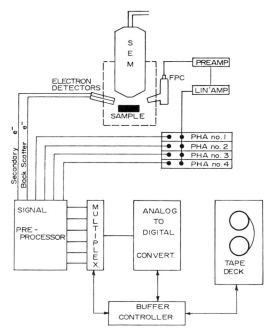

Fig. 1. Block diagram of SEM and digital magnetic tape recording system.

SPECIMEN HANDLING AND RECORDING

Specimen preparation

The powders were prepared for SEM image recording in the form of dispersions on aluminum-coated glass slides. The vacuum evaporated aluminum provides a conductive substrate having a perfectly flat surface on the scale of SEM resolution. No conductive overcoat was applied to the specimen. Furthermore, the secondary electron emission intensities from the aluminum are always less than from anywhere on the particles. Thus the particles are always seen as brighter than the background. The powders were dispersed in n-Hexane and transferred to the coated slides. The low surface tension of n-Hexane in contact with air proved to be very advantageous, the evaporation is fast and no residue is left (WEAST, 1968).

Recording parameters

Secondary electron images were recorded using a SEM voltage of 25 keV and 5×10^{-10} amp specimen current. A picture point density of 250 points per line and 250 lines was used for all images. The secondary signal is adjusted with the signal preprocessor to give a baseline of about -6 V and a peak on the particles of about $+8$ V.

Magnification selection is of obvious importance. One does not want to waste tape by recording too many points per particle. Conversely, too low a magnification does not adequately define the smaller particles in a given sample while too high a magnification setting tends to preclude measurement of the largest particles. Six images were recorded for each of the six specimens. Three were at $100\times$ while three were taken at $1000\times$ magnification.

The binary coded map technique used in this study has been described elsewhere (WHITE *et al.*, 1970a; WHITE *et al.*, 1970b). It involves choosing a signal intensity level and placing an asterisk in an array for each data point whose intensity is above the selected level. This array can then be printed and, assuming a correct level was chosen, the binary map will be a replica of what was seen on the oscilloscope at the time of recording. The information collected in this process is then analyzed to produce size and shape characteristics.

RESULTS AND DISCUSSION

Results of the particle size analysis are summarized in Figs. 2 and 3. Figure 2 is for the analysis of images taken at $100\times$ magnification; Figure 3 is for results of the analysis at $1000\times$. On the logarithmic scale is recorded the equivalent circular diameter of the particles, on the probability scale the cumulative frequency of the number of grains.

Under conditions of recording, a particle size truncation occurs at a lower limit of 0.8 μm in the case of the $1000\times$ images and about 3.0 μm for the $100\times$. In view of this truncation, the $1000\times$ results are more complete. Except for the uppermost portions of the curves (about 98% probability), the distributions are very nearly log-normal. The up–turn in the curves above 98% most likely results from the fact that particles which touch or overlap the boundary of the image are not included in the count.

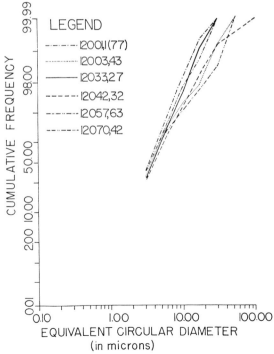

Fig. 2. Size distributions for the six samples generated from the $100\times$ images.

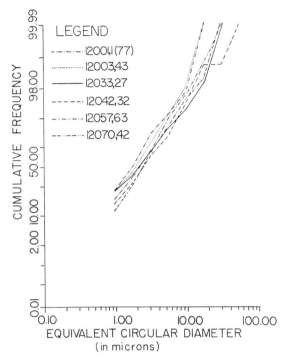

Fig. 3. Size distributions for the six samples generated from the 1000× images.

Thus, the largest particles tend not to be included in the curve and this bias has the observed effect on the shape of the distribution. Recordings at 5000× magnification showed the smallest grain size of the particles to be about 0.2 μm. The highest grain size determined with the light microscope of the six samples was around 400 μm.

DUKE *et al.* (1970) and RABINOWICZ (1970) mentioned that the lunar soil of Apollo 11 is deficient in grains smaller than 15 μm (based on weight fraction and not number

Table 1. Aspect ratios of 55 grains (equivalent circular diameter = 3.19–6.12 μm) from sample 12001,1(77)

$\dfrac{\text{major axis}}{\text{minor axis}}$	% of grains
1.00–1.29	16.4
1.30–1.59	16.4
1.60–1.89	21.8
1.90–2.19	5.4
2.20–2.49	7.3
2.50–2.79	9.1
2.80–3.09	7.3
3.10–3.39	5.5
3.40–3.69	1.8
3.70–3.99	1.8
4.00–4.29	1.8
4.30–4.59	3.6
7.25	1.8

count as used in this study). The counting of grains in the analyzed samples (Apollo 12) showed that more than the half of the grains are smaller than 10 μm.

A shape analysis has been made for 55 grains from sample 12001,1(77). The results are given in Table 1. The ratio of the major and minor axes (aspect ratio) of the least square fitted ellipse of each particle is a measure of the shape factor (MATSON *et al.*, 1970). A ratio of one indicates a circular-shaped particle. The larger the ratio the more elongated the particle. From Table 1 it is seen that the commonly occurring aspect ratio falls in the range 1.60–1.89, e.g., the grains are slightly to medium elongated.

With the exception of occasional round glass beads, the particles are typically quite angular. The angularity of the particles and the log-normal size distribution of these samples suggest mechanical comminution without subsequent sedimentary processes to sort and round the material.

Acknowledgments—This work was supported under NASA Grant NGR 39-009-(152) and through the Space Sciences and Engineering Laboratory of The Pennsylvania State University, operated under NASA Grant NGL-39-009-015.

REFERENCES

DUKE M. B., WOO C. C., BIRD M. L., SELLERS G. A., and FINKELMAN R. B. (1970) Lunar soil: Size distribution and mineralogical constituents. *Science* **167**, 648–650.

MATSON W. L., MCKINSTRY H. A., JOHNSON G. G., JR., WHITE E. W., and MCMILLAN R. E. (1970) Computer processing of SEM images by contour analyses. *Pattern Recognition* **2**, 303–312.

MCMILLAN R. E., JOHNSON G. G., JR., and WHITE E. W. (1969) Computer processing of binary maps of SEM images. Second Annual Scanning Electron Microscope Symposium, IITRI, 439–444.

RABINOWICZ E. (1970) Size distribution of lunar soil. *Nature* **228**, 1299.

WEAST R. C. (1968) *Handbook of chemistry and physics*, 49th edition. The Chemical Rubber Co.

WHITE E. W., GÖRZ H., JOHNSON G. G., JR., and MCMILLAN R. E. (1970a) Particle size distributions of particulate aluminas from computer processed SEM images. Third Annual Scanning Electron Microscope Symposium, IITRI, 57–64.

WHITE E. W., MAYBERRY K., JOHNSON G. G., JR. (1970b) Computer analysis of multi-channel SEM and X-ray images from fine particles. *Pattern Recognition*. (in press).

Proceedings of the Second Lunar Science Conference, Vol. 3, pp. 2027–2040
The M.I.T. Press, 1971.

Ultramicroscopic features in micron-sized lunar dust grains and cosmophysics

J. Borg and M. Maurette

Centre de Spectrométrie de Masse du C.N.R.S., 91-ORSAY, France

and

L. Durrieu and C. Jouret

Institut d'Optique Electronique du C.N.R.S., 31-TOULOUSE, France

(*Received* 23 *February* 1971; *accepted in revised form* 24 *March* 1971)

Abstract—Ultramicroscopic features in the finest lunar dust grains from the Apollo 11 and Apollo 12 missions have been studied by high voltage electron microscopy and compared to those observed in crushed fragments extracted from lunar and meteoritic rocks. Some striking features such as very high densities of nuclear particle tracks and amorphous coatings have been observed very frequently in the micron-sized lunar dust grains. These features have been tentatively used (1) to detect different types of low energy solar nuclear particles in view of deciphering the past history of solar activity; (2) to find some clues concerning the "fabric" of the lunar dust grains and breccias; (3) to get a better understanding of the optical and mechanical properties of lunar soils; and (4) to propose some applications of the present results to cosmophysics.

INTRODUCTION

THIS PAPER DESCRIBES a comparative electron microscope study of some radiation damage and habit and texture features observed in individual grains either extracted from 5 different size fractions of lunar dust samples 10084, 12032, 12070, and 12028 (55, 61, 62, 75, 98, 155, 203) or hand picked in lunar rocks 10046, 10047, and 12063 and in the Orgueil and Pesyanoe meteorites.

A preliminary account of this work was already published (BORG *et al.*, 1970a; BORG *et al.*, 1970b; DRAN *et al.*, 1970) or presented at the Houston Conference (BORG *et al.*, 1971). Some of our results were later confirmed by BARBER *et al.* (1971). The present paper contains new data as well as discussions where the ultramicroscopic features observed in the finest dust grains have been tentatively used (1) to detect different types of solar corpuscular radiations in view of tracing back the past history of solar activity; (2) to find some clues concerning the "fabric" of the lunar dust grains and breccias; (3) to get a better understanding of the optical and mechanical properties of the lunar soil for remote sensing purposes, and (4) to discuss some "cosmophysical" problems including the origin of meteorites enriched in solar type rare gas and the formation of solid bodies by accretion of dust particles in turbulent dust clouds surrounding young stars.

EXPERIMENTAL TECHNIQUES

The dust samples were separated into 5 size fractions (400 mesh residue, 400, 325, 200, and 100 mesh fractions) by sieving. Surface crystals from rocks 10047 and 12063 were extracted with a replicating tape. Individual grains were hand picked in internal chunks from the Orgueil and Pesyanoe meteorites and from lunar breccia 10046. The 400 mesh residue dust grains were dry deposited with a platinium loop on Fukami-Adachi type substrates; but the coarser dust and rock grains were crushed into small micron-sized fragments before being dispersed on the substrate. Before being dispersed, a fraction of the dust and rock grains, whether or not crushed, were chemically etched, heated or artificially irradiated as described elsewhere (Borg *et al.*, 1970a; Borg *et al.*, 1970b; Dran *et al.*, 1970). Some optical properties of the bulk dust samples were measured for orange light by Dollfus *et al.* (1971). Dr. R. Scott (oral communication) gave us some information concerning the bearing strength of various lunar soil samples which decreased in the following order at the Apollo 11, Apollo 12, and Surveyor VII landing sites: Apollo 11 > Apollo 12 > Surveyor VII.

We compared the results obtained by studying the same grains from sample 10084 with a 100 keV and a 1000 keV electron microscope, and we concluded that it is necessary to use a high voltage

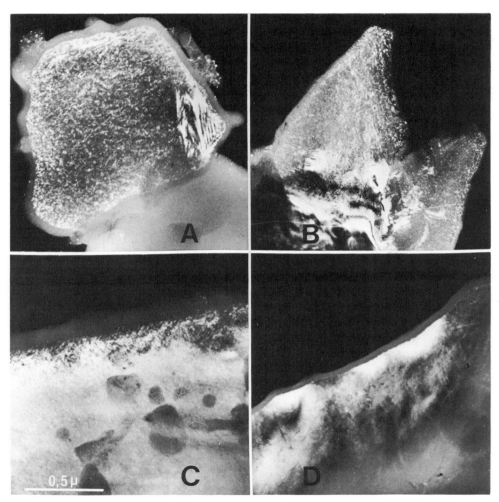

Fig. 1. 1 MeV dark field micrographs of various types of amorphous coatings observed either in sample 10084 (micrographs A and B) or in sample 12070 (micrographs C and D). In B and C we reported the most extreme variations in skin thicknesses we observed and in A and D the most frequently observed "average type" coatings.

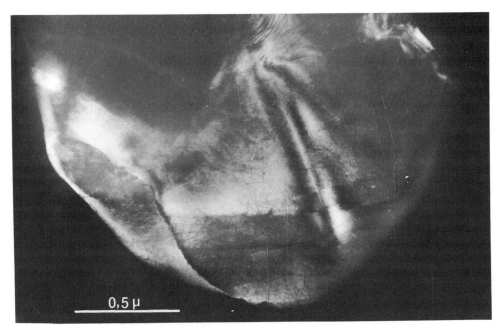

Fig. 2 1 MeV dark field micrograph of an Apollo 11 crushed dust grain containing a high density ($\sim 10^{11}$ tracks/cm²) of latent nuclear particle tracks appearing as the lines of dark contrast. The highest track densities (exceeding 10^{11} tracks/cm²) were observed in the most rounded grains generally surrounded with a coating of amorphous material.

electron microscope to study intergrain cementing and to see very faint contrast structures such as amorphous coatings (Fig. 1) and nonetched, or "latent," nuclear particle tracks (Fig. 2). Therefore the use of the 100 keV electron microscope was restricted to the study of the habit and size distribution of well-dispersed grains, to the observation of very slightly etched tracks in the dust particles (Fig. 3) and to the determination of the proportion of amorphous to crystalline grains.

Results for Lunar Dust

The following striking features, partially reported in Table 1, were generally observed in the finest and uncrushed grains extracted from the 400 *mesh dust residues* (1) the proportion of amorphous grains was generally smaller than 30% (column 2); (2) in 1 MeV dark field micrographs an important proportion of the crystalline grains —except those of ilmenite—were surrounded with a superficial layer of amorphous material (Fig. 1) whose thickness distribution is peaked at about 500 Å; (3) the electron diffraction patterns when taken below the amorphous coating reflected a good ordering of the grain lattice and indicated that the proportion of highly disordered crystalline grains was remarkably small ($\simeq 1\%$) in all dust samples; (4) about 90% of the crystalline grains contained very high densities of latent nuclear particle tracks (Fig. 2). The highest track densities ($\geq 10^{11}$ tracks/cm²) were observed in the most rounded grains showing a well developed and continuous amorphous coating (Fig. 1); in a very few of the coated grains the high track densities were hardly

Fig. 3. 100 keV bright field micrograph of an Apollo 11 grain, etched by using the "slight" etching conditions developed by Barber *et al.* (1971) This grain contains long but also short etched nuclear particle tracks appearing as the lines of white contrast. Very few tracks with length greater than 0.5 μ can be seen.

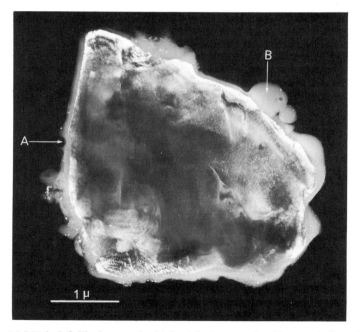

Fig. 4. 1 MeV dark field micrograph obtained during a study of intergrain boundaries. In A a small grain seems to have been lost from the coating surrounding the coarser grain and in B a definite grain surface cementing can be observed. It must be pointed out that the faintest contrast features such as amorphous coatings and latent tracks as well as grain cementings can only be easily observed with the 1 MeV microscope.

observable, only after taking several dark field micrographs of the same area with slightly different beam orientations; (5) approximately 10 to 20% of the crystalline grains showed etched tracks (Fig. 3) when using the slight etching conditions first developed by BARBER et al. (1971); however we observed nonetched latent tracks, in about 10% of the grains when using much stronger etching conditions, generally applied to enlarge the diameter of etched tracks at about 1000 Å, both in feldspar and pyroxene crystals, for scanning electron microscope observations (see Fig. 4 in BORG et al., 1970b); (6) small grains appeared frequently as cemented to the surface of coarser rounded and "coated" grains (Fig. 4); (7) heating the grains during 1 hour at 800°C under a vacuum caused the disappearance of the amorphous coatings and tracks and the growth of numerous "inclusions" (Fig. 5); (8) the coarseness of the size distribution in the 400 mesh residues varied erratically with the depth in core tube 12028 (Table 1, column 5); (9) the contrast structure of the latent tracks was generally complex (see for example the "interrupted" lines and those constituted of a central dark line edged by white lines in Fig. 2 of BORG et al., 1970b) and is currently analyzed because we believe that it should be "modulated" by the heat and shock histories of the grains and could help in deciphering such histories; (10) during the etching experiments in acid mixtures containing HF, the amorphous coatings on the grains disappeared.

Our other preliminary results bearing on a comparison of the various size fractions in dust samples 10084, 12070, 12032 are (1) the bulk uranium concentration ($\simeq 0.3$ ppm) is similar in the 400 mesh residue and in the 100 mesh fraction of sample 10084; (2) in sample 12032 stored energy measurements showed 2 exotherm peaks at about 300°C and 500°C with the 200 mesh grains but only a 300°C peak was observed with the 400 mesh residue; (3) no significant increase in the *external* etched

Table 1. Comparative studies of different size fractions of lunar dust samples.

Samples	400 mesh residues				200 mesh feldspar grains			Bulk fines	
	% amorphous	% coated	ρmax[1] ($\times 10^8$)	size[2]	% gradient[3]	ρext ($\times 10^8$)	ρint ($\times 10^8$)	$\log_{10}$ A[4]	$\log_{10}$ Pm[4]
10084	12	60	$\geq 10^3$		≤ 25	0.6–6	4–12	−1.12	−0.88
12032		15	$\geq 10^3$		$\simeq 15$	0.2–0.7	0.1–4	−0.84	−1.19
12070	20	40	$\geq 10^3$		≤ 15	1–7	1–16	−1.04	−0.95
12028,55	30	40	$\geq 10^3$	1.2			0.4–16	−0.99	−0.87
12028,61	20		$\simeq 10^3$	1.9					
12028,62	10	7	$\simeq 10^3$		≤ 25	0.04–1	0.1–0.5		
12028,75	5			0.82			0.2–16		
12028,98	4	5	$\leq 10^2$	0.73			2–25	−0.82	−1.04
12028,155							0.7–3		
12028,203	4	35	$\geq 10^3$	1.4	≤ 10	3–20	6–10	−0.92	−1.02
Breccia 10046	$\simeq 2$	< 1	$\geq 10^3$				0.5–2		

[1] If the track-producing particles are VH nuclei, then the maximum track density in the grains cannot exceed a value of about 10^{12} tracks/cm², which corresponds to artificial doses producing severe distortion in silicate lattices (SEITZ et al., 1970). [2] We reported the value of the ratio between the number of grains having sizes between 0.5 and 2 μ and 2 and 5 μ. [3] We give the proportions of 200 mesh feldspar grains showing a homogeneous edge zoned distribution of etched tracks where the surface track density decreases by a factor of about 2 to 3 on a depth $\sim 20 \mu$. These values are much smaller than those ($\sim 50\%$) obtained for glassy spherules and indicate most likely that feldspar grains were loaded with etched tracks when they were part of a rock surface, before being released in the dust. [4] These values have been measured for orange light by DOLLFUS et al. (1971).

Fig. 5. Dark field micrograph of an Apollo 11 grain, heated during 1 hour at 800°C in vacuum. This grain was probably loaded with a high concentration of rare gas and a high latent track density. Numerous new inclusions (gas bubbles or radiation induced phase transformations) have been produced during the annealing. Such "inclusions" should constitute a characteristic "extraterrestrial" signature for any dust grain, subsequently strongly heated during its accretion by the Earth or its "sintering" inside a chunk of meteorite.

track densities was observed with a scanning electron microscope when going from the 100 mesh to the 325 mesh grains, in samples 12070 and 12032; (4) by crushing the 5 different size fractions of sample 12070, we noted a marked increase in the probability of observing high latent track densities and partially broken coatings in micron-sized fragments, when going from the coarsest to the finest size fractions.

RESULTS FOR LUNAR AND METEORITIC ROCKS

None of the micron-sized fragments taken on the top surfaces of igneous rocks 10047 and 12063 showed the striking "lunar soil" pattern of amorphous coatings with high latent track densities exceeding 10^{11} tracks/cm^2 below. However in rock 12063 we saw about 5% of grains with high track densities of about 10^{10} tracks/cm^2. Finally these micron-sized fragments were constituted of about 100% of crystalline grains. By studying $\simeq 100$ individual grains in lunar breccia 10046 we found only 1 grain with a high latent track density and we observed no amorphous coating. Furthermore, small "inclusions" similar to those observed in dust grains heated under vacuum were frequently observed and the proportion of amorphous grains was much smaller ($\simeq 2\%$) than in sample 10084. These results support the idea that brecciation occurs by a heat sintering of the grains (DUKE *et al.*, 1970).

In meteoritic fragments, the latent tracks and coatings patterns were lacking when examining about 500 grains in Orgueil and 20 grains in Pesyanoe and we did not observe inclusions similar to those either developed in dust grains heated under vacuum or present in grains from breccia 10046. Furthermore the proportion of crystalline matter varied from about 10% in Orgueil to 100% in the dark parts of

Pesyanoe and a high proportion ($\simeq 10\%$) of the crystalline grains were highly disordered in Orgueil, but not in Pesyanoe. Therefore the Orgueil and Pesyanoe matrix were very different from the 400 mesh lunar dust residues which appear as a 2 component matrix made of a totally amorphous minor fraction mixed to a major fraction of well ordered crystalline grains.

ORIGINS OF THE AMORPHOUS COATINGS ON THE GRAINS AND STUDY OF THE ANCIENT SOLAR WIND

We have already demonstrated the amorphous character of the superficial coating on the grains and discussed some of its possible origins (DRAN *et al.*, 1970). We showed that it does not result from an interaction between the electron beam and the grain matter. Furthermore, it is unlikely a reaction layer produced by an atmospheric component because (1) the coating matter behaves similarly to the radiation damaged core of the tracks in being totally annealed when the grains are heated in vacuum at 800°C, during 1 hour; (2) any residence time effect in air or in nitrogen seems precluded by the erratic variation in the proportion of coated grains with the depth in core tube 12028 (Table 1, column 3); (3) comparative weathering (KELLER and HUANG, 1971) and exoelectron emission (BECKER and GAMMAGE, 1971) studies of lunar dust grains and crushed terrestrial basalts show that the surface of lunar dust grains is no more reactive to H_2O, $H_2O + CO_2$, and O_2 than terrestrial basalt as long as the grains are not heated; therefore a reaction layer origin for the dust grain coating is unlikely because we never observed such coating on crushed fragments extracted from lunar or meteoritic rocks or from oceanic basalts. Therefore we think that the amorphous coatings on the grains have a lunar origin. They could result from a vapor deposition on the grains (HAPKE *et al.*, 1971) or from a solar wind implantation producing either a sputtered coating (GOLD *et al.*, 1970) or a superficially metamictized layer (BORG *et al.*, 1970a; DRAN *et al.*, 1970).

By associating the annealing experiment just described, the striking similarity between the skin thicknesses and the expected ranges of solar wind type ions in solids and the straight relationship between the proportion of coated grains and the solar wind rare gas contents of the Apollo 11 and Apollo 12 soils, we infer that the amorphous coating is a solar wind metamictized layer. Thus, by studying variations in the coating thicknesses and in the proportion of coated grains with the depth in a "stratified" core tube it should be possible to detect any marked changes in the energy on intensity of the solar wind nuclei in the past (DRAN *et al.*, 1970). Such variations do exist (Fig. 1 and Table 1, column 3) but before relating them to changes in solar wind characteristics, one must study the growth of amorphous coatings produced by artificial solar wind type implantations in grains having different radiation stabilities.

With this solar wind origin for the coating, the much greater solar wind rare gas content in ilmenite grains could be attributed to their better radiation stability as reflected by the lack of a superficial coating. Therefore rare gas ions in ilmenite will not diffuse away as easily as those implanted in the amorphous layers of the more readily damaged silicate grains. Furthermore, the determination of the critical dose of solar wind type ions producing an amorphous coating in ilmenite grains will give an interesting *upper* limit for the exposure time of the grains in the solar wind.

Origins of the Latent Tracks and "Fabric"
of the Lunar Dust Grains

In natural samples very high track densities can only be produced by a bombardment of the grains in low energy solar cosmic rays or by the spontaneous fission of very heavy elements in high uranium bearing phases (WALKER, 1971). If the track densities in excess of 10^{10} tracks/cm² observed in 90% of the crystalline grains of samples 10084 were due to spontaneous fission, the average uranium content in the 400 Mesh residue of this sample should largely exceed that measured ($\simeq 0.3$ ppm) in the 100 Mesh fraction and in the bulk dust sample; furthermore, the α-recoil tracks stored in the grains should produce their metamictization. These 2 features are not observed (BORG *et al.*, 1971) and therefore the track producing nuclear particles are low energy heavy ions of solar origin.

These particles could be found either in the well-known solar flare cosmic rays or in a plausible "suprathermal" ion flux the proton component of which has been recently detected in space by FRANK (1970). Etched tracks studies in the Surveyor III glass filter (CROZAZ and WALKER, 1971; FLEISCHER *et al.*, 1971; PRICE *et al.*, 1971) indicate that the track production rate by solar flare VH nuclei is about 10^6 tracks/cm²/ year, in grains smaller than 10 microns. Therefore a relatively short exposure ($\geq 10^4$ years) of the grains on the surface of the Moon could apparently well account for the high track densities.

However, we favor a suprathermal ion origin for the tracks for the following reasons: (1) by supposing that VH ions are emitted simultaneously with the suprathermal protons discovered by FRANK, we deduce, by using the proton flux values quoted by FRANK and a hypothetical VH/p abundance ratio similar to that ($\simeq 10^{-5}$) measured in the solar flare and galactic cosmic rays, that the exposure times required to get track densities in excess of 10^{10} tracks/cm² with such ions are lower ($\geq 10^3$ years) than those deduced from the solar flare irradiation of the Surveyor III glass filter ($\geq 10^4$ years). Therefore they are more compatible with the minimum exposure age of the grains in the solar wind (≥ 300 years) deduced from saturation rare gas values (EBERHARDT *et al.*, 1970). Then our observation (BORG *et al.*, 1970a, 1971) of a correlation between the extent in growth of the amorphous coating and the track density below—the highest track densities are observed in the most rounded and coated grains—could be explained by assuming simply that the grains have been simultaneously implanted with solar wind and suprathermal nuclei during the *same* exposure on the lunar surface; (2) the distribution of the projected lengths of the etched tracks in Fig. 3 seems shorter than that expected for solar flare tracks; (3) pieces of evidence, to be discussed in the next paragraph, indicate that the variation of the high track density with the depth inside a grain is very steep. This variation is more compatible with a "pulse" shaped differential energy spectrum, similar to that reported by FRANK (1970) for suprathermal protons, than with kinetic energy spectrums similar to those measured recently for low energy solar flare cosmic rays by E. STONE (1971) and ARMSTRONG and KRIMIGIS (1970).

The evidence for a steep energy spectrum is (1) the marked increase in the probability of observing high track densities in broken fragments when the size of the initial grain decreases; this observation points out to a registration of the high track densities in a "micron-sized" superficial layer; (2) the values of the etched track densities observed by scanning electron microscopy, in the external surfaces of "strongly" etched grains extracted from the 325, 200, and 100 mesh fractions of samples 12070 and 10084, which do not depend on the grain size and are generally clustered in the range 10^8 to 10^9 tracks/cm² (Table 1, column 7). Such values are much lower than those ranging from 10^{10} to 10^{11} tracks/cm², observed by high voltage electron microscopy in the micron-sized regolith grains, either before (Borg et al., 1970a; Dran et al., 1970) or after (Barber et al., 1971) a slight chemical etching, 70 times weaker than that used for scanning electron microscope observations. This is indeed a stunning difference which could be explained by assuming that 90% of the crystalline grains in the 400 mesh residue from sample 10084 have had an "exotic" history, very different from that of the 325, 200, and 100 mesh grains. The finest regolith grains could be for example cosmic dust particles, exposed during long times in space to solar flare VH nuclei and subsequently accreted by the Moon. This accretion theory was first described by Gold (1971) and is advocated in the work of Barber et al. (1971). The other alternative is that the finest grains are similar to those extracted from the coarser fractions in being lunar rock degradation products (Shoemaker et al., 1970) irradiated with suprathermal ions (Borg et al., 1970a; 1970b; 1971). Then from the energy spectrum reported by Frank (1970) it can be expected that such ions have a "pulse" shaped spectrum. Therefore their maximum implantation depths in the grains would be strongly peaked at about 1 μ and they would be essentially registered in a micron-sized superficial layer in all the dust grains, whatever their size may be. Then after the "strong" etching required to see tracks with the scanning electron microscope, this superficial layer would have been etched out completely, thus leaving an internal surface showing only the smaller track densities due to low energy solar flare VH nuclei.

The validity of this suprathermal origin can be first criticized on the basis that the etched track studies in the Surveyor III glass filter show a "lack" of track producing particles with energy smaller than 0.5 MeV/amu. But these results could be as well explained by the large cone angle of the glass, preventing the observation of the tracks of particles with very short range even if they are very heavy, and by the complex multiple "micron-sized" layers overlaying the glass filter and made of MgF_2, silane, and dust grains (Hart, oral communication). Furthermore from track registration studies in minerals (Fleischer et al., 1967) it could be argued that suprathermal VH nuclei are not registered as etchable tracks in the dust grains. But preliminary results obtained by Peter (oral communication) show that this conclusion is no longer valid. Rare gas people could help greatly in checking the validity of these conclusions in verifying if the depth dependence of the rare gas content in ilmenite grains is compatible with a suprathermal spectrum in showing a marked drop at depths of about 1 to 2 μ. But at the present time the comparison of the meager rare gas data (Eberhardt et al., 1970; Kirsten et al., 1970) to the track densities allow us only to deduce that low energy solar VH nuclei are the most likely track producing particles.

We think that the finest regolith grains are not cosmic dust particles accreted by the Moon but lunar rock degradation products. This is supported by the correlation between the existence of a coating and that of high track densities below indicating that these 2 irradiations features have been most likely produced simultaneously. If these irradiations had occurred in space, it would be extremely difficult to explain how the very homogeneous coating surrounding the grains could have survived an impact on the Moon without being at least partially broken. However, Walker (oral communication) suggested that the grains could have first been irradiated in space, then shielded inside an "icy" cometary body which fell on the Moon thus releasing "unbroken" coated grains. But if this hypothesis is true a *marked* contamination of the lunar soil in chondritic material should have been discovered and this is not observed (Anders *et al.*, 1971; Borg *et al.*, 1971).

Coated Grains and the Mechanical Properties of the Regolith

If the amorphous coating on the silicate grains is composed of purely radiation-damaged material then it is likely to contain a very high stored energy. Such energy could be released when two grains come together in a mechanical collision. This could in turn melt the interface and produce a sticking of the grains—the new sintering mechanism based on stored energy release was kindly suggested to one of us (M. Maurette) by Dr. A. Turkevich during a discussion concerning the mechanical properties of the lunar dust. This would explain our observations where we see several small particles stuck to the surface of coarser and "coated" grains with an apparent continuity in the surface films. If this effect occurs, it might well (1) influence the formation of breccias; the relative scarcity of this type of rock at the Apollo 12 site would thus be linked with the fact that the proportion of coated grains is $\simeq$ 3 times less than in the Apollo 11 soil; and (2) explain the formation of a layer of micron-sized dust grains strongly cemented to some constituent parts of Surveyor III (Nickle, oral communication) by coated lunar dust grains, set in balistic motion by the Surveyor III or Apollo 12 LEM rocket exhausts and colliding "mechanically" with the Surveyor parts.

Grains showing such "cemented" structure should be more easily "anchored" in the dust blanket, therefore offering more resistance to penetration and turn over and thus we predict that the Apollo 11 soil, having a greater proportion of coated grains is more resistant to penetration than the Apollo 12 soil.

Relation of Microstructure to Optical Properties

The orange light albedo varies from 15% to 7% between different samples of fines (10084, 12070, 12032) and samples removed from different layers of the Apollo 12 core tube. As shown in Fig. 6 there is a strong correlation between the proportion of coated grains and the reduction in albedo. There seems to be no correlation with the proportion of glass in the samples. This lends support to the suggestion by Gold *et al.* (1970) that irradiation of the grains plays an important role in determining the optical properties. In this connection we wish to point out that it is not necessary to blacken the particles with color centers or cover them with a sputtered metallic coating. A simple amorphous layer on a silicate grain, composed of the same material as the grain itself, will have a smaller index of refraction and could act to trap incident radiation by internal reflection. This influence of the coating on the optical properties

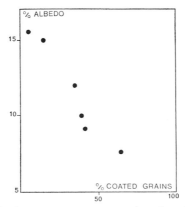

Fig. 6. This curve results from an attempt to correlate the microscopic characteristics of the 400 mesh residue grains, as viewed by high voltage electron microscopy, to the optical and mechanical properties of the bulk dust samples. The albedo varies clearly with the proportion of coated grains but not with that of amorphous grains.

of the grain is consistent with the observation that a "slight" etching of the lunar soil, which most likely dissolves the thin amorphous coating, also strongly increases the albedo of the sample (HAPKE *et al.*, 1971). On the other hand the albedo only slightly increases after heating the soil samples at 900°C, during 1 hour under vacuum (GOLD, oral communication). After such a treatment the amorphous coating disappears (BORG *et al.*, 1971). Therefore it would be argued that this coating has no influence on the optical properties of the lunar soil. However, we want to point out that the interpretation of the annealing experiment in term of albedo is very difficult for 2 reasons: (1) metamictized zircons recrystallize upon heating; however their index of refraction is still lower than that measured in non-metamictized crystals (PELLAS, 1965). Therefore the multidielectric structure of the finest dust grains where the most superficial layer has an index of refraction smaller than that below is probably preserved after the annealing, thus explaining why the optical properties of the soil cannot be strongly modified after such a treatment; (2) after the heating, aggregates of cemented grains similar to that reported in Fig. 5 have been produced in the 400 mesh residue which is therefore no longer comparable to the initial sample.

If radiation always plays the dominant role in reducing the albedo, then the high albedo of the highlands could be taken as evidence for less radiation processing of the regolith. Since this is contrary to much of what is believed, it is likely that other processes such as the redistribution of strongly absorbing ions such as Ti (CONEL and NASH, 1970) also influence the reduction of albedo. The average albedo at several landing sites on the lunar surface decreases in the following order: Apollo 11 < Apollo 12 < Apollo 14 < Surveyor VII (DOLLFUS, oral communication). Then if our correlation between the optical and mechanical properties of the grains and their ultramicroscopic features is well founded, we should find that the resistance to penetration at these 4 landing sites should decrease in the *inverse* order. This order corresponds to that suggested by F. SCOTT (oral communication) from totally independent evidence. Therefore the simple correlations that we have proposed to relate microscopic features to macroscopic properties seem to be partially justified.

Lunar Dust Grains and Cosmophysics

The sintering process we described to account for the dust grain cementing is different from other "radiation" sintering mechanisms previously proposed to explain the cohesion of lunar grains and based upon sputtering (Wehner *et al.*, 1964) and diffusion (Smoluchowski, 1965) because it can as well work with the condensed dust in the lunar soil as with extremely dispersed and turbulent dust clouds in which the grains have been superficially metamictized and can collide occasionally. Therefore it could be responsible for the first "accretion" step necessary to the development of a planetary system in dust nebulae surrounding young, T-Tauri type stars. Indeed in such nebulae all the necessary cementing "ingredients" are present: there is an ample supply of solar wind type ions which are injected at a rate 10^5 to 10^6 times higher than the present solar wind mass loss by the Sun, and marked turbulence in the dust clouds motions are observed. Therefore lunar type "microbreccias" could be formed during the first stage of the accretion step in an active volume where a "turn-over" mechanism is at work to continuously expose the grains in low energy solar nuclear particles.

Our thermal annealing experiment shows that if a cosmic dust grain is loaded with rare gas and tracks in space and subsequently accreted by the earth a characteristic extra-terrestrial signature could be still registered in the grain even if it gets strongly heated in the earth atmosphere. This signature will be constituted by a high density of inclusions, similar to those appearing in Fig. 5. These inclusions grow—and do not disappear—when the annealing temperature is increased and they should help in finding an extraterrestrial component in antarctica dust.

Finally, the study of the same thermally induced inclusions when associated with the observation of ultramicroscopic features in grains extracted from lunar breccias and meteorites should also give a new way of identifying those of the meteorites which result from the compaction of grains individually irradiated by low energy solar nuclear particles either in space or on the surface of a "lunar type" parent body. In such meteorites it could be expected that the grains contain either a typical lunar soil pattern of high track densities below the amorphous coatings or extended "thermal" inclusions if the sintering of the grains in the meteorite was triggered by an extensive metamorphism.

As a preliminary attempt to check this idea we examined about 500 grains extracted from the fine grained matrix of a carbonaceous type I chondrite, Orgueil, and about 20 grains from the dark part of a typical solar type gas-rich meteorite, Pesyanoe. The Orgueil meteorite is often described as a primitive object (Mason, 1964; Anders, 1970) resulting from the accretion of grains produced by a primordial "condensation" in the early solar system and the grains constituting the dark parts of a solar type gas-rich meteorite are considered by several investigators to have been individually irradiated in the "ancient" solar wind (Eberhardt *et al.*, 1965) and low energy solar flare cosmic rays (Lal and Rajan, 1969; Pellas *et al.*, 1969). None of the meteoritic grains showed the ultramicroscopic irradiation record observed frequently in the lunar dust grains, whether or not sintered into a "breccia." However, after a scanning electron microscope observation, Pellas (oral communication) measured high densities of etched tracks in $\simeq 30\%$ of the pyroxene grains in Pesyanoe, that he attributed to an individual irradiation of the grains in low energy solar flare cosmic rays. These etched track densities, ranging from 10^8 to 10^9 tracks/ cm², are quite similar to those in the lunar dust grains. Therefore they indicate that the lunar and Pesyanoe grains have had a similar "individual" irradiation history with low energy solar particles if the interpretation of PELLAS is correct. Then the non observation of "thermal" inclusions in Pesyanoe is puzzling because this meteorite has certainly been more metamorphized than the lunar dust grains constituting breccia 10046 and already loaded with such inclusions. Therefore, there is a difference between the ultramicroscopic irradiation record of meteoritic and lunar dust grains, considered as having been similarly irradiated by low energy solar cosmic rays, before their "sintering" into the meteoritic or lunar breccia. This difference has first to be ascertained by examining more grains both in gas-rich meteorites and in different types of lunar breccia; then its understanding will certainly give some clues concerning the origin of meteoritic matter considered either as very primitive or extremely enriched in solar type rare gas.

Acknowledgments—This work has been made possible by the generous cooperation of Professors G. DUPOUY and F. PERRIER from the Institut d'Optique Electronique du CNRS, Toulouse. The efficient help we received from R. SIRVIN and G. GONTRAN from the same Institution is gratefully acknowledged. One of us (M. MAURETTE) is deeply indebted to Drs. R. BERNAS, A. TURKEVICH, and R. M. WALKER for their very active and enthusiastic support and interest, and to Drs. E. ANDERS, E. STONE, and R. SCOTT for stimulating discussions. We also thank Dr. R. HUTCHINSON (British Museum, London) for the loan of meteoritic samples. We would also like to express our appreciation to J. CARO for assistance in manuscript preparation.

REFERENCES

ANDERS E. (1970) Meteorites and the Early Solar System. Univ. Chicago Rep. No. COO-382-107.

ANDERS E., LAUL J. C., KEAYS R. R., GANAPATHY R., and MORGAN J. W. (1971) Elements depleted on lunar surface: Implications for Origin of Moon and meteorite influx rate. Second Lunar Science Conference (unpublished proceedings).

ARMSTRONG T. P. and KRIMIGIS S. M. (1970) A statistical study of solar protons, alphas, and $Z > 3$ nuclei in 1967–68. Johns Hopkins Univ. Applied Physics Laboratory Preprint.

BARBER D. J., HUTCHEON I., and PRICE P. B. (1971) Extralunar dust in Apollo cores? *Science* **171**, 372–373.

BECKER K. and GAMMAGE R. B. (1971) Exoelectron emission and surface characteristics of lunar material. Second Lunar Science Conference (unpublished proceedings).

BORG J., DRAN J. C., DURRIEU L., JOURET C., and MAURETTE M. (1970a) High voltage electron microscope studies of extraterrestrial matter. 7th Int. Colloquium Corpuscular Photography and Visual Solid Detectors, Barcelona, paper No. 41.

BORG J., DRAN J. C., DURRIEU L., JOURET C., and MAURETTE M. (1970b) High voltage electron microscope studies of fossil nuclear particle tracks in extraterrestrial matter. *Earth Planet. Sci. Lett.* **8**, 379–386.

BORG J., DURRIEU L., JOURET C., and MAURETTE M. (1971) The ultramicroscopic irradiation record of micron-sized lunar dust grains. Second Lunar Science Conference (unpublished proceedings).

CONEL J. E and NASH D. B. (1970) Spectral reflectance and albedo of Apollo 11 lunar samples: effects of irradiation and vitrification and comparison with telescopic observations. *Proc. Apollo 11 Lunar Sci. Conf.*, Geochim. Cosmochim. Acta Suppl. 1, Vol. 3, pp. 2013–2023. Pergamon.

CROZAZ G., HAACK U., HAIR M., MAURETTE M., WALKER R. M., and WOOLUM D. (1970) Nuclear track studies of ancient solar radiations and dynamic lunar surface processes. *Proc. Apollo 11 Lunar Sci. Conf.*, Geochim. Cosmochim. Acta Suppl. 1, Vol. 3, pp. 2051–2080. Pergamon.

CROZAZ G. and WALKER R. M. (1971) Solar particle tracks in glass from the Surveyor III spacecraft. Second Lunar Science Conference (unpublished proceedings).

DOLLFUS A., GEAKE J. E., and TITULAER C. (1971) Polarimetric and photometric properties of Apollo lunar samples. Second Lunar Science Conference (unpublished proceedings).

DRAN J. C., DURRIEU L., JOURET C., and MAURETTE M. (1970) Habit and texture studies of lunar and meteoritic materials with a 1 MeV electron microscope. *Earth Planet. Sci. Lett.* **9**, 391–400.

DUKE M. B., WOO C. C., SELLERS G. A., BIRD M. L., and FINKELMAN R. (1970) Genesis of lunar soil at Tranquillity Base. *Proc. Apollo 11 Lunar Sci. Conf., Geochim. Cosmochim. Acta* Suppl. 1, Vol. 1, pp. 347–361. Pergamon.

EBERHARDT P., GEISS J., and GROGLER N. (1965) Further evidence on the origin of trapped gases in the meteorite Khor-Temiki. *J. Geophys. Res.* **70**, 4375.

EBERHARDT P., GEISS J., GRAF H., GROGLER N., KRAHENBUHL U., SCHWALLER H., SCHWARZMULLER J., and STETTLER J. (1970) Trapped solar wind noble gases, exposure age and K/Ar-age in Apollo 11 lunar fine material. *Proc. Apollo 11 Lunar Sci. Conf., Geochim. Cosmochim. Acta* Suppl. 1, Vol. 2, pp. 1037–1070. Pergamon.

FLEISCHER R. L., HART H. R., and COMSTOCK G. M. (1971) Very heavy solar cosmic rays: Energy spectrum and implications for lunar erosion. Second Lunar Science Conference (unpublished proceedings).

FLEISCHER R. L., HUBBARD E. L., PRICE P. B., and WALKER R. M. (1967) Criterion for track registration in solids. *Phys. Rev.* **156**, 353–357.

FRANK L. A. (1970) On the presence of low-energy protons ($5 \leq E \leq 50$ keV) in the interplanetary medium. *J. Geophys. Res.* **75**, 707–716.

GOLD T., CAMPBELL M. J., and O'LEARY B. T. (1970) Optical and high-frequency electrical properties of the lunar sample. *Proc. Apollo 11 Lunar Sci. Conf., Geochim. Cosmochim. Acta* Suppl. 1, Vol. 3, pp. 2149–2154. Pergamon.

GOLD T. (1971) The nature of the surface of the Moon. *Proc. Amer. Phil. Soc.* (in press).

HAPKE B. W., CASSIDY W. A., and WELLS E. N. (1971) The albedo of the Moon: evidence for vapor-phase deposition processes on the lunar surface. Second Lunar Science Conference (unpublished proceedings).

KELLER W. D. and HUANG W. H. (1971) Response of Apollo 12 lunar dust to reagents simulative of those in the weathering environment of the Earth. Second Lunar Science Conference (unpublished proceedings).

KIRSTEN T., MULLER O., STEINBRUNN F., and ZAHRINGER J. (1970) Study of distribution and variations of rare gases in lunar material by a microprobe technique. *Proc. Apollo 11 Lunar Sci. Conf., Geochim. Cosmochim. Acta*, Suppl. 1, Vol. 2, pp. 1331–1343. Pergamon.

LAL D. and RAJAN R. S. (1969) Observations relating to space irradiation of individual crystals of gas-rich meteorites. *Nature* **223**, 269–271.

MASON B. (1963) The carbonaceous chondrites. *Space Sci. Rev.* **1**, 621–646.

PELLAS P. (1965) Etude sur la recristallisation thermique des zircons metamictes. *Mém. Museum Nat. Hist. Nat.* **12**, 227–253.

PELLAS P., POUPEAU G., LORIN J. C., REEVES H., and AUDOUZE J. (1969) Primitive low energy particle irradiation of meteoritic crystals. *Nature* **223**, 272–274.

PRICE P. B., HUTCHEON I., COWSIK R., and BARBER D. J. (1971) Enhanced emission of iron nuclei in solar flares. Second Lunar Science Conference (unpublished proceedings).

SEITZ M., WITTELS M. C., MAURETTE M., WALKER R. M., and HECKMAN (1970) Mineral irradiations and implications to lunar and meteoritic materials. *Rad. Effects* **5**, 143–151.

SHOEMAKER E. M., HAIT M. H., SWANN G. A., SCHLEICHER D. L., SCHABER G. G., SUTTON R. L., DAHLEM D. H., GODDARD E. N., and WATERS A. C. (1970) *Proc. Apollo 11 Lunar Sci. Conf., Geochim. Cosmochim. Acta* Suppl. 1, Vol. 3, pp. 2399–2412. Pergamon.

SMOLUCHOWSKI R. (1966) Structure and Coherency of the lunar dust layer. *J. Geophys. Res.* **71**, 1569–1573.

STONE E. (1971) Low energy solar cosmic rays. *Phys. Rev. Lett.* (in press).

WALKER R. M (1971) Fossil track studies in extraterrestrial materials. *Rad. Effects* (in press).

WEHNER G. K., ROSENBERG D. L., and KENKNIGHT C. E. (1964) Investigation of sputtering effects on the Moon's surface. NASA Rep. No. CR 56292, N 64-24023.

Proceedings of the Second Lunar Science Conference, Vol. 3, pp. 2041–2047
The M.I.T. Press, 1971.

Morphology and petrostatistics of regular particles in Apollo 11 and Apollo 12 fines

GEORGE MUELLER

Institute of Molecular Evolution, University of Miami, Coral Gables,
Florida 33134

(*Received* 22 *February* 1971; *accepted in revised form* 31 *March* 1971)

Abstract—The regular particles of Apollo 11 and Apollo 12 fines and conglomerates are globules, spheroids, dumbbells, tear drops, rings, and crescents of glass and crystalline silicates. The forms can be derived through rotational elongation and flattening of freely suspended spray particles, which process is also indicated with the apparent centrifugation of denser phases towards the extremities of the zoned particles. Regular displacements of particles along alternate sides of termination of the long axes seem to indicate change of rotational momentum in the course of solidification of the sprays. The frequent sintered appendages suggest gentle settling of the still-hot particles on the lunar surface. Periodic reheating of particles is indicated by recrystallization, partially healed flaws, and fused edges of fragments. The distribution patterns of the diverse morphological types of particles from all the hitherto studied localities are similar to each other, throughout a diameter range between 0.5 and 200 μ. The ratio of elongate to globular particles increases with density, magnetism, darkening color, and increasing optical inhomogeneity of the particles.

The increase of elongation ratios with optical inhomogeneity of the particles seems to indicate that their rotation was caused by intensive radiation. This and other evidence seem to support the hypothesis that the bulk of lunar sprays was formed through decrepitation and subsequent melting of dust over the lunar surface, caused by unusually intensive solar flares.

INTRODUCTION

THIS PAPER DESCRIBES and interprets the structures and the statistical morphology of the regular particles. Sample 10086 of the fines and samples 10019 and 10091 of the conglomerates of Apollo 11, and Apollo 12 samples 12033 (the trench sample), 12001, 12032, 12037, and 12042 of the fines were observed under the Wild petrographic and Leitz phase contrast microscopes and AMR scanning microscope. Portions of the samples were sieved, washed, and sedimented in chloroform, gravitationally separated with bromoform, and mounted. The spheroidal particles were concentrated by repeated rollings on inclined paper sheets. Chemical composition of Apollo 11 samples was studied with the Philips electron microprobe, and the visible absorption spectra of the glasses was determined by the microspectrophotometer of STROHER and WOLKEN (1959).

OBSERVATIONS

The regular particles include globules, elongated forms, namely, prolate spheroids, cylindrical spheroids, dumbbells, and tear drops; flattened forms, namely, oblate spheroids, rings, and crescents with or without central films (Figs. 1, 2, and 3).

All the rare structures which were observed by FOX *et al.* (1970) and MUELLER and HINSCH (1970) among the Apollo 11 regular particles in fractional percentages

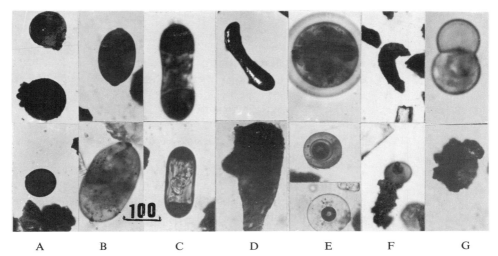

A   B   C   D   E   F   G

Fig. 1. Comparison of regular particles of Apollo 11, 10086 (top row) and Apollo 12, 12001 (bottom row). A. Globules with appendages. B. Prolate spheroids with knobs of nickel–iron. C. Dumbbell (top) and cylindrical spheroid (bottom) with symmetrical zones of brown glass. D. Distorted dumbbell with fused appendage (top), tear drop with fused appendage (bottom). E. Ring with central film (top), two rings with central films including bubbles (bottom). F. Free crescent (top), crescent with glass film and long sintered appendage (bottom). G. Smooth twin (top) rugged twin with parallel appendages (bottom).

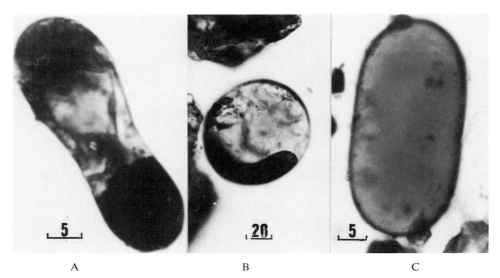

A   B   C

Fig. 2. Structures indicating centrifugation. A. Zoned dumbbell with thinner lobe occupied with dark glass, Apollo 12, 10086 B. Ellipsoid of pale yellow glass with the shortest axis vertical to the plane of the photomicrograph. Note droplet of dark brown glass squeezed into an equatorial position. Apollo 12, 12033. C. Displacement of pyroxene grains along alternate sides of the long axis of a cylindrical spheroid of brown glass, Apollo 11, 10086.

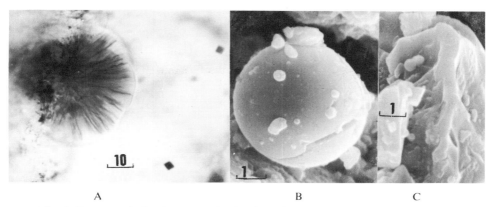

A   B   C

Fig. 3. Structures indicating events in the thermal history of particles. A. Part of a
globule of yellow glass with spherulite of an unidentified crystallite and octahedra of
spinel, a high temperature mineral, Apollo 12, 12001. B. Scanning electron photomicrograph of glass globule with plateaux-like circular protuberance and partially sintered fracture. Apollo 12, 12032. C. SEM of fragment of glass with indications of
incipient fusion of the edge, Apollo 12, 12032.

were also detected in approximately the same abundance in the Apollo 12 regular
particles.

The abundance of twinned regular particles (Fig. 1G) was found to increase from
about 0.1 % to 1.5 % of the total of regular particles, with the decrease of diameters
from 500 to 2 μ, in both Apollo 11 and Apollo 12 samples.

Approximately 0.4 % of the elongate particles of Apollo 11 fines, and 0.8 % of the
elongates of Apollo 12 fines, show zoning, which structure consists of concentrations
of orange and dark-colored glasses (Fig. 1C), further troilite and nickel–iron (Fig. 1B)
towards the extremities of the particles. Zonation may be restricted to only one
extremity, which is usually narrower than the opposite one (Fig. 2A).

The ellipsoid of flattened habit illustrated in Fig. 2B has a flattened droplet of
dark glass in the equatorial position, which tapers in a clockwise direction, and bulges
in an anticlockwise direction. Crystalline silicates symmetrically displaced on alternate
sides of the long axis of elongate particles have been observed in both Apollo 11 and
Apollo 12 fines (Fig. 2C).

The results of preliminary work with the electron probe indicate that the orange
to dark brown glass occupying the extremities of elongate or flattened particles is
exceptionally rich in titanium, and the microspectra of these glasses show a maximum
absorption feature at 340 μm, which is absent from the nonzoned glasses.

The surfaces of some of the particles are smooth. Other clearly recognizable
spheroidal particles in both the Apollo 11 and Apollo 12 samples have an uneven,
rugged surface (Fig. 1G, bottom). Approximately 40 % of the smooth and rugged
globular particles have a single appendage of sintered dust (Fig. 1A, top; 1F, bottom).
The appendages of elongate particles are usually of glass in optical continuity with the
rest of the particle (Fig. 1D). Most of the elongate particles with appendages are

distorted (Fig. 1D, top). Twinned particles with appendages in parallel position have been observed (Fig. 1G).

Fractured dumbbells are common but impact cratered or fractured particles of less vulnerable morphology are rare among the fines of Apollo 11 and Apollo 12. The markedly vesicular particles proved to contain about 5 times more surface craters than the particles of nonvesicular glass, indicating that a proportion of the craters may be scars of bubbles which burst when the particle was in a semisolid state. A very rare glass-spinel porphyry observed in the Apollo 11 fines by Fox *et al.* (1970) was also detected in the Apollo 12 fines (Fig. 3A). Partially sintered cracks (Fig. 3B) and indications for fusion of the edges of glassy fragments (Fig. 3C) have been observed.

Estimates of the number of regular particles per milligram in the diverse Apollo 11 and Apollo 12 samples are given in Table 1. It is shown in Table 2 that the ratio between markedly elongate and markedly globular particles remains within the same rather narrow limits throughout all the hitherto studied lunar localities. Counts of the diverse types of regular particles, on the basis of which Tables 2 and 3 have been computed, indicates the following generally applicable distributional trends: (A) Crystallized regular particles are predominantly single crystals of pyroxene, felspar, etc., with subordinate aggregates and porphyries; the rugged particles within each diameter range contain about a tenfold higher proportion of crystalline to glassy

Table 1. Estimates for the number of regular particles per mg of lunar fines

	10086	12033	12001	12037
Smooth particles	6,930	5,250	14,430	6,750
Rugged particles	2,200	3,050	2,750	1,350

The number of regular particles per mg of lunar fines was estimated on making 100 random microscopical counts of a total of 0.02 mg material, which represented 1% of the area of slides prepared from 2.00 mg lunar fines. The maximum likely error of these estimations is in the order of $\pm 10\%$. It appears from tentative estimations based on counts of 0.005 mg fines or ground conglomerates, that the Apollo 11, 10019 and 10091 conglomerates contain about 500 smooth and 2000 rugged particles per mg whereas the abundances of regular particles within the Apollo 12, 12032 and 12042 samples are well within a factor of 2 of those of Apollo 12, 12037.

Table 2. Comparison of ratios of elongate to globular* regular particles in Apollo 11 and 12 samples.

Diameter	A. 11 10086 (Fines)	A. 11 10091,3 (Conglomerate)	A. 12 12033 (Fines)	A. 12 12001 (Fines)	A. 12 12032 (Fines)	A. 12 12037 (Fines)	A. 12 12042 (Fines)
0.5–1	0.27	0.365	0.385	0.445	0.715	0.725	0.16
1–2	0.21	0.72	0.46	0.64	0.735	0.24	0.43
2–5	0.52	0.875	0.63	0.64	0.50	0.52	0.43
5–10	0.74	0.47	0.68	0.74	0.59	0.535	0.34
10–20	0.93	0.56	0.69	0.915	0.785	0.465	0.385
20–50	0.51	0.59	0.595	0.755			
50–100	0.43		0.51	0.49			
100–200	0.47			0.545			
200–500	0.56						

* Globular particle has eccentricity below 10%; elongate particle has eccentricity above 10%. Each of the values in Tables 2 and 3 is based on the counting of 100 smooth, globular particles, a single lobe of a fractured dumbbell counting as 0.5 elongate particle.

Table 3. Relation between densities and ratios of elongate to globular regular
particles of Apollo 11 and Apollo 12 fines

Sample	Range of diameters	Density below 2.85	Density above 2.85
A. 11 10085	10–105 m	0.47	0.68
A. 12 12033	44–105 m	0.31	0.79
A. 12 12001	44–105 m	0.45	0.59

particles than the smooth regular particles. (B) With the increase of diameters from
0.5 to 200 μ, the ratio between glassy and crystalline particles increases from approximately 0.001 to 0.8, and simultaneously the ratio of rugged to smooth particles
increases from about 0.01 to 2. (C) Elongated particles are about 15 times more
frequent than flattened particles. The only feature displaying a difference beyond
likely statistical error is that the Apollo 11 regular particles contain about 1% dumbbells, whereas the Apollo 12 fines contain approximately 2% dumbbells.

It was observed that in Apollo 11 and Apollo 12 fines the mean of the ratio
between elongate and globular particles increases with (A) density (Table 3), (B)
darkening shade, (C) magnetism, and (D) optical inhomogeneities, which have been
found to increase with density.

Some Theoretical Considerations

The morphology of the regular particles clearly indicates that they are sprays which
solidified in a freely suspended state. Intensive rotation could explain the development
of rotationally elongated and flattened forms, and the structures of zonation in which
darker and presumably denser material tended to centrifuge towards the marginal
extremities of the particles. Displacements of particles or glassy phases from alternate
sides of the long axes of some particles seem to indicate change of rotational momentum at the time of solidification of the spray.

The presence of sintered or fused appendages also suggest a low velocity settling
of the still hot particles on the lunar surface; cratering indicative of high velocity
impacts seems to be relatively rare. The spinel-glass porphyries suggest maximum
temperatures exceeding about 1800°C in the course of formation of some of the spray
particles; rise of temperature subsequent to solidification seems to be indicated by the
twinning, high proportion of recrystallized rugged, presumably old particles, and
further by the partially healed gushes and fused edges of fragments.

According to data given by Schmitt et al. (1970) and LSPET (1970), the Apollo 12
sample 12001 originates from the smoothest and least disturbed lunar surface of all
the samples which have been studied in the present work. It is interesting to note from
Table 1 that sample 12001 is the richest in smooth regular particles. The decrease of
recognizable regular particles, and smooth particles in particular in the other samples
included in Table 1, and in particular in the trench sample and the conglomerates,
suggest that the regular particles are generated on the actual lunar surface, and they
tend to erode on burial.

The considerable degree of similarity between the petrostatistics of regular
particles from the Apollo 11 and Apollo 12 sites, respectively, is interpreted to

preclude the possibility that the bulk of the particles formed through major volcanic eruptions or meteorite impacts. A general experience here on Earth suggests that each one of such a localized major event produces petrographically quite different sprays. Two spray-producing processes which may be of sufficiently uniform distribution to account for the uniformity of the petrostatistics of the regular particles have been proposed so far. These agencies are (A) hypervelocity impacts of minor and of micrometeorites according to CARTER and MACGREGOR (1971) and others; (B) unusually intensive solar flare activity proposed by MUELLER and HINSCH (1970).

Some of the hitherto known features of the lunar surface seem to be equally well explainable with either of the foregoing two hypotheses, whereas other features seem to point to the importance of solar flare effects. Thus GOLD (1969) interpreted the formation of glassy incrustations on the hollow lunar surfaces as due to fusion by unusually intensive solar flare activity which occurred not more than 100,000 years ago. It appears that if such exceptional flares are intensive enough to melt the hollow portions of the lunar surface, they would be also intensive enough to melt small particles resting on the lunar surface, from which heat could diffuse only through relatively narrow contacts with the rest of the surface. Such isolated particles would melt, some would remain on the surface, and the heat slowly diffusing through their narrow contacts would produce the appendages. Other particles would decrepitate and would be ejected a short distance over the lunar surface. The process has been partially simulated on focusing the beam of the electron microscope on fragments of lunar glass, resulting in their decrepitation and melting, under conditions which may be similar to the effects of a solar flare of unusual intensity (FOX et al., 1970).

The relatively higher value of elongation ratio of the denser, and usually optically less homogeneous, particles cannot be accounted for with any hypothesis based on spray formation through random meteorite impacts. In this connection it was demonstrated by RADZIEVSKY (1954) that meteorites may be destroyed in solar orbit by rotational bursting caused by irradiation of a surface with optical inhomogeneities. Finally, the structures indicating changes in rotational momentum in the course of the solidification of the particle are readily explained by continuing increase of rotational momentum through the effects of radiation.

According to the foregoing hypothesis of formation through solar flare activity of the bulk of the lunar sprays, it would be anticipated that the sprays would remain petrostatistically similar throughout all the basaltic zones of the lunar equatorial regions. The increase of viscosity of silicate melts with their increasing acidity may render the rotational subdivision of dumbbells into teardrops more difficult, and therefore regular particles of relatively larger diameters may be anticipated in those equatorial regions of the Moon (possibly some highlands) in which granitic rocks may predominate. The abundance of sprays would decrease towards the poles.

It was anticipated by MUELLER (1967) that the lunar surface would be mainly basaltic and that depletion of alkalies, sulfur, and organic volatiles would occur due to the effects of solar activity in vacuum. The foregoing hypothesis of solar flare origin of the bulk of the regular lunar particles seems to accentuate the effects of solar activity. It is probable that in the course of the past 4.5 eons-long history of the Moon, solar devolatilization may have had an effect to considerable depth of possible several kilometers along the present and past equatorial regions, due to the reworking of the lunar surface with the interplay between impacts, tectonic forces, volcanism, and erosion.

Systematical studies of the regular particles in the future may enable us to recognize and specify the distinct spray-generating agencies which produced the chondrules and microchondrules in meteorites; the various and varied terrestrial volcanic and impact sprays, the regular spray particles from the moon, and sprays from other celestial bodies of the solar system. The interpretation of petrostatistical features of regular particles may prove to be an alternative approach for the reconstruction of conditions which prevailed in the course of condensation and subsequent geological histories of celestial bodies.

Acknowledgments—The work was supported by NASA grant NAS 9-8101. It is contribution no. 188 of the Institute of Molecular Evolution. I would like to thank S. W. Fox of the Institute of Molecular Evolution for help and discussion, Gertrude W. Hinsch of the Institute of Molecular Evolution, and S. Moll of American Metals Research for scanning microscopy, J. J. Wolken of Carnegie-Mellon University, Pittsburgh, for the use of the microphotospectrometer, and the Philips Electronic Application Laboratory, Mt. Vernon, N.Y., for the use of the electron probe and scanning electron microscope.

REFERENCES

Carter J. M. and MacGregor I. D. (1970) Mineralogy, petrology and surface features of lunar samples 10062,35, 10069,30, and 10085,16. *Science* **167**, 661–663.

Fox S. W., Harada H., Hare P. E., Hinsch G., and Mueller G. (1970) Bio-organic compounds and glassy microparticles in lunar fines and other materials. *Science* **167**, 767–770.

Gold T. (1969) Apollo 11 observations of a remarkable glazing phenomenon on the lunar surface. *Science* **165**, 1345–1349.

LSPET (Lunar Sample Preliminary Examination Team) (1970) Preliminary examination of the lunar samples from Apollo 12. *Science* **167**, 1325–1339.

Mueller G. (1967) Mineral deposits of the moon. *Nature* **215**, 1149–1151.

Mueller G. and Hinsch G. W. (1970) Glassy particles in lunar fines. *Nature* **228**, 254–258.

Schmitt H. H., Lofgren G., Swahn G. A., and Simmons G. (1970) The Apollo 11 samples: Introduction. *Proc. Apollo* 11 *Lunar Sci. Conf., Geochim. Cosmochim. Acta* Suppl. 1, Vol. 1, pp. 1–54. Pergamon.

Radzievsky V. V. (1954) Ob odnom mechanismo raspada asteroidov i meteoritov (A mechanism for the disintegration of asteroids and meteorites). *Akad. Nauk. SSR Doklady* **97**, 49–52.

Stroher G. K. and Wolken J. J. (1959) A simplified microspectrophotometer. *Science* **130**, 1084–1088.

Proceedings of the Second Lunar Science Conference, Vol. 3, pp. 2049–2055
The M.I.T. Press, 1971.

Compositions, homogeneity, densities, and thermal history of lunar glass particles

Charles H. Greene, L. David Pye, Harrie J. Stevens, Daniel
E. Rase, and Herbert F. Kay*
State University of New York College of Ceramics at Alfred University,
Alfred, N.Y. 14802

(*Received* 24 *February* 1971; *accepted in revised form* 30 *March* 1971)

Abstract—It is known that the physical properties of a glass depend upon its chemical composition and thermal history. Glass spherules from the Apollo 12 mission were examined in regards to these factors. Chemical composition and homogeneity were determined by electron probe analysis. These results were useful in preparing synthetic lunar glasses. The thermal history of several spherules was inferred by measuring their density before and after heat treatment in their transformation range. Most spherules showed an increase in density; one showed a decrease. These results indicate that there are two distinct thermal histories associated with the formation of lunar glass spherules.

I. Introduction

The cooling of a melt to form a solid glass represents three distinct departures from equilibrium: (1) a failure to crystallize at the liquidus temperature; (2) a failure to evolve energy in the form of heat consistent with the melt temperature; (3) a failure to equilibrate dimensionally, consistent with the melt temperature.

These departures are a result of very high viscosities (10^7–10^{14} poise) associated with glass forming melts. The latter two occur specifically over a temperature interval known as the transformation region. The point at which dimensional non-equilibrium occurs is referred to as the transition temperature and is sensitive to the rate at which the melt is cooled. Faster cooling rates will give rise to higher transition temperatures. We thus arrive at the well known diagram (Fig. 1) which depicts dimensional changes that can occur as a melt is cooled to room temperature. These simple concepts form the basis of inferring the thermal history of a given glass, for by reheating this glass into its transformation region, the density of the glass will equilibrate in a direction (i.e. incease or decrease) dictated by its thermal history. Since most glasses (including those studied here) are dimensionally stable below their transformation region, this thermal history can be reconstructed by observing these density changes. This is true even though the glass in question may have been formed several billion years ago. In the main, the work reported here was an attempt to apply these concepts to the glass spherules retrieved by the Apollo 12 mission.

II. Experimental Procedures

(a) *Selection of glass spherules*

Glass particles were extracted from four half gram portions of lunar soil < 1 mm (samples 12070,38; 12057,61; 12001,74, and 12033,25). No reliable statistics on the relative abundances of

---

* Present address: Department of Physics, Bristol University, Bristol, England.

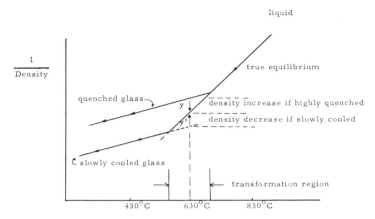

Fig. 1. Density-temperature relation of a glass formed by cooling from a melt.

glass particles in these four samples were compiled; however, sample 12001,74 appeared to contain more glass than the others. Only one glass spherule was found in sample 12033,25 which was taken 15 cm below the lunar surface. This suggests that the tiny glass spheres are more abundant on the surface than deeper in the lunar soil. Twelve of the largest and most perfect glass particles, free insofar as possible from surface defects, internal bubbles or solid inclusions, were selected for analysis by electron microprobe and density studies. Ten were spheres ranging from 0.1–0.4 mm in diameter, one was a fractured hemisphere and one was dumbbell shaped. These glass samples, labeled A–L, ranged in color from light amber through red brown to very dark brown.

(b) *Microprobe analysis*

Samples A–H were mounted in epoxy resin and studied by electron probe analysis. Before mounting samples A–D their densities were determined by the procedure outlined in section "d" below. Beam currents were kept to a minimum to avoid possible alteration in the density of the samples. The results of these analyses are shown in Table 1. Sample 9 is a synthetic composition melted for practice density determinations and heat treatment studies. The analyses of samples A, F, and G were repeated by other laboratories and sensibly confirmed.

In the first of these analyses, counts were made on six to eight points scattered at random over the exposed surface of the sample. These counts indicated that all of the glass samples except F

Table 1. Chemical composition of lunar glass spheres

Element	A wt.%	σ	B wt.%	σ	C wt.%	σ	D wt.%	σ	E wt.%	σ	F wt.%	σ	G wt.%	σ	H wt.%	σ	9 wt.%	σ
SiO_2	47.43	0.66	45.73	0.44	47.81	0.78	43.19	0.38	38.84	0.83	42.87	0.70	42.17	0.35	43.10	0.56	48.82	0.51
FeO	12.11	0.35	16.16	0.25	10.60	0.17	15.97	0.27	16.14	0.3	19.23	0.92	19.36	0.33	17.00	0.35	3.52	0.15
Al_2O_3	17.55	0.32	13.35	0.41	17.28	0.44	13.97	0.11	14.85	0.35	9.96		9.87		12.29	0.17	27.30	0.46
CaO	11.03	0.14	10.88	0.18	10.70	0.25	11.55	0.17	12.28	0.10	9.00	0.18	9.48	0.22	11.17	0.16	13.55	0.26
MgO	9.03	0.18	7.70	0.18	9.97	0.19	10.40	0.15	11.15	0.36	10.86	0.17	12.81	0.16	10.61	0.18	5.79	0.21
K_2O	0.14	0.03	0.08	0.02	0.34		0.05	0.01	0.05	0.02	0.16	0.03	0.06	0.02	0.07	0.02	0.02	0.02
TiO_2	2.01	0.02	3.25	0.07	1.87	0.06	2.99	0.07	2.94	0.07	4.3	0.1	3.50	0.10	3.35	0.07	1.68	0.04
Na_2O	0.2	0.06	0.2	0.03	0.5	0.07	0.1	0.03	0.05	0.02	0.3	0.02	0.1	0.05	0.1	0.03	0.07	0.04
Cr_2O_3	0.15	0.02	0.29	0.02	0.18	0.02	0.24	0.02	0.24	0.02	0.45	0.05	0.40	0.03	0.31	0.03	0.00	
MnO	0.17	0.01	0.24	0.01	0.15	0.02	0.22	0.02	0.24	0.01	0.25	0.01	0.26	0.02	0.24	0.03	0.01	0.01
NiO	0.05	0.01	0.06	0.01	0.04	0.01	0.04	0.02	0.05	0.02	0.07	0.02	0.07	0.01	0.05	0.01	0.02	0.01
P_2O_5	0.05	0.03	<0.05		0.07	0.02	<0.05		<0.05		0.11	0.04	<0.05		<0.05		<0.05	
Totals	99.92		97.94		99.51		98.72		96.83		97.56		98.08		98.29		101.38	

Table 2. Chemical analysis of special areas of sample F

Oxide	General Matrix	Area 1 outside	Area 1 inside	Area 2
Al_2O_3	10.08	19.87	3.82	22.87
SiO_2	42.20	68.52	88.60	36.42
K_2O	0.14	5.0	0.21	nil
CaO	9.12	0.67	nil	11.45
TiO_2	4.70	0.16	nil	0.49
FeO	20.52	0.46	nil	5.25

were reasonably homogeneous. Scans across sample F showed large fluctuations in the composition of the general matrix. In addition, small areas were found with unusual compositions. One area near the surface of the sample was high in silicon, aluminum and potassium. Adjacent to it was a very high silica area. Another small area high in aluminum and calcium but low in silicon was noted. These analyses are shown in Table 2. Also, tiny regions very high in iron were noted in sample F.

After the microprobe studies the eight lunar samples were polished with very fine alumina to remove the carbon coating and cut from the epoxy mount with a dental drill. Adhering resin was dissolved by heating for two hours in a sealed glass tube with thionyl chloride ($SOCl_2$) at 175°C. Tests with synthetic glass showed that this solvent did not attack glasses of this type and was very effective in dissolving the epoxy resin. Following the removal of samples A, B, C, and D, the densities were redetermined.

(c) Synthetic lunar glasses

For preliminary heat treatment and density experiments, homogeneous glasses similar to the samples from the moon were prepared by mixing a slurry of fine grained batch materials in a ball mill, drying thoroughly and melting in a silica tube. Iron was added as ferric oxide and reduced with wet hydrogen. A batch to yield glass with a composition similar to lunar samples A and C (47.5% SiO_2, 12% FeO, 17.3% Al_2O_3, 11% CaO, 10% MgO, and 2.2% TiO_2 by weight) was found to have a liquidus temperature of 1247 ± 2°C. Another batch to yield glass similar to lunar samples F and G (43.3% SiO_2, 19.8% FeO, 10.1% Al_2O_3, 9.7% CaO, 13.1% MgO 3.6% TiO_2, and 0.4% Cr_2O_3 by weight) was also prepared. This composition crystallized completely, however, when cooled rapidly in the one inch diameter melting tube. This indicates that samples F and G must have been cooled very rapidly or they would have crystallized. Such rapid cooling might have occurred by free radiation from such small particles.

(d) Density measurements

The densities of the glass particles were measured by suspending them in a mixture of a tetra-bromethane and either alpha-chloronaphthalene or methylene iodide. A mixture of these liquids were prepared so that the sample in question remained balanced at some temperature between 17°C and 33°C. In order to convert balancing temperatures to densities, the densities and thermal expansions of the various liquids were determined by a pycnometer method.

To minimize the effects of convection currents on the motion of the suspended glass particles and to facilitate thermal equilibrium of the liquid mixtures, density balances were made in a thin walled glass tube about 3 mm in diameter submerged in a well stirred thermostat. The temperatures of the liquids could be conveniently raised, lowered or held constant to a few thousandths of a degree by using a mercury-ether regulator.

Although it was possible to balance a sphere 0.3 mm in diameter to within ±0.005°C with this apparatus, difficulty was encountered with drift of the balance temperature over a period of days. Much of this difficulty was due to failure to wet the tiny samples completely so as to fill surface crevices and pores completely with liquid. This in turn gave rise to a slow dissolution of the air within the pores causing the apparent density of the glass to increase by as much as 0.002 gm/cc.

This difficulty was largely overcome by evacuating and baking the samples and then injecting the liquid mixture before admitting air. The samples were precleaned in hot (40°C) triply distilled acetone for thirty minutes and then dried at 150°C under vacuum (0.1 mm Hg) for thirty minutes.

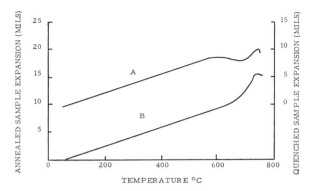

Fig. 2. Thermal expansion of synthetic moon glass M31B (similar in composition to spherules A and C); A—quenched glass, B—glass annealed at 700°C.

(e) *Heat treatment*

A crucial aspect of this investigation involved the selection of proper heat treatment temperatures for the lunar glass spherules which would allow a convenient measurement of their density changes with time. An improper selection might have erased their latent thermal histories. To aid in this selection, a thermal expansion measurement of the synthetic lunar glass similar to lunar sherules A and C was carried out. (Fig. 2) This result, coupled with other measurements of thermally induced density changes of this same glass, indicated that its transformation range is about 570–700°C. Consequently, an initial heat treatment temperature at 630°C was chosen for the lunar glass spherules.

The samples were heated in thin silica tubes at 630°C for increasing periods of time. After heat treatment at 630°C, each spherule was heated for ten hours at 650°C and again at 670°C. In three cases (spheres D, I, and K) a final heating for sixty hours at 630°C was given after the heat treatments at 650°C and 670°C.

III. DISCUSSION OF RESULTS

The densities determined are given in Table 3 and Figs. 3, 4, 5, and 6. It is evident that with two exceptions there is a general tendency for the density to increase with

Table 3. Densities of lunar glass samples before and after heating

Sample	(1)	(2)	1 hr 630°	3 hrs 630°	10 hrs 630°	60 hrs 630°	10 hrs 650°	10 hrs 670°
A	2.8787	2.8923	2.9127	2.9167	2.9202	2.9182	2.9187	(3)
B	2.9610	2.9685	2.9824	2.9902	2.9904	(3)		
C	2.8486	> 2.8494	(4)					
D	3.0042	3.0072	3.0072	3.0363	3.0204	3.0307	3.0218	3.0164
I	3.0145	(5)	3.0306	3.0315	3.0330	3.0352	3.0245	3.0185
J	2.7723		2.7664	2.7747	2.7764	2.7725	2.7730	2.7757
K	3.1899	(6)	3.1854	3.1836	3.1827	3.1912	3.1789	3.1755
L	2.9463	(7)	2.9659	2.9664	2.9715	3.0318	3.0273	3.0232

Notes:
(1) These measurements were made on the spherules before they were analyzed in the microprobe.
(2) After the electron microprobe experiments.
(3) Sample disintegrated, probably due to thermal shock.
(4) Sample C was lost before completion of the second density measurement.
(5) Samples I, J, K, and L were not analyzed by the electron microprobe.
(6) The high density of the last treatment (60 hours at 630°C) may be due to the breaking of the sample holder during treatment.
(7) During the 10 hour-630°C treatment "L" broke. The testing was then done on a fragment, apparently of a higher density.

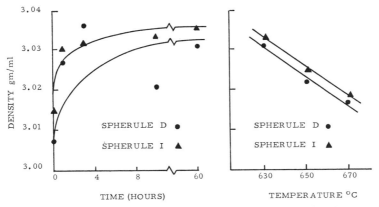

Fig. 3. Density-time-temperature relations of lunar glass spherules D and I.

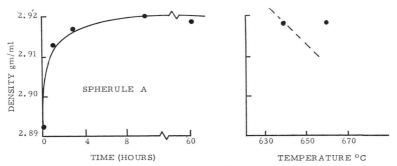

Fig. 4. Density-time-temperature relations of lunar glass spherule A.

time at 630°C corresponding to the interval y in Fig. 1. This behavior would be expected if they were quenched rapidly from a high temperature as they were formed on the moon. There is also a tendency for the densities of these samples to decrease as the temperature at which they are held for ten hours or more increases. This is the usual behavior of glass held in the transformation range for a sufficient time to bring it to equilibrium. The slopes of the equilibrium density temperature lines for samples D, I, K, and L are -0.00036, -0.00036, -0.00018, and -0.00022 gm/cc deg. These values are comparable to those found by RITLAND (1954) for an optical borosilicate crown glass. Therefore, it is concluded that these results are completely consistent with the theories of glass science and are not due to an unknown spurious effect.

The results for two samples, D and I, (Fig. 3) indicate that these lunar spherules were cooled very rapidly from a temperature above 670°C as they were formed.

Sample A (Fig. 4) comes rapidly to density equilibrium at 630° and the density does not decrease when it is held at 650°C. This indicates that 630°C is close to the upper end of the transformation range for this glass. The low initial density shows that it was quenched very rapidly indeed as it cooled originally on the moon. The rate of cooling must have been faster than in our experiments where the spherule in a thin silica capsule was cooled in air at room temperature.

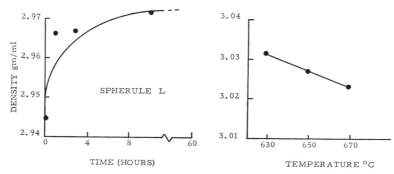

Fig. 5. Density-time-temperature relations of lunar glass spherule L.

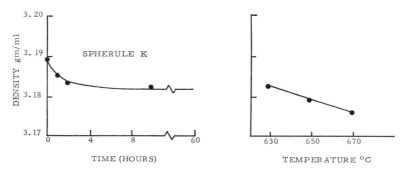

Fig. 6. Density-time-temperature relations of lunar glass spherule K.

Sample J showed erratic density changes suggesting that the transformation range for this glass is lower than 630°C so that the observed changes depend on cooling rate rather than the treatment temperature. However, the initial density of this sample is close to that observed after the heat treatments. This indicates that this spherule cooled after its formation on the moon at a rate comparable to the rate of cooling in the vitreous silica capsules in our laboratory.

The changes in the density of sample L (Fig. 5) are normal except that the fragment available for measurements after bringing the density to equilibrium was of a composition having a higher density than the average of the whole spherule.

Sample K is an interesting exception to the usual behavior (Fig. 6). Its decrease in density at 630°C, corresponding to the interval y' in Figure 1, indicates that it was cooled at a much slower rate than the other glass spherules. At present, the absolute differences in cooling rates can not be determined from our data. From the magnitude of the density changes, however, it is concluded that they are substantially different. There also exists the possibility that spherule K was simply reheated, i.e. annealed at some temperature around 610°C for an extended period. Again, our data are not complete enough to permit this distinction in thermal history. Collectively, our data indicate that there are important differences in the thermal histories of this type lunar glass and they must be accounted for in any theory attempting to explain their origins.

Acknowledgments—We are grateful to Dr. GEORGE DESBOROUGH of the U.S. Geological Survey in Denver for the electron microprobe analyses given in Table 1. We also acknowledge the help given us by Dr. WILLIAM KANE of the Corning Glass Works and Dr. RALPH GUTEMACHER of the Lawrence Radiation Laboratory in Livermore for confirming some of these analyses and for checking the homogeneity of the lunar glass.

We have had generous support from the faculty and staff of the New York State College of Ceramics including W. VOTAVO, G. CARTLEDGE, E. ORMSBY, L. HANKS, L. BURZYCKI, D. GREEN, D. CRONIN, R. PALMER, J. KNOX, D. EHMAN, D. PEDLEY, and Mrs. DORIS SNOWDEN.

The advice of Prof. W. G. LAWRENCE and Prof. WAYNE BROWNELL, and, particularly, the critical suggestions and micro-photographs of Prof. VAN DERCK FRECHÉTTE are appreciated.

REFERENCE

RITLAND H. N. (1954) Density Phenomena in the Transformation Range of a Borosilicate Crown Glass, *J. Amer. Ceram. Soc.* **37**, 370–378.

Proceedings of the Second Lunar Science Conference, Vol. 3, pp. 2057–2067
The M.I.T. Press, 1971.

Exoelectron emission and surface characteristics of lunar materials

RICHARD B. GAMMAGE and KLAUS BECKER

Health Physics Division, Oak Ridge National Laboratory, Oak Ridge, Tennessee 37830

(*Received* 23 *February* 1971; *accepted in revised form* 25 *March* 1971)

Abstract—The extreme radiation damaged and amorphous condition of the exterior region of regolith grains composing lunar surface and core-tube fines inhibits natural as well as artificial (γ radiation-induced) thermally stimulated emission of exoelectrons (TSEE). Removal of the interfering outer layer by etching permits observation of natural TSEE from deeper within. Particles removed from rock 12065 also show natural TSEE which varies with distance from the surface. Exoelectron traps responsible for peaks at about 600°C are concentrated in the plagioclase-rich grains. Artificial TSEE between 130° and 550°C, which can be induced by preheating in air between 300° and 700°C, is a sensitive diagnostic tool for studying surface oxidation and structural changes. Heat treatment under reducing conditions, or in vacuo, fails to develop the artificial TSEE responsiveness showing that some surface oxidation is first necessary. Among other reduced species, iron ions are involved in these oxidation processes, and adsorbed atmospheric gases (O_2) also seem to affect the TSEE characteristics.

INTRODUCTION

LOW ENERGY ELECTRONS are generally emitted from the thin surface layers of irradiated insulating materials (Fig. 1) at characteristic temperatures. This thermally stimulated emission of exoelectrons (TSEE) has been used to measure radiation doses and to study surface structure and solid state physical and chemical reactions. For a review see BECKER (1970a). BECKER (1970b) and GAMMAGE and BECKER (1970) showed that γ irradiated terrestrial rock and mineral samples (U.S. Geological Survey Standards) of basalt, granite, and olivine exhibited TSEE, which points to the common occurrence of TSEE.

The purpose of this study was two-fold: (1) to obtain information from "natural" TSEE about radiation and thermal histories. Such results would complement and expand upon the thermoluminescence measurements made by DALRYMPLE and DOELL (1970), HOYT *et al.* (1970), and others; (2) to use "artificial" TSEE (induced by γ radiation) to detect and follow degradation of, and reactions within, the lunar material as a function of temperature and ambient atmosphere. This can be done because of the extreme sensitivity of exoelectron emission to changing surface properties.

EXPERIMENTAL PROCEDURES

TSEE was measured by heating small quantities ($\sim$ 1–5 mg) of unirradiated or irradiated material inside a gasflow, water cooled, GM counter (Fig. 2) of the type previously described by OBERHOFER and ROBINSON (1970). The sample holders were graphite which was always annealed before readout.

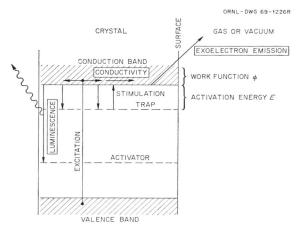

Fig. 1. Simplified schematic diagram of the three processes (stimulated luminescence, conductivity, and exoelectron emission) primarily associated with the release of trapped electrons in an ionic crystal during thermal or optical stimulation.

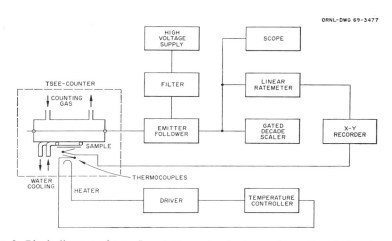

Fig. 2. Block diagram of a gasflow GM counter, heater, and counting electronics used for recording TSEE as a function of temperature during heating.

Heating was at a linear rate of between 1° and 5°C/sec (THORNGATE, 1969). TSEE peaks were recorded as a function of temperature up to about 650°C, where thermionic emission begins. It should be noted that the peak locations, as given in these graphs, are subject to some errors in the surface temperature measurement. The peak location also depends on the heating rate. The peak ratios vary somewhat with the dose level to which the samples have been exposed, resulting in a slightly different slope of the count vs. dose curve for peaks in different temperature ranges (Fig. 3). No saturation has been observed for doses up to 10^7 rad.

All the lunar samples were stored under atmospheric conditions. A conventional vacuum system was used for outgassing at elevated temperatures and for treatments in controlled atmospheres.

Using a ^{60}Co source, the samples were γ irradiated in air, in controlled atmospheres, or in vacuo.

RESULTS AND DISCUSSION

Standard U.S. Geological Survey rock and mineral samples

These materials were examined for both "natural" and "artificial" exoelectron activity. Several irradiated feldspars and olivines exhibited weak TSEE peaks between 130° and 600°C. A basalt (Fig. 4) with a high pyroxene content was unusual in that it did not show any TSEE unless it was previously oxidized by heating in air at 600°C.

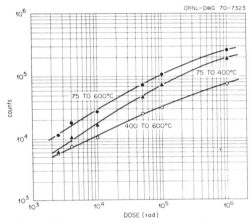

Fig. 3. Total artificial TSEE counts from the fines 12033,46, after "stabilizing" pre-treatment for 30 minutes at 650°C in air, as a function of ^{60}Co gamma radiation dose for all peaks between 75° and 600°C; for the low-temperature peaks (75° to 400°C) and for the high-temperature peaks (400° to 600°C).

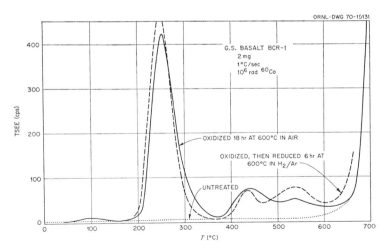

Fig. 4. TSEE of irradiated U.S. Geological Survey Standard basalt BCR-1 before modifying heat treatment; after oxidation at 600°C in air; and after an additional heating at 600°C in a reducing atmosphere (4% H_2, 96% Ar).

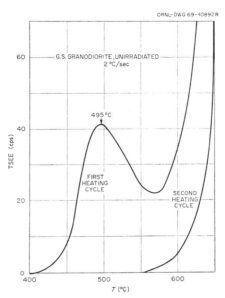

Fig. 5. TSEE from unirradiated, untreated U.S. Geological Survey Standard granodiorite GSP-1 during the first and second heating cycle.

The oxidation was not reversible by heating in a H_2/Ar mixture at 600°C. One material, a granodiorite, also exhibited a high-temperature TSEE peak (Fig. 5) without previous irradiation, presumably because of the integrated natural radiation effect in this material. It was, therefore, thought that a similar natural TSEE in lunar materials might be detectable.

Natural TSEE

The first results with surface fines (10065,20; 10087,5; 12070,18; and 12033,46) and core-tube fines (12025,13; 12028,13; and 12028,32) were disappointing because no TSEE peaks emerged. The integrated counts exceeded the background by only a few hundred counts. Even after exposure of the virgin materials to γ radiation doses of up to 10^6 rad, only a small increase in the count rate, principally around 450°C, was found. This indicates either that the radiation sensitivity of the surface layer was very small or that other factors were working to suppress the exoelectron emission.

Material from the bulk of an igneous crystalline rock (12065,85 and 12065,64, both being sections of a 2109 g pigeonite phorphyry rock of plagioclase, pigeonite, and ilmenite) was examined to test the possibility that buried solar wind in the surface might be interfering with TSEE. A rock fragment was fractured, scrapings taken from a fresh fracture surface and read out. This produced the first clear evidence of natural TSEE, with a peak around 450°C and considerable emission above 520°C, as shown in Fig. 6. Scrapings taken from the exterior surfaces of these rocks (surface and near surface material) showed only the emission above 520°C. This indicates that a partial bleaching has occurred for material in the surface regions. The TSEE spectra were adequately reproducible, and checking showed that possible spurious effects, such as light-induced or tribo-TSEE, were not responsible. It was found, however, that intense light can anneal the signal. With proper care the TSEE technique can prove useful in the determination of the dose distribution with depth in lunar rocks. One must work, however, with fracture, and not smooth sawn surfaces. Samples scraped from the smooth cut faces produced negative results; localized frictional heating during the cutting process most likely had already released any latent exoelectrons. There is additional evidence that severe localized frictional heating strongly modifies material in the region of the sawn faces; substantial modifications occur in the artificial TSEE spectrum, particularly with respect to the relative peak intensities.

Another technique for avoiding the possible interference of solar wind effects with TSEE is removal of the thin surface layer of regolith grains in which the solar wind particles are trapped. Indeed, etching of surface and core-tube fines with a solution of $2HF:1H_2SO_4:7H_2O$ for 30 seconds, followed by careful rinsing and drying, permitted natural TSEE to be seen (Fig. 7). The surface fines 12070,18 and the shallow core-tube sample 12025,13 (~ 1 cm depth) showed significant exoelectron emission only above 500°C. In contrast the core-tube sample, 12028,32, from a depth of 28 cm, showed three distinct lower temperature peaks and a higher integral number of counts. For the near surface material, the latent exoelectrons associated with the lower temperature peaks had probably been lost under the combined influence of diurnal temperature fluctuations and the bleaching effects of sunlight at the lunar surface. It was also shown by repeat experiments that etching itself did not induce any latent TSEE. These results complement the thermoluminescence measurements of Hoyt *et al.* (1971) who found marked thermal draining to have taken place in near surface material. The reproducibility of TSEE in these etching experiments, however, left much to be desired. It seems that the detection of natural exoelectrons from virgin regolith grains is not likely to become routinely easy.

The reasons were sought for the absence of natural TSEE from unmodified surface grains. It was determined that trapped solar wind was not directly to blame. The

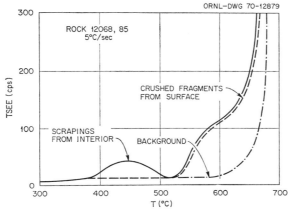

Fig. 6. Natural TSEE from freshly prepared fragments taken from the surface region or the interior of lunar rock 12068,85.

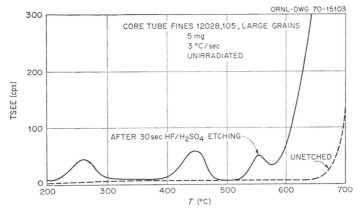

Fig. 7. Natural TSEE from the coarse grain fraction of an unetched sample of core-tube fines 12028,32–105 (depth 28 cm), and natural TSEE from the etched material.

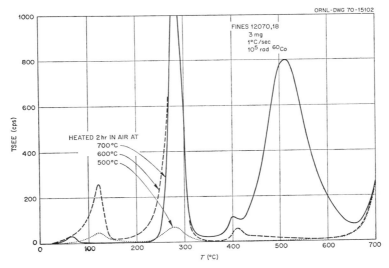

Fig. 8. Artificial (radiation-induced) TSEE for surface fines 12070,18 after preheating for 2 hours in air at various temperatures.

fines 12070,18 and 12033,46 were heated in a hard vacuum at 600°C to release most of the buried solar wind particles (EBERHARDT *et al.*, 1970) but no artificial TSEE could be induced by irradiation of these vacuum treated samples. The unmodified surfaces of regolith grains are in an extreme condition being non-stoichiometric, slightly oxygen-deficient, and glassy (HAPKE *et al.*, 1971). In addition, the micron-sized grains are characterized by exceedingly high latent track densities, some of $> 10^{11}/cm^2$ (BARBER *et al.*, 1971). In view of the highly defective nature, one might anticipate a high concentration of trapping centers and hence an appreciable exo-electron response. The cause of its absence or suppression is not entirely clear but is due to some condition peculiar to the skin, such as oxygen-deficiency or a preferentially reduced state.

Artificial TSEE

If instead of heating in vacuum, the fines are heated in air, at 300°C or higher, intense artificial TSEE can be induced. The TSEE spectra of Apollo 11 and 12 fines are very similar. Figure 8 shows an example of this induced activity for the temperatures of oxidation 500°, 600°, and 700°C. The peaks grow at different rates during prolonged heating in air (Fig. 9). One concludes that oxidation of the surface is a prerequisite for observing artificial TSEE. If during heat treatment the ambient atmosphere is reducing (4% H_2, 96% Ar), the artificial TSEE is suppressed completely. Also, it was discovered that if an oxidized sample is re-heated in a reducing atmosphere (Fig. 10), 95% of the artificial TSEE disappears. The process is almost completely reversible, which verifies the idea that an oxidized surface layer must be present. With increasing temperature of oxidation the artificial TSEE develops as follows: After prolonged heating at 300° or 400°C, a weak peak appears at ~ 600°C similar to that seen in Figs. 6 and 7 for rock and etched fines; oxidation between 500° and 650°C causes strong peaks between 130° and 300°C while the peak at ~ 600°C is eliminated, presumably be annealing of the trap; and fines oxidized at 700°C are characterized by a new, very strong peak at ~ 500°C. HAPKE *et al.* (1971) showed the following: The lunar fines are deficient in oxygen, complete oxidation bringing about a weight gain of 2%; at 500°C, oxygen begins to diffuse into the non-stoichiometric, siliceous glassy skin; and not until nearly 900°C do the bulk ferrous ions oxidize to ferric. In light of HAPKE's (1971) work it can be seen that our TSEE spectra are sensitive indicators of subtle changes occurring in the surface oxidation state between 300° and 700°C. Because TSEE is so sensitive to oxidation processes, various fines were periodically

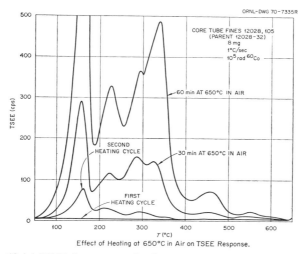

Fig. 9. Artificial TSEE from core-tube fines 12028,32 during the first heating cycle; during re-irradiation and re-heating of the same sample; and after prolonged heating of the sample at 650°C in air.

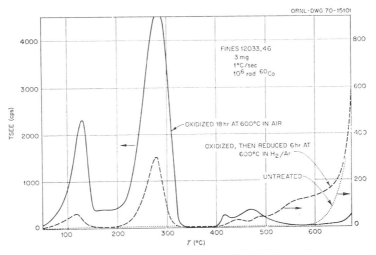

Fig. 10. Artificial TSEE from the untreated fines 12033,46; from the same sample after oxidation in air; and after additional reduction in H₂/Ar.

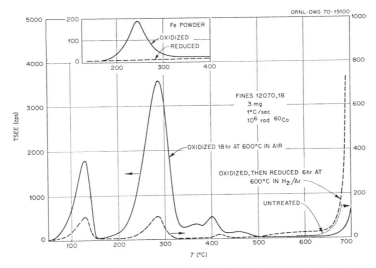

Fig. 11. Artificial TSEE from untreated, oxidized, and consecutively reduced surface fines 12070,18; and from analytical grade iron powder.

checked for evidence of oxidation during storage at room temperature in our laboratory; the results were always negative showing that the samples had not deteriorated via atmospheric oxidation.

In summary, it is likely that the solar wind, and allied effects, have produced a preferentially reduced state in the surface virgin fines from which no TSEE can occur (metals and semi-conductors are known to exhibit TSEE only if oxidized). Bulk material, which is free from the reducing effects of the solar wind, does not have to be heated and oxidized to develop a capacity for TSEE.

WEEKS *et al.* (1970) demonstrated by EPR methods that iron ions in a variety of chemical and magnetic states are the sources of intense electron resonance absorption in lunar samples, and that partially reduced Fe_2O_3 gives an EPR spectrum similar to that of the lunar fines. Parallel TSEE experiments with lunar fines and iron powder (Fig. 11) showed a similar, partial correspondence. In both materials, artificial TSEE can be induced by oxidation and destroyed by reduction. The TSEE spectra show some similarities in the peak locations, indicating that at least some of the artificial TSEE in lunar samples is closely related to oxidized surface pyroxenes or free iron. Also, the TSEE behavior of the iron containing terrestrial basalt (Fig. 3) bolsters this hypothesis since oxidation (presumably of surface iron ions in a reduced state) was necessary before exoelectron emission could be induced.

We have also attempted to identify the mineral components carrying the exoelectron traps which produce TSEE peaks in the vicinity of 600°C. A crude mineral separation, based on gross visual differences, was carried out with the coarsest grains of the surface fines 12070,18. Three components were obtained, namely, colorless fragments rich in feldspar (plagioclase), opaque dark fragments with a high Fe pyroxene content, and semitransparent colored glasses. All three components exhibited some artificial TSEE above 400°C as shown in Fig. 12. The plagioclase fraction, however, was the strongest emitter. Plagioclase in fines has been shown by DALRYMPLE and DOELL (1970) to be the primary carrier of thermoluminescence.

TSEE from defect centers in the plagioclase accounts primarily for the high-temperature peaks, whereas the low-temperature peaks (artificial TSEE) appear to be associated with oxidation products, including those of iron. With the latter, another effect—sorption, probably of oxygen—also seems to be involved. This assumption is supported by the fact that another TSEE emitter, ZnO, contains several peaks very similar to those of oxidized lunar fines (Fig. 13), and these have been attributed to

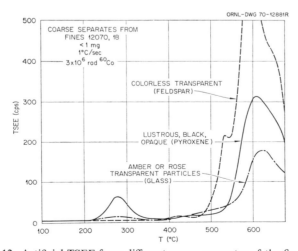

Fig. 12. Artificial TSEE from different coarse separates of the fines 12070,18.

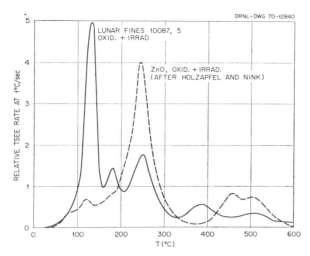

Fig. 13. Comparison of artificial TSEE in oxidized fines 10087,5 and in ZnO, normalized for the use of a TSEE reader with precision temperature calibration (in cooperation with PTB Braunschweig).

adsorbed oxygen by Holzapfel and Nink (1971). Additional support comes from the observation that the low-temperature peaks are considerably more pronounced for the finer fraction of the regolith grains with a higher surface area, which points to the involvement of surface centers.

Acknowledgments—The authors acknowledge valuable experimental assistance and stimulating suggestions by numerous colleagues, in particular, J. A. Auxier, R. K. Bennett, Jr., L. L. Hall, and R. A. Weeks, Oak Ridge. Research sponsored by NASA under Union Carbide Contract with the U.S. Atomic Energy Commission.

REFERENCES

Barber D. J., Hutcheon I. D., Price P. B., Rajan R. S., and Wenk H. R. (1971) Exotic particle tracks and lunar history. Second Lunar Science Conference (unpublished proceedings).

Becker K. (1970a) Principles of TSEE Dosimetry. *IAEA At. Energy Rev.* **8**, no. 1.

Becker K. (1970b) *Thermally Stimulated Exoelectron Emission from Terrestrial and Lunar Minerals.* ORNL Rep. no. TM–2869.

Dalrymple G. B. and Doell R. R. (1970) Thermoluminescence of lunar samples. *Science* **167**, 713–715.

Eberhardt P., Geiss J., Graf H., Grögler N., Krähenbühl U., Schwaller H., Schwarzmüller J., and Stettler A. (1970) Trapped solar wind noble gases, Kr^{81}/Kr exposure ages and K/Ar ages in Apollo 11 lunar material. *Science* **167**, 558–560.

Gammage R. B. and Becker K. (1970) *Research on Stimulated Exoelectron Emission from Lunar Materials.* ORNL Rep. no. TM-3131.

Hapke B. W., Cassidy W. A., Wells E. N., and Cohen A. J. (1971) Analysis of optical coatings on Apollo fines. Second Lunar Science Conference (unpublished proceedings).

Holzapfel G. and Nink R. (1970) Thermally and optically stimulated exoemission from ZnO. *Phys. Stat. Sol.* **3**, section A, no. 3, 181K–184K.

HOYT H. P., JR., KARDOS J. L., MIYAJIMA M., WALKER R. M., and ZIMMERMAN D. W. (1971) Thermo-luminescence and stored energy measurements of Apollo 12 samples. Second Lunar Science Conference (unpublished proceedings).

OBERHOFER M. and ROBINSON E. M. (1970) A new TSEE–GM–gasflow counter. *Kerntechnik* **12,** 34–38.

THORNGATE J. H. (1969) *An Instrument that Provides Linear Temperature Ramps for Use with TSEE Experiments.* ORNL Rep. no. TM–2687.

WEEKS R. A., CHATELAIN A., KOLOPUS J. L., KLINE D., and CASTLE J. G. (1970) Magnetic resonance properties of some lunar material. *Science* **167,** 704–707.

Proceedings of the Second Lunar Science Conference, Vol. 3, pp. 2069–2078
The M.I.T. Press, 1971.

Lunar glass I: Densification and relaxation studies

Rustum Roy,* Della M. Roy,* S. Kurtossy, and S. P. Faile
Materials Research Laboratory, The Pennsylvania State University
University Park, Pennsylvania 16802

(*Received* 23 *February* 1971; *accepted in revised form* 25 *March* 1971)

Abstract—Synthetic glasses simulating lunar glass compositions have been prepared and used as a model system for determining the magnitude of a pressure event by studying frozen-in density and refractive index changes. Subsequently, the same phenomena were investigated in actual lunar glasses. The pressure-temperature metastable equilibrium-refractive index dependence of the synthetic glasses (high titania, high iron, and anorthite) was determined in the range from 0 to 65 kb and 50°–525°C, and very substantial changes were observed. The time-temperature conditions needed to relax out these refractive index changes were determined, and most can be returned to their original value between 500°–600°C in approximately 1–2 hours.

Several sets of lunar glasses were investigated. On annealing, the index of refraction was found to increase for some chips (10085,49) from 1.662 to 1.675, in others it stayed almost constant, while in other fragments from the same sample it decreased from 1.673 to 1.660. In general, the magnitude of the refractive-index changes on annealing in some lunar glass samples is very small (and also complicated by compositional and local variations) compared to the major effects obtained for glasses densified at quite low static pressures. Some of our results are analogous to those obtained by others on feldspathic glasses. It is clear that the structure and hence density and refractive index of noncrystalline solids produced by shock events is not useful as a guide to the actual pressures reached, but probably reflects much more the effects of the latest high temperature event. Any index change measurements cannot exclude the possibility of a multiple impact history subsequent to shock melting and one should also consider these effects.

Introduction

Roy and Cohen (1961) first suggested that the densification of glasses could be used as a possible piezometer for the measurement and recording of pressures in the hundred kilobar region. It is obvious that such measurements and indications of pressure effects can be applied both to laboratory as well as to geologically provided samples. As the literature on shock metamorphism has grown over the last decade, we had expected to find evidence that the production of noncrystalline solids at high pressures gave rise to NCS phases with refractive indices very substantially greater than those of the glasses of the same composition made at atmospheric pressure. However, neither the natural nor laboratory-produced shock metamorphism samples have resulted in very large increases. From the earlier reports of De Carli and Jamieson (1959) to the more recent of Ahrens and Rosenberg (1968) the NCS phases resulting from laboratory-induced shock metamorphism were not substantially more dense than normally melted glasses, the high residual temperatures reached during the shock events having relaxed out the high pressure density. With the discovery that the Apollo 11 samples from the moon contained a large fraction of noncrystalline material, it was hoped that

---

* Also affiliated with the Departments of Geochemistry and Mineralogy and of Materials Science.

one would be able to possibly determine something of the pressure history of these materials if any refractive index changes could be observed and studied. Further, the magnitude and the kinetics of any relaxation effects in the refractive index values of such glasses could provide interesting clues to the structure of such glasses. The present preliminary report gives data on our research on both these aspects.

Preparation of Simulated Lunar Glasses

Prior to receipt of the Apollo 12 specimens, and in order to be able to have larger volumes of samples on which to base preliminary experiments, we have prepared a series of synthetic, typical glasses corresponding to pure anorthite, a moderate range $FeO + TiO_2$ content glass, and a high $FeO + TiO_2$ glass.

Glasses were prepared from the compositions shown in Table 1 by conventional mixing of oxides and carbonates, melting in platinum crucibles, regrinding, and rehomogenization. Glasses A and B were finally equilibrated at a low pO_2 by melting in molybdenum in a partial oxygen pressure corresponding to $10^{-12.5}$ at 1200°C and $10^{-11.2}$ at 1300°C respectively. Figure 1 shows these compositions on a summary plot of refractive index vs. composition for a large number of Apollo 11 and 12 glasses studied by Chao *et al.* (1970b). The glasses A and B were quite homogeneous in refractive index and were both dark brown in color.

Pressure-Temperature Effects on Synthetic Glasses

The experimental techniques used for these studies have been reported in a series of papers by Cohen and Roy (1962), Dachille and Roy (1962), and Myers *et al.* (1963), and will not be repeated here. All high pressure runs are made on prepelleted samples with an aspect ratio of roughly 10:1, surrounded by platinum disks. The temperatures reported are accurate to ±5° and the pressure in the opposed anvil system is generally accurate to ±5% of the value in this pressure range. However, in some of the runs made on the actual lunar samples, the volume of the sample was so small that the actual pressure variation is likely to be larger than this. To ensure homogeneity, sample preparation of the lunar glasses for the pressure apparatus was varied slightly from the usual technique because of the necessity of using small sized glass chips. The glass chips studied were placed within a measured amount of SiO_2 powder which would give the desired pellet thickness at the pressures used to prepare the samples for the pressure runs. This method proved quite successful when tested on simulated lunar glass, giving the refractive index of 1.773 for a simulated glass chip of the same size as the lunar fragment compared to a previous index of 1.775 for the

Table 1. Chemical composition (in wt.%) of simulated lunar glasses

Oxide	(A)	(B)	(C)
SiO_2	43	41	43.19
TiO_2	10	5	
Al_2O_3	10	17	36.65
FeO	19	12	
MnO	0.4	0.2	
MgO	6.5	9	
CaO	9.7	13.3	20.16
Na_2O	0.5	0.1	
K_2O	0.16	0.1	
N_D	1.699	1.648	1.580

COMPOSITION AND REFRACTIVE INDEX OF
LUNAR GLASSES OF IMPACT ORIGIN

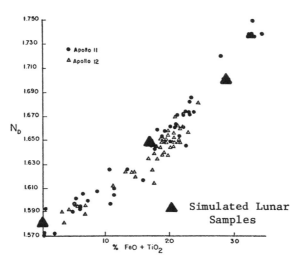

Fig. 1. Composition and refractive index of lunar glasses of impact origin (after CHAO
et al., 1970b).

normally prepared synthetic sample. The change of refractive index, as measured by
the Becké line method for Na_D, with pressure for the three main simulated lunar
glasses is summarized in Fig. 2. It should be noted that these are the refractive indices
of the glasses after quenching from the stated *p–t* conditions. These values should
represent the "equilibrium" value of the structural state of the particular glass at a
given temperature and pressure after release of elastic strain.

The meaning of such curves has been explained in the paper by COHEN and ROY
(1961, Fig. 4). Several interesting points may be noted in the data shown in Fig. 2.
First, it is remarkable that the relative change in the index of the three glasses is
roughly the same. Second, the magnitude of the change is surprisingly high in light of
the very moderate pressures involved. Indeed, although we originally planned to carry
the work to 200 kilobars, this appeared quite pointless in the light of the actually
observed changes in index. Third, the data indicate that the kinetics of the densification
process are a good indicator of the nature of the structural change taking place. In the
densification of silica SiO_2 glass, we have previously shown that this process is complete at temperatures in the vicinity of 500°C in a few milliseconds. However, in
certain other glasses the process takes as long as 20–30 hours to reach an equilibrium
value at these temperatures (see COHEN and ROY, 1962, on densification of soda-
magnesium-silica). It is interesting to note from Fig. 2 that there is a substantial
change between the values attained in 1.5 hours and those attained after 20 hours in
the simulated lunar glasses. In addition the data obtained for composition (A) show
that there is also a strong temperature dependence of the "equilibrium" refractive

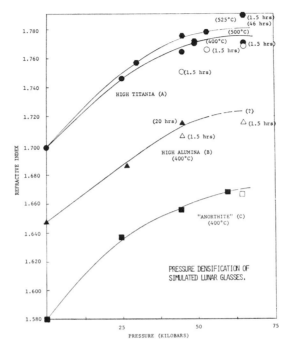

Fig. 2. Pressure densification of simulated lunar glasses.

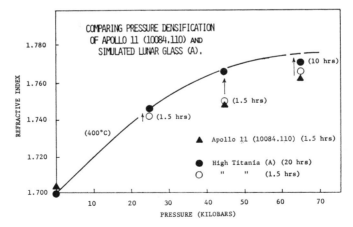

Fig. 3. Comparing pressure densification of Apollo 11 (10084,110) and simulated lunar glass (A).

Table 2. Comparison of densification data for Apollo 11 and simulated high titania lunar glass (A)

Glass type	Atm	45 kb	65 kb	Relaxed 600°C 1 hr
Apollo 11 (10084,110)	1.703	1.748	1.762	1.705
Simulated high titania (A)	1.699	1.750	1.766	1.705

index value attained at any particular pressure. The magnitude and direction of this change is quite similar to that observed for SiO_2.

The magnitude of index change, approximately 0.08 at 60 kb is also comparable to that reported by CHAO and BELL (1969), who showed a refractive index increase of approximately 0.04 for an An_{68} glass densified at 29.6 kb. These authors did not, however, report data on other types of glass compositions.

In Fig. 3 we show an analogous plot of refractive index vs. pressure of the changes induced in real lunar glass samples. (Individual chips were hand picked from a sampling of discrete glass particles from sample 10084,110.) The magnitude and sign of the refractive index changes are very similar to those of simulated lunar glass compositions. Indeed, it is striking how closely the real and the simulated samples follow each other in both absolute magnitude of the refractive index and its dependence on pressure. This comparison is brought out in Fig. 3 and even more sharply in Table 2. The above results seem to indicate that certain lunar glasses if produced by shock metamorphism or other processes behave quite normally with respect to their pressure-temperature history. In other words, they appear to show no evidence for remanent changes in refractive index which would be caused by transient or static high pressures in excess of a few kilobars.

ANNEALING STUDIES

Before we proceed to a discussion of the densification results, we will report on the relaxation of refractive index as a function of temperature and time in the densified simulated lunar compositions as well as of Apollo 11 (10085,49) specimens. Figure 4 shows the rate of change of refractive index of the simulated lunar glasses B and C as a function of time at different temperatures. It will be noted that most of the "excess" refractive index is annealed out within a period of 1 or 2 hours at temperatures in the vicinity of 500°–600°C. However, even at temperatures in the order of 160°–200°C there appears to be a substantial change in a matter of hours in compositions B and C, although A showed no change at 250°C, suggesting that the temperatures attained in the lunar day *may* be sufficient to permit relaxation. This is a rather unexpected result and may be one reason why no permanent densification has been found. Another factor to consider is that only a small fraction of the maximum shocked density is apparently retained in naturally and experimentally shocked glasses (BELL and CHAO, 1970). Recently BELL and CHAO investigated naturally densified glasses from the Ries Crater and synthetically shock-wave produced NCS glasses from An_{67} crystals. Their data for the latter glasses show greater remanent densification of the glass obtained at 285 kb than for the glass shocked at 325 kb, indicating the importance of the release adiabat over the peak pressure (FRENCH and SHORT, editors,

1968). Though the higher shock pressure gave a lower final densification, the rate of change of index on annealing after the initial drop is lower, even when annealing is carried out at a high temperature. The naturally densified feldspar glasses required even shorter annealing times. Our data in Fig. 4, glass C, also point to an interesting feature: that the rate of change of index on annealing is smaller if the glass has previously been allowed to "equilibrate" at the high pressure for the longer times.

Figure 5 shows the data on the simulated lunar glass A, which again shows a very rapid decrease of refractive index with time, and on the same figure we also show the results obtained on various lunar chips (10085,49) obtained from the glass lining of pit craters. It will be noted that the behavior is quite different for the lunar glasses in that they show only very small changes in index. Some show instead of the expected *decrease* of refractive index, a very small *increase* in refractive index at these temperatures (bottom of Fig. 5). In one case a decrease in refractive index was observed when a glass chip (10085,49) went from a value of 1.673 to 1.660 after annealing at 580°C for two hours. The data have been obtained principally in sealed platinum systems to avoid changes in oxidation state. The glasses have been compared under various oxidizing conditions in order to estimate or avoid changes in refractive index due to changes in valence states. It appears from the data on simulated glasses that oxidation and reduction play a minor role in the change of refractive indices; this

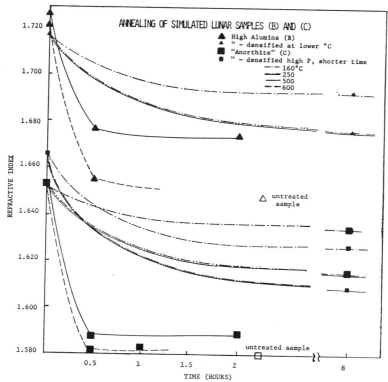

Fig. 4. Annealing of simulated lunar samples (B) and (C).

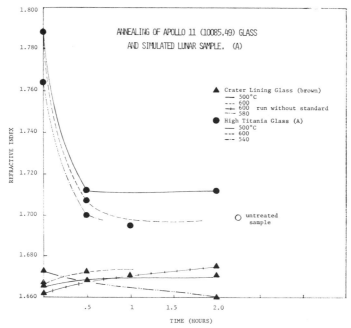

Fig. 5. Annealing of Apollo 11 (10085,49) glass and simulated lunar sample (A).

source of possible complication is, however, important in discussing the changes in the real lunar samples since the changes involved are so small. See Fig. 4, sample C.

DISCUSSION OF DATA

Analysis and interpretation of the data are far from complete. However, some of the results already suggest certain conclusions. VON ENGELHARDT *et al.* (1970a) have earlier reported the increase on annealing in refractive index of the lunar glasses of compositions close to the plagioclases. They noted that *some* of the lunar diaplectic plagioclase glasses (as well as glasses from terrestrial impact craters) actually increased their refractive index upon annealing. BELL and CHAO (1970), however, reported a negative index change on annealing for densified glasses from terrestrial impact and experimental shock-wave events. The maximum change in index of refraction for one impactite (An_{22}) was 0.0076. Later, CHAO *et al.* (1970a) noted that in a thetomorphic plagioclase glass from Apollo 11 the refractive index decreased from 1.5688 to 1.5674 upon annealing. Such differences in the index of refraction on annealing of various thetomorphic glasses suggest variations in the pressures reached during shock vitrification. (FRENCH and SHORT, 1968, p. 509). The decrease in the refractive index reported by CHAO for a lunar feldspathic glass is much less than that of 0.013 observed by us for a single dark brown glass chip from (10085,49) (bottom of Fig. 5). In addition, some of our data in Fig. 5 for pit-crater chips from the same sample show that in certain lunar glasses high in iron and titania there is virtually no change in refractive index, or only a minor increase. Two additional papers in the present proceedings also describe increases in density and/or refractive index on annealing lunar glasses: GREENE *et al.* (1971) reported a density increase of ~ 0.02 for two glasses annealed at 630°, and COOPER *et al.* (1971) described refractive index increases up to 0.0074 on annealing one set of glasses and a density increase of up to 0.033 on another fragment.

The large decrease in the index of refraction (in 10085,49) for short annealing indicates that the temperatures during the latest shock event must have been low and suggests a history of multiple high-velocity impact. Other lunar investigators have found evidence of microcratering in the glass fines. Thus, though one would generally expect relatively larger index changes on annealing for thetomorphic glass than for pit crater glass, a product of shock induced melting, one should consider the potentially more complex history of the actual lunar samples.

No effort will be made to review the very substantial literature on static pressure densification of common silicate and borate glasses which has grown up since 1960. Suffice it to report that there is now general agreement that a noncrystalline phase appears to attain a metastable equilibrium configuration (and hence density and refractive index), which is a continuous function of temperature and pressure. Reasonable values for these effects in a very wide range of compositions are now available for pressures up to roughly 100 kb. Likewise kinetic data are available both for the densification and the relaxation processes on many glasses. Extreme care must be exercised in measurement and interpretation of all these data, since we are dealing with delicate balances between competing kinetics of reactions among various metastable states.

It is now abundantly clear that the terrestrial and lunar noncrystalline solids (NCS), generally thought to be the product of shock impacts, do *not* show remanent effects which would place them on the same scale as the static data. Indeed it is generally unwise to attempt to relate changes in solids caused by shock pressures quantitatively to those produced by static pressures (ROY, 1969). Instead, one may perhaps attempt to understand qualitatively the structural changes in NCS produced by various means. Thus one of the major puzzles is the frequent observation of lunar NCS *increasing* their refractive indices very substantially on annealing.

Increases of such magnitude in the refractive index of a normal glass upon annealing are extremely rare. Recently RINDONE (1965) reported a similar finding in some high index high titania glasses. He ascribed such changes to a change in coordination of Ti^{4+} from four to six. Although two of our samples do contain considerable Ti, this obviously cannot be the correct explanation since the near plagioclase-compositions reported on by VON ENGELHARDT (1970a) show the same phenomenon.

One possibility suggested by our own work, by laser-Raman spectroscopy on some of the same lunar samples, is the incipience of crystallization (WHITE *et al.*, 1971). Impact glasses are known to contain crystalline inclusions (O'KEEFE, 1963) and certainly these lunar NCS show a wide range of not only compositional but structural inhomogeneity. It is conceivable that phase separation and subsequent growth to sub-x-ray detection size could cause such refractive index changes.

It is quite obvious that lunar glasses contain little record of the pressures reached in various events on the lunar surface as part of their permanent refractive index. Indeed, the puzzle that now confronts the crystal chemist is *how* the dynamic high pressure gives rise to a noncrystalline solid, i.e., how shock amorphization works on an atomic scale. One route is by the process of low-temperature, stepwise reaction of a high pressure phase to the equilibrium phase whereby, for example, DACHILLE *et al.* (1963) converted stishovite to a NCS by low-temperature annealing. The observation of the effect of an increase of refractive index upon annealing would suggest alternatively that the process responsible for the destruction of crystallinity is not the high pressure part of the shock cycle. It is possible to conceive of the destruction of a structure which does not undergo a phase transition by the rarefaction wave as the lattice is subjected to the high negative pressure expansion of the lattice and propagation of all surface defects and dislocations throughout the crystal. Such disorder cannot subsequently be reversed in the compression part of the cycle—glasses typically do not crystallize under shock pressures. Of course, the destruction of crystallinity may also be ascribed merely to excessive heating in the shock wave. Preliminary laser experiments (KNOX and WEBER, 1971) suggest that there are threshold energies both for melting and for the development of the globular morphologies observed. In any case it would appear that the thermal or thermal-simulating processes dominate the thermodynamic environment which produced the lunar glasses.

Acknowledgements—Support for this research has come from NASA Grant 39-009-(155) as well as from NASA Institutional Grant to The Pennsylvania State University SSEL NGR-39-009-015.

REFERENCES

AHRENS T. J. and ROSENBERG J. T. (1968) Shock metamorphism: Experiments on quartz and plagioclase. In *Shock Metamorphism of Natural Materials* (editors B. M. French and N. M. Short), pp. 59–82. Mono Book.

BELL P. M. and CHAO E. C. T. (1970) Annealing experiments with naturally and experimentally shocked feldspar glasses. *Carnegie Inst. Wash. Yearb.* **68,** 336–339.

CHAO E. C. T. and BELL P. M. (1969) Annealing characteristics of dense feldspar glass. *Carnegie Inst. Wash. Yearb.* **67,** 126–130.

CHAO E. C. T., JAMES O. B., MINKIN J. A., BOREMAN J. A., JACKSON E. D., and RALEIGH C. B. (1970a) Petrology of unshocked crystalline rocks and shock effects in lunar rocks and minerals. *Science* **167,** 644–647.

CHAO E. C. T., BOREMAN J. A., MINKIN J. A., and JAMES O. B. (1970b) Lunar glasses of impact origin; physical and chemical characteristics and geological implications. *J. Geophys. Res.* **75,** 7445–7479.

COHEN H. M. and ROY R. (1961) Effects of ultrahigh pressures on glass. *J. Amer. Ceram. Soc.* **44,** 523–524.

COHEN H. M. and ROY R. (1962) Effects of high pressure on glass. In *Physics and Chemistry of High Pressures*, pp. 133–139. Soc. Chem. Industry, London.

COOPER A. R., VARSHNEYA A. K., SWIFT J., and YEN F. (1971) Properties of lunar glasses. Second Lunar Science Conference (unpublished proceedings).

DACHILLE F. and ROY R. (1962) Modifications of opposed anvil devices. In *Physics and Chemistry of High Pressures*, pp 77–84. Soc. Chem. Industry, London.

DACHILLE F., ZETO ROBERT J., and ROY R. (1963) Coesite and stishovite: Stepwise reversal transformations. *Science* **140,** 991–993.

DE CARLI P. S. and JAMIESON J. C. (1959) Formation of an amorphous form of quartz under shock conditions. *J. Chem. Phys.* **31,** 1675–1676.

ENGELHARDT W. VON, ARNDT J., MULLER W. F., and STOFFLER D. (1970) Shock metamorphism in lunar samples. *Science* **167,** 669–670.

FRENCH B. M. and SHORT N. M., editors (1968) *Shock Metamorphism of Natural Materials*. Mono.

GREENE C. H., PYE L. D., STEVENS H. J., RASE D. E., and KAY H. F. (1971) Compositions, homogeneity, densities and thermal history of lunar glass particles. Second Lunar Science Conference (unpublished proceedings).

KNOX B. E. and WEBER J. N. (1971) Glass formation by pulsed laser interaction with lunar surface materials and synthetics. Second Lunar Science Conference (unpublished proceedings).

MYERS M., DACHILLE F., and ROY R. (1963) Contributions to calibration of high pressure systems from studies in opposed-anvil apparatus. In *High Pressure Measurements* (editors A. A. Giardini and E. C. Lloyd), pp. 17–33. Butterworths.

O'KEEFE J. A. (1963) The origin of tektites. In *Tektites* (editor J. A. O'Keefe), p. 175. Univ. of Chicago Press.

RINDONE G. E. (1965) Unusually large refractive index changes in titania glasses with heat treatments. In *Proceedings of the 7th International Congress on Glass, Brussels, Belgium* (editors Emile Plumet and R. Chambon), paper no. 103. Gordon and Breach.

ROY R., and COHEN H. M. (1961) Effects of high pressure on glass: A possible piezometer for the 100 kilobar regions. *Nature* **190,** 798–799.

ROY R. (1969) Neglected role of metastability in high pressure research. *Reactivity of Solids* (editors J. W. Mitchell, R. C. DeVries, R. W. Roberts, and P. Cannon), pp. 777–788. John Wiley.

WHITE W. B., WHITE E. W., GORZ H., HENISCH H. K., FABEL G. W., and ROY R. (1971) Examination of lunar glass by optical, raman, and X-ray emission spectroscopy, and by electrical measurements. Second Lunar Science Conference (unpublished proceedings).

Proceedings of the Second Lunar Science Conference, Vol. 3, pp. 2079–2082
The M.I.T. Press, 1971.

Neutron diffraction study of lunar materials

STANLEY J. PICKART and HARVEY ALPERIN
Solid State Division, Naval Ordnance Laboratory, Silver Spring, Maryland 20910

(Received 20 February 1971; accepted in revised form 30 March 1971)

Abstract—Neutron diffraction studies were carried out on Apollo 12 samples 12070,119; 12071,6; and 12008,7 at 300,77 and 4.2 K. Evidence of a magnetic ordering was observed in sample 12071,6 and is tentatively attributed to an iron-rich pyroxene phase. From the relative intensities of the observed d-spacings, it is concluded that the first two samples contain plagioclase and pyroxene in roughly equal proportions, while the latter is predominantly pyroxene, with some olivine.

INTRODUCTION

THIS WORK describes neutron diffraction measurements taken on three Apollo 12 lunar samples in the form of rock chips and fines. Although the primary purpose of the study (investigation of the antiferromagnetic structure of lunar ilmenite) was thwarted because of the lack of clear evidence for $FeTiO_3$, an indication of antiferromagnetic ordering was observed in one sample, and is attributed to an iron-rich pyroxene phase. Some qualitative information about relative sample compositions was also obtained.

Neutron diffraction analysis is a valuable tool in magnetic studies because the interaction between the neutron spin and unpaired atomic electron gives rise to an additional, magnetic intensity component in the presence of ordered spin lattices. Moreover, information about phase constitution complementary to X-ray analyses is obtained because the neutron-nuclear scattering amplitudes vary widely across the periodic table, making lighter atoms and ordering effects more visible in many cases. A further pertinent fact is that the neutron intensities are more representative of the *volume* composition, because of the low inherent absorption. These advantages are offset somewhat by the relatively poor resolution required at least for samples of small size; *d*-spacings are difficult to measure (or resolve) to better than about 1%.

EXPERIMENTAL TECHNIQUES

Method

The present measurements were taken at 1.38 A for the most part, using the (111) planes of a deformed Ge crystal as monochromator, with 20′ collimation throughout. Some additional runs were taken at 2.38 A with a pyrolytic graphite monochromator to improve resolution. Limited checks for the presence of single crystal grains were made, with negative results. Data were taken at the NBSR reactor facility, using a computer-controlled spectrometer control system (ALPERIN and PRINCE, 1970).

Samples

The samples studied are listed in Table I, along with their description and the experimental conditions under which measured. In toto, some 35 diffraction runs were taken, many of them repetitions. This procedure was followed to improve data quality; data collation can readily be done with the on-line control system.

Table 1. Disposition of samples

Sample No.	Weight	Form	Temperature of study
12070,119	0.97 g	fines	300, 77, 4.2 K
12071,6	2.07 g	3 chips and incidental fines	300, 4.2 K
12008,7	1.409 g	single chip	300 K

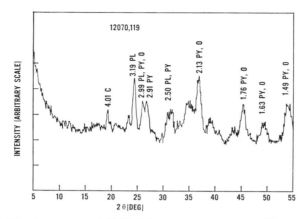

Fig. 1. Diffraction pattern of 12070,119 at room temperature. The wavelength in this and the following figures is 1.38 A. Only strong lines are identified. Symbols: PL = plagioclase (anorthite, albite, sanidine), PY = pyroxene (pigeonite, angite), O = olivine (forsterite), C = cristobalite.

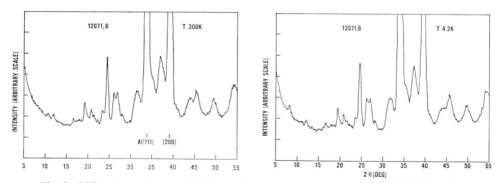

Fig. 2. Diffraction patterns of 12071,6 at (a) room temperature and (b) 4.2 K. Note the additional small peak near 8° in the second pattern.

Results

Diffraction patterns (collated data) from the three samples are displayed in Figs. 1–3. Low temperature data for sample 12070,119 were identical to room temperature and are not shown. Apart from the strong lines due to the Al sample holder (a $\frac{1}{8}''$ diameter V cylinder was used for the fines) it is apparent that samples 12070,119 and 12071,6 are fairly close in composition while 12008,7 is quite different. The low temperature pattern of 12071,6 (Fig. 2b) is identical to the room temperature one within statistics except for a weak but unmistakable additional peak at low angles. This is attributed to a magnetic ordering, as discussed below.

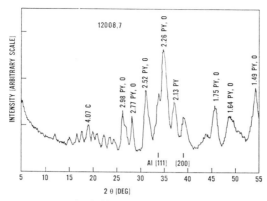

Fig. 3. Diffraction pattern of 12008,7 at room temperature. Symbols as in Fig. 1. Decreased plagioclase is inferred from near absence of strong reflection at 3.19 A, increased olivine and pyroxene from spacings at 2.77, 2.52, and 2.26.

DISCUSSION

Mineralogy

We looked for the presence of all minerals found by LSPET (1970). On the basis of the relative intensities of the observed interplanar d-spacings, we can make the following tentative identifications: (a) Samples 12070,119 and 12071,6 are primarily plagioclases (anorthite, sanidine) and pyroxenes (pigeonite and augite) in roughly equal proportions. Quartz, cristobalite, and olivine (forsterite) are possibly present. Ilmenite and troilite are not visible, with an upper limit of 2–3% by volume estimated from the signal-to-noise ratio. (b) Sample 12008,7 is predominantly pyroxenes and olivine. Some evidence is seen for spinel, quartz, plagioclase, and cristobalite. Ilmenite is possibly present.

It is difficult to make these estimates more than qualitative for two reasons. The first is the aforementioned low resolution (20′ in our case) which does not display the line splittings of the various triclinic, monoclinic and orthorhombic forms of plagioclase and pyroxene. This is obvious from the diffraction patterns, where narrow lines, comparable to the resolution width, contrast with broad, clearly unresolved ones. A second problem is the difference of relative intensities between neutrons and X-rays; not many complex silicate minerals have been studied with neutrons, and while the neutron scattering amplitudes are qualitatively similar for most of the mineral phases cited, some differences will occur, particularly at high angles because of the absence of a form factor correction for neutrons.

While these estimates are admittedly qualitative, and certainly do not provide any unexpected information, nevertheless they are felt to be useful because they demonstrate the possibility of using neutron diffraction as an analytical tool in lunar sample studies. The complementary advantages to X-rays, as mentioned earlier, are the different relative visibility of atoms and the sensitivity of the scattering to volume. Samples of 1 gm size are unusually small, even for single phase material studied at high flux reactors, and it was by no means clear at the outset that useful information could be obtained.

Magnetic ordering

The observation of a peak attributable to magnetic ordering in sample 12071,6 is intriguing because all known terrestrial antiferromagnets so far studied by neutron diffraction can be eliminated as its origin. The known candidates are ilmenite (SHIRANE *et al.*, 1959), troilite (ANDRESEN, 1960) and olivine (fayalite) (SANTORO *et al.*, 1966). None of these shows a strong magnetic reflection with the measured *d*-spacing, 9.18A.

However, this line indexes quite well as the (200) or (010) reflection of an ortho-rhombic synthetic iron silicate, $FeSiO_3$, which is known to be antiferromagnetic (SAWAOKA *et al.*, 1968), raising the possibility that the ordering is due to an iron-rich pyroxene (eulite ?). An anomaly in the susceptibility of Apollo 11 lunar material has also been attributed to this phase by NAGATA *et al.* (1970). Of course, not much can be said about the magnetic structure from these limited data, other than that such a reflection could arise if the periodicity of the antiparallel Fe^{2+} spins were a/2 or b, with their direction along c.

CONCLUSION

Although the analyses presented here are qualitative, we believe the above results demonstrate the utility of neutron diffraction for the identification of mineral phases and magnetic ordering in lunar materials. More quantitative analyses could be made if the isolated, synthetic minerals were studied or if larger samples allowing better resolution were available. Should subsequent study on terrestrial material confirm the proposed magnetic ordering, it will probably be the only antiferromagnetic structure first determined from lunar material.

REFERENCES

ALPERIN H. A. and PRINCE E. (1970) A time-shared computer system for diffractometer control. *J. Res. Nat. Bur. Stand. (U.S.).* **74C**, 89–95.
ANDRESEN A. (1960) A neutron diffraction study of FeS. *Acta Chem. Scand.* **14**, 919–920.
LSPET (Lunar Sample Preliminary Examination Team) (1970) Preliminary examination of the lunar samples from Apollo 12. *Science* **167**, 1325–39.
NAGATA T., ISHIKAWA Y., KINOSHITA H., KONO M., SYONO Y., and FISHER R. M. (1970) Magnetic properties of lunar crystalline rock and fines. *Science* **167**, 703–704.
SANTORO R. P., NEWNHAM R. E., and NOMURA S. (1966) Magnetic properties of Mn_2SiO_4 and Fe_2SiO_4. *J. Phys. Chem. Solids* **27**, 655–66.
SAWAOKA A., MIYAHARA M., and AKIMOTO S. (1968) Magnetic properties of metasilicates and metagermanates with pyroxene structure. *J. Phys. Soc. Japan* **25**, 1253–57.
SHIRANE G., PICKART S. J., NATHANS R., and ISHIKAWA Y. (1959) Neutron diffraction study of antiferromagnetic $FeTiO_3$ and its solid solutions with α-Fe_2O_3. *J. Phys. Chem. Solids* **10**, 35–43.

Proceedings of the Second Lunar Science Conference, Vol. 3, pp. 2083–2092
The M.I.T. Press, 1971.

Auger electron spectroscopy of lunar materials

Gary L. Connell, Richard F. Schneidmiller, Paul Kraatz,
and Y. P. Gupta

Northrop Corporate Laboratories, Hawthorne, California 90250

(*Received* 22 *February* 1971; *accepted in revised form* 31 *March* 1971)

Abstract—Lunar samples 12002,172 and 12021,18 have been examined by Auger electron spectroscopy (AES). Standard techniques were used, utilizing a 1450 eV, 20 μA primary electron beam to produce the Auger electrons which have discrete energies, characteristic of the elements from which they arise. The elements present in the first few atomic layers (10–15 Å) in various areas were found to be Al, Si, S, Cl, K, C, Ca, Ti, O, and Fe. Controlled in situ sputter cleaning by 340 V Ar ions did not reveal any additional features, except the expected removal of S, Cl, and C contaminants. In addition, the first 1 to 3 μ were examined by a nondispersive X-ray spectrometer in a scanning electron microscope. This revealed the following elements: Al, Si, K, Ca, Ti, Fe, Mg, Na, Cr, Zr, Y, Sr, Ni, S, Cl, and P. Bulk analysis (20–30 μ depth) by X-ray fluorescence (for $Z \leq 36$) found the material to be composed of Al, Si, K, Ca, Ti, Fe, Mn, Cr, P, and Cl. From these elemental analyses, it is reasonable to postulate that both of the lunar rock samples comprise some or all of the following mineral species: plagioclase feldspar (probably near the calcic end of the series), pyroxenes of the enstatite-hypersthene series or the diopside-hedenbergite series, olivine, ilmenite, and possibly chromite and manganese garnet (spessartite).

Introduction

In many laboratories of the world numerous studies of the physical, chemical, and geological properties of the lunar materials from Apollo 11 and Apollo 12 have been in progress. However, aside from the present investigation of the lunar material by Auger electron spectroscopy, most studies are concerned with the investigation of the bulk properties. On the other hand, Auger electron spectroscopy (AES) is used for determining the surface properties, particularly the elemental distribution in the first few atomic layers (10–15 Å).

The atoms in a solid, when bombarded with electrons or photons of a sufficient energy, can be ionized. The atom in the ionized state has higher energy and consequently is unstable relative to its ground state. The excess energy of the ionized atoms can be released by emission of electromagnetic radiation (such as characteristic X-rays) or Auger electrons. The former process is called fluorescence and forms the basis for emission spectroscopy such as X-ray fluorescence and electron microprobe analyses.

Auger electron spectroscopy utilizes emission of Auger electrons from the ionized atoms during their relaxation. The Auger electron emission is generally more efficient for elements of atomic number less than 20, while fluorescence is more probable for elements of higher atomic numbers (although, strictly speaking, X-ray production is negligible for ionization by primary electron beam energies of 1 to 2 keV). Auger electrons comprise a small part of the total secondary electrons emitted from a surface under bombardment by electrons, photons, or ions. This is so because the Auger electron yield is much smaller than the total flux of secondary electrons. Thus, in the energy distribution of secondary electrons, Auger electrons usually appear as minute peaks of characteristic energies. The magnitude of these peaks may be so small as to be masked by the overall secondary electron yield. This condition, therefore, necessitates techniques with very high sensitivity for detecting Auger electrons. A technique commonly used for detecting the Auger peaks is to electronically differentiate the secondary electron distribution function with respect to energy. In so doing, the large uniform background, due to secondary electrons, contributes only minimally to the derivative distribution except in the range up to about 100 eV, where the secondary electron distribution changes rapidly. In the derivative distribution, Auger electron peaks appear as highly enhanced inflections or doublets. Moreover, only those Auger electrons that come from the first few atomic layers are detected since the range or escape depth of Auger electrons is very small, because they must not only escape, but they must escape with no loss of energy in order to be identified as characteristic Auger electrons.

The procedure of obtaining the Auger spectrum from a given specimen as well as a general review of this technique are described elsewhere (CONNELL and GUPTA, 1971; HARRIS, 1968; WEBER and PERIA, 1967). Figure 1 shows typical curves of $N(E)$ versus E and $dN(E)/dE$ versus E, where $N(E)$ and $dN(E)/dE$ are, respectively, the secondary electron distribution function and the derivative of the distribution $N(E)$ with respect to energy E. The Auger spectra of the lighter elements are very simple in structure. For heavier elements, increased spectral complexity and decreased Auger yield cause some difficulties in the interpretation of an Auger spectrum and sometimes even in the identification of peaks from these elements. The sensitivity of AES is, however, very high. In most cases, bulk concentrations of more than 100 parts per million may be detected (WEBER, private communication).

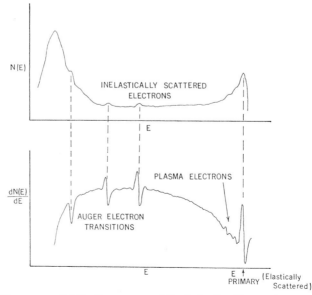

Fig. 1. The energy distribution curve and its derivative (Auger electron spectra).

Optical Microscopy and Physical Description

In referencing Auger spectra and in assisting other studies to be identified with the specific sides of the specimen, the following nomenclature was used: For specimen 12002,172, the sides 1, 2, and 4 are the outer original surfaces of the sample, while side 3 is an inner cut surface having a fracture band across the upper half inch of the micrograph. For the lunar sample 12021,18 the sides 1, 2, and 3 are the "original" surfaces (sample chipped from surface of 12021).

Sample 12002,172

Approximate mass: 1.30 gm; approximate dimensions: 1 cm × 1 cm × 1 cm. Consists of interlocking, subhedral to euhedral, grayish brown plagioclase feldspar laths with interstitial greenish yellow pyroxene and subhedral to euhedral yellowish green olivine. Grain size: minimum, approximately 0.1 mm; maximum, approximately 1 mm; average, approximately 0.7 mm. Olivine averages 0.5 to 1.0 mm. Doleritic texture with many voids averaging 1 to 2 mm.

Sample 12021,18

Approximate mass: 1.16 gm; approximate dimensions: 1 cm × 1 cm × 0.6 cm. Consists of subhedral to euhedral colorless plagioclase feldspar laths with interstitial reddish brown pyroxene and minor olivine, ilmenite, and garnet. Grain size: minimum, approximately 0.5 mm; maximum, 2 mm +; average, 1.0 mm. Ilmenite averages about 0.5 mm. Doleritic texture with few voids, averaging 1.0 mm. Several intergranular fractures, showing openings of 0.2 to 0.3 mm width. No visible mineralogic alteration along fractures. All grain sizes given are approximate, since they were determined by comparison with a millimeter scale held in the field of view. Mineral identifications are all tentative, since thin sections were not available. Study of thin sections and the use of a micrometer ocular can verify both grain size measurements and mineral identification.

Auger Electron Spectroscopy

The lunar specimen was placed in a stainless steel sample holder. Prior to use, the holder was cleaned in acetone, methyl alcohol, and transene, and then vacuum baked for 1.5 hours. Each of the four sides of the sample holder have an 8 mm hole through which the electron beam could probe the sample. The holder was attached to a crystal manipulator which allowed movement of the specimen in x, y, and z directions, as well as tilt and continuously variable rotation through 360°.

After placing the specimen in the vacuum chamber, the system pressure was reduced to less than 3×10^{-9} torr and maintained at this pressure throughout the examination.

Initially, an Auger spectrum of the sample holder was taken. This revealed the following elements: P, S, C, Cr, O, Fe, and Ni. Then the lunar specimens were examined before and after in situ sputter cleaning by 340 V Ar ions, as shown in Fig. 2. Figure 3 shows a spectrum from an area that was not sputter cleaned. The

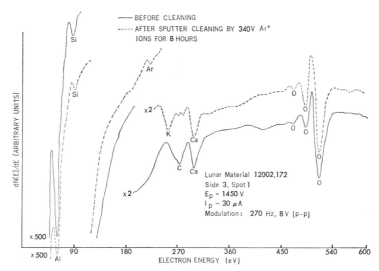

Fig. 2. Auger electron spectra of lunar sample 12002,172, side 3, spot 1, before and after in situ sputter cleaning.

results of all AES analyses performed on the two specimens are summarized in Tables 1 and 2, respectively. The Auger spectra are identified by the sample number, side number, and spot number. Generally, the spectra were obtained up to about 600 V unless transitions were detected beyond that value. This restriction on the range of Auger electron energies covered is due to the limited capability of the primary electron gun (1450 V) in our present Auger spectrometer, since the maximum detection efficiency of an Auger peak of specific energy E occurs at a primary electron beam energy of $2E$ to $5E$ (PERIA, 1968).

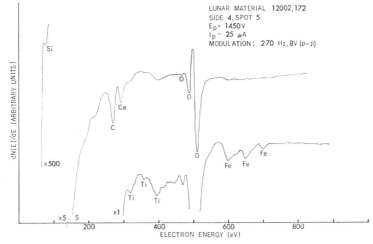

Fig. 3. Auger electron spectra of lunar sample 12002,172, side 4, spot 5.

Table 1. Lunar material 12021,18.

Figure	Side	Spot	Elements detected	Comments
29	3	1	Si, S, K, C, Ca, O	
30	3	2	Si, K, C, Ca, Ti, O	
31	3	2	Si, C, Ca, Ti, O	
32	3	3	Si, K, C, Ca, Ti, O	
33	3	4	Si, S, C, Ca, O	
34	3	5	Si, S, Cl, K, C, Ca, O	
35	3	6	Si, S, C, Ca, O	
36	3	7	Si, S, K, C, Ca, Ti, O	
37	3	8	Si, S, Cl, K, C, Ca, Ti, O	
38	3	10	Si, S, K, C, Ca, Ti, O	
39	3	10	Si, S, Cl, Ar, K, C, Ca, Ti, O	4 hrs sputter cleaned
40	3	10	Si, S, Ar, K, C, Ca, Ti, O	8 hrs sputter cleaned
41	3	10	Si, S, Cl, Ar, K, C, Ca, Ti, O	12 hrs sputter cleaned
42	3	11	Si, S, Cl, Ar, K, C, Ca, O	Examined after 12 hrs sputter cleaning of spot 10, side 3
43	3	12	Si, S, Ar, C, Ca, O	Examined after 12 hrs sputter cleaning of spot 10, side 3

The following elements were identified in these studies: Al, Si, K, Ca, Ti, O, and Fe. S, Cl, and C are also indicated and could probably be contaminants. However, S, Cl, and C were not always removed by sputter cleaning. The Ar transitions detected are caused by implantations of these ions on the surface during sputter cleaning, since these peaks have also been detected on sputter-cleaned single crystal silicon and other materials examined. Mg and Na were not detected in the present AES studies, since their major KLL Auger transitions occur at 1000 V (Na), and 1150 V and 1200 V (Mg), where the appearance of plasma electrons (since the primary electron beam energy was 1450 V), caused excessive noise and where the K-shell ionization cross sections would be low. Alternatively, the LM Auger electron transitions for Na and Mg occur in the range of 40–50 V. Since the background in this range rises rapidly, detection is less certain.

Table 2. Lunar material 12002,172.

Figure	Side	Spot	Elements detected	Comments
11	3	1	Al, Si, Cl, C, Ca, O	
12	3	2	Al, Si, C, Ca, O	
13	3	3	Si, K, C, Ca, O	
14	1	1	Al, Si, Cl, C, Ca, O	
15	3	1	Al, Si, C, Ca, O	Taken before beginning sputter cleaning
16	3	1	Si, K, C, Ca, O	4 hrs sputter cleaned
17	3	1	Al, Si, Ar, K, Ca, O	5 hrs sputter cleaned
18	3	1	K, C, Ca, O	5 hrs sputter cleaned
19	3	1	Si, S, Cl, Ar, Ca, Ti, O	20.5 hrs sputter cleaned
20	3	1	K, C, Ca, Ti, O	20.5 hrs sputter cleaned
21	2	1	S, Cl, C, Ca	
22	2	3	Si, S, C, Ca, O	
23	2	4	Si, S, K, C, Ca, O	
24	2	4	S, Cl, C, Ca, O	
25	4	2	Si, Cl, C, Ca, Ti, O	
26	4	3	Si, S, C, Ca, Ti, O	
27	4	4	S, C, Ca, Ti, O	
28	4	5	Si, S, C, Ca, Ti, O, Fe	

A major problem encountered was charge buildup on the sample surface during electron bombardment. This is not surprising, since most silicate minerals are poor conductors. For this reason certain spots could not be examined. As the samples were sputter cleaned at a rate of 5–10 Å/hr, the problem of charging became worse, causing the appearance of anomalous "peaks" as well as shifting the true Auger electron peaks. An attempt was made to always reference the C peak to its normal value (270 V) in order to compensate for the potential shift. In many cases the Si transition was observed at 80 V, rather than at 90 V obtained from single-crystal Si. This ambiguity is due to the chemical shift of the electronic levels (oxidation reduces the energy of the major [LMM] transition).

A final experimental problem concerns the beam size and the uncertainty of its location. The beam diameter was about 1 mm. The grain size in the lunar material is generally less than this. Thus, more than one single-crystal mineral phase was being examined at each spot.

SCANNING ELECTRON MICROSCOPY

Surface microstructural features of the lunar samples were examined by scanning electron microscopy (SEM), a technique especially useful in looking directly at rough or irregular surfaces. The SEM used in this study was a JEOLCO JSM-2, with a Princeton Gamma-Tech nondispersive X-ray spectrometer attachment for determining elemental analysis both on areas containing groups of crystals as well as on individual crystals.

The results of the nondispersive X-ray spectrometry are given in Tables 3 and 4. Micrographs of these areas were also obtained, but they will not be reported here.

From the data in Table 3 it is evident that the major elements in sample 12002,172

Table 3. Nondispersive X-ray spectroscopic analysis of sample 12002,172.

	Side	Area	Si	Ca	Fe	Ti	Al	Mg	Na	K	Cr	Zr*	Y*	Sr*	Ni	Cu	Description
A	1	1	×	++	××	——	—	—	—	—	—					—	Center
B	3	1	×	++	××	——	—	—	—	—	—					—	Center
C	1	2	××	++	×	—	—	—	—		—	——		—	—□	—	□ Large crystal
D	1	2	++	+	××	—		—			—	—		——	—	—	Yellow-green crystal
E	1	2	++	—	××	—	——		——	—	++		—	—	—	—	Area between large voids
F	1	2	+	+	××	—			—	—	——			—□	——	Large crystal
G	1	3	××	+	++	——		—			—	—		—	—	—	Large crystal
H	1	4	××	+	××	——	—	—		—	——		—	—	—	——	Black area
J	1	5	××	+	××	——			—		—		—		——	Corner
K	3	5	++	+	××	—	—	—	—		—			—	—	—	Fracture area
L	1	6	+	+	××	—	—	—	——	—	—		—	——	Large crystal

Symbols	Intensity		Instrumental settings		
				YV/IN	Display
××	Strongest (100)				
×	Strong				
++	Medium		Normal	0.5	5 × 10³
+	Weak		□	0.5	10³
——	Very weak		△	0.2	10³
—	Very very weak		*	0.1	10³

Table 4. Nondispersive X-ray spectroscopic analysis of sample 12021,18.

	Side	Area	Si	Ca	Fe	Al	Al	Mg	Na	K	Cr	Zr*	Y*	Sr*	Ni	Cu	Description
A	1	1	×	+	××	—		—			—	—	—		—△	— —	Center
B	3	1	++	+	××	—	—	—	—	—	—	—	—	—		— —	Center
C	3		++	+	××	—		—	—□	—□	—	—				+	Sponge-like structure
D	3		+	+	××	—□	—	—□	—	—□	+				—□	+	Large black crystal
E	loose chip		××		××			—	—□	—		—□	—			— —	Yellow-green crystal
F	loose chip		××	+	×	++			—		—△	—△		—		—△	Reddish-brown cryst W (on carbon substrate)
G	3	3	+		××			—		—	—□			—		+	Growth step structure
H	loose chip		+		××	××	—	—	—			—		—		—△	Black crystal (on carbon substrate)

Symbols	Intensity		Instrumental settings		
				YV/IN	Display
××	Strongest (100)				
×	Strong				
++	Medium		Normal	0.5	5 × 10³
+	Weak		□	0.5	10³
— —	Very weak		△	0.2	10³
—	Very very weak		*	0.1	10³

are iron, calcium, and silicon. Minor elements are titanium, aluminum, magnesium, sodium, potassium, chromium, copper, nickel, zirconium, yttrium, and strontium. Similarly, from Table 4, the major elements for sample 12021,18 are iron, calcium, and silicon, and the minor elements are titanium, aluminum, magnesium, sodium, potassium, chromium, copper, nickel, zirconium, yttrium, and strontium. Thus, the two samples, although from two different lunar rocks, have the same elemental composition. This elemental composition is similar to that of terrestrial basalt material.

Before any conclusions can be made from the results of this portion of our study, several remarks should be made. First, in using the nondispersive X-ray spectrometer with the SEM, an iron peak is always present in the spectrum obtained, due to internal scatter of X-rays from iron components within the SEM. Secondly, it is important to remember that the sample holder was made from 2024 Al (4.5% Cu, 1.5% Mg, 0.6% Mn, and the balance Al) and that it is possible (although not too probable) that some of the elements detected could have originated from this aluminum alloy sample holder. And, finally, it should be noted that the results from the nondispersive X-ray spectrometer are semiquantitative in nature, due in part to the rough nature of the sample.

The following can be concluded for lunar sample 12002,172 from the result tabulated in Table 3. It can be seen that the elemental composition of the center of side 1, the original outer rock surface (A), is essentially the same as that from the central area on side 3 (B), an inner cut surface. The only difference is in the minor elements present, which is not significant. This indicates that the elemental composition of this sample has not changed significantly from the original outer surface to this depth, a distance of about 5 mm.

Some reasonable postulations can be made as to the major mineral species in each area or crystal on which an elemental analysis was made as listed in Table 3. Pyroxene (Mg, Fe) SiO_3 or Ca (Mg, Si) $AlSi_2O_8$ is the most reasonable major mineral species in A and B; plagioclase feldspar (Ca, Na) (Al, Si) $AlSi_2O_8$ in C and G; olivine (Mg, Fe)$_2$ SiO_4 in D; pyroxene and olivine in H, J, and K; chromite (Mg, Fe) Cr_2O_4 and ilmenite $FeTiO_3$ in F and L; and chromite and silicates in E. The same type of postulations can be made for the areas and crystals in Table 4. Thus, pyroxene is the most reasonable mineral species to satisfy the elemental analysis of A, B, C, and F; chromite for D, olivine for E and G; and finally, ilmenite for H. The presence of copper in both tables must be due either to the sample holder or to internal scattering effects (as in the case of iron), since it was not detected by either X-ray fluorescence or electron microprobe analysis. The determination of mineral species is further complicated by the fact that the spectrum obtained was not only from the major mineral phase present, but also from the surrounding mineral matrix. Thus, the spectrum for the major minerals is slightly obscured due to the presence of this background effect. This is not true for the crystals mounted on the carbon substrates (E, F, and H) since there was no background effect and the spectrum obtained is very close to that expected from the individual minerals.

The SEM has proved to be valuable for the determination of the microstructure of the lunar samples, and when combined with optical examination, it makes it possible to determine the location of areas under examination. The use of the non-dispersive X-ray spectrometer in the SEM provides a means of identifying the major mineral or minerals present in the area being examined, with a reasonable degree of certainty.

There are several techniques which could be used to improve the experimental procedure which would yield even more information in any future studies. One of the ways would be to coat the lunar samples either with gold or an antistatic spray to eliminate the charging effects. This would make it possible to obtain both higher resolution and higher magnification micrographs resulting in a higher yield of information. The examination of petrographic thin sections in the SEM in conjunction with optical examination would yield much more definitive information on the mineral species present.

X-RAY FLUORESCENCE SPECTROSCOPY

X-ray fluorescence analyses were performed on each sample and on a terrestrial basalt of known composition. A GE XRD-6 vacuum spectrometer was used for these analyses. The principles and techniques for X-ray fluorescence analysis are described elsewhere (CULLITY, 1959).

The measurements reported here are entirely qualitative since no suitable standards were available and the sample surfaces could not be properly prepared (ground flat). The purpose of this study was to provide an elemental analysis representative of the bulk of each sample for comparison with the Auger analysis of sample surfaces. In each case, an entire side of a lunar sample (approximate area 1 cm²) was irradiated at different times with X-rays from a tungsten target and a chromium target (GE type EA 75 dual target tube) operated at 50 kV and 20 mA. The maximum effective depth of penetration of the primary X-rays is approximately 20 to 30 microns (200,000 to 300,000 Å).

In order to detect elements of interest in the lunar samples, two combinations of primary radiation source (tube target) and analyzing crystal were utilized to scan each sample. For longer wavelengths emitted by the lighter elements (atomic numbers 13 through 22), the primary radiation source was chromium (unfiltered) and the analyzing crystal was Pentaerythrital (PET), which has a lattice spacing (d) of 8.750 Å. For shorter wavelengths emitted by elements of atomic number 22 and above, the primary radiation source was tungsten (unfiltered) and the analyzing crystal was lithium fluoride (LiF), having a lattice spacing (d) of 4.0267 Å. The counter tube used was a gas flow proportional type utilizing a 90% argon, 10% helium mixture at 3 psi, a flow rate of approximately 1 cm^3/sec, and a potential of 1750 volts. This tube is most efficient for radiation of wavelength greater than 1.0 Å, but becomes inefficient for radiation of shorter wavelengths. Thus, an analysis of elements only up to atomic number 36 was performed.

From this analysis we conclude that the major elements present in the lunar sample 12002,172 are iron, calcium, silicon, and aluminum. Minor elements are titanium, manganese, chromium, chlorine, sulfur, phosphorus, and possibly potassium, although this last element is also present in the sample holder-set screw assembly. Similarly, lunar sample 12021,18 contains as major elements iron, calcium, silicon, and possibly aluminum and minor elements titanium, manganese, chromium, chlorine, sulfure, phosphorous, and again, possibly potassium. Clearly, the elemental compositions of both the lunar samples are very similar to each other. The elements present in the set screw of the sample holder are iron and minor amounts of manganese and chromium. Apparent impurities and contaminants in the sample holder-set screw combination include calcium, potassium, sulfur, and chlorine.

Summary of Results

From these elemental analyses, it is reasonable to postulate that both of the lunar rock samples comprise some or all of the following mineral species: plagioclase feldspar (Ca, Na) (Al, Si) AlSi$_2$O$_8$ (probably near the calcic end of the series), pyroxenes of the enstatite-hypersthene series (Mg, Fe) SiO$_3$ or the diopside-hedenbergite series Ca (Mg, Fe) Si$_2$O$_6$, olivine (Mg, Fe)$_2$ SiO$_4$, ilmenite FeTiO$_3$, and possible chromite (Mg, Fe) Cr$_2$O$_4$ and manganese garnet (spessartite) Mn$_3$Al$_2$ (SiO$_4$)$_3$. The chlorine, sulfur, carbon, and phosphorous are undoubtedly contaminants.

Clearly, elemental analyses such as these give little specific information on the mineralogic nature of the rocks themselves. Conjunct studies of petrographic thin sections would yield far more precise information on the petrology of the lunar materials.

Despite some experimental problems it appears that Auger electron spectroscopy is a suitable analytical method for elemental analysis. This technique may also be promising for remote determinations (such as unmanned probes). It is the only method currently being used for revealing the near surface (10–15 Å) composition of Apollo 12 material. Controlled sputter cleaning did not reveal any additional features, except the expected removal of surface contaminants. Although the present AES analysis was qualitative, it does appear feasible that under more carefully controlled experimental conditions, and through measurement of the $N(E)$ versus E curve, the ratios of Auger electron peak areas for different elements would be meaningful, and valid conclusions could be drawn concerning elemental segregation and distribution. A better experimental approach would be to use an electron beam of more than 3 keV and smaller spot size. If this could be done in a SEM, simultaneously with visual observations and nondispersive X-ray analysis, an ideal situation would be achieved with a resulting multitude of experimental information.

REFERENCES

CONNELL G. L., and GUPTA Y. P. (1971) Auger electron spectroscopy. *ASTM Materials Research and Standards* **11**, 8–13.

CULLITY, B. D. (1959) *Elements of x-Ray Diffraction*, 2nd edition, Addison-Wesley.

HARRIS, L. A. (1968) Secondary electron spectroscopy. *Ind. Res.* **10**, 53–56.

HARRIS, L. A. (1968) Some observations of surface segregation by Auger electron emission. *J. Appl. Phys.* **39**, 1428–1431.

PERIA, W. T. (1968) Physics of electron-electron and electron-photon interaction. A. F. Tech. Rep. No. AFAL–TR–69–177.

WEBER R. E., and PERIA W. T. (1967) Use of LEED Apparatus for the Detection and Identification of Surface Contaminants. *J. Appl. Phys.* **38**, 4355–4358.

Proceedings of the Second Lunar Science Conference, Vol. 3, pp. 2093–2102
The M.I.T. Press, 1971.

Some results from the Apollo 12 Suprathermal Ion Detector

J. W. Freeman, Jr., H. K. Hills, and M. A. Fenner
Department of Space Science, Rice University, Houston, Texas 77001

(*Received* 22 *February* 1971; *accepted in revised form* 29 *March* 1971)

Abstract—The Apollo 12 ALSEP Suprathermal Ion Detector has repeatedly detected clouds of positive ions of energies ranging from tens of electron-volts to several thousand electron-volts. The majority of these events are considered to be natural events typical of the ambient lunar surface environment; however, one such event was caused by the impact of the Apollo 13 S-IVB stage. This plasma cloud (or clouds) was also detected by another experiment, the Apollo 12 ALSEP Solar Wind Spectrometer. The data from the Suprathermal Ion Detector indicates that ions in the mass range 55 to 130 amu/q were present in the cloud. The ion energies observed ranged from 10 to 500 electron-volts/unit charge. The maximum integral ion flux from the near vertical direction in which the Suprathermal Ion Detector looks was 3×10^7 ions/cm²-sec-ster; however, the Solar Wind Spectrometer reported a still larger flux from horizontal directions. This paper contains further details of this event as well as a brief list of some natural phenomena observed.

Introduction

The Apollo 12 ALSEP 1 Suprathermal Ion Detector Experiment (SIDE) was designed to detect possible lunar ionospheric phenomena near the lunar surface. Data from this experiment have been varied and highly complex. Consequently, a substantial portion of the analysis effort during the first year has been spent sifting the observations to establish categories for the phenomena observed, especially those of a repetitive or diurnal character. Additional attention has been given to a few singular events such as the Apollo 13 S-IVB impact, the Apollo 12 LM impact, the lunar eclipses, and solar flare-induced events.

A unifying feature found between certain diurnal phenomena, such as clouds of suprathermal ions often seen before and after lunar sunrise and sunset, and certain nonrepetitive phenomena is the apparent suggestion of an ion acceleration mechanism operative near the lunar surface. We have found the S-IVB impact event to be particularly useful in attempting to study this acceleration mechanism. For this reason, the principal emphasis in this paper will be placed upon this event.

The SIDE consists of two detectors which accept positive ions from a direction 15 deg west of vertical. Their fields of view are approximately square and 6 deg on a side. The first, the Total Ion Detector (TID), measures the differential energy spectrum of positive ions in twenty steps from 10 eV up to 3500 eV; the second, the Mass Analyzer Detector (MAD), determines the mass per unit charge of positive ions at each of the six energy levels 48.6, 16.2, 5.4, 1.8, 0.6, and 0.2 eV. The mass range is selectable and that being scanned during the S-IVB impact event covered 10 to 130 amu/unit charge in ten channels. A channel-electron-multiplier operated in the pulse mode serves as the sensor for each detector.

The SIDE instrument package has its main ground line tied to a 24-inch diameter spider web-shaped electrode through a stepped voltage supply. The electrode (called the *ground plane*) is placed on the lunar surface beneath the instrument by the deploying astronaut. At the time of the S-IVB impact this ground plane stepping voltage had been halted with the SIDE 16.2 V negative with respect to the ground plane. Further details of the SIDE are given by Freeman *et al.* (1970).

The S-IVB stage of the Apollo 13 launch vehicle impacted the lunar surface at a point approximately 139 km west of the Apollo 12 site. Twenty seconds later the Apollo 12 SIDE reported a unique, intense burst of suprathermal positive ions. The characteristics of this cloud of ions as indicated by the total ion energy spectrometer and mass analyzer components of the SIDE were as follows: (1) The maximum total flux encountered was approximately 3×10^7 ions/cm^2-sec-ster. (2) The ions had energies principally in the range 10 to 500 eV. (3) At least during the early portion of the event the majority of the ions had mass per unit charge in the 66 to 90 amu/q range. (4) The duration of the burst, as seen in the vertical flux measured by the SIDE, was seven minutes, with the peak flux occurring some 75 seconds after the impact.

There can be little doubt that this burst of ions is associated with the S-IVB impact. A study has been made of seven other lunar orbits for which data from this presunrise period are available, and no bursts of comparable magnitude are found at or near this location in lunar orbit. The Apollo 12 Solar Wind Spectrometer also reported an unusual burst of ions at this time, and a study of the combined data from the two experiments is reported in a paper by Snyder *et al.* (1971) presented at the Second Lunar Science Conference.

At impact (Saturn Flight Evaluation Working Group 1970) the S-IVB had a mass of 1.34×10^4 Kg (Stage dry weight—all residual propellants assumed dissipated) and velocity of 2.58 km/sec, thus its kinetic energy was 4.5×10^{10} joules. The direction at impact was nearly vertical.

The SIDE S-IVB Impact Data

Figure 1 is a time profile of the event showing the count-rate of each detector as well as the instantaneous energy and mass level for the MAD. It can be seen that the first three TID energy spectra after the impact show an initial burst of ions arriving at the ALSEP within two minutes. This is followed by a drop in intensity at about 0111 U.T. which builds slowly to a second broad peak in the flux ending at around 0117 U.T.

The TID shows a rather interesting energy vs time profile (see Fig. 2). The initial burst is characterized by ions in the 50 to 70 eV energy range while during the lull between 0111 and 0114 U.T. the energy at maximum flux falls to the 10 or 20 eV energy step. It then builds gradually to energies as high as 500 eV during the later part of the event. A careful comparison with the energies of the maximum ion flux seen by the vertical looking cup in the Solar Wind Spectrometer indicates that the SIDE ground plane is probably effective in causing ions entering the SIDE to be preaccelerated by the 16 volts on the ground plane, hence, the ions seen between

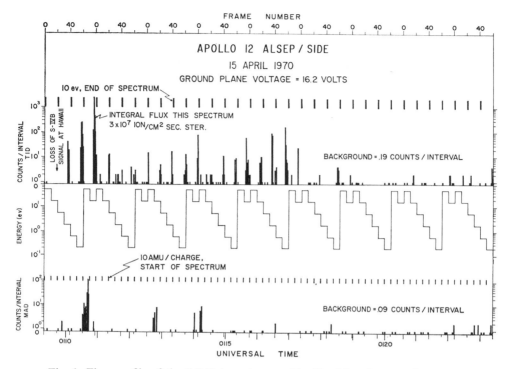

Fig. 1. Time profile of the S-IVB impact event. The Total Ion Detector (TID) and Mass Analyzer Detector (MAD) responses are the raw sensor counts per 1.13 second accumulation interval. The TID energy settings cycle repeatedly through 20 steps, from 3500 eV down to 10 eV. The 10 eV frames are indicated by the vertical marks above the TID response histogram. The line step function above the MAD response histogram gives the instantaneous energy setting for the MAD and the vertical marks between this and the MAD response histogram indicate the lowest mass per unit charge range accepted by the MAD. The abscissa across the top is the SIDE frame number. Each frame corresponds to approximately one accumulation interval. Since this is the natural time interval for the instrument it is repeated on the lower abscissa with Universal Time in minutes overlayed. We have ignored the 1.323 second signal delay time from the moon.

0111 and 0114 U.T. may have been originally of thermal energies. Likewise, the other energies indicated should be reduced by about 16 eV.

Referring to the lower histogram in Fig. 1, a mass spectrum is obtained by the MAD around 0110:40 U.T. Flux temporal variations as indicated by the TID preclude the deduction of a correct mass spectrum, however, it is significant that a large fraction of the ions are seen in the high mass range. This indicates the ions can not be solely solar wind ions but rather must include ions created from impact generated gas. If the flux is considered stationary during the mass spectrum the majority of the ions had mass per unit charge in the range approximately 66 to 90 amu/q. The average mass spectrum over the first 10 minutes of the event is given

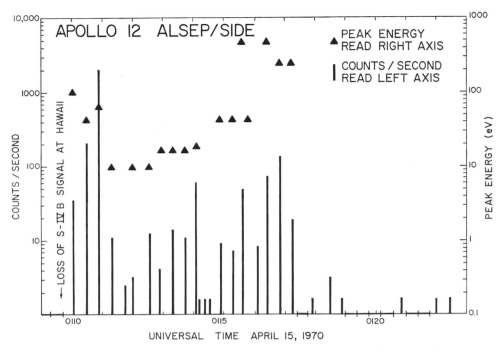

Fig. 2. TID data covering the same time period as in Fig. 1, showing the maximum TID counting rate in each 20-step spectrum, together with the energy at which the maximum counting rate appeared.

in Fig. 3. The mass per unit charge ranges indicated for the large fluxes are consistent with the assumption that these ions came from vaporized lunar surface material.

Discussion

We now address ourselves to the question of why an impact event such as this should result in the production of suprathermal ions detectable at the ALSEP site.

Consider first the possibility that these ions were not created at the instant of the impact or in the impact region but arose from the ionization of a neutral gas cloud. We can estimate the required neutral atom concentration if we assume that the neutral gas cloud from the impact expanded into the solar wind and sunlight and was then partially ionized by charge exchange and photoionization. These processes are found to have comparable ion production rates if we take the cross-sections for ^{40}Ar and the photon flux as given by Bernstein *et al.* (1963), with a solar wind flux of 2×10^8/cm² sec. Then, using a 5.5 second mean residence time for ions in the cloud, we arrive at an order of magnitude estimate of 5×10^7 atoms/cm³. (If additional ionization mechanisms are operative this number will be correspondingly lower.) This implies a mean free path for interaction between the solar wind and the neutral gas cloud of the order of several hundred kilometers.

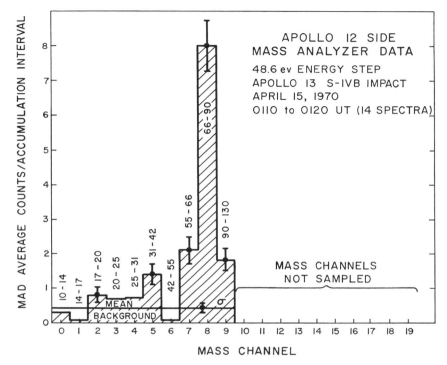

Fig. 3. Mass spectrum at 48.6 eV, showing the average counts per accumulation interval over the 10 minutes of the major part of the event shown in Fig. 1. The mass per unit charge range is indicated above each channel.

Figure 4 is a drawing of the event as seen in cross-section through the moon's equatorial plane. Taking the time from ionization to ion detection to be zero and therefore the time to first detection as an upper limit on the time-of-flight for the neutrals from the impact point up to the solar wind we can establish a lower limit on the neutral particle temperature of the order of $10^{4\circ}$K. When this is combined with the calculated neutral number density we find a neutral gas pressure of the order of 10^{-4} dynes/cm^2; a number more than 3 orders of magnitude larger than that for the solar wind. Thus, it would appear possible for the expanding gas cloud to easily deflect the solar wind if a suitable interaction mechanism exists.

ALFVÉN (1954) and more recently LEHNERT (1970) have drawn attention to effects in cosmical plasmas interacting with a neutral gas in the presence of a magnetic field whereby the plasma and neutral gas are not able to interpenetrate. Laboratory experiments by the Stockholm group (LEHNERT et al., 1969) demonstrate that interpenetration is inhibited, so long as the plasma temperature and energy input remain above certain critical values. These critical values are apparently related to the ionization potential of the neutral gas. Such a mechanism obviates the application of the classical interaction cross-section to this problem and a strong interaction between

TO SUN

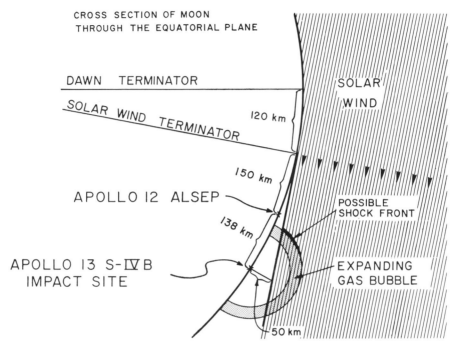

Fig. 4. An equatorial cross sectional view of the moon (not to scale) showing the location of the impact and ALSEP sites with respect to the dawn and solar wind terminators. A 5 deg aberration angle has been assumed for the solar wind.

the impact gas cloud and the solar wind may be possible giving rise to a temporary shock front at the surface of the gas bubble as illustrated in Fig. 4.

Next consider the fate of charge exchange or photoionization created ions in the undeviated solar wind. MANKA and MICHEL (1970) have studied this problem. They considered the effect of the $\vec{V}_{SW} \times \vec{B}_{SW}$ solar wind electric field on lunar atmospheric ^{40}Ar ions. They showed that the $\vec{E}_{SW} \times \vec{B}_{SW}/B_{SW}^2$ cycloidal motion will drive the ions either northward or southward initially if, as is most often the case, the interplanetary magnetic field lies in the ecliptic plane. We cannot rule out, however, that the interplanetary field could have been pointing out of the ecliptic plane at the time of this event. A further important point is that thermal ions picked up by the solar wind electric field in this manner will very quickly acquire a component of velocity in the direction of solar wind flow. For ^{40}Ar ions the cyclotron radius for the cycloidal motion is of the order of 10 lunar radii. For these reasons it is unlikely that a source *downwind* of the ALSEP station would produce ions that could be detected by detectors whose look axes lie in the ecliptic plane if $\vec{E}_{SW} \times \vec{B}_{SW}/B_{SW}^2$ acceleration were

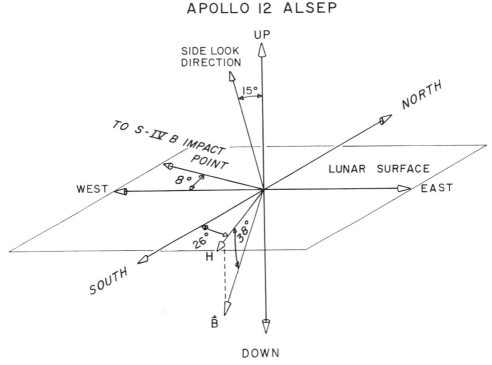

Fig. 5. A diagram showing the relationship between the SIDE look direction, the direction to the impact point, and the steady local magnetic field measured by the ALSEP magnetometer.

the only operative mechanism. Unfortunately, there is no data from the Apollo 12 ALSEP magnetometer for the time period of the S-IVB impact. Hence, ion motions influenced by the local magnetic field or the solar wind field cannot be predicted with any degree of certainty. However, the Apollo 12 ALSEP magnetometer has shown that there is a steady magnetic field, of local origin, of the order 36 gammas at the ALSEP site (DYAL *et al.*, 1970). Figure 5 shows the orientation of this field with respect to the SIDE detectors and the direction to the S-IVB impact point. This field could be that required in the mechanism described by Lehnert to inhibit inter-penetration.

We must also examine the hypothesis that the ions were created and/or energized at the impact. We are not in a position to discuss this possibility at this time; however, we note in passing that the time duration and time profile of the event would seem to argue against this, at least as an explanation for the late arriving ions. If the emission were impulsive the late ions would have traveled many lunar radii before arriving at the ALSEP site.

In considering the acceleration of the ions from this impact event, we should note that during a separate event the Apollo 12 LM ascent stage impacted about 79 km east southeast of ALSEP 1, and that the SIDE subsequently detected fluxes of ions of energy 250 eV and 500 eV, with intensities an order of magnitude less than in the S-IVB event. In view of this, it will be of great interest to observe the features of the Apollo 14 LM impact with both the ALSEP 1 and ALSEP 4 (Apollo 14) instruments simultaneously.

We hope that the study of these man-made impact events will aid in the understanding of numerous other natural events.

Diurnal Phenomena

Figure 6 shows the lunar orbit in relation to the earth magnetosphere and indicates the directions in which the SIDE detectors look at various points throughout the orbit. Also indicated are regions where ion events are observed with repetition from one lunar cycle to the next. There is, however, a considerable amount of variation in the activity seen in these regions from cycle to cycle. It is evident from the figure that, when leaving the magnetospheric tail, the SIDE can see ions flowing downstream in the magnetosheath region. Further discussion of these numerous additional phenomena reported by the SIDE is beyond the scope of this brief report; however, Table 1 enumerates the principal classes of events observed.

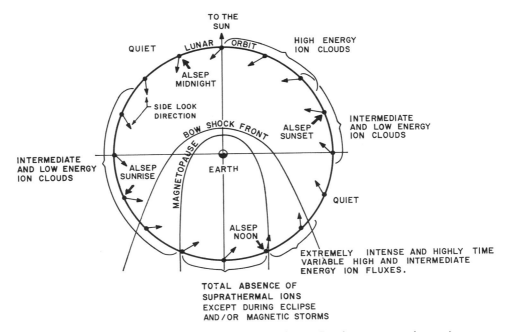

Fig. 6. A diagram showing the look directions of the SIDE detectors at various points throughout the moon's orbit. An approximate indication of the ion event activity as seen repetitively on a lunar day-night basis is also given.

Table 1. Principal categories of ALSEP 1 SIDE data.

Singular Events

1. S–IVB and LM impacts
2. 2 Partial eclipses
3. Solar flare effects outside the magnetosphere
4. Geomagnetic storms

Diurnal Phenomena

1. Medium energy ion clouds near sunrise and sunset
2. Energetic ions thought to be from earth's bow shock; found during lunar evening and night
3. Lunar surface temperature observations
4. Ion fluxes during magnetopause, magnetosheath, and bow shock crossings
5. Absence of fluxes in magnetotail except during a magnetic storm

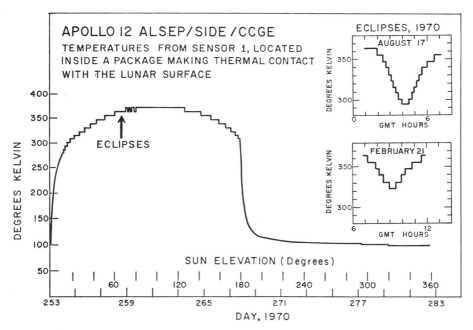

Fig. 7. Typical temperatures recorded during a lunar cycle by a sensor in a package resting on the lunar soil. No corrections have been attempted to convert these measurements to actual surface soil temperatures. The insets show temperatures recorded during the two eclipses experienced by ALSEP 1 in 1970.

Lunar Surface Temperature

The Cold Cathode Gauge Experiment (CCGE), whose electronics is incorporated into the SIDE, has its sensor in a separate small package connected to the SIDE by a four-foot cable. This package, which contains the SIDE/CCGE temperature Sensor 1, rests on the lunar soil and has no active thermal control. Thus the measured temperature is an indication of the lunar surface temperature. Figure 7 shows typical temperatures measured by sensor 1 during a lunar day-night cycle. These are not actual lunar surface temperatures, and no corrections have been made to obtain actual surface temperatures, but the required corrections are presumed to be small. The insets show the temperature variations recorded during the two eclipses which occurred in 1970.

Acknowledgments—It is a pleasure to acknowledge helpful discussions with Drs. A. J. DESSLER, HANS BALSIGER, and CONWAY SNYDER. We also thank Mr. TOM WILSON, who supplied information on the S-IVB impact, and Mrs. NITA JONES, who assisted with the data preparation. Special thanks are due to the Rice University Space Science Facility staff, the NASA/MSC Apollo Lunar Surface Experiments Office staff, and astronauts PETE CONRAD and AL BEAN for helping to make possible a successful experiment.

   This research was supported under NASA Contract NAS 9-5911.

REFERENCES

ALFVÉN H. (1954) *On the Origin of the Solar System*, Oxford, Clarendon Press.

BERNSTEIN W., FREDRICKS R. W., VOGL J. L., and FOWLER W. A. (1963) The lunar atmosphere and the solar wind. *Icarus* **2**, 233–248.

DYAL P., PARKIN C. W., and SONETT C. (1970) Apollo 12 magnetometer: Measurement of a steady magnetic field at the surface of the moon. *Science* **169**, 762–764.

FREEMAN J. W., JR., BALSIGER H., and HILLS H. K. (1970) The suprathermal ion detector experiment. Apollo 12 Preliminary Science Report, NASA SP-235.

LEHNERT B. (1970) Minimum temperature and power effect of cosmical plasmas interacting with neutral gas. Royal Institute of Technology, Division of Plasma Physics, Report 70–11, Stockholm, Sweden.

LEHNERT B., BERGSTRÖM J., HOLMBERG S., and TENNFORS E. (1969) Experiments on the minimum power input of a rotating plasma. Royal Institute of Technology, Division of Plasma Physics, Report 69–34, Stockholm, Sweden.

MANKA R. H. and MICHEL F. C. (1970) Lunar atmosphere as a source of argon-40 and other lunar surface elements. *Science* **169**, 278–280.

SATURN FLIGHT EVALUATION WORKING GROUP (1970) Saturn V launch vehicle flight evaluation report AS-508 Apollo 13 Mission. George C. Marshall Space Flight Center Report MPR–SAT–Fe–70–2, June 20.

SNYDER C. W., CLAY D. R., and NEUGEBAUER M. (1971) An impact-generated plasma cloud on the moon. Second Lunar Science Conference (unpublished proceedings).

Proceedings of the Second Lunar Science Conference, Vol. 3, pp. 2103–2123
The M.I.T. Press, 1971.

Mössbauer instrumental analysis of Apollo 12 lunar rock and soil samples

C. L. HERZENBERG, R. B. MOLER, and D. L. RILEY

IIT Research Institute, Chicago, Illinois, 60616

(*Received* 22 *February* 1971; *accepted in revised form* 31 *March* 1971)

Abstract—Nuclear gamma resonance measurements show that in Apollo 12 returned lunar samples iron is generally abundant, and it has been identified specifically only in the ferrous and to a minor extent in the metallic state. Substantial differences among the spectra of the lunar crystalline rocks indicate considerable diversity in the phase distribution of iron and in the modal mineralogy. Two classes of crystalline rocks can be distinguished on the basis of their Mössbauer spectra. Analyses indicate the Apollo 12 crystalline rocks contain less ilmenite and more olivine than the Apollo 11 rocks.

Spectra of lunar soils returned on Apollo 12 exhibit quite remarkable similarity, but are notably different from the characteristic Apollo 11 soil spectra, suggesting that the nuclear gamma resonance spectra of lunar soils provide regional soil signatures. In the Apollo 12 soils, ilmenite and metallic iron are less abundant, and olivine is somewhat more abundant, than in the Apollo 11 soils. Evidence for a resonance associated with the "kreep" component has been found.

The unusual lunar microbreccia 12013 exhibits resonances indicating the presence of the major lunar minerals, with both olivine and ilmenite less abundant than in any other Apollo 12 sample examined. In addition, anomalous features are present which appear to be associated with the "kreep" component in the dark lithology and with a history of intense shock.

INTRODUCTION

THIS PAPER reports Mössbauer instrumental analyses for the second sampling of lunar material which has been returned to earth, and constitutes a sequel to our earlier major report on Mössbauer spectrometry of the first returned lunar samples (HERZENBERG and RILEY, 1970a).

In the studies discussed here, we have examined the nuclear gamma resonance of the nuclide Fe^{57} which is naturally present in the lunar material. The particular Apollo 12 lunar samples studied, their characteristics, and the techniques employed in their study are summarized in Table 1. Techniques, equipment, and methods of data analysis in use have already been described in detail in our earlier papers (HERZENBERG and RILEY, 1970a, 1970b; HERZENBERG *et al.*, 1971b).

ANALYSIS

In this, as in our previous studies, each individual spectral measurement has been accompanied by calibration measurements, including measurements on geochemical standards as well as spectrometer calibration standards (HERZENBERG and RILEY, 1970a). Laboratory and data analysis procedures which have been applied are based on and are similar to those previously developed and used in other analytical applications of Mössbauer spectrometry (SPRENKEL-SEGEL and HANNA, 1964; MUIR, 1968; LERMAN *et al.*, 1968; HERZENBERG, 1969; HERR and SKERRA, 1969; DE VOE and SPIJKERMAN, 1970; HERZENBERG and RILEY, 1970b) Previous analytic applications have included elemental abundance studies (GOLDANSKII *et al.*, 1968; DE VOE and SPIJKERMAN, 1970; HERZENBERG, 1969;

Table 1. Apollo 12 lunar samples analyzed by Mössbauer spectrometry

Samples Studied	Original material	Form of sample	Mass (grams)	Areal density (mg/cm²) (nominal)	Sample containment*	Mössbauer spectrometry mode
12032,63	fines	powder	0.21	43.6	special powder compression holder	transmission
12033,93	fines	powder	0.200	41.5	special powder compression holder	transmission
12037,12	fines	powder	0.2003	41.5	special powder compression holder	transmission
12013,17	microbreccia	powder	0.0372	40	adapted powder compression holder	transmission
12013,7	microbreccia	thin section	ND	STD	polished standard thin section	scattering
12002,168	crystalline rock	powder	0.200	41.5	special powder compression holder	transmission
12004,52	crystalline rock	powder	0.2	41.5	special powder compression holder	transmission
12004,48	crystalline rock	thin section	ND	40	uncovered on beryllium disc	transmission
12004,35	crystalline rock	chip	0.9	N.A.	in nitrogen filled plastic bag	scattering
12020,17	crystalline rock	powder	0.200	41.5	special powder compression holder	transmission
12020,56	crystalline rock	probe mount	ND	40	uncovered on beryllium disc	transmission and scattering
12020,44	crystalline rock	chip	1.13	N.A.	in nitrogen filled plastic bag	scattering
12063,100	crystalline rock	powder	0.200	41.5	special powder compression holder	transmission
12063,114	crystalline rock	probe mount	ND	40	uncovered on beryllium disc	transmission and scattering
12063,58	crystalline rock	chip	1.05	N.A.	in nitrogen filled plastic bag	scattering

* More detailed descriptions are given in Herzenberg and Riley (1970a).
N.A.: not applicable; ND: data not supplied; STD: standard thickness.

Herzenberg and Riley, 1970b); distribution of iron among oxidation states (Lerman et al., 1968; Hogg and Meads, 1970; Hogarth et al., 1970) and studies involving the quantitative determination of the relative amounts of iron in dissimilar phases (Muir, 1968; Lerman et al., 1968; Herzenberg, 1969; Sprenkel-Segel, 1970). Semiquantitative total iron abundance estimates have been obtained by comparison of the total resonant absorption exhibited by lunar samples with the total resonant absorption exhibited by standard calibration absorbers of known iron abundance and known areal density of iron in accompanying calibration measurements (Herzenberg, 1969; Herzenberg and Riley, 1970b). Individual mineral contributions in the spectra have been identified on the basis of their line locations. Semiquantitative determination of the relative amounts of iron in different mineral phases has proceeded using techniques among those developed for quantitative determination of the relative amounts of iron in dissimilar phases in multiphase mixtures (Sprenkel-Segel and Hanna, 1964; Muir, 1968). Areal analysis procedures for obtaining the relative resonant absorption from different phases are similar to those used in our previous work (Herzenberg and Riley, 1970a), and are based on techniques employed in previous studies by other groups (Muir, 1968; Herr and Skerra, 1969; Lerman et al., 1968). In particular, the relative amount of resonant absorption attributed to an individual mineral phase in a spectrum has been determined as proportional to the summed absorption areas associated with the resonance lines corresponding to that mineral phase, corrected as necessary for saturation effects. (In the cases of phases for which not all of the individual lines are separately easily distinguishable, the contributions associated with obscured lines have been obtained from ideal or separately determined intensity ratios; e.g., in evaluating the inner line contributions of the magnetically ordered spectra of metallic iron and troilite, 3:2:1 ratios have been employed.) The relative resonant absorptions have been used directly to provide the numerical estimates of the phase distribution of iron in the mineral phases present in the sample. This procedure should provide correct first order estimates, and corresponds to the assumption of equal resonant (recoil-free) fractions in each contributing coordination site and phase. While on the basis of theoretical calculations and a limited number of experimental studies the resonant fractions for iron in the contributing phases are expected to be reasonably similar (GAY et al., 1970; HERZENBERG and RILEY, 1970a), this approximation is expected to be correct only to within about 20% (HERZENBERG and RILEY, 1970a; GAY et al., 1970; HERR and SKERRA, 1969; SPRENKEL-SEGEL, 1970). We consider that this approach remains the procedure of choice until accurate recoil-free fractions have been measured directly for all of the contributing phases in these samples.

We have also used the Mössbauer results to estimate semiquantitatively the relative amounts (weight percentages) of iron-containing minerals in the lunar samples. The following procedure has been used: the total iron abundance (wt. %) in the sample (in some cases obtained initially from total Mössbauer resonant absorption measurements; later values used were from reported chemical analyses) was partitioned according to the phase distribution of iron (obtained by areal analyses of Mössbauer spectra) to yield, as an intermediate result, the wt. % of iron in the sample which is associated with each iron-bearing mineral phase. Using compositional data for individual phases (LSPET, 1970; Apollo 12 Lunar Sample Information Catalog, 1970), the wt. % of each iron-bearing mineral phase present in the sample was calculated from the wt. % iron in the sample corresponding to this phase. Previous studies on the resonant absorption characteristics of actual and simulated lunar samples (HERZENBERG and RILEY, 1970a) verify that these procedures permit a reasonably satisfactory semiquantitative analysis of lunar sample composition.

EXPERIMENTAL RESULTS

Lunar crystalline rocks

Figure 1 is a Mössbauer hyperfine absorption spectrum measured in transmission for sawcuttings from the fine grained microgabbro 12063. The general similarity of this spectrum to the spectra of lunar samples returned on Apollo 11 (HERZENBERG and RILEY, 1970a; HERZENBERG et al., 1971b; HOUSLEY et al., 1970) is evident.

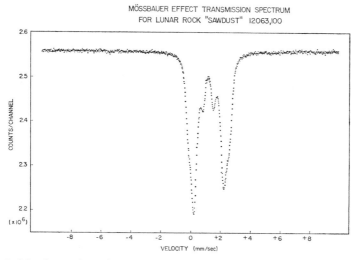

MÖSSBAUER EFFECT TRANSMISSION SPECTRUM
FOR LUNAR ROCK "SAWDUST" 12063,100

Fig. 1. Mössbauer hyperfine spectrum for lunar crystalline rock 12063 (powder specimen measured at room temperature).

Table 2. Phase distribution of Iron in Apollo 12 lunar crystalline rocks from Mössbauer
instrumental analysis*

Mineral phase	12002	12004	12020	12063
Metallic iron	0.9	0.4	0.5	0.8
Troilite	0.5	0.9	0.5	0.7
Ilmenite	7.2	6.3	5.9	13.7
Pyroxene[a]	65.8	68.3	75.0	70.6
Olivine	25.6	24.1	18.1	14.2
Total silicate (Fe^{2+} in pyroxenes, olivines, and glasses)	91.4	92.4	93.1	84.8

* % total resonant absorption associated with each mineral phase is tabulated. Tabulated data are based primarily on results obtained for transmission measurements in thin sections (probe mounts), except for the case of 12002, for which only a powder sample was available. However these results have been supplemented by data from transmission measurements on powder samples and scattering measurements on rock chips and thin sections.
[a] Includes small contributions from absorption in spinels and feldspars.

From the total resonant absorption, we estimate an approximate total iron abundance in this sample of 18 $\pm$ 3 wt. %, slightly higher than the 16.7 wt. % reported from chemical measurements (Apollo 12 Lunar Sample Information Catalog, 1970). The phase distribution of iron, and the inferred weight percentages of minerals in this sample, are given in Tables 2 and 3.

The compositional data we obtain are in reasonably good agreement with what would be expected on the basis of the report of Taylor et al. (1970), although we detect somewhat less iron in olivine than would be expected on the basis of their reported modes. There is no evidence for ferric iron or any unusual mineral phase in this sample.

Rock 12063 is found to be more nearly similar in its Mössbauer spectrum and in the phase distribution of iron derived therefrom to Apollo 11 material than any other Apollo 12 sample we have investigated. It appears to be especially similar in these respects to the Apollo 11 medium grained ophitic ilmenite basalt 10003 (Herzenberg and Riley, 1970a), the chief differences being stronger olivine resonances and less intense ilmenite resonances for this Apollo 12 rock. The similarity between the Apollo 11 rock 10003 and Apollo 12 crystalline rocks, in particular 12063, and similar ophitic basalts, has also been noted on the basis of other measurements (LSPET, 1970; Warner and Anderson, 1971).

Table 3. Weight percentages of minerals in Apollo 12 crystalline rocks from
Mössbauer instrumental analysis*

Mineral phase	12002	12004	12020	12063
Metallic iron	0.15	0.07	0.08	0.13
Troilite	0.13	0.24	0.13	0.18
Ilmenite	3.3	2.9	2.7	6.2
Pyroxene[a]	53.9	55.3	60.3	57.2
Olivine	18.2	18.2	13.5	8.8
Feldspar[b]	24.3	23.3	23.3	27.5

* Based on entries in Table 2, in conjunction with independently reported iron abundances and compositional data.
[a] Includes a small contribution from spinels; this appears to be appreciable only for the case of rock 12004, for which it may amount to a few wt.%.
[b] Obtained by difference.

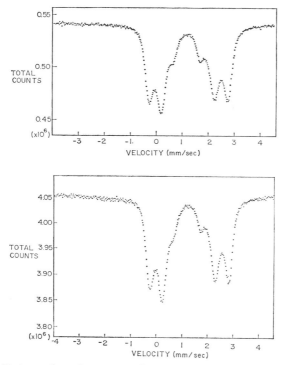

Fig. 2. Mössbauer hyperfine spectra for two samples of Apollo 12 lunar crystalline rock 12063 (measured at 80°K); upper spectrum: powder specimen; lower spectrum: probe mount.

We show in Fig. 2 spectra measured at 80°K for a powder sample and a probe mount sample of rock 12063. These separately measured spectra for different samples of the same rock are clearly very closely similar. It is evident that essentially concordant results are obtained by these two experimental approaches. Minor differences between these spectra, and differences among these and similar spectra measured on a separate instrument in scattering for the probe mount and the chip from 12063, are attributable to absorption by iron impurities in substrates and detector windows, differential penetration and saturation effects, and other minor but known effects. The similarity in results obtainable using these different types of sample and different modes of spectrometry testify to the significance and consistency of the results.

Figure 3 shows a Mössbauer hyperfine absorption spectrum measured in transmission for the powder sample associated with rock 12002, an olivine dolerite. Figure 4 shows the inner portion of a spectrum measured at 80°K, together with a computer fit. From the total resonant absorption observed for this sample, we estimate the total abundance of iron as 17 wt.%, in good agreement with other reports (WILLIS et al., 1971). On the basis of 8-line (25 free parameter) computer fits to room temperature spectra, and 6-line (19 free parameter) computer fits to spectra measured at liquid nitrogen temperature, as well as additional analyses, we obtain the analytic results

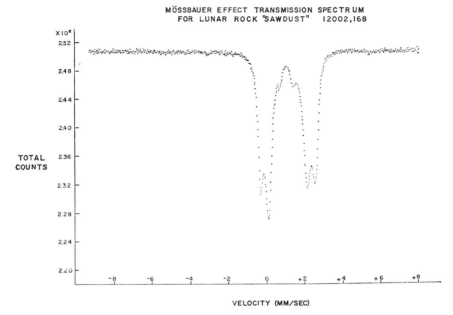

Fig. 3. Mössbauer hyperfine spectrum for lunar crystalline rock 12002 (powder specimen measured at room temperature).

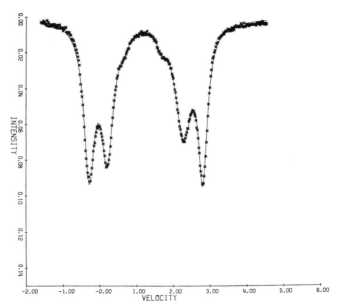

Fig. 4. Mössbauer hyperfine spectrum for lunar crystalline rock 12002 (powder specimen measured at 80°K), with least-squares computer fit.

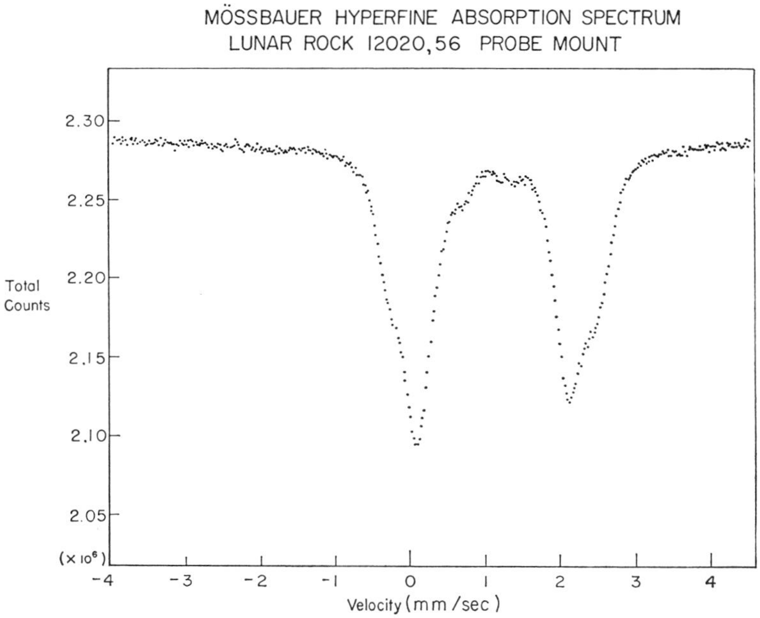

MÖSSBAUER HYPERFINE ABSORPTION SPECTRUM
LUNAR ROCK 12020,56 PROBE MOUNT

Fig. 5. Mössbauer hyperfine spectrum for lunar crystalline rock 12020 (probe mount measured at room temperature).

in Tables 2 and 3. It is clear that this sample contains an unusually high abundance of olivine, as reported previously (PAPANASTASSIOU and WASSERBURG, 1970). There is no evidence for ferric iron or any unidentified iron-bearing mineral phases in significant amounts in this specimen.

Figure 5 shows a spectrum measured in transmission through a probe mount from rock 12020, an olivine basalt. This spectrum indicates that this rock has considerably less ilmenite, and a higher proportion of iron in the silicate phases, than any other lunar crystalline rock we have examined. There is no evidence for ferric iron or any unusual mineral phase in this rock.

Significant compositional differences exist among the different samples associated with this rock, as can be seen by comparing Fig. 5 with Fig. 6, which is a spectrum measured for the powder sample associated with rock 12020. It is obvious from these spectra the powder contains more ilmenite, and it appears to have considerably more olivine (and possibly more disordered pyroxene) that the probe mount. Our rough estimate of the total abundance of iron for the powder (18 wt. %) is higher than the corresponding value for the probe mount; chemically determined values for the abundance of iron reported for this rock (typically 16.6 wt. %) (Apollo 12 Lunar Sample Information Catalog, 1970; WAKITA et al., 1971) are intermediate. Scattering measurements on the chip from this rock give spectra intermediate between the spectra produced by the thin section and the powder sample. These observations support the view that the variations observed in the case of these particular separate samples represent actual inhomogeneities in the rock itself. We note in this connection

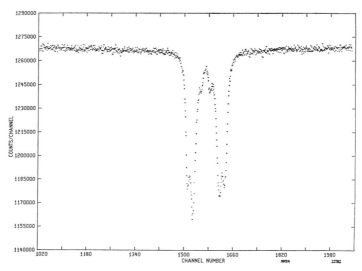

Fig. 6. Mössbauer hyperfine spectrum for lunar crystalline rock 12020 (powder specimen measured at room temperature).

that two different samples of an Apollo 11 rock (10058) exhibited sampling variations in Mössbauer studies by Housley et al. (1970).

Figure 7 shows a Mössbauer hyperfine spectrum measured at 80°K in transmission through a powder sample associated with rock 12004, an olivine basalt. Analytic data are given in Tables 2 and 3. We observe small intensity asymmetries of the major peak regions in spectra of this rock, the origin of which is not yet known with assurance, but which we tentatively associate with paramagnetic spectral contributions from the spinels which have been reported present in this rock (Simpson and Bowie, 1971; Banerjee et al., 1967).

Several general observations can be made in regard to the results derived from these measurements conducted on the Apollo 12 crystalline rocks. The Apollo 12 crystalline rock spectra are qualitatively fairly similar to the spectra of Apollo 11 crystalline rocks. The major features of the Apollo 12 crystalline rock spectra are readily accounted for in terms of contributions by lunar mineral phases previously identified in Apollo 11 crystalline rocks. Our identifications and semi-quantitative analytic results for these phases are indicated in Tables 2 and 3. However, the spectra characteristic of Apollo 12 rocks are generally easily distinguished from the spectra typical of Apollo 11 crystalline rocks (Herzenberg and Riley, 1970; Housley et al., 1970), because of the quantitative differences in the composition of crystalline rocks from the two sites.

On the basis of the Mössbauer spectra, we can state that the Apollo 12 crystalline rocks which we have studied systematically contain considerably less ilmenite and more olivine than the Apollo 11 crystalline rocks. The ilmenite content and the olivine content, as determined from the Mössbauer spectra, show variations of the order of a factor of 2 among the Apollo 12 crystalline rocks which we have examined

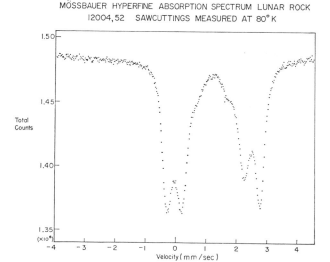

Fig. 7. Mössbauer hyperfine spectrum for lunar crystalline rock 12004 (powder specimen measured at 80°K).

(Tables 2 and 3). From the varied spectra produced by these Apollo 12 crystalline rocks it may be concluded that these Apollo 12 crystalline rocks exhibit a much greater diversity than the Apollo 11 rocks which we studied. This diversity in the spectra is indicative of a much greater range in the modal mineralogy, as would be expected on the basis of the observations of LSPET (1970).

On the basis of the Mössbauer spectra and the derived analytical data (Tables 2 and 3), the Apollo 12 lunar crystalline rocks may be separated into two classes (HERZENBERG et al., 1971a). The type 1 rocks (12002, 12004, and 12020) all have a similar ilmenite content, which is about a factor of 2 below the ilmenite content of the type 2 rock (12063). It is of interest that this classification of a limited number of rocks on the basis of their Mössbauer spectra agrees with the classification of crystalline rocks proposed by WARNER and ANDERSON (1971). Furthermore, the variation of ilmenite content which we observe with this classification is in accord with the observations of WAKITA et al. (1971) on titanium enrichment and depletion in these rock types. The classification of these crystalline rocks on the basis of their Mössbauer spectra is also in accord with the petrographic classification associated with the stratigraphic sequence proposed by SCHMITT and SUTTON (1971).

We also note that our results agree with the more detailed observations (WARNER and ANDERSON, 1971) that the Apollo 12 type 2 rocks are more similar to the type B crystalline rocks returned by Apollo 11; whereas the type 1 rocks are found only among the Apollo 12 return.

The weight percentages of ferromagnetic metallic iron and troilite in the Apollo 12 crystalline rocks (Table 3) are in the same general range as values obtained from Mössbauer studies on Apollo 11 crystalline rocks (HOUSLEY et al., 1970). These values are also reasonably close to preliminary values determined from Mössbauer studies on another Apollo 12 crystalline rock (Housley *et al.*, 1971). However, our estimates of metallic iron abundances in the crystalline rocks using Mössbauer methods appear to be higher (about a factor of 2) than the average of some preliminary estimates for metallic iron obtained from studies of the magnetic properties of various Apollo 12 rocks (Nagata *et al.*, 1971; Hargraves and Dorety, 1971; Pearce *et al.*, 1971). Further work will be required to resolve this discrepancy.

Lunar microbreccia 12013

The microbreccia 12013 is a heterogeneous rock, with three distinct lithologies; and contains a dark portion having "kreep" composition (James, 1971; Meyer *et al.*, 1971), in addition to a granitic and other components. The rock 12013 spectra which we observe may therefore be expected to exhibit the major features of the spectra of crystalline "kreep," superimposed by features originating from other portions of this rock.

Measurements have been conducted in scattering and transmission on two spectrometers for two samples of this rock: an ordinary uncovered polished thin section, and a sample of approximately 35 mg of sawdust. A Mössbauer hyperfine absorption spectrum of the sawdust is shown in Fig. 8. The spectra of this rock, while presenting the general pattern familiar for lunar samples, also exhibit distinctive differences from the spectra of all other lunar samples which we have investigated.

The spectrum of this rock exhibits peaks which are characteristic of the presence of Fe^{2+} in pyroxenes, olivine, and ilmenite. Spectral analysis indicates that this sample contains less olivine than the Apollo 12 crystalline rocks or any of the Apollo 12 fines we have examined. We estimate that 2.5 wt. % or less of the sample is olivine.

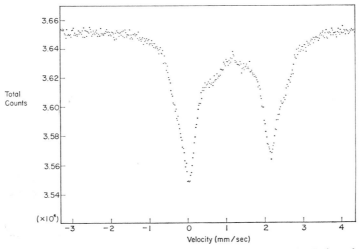

Fig. 8. Mössbauer hyperfine spectrum for lunar rock 12013 (powder specimen measured at room temperature).

Ilmenite accounts for less than 2.6 wt. % of this rock 12013, our best estimate being 0.9 wt. %. Ilmenite is thus clearly less abundant in 12013 than in the crystalline rocks (Table 3), and appears also to be less abundant than in the fines. The low abundances of ilmenite and olivine and also the magnitudes of certain anomalous features in the spectra of rock 12013, indicate that this breccia could not have been formed from local Apollo 12 soils or fragments of crystalline rocks such as we have examined, and suggest a foreign origin.

Evidence for the presence of metallic iron and troilite was sought specifically. In spectra measured over a higher velocity range for rock 12013, we find characteristic resonances indicating the presence of a small amount of metallic iron, confirming the previous observations of LAUL *et al.* (1970). Using a value of 8 wt. % for the total abundance of iron (MORGAN and EHMANN, 1970), we conclude that 0.35 wt. % of the sample is ferromagnetic metallic iron. This is somewhat larger than the amount of metallic iron found in the Apollo 12 crystalline rocks (Table 3), and more nearly similar to the amount found in the fines (Table 5). We also detect resonances characteristic of iron in troilite, and on the basis of 8 wt. % total iron, estimate that approximately 0.17 wt. % troilite is present in this sample. It appears from this result that the amount of metallic iron in this sample is somewhat larger than the amount of troilite, a result at variance with previous observations on other samples of this rock (DRAKE *et al.*, 1970).

Rock 12013 spectra also exhibit certain quite unusual features. In order to emphasize these unusual characteristics, in Fig. 9 a spectrum of rock 12013 is compared with the spectra of two crystalline rocks returned on the Apollo 12 mission.

After absorption attributable to the ferrous ion in the usual silicate environments and in ilmenite has been accounted for, an excess absorption is still present in the low velocity region, an appreciable amount of additional absorption is present in the spectral range between the major absorption regions, and a small additional absorption is present just beyond the major absorption regions. These features of the rock 12013 spectra appear to originate from a combination of absorption contributions in the low velocity region together with a broad diffuse troughlike absorption component extending between about −1 and +3 mm/sec.

While one must consider the possibility that some absorption features might originate from contaminants, previous research has shown that the major contaminant in this particular sample cannot be iron (MORGAN and EHMANN, 1970). At present we have no evidence that would not favor the interpretation of these unusual resonant features as intrinsic absorption characteristics of rock 12013; this conclusion is substantiated by the fact that similar spectra are observed under very different experimental conditions for the two separate samples of this rock.

We note that some of the excess resonant absorption exhibited by rock 12013 appears at low velocities in the same general region in which excess absorption has been detected in all the Apollo 12 fines so far examined. As discussed later in the context of the fines, we believe that a portion of this anomalous absorption in the low velocity region may originate from a mineral phase in the "kreep" component.

The additional absorption present in the region between the major peaks is visible in Fig. 9, which also makes it apparent that this cannot be accounted for by the

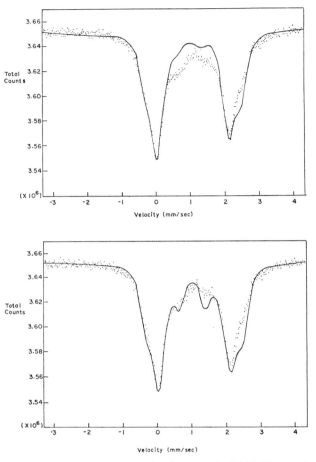

MÖSSBAUER SPECTRUM OF LUNAR ROCK 12013 COMPARED WITH
LUNAR CRYSTALLINE ROCK SPECTRA

Fig. 9. Mössbauer hyperfine spectra for lunar rock 12013 (data points) compared
with Apollo 12 crystalline rock spectra; upper curve 12020, lower curve, 12063.

absorption lines previously observed and identified in the crystalline rock spectra.
The characteristics of additional spectra measured at 80°K rule out the possibility
that the excess absorption in this region and in the low velocity region of the spectrum
could be attributable to chromite and ulvöspinel resonances; the low abundance of
chromium (LSPET, 1970) and the rarity of ulvöspinel in this rock (MEYER *et al.*,
1971) substantiate this conclusion. The low temperature spectra also appear to rule
out the possibility that this absorption could originate from other magnetically
ordered iron titanium oxide minerals (HERZENBERG and RILEY, 1970a, 1970b).
Finally, we wish to point out that absorption in this wide region has been observed
in the case of a previously studied intensely shocked terrestrial basalt (HERZENBERG

and RILEY, 1970b). Spectra of a terrestrial basalt shocked to in excess of 500 kilobars, when compared to spectra of the corresponding unshocked basalt, exhibit a pronounced excess of absorption throughout the entire broad region between the olivine peaks; in addition, an excess absorption in the vicinity of the low velocity paramagnetic peaks was observed for the intensely shocked sample of terrestrial basalt. Spectral characteristics at 80°K are also similar. While the available comparative evidence is at present rather limited, it suggests that certain of the observed features in the rock 12013 spectrum may be associated with a history of intense shock exposure. This conclusion is consistent with previous independent observations on the petrology of this specimen indicative of intense shock exposure (JAMES, 1970).

We suggest that it would be useful to re-examine other lunar samples for these unusual spectral features, which might have gone undetected if they were comparatively weak, rather than prominent features as in 12013. It may be anticipated that similar features will be observable in spectra of "kreep" type rocks returned in future lunar missions.

Lunar fines

Three separate specimens of lunar fines, each from a different documented soil sample from the Apollo 12 return, have been studied. The spectra for the three distinct samples of Apollo 12 fines (Fig. 10) are closely similar to each other. They are also similar to spectra of other Apollo 12 fines samples (HOUSLEY *et al.*, 1971; BANCROFT *et al.*, 1971; DUCHESNE *et al.*, 1971). However, minor differences among the spectra of these lunar soils do exist, which is consistent with other observations that the regolith at the Apollo 12 landing area is not homogeneous. That the Mössbauer effect spectra of the various Apollo 12 soil samples are very nearly identical is a surprising result in view of the fact that the three samples are visually distinct, 12033 in particular being a notably distinctive light colored soil. Furthermore, they were collected separately, at separate sites, up to 240 meters apart on the surface of the moon.

In our previous study of Apollo 11 samples, we found that all the Apollo 11 soil spectra were virtually indistinguishable (HERZENBERG and RILEY, 1970a) Comparing the spectra (Figs. 10–12) and analytical results (Tables 4 and 5) for the Apollo 12 soils with those of the Apollo 11 soils shows them to be distinctly different, a result which is in agreement with independent mineralogical and petrological modal analyses (BUTLER *et al.*, 1970). We observe that at each landing site the local soils exhibit remarkably uniform Mössbauer spectra and yet the soils from the two sites are easily distinguished. It appears that Mössbauer spectra of local surface and near-surface soils are insensitive to local geological events of the type proposed to account for the observed distribution of crystalline rocks and breccias (WARNER and ANDERSON, 1971) or of the detailed local geological history of the regolith, and hence we may be observing something close to regional characteristics in the Mössbauer spectra of the lunar soils (HERZENBERG *et al* , 1971a).

We suggest the possibility that the gross phase distribution of iron in lunar soils (as indicated by the salient features of their Mössbauer spectra) may represent a regional characteristic, nearly constant for the near-surface regolith over regions

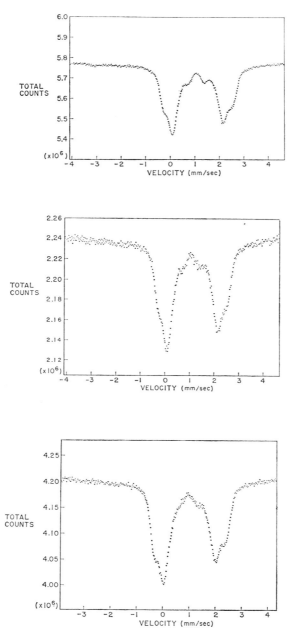

Fig. 10. Mössbauer spectra for three lunar soils from the Apollo 12 return compared: 12032 (upper); 12033 (middle); 12037 (lower).

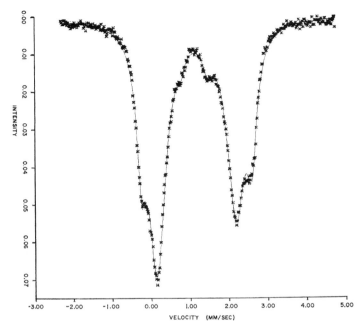

Fig. 11. Mössbauer hyperfine spectrum for lunar soil 12037 (measured at room temperature), with least-squares computer fit.

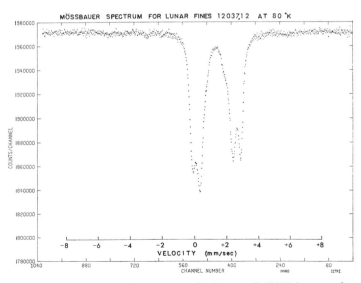

Fig. 12. Mössbauer hyperfine spectrum for lunar soil 12037 (measured at 80°K).

Table 4. Phase distribution of iron in Apollo 12 fines*

Mineral phase	12032,63	12033,93	12037,12
Metallic iron	2.7	2.4	2.4
Troilite	0.8	0.8	1.3
Ilmenite	6.1	5.7	6.1
Pyroxene[a]	51.2	51.3	45.4
Glass[a,b]	22.4	23.9	25.7
Olivine[a]	12.0	10.5	12.7
Anomalous resonance[c]	4.7	5.4	6.3
Total silicate (Fe^{2+} in pyroxenes, olivines, and glasses)	85.6	85.7	83.8

* % total resonant absorption associated with each mineral phase is tabulated.

[a] Silicate separation into olivine, pyroxene, and glass contributions are based on room temperature and (for 12037) 80°K measurements, and comparison with both Apollo 11 fines and simulated lunar sample spectra.

[b] Entry corresponds to materials vitreous on a molecular scale, and does not include devitrification products.

[c] This represents an absorption excess in the low velocity spectral region, discussed further in text.

corresponding in size to distances of perhaps the order of a kilometer or more, but showing considerable variation over distances comparable to the separation of the landing sites, and to the dimensions of the major visual features of the lunar nearside.

Total resonant absorption measurements have permitted rough estimates of the total iron abundance in the three soils. These results, which are listed in Table 5, are in reasonable agreement with other recently reported measurements (Wakita et al., 1971). Since total iron abundance values obtained from total resonant absorption measurements are subject to greater error than are the phase distribution percentages, we have used reported total iron abundances (LSPET, 1970; Wakita et al., 1971) to derive the other analytic results for these soils given in Tables 4 and 5.

From analyses of the Mössbauer spectra we calculated that about 85% of the iron is ferrous iron in iron-bearing silicate phases (pyroxenes, glasses, and olivine). Our spectra indicate that olivine is somewhat more abundant in the Apollo 12 fines than in the Apollo 11 fines (Herzenberg and Riley, 1970a). While it is difficult using this technique to obtain an accurate separation of the contributions of iron in the pyroxenes and iron in the lunar glasses, it appears that the contribution from iron in the pyroxenes dominates that from iron in glasses in these Apollo 12 fines, in contrast to the situation for Apollo 11 fines (Herzenberg and Riley, 1970a).

Table 5. Apollo 12 fines selected compositional data obtained from Mössbauer instrumental analysis.*

	12032	12033	12037
Metallic iron	0.32	0.30	0.26
Troilite	0.15	0.14	0.25
Ilmenite	1.9	1.7	2.0
Total iron content	12	12	10

* Weight percentages are tabulated. Average elemental iron abundances reported at the Apollo 12 conference rather than our determinations tabulated here have been used in calculating weight percentages of individual phases.

We find that the Apollo 12 soils are reduced in ilmenite content by about a factor of 3 relative to the Apollo 11 soils, which is in correspondence with recent observations on the relative abundances of titanium in Apollo 12 and Apollo 11 soils (WAKITA et al., 1971).

While the ilmenite content in Apollo 12 fines is closer to that of Apollo 12 type 1 than type 2 rocks, it is actually lower than for either type of crystalline rock. This reduced ilmenite content of the fines is evidence for the presence of a significant quantity of a foreign component (of yet lower ilmenite content) in the local soils in addition to comminuted local crystalline rocks.

Our results indicate that the Apollo 12 soils contain less magnetic material (that is, a smaller abundance of magnetically ordered mineral phases) than do the Apollo 11 soils. We find that, while the dominant ferromagnetic mineral is again metallic iron, the Apollo 12 soils are lower in ferromagnetic metallic iron (α-iron) content by approximately a factor of 2 relative to Apollo 11 soils, in agreement with the preliminary results reported by PEARCE et al. (1971). Our values, however, are somewhat lower than the values reported by NAGATA et al. (1971), HOUSLEY et al. (1971), and BANCROFT et al. (1971), for other fines. As was observed previously for Apollo 11 samples (HOUSLEY et al., 1970), we find that the ratio of metallic iron to troilite iron is greater in the cases of the fines (Tables 4 and 5) than in the cases of the crystalline rocks (Tables 2 and 3). The actual value of this ratio for the Apollo 12 fines is approximately equal to the corresponding ratio for Apollo 11 fines (HOUSLEY et al., 1970), in agreement with the observations of other groups (BANCROFT et al., 1971).

The reduced metallic iron content of the Apollo 12 soils is accompanied by a reduced content of vitreous material (Table 4). This suggests that much of the metallic iron in these Apollo 12 soils is associated with the glassy fraction, in agreement with the results of other groups (HOUSLEY et al., 1971), just as was observed for Apollo 11 samples. The relatively low abundance of glasses with metallic inclusions in the Apollo 12 soils relative to Apollo 11 soils suggests the possibility that comminutive processes other than meteorite impact have played an important role in the production of the local regolith.

For all the Apollo 12 soils, we find that the amount of resonant absorption ascribed to Fe^{2+} in glasses is considerably smaller than we found for the Apollo 11 soils. Analysis of the contribution of Fe^{2+} in glasses has been based on the recognition and separation of spectral contributions which appear to be characteristic of vitreous (HERZENBERG and RILEY, 1970a; SULLIVAN et al., 1971) or slightly devitrified (HOUSLEY et al., 1970; HERZENBERG and RILEY, 1970a) basaltic glasses. It should be recognized that we are therefore distinguishing here materials which appear vitreous or exhibit only incipient crystallization on a molecular scale. Thus, soil fractions categorized as glasses using other techniques (McKAY et al., 1971) may in some cases be expected to include material classified under devitrification products using this technique. Devitrification of lunar glasses has been observed using other techniques (WHITE et al., 1971; LOFGREN et al., 1971).

In addition to the usual spectral features which are present in the spectra of the Apollo 12 crystalline rocks, all of the Apollo 12 soil samples exhibit an additional characteristic absorption, accounting for about 5% of the total absorption. This additional absorption occurs in the low velocity region occupied by the most prominent manifest peak in the spectra. An asymmetry in intensity of the quadrupole doublets originating from the major phases could produce excess absorption here; however, we do not find an asymmetry of sufficient magnitude in the spectra of the major phases present in the crystalline rocks or the Apollo 11 fines (HERZENBERG and RILEY, 1970a) to account for this; and a highly asymmetric spectrum would be required for Fe^{2+} in any minor phase to produce it. Absorption in this general region can be produced by paramagnetic metallic iron, low spin ferrous iron, and ferric iron. Independent evidence indicates that ferric iron is absent from Apollo 12 fines (ANNELL, 1971; DUCHESNE et al., 1971), and low spin ferrous iron has so far not been identified in any lunar mineral phase. It seems probable that this absorption is produced by a form of metallic iron, probably extremely finely divided superparamagnetic metallic iron (HERZENBERG and RILEY, 1970a), as has already been suggested (DUCHESNE et al., 1971).

In the cases of the crystalline rocks, substantially all absorption in this region of the spectrum has been accounted for without residual anomaly. Hence this excess characteristic absorption detected in the Apollo 12 fines, whatever its identity, appears to be specific to the fines. As it has already been established from independent studies that the Apollo 12 soils contains a large proportion of an admixed foreign component having a very different composition from that observed among the larger igneous rock samples (HUBBARD et al., 1971; MEYER et al., 1971), it is not unreasonable to suggest that the anomalous resonant absorption detected in the Apollo 12 fines is associated with this foreign "kreep" component (HERZENBERG et al., 1971a). Evidently the Mössbauer spectra of "kreep" composition materials must include other features in addition to the anomalous ones. In the case of "kreep" glasses, which are present in lunar fines (MEYER et al., 1971), these might be expected to include previously studied features of lunar glasses, as well as features of the mineralogically more complex "kreep" bearing mineral phases present in rock 12013 (MEYER et al., 1971), which is compatible with our observations. In any event, it is clear that a further understanding of the nature and origin of this exotic contribution to the regolith will be forthcoming from a more detailed study of the Mössbauer resonance spectra of "kreep" bearing materials.

SUMMARY AND CONCLUSIONS

All of the lunar samples from the Apollo 12 return on which we have performed measurements and analyses contain an appreciable abundance of iron, typically higher for the crystalline rocks and somewhat lower for the fines and microbreccia than was the case for Apollo 11 samples. The iron is predominantly in the ferrous state, and is located mainly in pyroxenes, olivines, lunar glasses, and ilmenite. Small amounts of iron are also detected in metallic phases and troilite and the Mössbauer spectra indicate that metallic iron is the major magnetic constituent in the Apollo 12 samples.

Apollo 12 crystalline rocks exhibit considerable diversity, but in all cases studied show appreciably lower ilmenite content, and higher olivine content than Apollo 11 crystalline rocks. Two major classes of Apollo 12 crystalline rocks may be distinguished on the basis of their Mössbauer spectra.

Measurements on the unusual microbreccia 12013 show in addition to resonances detected in other samples, unusual spectral features not previously reported in lunar samples. Some of these features appear to be associated with the "kreep" component in the dark lithology, and some are suggestive of having originated from exposure to intense shock.

Apollo 12 lunar fines spectra also exhibit one of these unusual spectral features which we tentatively associate with the exotic "kreep" component in the fines. Our measurements show that the Apollo 12 fines contain more olivine and only one third as much ilmenite as the Apollo 11 fines. We find very different spectra for Apollo 11 and Apollo 12 fines, but a remarkable similarity in the spectra of all fines from a given site, which suggests that the nuclear gamma resonant spectra can provide regional soil signatures for lunar soils.

Acknowledgments—We are very grateful for the generous help of several of our colleagues at IIT Research Institute, including J. W. MANDLER, R. S. RYSKIEWICZ, E. H. MALONE, and especially S. I. BAKER for assistance in special computer program development. We wish also to thank a number of other lunar sample investigators for helpful discussions and communicating the results of their studies on Apollo 12 samples prior to publication. We are particularly grateful to R. A. SCHMITT for allowing us to study a sample in his custody. This research was performed with the support of NASA Contract NAS9-8083.

REFERENCES

ANNELL C. S., CARRON M. K., CHRISTIAN R. P., CUTTITA F., DWORNIK E. J., HELZ A. W., and LIGON D. T., JR. (1971) Chemical and spectrographic analyses of lunar samples from Apollo 12 mission. Second Lunar Science Conference (unpublished proceedings).

Apollo 12 Lunar Sample Information Catalog (1970) Lunar Receiving Laboratory, Science and Applications Directorate (MSC-01512), Manned Spacecraft Center, NASA, Houston, Texas.

BANCROFT G. M., BOWN M. G., GAY P., MUIR I. D., and WILLIAMS P. G. L. (1971) Mineralogical and petrographic investigation of some Apollo 12 samples. Second Lunar Science Conference (unpublished proceedings).

BANERJEE S. K., O'REILLY W., GIBB T. C., and GREENWOOD N. N. (1967) The behavior of ferrous ions in iron-titanium spinels. *J. Phys. Chem. Solids* **28**, 1323–1335.

BUTLER J. C., KING E. A., and CARMAN M. F. (1970) Grain size and modal analyses of lunar regolith material returned by Apollo 11 and 12. *Geol. Soc. Amer. Abs.* **2**, 512.

DE VOE J. R. and SPIJKERMAN J. J. (1970) Mössbauer spectrometry. *Anal. Chem.* **42**, 366R–388R.

DRAKE M. J., MCCALLUM I. S., MCKAY G. A., and WEILL D. F. (1970) Mineralogy and petrology of Apollo 12 sample No. 12013: a progress report. *Earth Planet. Sci. Lett.* **9**, 103–123.

DUCHESNE J., DEPIREUX J., GERARD A., GRANDJEAN F., and READ M. (1971) A study by electronic paramagnetic resonance and Mössbauer spectroscopy of some lunar samples collected by Apollo 12. Second Lunar Science Conference (unpublished proceedings).

GAY P., BANCROFT G. M., and BOWN M. G. (1970) Diffraction and Mössbauer studies of minerals from lunar soils and rocks, in *Proc. Apollo 11 Lunar Sci. Conf., Geochim. Cosmochim. Acta* Suppl. 1, Vol. 1, pp. 481–497. Pergamon.

GOLDANSKII V. I., DOLENKO A. V., EGIAZAROV B. G., ZAPOROZHETS V. M, ISAKOV L. M., MAKAROV E. F., TRUKHTANOV V. A., and CHUPROVA I. D. (1968) The devices and methods of analysis of ores and minerals for Sn and Fe based on Mössbauer effect and their practical applications. Proceedings of the Institute of Nuclear Geophysics and Geochemistry, *Nuclear Geophysics*, issue 3, pp. 220–225.

HARGRAVES R. B. and DORETY N. (1971) Magnetic properties of some lunar crystalline rocks returned by Apollo 11 and Apollo 12. Second Lunar Science Conference (unpublished proceedings).

Herr W. and Skerra B. (1969) Mössbauer spectroscopy applied to the classification of stone meteorites. In *Meteorite Research* (ed. P. Millman). North-Holland.

Herzenberg C. L. (1969) Mössbauer spectrometry as an instrumental technique for determinative mineralogy. In *Mössbauer Effect Methodology* (ed. I. J. Gruverman), Vol. 5. Plenum.

Herzenberg C. L. and Riley D. L. (1970a) Analysis of first returned lunar samples by Mössbauer spectrometry, in *Proc. Apollo 11 Lunar Sci. Conf.*, *Geochim. Cosmochim. Acta* Suppl. 1, Vol. 3, pp. 2221–2241. Pergamon.

Herzenberg C. L. and Riley D. L. (1970b) Current applications of Mössbauer spectrometry in geochemistry. In *Developments in Applied Spectroscopy* (ed. E. L. Grove), Vol. 8, pp. 277–291. Plenum.

Herzenberg C. L., Moler R. B., and Riley D. L. (1971a) Preliminary results from Mössbauer instrumental analysis of Apollo 12 lunar rock and soil samples. Second Lunar Science Conference (unpublished proceedings).

Herzenberg C. L., Riley D. L., and Moler R. B. (1971b) The application of Mössbauer spectrometry to lunar and terrestrial rock samples. To be published in *Applications of Low Energy X- and Gamma Rays* (ed. C. A. Ziegler). Gordon and Breach.

Hogarth D. D., Brown F. F., and Pritchard A. M. (1970) Biabsorption, Mössbauer spectra, and chemical investigation of five phlogopite samples from Quebec. *Can. Mineral.* **10**, 710–722.

Hogg C. S. and Meads R. E. (1970) The Mössbauer spectra of several micas and related minerals. *Min. Mag.* **37**, 606–614.

Housley R. M., Blander M., Abdel-Gawad M., Grant R. W., and Muir A. H., Jr (1970) Mössbauer spectroscopy of Apollo 11 samples, in *Proc. Apollo 11 Lunar Sci. Conf.*, *Geochim. Cosmochim. Acta* Suppl. 1, Vol. 3, pp. 2251–2268. Pergamon.

Housley R. M., Grant R. W., Muir A. H., Jr., Blander M., and Abdel-Gawad M. (1971) Mössbauer studies of Apollo 12 samples. Second Lunar Science Conference (unpublished proceedings).

Hubbard N. J., Gast P. W., and Meyer C. (1971) The origin of the lunar soil based on REE, K, Rb, Ba, Sr, P and $Sr^{87/86}$ data. Second Lunar Science Conference (unpublished proceedings).

James O. B. (1971) Petrology of lunar microbreccia 12013,6. Manuscript submitted for publication (personal communication).

Laul J. C., Keays R. R., Ganapathy R., and Anders E. (1970) Abundance of 14 trace elements in lunar rock 12013,10. *Earth Planet. Sci. Lett.* **9**, 211–215.

Lerman A., Stiller M., and Hermon E. (1968) Mössbauer quantitative analysis of Fe^{+3}/Fe^{+2} ratios in some phosphate and oxide mixtures: possibilities and limitations. *Earth Planet. Sci. Lett.* **3**, 409–416.

Lofgren G. (1971) Devitrified glass fragments from Apollo 11 and Apollo 12. Second Lunar Science Conference (unpublished proceedings).

LSPET (Lunar Sample Preliminary Examination Team) (1970) Preliminary examination of lunar samples from Apollo 12. *Science* **167**, 1325–1339.

McKay D., Morrison D., Lindsay J., and Ladle G. (1971) Apollo 12 soil and breccia. Second Lunar Science Conference (unpublished proceedings).

Meyer C., Aitken F. K., Brett R., McKay D. S., and Morrison D. A. (1971) Rock fragments and glasses rich in K, REE, P in Apollo 12 soils: their mineralogy and origin. Second Lunar Science Conference (unpublished proceedings).

Morgan J. W. and Ehmann W. D. (1970) Lunar rock 12013: O, Si, Al and Fe abundances. *Earth Planet. Sci. Lett.* **9**, 164–168.

Muir A. H., Jr. (1968) Analysis of complex Mössbauer spectra by stripping techniques. In *Mössbauer Effect Methodology* (ed. I. J. Gruverman), Vol. 4, pp. 75–102. Plenum.

Nagata T., Fisher R. M., Schwerer F. C., Fuller M. D., and Dunn J. R. (1971) Magnetic properties and remnant magnetization of Apollo 12 lunar materials and Apollo 11 lunar microbreccia. Second Lunar Science Conference (unpublished proceedings).

Papanastassiou D. A. and Wasserburg G. J. (1970) Rb-Sr ages from the Ocean of Storms. *Earth Planet. Sci. Lett.* **8**, 269–278.

PEARCE G. W., STRANGWAY D. W., and LARSON E. E. (1971) Magnetism of two Apollo 12 igneous rocks. Second Lunar Science Conference (unpublished proceedings).

SCHMITT H. H. and SUTTON R. L. (1971) Stratigraphic sequence for samples returned by Apollo missions 11 and 12. Second Lunar Science Conference (unpublished proceedings).

SCHNETZLER C. C., PHILPOTTS J. A., and BOTTINO M. L. (1970) Li, K, Rb, Sr, Ba, and rare-earth concentrations, and Rb-Sr age of lunar rock 12013. *Earth Planet. Sci. Lett.* **9**, 185–192.

SIMPSON P. R. and BOWIE S. H. U. (1971) Opaque phases in Apollo 12 samples. Second Lunar Science Conference (unpublished proceedings).

SPRENKEL-SEGEL E. L. and HANNA S. S. (1964) Mössbauer analysis of iron in stone meteorites. *Geochim. Cosmochim. Acta* **28**, 1913–1931.

SPRENKEL-SEGEL E. L. (1970) Recoilless resonance spectroscopy of meteoritic iron oxides. *J. Geophys. Res.* **75**, 6618–6630.

SULLIVAN S., THORPE A. N., ALEXANDER C. C., SENFTLE F., and DWORNIK E. (1971) Magnetic properties of individual glass fragments and microscopic spherules, Apollo 12 lunar sample. Second Lunar Science Conference (unpublished proceedings).

TAYLOR L. A., KULLERUD G., and BRYAN W. B. (1970) Apollo 12 sample 12063,9. *Trans. Amer. Geophys. Union* **51**, 583.

WAKITA H., REY P., and SCHMITT R. A. (1971) Abundances of the 14 rare earth elements plus 22 major, minor, and trace elements in ten Apollo 12 rock and soil samples. Second Lunar Science Conference (unpublished proceedings).

WARNER J. L. and ANDERSON D. H. (1971) Lunar crystalline rocks: Petrology, geology and origin. Second Lunar Science Conference (unpublished proceedings).

WHITE W. B., WHITE E. W., GÖRZ H., HENISCH H. K., FABEL G. W., and ROY R. (1971) Examination of lunar glass by optical, raman, and X-ray emission spectroscopy, and by electrical measurements. Second Lunar Science Conference (unpublished proceedings).

WILLIS J. P., AHRENS L. H., DANCHIN R. V., ERLANK A. J., GURNEY J. J., HOFMEYR P. K., MCCARTHY T. S., and ORREN M. J. (1971) Some interelement relationships between lunar rocks and fines, and stony meteorites. Second Lunar Science Conference (unpublished proceedings).

Proceedings of the Second Lunar Science Conference, Vol. 3, pp. 2125–2136
The M.I.T. Press, 1971.

Mössbauer studies of Apollo 12 samples

R. M. Housley, R. W. Grant, A. H. Muir, Jr., M. Blander,
and M. Abdel-Gawad

Science Center, North American Rockwell Corporation Thousand Oaks,
California 91360

(*Received* 22 *February* 1971; *accepted in revised form* 29 *March* 1971)

Abstract—We have analyzed Mössbauer data on Apollo 12 fines samples 12042,38 and 12001,117, core tube samples 12025,15; 12025,42; 12028,88; and 12028,113; and rocks 12038,47, 12052,16, and 12063,59. We find that the dark gray fine-grained material from widely separated surface locations and different depths contains a nearly uniform distribution of major Fe containing phases suggesting either that extensive mixing has taken place or less likely that all these samples were derived from similar rocks. The phase composition of the separated mineral and rock fragments indicates that rocks fairly rich in olivine and poor in ilmenite have made the major contributions to the fines. The fines contain about 0.5–1 wt. % metallic Fe which greatly exceeds the amount present in the igneous rocks. Most of this Fe metal is enclosed in glassy material and about half of it is in particles less than 40 Å in diameter. These small Fe particles may be partially responsible for the dark color of much of the glass. This fine Fe metal probably results from reduction of Fe^{2+} during impact events. Although the ~ 0.1 wt. % metal content of the rocks is sufficient to contain all the Ni, thermodynamic analysis indicates that a large fraction may have been in olivine at high temperatures (1100–1200°C).

Introduction

In this paper we discuss our analysis of Mössbauer spectra of Apollo 12 bulk fines 12042,38 and 12001,117; the core tube samples 12025,15; 12025,42; 12028,88 and 12028,113; and the rocks 12038,47; 12052,16 and 12063,59. Resonance absorption resulting from ilmenite, pyroxene, olivine, glass, troilite, and metallic Fe has been distinguished. Several problems were studied which advantageously utilize the ability of Mössbauer spectroscopy to easily provide accurate relative abundance data for Fe bearing phases in complex fine grained mixtures.

In the submillimeter core tube and bulk fines samples we compared the phase composition from sample to sample and compared these compositions with those of the local igneous rocks. These studies have contributed to understanding the extent of mixing in the lunar regolith and place some constraints on the types of igneous rocks which may have been the source material for the fines. We have also used Mössbauer spectroscopy to study the size distribution and association of Fe metal in the lunar fines and rocks with the object of evaluating the contributions to the ~ 0.5–1.0 wt. % metallic Fe found in the fines from igneous rocks, from iron meteorites, and from Fe^{2+} reduced during impact events. In this case our technique allows a separate semi-quantitative measure of the contribution from extremely small Fe particles < 40 Å in diameter which appear to be paramagnetic. The FeS content was quantitatively determined in the same measurements.

In support of our Mössbauer measurements and to document our samples, we have examined grain mounts of all our samples by optical microscopy in reflected light, of a number of samples in transmitted light, and have used other techniques such as microprobe analysis when appropriate. One unusual fragment from the fines 12038,42 having a fan-like texture and the composition of an anothosite gabbro is described.

APPARATUS

Mössbauer spectra were obtained and analyzed using standard techniques. Two electromechanical drive systems and two programmed mechanical drive systems were used. All runs were made with the samples as absorbers using Co^{57} in Cu sources at room temperature. For all spectra, zero velocity is referenced to metallic Fe at room temperature.

SAMPLE DESCRIPTION AND MICROSCOPIC OBSERVATIONS

The igneous rocks which we examined consist essentially of clinopyroxenes, calcic plagioclase, olivine, and ilmenite. Minor constituents include chromite, ulvö-spinel, troilite, and metallic Fe. The samples of fines consist predominantly of glass, fragments of breccia, igneous rocks, and individual mineral grains.

Our observations generally agree with the petrographic descriptions given by LSPET (1970a). We report here, however, a few selected observations which are significant.

In our Apollo 12 igneous rocks troilite and metallic Fe occur mostly independently; Fe globules or crystals within troilite, which characterize the Apollo 11 samples are rather rare. Similar observations were made by TAYLOR et al. (1971) and RAMDOHR et al. (1971).

In sample 12063,59 (ophitic basalt, WARNER and ANDERSON, 1971) the ilmenite shows local intense deformation with partial recrystallization. This is in agreement with similar observations by TAYLOR et al. (1971). RAMDOHR et al. (1971) however found no sigmoidally deformed twin lamellae and concluded that their samples were insufficiently shocked to affect the ilmenite.

Sample 12038,47 (ophitic basalt, WARNER and ANDERSON, 1971) shows what appears to be subsolidus breakdown of ulvöspinel into ilmenite and metallic Fe. The breakdown is more evident along fractures (Fig. 1a). Similar observations were made by RAMDOHR et al. (1971) in samples 12018 and 12063 and by HAGGERTY and MEYER (1970) in sample 12052. This feature thus appears to be common in the Apollo 12 rocks and must account for part of the metallic Fe of the igneous rocks.

In sample 12042,38 (fines) we found an unusual fragment with a fan-shaped texture which resembles some pyroxene or devitrified glass chondrule fragments (Fig. 1b). In thin section the grain has first order yellow interference colors, straight undulatory extinction parallel to the fibers and is devoid of opaque minerals. The average chemical composition obtained with a defocused microprobe beam is given in Table 1 and is that of anorthositic gabbro.

HOMOGENEITY OF FINES

Although the regolith at the Apollo 12 site was generally dark gray in color, a persistent lighter colored layer was noted by the astronauts (LSPET, 1970b). Generally buried under several centimeters of darker material, it was occasionally exposed at the surface and was buried more deeply near the rims of some craters. In the double core tube from which our samples came, a light layer was found in the bottom from 39–41 cm in depth and an unusual layer of coarse olivine and glass was found at 13–15 cm depth.

To obtain information on the variation of phase compositions within the gray fine grained material which is predominant at the site, we have compared spectra from

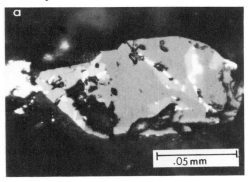

Fig. 1. (a) Ulvöspinel (gray) appears to break down into ilmenite (light gray) and metallic Fe (white) particularly along fractures and grain boundaries (sample 12038,47). Polished section. (b) Fan-shaped fragment in lunar fines. The fragment with a fibrous texture resembling some fan-like pyroxene chondrules has an average composition of anorthositic gabbro, (sample 12042,38). Thin section, crossed nicols.

Table 1. Microprobe analysis of unusual fragment
in lunar fines 12042,38.

	wt. %
SiO_2	44.8
Al_2O_3	23.8
CaO	14.3
FeO	5.1
MgO	10.6
TiO_2	0.2
Na_2O	trace
Total	98.8

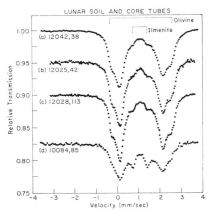

Fig. 2. Room temperature Mössbauer spectra of Apollo 12 dust (a), core tube sample from ~ 5 cm (b), core tube sample from ~ 30 cm (c), and Apollo 11 dust (d). (Zero velocity is referenced to metallic Fe at room temperature.)

the four core tube samples 12025,15; 12025,42; 12028,88; and 12028,113 from the approximate depths of 0.5, 5, 19, and 30 cm, respectively and the surface samples 12042,38 and 12001,117. The core tube samples are all from dark gray regions although 12028,88 was from a visually somewhat lighter layer (LSPET, 1970a). Sample 12042 obtained about 75 m away from the double core tube location was collected because it contained many cohesive clots of material making the texture clearly distinct from other areas. No mention of light colored material in the vicinity was made (LSPET, 1970a). Sample 12001 was the fine material in the selected sample container. This material was collected during the first EVA on a traverse roughly 300 meters away from the double core tube location (LSPET, 1970a).

All six fines samples yielded very similar spectra, quite different from spectra of Apollo 11 fines, indicating a close similarity in phase composition among themselves and a distinct difference from Apollo 11 type fines. Spectra of the fines 12042,38, two of the core tube samples and the Apollo 11 fines 10084,85 are compared in Fig. 2. The data on these six samples suggest that the regolith at the Apollo 12 site is fairly homogeneous with respect to major phases except for a modest number of visually distinct regions. This most probably means that the regolith has, for the most part, been fairly well mixed and that the visually distinct layers are the result of isolated events late in lunar history. The alternative possibility that the dark fines all formed from similar rocks does not seem as likely in view of the variety of rocks which were returned from the Apollo 12 site.

Local Rock Content of Fines

About 1.37 g of 12042,38 was separated into size fractions mainly by dry sieving assisted when necessary by ultrasonic agitation in acetone. The size distribution is very similar to that of Apollo 11 material (Duke *et al.*, 1970). The 44–149 μm material weighing 357 mg was further separated into heavy (108 mg) and light (243 mg) fractions with methylene iodide (sp.g. 3.3). Microscopic examinations confirmed the

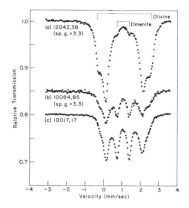

Fig. 3. Room temperature Mössbauer spectra of the heavy fraction of the Apollo 12 dust (a), the Apollo 11 dust (b), and a representative type A Apollo 11 rock (c). (Zero velocity is referenced to metallic Fe at room temperature.)

expectation that grains of major Fe containing mineral phases would be in the heavy fraction with only an occasional glass spherule being seen. The light fraction consisted mostly of frequently vesicular glass, microbreccia fragments and plagioclase with moderate amounts of adhering heavier phases.

Figure 3 shows the spectrum of the heavy fraction described above, which we believe is characteristic of the mineral grains in the fines no longer obscured by the Fe containing glass. Also shown are spectra of a similar heavy separate from the Apollo soil 10084,85 and of Apollo 11 rock 10017 which it closely resembles. Differences in overall area are largely caused by differences in sample thickness. Since most rocks from the Apollo 11 site are quite similar with respect to major phases, this close resemblance of spectra supports the view (HOUSLEY et al., 1970) that the majority of mineral fragments in the regolith material over the mare regions come from the local rocks. It seems probable, therefore, that the spectrum of the 12042,38 heavy fraction is indicative of the phase composition of a suitably weighted average of the local Apollo 12 rocks.

Figure 4 shows spectra of the three Apollo 12 rocks which we have studied. These spectra and those of the heavy fraction of the fines 12042,38 have been analyzed both by least squares fitting, using necessary constraints on line positions and widths, and by spectrum stripping using terrestrial mineral standards (MUIR, 1968). The contribution of olivine is difficult to quantify because of strong overlap with the pyroxene M1 site lines. This necessitates introduction of a model to fit the data by constraining either the positions or widths of the pyroxene and olivine lines, or both. Good fits can be obtained with a range of physically reasonable choices, yielding significantly different olivine-pyroxene ratios. Therefore, we do not report quantitative results for olivine and pyroxene separately. The ratios of absorption resulting from Fe in ilmenite to that resulting from total silicate Fe obtained by the two methods are in good agreement, and a weighted average is presented in Table 2. The indicated uncertainties

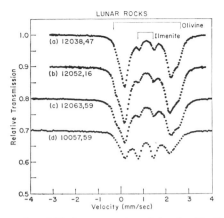

Fig. 4. Room temperature Mössbauer spectra of Apollo 12 olivine basalts 12038 (a), 12052 (b), 12063 (c), and Apollo 11 basalt (type A) 10057 (d). (Zero velocity is referenced to metallic Fe at room temperature.)

Table 2. Ratio of absorption area for Fe in ilmenite to Fe in total silicates

Sample	Sample description	$(Fe)_{ilmenite}/(Fe)_{silicates}$
12042,38	fines	0.04 ± 0.02
12042,38,H	fines (sp.g. $>$ 3.3)	0.04 ± 0.02
12038,47	olivine basalt	0.11 ± 0.02
12052,16	olivine basalt	0.10 ± 0.02
12063,59	olivine basalt	0.15 ± 0.02
10017,17	basalt	0.6 ± 0.05

are estimated absolute uncertainties; the uncertainty for relative comparisons of the tabulated ratios should be considerably smaller. The mineral grains in the fines are substantially lower in ilmenite and higher in olivine than any of the rocks we studied with rock 12052,16 being nearest to the fines in composition. The fines phase composition resembles more closely those of rocks 12002, 12004, and 12020 which were studied by HERZENBERG *et al.* (1971). Together these six rocks span a range of major phase compositions which bracket that of the mineral grains in the fines, making it possible to derive the fines in large part from a suitable mixture of them.

METALLIC Fe–Ni IN FINES

In this section we discuss observations of the abundance, association and size distribution of metallic Fe in our samples. As with the Apollo 11 samples, there is much more metal (mostly associated with the glass) in the lunar fines than in the igneous rocks. In addition, a large fraction of the metallic Fe is in very small <40 Å diameter particles.

In the course of our measurements we observed that the absorption resulting from ferromagnetic Fe metal increased more on cooling the sample to 77°K than could be explained by the change in recoil free fraction f alone. This suggested that some of the Fe metal was present as very small particles which behaved superparamagnetically at room temperature, but ferromagnetically at 77°K.

Recently HERZENBERG (1971) called attention to excess area amounting to a few % of the total which was found in the low velocity region of spectra of lunar fines. We confirmed this observation and found from analysis of spectra of our heavy liquid separates that the excess absorption was associated with the glassy material which also contains the majority of the Fe metal. This suggests that this excess absorption also is associated with superparamagnetic Fe. HAFNER et al. (1971) also noticed excess absorption near zero velocity in spectra of the lunar fines, which they attributed to superparamagnetic Fe.

Spectra were taken using a Cu source $f = 0.71$ (HOUSLEY et al., 1964a) and were corrected for background (~ 10–15%) in a standard manner (HOUSLEY et al., 1964b). The low energy (most negative) lines of ferromagnetic Fe and FeS are far separated from other absorption lines and fairly well resolved from each other (HOUSLEY et al., 1970). The absorption area attributed to ferromagnetic Fe was 4 times the area measured for the low velocity Fe line by summing data points in the velocity range about -6.75 to -4.75 mm/sec (all velocities are relative to a zero at the center of a metallic Fe pattern). The absorption area attributed to Fe in FeS was similarly 4 times the area in the range -4.65 to -3.55 mm/sec.

The area due to superparamagnetic Fe was taken to be the difference between the area obtained by summing over the range -1.13 to $+1.10$ mm/sec and that obtained over the range $+1.13$ to 3.36 mm/sec. The dividing line between ranges is midway between the peaks of ilmenite so that both ranges are expected to have equal contributions from Fe^{2+} and in addition both ranges contain equal contributions from ferromagnetic Fe.

Observed differences could a priori be attributed to a bad choice of ranges, Fe^{3+}, f anisotropy, or preferred orientation of grains as well as superparamagnetic Fe. Differences obtained for all our lunar rocks are small and in some cases negative. Thus it appears that f anisotropy and/or preferred orientation generally leads to a small negative effect. The high abundance of free Fe metal in the lunar glass makes Fe^{3+} very unlikely. As another test to verify that the observed area difference was due to superparamagnetic Fe we compared this difference and the ferromagnetic component in the light fraction of 10084,85 at room temperature with similar data taken at 77°K. The difference near zero velocity decreased on cooling and the ferromagnetic area increased a corresponding amount. This is the type of change expected to be caused by particles in a certain size range passing through a critical range of reorientation times because of the temperature change and supports the assumption that the excess area is caused by superparamagnetic very fine grained metallic Fe.

Our areas have been converted into absolute amounts of ferromagnetic Fe, superparamagnetic Fe, and FeS using $f_{FeS} = 0.82$ and $f_{Fe} = 0.80$ at room temperature (HOUSLEY et al., 1970), $f_{Fe} = 0.92$ at 77°K and taking the maximum resonant cross section as 2.56×10^{-18} cm² corresponding to an internal conversion coefficient of 8.2. The results are presented in Table 3 with error limits corresponding to one standard deviation determined from counting statistics. The possible systematic errors discussed up to now might necessitate increases in the superparamagnetic Fe of up to about 0.1 wt. %. Saturation effects resulting from the finite absorber thickness also cause our values for superparamagnetic Fe to be low by about 10%.

Table 3. Metallic Fe and FeS content of samples*

Sample	Ferromagnetic Fe wt.%	Superparamagnetic wt.%	FeS wt.%
77°K			
10084,85 (sp.g. < 3.3)	0.61 ± 0.04	0.29 ± 0.02	—
Room Temperature			
10084,85 (sp.g. < 3.3)	0.51 ± 0.04	0.39 ± 0.02	—
12042,38 (sp.g. < 3.3)	0.34 ± 0.05	0.17 ± 0.02	0.36 ± 0.06
12042,38 (sp.g. > 3.3)	0.04 ± 0.07	—	0.10 ± 0.08
12025,15	0.38 ± 0.06	0.35 ± 0.02	0.23 ± 0.07
12025,42	0.37 ± 0.06	0.35 ± 0.02	0.42 ± 0.06
12028,88	0.39 ± 0.08	0.22 ± 0.02	0.36 ± 0.08
12028,113	0.31 ± 0.05	0.43 ± 0.01	0.29 ± 0.06
12038,47	0.07 ± 0.03	0.04 ± 0.01	0.23 ± 0.03
12052,16	0.11 ± 0.06	0.03 ± 0.01	0.23 ± 0.06

* The error limits are one standard deviation due to counting statistics only. Because of possible systematic errors discussed in the text the values, particularly those for superparamagnetic Fe, should more likely be regarded as lower limits.

To appear completely ferromagnetic in an Fe^{57} Mössbauer measurement, the reorientation times for the magnetic moments of the Fe particles must be longer than about 10^{-7} sec (van der Woude and Dekker, 1965). Aharoni (1969) has suggested that the reorientation rate γ of spherical particles should be given by

$$\frac{1}{\tau} = \frac{\gamma_0 K}{M_s} \alpha^{1/2} e^{-\alpha} \tag{1}$$

where γ_0 is the gyromagnetic ratio, K the magnetocrystalline anisotropy, M_s the saturation magnetization and

$$\alpha = \frac{KV}{kT},$$

where V is the particle volume and kT the absolute temperature in energy units. Using constants measured for pure Fe in the above expressions one finds that the transition between ferromagnetic and superparamagnetic behavior starts at a diameter of about 40 Å for spherical particles at room temperature and at a somewhat smaller size at 77°K.

This division, however, is not sharp. In the range of reorientation times between 10^{-7} and 10^{-9} sec considerable intensity would lie between our paramagnetic and ferromagnetic regions (van der Woude and Dekker, 1965). The maximum value assumed by expression (1) for Fe at room temperature is about 10^8. Therefore, a substantial fraction of the area caused by small Fe particles will probably be lost in the analysis. This is another reason why the values presented for superparamagnetic Fe in Table 3 may underestimate the actual amount of fine grained Fe present. The spectra expected for fine grained Fe are similar to the ones obtained by Lindquist *et al.* (1968) for Fe in fine Ni particles.

Metallic Fe particles in the diameter range 26–110 Å were found in the fines by Runcorn *et al.* (1970) from magnetic measurements, while Nagata *et al.* (1970) interpreted their measurements as evidence of particles with diameters ranging from 70–130 Å.

The abundance of Fe metal in the fines and its strong association with glassy material rather than mineral fragments indicate that it either is of meteoritic origin directly or that it results from reductive heating of the lunar material during impact events (HOUSLEY et al., 1970). Surface films produced by vapor phase transport can at most make a small contribution to the total Fe metal since they would be equally probable on the mineral grains. The abundance of fine Fe metal argues for reduction as the most probable origin. This may have been assisted by the solar wind hydrogen which saturates most particle surfaces (HOUSLEY et al., 1970). The presence of phases containing Ti^{4+} which tends to reduce to Ti^{3+} at high temperatures may also have been important to the reduction process (GILLES et al., 1967, 1968). GOLDSTEIN and YAKOWITZ (1971) have also argued on the basis of composition data that most of the larger metal grains enclosed in other particles of the Apollo 12 soil are not of meteoritic origin.

It is interesting to note that the slightly lighter gray core tube sample 12028,88 contains significantly less superparamagnetic Fe than the others consistent with the view that superparamagnetic Fe plays an important role in the darkening of lunar fines.

METALLIC Fe–Ni IN ROCKS

From the overall mineralogy of the Apollo 12 rocks, the presence of about 0.1 wt.% metallic Fe, and the absence of any similar amount of Fe^{3+} it must be concluded that they, like the Apollo 11 rocks crystallized at a P_{O_2} lower than that of the quartz-fayalite-iron buffer.

The total Ni metal content in rock 12052 reported by LSPET (1970a) is 32 ppm. If we were to assume that all of this is in the metal phase we would expect an average Ni content of $\sim 3\%$ in that phase. CHAMPNESS et al. have analyzed two metal grains from rock 12052 finding Ni contents of 1.9% and 2.3%.

Despite this agreement there is good reason to believe, based upon the analyses of Fe–Ni grains reported by REID et al. (1970) and BRETT et al. (1971), that a significant fraction of the Ni metal in some rocks is in olivine, or at least was at high temperatures. The relevant chemical equilibrium,

$$Fe(metal) + \tfrac{1}{2}Ni_2SiO_4(olivine) \leftrightharpoons \tfrac{1}{2}Fe_2SiO_4(olivine) + Ni(metal), \qquad (2)$$

is independent of the properties of the remaining liquid including the P_{O_2}. The solutions of these constituents in olivine or metal appear to be ideal (BUSECK and GOLDSTEIN, 1969; NAFZIGER and MUAN, 1967; CAMPBELL and ROEDER, 1968; and KAUFMAN and COHEN, 1956). Consequently,

$$K \cong \frac{N_{Ni}(m)}{N_{Fe}(m)} \frac{N_{Fe^{2+}}(ol)}{N_{Ni^{2+}}(ol)} = \frac{W_{Ni}(m)}{W_{Fe}(m)} \frac{W_{Fe^{2+}}(ol)}{W_{Ni^{2+}}(ol)} \qquad (3)$$

where N is mole (or ion) fraction, W is weight fraction, m designates metal, and ol, olivine. Values of K may be calculated from known thermodynamic data and are given in Table 4.

From these equilibrium constants one sees that olivine of composition Fa_{30} in equilibrium with metal containing 30% Ni, a value which is frequently exceeded in

Table 4. Equilibrium constants for the reaction Fe $+ \frac{1}{2}Ni_2SiO_4 \rightleftharpoons$
$Ni + \frac{1}{2}Fe_2SiO_4$

T°C	K*	K**,†
800	(616)	(319)
900	481	(264)
1000	391	(225)
1100	327	(189)
1200	(280)	(168)

( ) extrapolated
* TAYLOR and SCHMALZREID (1964); ** CAMPBELL and
ROEDER (1968); † ROBIE and WALDBAUM (1968).

the very earliest metal particles to form (REID *et al.*, 1970), at 1200°C will contain about 500 ppm Ni. Less than 20% olivine of this composition would accommodate all the Ni in most of the rocks and only 6% would be required for rock 12052. Consequently, a large fraction of the Ni is expected to be in olivine at high temperature. This fraction is dependent upon P_{O_2} only to the extent that P_{O_2} controls the Fe^{2+} content of the olivine. Given the Fe^{2+} content the calculations can be made from the known equilibrium constants for Eq. 2. As the temperature decreases more metal forms and the increasing ratio of metal to olivine and decreasing value of the equilibrium constant will lead to transfer of Ni from olivine to metal as long as re-equilibration can take place. The change in K over the temperature range over which re-equilibration could occur is only about a factor of two so the early Ni concentration must have been at least half as large as the maximum presently measured value.

It seems likely that olivine accumulation in any series of cogenetic rocks will also lead to enrichment in Ni. Using the Ni values reported by LSPET (1970a) the cogenetic series proposed by KUSHIRO *et al.* (1971) show this trend.

SUMMARY

(1) Our results on fine grained gray fines from four different depths in the core tube and from two widely separated surface localities suggest either that they have all been derived locally from the same distribution of rock types, or more probably that for the most part the lunar regolith has been fairly well mixed. This would in turn imply that the stratification observed in the core tube samples resulted from relatively recent events. The phase compositions of the dark fine grained Apollo 12 fines are characteristic of the vicinity of the landing site and are distinctly different from that of the Apollo 11 fines. (2) The phase composition of the mineral and rock fragments (glassy material excluded) in the dark gray Apollo 12 fines can be matched with suitable mixtures of the returned rocks, rocks relatively rich in olivine and poor in ilmenite making the major contributions. (3) Both Apollo 11 and Apollo 12 fines contain much more Fe metal than the igneous rocks at the sites. Most of this metal is associated with the glassy material and a substantial fraction of it is present as particles $\leqslant 40$ Å in diameter, suggesting that most of this excess Fe metal may have resulted from the reduction of Fe^{2+} during impact events. In view of our results, it seems worthwhile to investigate the possibility that fine grained Fe metal may make a significant contribution to the optical properties of lunar glasses. (4) Thermodynamic

analysis of the distribution of Fe and Ni between olivine and metal leads to the conclusion that much of the Ni in many lunar rocks has been in the olivine phase at high temperatures.

Acknowledgments—The technical assistance provided by W. B. NORRIS and K. G. RASMUSSEN is gratefully acknowledged. We have profited from discussions with Dr. C. L. HERZENBERG and Prof. S. S. HAFNER. This work was largely supported by NASA Contract NAS 9-10208.

REFERENCES

AHARONI A. (1969) Effect of a magnetic field on the superparamagnetic relaxation time. *Phys. Rev.* **177**, 793–796.

BRETT R., BUTLER P. JR., MEYER C. JR., REID A. M., TAKEDA H., and WILLIAMS R. J. (1971) Apollo 12 igneous rocks 12004, 12008, 12009, and 12022: A mineralogical and petrological study. Second Lunar Science Conference (unpublished proceedings).

BUSECK P. R. and GOLDSTEIN J. I. (1969) Olivine compositions and cooling rates of pallasitic meteorites. *Bull. Geol. Soc. Amer.* **80**, 2141–2158.

CAMPBELL F. E. and ROEDER R. (1968) The stability of olivine and pyroxene in the Ni-Mg-Si-O system. *Amer. Mineral.* **53**, 257–268.

CHAMPNESS P. E., DUNHAM A. C., GIBB F. G. F., GILES H. N., MACKENZIE W. S., STUMPFL E. F., and ZUSSMAN J. (1971). Some aspects of the mineralogy and petrology of Apollo 12 rocks, Second Lunar Science Conference (unpublished proceedings).

DUKE M. B., WOO C. C., SELLARS G. A., BIRD M. L., and FINKELMAN R. B. (1970) Genesis of lunar soil at tranquillity base. *Proc. Apollo 11 Lunar Sci. Conf., Geochim. Cosmochim. Acta* Suppl. 1, Vol. 1, pp. 347–361. Pergamon.

GILLES P. W., CARLSON K. D., FRANZEN H. F., and WAHLBECK P. G. (1967) High temperature vaporization and thermodynamics of the titanium oxides. I. Vaporization characteristics of the crystalline phases. *J. Chem. Phys.* **46**, 2461–2465.

GILLES P. W., FRANZEN H. F., STONE G. D., and WAHLBECK P. G. (1968) High temperature vaporization and thermodynamics of the titanium oxides. III. Vaporization characteristics of the liquid phase. *J. Chem. Phys.* **48**, 1938–1941.

GOLDSTEIN J. I. and YAKOWITZ H. (1971). Metal particles and inclusions in Apollo 12 lunar soil. Second Lunar Science Conference (unpublished proceedings).

HAFNER S. S. (1971) State and location of iron in Apollo 11 samples. In *Mössbauer Effect Methodology* (edited by I. J. Gruverman), Vol. 6, pp. 193–207. Plenum Press.

HAGGERTY S. E. and MEYER H. O. A. (1970) Apollo 12: Opaque Oxides. *Earth Planet. Sci. Lett.* **9**, 379–387.

HERZENBERG C. L. (1970) Private communication.

HERZENBERG C. L., MOLER R. B., and RILEY D. L. (1971). Mössbauer spectrometry of Apollo 12 lunar rock and soil samples. Second Lunar Science Conference (unpublished proceedings).

HOUSLEY R. M., DASH J. G., and NUSSBAUM R. H. (1964a) Mean-square displacement of dilute iron impurity atoms in high-purity beryllium and copper. *Phys. Rev.* **136**, A464–A466.

HOUSLEY R. M., ERICKSON, N. E., and DASH J. G. (1964b) Measurement of recoil-free fractions in studies of the Mössbauer effect. *Nucl. Instr. Methods* **27**, 29–37.

HOUSLEY R. M., BLANDER M., ABDEL-GAWAD M., GRANT R. W., and MUIR A. H., JR. (1970) Mössbauer spectroscopy of Apollo 11 samples. *Proc. Apollo 11 Lunar Sci. Conf., Geochim. Cosmochim. Acta* Suppl. 1, Vol. 3, pp. 2251–2268. Pergamon.

KAUFMAN L. and COHEN M. (1956) The martensitic transformation in the iron-nickel system. *Am. Inst. Min. Met. Engrs., Trans.*, Vol. 206, pp. 1393–1401.

LINDQUIST R. H., CONSTABARIS G., KÜNDIG W., and PORTIS A. M. (1968) Mössbauer spectra of ^{57}Fe in superparamagnetic nickel. *J. Appl. Phys.* **39**, 1001–1003.

LSPET (Lunar Sample Preliminary Examination Team) (1970a) Apollo 12 lunar sample information. NASA TR R-353.

LSPET (Lunar Sample Preliminary Examination Team) (1970b) Preliminary examination of lunar samples from Apollo 12. *Science* **167**, 1325–1339.

MUIR A. H., JR. (1968) Analysis of complex Mössbauer spectra by stripping techniques. In *Mössbauer Effect Methodology* (edited by I. J. Gruverman), Vol. 4, pp. 75–101. Plenum Press.

NAFZIGER R. H. and MUAN A. (1967) Equilibrium phase compositions and pyroxenes in the system MgO—"FeO"—SiO_2. *Amer. Mineral.* **52**, 1364–1385.

NAGATA T., ISHIKAWA Y., KINOSHITO H., KONO M., SYONO Y., and FISHER R. M. (1970) Magnetic properties and natural remanent magnetization of lunar materials. *Proc. Apollo 11 Lunar Sci. Conf.*, *Geochim. Cosmochim. Acta* Suppl. 1, Vol. 3, pp. 2325–2340. Pergamon.

RAMDOHR P., EL GORESY A., and TAYLOR L. A. (1971) The opaque minerals in the lunar rocks from Oceanus Procellarum. Second Lunar Science Conference (unpublished proceedings).

REID A. M., MEYER C. JR., HARMON R. S., and BRETT R. (1970) Metal grains in Apollo 12 igneous rocks. *Earth Planet. Sci. Lett.* **9**, 1–5.

ROBIE R. A. and WALDBAUM D. R. (1968) Thermodynamic properties of minerals and related substances at 298.15°K (25.0°C) and one atmosphere (1.013 bars) pressure and at higher temperatures. *Geol. Survey Bull.* 1259, U.S. Govt. Printing Office, Washington, D.C.

RUNCORN S. K., COLLINSON D. W., O'REILLY W., BATTEY M. H., STEPHENSON A., JONES J. M., MANSON A. J., and READMAN P. W. (1970) Magnetic properties of Apollo 11 lunar samples. *Proc. Apollo 11 Lunar Sci. Conf.*, *Geochim. Cosmochim. Acta* Suppl. Vol. 3, pp. 2369–2387. Pergamon.

TAYLOR L. A., KULLERUD G., and BRYAN W. B. (1971) Mineralogy of two Apollo 12 samples. Second Lunar Science Conference (unpublished proceedings).

TAYLOR R. W., and SCHMALZREID H. (1964) The free energy of formation of some titanates, silicates and magnesium aluminate from measurements made with galvanic cells involving solid electrolytes. *J. Phys. Chem.* **68**, 2444–2449.

VAN DER WOUDE F. and DEKKER A. J. (1965) The relation between magnetic properties and the shape of Mössbauer spectra. *Phys. Stat. Solidi* **9**, 775–786.

WARNER J. L. and ANDERSON D. H. (1971) Lunar crystalline rocks—petrography, geology, and origin. Second Lunar Science Conference (unpublished proceedings).

Proceedings of the Second Lunar Science Conference, Vol. 3, pp. 2137–2151
The M.I.T. Press, 1971.

Infrared vibrational spectroscopic studies of minerals from Apollo 11 and Apollo 12 lunar samples

Patricia A. Estep, John J. Kovach, and Clarence Karr, Jr.
Morgantown Energy Research Center, U.S. Department of the Interior,
Bureau of Mines, Morgantown, West Virginia 26505

(*Received* 20 *February* 1971; *accepted in revised form* 31 *March* 1971)

Abstract—Infrared vibrational spectral correlations, derived from terrestrial and synthetic minerals, were used to characterize structures of the predominant lunar silicate minerals isolated from crystalline rocks and dusts. Absorption bands in the low-frequency region (400–180 cm^{-1}) were used to determine specific compositions for isolated pyroxene (Fs$_{21}$ to Fs$_{37}$), plagioclase (An$_{81}$ to An$_{100}$), and olivine (Fa$_{28}$ to Fa$_{34}$). For each of these predominant minerals, spectral similarities for separates from both rocks and dusts were observed. Further correlations from the fundamental vibrations of silicate SiO$_4$ tetrahedra were used to determine basic compositions for bulk samples from the dusts. The distinctive lunar basaltic spectra, predominant in pyroxene, matched better with some ocean tholeiitic basalts than with tektites, meteorites, or any other terrestrial rock type, but in no case was a good comparison obtained.

Iron composition variations in lunar pyroxenes produced spectral changes that were color related and may be correlatable with distribution of cations over the nonequivalent octahedrally coordinated sites. Colors of lunar glass were also observed to be composition dependent, and infrared spectral evidence is given to support the origin of light glass from plagioclase and dark glass from pyroxenes. Spectra of lunar glass were markedly different from those of tektites.

Introduction

This paper reports the first application of infrared vibrational spectroscopy to the analysis of lunar samples. The potential of this method for the structural analysis of rocks and minerals that could be found on the lunar surface was previously demonstrated by Lyon (1963) for the mid-infrared region to 400 cm^{-1}. The recently developed easy accessibility of the low-frequency far-infrared region to 30 cm^{-1} has further contributed to the usefulness of infrared spectroscopy in determining specific molecular structure of minerals. We have characterized the structures of the predominant lunar silicate minerals of pyroxene, plagioclase, and olivine, isolated from both rocks and dusts, through use of infrared spectral-structural correlations derived from terrestrial and synthetic minerals. Cation substitutions in these silicates were determined from data obtained in the low-frequency vibration region. We used spectra of the separated minerals and further correlations from the fundamental vibrations of silicate SiO$_4$ tetrahedra to classify bulk compositions of dust sieved fractions, composite grains, and glass particles. Cation ordering in some lunar pyroxene and plagioclase separates was indicated from changes in infrared absorption band shapes, resolution, and intensities. Such information on atomic distributions over the structural sites is useful in deducing formation conditions for these lunar minerals and can contribute to reconstructing the moon's early history.

EXPERIMENTAL TECHNOLOGY

Mineral separations

All lunar samples were handled, and pellets were prepared in a glove box purged with dry nitrogen. Available for our infrared studies were one Apollo 11 dust (10085,46) and six Apollo 12 samples: three crystalline rocks (12018,26, 12020,26, 12021,24) and three lunar dusts (12001,60, 12057,57, 12070,24). The dusts were dry-sieved to obtain mineral separates of a size suitable for microscopic isolation ($10–60\times$) and infrared analysis. Grain size distribution data for the four dusts are given in Table 1. For the three Apollo 12 dusts, a statistical two-way contingency test of the data (conducted at the 95% confidence level) showed that dusts 12070,24 (contingency sample) and 12057,57 (documented sample) (both appearing to be bimodal) have the same distribution, but their distributions are different from that of dust 12001,60 (selected sample). Dust mineral grains were microscopically separated only from the $+100$ mesh sieved fractions and ranged from 1600 to 149 μm for the Apollo 12 dusts. A single grain of about 1300 μm was necessary to give the 1 milligram of sample required for a good mid-infrared spectrum. A typical dust grain size of 250 μm required 50 combined grains. Data from a number of selections, relating number and size of grains required for 1 milligram, were plotted and used as working curves to facilitate subsequent microscopic isolations. Crystalline fragments chipped from rock samples gave single phases that were typically 300 μm. About 40 of these were required for a good mid-infrared spectrum.

Pellet preparation

Isolated mineral grains were placed in a mullite mortar fitted with a specially designed stainless steel funneled cylinder. A 3 mm diameter plunger was tapped down through the cylinder onto the confined sample, producing completely recoverable crushed fragments directly on the mortar surface. The cylinder assembly was removed, and further hand grinding was continued in the same mortar in order to reduce particles to a size suitable for a good infrared spectrum. To prepare a pellet for the mid-infrared region (4000–200 cm^{-1}), 500 milligrams of powdered cesium iodide (Harshaw Chemical Co.) were added directly to the preground sample in the mortar, and the mixture was blended for 5 minutes. This mixture was triple-pressed in a die at 23,000 lb total, with a 15-minute total press time. The resulting 13 mm diameter $\times$ 0.8 mm thick pellet was placed in a cell specially constructed to exclude air, and scanned on a Perkin–Elmer 621 grating spectrophotometer, purged with dry air. A comparison of spectra obtained for pellets mounted in this special holder with those of pellets exposed to the atmosphere in a regular pellet holder showed no detectable structural changes. Therefore, all subsequent cesium iodide pellets were scanned exposed to dry air. Pellets for far-infrared spectra were prepared in a similar manner, using 4 milligrams of sample and 150 milligrams of polyethylene (Uvasol, E. Merck AG) as a pellet matrix. Th 1-inch diameter pellets were scanned on a Perkin–Elmer FIS-3 vacuum spectrophotometer, covering the range 400 to 30 cm^{-1}.

RESULTS AND DISCUSSION

Structure determination of mineral separates

Pyroxene, plagioclase, olivine, and ilmenite (in order of decreasing abundance) were isolated from the lunar rocks and dusts and these were readily identified from

Table 1. Grain size distribution for lunar dusts.

Screen size, mesh	Grain size, μm	Weight % of total dust			
		12001,60	12057,57	12070,24	10085,46
$+100$	>149	29.93	27.56	21.73	91.5
$+200$ -100	149–74	23.80	28.09	24.10	n.d.*
$+325$ -200	74–44	18.87	18.05	19.46	n.d.
$+400$ -325	44–37	17.08	5.37	5.91	n.d.
-400	<37	10.32	20.93	28.80	n.d.

* n.d. = not determined: the $+100$ mesh fraction from dust 10085,46 (fines to coarse fines) contained larger grains (7000–149 μm) than those for the three other dusts.

Table 2. Infrared absorption bands of mineral separates from lunar samples.

Mineral separate	Derived composition	Sample source	Frequencies,* cm^{-1}
Pyroxene (yellow transparent angular fragments)	Fs_{22}	12021,24	240(w), 287(w), 310(w), 339(mw), <u>393</u>(mw), 412(sh), 450(sh), 475(sh), 502(s), 545(sh), 638(mw), 669(w), 728(w), 882(s), 955(s), 1022(sh), 1055(s), 1130(sh)
Pyroxene (dark amber transparent angular fragments)	Fs_{37}	12021,24	235(w), 285(w), 321(mw), <u>380</u>(w), 483(s) 535(sh), 628(mw), 658(w), 722(w), 875(s), 917(w), 960(s), 1053(s)
Plagioclase (single chalky-white grain)	An_{100}	10085,46	144(w), 166(vw), 179(vw), 192(vw), 209(w), <u>237</u>(mw), 287(vw), 305(w), 318(w), 353(w), 380(w), 390(m), 400(w), 430(w), 455(w), 468(w), 483(w), 538(w), 577(m), 602(w), 622(m), 665(w), 680(w), 697(w), 728(w), 757(w), 926(s), 945(w), 965(w), 983(w), 1016(s), 1082(s), 1137(s)
Olivine (yellow transparent equant grains)	Fa_{33}	12018,26	283(w), 352(m), <u>398</u>(m), 493(s), 510(s), 590(m), 832(w), 880(s), 938(w), 972(s), 1065(sh)
Ilmenite (black lustrous grains)		10085,46	285(m), 320(w), 365(w), 440(mw), 522(s), 675(mw)

* Analytical frequencies used to derive chemical compositions are underlined.
s = strong; m = medium; w = weak; sh = shoulder.

their distinctive infrared spectra. Table 2 lists absorption band frequencies for some separates of each of these minerals, and Fig. 1 gives examples of spectra for the predominant lunar silicates. Absorption bands appearing in the infrared spectrum of a silicate structure are commonly assigned to vibrations of SiO_4 tetrahedra, to octahedra or tetrahedra of substituted cations (metal-oxygen bonds), and to lattice vibrations. The frequencies of all these absorption bands can be affected by changes in cation substitution through changes in bond distances and bond force constants. This sensitivity to small changes in molecular structure allowed us to determine some specific cation substitutions for each of the predominant lunar silicate minerals. Correlation curves based on composition dependent frequency shifts were derived from a number of synthetic and terrestrial standards for each of the silicate classes. These were then applied in a determination of specific chemical composition for individual lunar separates after the silicate molecular structure was established from overall spectral features. The absorption bands selected for use in the cation determinative curves were all from the low-frequency region 400–180 cm^{-1} and are shown starred in Fig. 1. Although these varied from medium to weak in absorption intensities, they were selected as the most suitable compromise between maximum frequency shifts with compositional changes and best fits of the data to the determinative curves.

Pyroxenes. Pyroxenes were isolated as colored transparent angular fragments from the rocks and dusts listed in Table 3. All spectra were similar to that shown in Fig. 1, curve (b) in band shapes and relative intensities. These pyroxene spectra more closely matched those of a series of seven synthetic pigeonites that we studied in the composition range $Wo_{10}En_{75}Fs_{15}$ to $Wo_{10}En_{30}Fs_{60}$ (Tem-Pres Research Division,

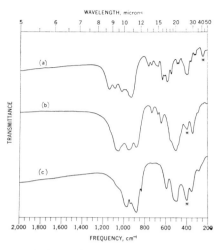

Fig. 1. Infrared spectra of the predominant silicate minerals in lunar samples (a) plagioclase (An_{100}) from dust 10085,46, single chalky-white grain, 2800×1400 μm; (b) pyroxene (Fs_{24}) phenocrysts from rock 12021,24, 15 yellow transparent angular fragments, avg. 600 μm; (c) olivine (Fa_{32}) from rock 12018,26, 134 yellow transparent equant grains, avg. 170 μm

The Carborundum Co., State College, Pennsylvania) than those of any other pyroxenes with which we compared them. Spectra of terrestrial calcic clinopyroxenes, e.g., a series falling in composition along the diopside-hedenbergite tie line (samples from G. M. Bancroft), did not compare with spectra of the lunar pyroxenes that we isolated. Similarly, spectra of a series of terrestrial aluminous augites and salites from metamorphic rocks with extensive and varied cation substitutions (Wo_{43} to Wo_{56}, with 2.2 to 13.5% Al_2O_3, A. T. RAO, Andhra University, Waltair, India) (RAO, 1969) all differed markedly from those of our isolated lunar pyroxenes. A volcanic augite from Kakanui, New Zealand ($Wo_{30}En_{45}Fs_{10}$) also with substantial cation substitution and showing a more band-broadened spectrum than the metamorphic clinopyroxenes because of its disorder (HAFNER and VIRGO 1970) did not match well with our isolated lunar pyroxenes. Spectra of the calcium poor clinopyroxenes, clinoenstatite (synthetic and natural) and clinohypersthene (synthetic), as well as spectra of orthopyroxenes from enstatite to hypersthene ($Fs_{14.5}$ to $Fs_{85.9}$, samples studied by BANCROFT *et al.*, 1967; and others by LYON, 1963), did not compare well with those of our isolated lunar pyroxenes.

   The chemical inhomogeneity of pyroxenes due to compositional zoning has been described by BENCE *et al.* (1970) for rock 12021 and by other workers for lunar pyroxenes. Since it was necessary to combine several grains for a single infrared analysis, the spectra represent the predominant composition for the pyroxene separates and thus yield modal information. Although other workers have reported the identification of pyroxene in the augite range, we tentatively conclude that the predominant compositions of our pyroxene separates lie in the pigeonite to subcalcic augite range, based on the above described spectral comparisons. Studies on synthetic

Table 3. Analyses of mineral separates from lunar rocks and dusts.

Mineral isolated	Sample*	Description of mineral separates†	Number of analyses	Analytical frequencies cm^{-1}	Derived composition
Pyroxene	12018,26	Transparent light to dark amber angular fragments, 185 μm; white granular opaque grain, composited with olivine, 350 μm	3	392–385	Fs_{23}–Fs_{30}
	12021,24	Yellow to dark amber angular fragments, 650–250 μm	9	393–380	Fs_{22}–Fs_{37}
	12070,24	Dark yellow to reddish-amber angular fragments, 350–200 μm	7	394–382	Fs_{21}–Fs_{34}
	10085,46	Medium to dark amber angular fragments, 300–150 μm	4	388–385	Fs_{27}–Fs_{30}
Plagioclase	12018,26	Colorless transparent angular fragments, 160 μm	1	228	An_{85}
	12021,24	Colorless transparent angular fragments, 450–320 μm	5	230–235	An_{89}–An_{98}
	12070,24	Colorless transparent angular fragments, 520–200 μm; light to dark grey opaque grains, 580–200 μm	7	228–232	An_{85}–An_{92}
	10085,46	Chalky-white grain 2800 $\times$ 1400 μm; dark grey granular grain, 1500 μm; colorless transparent angular fragments, 250–175 μm	6	226–237	An_{81}–An_{100}
Olivine	12018,26	Yellow transparent equant grains, 400–170 μm; white granular opaque grains, composited with pyroxene, 350 μm	3	398	Fa_{33}
	12020,26	Yellow transparent equant grains, 570 μm	1	400	Fa_{29}
	12070,24	Light yellow transparent angular fragments, 470–270 μm; brown translucent grain, 1000 $\times$ 1000 μm brown opaque blocky grain, 1000 $\times$ 1000 μm	4	401–397	Fa_{28}–Fa_{34}
	10085,46	Grey fine-grained blocky particle, composited with pyroxene, 1500 $\times$ 1000 μm	1	400	Fa_{29}
Ilmenite	12018,26	Black lustrous grains, 100 μm	1		
	12021,24	Black lustrous grains, 100 μm	1		
	10085,46	Black lustrous grains in aggregates with colorless transparent plagioclase and amber transparent pyroxene, 100 μm	1		

* Lunar samples were classified as follows: rock 12018,26, olivine dolerite; rock 12020,26, olivine basalt; rock 12021,24, pigeonite dolerite to porphyritic gabbro with variolitic texture; dust 12001,60, selected sample; dust 12057,57, documented sample; dust 12070,24, contingency sample; dust 10085,46, fines to coarse fines.

† Sizes are given as weighted average particle sizes for the analyses.

augites and sub-calcic augites are in progress to determine the effects on infrared spectra of systematic changes in calcium content and aluminum substitution into these structures. An example of the closely matching general features of lunar pyroxene spectra compared with those of the synthetic pigeonites is shown in Fig. 2. Curve (c) is the spectrum of dark amber pyroxene fragments isolated from rock 12018,26, and these appear to be intermediate in band shapes, resolution, and frequencies between the synthetic pigeonites with iron contents of Fs_{30} and Fs_{38}. We observed that as the color of lunar pyroxenes progressively deepened from yellow to dark amber, nearly all absorption bands systematically shifted to lower frequencies (Table 2). Studies on the closely matching synthetic pigeonites showed that these frequency decreases are

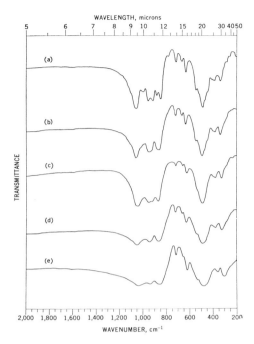

Fig. 2. Spectral comparison of lunar pyroxene with synthetic pigeonites (a) synthetic $Wo_{10}En_{75}Fs_{15}$; (b) synthetic $Wo_{10}En_{60}Fs_{30}$; (c) lunar pyroxene (Fs_{32}) from rock 12018, 26, 104 dark amber fragments; (d) synthetic $Wo_{10}En_{52}Fs_{38}$; (e) synthetic $Wo_{10}En_{30}Fs_{60}$.

related to increasing Fe^{2+} content. Decrease in bond energies in this silicate structure results from replacement of smaller (0.72 Å) and lighter Mg^{2+} ions by larger (0.77 Å) and heavier Fe^{2+} ions. We selected the absorption band shifting from 400 to 368 cm^{-1} in synthetic pigeonites Fs_{15} to Fs_{60} (Fig. 2), for use in a cation determinative curve, shown in Fig. 3. The observed range of 394 to 380 cm^{-1} for this absorption band in lunar pyroxene separates indicated Fe^{2+} compositions in the range Fs_{21} to Fs_{37}. Specific frequency ranges observed for various colors of lunar pyroxenes are shown in Fig. 3 to demonstrate the correlation of color with Fe^{2+} content. We studied this same analytical absorption band in terrestrial orthopyroxene spectra and observed the same correlation between Fe^{2+} content and frequencies. The decrease in Ca^{2+} content and the change to orthorhombic crystal system shifted frequencies for the analytical absorption band very little from those of the monoclinic pyroxenes.

In addition to frequency shifts with increasing Fe^{2+} content, we observed in spectra of both lunar clinopyroxenes and the series of synthetic clinopyroxenes (Fig. 2) that there is progressive band broadening and a systematic decrease of absorption intensities with increasing Fe^{2+} content (yellow to amber for lunar pyroxenes). These effects could be due to the increasing nonequivalence of neighboring chains in the pyroxene structure, known to occur with increasing Fe^{2+} substitution (MORIMOTO, 1960). Mössbauer data for terrestrial orthopyroxenes (BANCROFT *et al.*, 1967) and lunar clinopyroxenes (HAFNER and VIRGO, 1970) show that increasing Fe^{2+} content is accompanied by a more disordered Fe^{2+} distribution over the nonequivalent octa-hedrally coordinated sites, M1 and M2. If the observed broadening and intensity decrease of infrared absorption bands in clinopyroxenes is related to this same phenomenon, then the well-resolved spectrum obtained for the iron-poor yellow lunar pyroxenes (e.g., curve b, Fig. 1) suggests a relatively ordered structure. From their Mössbauer studies, HAFNER and VIRGO (1970) have interpreted cation ordering in lunar pyroxenes as an indication of slow cooling at relatively low equilibrium tempera-tures. The synthetic clinopyroxenes, soaked up to 288 hours at 850°C and slow-cooled, are expected to have achieved equilibrated distributions of cations over the non-equivalent sites. Further support from infrared data for ordering in lunar pyroxenes was obtained by comparison with a terrestrial volcanic pigeonite

$$Ca_{0.29}Na_{0.02}K_{0.01}Mn_{0.02}Mg_{0.92}Fe^{2+}_{0.74}Al_{0.06}Si_{1.94}O_6$$

(Hakone Volcano, Japan, No. 101824) determined to be appreciably ordered from Mössbauer studies (BANCROFT and BURNS, 1967). The spectrum of this pigeonite compares with lunar pyroxene spectra in frequencies, overall band shapes, and resolu-tion nearly as well as the synthetic pigeonites. All other terrestrial pigeonite samples that we studied, presumably with more rapid cooling histories, exhibited band-broadened spectra, suggesting considerable disorder.

Thus the infrared comparisons that we made indicate that the lunar pyroxenes isolated were monoclinic, calcium-poor, ranging in iron composition Fs_{21} to Fs_{37},

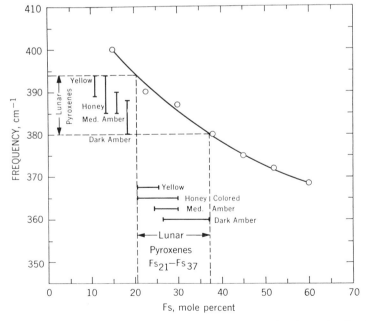

Fig. 3. Determinative curve for synthetic pyroxenes, for the absorption band shifting 400 to 368 cm^{-1} for $Wo_{10}En_{75}Fs_{15}$ to $Wo_{10}En_{30}Fs_{60}$.

probably more ordered for the iron-poor pyroxenes, and similar in structure for samples from both rocks and dusts.

Plagioclase Feldspars. Plagioclase feldspars isolated from the rocks and dusts listed in Table 3 varied typically from colorless transparent fragments to white-grey opaque grains. The spectrum of a single chalky-white, black-flecked anorthite grain (An_{100}) isolated from the Apollo 11 lunar dust is shown in Fig. 1, curve (a). Its unusually well-resolved spectral features in both the mid- and far-infrared regions (to 30 cm^{-1}) indicated that this particular lunar plagioclase sample was more highly ordered (in Si/Al distribution) than most of the lunar plagioclase, or that in any of the terrestrial anorthosites or anorthite specimens that we studied. Previous systematic studies of the variation of infrared spectra of plagioclase feldspar with composition (Thompson, 1967; Angino, 1969) and data from this laboratory on analyzed terrestrial plagioclase samples indicated that the absorption band shown starred in curve (a), Fig. 1, was a good choice for a determinative curve. This band shifts linearly from 185 to 235 cm^{-1} (An_5 to An_{95}) with increasing Ca^{2+} content of the plagioclase ($v_{cm}^{-1} = 180.35 + 0.5637An_{mole\ \%}$). The shift does not reflect simple mass or ionic radii effects and may be related to increasing substitution of Al^{3+} in the more calcic plagioclase (Angino, 1969). Observed frequencies of 226 to 237 cm^{-1} for the lunar plagioclase separates were applied to the determinative curve, and these indicated bytownite to pure anorthite, An_{81} to An_{100}.

We compared spectra of several anorthositic and gabbroic anorthositic grains isolated from dust 12070,24 with those from three types of terrestrial anorthosites; massive Adirondack (An_{44} to An_{56}, Y. W. Isachsen, University of the State of New York); stratiform Bushveld (An_{95}, J. Ferguson, University of Witwatersrand, South Africa); and Group III anorthosites from West Greenland (An_{47} to An_{96}, B. F. Windley. University of Leicester England). The latter have been suggested to be the first terrestrial rocks to have any marked affinities with those on the moon (Windley, 1970). The Bushveld (An_{95}) and some Greenland (An_{92} to An_{96}) anorthosite spectra compared favorably with lunar anorthosite spectra because of their high anorthite contents. We also obtained a good match of lunar anorthositic spectra with those of two eucrites, Sioux County, Nebraska and Pasamonte, New Mexico, that were both predominant in calcic plagioclase (An_{95}).

Olivines. Olivines with varied morphologies were isolated from the lunar rocks and dusts listed in Table 3. Their infrared spectra, e.g. curve (c), Fig. 1, showed them to be in the forsterite-fayalite olivine series. Spectra were similar for separates from both rocks and dusts. As with pyroxenes, frequencies of olivine bands are composition dependent (Duke and Stephens, 1964; Burns and Huggins, 1970) and decrease as Fe^{2+} content increases. We selected for use in a cation determinative curve the low-frequency absorption band shown starred in curve (c), Fig. 1. The correlation curve ($v_{cm}^{-1} = 419 - 0.653Fa_{mole\ \%}$) was determined from a series of nine synthetic olivines in the Fe-Mg series (Tem-Pres Research), in which this absorption band shifted linearly from 418 to 356 cm^{-1} for Fa_0 to Fa_{100}. For lunar olivine separates this band ranged from 401 to 397 cm^{-1}, indicating a narrow composition range of Fa_{28} to Fa_{34} from the correlation plot. We compared lunar olivine compositions with those of a series of 32 chondrites that we examined, which varied from Fa_{14} to Fa_{29} (410 to 400 cm^{-1}) in agreement with values reported for most chondrites by MASON (1962). Only six other chondrites that we studied, with frequencies in the range 395 to 387 cm^{-1}, had higher fayalite compositions of Fa$_{37}$ to Fa$_{49}$. Thus the fayalite content of the lunar olivines that we isolated comprised a narrower composition range than that for the meteorites that we studied.

Ilmenite. Ilmenite occurred as fine grains in rocks and dusts (Table 3) and was typically isolated in fragments averaging 100 μm, still associated with trace amounts of unidentified silicates (1093, 1065, and 1050 cm^{-1}). A typical isolation required 324 grains to obtain enough for a mid-infrared spectrum. We have observed significant variations in terrestrial ilmenite spectra for both natural and synthetically prepared samples, with frequency shifts up to 30 cm^{-1}. These infrared differences may be related to subtle structure variations, and further characterization of the lunar ilmenite structure beyond a straightforward identification may be possible from a thorough study of these effects.

Determination of bulk composition of composite lunar samples

We obtained infrared spectra of the lunar dust composite samples listed in Table 4 for comparison with a variety of samples. To facilitate these comparisons we used the stretching vibrations of Si—O bonds in silicate SiO$_4$ tetrahedra, appearing in the

Table 4. Infrared absorption data for composite samples from lunar dusts

	Frequency, cm^{-1}	
Sample description	Si—O stretching region	Si—O—Si bending region
Dust 12001,60		
+200 −100 Mesh sieved fraction	985	487
+325 −200 Mesh sieved fraction	990	485
+400 −325 Mesh sieved fraction	1000	485
−400 Mesh residue	980	470
Dust 12057,57		
+200 −100 Mesh sieved fraction	980	480
+325 −200 Mesh sieved fraction	980	480
+400 −325 Mesh sieved fraction	990	485
−400 Mesh residue	1000	485
Dust 12070,24		
8 Individual basaltic grains		
from +100 mesh sieved fraction	1005–990	495–480
+200 −100 Mesh sieved fraction	985	472
+325 −200 Mesh sieved fraction	985	475
+400 −325 Mesh sieved fraction	987	473
−400 Mesh residue	990	467
Dust 10085,46		
20 Individual basaltic grains		
from +100 mesh sieved fraction	1005–975	500–475
−100 Mesh sieved fraction	990	492
Lunar glass		
Dust 10085,46, light green	1000	467
Dust 10085,46, reddish-brown	972	485
Dust 10085,46, dark brown	970	478
Dust 10085,46, dark brown	965	480
Dust 10085,46, dark brown, vesicular	960	480
Dust 12070,24, light green	990	468
Dust 12070,24, reddish-brown beads	970	480

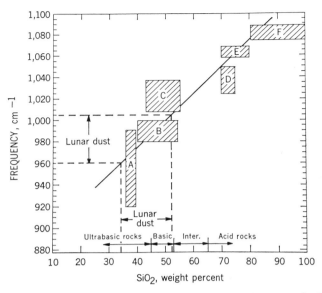

Fig. 4. Variation of the Si—O stretching frequency with SiO₂ content for igneous silicate
rocks and glasses.

A = 42 chondritic meteorites, using %SiO₂ range from UREY (1953).
B = (a) 15 basalts, mostly tholeiitic: ocean basalt from the East Pacific Rise (Amph 3M,
S. R. Hart, Carnegie Institution of Washington); Puerto Rico Trench basalt (No. 5,
S. R. Hart); and other Atlantic and Pacific ocean basalts (W. W. Schneider and P. B.
Helms, Scripps Institution of Oceanography). (b) 8 synthetic pseudo-lunar glasses
(54–46 %SiO₂, C. H. Greene, Alfred University). (c) Serpentine (TC-5, D. E. Fogelson,
USBM, Minneapolis, Minnesota). (d) 2 eucrites: Sioux County, Nebraska, and Pasa-
monte, New Mexico (C. B. Moore, Arizona State University), using %SiO₂ range from
UREY (1953). (e) Laminated gabbro from the Romanche Trench (NMNH 110753, V. T.
Bowen, Smithsonian Institution) and Fe–Ti rich gabbro (No. 27, Adirondacks, R. B.
                    Hargraves, Princeton University).
C = 23 basalts, mostly alkali-rich; 7 from the Afar Triangle, Ethiopia (F. Barberi,
University of Pisa, Italy); Greenland Disco basalt (No. 53479, W. G. Melson, Smithson-
ian Institution); Puerto Rico Trench basalts (No. 14, S. R. Hart), (F. B. Wooding, Woods
Hole Oceanographic Institution); and other Atlantic and Pacific ocean basalts (AD4-1,
                    S. R. Hart), (W. W. Schneider), (P. B. Helms).
D = 9 acid rocks and glasses, including 7 simulated lunar rocks of rhyolites, dacite,
                    granodiorite, pumice, tuff, and obsidian (D. E. Fogelson).
E = 24 tektites, including Javanites, Australites, Indochinites, Moldavites, and
                    Philippinites, using %SiO₂ range from MASON (1962).
F = 7 high-silica glasses, commercial products.

frequency range of 1200 to 800 cm⁻¹, for classification of bulk compositions. These
intense vibrations are highly sensitive to changes in bond force constants, and LYON
(1965) has shown that their frequency shifts can be correlated with SiO₂ content.
Figure 4 shows this general correlation applied to a wide range of bulk compositions,
and the plot includes 133 samples of terrestrial rocks, synthetic glasses, tektites, and
meteorites. The correlation was found to apply to samples that are highly crystalline,

glassy, fine-grained, or coarse-grained. Samples with high SiO_2 content, such as acid rocks, tektites, and synthetic glasses (blocks D, E, F), are found at higher frequencies, while low SiO_2 content samples such as meteorites, basalts, gabbros, and other basic rocks (blocks A, B, C) appear at considerably lower frequencies. This shift of the Si—O stretching vibration reflects both chemical and mineralogical compositions, showing a decrease in frequency as the Si/O mole ratio decreases (0.5 to 0.25) in going from tectosilicates to nesosilicates (LAUNER, 1952; SAKSENA, 1959). The frequency decrease reflects a decrease in the "degree of polymerization" of SiO_4 tetrahedra in a silicate lattice and consequently, decreasing bond energies. For example, in Fig. 1 the center of gravity for the Si—O stretching vibration is seen to decrease from 1010 cm^{-1} to 980 cm^{-1} to 930 cm^{-1} for plagioclase (tectosilicate), pyroxene (inosilicate), and olivine (nesosilicate), respectively. Substitution of cations into a silicate structure, whether replacement for Si in the tetrahedra or substitution into octahedral sites, results in lower Si—O stretching frequencies (MILKEY, 1960; STUBICAN and ROY, 1961). Heavy cations such as Fe^{2+} and Ti found in basalts and gabbros decrease the Si—O stretching frequency more than lighter cations.

Dust grains and sieved fractions. We used this frequency-silica content correlation and comparisons of other overall spectral features to classify compositions of composite samples from the lunar dusts. Si—O stretching frequencies for individual dust grains and dust sieved fractions ranged from 1005 to 975 cm^{-1} (Table 4), and these indicated basic to ultrabasic compositions of 52 to 40% SiO_2, from the correlation plot in Fig. 4. Although gross spectral features classify the lunar samples as basaltic, in no case was an exact comparison obtained with spectra of any of the large number of terrestrial rock types with which we compared them. The distinctive lunar basaltic spectrum, as shown, for example, in Fig. 5, curve (a), compared more favorably in general features with that of some ocean tholeiitic basalts (curve b) than with most samples. It was markedly different from that of olivine-rich chondrites (curve c), tektites (curve d), and plagioclase-rich eucrites.

Isolated basaltic grains ranged from microgranular grey aggregates to dark grey-brown blocky fragments and their spectra were similar to each other and also to those of the dust sieved fractions as shown for example in Fig. 6, curves (b) and (c), for dust 12070,24. These unique basalt-type spectra appear to be dominated by pyroxenes, as shown in Fig. 6 by comparison with the spectrum of a pyroxene (Fs_{27}) (curve a) isolated from the same dust. The slight shifts of the medium intensity absorption band in the Si—O—Si bending region to lower frequencies (500 to 467 cm^{-1}) in some dust sieved fractions (see Table 4) may be associated with increasing iron content, as observed for the lunar pyroxene separates.

Glasses. Glasses were isolated from the lunar dusts listed in Table 4 as beads (700 to 200 μm) and irregular fragments. Si—O stretching frequencies in the range 1000 to 960 cm^{-1} similarly classified these as basic compositions from the correlation plot in Fig. 4 (50 to 34% SiO_2). For Apollo 11 glasses, a trend of increasing depth of color with increasing iron substitution and decreasing SiO_2 content has been reported (ANDERSON *et al.*, 1970; VON ENGLEHARDT *et al.*, 1970), and the data in Table 4 show this correlation. For the glasses isolated from the Apollo 11 sample, the decrease of the Si—O stretching frequency from 1000 cm^{-1} for light-colored glass to 960 cm^{-1} for

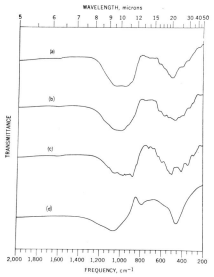

Fig. 5. Infrared spectra of (a) lunar basalt from dust 12070,24, 2 light grey micro-granular aggregates, 1000 and 600 μm; (b) ocean tholeiitic basalt from the Pacific Antarctic Ridge (No. 21–7–102B, T. E. Simkin, Smithsonian Institution); (c) chondritic meteorite, Plainview, Texas; (d) Javanite, Sangiran Dome, Central Java.

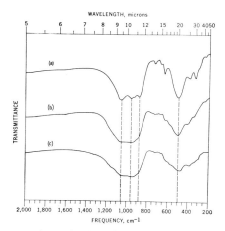

Fig. 6. Infrared spectra of samples from lunar dust 12070,24. (a) pyroxene (Fs_{27}), 28 transparent medium amber angular fragments, avg. 300 μm; (b) microgranular aggregates, 11 light-grey aggregates, 1700 to 400 μm; (c) dust sieved fraction, +325, −200 mesh.

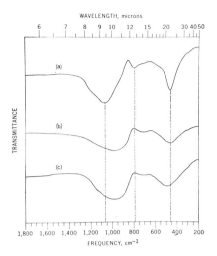

Fig. 7. Spectral comparison of lunar glass with a tektite and a pseudo-lunar glass. (a) Indochinite, Khon Kaen, Thailand; (b) pseudo-lunar glass No. M-17 (C. H. Greene, 45% SiO_2, 15% CaO, 14% Al_2O_3, 10.3% FeO, 0.8% Fe_2O_3, 8% TiO_2, 4% MgO, and 0.08% Na_2O); (c) lunar glass from dust 10085,46, single dark brown vesicular fragment, 1500 × 1000 μm.

dark-colored glass indicates decreasing SiO_2 content from Fig. 4 (50 to 34%) and is presumably associated with increasing iron substitution. This color correlation also applied in the case of the two Apollo 12 glasses, even though fine structure appearing on the major silicate absorption bands suggested some devitrification. The Si—O stretching frequency for light-colored glass (1000 to 990 cm^{-1}) approximates that of lunar plagioclase (1010 cm^{-1}), while that of the dark-colored glass (972 to 960 cm^{-1}) approximates that of lunar pyroxenes (980 to 970 cm^{-1}), supporting the belief that these two types of glass are formed from these minerals. We compared spectra of the lunar glasses with those of 24 tektites and a suite of eight pseudo-lunar artificial glasses synthesized by Greene (Fig. 4) to approximate the surface composition of Mare Tranquillitatis (TURKEVICH et al., 1969). The examples in Fig. 7 demonstrate the better match in frequencies and band shapes of lunar glass spectra with those of the pseudo-lunar glasses than with those of tektites. The higher frequencies for the Si—O stretching vibrations in tektites (1070 to 1055 cm^{-1}) reflect their higher SiO_2 content (70 to 80%), as compared with that of the pseudo-lunar glasses that we studied (54 to 46% SiO_2, 1000 to 980 cm^{-1}) (See Fig. 4). Frequencies for the three predominant absorption bands in the 24 tektites that we studied did not vary more than 15 cm^{-1}.

Acknowledgements—We thank Edward E. Childers and Arthur L. Hiser of this laboratory for isolating and obtaining spectra on terrestrial minerals and rocks, and John J. Renton, West Virginia University, for authenticating structures of terrestrial minerals by X-ray diffraction. We are grateful to R. G. Burns, Massachusetts Institute of Technology, G. M. Bancroft, University of Western Ontario,

Canada, J. S. White, Smithsonian Institution, and J. D. Stevens, Kennecott Copper Corporation, for supplying pyroxene samples; to F. C. Dehn, PPG Industries, for terrestrial ilmenite samples; to W. R. Riedel, Scripps Institution of Oceanography, for ocean basalts; to C. H. R. von Koenigswald, Forschungs-Institut Senckenberg, Germany, D. R. Chapman, NASA, Ames Research Center, and C. B. Moore, Arizona State University, for tektite samples; to E. J. Olsen, Chicago Natural History Museum, for rare meteorite samples; and to many others (as noted in the text) who generously supplied analyzed samples for our comparisons. This research was sponsored by NASA under Contract no. T–88614.

REFERENCES

ANDERSON O. L., SCHOLZ C., SOGA N., WARREN N., and SCHREIBER E. (1970) Elastic properties of a micro-breccia, igneous rock and lunar fines from Apollo 11 mission. *Proc. Apollo 11 Lunar Sci. Conf., Geochim. Cosmochim. Acta* Suppl. 1, Vol. 3, pp. 1959–1973. Pergamon.

ANGINO E. E. (1969) Far infrared absorption spectra of plagioclase feldspars. *Kansas Geol. Surv. Bull.* Pt. 1, **194,** 9–11.

BANCROFT G. M. and BURNS R. G. (1967) Distribution of iron cations in a volcanic pigeonite by Mössbauer spectroscopy. *Earth Planet. Sci. Lett.* **3,** 125–127.

BANCROFT G. M., BURNS R. G., and HOWIE R. A. (1967) Determination of the cation distribution in the orthopyroxene series by the Mössbauer effect. *Nature* **213,** 1221–1223.

BENCE A. E., PAPIKE J. J., and PREWITT C. T. (1970) Apollo 12 clinopyroxenes: Chemical trends. *Earth Planet. Sci. Lett.* **8,** 393–399.

BURNS R. G. and HUGGINS F. E. (1970) Cation determinative curves and evidence of ordering in Mg—Fe—Mn olivines from vibrational spectra. Pres. Nov. 3, 1970, Geol. Soc. Amer. Symp. on Crystal Chem. of Silicates, Milwaukee, Wisconsin, Amer. Mineral. In preparation

DUKE D. A. and STEPHENS J. D. (1964) Infrared investigation of the olivine group minerals. *Amer. Mineral.* **49,** 1388–1406.

HAFNER S. S. and VIRGO D. (1970) Temperature-dependent cation distribution in lunar and terrestrial pyroxenes. *Proc. Apollo 11 Lunar Sci. Conf., Geochim. Cosmochim. Acta* Suppl. 1, Vol. 3, pp. 2183–2198. Pergamon.

LAUNER P. J. (1952) Regularities in the infrared absorption spectra of silicate minerals. *Amer. Mineral.* **37,** 764–784.

LYON R. J. P. (1963) Evaluation of infrared spectrophotometry for compositional analysis of lunar and planetary soils. Stanford Res. Inst., Final Rept. under contract NASr-49(04), Pub. by NASA as Tech. Note D-1871.

LYON R. J. P. (1965) Analysis of rocks by spectral infrared emission (8 to 25 microns). *Econ. Geol.* **60,** 715–736.

MASON B. (1962) *Meteorites.* John Wiley, pp. 62, 201.

MILKEY R. G. (1960) Infrared spectra of some tectosilicates. *Amer. Mineral.* **45,** 990–1007.

MORIMOTO N., APPLEMAN D. E., and EVANS H. T., JR. (1960) The crystal structures of clinoenstatite and pigeonite. *Z. Kristallogr.* **114,** 120–147.

RAO K. S. R., RAO A. T., and SRIRAMADAS A. (1969) Porphyritic plagioclase-hornblende-pyroxene granulite from charnockitic rocks of Chipurupalli, Visakhapatnam District, Andhra Pradesh, South India. *Mineral. Mag.* **37,** 497–503.

SAKSENA B. D. (1959) Classification of silicate structures from infrared studies. Proc. of Symp. on Raman & Infrared Spectroscopy, Nainital, 93–102.

STUBICAN B. and ROY R. (1961) Infrared spectra of layer-structure silicates. *J. Amer. Ceram. Soc.* **44,** 625–627.

THOMPSON C. S. (1967) Determination of the composition of plagioclase feldspars by means of infrared spectroscopy. Ph.D. thesis, University of Utah.

TURKEVICH E. J., FRANZGROTE J., and PATTERSON J. H. (1969) Chemical composition of the lunar surface in mare tranquillitatis. *Science* **165,** 277–279.

UREY H. C. and CRAIG H. (1953) The composition of the stone meteorites and the origin of the meteorites. *Geochim. Cosmochim. Acta* **4,** 36–82.

VIRGO D., HAFNER S. S., and WARBURTON D. (1971) Cation distribution studies in clinopyroxenes, olivines and feldspars using Mössbauer spectroscopy of ^{57}Fe. Second Lunar Science Conference (unpublished proceedings).

VON ENGELHARDT W., ARNDT J., MÜLLER W. F., and STÖFFLER D. (1970) Shock metamorphism in lunar samples. *Science* **167,** 669–670.

WINDLEY B. F. (1970) Anorthosites in the early crust of the earth and on the moon. *Nature* **226,** 333–335.

Proceedings of the Second Lunar Science Conference, Vol. 3, pp. 2153–2164
The M.I.T. Press, 1971.

Microchemical, microphysical, and adhesive properties of lunar material, II

J. J. GROSSMAN, J. A. RYAN, N. R. MUKHERJEE, and M. W. WEGNER

Space Sciences Department, McDonnell Douglas Astronautics Company—West
Huntington Beach, California 92647

(Received 22 February 1971; accepted in revised form 5 April 1971)

Abstract—A vacuum pumped Apollo lunar sample, 12001,118, was transported successfully from the Lunar Receiving Laboratory to our laboratories for gas exposure experiments. Disruptions in oxygen and in water vapor have been observed. Etching techniques of the defect structures have been further developed and both dislocations and tracks have been found in close association. Tracks have been found in olivine, pyroxenes, and possibly in ilmenite. Track gradients show that the (E, S, T) face of 12052,43 has been exposed to the solar wind. The net adhesion force of a UHV fractured specimen of the rock fractured was too small to be measured. Measurement of electrostatic charge distributions indicated the specimen had a net positive average charge of about 10^6 charges cm^{-2}. Exposure to dry N_2 did not effect the charge density. Both air and ultraviolet discharged the fractured faces. The unusual electrostatic adhesion processes are expected for lunar materials in the absence of a tenuous lunar atmosphere.

INTRODUCTION

THE OBJECTIVES OF THIS program are to obtain an improved understanding of solid state phenomena and to investigate the nature and relative effectiveness of mechanisms which may produce agglomeration and disruption of lunar material. In this paper we summarize results of three classes of experiments on Apollo 12 lunar material, namely, the effects of exposure of a vacuum sample to various gases, the microanalysis of particles and particle interfaces, and adhesion measurements in an ultrahigh vacuum system.

EXPERIMENTAL METHODS

Gas exposure

Lunar soil sample 12001,118 was kept under vacuum continuously from its collection on the lunar surface in November 1969 to its delivery to our laboratories where portions were placed in reaction tubes. The sequence involved sealing the sample in an aluminum container (performed in Chamber F-201 of the Lunar Receiving Laboratory) and storage until November 1970 when the container was opened at ultrahigh vacuum in F-201 and subsequently placed in a specially designed (by us) steel matchbox sample container. This was maintained at UHV by ion pumping in a NASA shipping container. The sealed shipping container, with a pressure in the mid 10^{-8} torr range, was transported to our laboratory, connected to our ultrahigh vacuum micromanipulator and sample transfer system and evacuated to the low 10^{-9} torr range. The shipping/storage chamber was then opened, the matchbox moved into the micromanipulator chamber, and the sample transferred to the gas reaction tubes (GROSSMAN *et al.* 1970b). After closing the matchbox, it was returned to the vacuum pumped NASA Storage Chamber which was also closed, Fig. 1. All gas reaction tubes were pinch-off vacuum-sealed, our chamber opened, removed and connected to the UHV gas manifold, and the gas reaction experiments started.

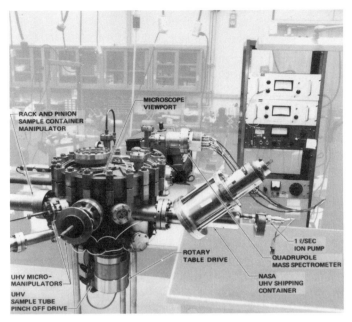

Fig. 1. NASA vacuum pumped shipping and storage container connected to MDAC-West UHV micromanipulator chamber. The container is closed, storing the remainder of lunar sample 12001,118 after vacuum transferring a part of the sample to gas exposure cells.

Microchemical and microphysical analysis and adhesion studies

The procedures were essentially the same as for Apollo 11 samples (GROSSMAN *et al.*, 1970b).

EXPERIMENTAL RESULTS

Gas exposure

To date, one sample tube containing two 2.5 mm particles, one 1 mm bonded pair, and some fines smaller than 200 μm have been exposed at 300°K successively to (a) pure O_2 (impurity < 5 ppm) at 0.9 torr; (b) pure H_2O at 1 and 12 torr; and (c) a mixture of O_2 + 2.8 % H_2O at 760 torr. After eight hours of exposure under condition (a), no perceptible change was observed. However, after eighteen hours, the bonded pair was found to be disrupted. Under each of conditions (b) and (c) several small crystals were found to be disrupted from the two 2.5 mm particles after several hours of exposure. The experiment is being continued for exposure of original UHV samples to dry N_2, organic acid, and organic base.

The disrupted surfaces of the lunar soil particles (Apollo 11 sample number 10084,93; GROSSMAN *et al.*, 1970b, Fig. 2(b) and (d) were examined both with a light microscope and a scanning electron microscope (SEM). A number of stereophotographs were taken with the SEM while viewing the samples normal to these disrupted areas. For illustrative purposes, one photograph of a stereo pair at 90° orientation of the former and one at 30° orientation of the latter are shown in Figs. 2 and 3. A detailed examination reveals the following: (a) Each disrupted area is sufficiently porous permitting interparticle diffusion of a gas; (b) the shape of individual small crystals and protrusions on larger plate-like crystals suggests that some deposition and agglomeration have taken place from a liquid spray; (c) no large scale chemical reaction is apparent; (d) signatures of material fracture which are expected to remain after disruption of strong bonding are not observed at these magnifications; and (e) the presence of small crystals is probably due to comminution and rebonding processes.

High magnification photographs, such as the one in Fig. 3, show another interesting phenomenon. Droplets of small splash particles, which were formed presumably by liquid spray, appear to be comparatively less abundant in the area where the disrupted crystal contacted the larger particle. Further evidence of such deposits is found at low magnification on the surface of the disrupted crystal (shaped like a rice grain). Although this evidence suggests that some of the spray deposited material occurred while the gas disrupted crystal was attached to the larger particle, the argument is weakened by the deposition asymmetry which is within 2σ of the expected statistical variance. Prior to gas exposure, the particles were also subjected to considerable mechanical motions, such as lateral motion and roll by magnetically controlled positioners (glass-coated iron wire of about 0.5 mm diameter and 3 mm long), for manipulating the particles in the field of view of the microscope. Some of the

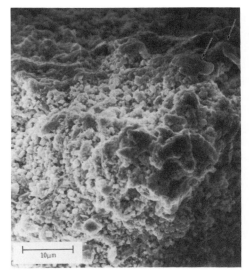

Fig. 2. Scanning electron micrograph (from stereo pair) viewed normal to bonding plane of two disrupted crystals, one right of center over horseshoe area and the second above this area and further to the right. Arrows point to examples of deposits attributed to liquid spray. (See GROSSMAN *et al.* (1970b), Fig. 2a for the disrupted crystals.)

Fig. 3. Scanning Electron Micrograph viewed 30° from normal of plan from which rice grain shaped crystal disrupted (see GROSSMAN *et al.* (1970b), Fig. 2c for the bonded rice grain shaped crystal). The crystal was adhering in an end on position in the cavity in the lower central region of the photomicrograph. Arrows point to examples of deposits attributed to liquid spray.

particles which appeared bonded at first, separated during translation. A few small crystals fell off from larger particles, but the crystals which were bonded in the areas shown in Figs. 2 and 3 remained in position without change of orientation. They disrupted only upon exposure to gas.

Microchemical and microphysical analysis

Two samples from Apollo 12 lunar rock chip 12052,43 were mounted in an epoxy resin. A low vacuum was used during monitoring to better fill the pores, vugs, and vesicles with resin. The samples were then polished, the final critical polishing being done with 0.05 μm Al_2O_3 to obtain a scratch-free, minimum relief surface. Next, they were etched with etchants to accommodate the different etch rates of the lunar mineral species. These etchants, chosen in order of increasing etch-rate are as follows: H_2SiF_6:HCl:citric acid (1:1:1); HBF_4:HCl:citric acid (1:1:1); HF:HCl:H_2O (1:3:8); and HF:HCl:citric acid (1:1:1). Finally, residues which remained after etching, produced in part by the greater reactivity of the polish-damaged surface layer, were removed using suitable chemical complexing agents.

The smaller section of 12052,43, designated 2, was mounted in epoxy and oriented so that the normal to the polished surface was the body diagonal penetrating the NASA orientation cube at the corner intersection of the E, S, and T planes. An adjacent flat chip from the remaining long section, designated 43,1,1 was also mounted in epoxy and polished, exposing the outer surface of the rock. Sample 12052,43,2 was etched with the last three of the series of four etchants after the second polishing. There was no repolishing between the etching steps.

Dislocations can be distinguished from tracks by their crystallographically oriented pit edges as compared to random orientation of elongated-track directions. In addition, dislocation are also found polygonized and associated with twin boundaries. In olivine of sample 12052,43,2 (Fig. 4) the HBF_4:HCl:citric acid produced crystallographically oriented dislocation etch figures as well as numerous tracks whose elongation directions were randomly oriented with respect to the crystallographic axis. The rectangular dislocation etch figures are characteristic of the (100) face (MAURETTE, 1966; YOUNG, 1966; WEGNER and GROSSMAN, 1969). The dislocation density is 7×10^6 cm^{-2} and the track density is 1.8×10^7 cm^{-2} at a depth of approximately 100 μm. Slight etching of plagioclase laths and some glassy areas was observed.

The second etching step, with HF:HCl:H_2O, caused deep etching of olivine and etching of the glassy matrix, plagioclase laths, and skeleton crystals. Plagioclase exhibited dislocation etch pits, some polygonized dislocations, defects at twin boundaries, and tracks. Ilmenite crystals were not etched.

The last etching, HF:HCl:citric acid (1:1:1), produced exsolution striations in pyroxenes and tracks in pyroxenes and, possibly, ilmenite. Tracks in pyroxenes appeared as round dots in the light microscope (Fig. 5). Other tracks were found on the edges of the two pyroxene crystals (Fig. 6) and possibly in ilmenite near its interface with chromite crystals (Fig. 7). The shape, density, and orientation of these tracks, similar in both pyroxenes and ilmenite, are similar to heavy cosmic ray tracks described by FLEISCHER *et al.* (1967, 1971), BARBER *et al.* (1971), and BHANDARI *et al.* (1971). In other lunar pyroxene crystals (Fig. 6) they resemble tracks in terrestrial

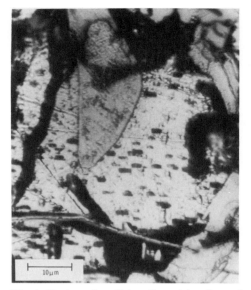

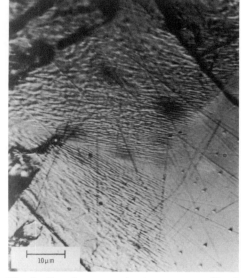

Fig. 4. Rectangular dislocation etch pits and conical tracks in olivine (12052,43,2).

Fig. 5. Cosmic-ray tracks in pyroxene (dots) (12052,43,2).

pyroxenes as described by CONDIE *et al.* (1969). The ilmenite-bordered chromium-spinel crystal has a composition similar to that found in 12013,10, area 24 (LUNATIC ASYLUM 1970). Possible surface radiation damage in pyroxene and ilmenite has also been observed by other investigators (CHRISTIE *et al.*, 1971).

Another pyroxene crystal with an hourglass structure visible in polarized light produced dense etching striations in the outer portion (Fig. 8). EMX analysis showed higher Ca, lower Mg, and slightly less Fe and Si in the etched region as compared to the central portion (CHRISTIE *et al.*, 1971). A pyroxene crystal which penetrated the exterior surface of the rock provided a track density gradient to a depth of 70 μm into the surface. The density gradient corresponded to that reported by FLEISCHER *et al.* (1970), but the net density was a factor of three less. This suggests that the (E, S, T) face of 12052 was exposed directly to space and the solar wind.

The densities of crystallographically aligned dislocation etch figures and randomly oriented conical tracks varied with the crystal orientation (CONDIE *et al.*, 1969; WEGNER and GROSSMAN, 1969; BARBER *et al.*, 1971). In section 12052,43,1,1 the olivine (100) and (010) surfaces (Fig. 9) were first identified by their characteristic etch pit shapes (GROSSMAN *et al.*, 1970a; WEGNER and GROSSMAN, 1969; YOUNG, 1966) and later confirmed by EMX composition analysis (Table 1). Although the chemical composition of the four olivines is quite similar, their reaction with the same etchant (H_2SiF_6 complex) varied; two of the crystals were more heavily etched and acquired a brown discoloration. This could be caused by differences in crystal orientation and those element contents in the mineral which are known to modify the crystal-free energy and thereby change its reactivity. The specific source of the staining has not been identified.

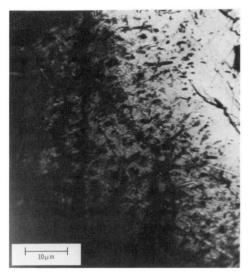

Fig. 6. Tracks in edge of pyroxene crystal (12052,43,2).

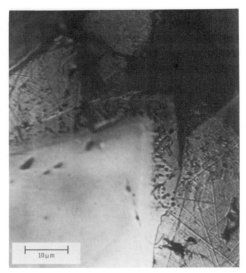

Fig. 7. Possible tracks in ilmenite bordering a chromite crystal (12052,43,2). Similar tracks are found in adjoining pyroxene crystals.

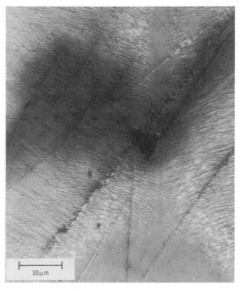

Fig. 8. Etched striations in high calcic phase of pyroxene crystal (12052,43,2).

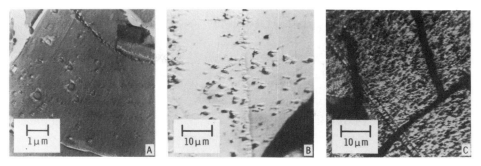

Fig. 9. Dislocation etch pits and tracks in polished olivine crystals. Approximate orientations are (a) (100) face; (b) (010) face; (c) (001) face (12052,43,1,1).

Table 1. EMX analysis of olivines in 12052,1,1

| Area | Wt. % of Element | | | |
	1	2A	2B	3
Mg	20.97	19.53	19.53	18.97
Si	17.52	16.97	17.16	17.24
Fe	20.50	22.21	22.39	23.32
Mn	0.21	0.24	0.26	0.22
Co	0.41	0.46	0.44	0.44
Ti	0.07	0.08	0.09	0.09
Cr	0.27	0.22	0.21	0.21
O	39.53 +	38.55 +	38.83	38.80
TOTAL	99.57	98.26	98.91	99.29

Microcracks at interfaces of mineral species were observed by optical microscopy and scanning electron microscopy. An example of an olivine crystal with microcracks at the ilmenite inclusion in sample 12052,43,1,1 is shown in Fig. 10. In Fig. 11a electron micrographs of terrestrial twinned albite, cleaved and etched (WEGNER and GROSSMAN, 1969) are compared in Fig. 11b to polished and etched twinned lunar plagioclase in 10058,40,2, both exhibiting similar etch figure shapes, depending on twinning angles. Figure 11c shows dislocation etch pits of the polished lunar plagioclase at high magnification.

Adhesion studies

Two successful fractures of 12052,43 were made at ultrahigh vacuum (10^{-9} to 10^{-10} torr). Attempts to measure touch contact adhesion between the segments revealed that the net adhesion force was less than that measurable by the system ($< 10^2$ dynes). However, indications of electrostatic charging were observed (specifically in adherence of fragments to the metal pieces associated with the fractures, and particle-alignment changes as the rock fragments were brought toward contact). The observed behavior was similar to that of the Apollo 11 samples and terrestrial rocks, but the adhesion force was orders of magnitude lower than the adhesion force measured for various terrestrial minerals studied previously.

The smaller section from the first fracture of 12052,43 was mounted on the UHV rotary feedthrough of the Cartesian cross assembly for measuring electrostatic charge (GROSSMAN, 1969). Basically, a chisel and anvil are disposed along a line perpendicular to the rotary feedthrough. A 630 μm electrometer probe wire penetrates a ground-plane which is parallel to the fractured face. This probe translates radially along the third Cartesian axis. The surface can be mapped by measuring either the charge or, more sensitively, the current, produced by scanning the surface at different rotation rates. The fracture direction was along one diagonal of the rectangular sample, but the fracture face penetrated the surface in approximately a circular arc centered near the chisel penetration (Fig. 12). The net charge of this face was found to be positive from the positive/negative current peak pair found as the electrometer probe traversed this fracture edge. In addition, a second positively charged section was found near the center of the arc, at a point just above the fracture arc. The average charge density in this region was estimated to be 5×10^6 charges/cm^2, which is five orders of magnitude less than that found with single crystal terrestrial rocks and at least a factor of 10 lower than coarse-grained granite.

When the vacuum chamber pressure was raised to that of the atmosphere with air, the effect of gas adsorption was observed as a change in relative (current) peak heights. After the sample was exposed to the laboratory atmosphere for approximately one week, the chamber was pumped to 10^{-9} torr. We then found the charge had decreased irreversibly by a factor of 20, which was very near the detectivity limit of the

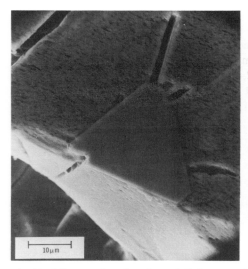

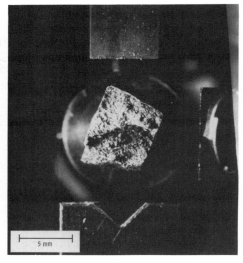

Fig. 10. Microcracks at interface of olivine crystal and ilmenite inclusion (12052,43,1,1).

Fig. 12. Fracture face of lunar sample 12052,43,1,2,1,1. After measuring the electrostatic charge distribution, the specimen is approximately 5 mm wide. Present orientation is approximately a 150° clockwise rotation from fracture position.

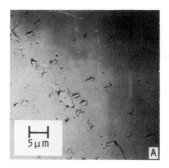  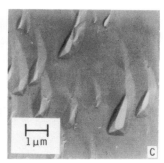

Fig. 11. Dislocation etch pits and twin boundaries in (a) Madagascar albite, cleaved (001) and (b) polished lunar plagioclase (10058,40,2). (c) is the same as (b) with higher magnification, showing pointed bottom edge dislocation etch pits in polished lunar plagioclase.

system. The same sample was fractured a second time from the opposite direction and again the fracture plane penetrated the crystal surface, but this time as a straight edge. A negative/positive peak pair, associated with the fracture edge, was found at UHV. This peak pair remained stable when the pressure was raised to that of the atmosphere with dry nitrogen and the chamber then repumped to 10^{-9} torr. The sample was then irradiated with UV light through a sapphire window, and the charge decreased a factor of 30 below background in four hours.

DISCUSSION

The Apollo 12 rock chip 12052,43 is an olivine basalt, consisting mainly of plagioclase, pyroxenes, olivine, ilmenite, and a chromium-spinel in a glassy matrix. Single crystals of some species are as long as 1 mm and 0.5 mm wide. The pyroxenes and olivine are mostly equant and in some cases rounded, the plagioclase occurs in long radial laths and as skeleton crystals, and ilmenite plates intersect the surface parallel to plagioclase laths. A band of ilmenite surrounding a chromium-spinel core was found. The rock is very porous, with numerous craters lined with glass splashes and glass rims. Microfractures or glassy deposits give the surface a "sugary" effect. Single crystals, visible in the light microscope, exhibit deformation features, such as bent twinning bands, fractures, and displacement. Few spheres were visible on the surface but some were found embedded in the rock after it was sectioned.

The chemical etch-pit method was employed to reveal the processes that modified the structure of lunar rocks. The relative simplicity of the method, which allows study of the structure of bulk cleaved or polished crystals in depth by successive polishing/etching steps, has some advantage over, and complements, studies of the rocks by thin section and transmission electron microscopy (CHRISTIE et al., 1970). Chemical etching brought out a variety of crystal structural features. These include twinning, banding, exsolution with twinning, growth bands, and skeletal structures. Deformation features, as kink bands, subgrain low angle tilt boundaries, and microfractures were also found. Dislocations and cosmic-ray tracks (ratio 1:3 and less) were found in pyroxenes, olivine, plagioclase, and possibly in ilmenite in the interior of 12052,43.

Other minerals in the rock, such as apatite, crystobalite, and others were not examined in detail.

The most interesting accomplishments of the chemical etching studies of the Apollo 11 and 12 lunar rocks were the findings of both dislocations and fossil tracks in olivine, tracks in pyroxene which appeared similar to heavy cosmic-ray tracks, possible tracks in ilmenite, and the variations in the pyroxene composition exhibited by etched striations in the highly calcic augite. These results show that the lunar rocks examined were subject to deformation melting and high temperature recrystallization; to slow cooling resulting in exsolution, chemical differentiation, and fractionation; and to crystal structure damage from cosmic rays, fissionable compounds, and meteoritic impacts. The types of dislocation observed indicate that relatively little mechanical deformation has occurred in lunar rock 12052,43. Variations in thermal expansion of crystalline inclusions may have induced microcracks which in turn could have contributed to the fracturing of the lunar rocks.

In an attempt to assess the causes of the bonding, elemental analyses of the gas-disrupted areas and their surroundings were made by electron microprobe. A search was made not only for the major elements but also for those minor elements which are known to form hydrogen bonding or hydrides in the solid state, because it was thought that hydrogen atoms derived at the lunar surface from solar wind and solar cosmic-ray protons might form hydrogen bonding in conjunction with these elements. There were no distinctive differences in elemental composition between the gas-disrupted areas and their surroundings. Nevertheless, in some cases atomic bonding might still exist. If so, the contact areas for bonding of the gas-disrupted particles are few, and the linear dimension of each area is smaller than about 0.1 μm, the limit of linear resolution in Fig. 3.

The net positive charge found on the UHV fractured face of lunar rock chip 12052,43 is what would be expected from strain-free fracture of a homogeneous dielectric with a positive electron work function. Both the probe and probe distance used are large compared to the average grain size, and, therefore, average out local inhomogeneous electric fields previously observed on Apollo 11 samples by the orientation effect on 100 μm and smaller particles. We conclude, therefore, that the lack of adhesion of fractured lunar rocks on a macroscopic scale is due both to the statistical local charge averaging to zero over the rough surface and the repulsive force produced by work function effects when creating a fresh surface in UHV.

A general conclusion is that samples returned by Apollo 11 and 12 missions are susceptible to disruption when exposed to oxygen and water vapor. When the interior surface of the fractured rock was exposed to nitrogen, no change in the macroscopic charge distribution was observed. This can be used as direct evidence that the bulk properties of lunar rocks are not affected by processing in dry nitrogen. However, some surface properties, such as irreversible adsorption of the lunar fines, may be affected and require investigation in these special cases.

A discussion of electrostatic adhesion forces (greater than Van der Waals' forces) stabilized by solar ultraviolet irradiation was recently presented (GROSSMAN, 1970). Briefly, disturbed lunar soil particles which randomly contact new particles will undergo induced photoelectric and photoconductive charge redistribution until a steady state charge distribution is established. The average thickness of the charged layer (the Debye length) and, therefore, the force of adhesion depends on the particle work functions, solar flux, the temperature, the UV adsorption coefficients, the dielectric constants and the conductivity. Preliminary results (GUPTA, 1970) suggest that surfaces of lunar material damaged by solar cosmic rays have higher conductivity than terrestrial particles. We conclude tentatively that the unusual cohesive properties of lunar soil are due to a combination of electrostatic effects partly attributable to the absence of a lunar atmosphere.

Acknowledgements—We thank W. M. HANSEN, A. D. PINKUL, E. GONZALES, and Dr. N. N. GREENMAN for their valuable contribution to the program; E. L. MILLER for scanning electron microscopy and microprobe analysis; and A. PHILLIPS and Mrs. B. TOOPER for electron microscopy. This program was supported by NASA Contract NAS 9–8082.

REFERENCES

BARBER D. J., HUTCHEON I., and PRICE P. B. (1971) Extralunar dust in Apollo cores. *Science* **171**, 372–374.

BARBER D. J., HUTCHEON I. D., PRICE P. B., and WENK R. (1971) Exotic particle tracks and lunar history. Second Lunar Science Conference (unpublished proceedings).

BHANDARI N., BHAT S., LAL D., RAJAGOPOLAN G., TAMHANE A. S., and VENKATAVARADAN V. S. (1971) Fossil track studies in lunar materials, I–III. Second Lunar Science Conference (unpublished proceedings).

CHRISTIE J. M., FISHER R. M., GRIGGS D. T., HEUER A. H., LALLY J. S., and RADCLIFFE S. V. (1971) Comparative electron petrography of Apollo 11, Apollo 12, and terrestrial rocks. Second Lunar Science Conference (unpublished proceedings).

CONDIE K. C., KUO C. S., WALKER R. M., and MURTHY R. V. (1969) Uranium distribution in separated clinopyroxenes from four eclogites. *Science* **165**, 57.

FLEISCHER R. L., PRICE P. B., WALKER R. M., and MAURETTE M. (1967) Origins of fossil charged-particle tracks in meteorites. *J. Geophys. Res.* **72**, 331–353.

FLEISCHER R. L., HAINES E. L., HANNEMAN R. E., HART H. R., JR., KASPER J. S., LIFSHIN E., WOODS R. T., and PRICE P. B. (1970a) Particle track, X-ray, thermal, and mass spectrometric studies of lunar paterials. Science **167**, 568–571.

FLEISCHER R. L., HAINES E. L., HART H. R. JR., WOODS R. T., and COMSTOCK G. M. (1970b) The particle track record of the Sea of Tranquility. *Proc. Apollo 11 Lunar Sci. Conf.*, Geochim. Cosmochim. Acta Suppl. 1, Vol. 3, pp. 2103–2120. Pergamon.

FLEISCHER R. L., HART H. R., and COMSTOCK G. M. (1971) Very heavy solar cosmic rays: Energy spectrum and implications for lunar erosion. Second Lunar Science Conference (unpublished proceedings).

GROSSMAN J. J. (1969) Electrostatic charge distribution on ultrahigh vacuum cleaved silicates. *J. Vac. Sci. Tech.* **6**, 233–236.

GROSSMAN J. J., RYAN J. A., and WEGNER M. R. (1970a) Experimental investigation of ultrahigh vacuum adhesion as related to the lunar surface. Final Contract Report, Fifth Year Summary, 1 January 1969 through 19 June 1970, NAS7–307.

GROSSMAN J. J., RYAN J. A., MUKHERJEE N. R., and WEGNER M. W. (1970b) Microchemical, microphysical, and adhesive properties of lunar material. *Proc. Apollo 11 Lunar Sci. Conf.,* Geochim. Cosmochim. Acta Suppl. 1, Vol. 3, pp. 2171–2181. Pergamon.

GROSSMAN J. J. (1970) Lunar soil adhesion due to electrostatic forces stabilized by solar radiation. *Trans. Am. Geophys. Union* **51(11)**, 772.

GUPTA Y. P. (1970) Private communication.

LUNATIC ASYLUM (1970) Mineralogic and isotropic investigations on lunar rock 12013. *Earth Planet. Sci. Lett.* **9**, 137–163.

Maurette M. (1966) Etude des traces d'ions lourds doris les mineraux naturels d'origine terrestre et extra-terrestre. *Bull. Soc. Franc. Miner. Crist.* **89,** 41–79.

Wegner M. W. and Grossman J. J. (1969) Dislocation etching of naturally deformed olivine. *Trans. Amer. Geophys. Union* **50,** 676.

Young C., III (1966) Applications of dislocation theory to upper-mantle deformation. Ph.D. dissertation, Stanford University.

Proceedings of the Second Lunar Science Conference, Vol. 3, pp. 2165–2172
The M.I.T. Press, 1971.

Pressure-volume properties of two Apollo 12 basalts

D. R. Stephens and E. M. Lilley

Lawrence Radiation Laboratory, University of California,
Livermore, California 94550

(*Received* 18 *February* 1971; *accepted* 16 *March* 1971)

Abstract—Two porous basalts from Apollo 12 (12002,180,181 and 12022,62) were studied under hydrostatic pressure to 10 kbar in a fluid pressure medium and to 40 kbar in a quasihydrostatic medium. Compressibilities by the two methods are in excellent agreement. The compressibilities of the Apollo 12 rocks are qualitatively similar to those of Apollo 11 basalts; volume compressibilities are initially very high (about 20 Mbar^{-1}) and decrease very rapidly to about 6–7 kbar. Beyond this, the compressibility decrease is much more gradual. Porosity persists to 40 kbar in the Apollo 12 basalts, as it did in the Apollo 11 samples. Volume changes with pressure are less for the Apollo 12 rock than for the Apollo 11 rocks.

Introduction

As part of a program to investigate the high-pressure properties of lunar materials, we studied three samples of two basalts from the Apollo 12 returned lunar samples: 12002,180,181 and 12022,62. The PV data can be used to calculate the lunar density as a function of the lunar radius. The results can be used in calculating impact processes, and basic information on lunar surface material is obtained.

In previous work on Apollo 11 rocks (Stephens and Lilley, 1970a), our quasi-hydrostatic data to 40 kbar agreed with, in general, but showed some interesting differences from, hydrostatic strain data taken by Anderson *et al.* (1970). To determine if this was a real effect due to the two different techniques or was just due to some-what different samples of the same rock, we examined our samples to 10 kbar in a hydrostatic environment and then ran the same samples to 35–40 kbar in the quasi-hydrostatic system (see below). Since the same samples were used in both methods, the data should be directly comparable, barring any irreversible effects due to the initial hydrostatic pressurization.

Experimental Techniques

The samples as received were approximate parallepepeds; rock 12022 had approximate dimensions of $1.0 \times 1.1 \times 1.1$ cm, and the rock 12002 specimens were approximately $0.7 \times 0.7 \times 1.6$ cm. Both rocks were appreciably vesicular. We prepared the samples for hydrostatic studies by two methods. In the first method, the samples were sputtered with 0.005 cm of copper before being electroplated with an additional 0.015 cm. Thus, the jacket should have been intimately bonded to the rock. In the second method, the sample was covered by a 0.007 mm thick, semiflexible, epoxy film, by the technique described by Schock and Duba (1970). Uncoated Constantan-foil strain gauges were then bonded to the surface in three mutually perpendicular directions. The correction for the pressure effect on the gauge resistance was taken from Brace (1964) as +0.055 Mbar^{-1} in strain.

The accuracy of compressibilities measured by the strain-gauge technique is believed to be ±2% (Brace, 1964; Schock and Duba, 1970). The four sample runs made in this work show that the

β_L (linear compressibility) for a given run fit a smooth curve to about $\pm 5\%$, while all β_L measured for a given rock agree on the average to $\pm 10\%$. We believe these variations are primarily due to the effect of anisotropic sampling of the porosity within the rock.

After the hydrostatic tests, the jacketing material was removed and the samples were potted in indium in right circular cylindrical geometry. The encapsulated cylinder was then pressurized to 35–40 kbar in lead by our previously described technique (Stephens and Lilley, 1970b). Volume compressibilities by this method for these lunar samples are accurate to $\pm 5\%$ below 20 kbar and $\pm 10\%$ at higher pressures.

Results and Discussion

Linear compressibilities (β_L)

Representative data for linear compressibilities are shown in Fig. 1 for sample 12002,180, in an epoxy jacket. (The directions T, S, E refer to orientation of the rock as given by the Curator's Office, NASA-MSC.) Linear compressibilities are appreciably different at low pressures and appear to converge toward a common value as pressure is increased. This is also seen in the data of Anderson *et al.* (1970).

Linear compressibilities for the two 12002 samples and two runs for the 12022 sample are shown in Figs. 2 and 3. We first measured rock 12022,62 in a copper jacket to 6 kbar, then removed the copper and ran the rock in an epoxy jacket to 10 kbar. Note that β_E was one of the most compressible directions in the first run, whereas it was the least compressible in the second run. Variations in the linear compressibility in the T-direction for sample 12002,181 were of the same order as the

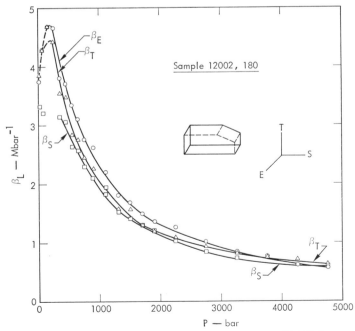

Fig. 1. Linear compressibilities versus pressure for sample 12002,180, in epoxy jacket. T, E, S refer to top, east, and south directions for this rock, as reported by the Curator's Office, NASA-MSC.

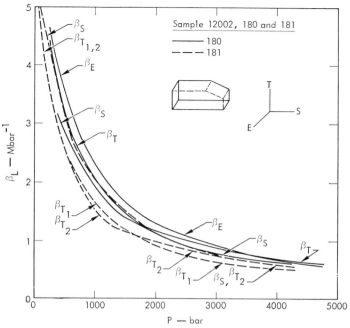

Fig. 2. Linear compressibilities versus pressure for rock 12002. Sample 180 was run
in an epoxy jacket, and sample 181 was encapsulated in copper.

variation in all three directions for sample 12002,180. Since the effect of porosity is
dominant in determining the compressibilities of these rocks, it appears that the
linear compressibility differences are due more to random sampling of differing
amounts of porosity in the rocks by the placement of the strain gauges rather than due
to any gross intrinsic anisotropies.

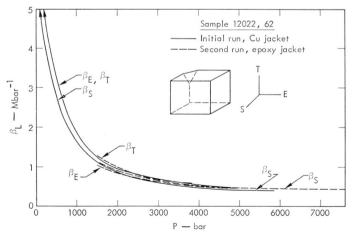

Fig. 3. Linear compressibilities versus pressure for rock 12022.

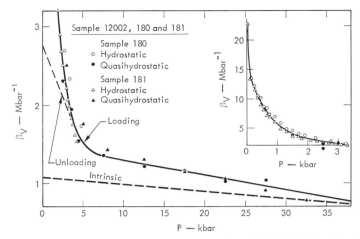

Fig. 4. Volume compressibilities versus pressure for rock 12002. Unloading data points
are not shown.

Volume compressibilities (β_V)

Volume compressibilities as a function of pressure are shown in Figs. 4 and 5. Hydrostatic and quasihydrostatic data are shown along with Voigt-Reuss-Hill (Hill, 1952) calculations of the intrinsic compressibilities of rocks 12002 and 12022, based on the mineralogic analyses (L,SPET 1970) for these rocks and compressibility data for minerals (Birch, 1966; Chung, 1970). Note that the 40-kbar quasihydrostatic data are in excellent agreement with the hydrostatic results. The hydrostatic volume compressibilities were determined as the sum of the linear compressibilities—that is, $\beta_T + \beta_E + \beta_S$.

The data for both Apollo 12 basalts are in qualitative agreement with the data of Anderson *et al.* (1970) and our data for rock 10017, in that the initially very high compressibility decreases rapidly to 2–7 kbar, followed by a much more gradual decrease with pressure. This behavior is consistent with Walsh's model (Walsh, 1965) of porosity in the form of narrow cracks. Narrow cracks have a large initial effect upon β but are closed by a pressure of a few kilobars, after which they have little influence. Ellipsoidal and spherical pores have a correspondingly lesser effect upon β and are closed at much higher pressures.

Thus, in both the Apollo 11 and Apollo 12 basalts, the presence of narrow cracks is inferred by the rapid decrease in β_{vol} at low pressures. The more gradual decrease in β at higher pressures (while still greater than the intrinsic β) is believed due to the effect of remaining ellipsoidal and spherical pores. A comparison of the experimental versus intrinsic compressibilities suggests that porosity persists in rock 12022 to 40 kbar, whereas porosity was almost completely eliminated in rock 12002 at 38–40 kbar.

The differences in the different sets of data are that the inflection point in the β_{vol} versus P was observed by Anderson *et al.* (1970) at about 2 kbar for 10017, at about 6 kbar in our work for the same rock, and at about 6–7 kbar in this study of rocks

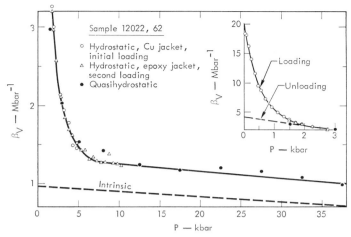

Fig. 5. Volume compressibilities versus pressure for rock 12022. Unloading data points are not shown.

12002 and 12022. Initial compressibilities were 8–10 Mbar^{-1} for 10017 and over 20 Mbar^{-1} for these Apollo 12 basalts.

Our data may also be compared to compressibilities calculated from elastic wave data measured as a function of pressure for the same rock. ANDERSON et al. (1970) measured compressional and shear velocities for rock 10017, and WANG et al. (1971) obtained elastic wave data for rocks 12002 and 12022. Compressibilities calculated in this way are about one-half our directly measured values. For example, for rock 12002 at 6.5 kbar, the data of WANG et al. (1971) suggest a β_{vol} of 0.80 Mbar^{-1}, compared to our value of 1.40 Mbar^{-1}. The reason for this discrepancy is believed to be due to the inhomogeneities in the rock (such as porosity) that influence the static compressibilities more than the 1 MHz dynamic measurements where the wave may effectively average-out the presence of a finite quantity of cracks or pores. SIMMONS and BRACE (1965) showed that large differences exist at low pressures between static and dynamic measurements; they attributed this to the presence of narrow cracks in the rocks. The rocks they studied were not vesicular, whereas the lunar rocks contain both narrow cracks and spherical pores. SIMMONS and BRACE (1965) observed that the agreement between the two measurements became good at a few kilobars, because of the closing of the narrow cracks. Since higher pressures are required to close spherical pores, it is expected that the differences in static and dynamic measurements for vesicular rocks would extend to higher pressures, which indeed seems to be the case of the lunar rocks.

It appears from the initial compressibilities that the Apollo 12 basalts studied contained more narrow cracks than did rock 10017. At high pressures, rock 10017 is more compressible than the Apollo 12 basalts (shown in both our work and in ANDERSON et al. 1970), indicating that the Apollo 11 basalt contained more spherical and elliposidal porosity. The net effect is that the volume change with pressure is considerably greater for rock 10017 than for either 12002 or 12022 as is shown in Fig. 6, indicating that rock 10017 was more porous than the Apollo 12 rocks. In addition,

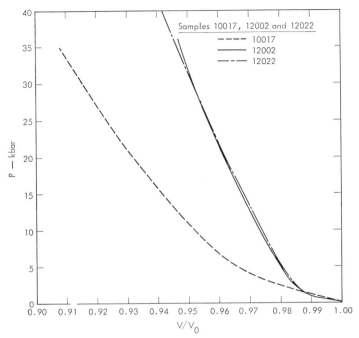

Fig. 6. Loading relative volume (V/V_0) versus pressure for rocks 10017, 12002, and 12022.

the significant decrease in compressibility upon unloading from 40 kbar, which was observed for rock 10017, was not observed for rocks 12002 and 12022, which again suggests that the Apollo 12 basalts contained less total porosity. The irreversible compaction after pressurization to 40 kbar and release to 1 atm are also consistent with this interpretation; permanent compaction of the Apollo 12 rocks was less than 1%, as compared to 5% for the Apollo 11 basalt.

The data in Fig. 6 show that the net volume changes of rocks 12002 and 12022 are quite similar, notwithstanding their somewhat different mineralogy and intrinsic compressibilities (see also Table 1 for compressibility and relative volume data for these rocks). In fact, the differences between the two rocks are no greater than the differences we have observed in samples cut from the same block of porous terrestrial basalts (STEPHENS and LILLEY, 1970b). It appears that the large effect of porosity tends to exceed the intrinsic effects of the mineralogy differences. Perhaps the similar compressibilities in rocks 12002 and 12022, suggesting a similar porosity distribution and content, in turn suggest that the mechanics of solidification of rocks 12002 and 12022 were related.

CONCLUSIONS

Linear compressibilities varied with direction for rock 12002 and 12022, but intrinsic anisotropy cannot be demonstrated. All linear compressibilities agree within ±10% for a given rock. Volume compressibilities determined by hydrostatic and

Table 1. Loading-unloading pressure-volume and compressibility data. Initial densities obtained directly from PV measurement technique.

P (kbar)	Rock 12002, ρ_0 = 3.294 g/cm³ V/V_0 Loading	Unloading	β_V(Mbar⁻¹) Loading	Unloading	Rock 12022, ρ_0 = 3.195 g/cm³ V/V_0 Loading	Unloading	β_V(Mbar⁻¹) Loading	Unloading
0	1.000	0.99025	24	2.82	1.0000	0.99234	20	4.27
0.05	0.99887		18		0.99903		18.6	
0.1	0.99812		14.2		0.99813		17.3	
0.2	0.99691	0.98969	12.4		0.99653		14.9	
0.3	0.99574	0.98941	11.1		0.99516		12.8	
0.5	0.99375	0.98887	9.0	2.66	0.99299		9.35	3.76
0.7	0.99213	0.98835	7.44		0.99136		7.35	
1	0.99018	0.98758	5.78	2.49	0.98952	0.98863	5.37	3.32
2	0.98538	0.98521	3.21	2.18	0.98577	0.98572	2.84	2.63
3	0.98332	0.98314	2.32	1.92	0.98344		2.00	
4	0.98135	0.98134	1.74	1.69	0.98167		1.63	
6	0.97832		1.42		0.97879		1.36	
8	0.97564		1.34		0.97624		1.27	
10	0.97306		1.30		0.97378		1.25	
12	0.97057		1.26		0.97136		1.23	
14	0.96816		1.22		0.96899		1.21	
16	0.96584		1.18		0.96665		1.19	
18	0.96360		1.14		0.96436		1.17	
20	0.96144		1.11		0.96211		1.16	
22	0.95934		1.07		0.95990		1.14	
24	0.95732		1.03		0.95773		1.12	
26	0.95539		0.99		0.95559		1.10	
28	0.95353		0.95		0.95350		1.08	
30	0.95175		0.92		0.95144		1.07	
32	0.95004		0.88		0.94942		1.05	
34	0.94841		0.84		0.94745		1.03	
36	0.94685		0.80		0.94550		1.01	
38					0.94360		0.99	
40					0.94174		0.97	

quasihydrostatic methods are in excellent agreement. Rocks 12002 and 12022 both are porous; and it can be inferred that the porosity is in the form of narrow cracks, as well as ellipsoidal or spherical pores, or both. This porosity influences the measured compressibilities to 40 kbar.

Volume compressibilities of the Apollo 12 rocks are qualitatively similar to that of the Apollo 11 basalt—that is, initially very high compressibilities, decreasing very rapidly with pressure to approximately 6–7 kbar, followed by a more gradual decrease toward higher pressures.

Both of the Apollo 12 basalts seem to contain more narrow cracks and fewer spherical pores than the Apollo 11 basalt 10017 contains.

The volume changes with pressure for rocks 12002 and 12022 are in reasonably good agreement with each other; both are less compressible than the Apollo 11 basalt at high pressure. There is also less irreversible compaction of the Apollo 12 rocks upon unloading from high pressure, confirming the conclusion that the Apollo 12 basalts have more narrow cracks and fewer pores than does the Apollo 11 basalt 10017.

Acknowledgments—This work was performed under the auspices of The U.S. Atomic Energy Commission.

REFERENCES

Anderson O., Scholz C., Soga N., Warren N., and Schreiber E. (1970) Elastic properties of a micro-breccia, igneous rock and lunar fines from Apollo 11 mission. *Proc. Apollo 11 Lunar Sci. Conf., Geochim. Cosmochim. Acta* Suppl. 1, Vol. 3, pp. 1959–1973. Pergamon.

Birch F. P. (1966) Compressibility, elastic constants. In *Handbook of Physical Constants* (editor S. P. Clark, Jr.), Geol. Soc. Amer. Mem. 97, Sec. 7.

Brace W. F. (1964) Effect of pressure on electric-resistance strain gauges. *Exptl. Mech.* **4**, 212–216.

Chung D. H. (1970) Effects of iron/magnesium ratio on P- and S-wave velocities in olivine. *J. Geophys. Res.* **75**, 7353–7361.

Hill R. (1952) The elastic behavior of a crystalline aggregate. *Proc. Phys. Soc. London* **A65**, 349–354.

LSPET (Lunar Sample Preliminary Examination Team) (1970) Lunar sample information catalog, Apollo 12. MSC-01512.

Schock R. N. and Duba A. (1970) Precise determination of the deformational properties of rocks to 4 kbar confining pressure. *Trans. Amer. Geophys. Union* **51**, 827.

Simmons G. and Brace W. F. (1965) Comparison of static and dynamic measurements of compressibility of rocks. *J. Geophys. Res.* **70**, 5649–5656.

Stephens D. R. and Lilley E. M. (1970a) Loading and unloading pressure-volume curves to 40 kbar for lunar crystalline rock, microbreccia and fines. *Proc. Apollo 11 Lunar Sci. Conf., Geochim. Cosmochim. Acta* Suppl. 1, Vol. 3, pp. 2427–2434. Pergamon.

Stephens D. R. and Lilley E. M. (1970b) Pressure-volume equation of state of consolidated and fractured rocks to 40 kbar. *Int. J. Rock Mech. Min. Sci.* **7**, 257–296.

Walsh J. G. (1965) The effects of cracks on the compressibility of rocks. *J. Geophys. Res.* **70**, 381–389.

Wang H., Todd T., Weidner D., and Simmons G. (1971) Elastic properties of Apollo 12 rocks. Second Lunar Science Conference (unpublished proceedings).

Proceedings of the Second Lunar Science Conference, Vol. 3, pp. 2173–2181
The M.I.T. Press, 1971.

Some physical properties of Apollo 12 lunar samples

T. Gold, B. T. O'Leary, and M. Campbell

Center for Radiophysics and Space Research, Cornell University,
Ithaca, New York 14850

(*Received* 22 *February* 1971; *accepted in revised form* 31 *March* 1971)

Abstract—The size distribution of the lunar fines is measured, and small but significant differences are found between the Apollo 11 and Apollo 12 samples as well as among the Apollo 12 core samples. The observed differences in grain size distribution in the core samples are related to surface transportation processes, and the importance of a sedimentation process versus meteoritic impact "gardening" of the mare grounds is discussed. The optical and the radio frequency electrical properties are measured and are also found to differ only slightly from Apollo 11 results.

Apollo 12 Grain Size Analysis

The Apollo 12 lunar fines were subjected to similar grain size analysis to that carried out for the Apollo 11 sample (Gold *et al.*, 1970). The general appearance and the appearance under the microscope of all samples of fines are rather similar, and the measured optical properties also show only small but significant differences. Although this type of uniformity was expected as a consequence of ground-based optical observations of the moon (Hapke, 1968), it nevertheless has to be emphasized as a remarkable conclusion.

The particle size distribution has been determined by two methods: electron microscopy and sedimentation rate in a column of water. The first was described in the Apollo 11 report (Gold *et al.*, 1970) and is of greatest value for particle sizes ranging down from 10 microns to less than 0.1 micron; it utilizes scanning electron micrographs of small "sections" of powder. The second method utilizes a sedimentation column which has been improved and perfected more recently.

The water sedimentation column consists of a vertical pipe 70.9 cm long, terminating below in a cubical box of optical glass plate. A photographic flash gun is imaged through a large aperture lens with focus just below the point of entry of the tube. Flash synchronized photographs are taken in a viewing direction perpendicular to the direction of the light. Stray and multiply scattered light is carefully excluded, and as a result the light scattered by a particle as small as 1 micron gives a perfectly recordable image. The water column is heated at the top and the temperature distribution along it is carefully controlled so that no thermal convection can set in. The particle sizes are deduced by Stokes' Law assuming them to be spherical. While this is of course not accurate, the optical and electron microscope examination showed the particles to be on the whole rather compact shapes, making this error rather small. Freedom from disturbing convection in the column is demonstrated by taking the photographs in pairs with a short duration in between, showing that each group of particles has settled a distance in that short time appropriate to its settling time from the top.

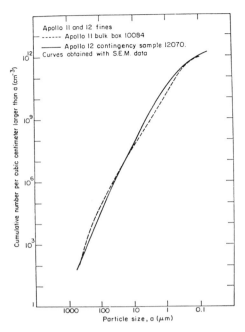

Fig. 1. The cumulative particle size distribution for the Apollo 11 and Apollo 12 bulk fines, determined from electron microscope data.

For an absolute measurement this method would perhaps not be sufficiently accurate, both for reasons of the particle shapes and perhaps also their unknown densities. For a comparison the method is very good, and it is much easier to accumulate good statistics than by the method of counting particles under the microscope.

Figure 1 compares the small-size particle size distribution of the Apollo 11 bulk box with that of the Apollo 12 contingency sample; the data, obtained by electron microscopy, are plotted as the cumulative number, per cubic centimeter, of particles larger in size than the abscissa value. A porosity of 0.5 is assumed and the number of particles counted is about 2000 in each case. The two curves are very similar, showing greatest divergence at particle sizes of a few microns; the difference, which amount to less than a factor 2.5, is probably real. Its significance is shown a little more clearly in Fig. 2 in which the differential rather than cumulated particle density is plotted.

The Apollo 12 contingency sample and three core samples (from cores 12025 and 12028) have been analyzed by the sedimentation column method, and the comparisons are shown on Figs. 3 and 4. From these curves it would appear that the surface sample from Apollo 12 is slightly coarser grained than that from Apollo 11. Among the core samples there is also a variation in the grain size distribution, with the deeper samples being somewhat richer in small particles than the surface and close subsurface ones. In particular the sample taken from a trench 15 cm deep (sample 12033) is significantly different in appearance from most others, and the size distribution analysis shows this one to possess a much larger proportion of small particles.

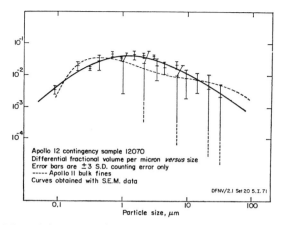

Fig. 2. The differential particle volume distribution for the Apollo 11 and Apollo 12 bulk fines, determined from electron microscope data.

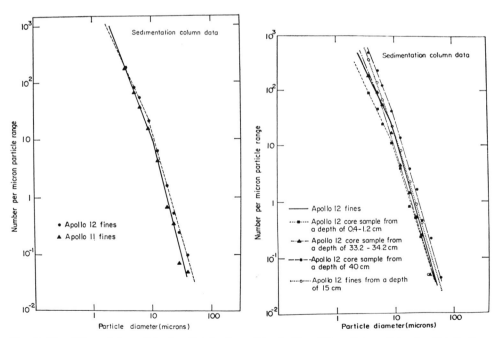

Fig. 3. The differential particle size distribution for the Apollo 11 and Apollo 12 bulk fines, determined by the sedimentation column method.

Fig. 4. The differential particle size distribution for the Apollo 12 bulk and core samples, determined by the sedimentation column method.

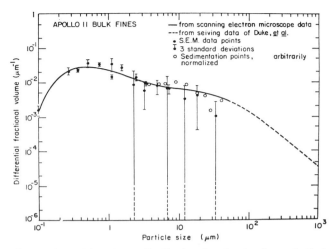

Fig. 5. Differential particle volume distribution for the Apollo 11 bulk fines. Curve
fits the electron microscope data, sedimentation data are also shown.

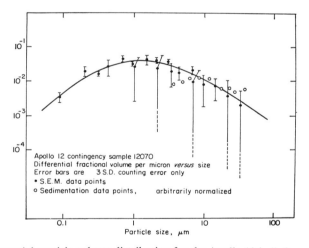

Fig. 6. Differential particle volume distribution for the Apollo 12 bulk fines. Curve fits
the electron microscope data, sedimentation data are also shown.

Figures 5 and 6 compare grain size analysis data obtained by the two different methods.

The fact that the grain size distribution in the core sample shows significant differences within tens of centimeters variation of depth requires comment. Differences over intervals of some centimeters in the core sample are also seen in the albedo (note color differences reported by the Lunar Sample Preliminary Examination Team, 1970), and very striking chemical differences have been reported (E. Anders, 1971). One has to discuss how sharply defined layers or other local configurations could be preserved despite the fact that some plowing of the ground by meteoritic impact must be taking place.

A material of different grain size, albedo or chemical composition could be derived either from a sufficiently distant or deep crater for this material not to have been previously well mixed by meteorite impact, or it could be material that is different as a result of contamination with some direct meteoritic infall. But it is not enough to account for possible sources of such different material. One must also understand firstly how it can have been deposited without excessive mixing, and secondly how it can have avoided being mixed by the plowing over which meteorites must be causing on the lunar surface.

The deposition of the material must be gentle and it cannot have reached its present position by being flung there on ballistic trajectories from a distant and deep crater. A layer some centimeters thick could not be deposited from such ballistic trajectories without mixing with a layer very many times its own thickness. The material seen in the core must thus have reached its position by a surface transportation process resulting in a sufficiently gentle sedimentation to avoid mixing. Secondly, in order to preserve such layers, one has to suppose that further sedimentation has taken place so that the overburden can protect the layer from meteorite plowing. If the rate of the meteorite plowing process were known, one could conclude what the rate of deposition has to be to have a significant probability that a layer at a given depth would be seen preserved. It is quite clear that even a single example of a very inhomogeneous core demonstrates that the ground has not been turned over hundreds of times to these depths, as had been calculated from estimates of the meteoritic infall rate. The mare ground seems to be subject to a sedimentation process much more than to a "gardening" process.

DIELECTRIC CONSTANT MEASUREMENT

The measurements of the high frequency electrical properties at 450 MHz were made by the same methods employed for the Apollo 11 samples (GOLD et al., 1970; CAMPBELL and ULRICHS, 1969). Moisture effects were avoided by having solid samples cut dry in laboratory atmosphere; as a precaution these were vacuum baked at 120°C for two days. Powder samples were stored in a dessicator with a large excess of anhydrous silica gel. The results do not disclose any marked difference in the dielectric constant of powder material from site to site. In Fig. 7 the dielectric constant measurements, as a function of bulk powder density, are shown for two Apollo 12 sites—one at a depth of 15 cm below the surface—as well as for the Apollo 11 bulk sample. The two Apollo 12 samples were chosen for their contrasting physical appearances, sample 12033 being much lighter in color and finer in texture than sample 12070. The variation of dielectric constant with density follows the Rayleigh formula (CAMPBELL and ULRICHS, 1969) in all cases and, indeed a single such curve fits all the data within $\pm 1\%$ excepting only the highest density point of sample 12070. The ground-based radar determinations of the dielectric constant (see EVANS and HAGFORS, 1968) are in complete accord with these measurements if one assumes a density of about 1.7 g cm^{-3} for the soil at a depth of 20 cm, an assumption which does no violence to the known properties of the soil.

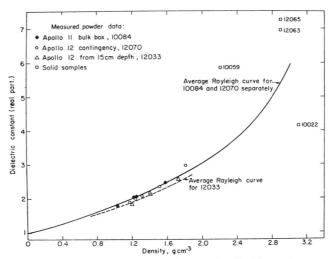

Fig. 7. Dielectric constant measurements for two Apollo 12 powder samples and the Apollo 11 bulk sample, as a function of bulk powder density. Dielectric constant vs. density points for four solid lunar rocks are also shown.

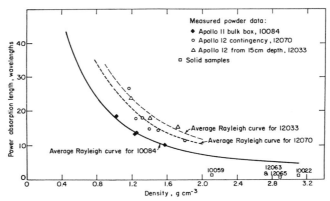

Fig. 8. The variation with density of the absorption length in two Apollo 12 powder samples and the Apollo 11 bulk sample. Points for four solid rocks are also shown.

Also shown on Fig. 7 are dielectric constant—density points for four solid lunar rocks, two each from Apollo 11 and Apollo 12. The latter pair, 12063 and 12065, are very similar petrologically and lie closely adjacent in the figure. Some allowance should be made for the porosity ($\sim 15\%$) of sample 10022 but this cannot greatly change the scatter of the points corresponding to this small but not atypical selection of rocks. None of the four solid rocks, nor any mixture of them, could be ground to a powder with the electrical properties (dielectric constant *and* loss tangent) of the dust samples, a conclusion in which we concur with the mineralogists.

Figure 8 shows in a similar way the variations with density of the absorption length in the powder samples, with points for the four solid rocks in addition. Again, assuming plausible densities for the powder at depths of a few centimeters, the data agree with prior ground-based radiothermal observations by KROTIKOV and TROITSKY (1963) and others.

OPTICAL PROPERTIES

The optical reflectivity and polarization of the Apollo 12 soil sample were measured as a function of phase angle with the same instrument and in the same manner

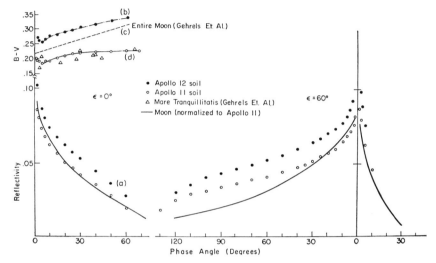

Fig. 9. (a) Reflectivity of the Apollo 11 and Apollo 12 soil vs. phase angle at 0.56 μm wavelength for viewing angles ε = 0 deg and 60 deg. (b) Color index B–V of the powder samples vs. phase angle for ε = 0 deg. Also plotted are (c) the reddening function of the entire moon, as determined by GEHRELS et al. (4), and (d) B–V values for a region of Mare Tranquillitatis.

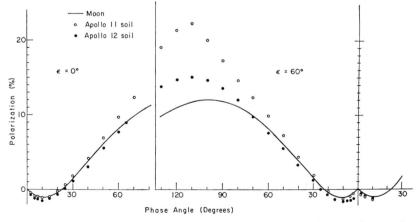

Fig. 10. The polarization of the Apollo 11 and Apollo 12 powders vs. phase angle at 0.56 μm wavelength for viewing angles ε = 0 deg and 60 deg.

as done previously for the Apollo 11 samples (O'Leary and Briggs, 1970). Both Apollo 11 and 12 samples were prepared by gradually dropping the fine-grained soil from a height of about 2 cm onto a sample tray.

Figures 9 and 10 indicate the dependence of reflectivity and polarization on phase angle for two viewing angles, ε, of 0° and 60°, as measured from the normal to the surface of the sample. While the Apollo 11 and 12 samples have similar photometric curves, the Apollo 12 sample is noticeably brighter than Apollo 11 (Fig. 9). The curves labeled "Moon" are taken from Hapke (1968) and normalized to the normal albedo of the Apollo 11 sample. The Apollo 12 soil has a normal albedo at 0.56 μm wavelength of 0.125 ± 0.003 as compared with 0.102 ± 0.003 for the Apollo 11 sample. Moreover, the Apollo 12 soil is redder than both the Apollo soil and the mean value for the moon (Gehrels et al., 1964). Finally, the Apollo 12 soil shows greater reddening with phase angle than the Apollo 11 soil. At $\varepsilon = 60°$, the photometric functions of both the Apollo 11 and 12 soils indicate a flattening toward larger phase angles compared with the lunar curve. The difference can probably be attributed to large scale roughness of the lunar surface as observed from the earth.

In Fig. 10 the polarization of the Apollo 12 soil is very similar to that of the moon as a whole (Hapke, 1968). However, for $\varepsilon = 60°$, both samples show peaks in polarization at greater phase angles than for the moon (Pellicori, 1969). The maximum polarization from the Apollo 12 sample is in good agreement with earth-based observations, while that of Apollo 11 is anomalously high. The interpretation of these data is somewhat uncertain, however, because of such factors as compaction, interaction with moisture and relative quantities of surface and subsurface soil contained in a given sample.

A study of the dependence of polarization and reflectivity on the degree of compaction, along with spectrophotometry of Apollo 12 soil and rocks, will be reported elsewhere (Briggs and O'Leary, in preparation).

Acknowledgments—We wish to thank Dr. Elizabeth Bilson for assistance with the sedimentation column work, Mr. Frank Briggs for assistance with optical studies, and Miss Joan Winters for assistance with electron microscope particle size counts. Mr. H. J. Eckelmann and Mr. S. M. Colbert helped in the design of the sedimentation column.

We wish to acknowledge gratefully the assistance given us by the Instrumental Analysis Research Department of the Corning Glass Works with scanning electron microscope work.

Work on lunar samples was carried out under NASA Contract NAS 9-8018.

References

Anders E., Laul J. C., Keays R. R., Ganapathy R., and Morgan J. W. (1971) Elements depleted on the lunar surface: Implications for origin and meteorite influx rate. Second Lunar Science Conference (unpublished papers).

Briggs F. and O'Leary B., in preparation.

Campbell M. J. and Ulrichs J. (1969) Electrical properties of rocks and their significance for lunar radar observations. *J. Geophys. Res.* **74,** 5867–5881.

Evans J. V. and Hagfors T. (1968) *Radar Astronomy*, McGraw-Hill.

Gehrels T., Coffeen T., and Owings, D. (1964) Wavelength dependence of polarization, III. The lunar surface, *Astron. J.* **69,** 826–852.

GOLD T., CAMPBELL M. J., and O'LEARY B. T. (1970) Optical and high-frequency electrical properties of the lunar sample. *Proc. Apollo 11 Lunar Sci. Conf., Geochim. Cosmochim. Acta* Suppl. 1, Vol. 3, pp. 2149–2154. Pergamon.

KROTIKOV V. D. and TROITSKY V. S. (1963) Radio emission and the nature of the moon. *Usp. Fiz. Nauk* **81,** 589–639.

LSPET (Lunar Sample Preliminary Examination Team) (1970) Preliminary examination of the lunar samples from Apollo 12. *Science* **167,** 1325–1339.

O'LEARY B. and BRIGGS F. (1970) Optical properties of Apollo 11 moon samples. *J. Geophys. Res.* **75,** 32, 6532–6538.

PELLICORI S. F. (1969) Wavelength dependence of polarization, XIX. Comparison of the lunar surface with laboratory samples. *Astron. J.* **74,** 1066–1072.

Proceedings of the Second Lunar Science Conference, Vol. 3, pp. 2183–2195
The M.I.T. Press, 1971.

Optical properties of mineral separates, glass, and anorthositic fragments from Apollo mare samples

John B. Adams

Caribbean Research Institute, College of the Virgin Islands, St. Croix 00820

and

Thomas B. McCord

Planetary Astronomy Laboratory Department of Earth and Planetary Sciences
Massachusetts Institute of Technology, Cambridge, Massachusetts 02139

(*Received* 22 *February* 1971, *accepted* 29 *March* 1971)

Abstract—Visible and near-infrared spectral reflectivity measurements of mineral separates from an Apollo 12 basalt demonstrate that pyroxene absorption bands dominate the curves of mare rocks and soil. Plagioclase, ilmenite, olivine, and other minerals have relatively little effect on the shapes of the reflectivity curves, although the proportions of feldspar and of opaques can affect albedo. By adding artificial glass back to the basalt from which it was made, it is shown that progressive vitrification of ilmenite-rich mare rocks causes darkening and masking of the pyroxene absorption bands, without imparting any of the (weak) band structure of the glass. Anorthositic lithic fragments separated from Apollo 11 soil have reflectivity curves that are dominated by low-Ca pyroxene, whereas telescopic curves of the lunar highlands show a band that indicates pyroxene of the same average composition as occurs at the Apollo 11 and Apollo 12 sites.

Introduction

This paper presents the results of laboratory measurements of the spectral reflectivity of Apollo 12 samples and a comparison of the results with Apollo 11 samples and with telescopic measurements of the lunar surface. An interpretation of the telescopic data, taking into account the laboratory studies of lunar samples, is published separately (Adams and McCord, 1971).

The following Apollo 12 samples were examined: fines 12042,41 and 12070,111; samples from the double core tube 12025,25; 12025,50; and 12028,97; and rocks 12053,23; 12053,29; 12053,30; and 12063,75; 12063,79; 12063,82. The core-tube samples are, respectively, from the following depths (uncorrected for compaction): 0.4–1.2 cm, 6.0–7.0 cm, 19.7–20.8 cm. The two rocks each consist of chips from top, interior, and bottom surfaces. Reflectivity measurements were made with a Beckman DK2–A ratio-recording spectrophotometer. The instrument and sample-handling procedures are discussed by Adams and McCord (1970). In the present study all samples were handled in air after it was determined that no changes in reflectivity resulted from exposure to dry air at room temperatures.

In our reports on the Apollo 11 samples (Adams and Jones, 1970 and Adams and McCord, 1970), we called attention to the differences between the spectral reflectivity curves for crystalline rocks and those for breccias and fines. The rock curves have well-developed absorption bands, whereas the breccias and the fines have only faint vestiges of bands. The main absorption bands in the rocks were attributed to Fe^{2+} in

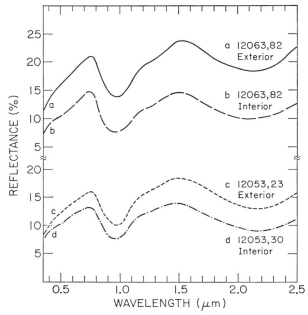

Fig. 1. Spectral reflectivity curves of exterior and interior surfaces of chips of Apollo 12 basalts 12063 and 12053. All measurements are relative to MgO.

pyroxene. The progressive degradation of these bands from rocks to breccias to fines was correlated with an increase in the percentage of dark glass.

The Apollo 12 samples are generally similar to those from the Apollo 11 site in their optical properties. The rocks (Fig. 1) exhibit strong absorption bands whereas the surface fines (Fig. 2) have weak bands. Our samples of Apollo 12 fines included light and dark material from the double core tube. These samples provided important additional evidence on the factors controlling the strength of the optical absorption bands. We have obtained further information on the origin of the bands in the Apollo 12 materials by analyzing mineral separates (Fig. 3). We have, in addition, fused Apollo 12 crystalline rock and investigated the optical properties of the glass and of glass-crystal mixes (Fig. 4). The results of these analyses lead to a consistent explanation for the main optical properties of the lunar samples at the Apollo 11 and 12 sites and correlate well with the telescopic measurements.

LABORATORY RESULTS

The optical properties of the moon as seen from an earth-based telescope are dominated by the fine soil at the surface. The Apollo 11 and 12 soils are made up of a complex assortment of silicate minerals, oxides, and glasses, with very minor sulfides and metals. To understand the optical properties of the bulk soil it is useful to start with the properties of individual mineral species and work toward multiphase assemblages.

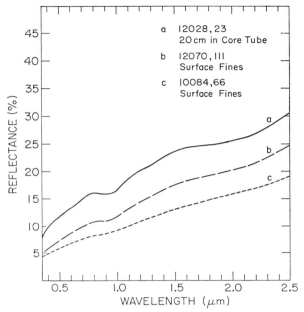

Fig. 2. Spectral reflectivity of Apollo 11 surface fines, Apollo 12 surface fines, and Apollo 12 fines from 20 cm deep in the double-core tube.

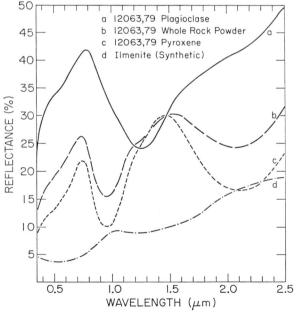

Fig. 3. Spectral reflectivity of Apollo 12 basalt powder 12063, and plagioclase and pyroxene separates from the same rock. Ilmenite is a synthetic sample.

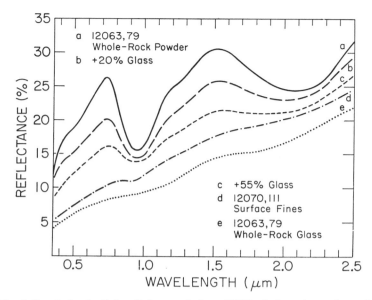

Fig. 4. Spectral reflectivity of glass made from 12063 whole-rock powder; mixtures of rock powder plus 20% glass and plus 55% glass; compared with curve of Apollo 12 surface fines.

Mineral separates were made from 0.5 gms of rock 12063 using a combination of magnetic techniques and hand-picking under the microscope. The rock consists of 51% pyroxene and 27% plagioclase, with 8% olivine and 11% opaques, mostly ilmenite (WARNER, 1970). It was possible to obtain only nearly pure separates of pyroxene and of feldspar owing to the limited amount of starting material. Figure 3 shows spectral reflectivity curves for the whole rock, the plagioclase separate, and the pyroxene separate, all sized to particles between 125 μm and 500 μm. Also shown for reference is a curve for synthetic ilmenite powder ($< 44\ \mu$m).

The pyroxene has two strong absorption bands at 0.95 μm and at 2.1 μm. These bands are produced by Fe^{2+} on a highly distorted (M2) octahedral site, and their assignments have been treated in detail elsewhere (BANCROFT and BURNS, 1967 and ADAMS and McCORD, 1970). A weak Ti^{3+} band occurs at 0.5 μm, and there is the suggestion of a band at 1.25 μm, which probably arises from Fe^{2+} on the M1 site.

The curve for plagioclase is characterized by a strong Fe^{2+} band centered at 1.25 μm and by strong absorption at the blue end of the spectrum. Comparison with curves of terrestrial calcic feldspars suggests that our lunar specimen still includes a component of pyroxene. This is seen in the flattening at 0.5 μm, the shallow depression near 1 μm, and in the faint band at 2.1 μm. Tiny inclusions (seen under the microscope) of the strongly absorbing pyroxene in the relatively clear plagioclase account for the above features.

To conserve sample, no attempt was made to separate ilmenite from rock 12063. The main spectral features are illustrated by a synthetic stoichiometric ilmenite. The band at 0.5 μm is assigned to Ti^{3+}, and the broad depression between 1.0 μm and 1.5 μm is assigned to Fe^{2+}. Note that the curve in Fig. 3 is from a < 44 μm powder that allows light to pass through individual particles without being totally absorbed. For larger particle sizes, the absorption bands become less distinct and the overall reflectivity decreases.

Olivine, which makes up < 10% of rock 12063 was not separated—again, to conserve sample. The optical properties of olivine are well known (ADAMS, 1968), and we would expect a broad Fe^{2+} band at 1.02 μm.

It is evident from Fig. 3 that the curve of the whole rock is overwhelmingly dominated by the spectral features of the pyroxene. Only the weak band at 1.25 μm in the rock curve is derived largely from plagioclase, as was suggested by CONEL and NASH (1970), although it overlaps the Fe^{2+} bands in pyroxene and in ilmenite, as was pointed out by ADAMS and MCCORD (1970). Ilmenite, olivine, and other minor minerals do not contribute significantly to the spectral reflectivity of rock 12063. Rock chips (Fig. 1), although slightly darker than the whole rock powders, show the same main spectral features. All curves are dominated by pyroxene. Freshly broken surfaces of the two Apollo 12 samples are slightly darker and have flatter curves than those for the natural outer surfaces of the specimens. We found a similar relationship for Apollo 11 rock samples. The natural outer surfaces of the rocks are highly microfractured. These optical discontinuities cause more light to be reflected, in the same way that crushing the rock to progressively finer particle sizes increases the albedo (ADAMS and FILICE, 1967). Although the exterior surfaces also contain glassy pits, there is no apparent effect on the spectral curves from the small amount of glass. Significantly, the "space weathering" of the rocks does not shift the positions of the absorption bands.

The spectral curves of the fines (Fig. 2) are different from those of the rocks. The two pyroxene bands are present, but they are very weak, and the integral reflectivity is about one-half that of the rock powder. Of the five Apollo 12 samples of fines that we measured, all had identical spectral curves except the light gray fines from 20 cm in the core tube. The other four samples available to us were from the surface or within the top 7 cm of the core tube, and have a curve given by 12070,111 in Fig. 2. Our 20 cm core-tube sample has a higher overall reflectivity and stronger absorption bands than the surface fines. NASH and CONEL (1971) reported on a different set of samples from the core tube and showed that albedo does not vary as a simple function of depth. They found, however, that the brighter materials have deeper absorption bands.

The most striking difference between lunar soil and rocks is the presence of abundant glass in the soil. Although the glass is highly varied in composition and therefore in color and refractive index, the most abundant type is a dark reddish brown.

We are still separating glasses from the lunar soil in an attempt to make a direct measurement of their optical properties. Separation of a sufficient quantity of glass is, however, a very slow procedure. Meanwhile, we have made artificial glass from crystalline rock (12063). The spectral curve for this glass and for mixtures of crystalline rock powder and glass powder are shown in Fig. 4.

Glass was made from 40 mg of the whole-rock powder of rock 12063. The charge was held in a platinum tube at 1300°C for 1.5 hours. The furnace was purged with dry N_2 during the run to prevent oxidation of the sample. The fused product is very dark brown. This glass was crushed and the powder was measured with the spectro-photometer. The curve for the glass has two broad absorption bands at approximately 1.1 μm and 1.9 μm. These bands are from Fe^{2+} on highly distorted sites. The curve and the bands are closely similar to those of terrestrial basaltic glasses that we have measured in our laboratory. The results are also in agreement with CONEL and NASH (1970) (see also CONEL, 1970).

To simulate the production of lunar soil by partial vitrification of crystalline rock, we added glass powder back to the same rock powder from which the glass was made (Fig. 4). Three points are significant: (1) The addition of dark glass lowers the albedo of the overall powder. (2) The absorption bands of the glass (1.1 μm and 1.9 μm) do not appear when the glass is mixed with the crystalline powder. (3) The absorption bands in the curve of the rock powder are progressively weakened as more glass is added. The disappearance of the glass bands in the mixed powder is not surprising in view of the absence of ilmenite (and plagioclase) bands in the rock-powder curve (Fig. 3). The weakening of the pyroxene bands in the rock-powder curve with the addition of glass is caused by the overall darkening of the mix, which leads to a lessening of differential absorption. The same effect can be produced by adding carbon-black or any other very dark material.

Figure 4 also shows the curve for Apollo 12 surface fines. We conclude that the curves of the natural mare soils can be explained in terms of crystalline rock powders (in which pyroxene dominates the optical properties), that have been partially melted to yield a mixture of dark glass and crystalline phases. As more glass is produced by micrometeoroid bombardment at the lunar surface, the soil should become darker and the pyroxene bands become less distinct. There is evidence for this effect in Fig. 2. The 20 cm core tube sample contains about 10% glass (our estimate), the Apollo 12 surface fines have about 20% glass (LSPET 1970), and the Apollo 11 soil (10084) contains approximately 50% glass (LSPET 1969). These curves show that as the glass content increases, the albedo decreases and the pyroxene bands become fainter.

Although our experiments with artificial glass illustrate the importance of glass for the lunar optical properties, there are differences between our laboratory mix and the natural lunar fines. Notably, our crushed glass consists wholly of chips and splinters that transmit more light than the spherical or equant grains of lunar glass which trap light by multiple internal reflections. Even very small (< 20 μm) spheres of the lunar brown glass are dark under the microscope. When we grind the artificial glass to particle sizes approaching the lunar material, the albedo is too high owing to the decreased mean path length of light in the irregular grains. The most important difference, however, is that none of the dark glass sticks to the other mineral grains. Therefore, a greater surface area of light particles is exposed in the laboratory mix than in the natural mare soil. The lack of sticking and the marked difference in particle shape may explain why the addition of 20% artificial glass does not reduce the albedo to that of the Apollo 12 soil containing 20% natural glass. It may also explain why CONEL and NASH (1970) found a higher albedo for their artificial glass than for the Apollo 11 soil. We also note that rock 12063 from which we made glass contains about

10 % ilmenite. This percentage may not be representative of the ilmenite content of the average local basalt. Higher concentrations of ilmenite will yield darker glasses.

GOLD et al. (1970) and O'LEARY and BRIGGS (1970) reported that sputter-deposited metal coatings on Apollo 11 soil particles cause the low albedo of the soil. The presence of ubiquitous metal coatings, however, is contradicted by other evidence from electron microscopy, Mössbauer studies, magnetic properties, and electrical properties. (See, for example, GOLDSTEIN et al., 1970; MCKAY et al., 1970; FRONDEL et al., 1970; HERZENBERG and RILEY, 1970; STRANGWAY et al., 1970; GOLD et al., 1970.)

HAPKE et al. (1970) drew attention to possible sputter-deposited opaque coatings on Apollo 11 fines, and HAPKE et al. (1971) presented evidence, based on an acid-leaching technique, for impact-produced vapor-deposited glassy coatings (approximately 2 μm thick) on the Apollo 12 fines. It is well known that glass cements particles in the breccias and soils, and that glass partially or even completely coats some grains. The occurrence and formation of glass, however, are highly complex and do not fit a simple vapor deposition model (MCKAY et al., 1970; MCKAY et al., 1971; FREDRIKSSON et al., 1970). We agree with HAPKE et al. (1970) that dark glass in the mare soil lowers the albedo; however, it is important to emphasize that glass does not coat all grains and, in fact, occurs in many forms such as local splashes, interstitial "cement," spherules, and irregular fragments. The ability of glass to stick to other particles strongly affects the optical properties as we pointed out earlier. In the mare soils sticking apparently has occurred over a range of temperatures, from the softening point of a glass through the liquid and vapor phases.

Using high voltage transmission electron microscopy, BORG et al. (1971) found amorphous rinds 0.1 μm thick on 1 μm diameter particles of Apollo 12 soil. These rinds appear to be a radiation-damaged outer portion of the crystalline material rather than a surface-deposited layer. BORG et al. (1971) also reported that the damaged particles have a lower albedo than the undamaged ones. A definitive test, however, requires classifying and separating otherwise similar 1 μm lunar soil particles according to whether they have damaged outer layers or not. Albedo measurements of the two classes of particles would require the separation of at least 50 mg of each class. This is a formidable task in view of the difficulties in handling (and classifying) such small particles. Until such separates can be made, without bias as to percentages of mineral and glass species, or until albedo can be measured on single 1 μ particles, the reported effect of the damaged layers on albedo must remain in doubt. If the 0.1 μm rinds do lower the albedo, it must be determined that such damaged layers also occur for the wide size range of larger particles before a generalization can be made about the lunar soil as a whole. We note that the exteriors of rocks are in fact brighter than their interiors (Fig. 1). Any darkening effect on the rocks is lost in the brightening due to microbrecciation. In like manner, any darkening of the soil by radiation damage is overwhelmed by the readily visible darkening due to the production of glass.

Based on the existing data we do not rule out the possibility that radiation damage lowers the albedo of lunar soil, but further work is necessary to determine the magnitude of any effects. Our results indicate that such effects on albedo must be minor, if they occur at all, and that they are not required to explain the optical properties of the mare soil samples.

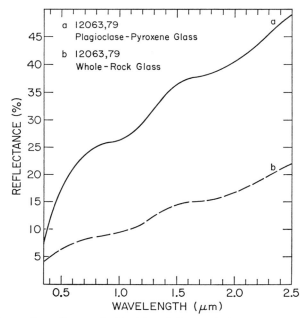

Fig. 5. Comparison of spectral reflectivity of glass made from Apollo 12 basalt (12063) containing about 10% ilmenite (lower curve), and glass made from a 1:1 by weight mixture of plagioclase and pyroxene from the same rock (upper curve).

In our study of the Apollo 11 soil (Adams and Jones, 1970; Adams and McCord, 1970) we proposed that the strong optical absorption in the glass was caused by iron and titanium that had been derived largely from ilmenite. Our glass experiments with Apollo 12 samples support this idea. We used portions of our mineral separates of plagioclase and of pyroxene to prepare a mixture that simulated a mare basalt that was free of ilmenite. Pure plagioclase and pure pyroxene powders were mixed 1:1 by weight and fused in a platinum tube, as previously described. The resulting glass is light tan, as contrasted with the very dark brown of the glass derived from the whole rock. Spectral reflectance curves of the two types of glass are shown in Fig. 5. The plagioclase-pyroxene glass has a higher overall reflectivity than the whole-rock powder (12063).

We conclude that ilmenite in the mare basalts is essential to the production of the dark glass, which, in turn, accounts primarily for the low albedo of the soil. Rocks without ilmenite (or other opaque phases) would be expected to undergo little or no darkening at the lunar surface as a result of vitrification.

ANORTHOSITIC ROCKS COMPARED WITH LUNAR HIGHLANDS

We turn now to the feldspathic component of the Apollo 11 and 12 soils. We separated anothositic fragments from the Apollo 11 soil and measured the spectral reflectivity (Fig. 6). We took care to exclude all pieces of coarse basalt from our

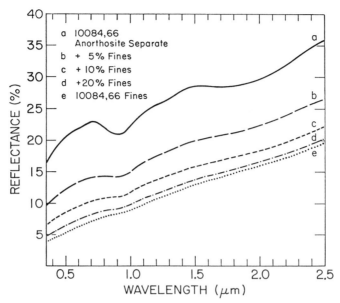

Fig. 6. Spectral reflectivity of anorthositic lithic fragments from Apollo 11 soil; compared with curves of the same material plus 5%, 10%, and 20% of fines from Apollo 11 soil (10084).

anorthositic sample, following the criteria of Wood *et al.* (1970). There are two striking features of the curve for the anorthositic separate: (1) The curve is dominated by pyroxene features rather than by the plagioclase, even though pyroxene makes up $< 10\%$ of the sample. This can be understood in terms of our previous discussion of mineral separates (Fig. 3). (2) The pyroxene bands are at 0.91 μm and at 1.8 μm. These band frequencies correspond to pigeonite or orthopyroxene. Low calcium pyroxenes are characteristic of the anorthositic fragments as has been verified by optical and microprobe analyses (Wood *et al.*, 1970).

We did not have a large enough sample of Apollo 12 soil to separate out the "foreign" feldspathic component. However, reports on the mineralogy of, for example, "Luny Rock 1" (10085, Albee *et al.*, 1970), rock 12013 (Drake *et al.*, 1970), "KREEP" (Gast and Hubbard, 1971), "Norite" (Wood *et al.*, 1971) refer consistently to feldspathic rock containing minor low-calcium pyroxene. Although we have not yet made direct measurements on these Apollo 12 feldspathic rocks, we expect, based on the mineralogy, that the reflectivity curves will be very similar to the Apollo 11 anorthositic separate (Fig. 6). The band positions should be very near 0.91 μm and 1.8 μm.

The Apollo 11 and 12 feldspathic rocks have far less ilmenite (typically $\leqslant 5\%$) than the basalts. Anorthositic glasses from the Apollo 11 soil have about the same albedo as the crystalline fragments, and we expect that, in general, the Apollo 11 and 12 feldspathic materials alone darken little, if at all, by impact vitrification at the lunar surface. If comparable feldspathic rocks comprise the lunar highlands (Wood *et al.*,

1970), an interesting problem arises as to how highland bright craters and rays with albedo ~ 0.25 darken with time to an albedo of ~ 0.13. ADAMS and MCCORD (1971) pointed out that the optical properties of the highlands may imply contamination by a few % of the dark mare soil. We will not review here the arguments supporting the contamination hypothesis. We have, however, added Apollo 11 fines back to our Apollo 11 anorthositic separate to observe the change in spectral reflectivity. Figure 6 illustrates that addition of 5% to 10% of the dark fines degrades the absorption bands of the anorthositic material and depresses the albedo.

There are two important differences between the laboratory curve for the Apollo 11 anorthositic rock and the telescopic curves of the lunar highlands (ADAMS and MCCORD, 1970; MCCORD *et al.*, 1971; and Figs. 7 and 8): The pyroxene band is (1) faint to absent for most of the highlands, (2) except for bright craters and rays where the band is at 0.95 μm (rather than at 0.91 for the anorthositic sample).

ADAMS and MCCORD (1971) presented evidence that the bright craters and rays have a higher crystal/glass ratio than the surrounding areas. The absorption band at 0.95 μm indicates that highland "rocky" areas have an average pyroxene composition similar to that found in the mare basalts, in contrast to the low calcium pyroxene of the feldspathic rocks recovered from the Apollo 11 and 12 sites.

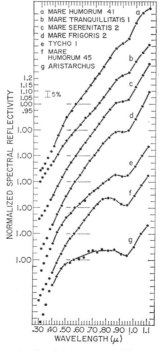

Fig. 7. Normalized spectral reflectivity curves of 18 km-diameter areas on the lunar surface (McCord *et al.*, 1971). (a) highland area near edge of Mare Humorum, (b) dark mare area (Tranquillitatis), (c) standard mare area (Serenitatis), (d) mare area (Frigoris), (e) highland bright crater (floor of Tycho), (f) mare bright crater (Mare Humorum), (g) mare bright crater (Aristarchus).

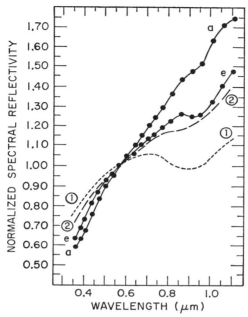

Fig. 8. Normalized spectral reflectivity curves of (1) anorthositic separate from Apollo 11 soil, (2) anorthositic material plus 10% fines from Apollo 11 soil, and telescopic curves of (a) highland area, and (e) highland bright crater (Tycho).

The telescopic curves of "background" highland areas (other than bright craters and rays) have only a slight change of slope in the 0.9 μm to 1 μm region. The alteration of highland bright crater and ray material to "background" soil is accompanied by an almost complete degradation of the pyroxene band. Contamination by mare fines (or any other dark material), although it lowers the albedo, does not cause this much degradation of the pyroxene band (see Fig. 8). It is unlikely that the highland regolith is devoid of a pyroxene component, as the pyroxene band appears wherever (subsurface) bright-crater material is exposed. The disappearance of the band, instead, may be due to disordering of the pyroxene structure by extensive impact melting and shock alteration of the soil. Complete melting of the soil, however, probably would produce the weak Fe^{2+} band at 1.1 μm, which is not observed.

Acknowledgments—We thank Professor DAVID WONES of M.I.T. for making possible the vitrification experiments. Mr. JEAN LARSEN and Miss CLAUDIA GELLERT assisted with the laboratory measurements at C.R.I. This work was supported by NASA grants and contracts (NGR-22-009-350, NGR-52-083-003, and NAS 9-9478).

REFERENCES

ADAMS J. B. (1968) Lunar and martian surfaces: Petrologic significance of absorption bands in the near-infrared. *Science* **159**, 1453–1455.
ADAMS J. B. and FILICE A. L. (1967) Spectral reflectance 0.4 to 2.0 microns of silicate rock powders. *J. Geophys. Res.* **72**, 5705–5715.
ADAMS J. B. and JONES R. L. (1970) Spectral reflectivity of lunar samples. *Science* **167**, 737–739.
ADAMS J. B. and McCORD T. B. (1970) Remote sensing of lunar surface mineralogy: Implications from visible and near-infrared reflectivity of Apollo 11 samples. *Proc. Apollo 11 Lunar Sci. Conf.*, *Geochim. Cosmochim. Acta* Suppl. 1, Vol. 3, pp. 1937–1945. Pergamon.

Adams J. B. and McCord T. B. (1971) Alteration of lunar optical properties: Age and composition effects. *Science* **171**, 567–571.

Albee A. L., Burnett D. S., Chodos A. A., Eugster J. C., Huneke D. A., Papanastassiou F. A., Podosek G., Price Russ, III, Sans H. G., Tera F., and Wasserburg G. J. Ages, irradiation history, and chemical composition of lunar rocks from the Sea of Tranquillity. *Science* **167**, 463–466.

Bancroft G. M. and Burns R. G. (1967) Interpretation of the electronic spectra of iron in pyroxenes. *Amer. Mineral.* **52**, 1278–1287.

Borg J., Durrieu L., Jouret C., and Maurette M. (1971) The ultramicroscopic irradiation record of micron-sized lunar dust grains. Second Lunar Science Conference (unpublished proceedings).

Conel J. E. (1970) Coloring of synthetic and natural lunar glass by titanium and iron. *Jet Propulsion Laboratory Space Programs Summary* 37–62. Vol. 3, pp. 26–31.

Conel J. E. and Nash D. B. (1970) Spectral reflectance and albedo of Apollo 11 lunar samples: Effects of irradiation and vitrification and comparison with telescopic observations. *Proc. Apollo 11 Lunar Sci. Conf., Geochim. Cosmochim. Acta* Suppl. 1, Vol. 3, pp. 2013–2023. Pergamon.

Drake M. J., McCallum I. S., McKay G. A., and Weil D. F. (1970) Mineralogy and petrology of Apollo 12 sample no. 12013: A progress report. *Earth Planet. Sci. Lett.* **9**, 103–123.

Fredriksson K., Nelen J., and Melson W. G. (1970) Petrography and origin of lunar breccias and glasses. *Proc. Apollo 11 Lunar Sci. Conf., Geochim. Cosmochim. Acta* Suppl. 1, Vol. 1, pp. 419–432. Pergamon.

Frondel C., Klein C., Jr. Ito J., and Drake J. C. (1970) Mineralogical and chemical studies of Apollo 11 lunar fines and selected rocks. *Proc. Apollo 11 Lunar Sci. Conf., Geochim. Cosmochim. Acta* Suppl. 1, Vol. 1, pp. 445–474. Pergamon.

Gast P. W. and Hubbard N. J. (1971) Rare earth abundances in soil and rocks from the Ocean of Storms. Second Lunar Science Conference (unpublished proceedings).

Gold T., Campbell M. J., and O'Leary B. T. (1970) Optical and high-frequency electrical properties of the lunar sample. *Proc. Apollo 11 Lunar Sci. Conf., Geochim. Cosmochim. Acta* Suppl. 1, Vol. 3, pp. 2149–2154. Pergamon.

Goldstein J. I., Henderson E. P., and Yakowitz H. (1970) Investigation of lunar metal particles. *Proc. Apollo 11 Lunar Sci. Conf., Geochim. Cosmochim. Acta* Suppl. 1, Vol. 1, pp. 499–512. Pergamon.

Hapke B. W., Cohen A. J., Cassidy W. A., and Wells E. N. (1970) Solar radiation effects on the optical properties of Apollo 11 samples. *Proc. Apollo 11 Lunar Sci. Conf., Geochim. Cosmochim. Acta* Suppl. 1, Vol. 3, pp. 2199–2212. Pergamon.

Hapke B. W., Cassidy W. A., and Wells E. N. (1971) The albedo of the Moon: Evidence for vapor-phase deposition processes on the lunar surface. Second Lunar Science Conference (unpublished proceedings).

Herzenberg C. L. and Riley D. L. (1970) Analysis of first returned lunar samples by Mössbauer spectrometry. *Proc. Apollo 11 Lunar Sci. Conf., Geochim. Cosmochim. Acta* Suppl. 1, Vol. 3, pp. 2221–2241. Pergamon.

LSPET (Lunar Sample Preliminary Examination Team) (1969) Preliminary examination of the lunar samples from Apollo 11. *Science* **165**, 1211–1227.

LSPET (Lunar Sample Preliminary Examination Team) (1970) Preliminary examination of the lunar samples from Apollo 12. *Science* **167**, 1325–1339.

McCord T. B., Charett M., Johnson T. V., Lebofsky L., and Pieters C. (1971) *Lunar Spectral Types*. In preparation.

McKay D. S., Greenwood W. R., and Morrison D. A. (1970) Origin of small lunar particles and breccia from the Apollo 11 site. *Proc. Apollo 11 Lunar Sci. Conf., Geochim. Cosmochim. Acta* Suppl. 1, Vol. 1, pp. 673–694. Pergamon.

McKay D., Morrison D., Lindsay J., and Ladle G. (1971) Second Lunar Science Conference (unpublished proceedings).

Nash D. B. and Conel J. E. (1971) Luminescence and reflectance of Apollo 12 samples. Second Lunar Science Conference (unpublished proceedings).

O'Leary B. T. and Briggs F. (1970) Optical properties of Apollo 11 moon samples. *J. Geophys. Res.* **75**, 6532.

STRANGWAY D. W., LARSON E. E., and PEARCE G. W. (1970) Magnetic studies of lunar samples—breccia and fines. *Proc. Apollo 11 Lunar Sci. Conf., Geochim. Cosmochim. Acta* Suppl. 1, Vol. 3, pp. 2435–2451. Pergamon.

WARNER J. (compiler) (1970) Apollo 12 Lunar-sample information. *NASA Technical Report R–353.*

WOOD J. A., DICKEY J. S., MARVIN U. B., and POWELL B. N. (1970) Lunar anorthosites and a geophysical model of the moon. *Proc. Apollo 11 Lunar Sci. Conf., Geochim. Cosmochim. Acta* Suppl. 1, Vol. 1, pp. 965–988. Pergamon.

WOOD J. A., MARVIN U. B., REID J. B., TAYLOR G. J., BOWER J. F., POWELL B. N., and DICKEY J. S., JR. (1971) Relative proportions of rock types, and nature of the light-colored lithic fragments in Apollo 12 soil samples. Second Lunar Science Conference (unpublished proceedings).

Proceedings of the Second Lunar Science Conference, Vol. 3, pp. 2197–2202
The M.I.T. Press, 1971.

Spectral directional reflectance of lunar fines as a function of bulk density

R. C. Birkebak and C. J. Cremers

University of Kentucky, Lexington, Kentucky 40506

and

J. P. Dawson

Scientific Specialties Corp., Houston, Texas

(Received 9 February 1971; accepted in revised form 5 April 1971)

Abstract—The spectral directional reflectance was measured for the lunar fines (samples 10084, 12001, 12070) as a function of bulk density from 1300 to 1800 kg/m³. The reflectance of the sample increased by as much as 40% with a density change from 1300 to 1800 kg/m³. Sample 12001 when examined under vacuum, before exposure to the atmosphere, showed a relatively strong absorption band at 0.62 μm which became much weaker after exposure to the atmosphere.

Introduction

Among the thermophysical properties needed by the heat transfer specialist for energy balance calculations on equipment on the lunar surface is the spectral directional reflectance. The surface may become more compacted around equipment on the moon due to intentional packing or because of astronaut activities. This packing of the surface causes a change in material density which will affect the thermophysical properties. The spectral directional reflectance of lunar fines is a function of angle of illumination and material bulk density.

A second reason for studying reflectance as a function of density is that none of the other investigators of Apollo samples, to our knowledge, state the density of the fines in their experiments. Some normalize their results to compare with telescopic measurements, but these results are not useful for heat balance calculations. We believe the variation on the magnitude of the Apollo 11 results may be caused by different densities used in the experiments.

When a fines sample is prepared by simple pouring it into a container and carefully leveling the surface, a bulk or material density of approximately 1300 kg/m³ is obtained. During the Apollo 11 and 12 missions core tube samples were obtained. Our bulk densities were selected according to the reported bulk densities of these core tube samples. The Apollo 12 Preliminary Science Report (1970) reported average in situ bulk densities of 1640 ± 40 kg/m³ for Apollo 11 and an average in situ bulk density 1800 ± 200 kg/m³ for Apollo 12.

In our study the bulk densities used were approximately 1300, 1400, 1600, and 1800 kg/m³. The directional reflectances were measured for a wavelength range from 0.55 to 2.2 μm and for angles of illumination of 10, 20, 30, 45, and 60 degrees. Studies were made on lunar fines, Apollo 11 sample 10084,68, and Apollo 12 samples 12001,19 and 12070,125. The only vacuum-packaged sample was 12001. Complete measurements are presented for sample 10084,68 and a partial set for samples 12070 and 12001.

The *directional reflectance* is defined as the ratio of the hemispherically reflected energy to the energy in the incident beam. The energy absorbed by lunar fines is easily calculated by subtracting the directional reflectance from unity.

Measurement Techniques

The center-mounted sample integrating sphere reflectometer measurement technique employed is described in detail in Birkebak *et al.* (1970). The directional reflectance used in this paper refers to the condition that the surface is illuminated by incident radiation contained in a solid angle $\Delta\Omega_i$, oriented at a specific angle 'Ψ' relative to the surface normal. The energy reflected is collected over the hemispherical space above the surface.

Sample Preparation

The Apollo 12 samples were packaged in special sample handling systems described by Birkebak *et al.* (1971). We had no control over the "as received" condition of these samples. Because of settling and some sample spillage during transportation from the Lunar Receiving Laboratory at NASA to our laboratory, we were able only to calculate density values which give a lower bound on the "as received" condition. The estimated "as received" bulk density of sample 12001 is 1630 kg/m³. Sample 12070 was repackaged so that densities up to 1800 kg/m³ could be obtained. Time allowed only the "as received" bulk density measurement on sample 12001 because of its use in thermal conductivity measurements.

The sample holder consisted of a Teflon cup 25 mm in diameter and approximately 3 mm in depth. Samples were measured out to the proper weight corresponding to the desired density and then carefully poured into the sample holder. To achieve a level surface the fines were packed by use of a vibrating tool held on the holder edge. Initial smoothing and packing of the surface is achieved with a stainless steel spatula.

Results

One Apollo 11 and two Apollo 12 samples were investigated. Apollo sample 12001 was packaged under vacuum conditions at LRL and could not be manipulated without destroying the vacuum environment to which it had always been exposed. Sample 12001 will be discussed first, then samples 10084 and 12070, for which multiple-density measurements were obtained. Sample 12070 was packaged and shipped in the special handling system under dry nitrogen.

Sample 12001,19

Ten and one-half grams of sample under vacuum were received in the special vacuum handling system which was connected directly to our vacuum integrating sphere system. The integrating sphere system was pumped to 10^{-6} torr and no increase in system pressure was noted when the transfer apparatus valve was opened to the system. Upon inspection of the sample it was noted that it had been disturbed; that is, in transit some of the sample had spilled from the sample holding cup into the handling system. Further inspection revealed that some small amount of lunar fines on the lip of the sample holding cup had prevented the sample cover from closing, thereby, allowing a small amount of sample to be vibrated out of the sample cup. The "as received" sample surface was not flat or level with the top of the sample holder cup but was inclined slightly with one chordal area of the sample depressed relative to its surroundings. The appearance suggested that the sample had slid back and forth in the holder.

Two events prevented us from obtaining absolute reflectances for this sample, and therefore only a normalized reflectance is presented. First, after the initial opening of the handling system gate valve, lunar fines from the spillage coated the o-ring seals thereby preventing a vacuum seal when the valve was closed. Our test procedure for setting the various angles of illumination required the blanking off of the handling system and the return to atmospheric pressure of the integrating sphere prior to setting of a new angle. The leakage caused by the contaminated o-ring seals eventually lead to exposure of the lunar fines to dry nitrogen and the atmosphere.

The second system failure, which unfortunately occurred simultaneously with the leakage, was that of the magnesium oxide coating on the interior of the integrating sphere reflectometer. The system had been refurbished but the MgO coating was not of sufficient thickness to be a diffusely reflecting surface at all wavelengths. This was truly unfortunate since this sample was our only true vacuum material tested.

Even with the above mishaps we were able to obtain some interesting and hopefully useful results. Because our 100% line measurement was in error, we have normalized our reflectance results in Fig. 1 with the measure sample reflectance at 1 μm. The results below this wavelength should give a true normalized curve. Of the three Apollo lunar fines tested, this sample, 12001,19, was the only vacuum packaged thermophysical property sample and the only one that showed more structure to the reflectance curve and a distinct absorption band. A very sharp band occurred between 0.60 and 0.64 μm. The origin of this band is unknown at this time. The depth of the absorption band is much greater than the probable error in our measurements of 2%. Our measurements indicate that upon exposure to dry nitrogen and atmospheric

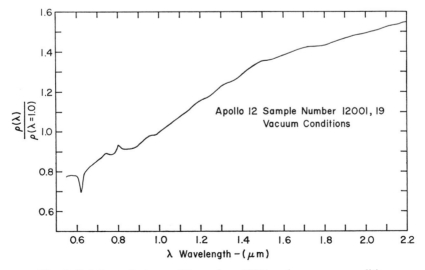

Fig. 1. Relative reflectance of lunar fines 12001 under vacuum conditions.

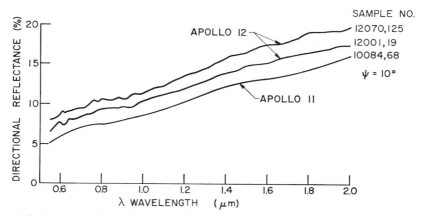

Fig. 2. Comparison of the spectral reflectance of Apollo 11 and Apollo 12 fines.

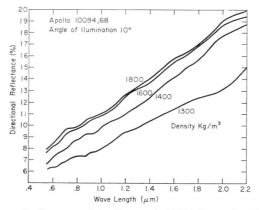

Fig. 3. Spectral reflectance of Apollo sample 10084,68 as a function of density.

air this band decreases in magnitude with time. It has been reported that the material is deficient in oxygen and the decrease in the absorption band is probably associated with chemical changes that have occurred. In Fig. 2 the results of reflectance vs. wavelength after exposure and repackaging of the sample to a density of 1300 kg/m³ is compared to our other lunar fines samples. However, no further tests for density effects have been conducted since the sample has been in use for thermal conductivity measurements.

Sample 10084,68

The results for this Apollo 11 sample for a density of approximately 1300 kg/m³ have been previously reported by Birkebak *et al.* (1970). A typical reflectance curve is shown in Fig. 2. In our new series of tests, densities of approximately 1400 (actually 1398.5 kg/m³), 1600, and 1800 kg/m³ were used. The highest density achieved by our method of packing was estimated to be between 1850 to 1900 kg/m³. The dependence of directional reflectance on density is shown in Fig. 3. In general, the reflectance increases with density for all wavelengths with an increase of up to 40% from the smallest to the largest density. As the surface becomes more compacted, that is a greater number of particles per unit volume, the void fraction decreases, and hence fewer number of cavities to trap and absorb radiation are present. Therefore, the reflectance increases for the sample. The reflectance increases more rapidly for the smaller densities changes. From our results it can be seen that we are approaching the limit where the porosity of the sample does not affect the reflectance values, that is, the material behaves as if it were a solid material.

The effect of angle illumination on reflectance for sample 10084,68 with a density

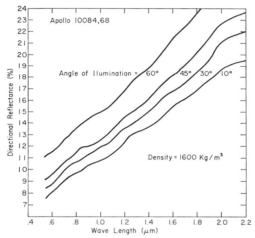

Fig. 4. Spectral reflectance of Apollo sample 10084,68 as a function of angle of illumination.

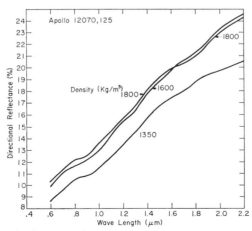

Fig. 5. Spectral reflectance of Apollo sample 12070,125 as a function of density.

of 1600 kg/m³ is shown in Fig. 4. Additional data of this type for the other densities are given in BIRKEBAK et al. (1971).

Sample 12070,125

Results for this sample were obtained for the "as received" density estimated to be 1350 kg/m³ and for densities of 1600 and 1800 kg/m³. Figure 5 presents the results obtained for these densities. The "as received" sample was run under vacuum condition 5×10^{-6} torr. Additional results for other angles of illumination and tabulated values for these measurements, are found in BIRKEBAK et al. (1971).

Again the effects of density change is the same as for sample 10084; for the middle and highest densities the increase in reflectance is small as compared to the lowest and middle densities. The reflectance for the "as received" condition is shown in Fig. 2. The directional reflectance as a function of angle of illumination is similar to that found for sample 10084,68.

Acknowledgments—We thank BLAKE NEVILLE, Jr. for his assistance and support with the design and construction of the equipment. We also want to thank TODD BIRKEBAK for taking and reducing some of the data. Support by NASA Contract NAS 9-8098 is gratefully acknowledged.

REFERENCES

BIRKEBAK R. C., CREMERS C. J., and DAWSON J. P. (1970) Directional spectral and total reflectance of lunar material. *Proc. Apollo 11 Lunar Sci. Conf.*, *Geochim. Cosmochim. Acta* Suppl. 1, Vol. 3, pp. 1993–2000. Pergamon.

BIRKEBAK R. C., CREMERS C. J., and DAWSON J. P. (1971) Thermophysical properties of lunar material (1971) TR-2, High Temperature and Thermal Radiation Laboratory, University of Kentucky, Lexington, Kentucky.

LSPET (Lunar Sample Preliminary Examination Team) (1970) Preliminary examination of lunar samples from Apollo 12. *Science* **167,** 1325–1339.

SCOTT R. F., CARRIER W. D., COSTES N. C., and MITCHELL J. K. (1970) Mechanical properties of the lunar regolith. *Apollo* 12 *Preliminary Science Report*, NASA SP-235, pp. 61–182.

Proceedings of the Second Lunar Science Conference, Vol. 3, pp. 2203–2211
The M.I.T. Press, 1971.

Far infrared properties of lunar rock

P. A. Ade, J. A. Bastin, A. C. Marston, S. J. Pandya,
and E. Puplett

Physics Department, Queen Mary College, Mile End Road, London E.1, England

(*Received* 16 *March* 1971; *accepted in revised form* 13 *April* 1971)

Abstract—Measurements of the dielectric constant of Apollo 11 and Apollo 12 samples at 8.9×10^{11} Hz give values in the range 6.2–7.2. A breccia sample has a dielectric constant of 3.7 ± 0.2. From 200 μm to 2000 μm the absorption spectrum of the fines has been investigated in detail and shows no sharp spectral features. The logarithmic slope of the absorption coefficient as a function of wavelength in this range varies monotonically from -3.3 ± 0.3 at 200 μm to -1.1 ± 0.3 at 2000 μm. The absolute value of the absorption coefficient at 2000 μm together with earth based measurements of the lunar brightness temperature at this and nearby wavelengths implies a conductivity for the top few centimetres of the lunar regolith of $8 \pm 4 \times 10^{-5}$ Jm^{-1} K^{-1}s^{-1}: a value over an order of magnitude less than that found from laboratory measurements with lunar fines. Also described are applications of these results to the determination of average surface temperatures of various regions of the lunar disc and the use of the moon as a calibrating source in submillimetre astronomy.

Far Infrared Properties of Lunar Rock

The study of the interaction of far infrared electromagnetic radiation with lunar rock has several applications. In this wavelength range a knowledge of the reflection coefficient is used to relate brightness temperatures, measured at various points of the lunar disc, to the corresponding surface and subsurface temperatures of the regolith. The reflection coefficient is also required in interpretations of relative and absolute intensities of radar reflection observations. The far infrared electromagnetic absorption coefficient depends on the lattice modes of vibration of the crystalline state of the rock and is thus capable of giving structural information; the theory of the subject in the case of polycrystalline agglomerates being still largely undeveloped. A comparison of the microwave brightness temperature measurements for the moon with laboratory measurements of lunar rock has been used to determine the thermal conductivity of the top surface of the regolith in situ on the lunar surface. The measurements of this coefficient are also of interest in determining the magnitude of the radiative term in the heat transfer process through lunar rock. In addition to reflection measurements, the absorption coefficient is also required to relate lunar brightness temperature measurements to thermodynamic temperatures at specific depths within the top layers of the regolith. The absorption coefficient and reflection coefficient may together be used to determine the loss tangent for the rock material. It is of interest to compare determinations of this parameter at far infrared frequencies with the extrapolation of measurements at lower frequencies where the lower attenuation may make possible the electrical prospecting of subsurface lunar layers.

Measurements of dielectric constant at 8.9×10^{11} Hz

The dielectric constant was determined from measurement of the reflection coefficient using rectangular cut blocks of lunar rock with, as a standard in each case, a polished brass block cut to the same shape as the sample under investigation. The arrangement is shown in Fig. 1. A cyanogen laser giving a high intensity at 338 μm was used as a source. The power was sufficiently intense for a Golay pneumatic cell to detect the reflected radiation with a high signal to noise ratio. The sample was placed on table E with the brass block resting on it, the two blocks having corresponding vertical faces coplanar. The laser beam, the four normals to the vertical rock faces and the detector were all in the same horizontal plane. The table was rotated by a motor at a constant rate so that reflections from four of the six faces were recorded automatically and at equal intervals of time. After several cycles, the table was lowered by exactly the height of the block and similar measurements were made with the brass block. Pen recordings of a series of reflections from both blocks are shown in Fig. 2. It will be noticed that there is some, although by no means an exact, correlation between the ratios of intensities of corresponding faces of the two blocks. This lack of correlation became more apparent as the distance between the sample and the detector was increased, the reflection from the brass surfaces being far more uniform in intensity. The effect was accounted for by slight concave or convex profiles of the rock samples. In addition, smaller scale size deviations from a perfect plane (saw-cuts on the sample surface) were found to be the cause of deviations in the reflected intensity from face to face especially at large angles of incidence. For this reason, in the case of one sample only (12063,96), the surface of the faces was ground flat to about 5 μm. This resulted in a fractional mass loss of about 0.008 of the sample but enabled far more consistent measurements to be made. Figure 3 shows the reflections in the case of this sample after the faces had been smoothed. It is clear that this gave much more uniformity in the fractional reflection. It has also lead to a somewhat higher value of the dielectric constant. For this smoothed sample, our reflection measurements now give a value of the dielectric constant of 6.25 ± 0.2 whereas, with the unsmoothed sample (as reported by BASTIN *et al.*, 1971), a value of 5.8 ± 0.4 was obtained. The sample is, however, somewhat unusual in being quite remarkably homogeneous. In the case of the breccia, for example, a conspicuously large inclusion composed mostly of ilmenite (diameter ~ 4 mm), gave a noticeably increased reflection, and other smaller inclusions were found to exhibit the same effect.

Using a unidirectional grid as an analyser, the laser beam was found to be very highly polarized with the electric vector parallel to the plane of incidence (p). An additional grid polarizer was used to make the polarization virtually complete (i.e., such that the s component was less than the noise level in the detector). Measurements of the reflection coefficient were made for a range of angles of incidence from $20°$ to $85°$. The plane of polarization was then rotated through an angle of $\pi/2$ and the measurements were repeated in the s mode. Results for samples 12063,96 and 10065,30 are shown in Figs. 4 and 5. In the case of the breccia sample, random orientations of the

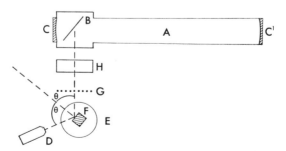

Fig. 1. Determination of reflection coefficient of lunar sample in block form. A is a cyanogen laser with reflecting mirrors CC' and beam splitter B. H rotates the beam of polarization by $\pi/2$ and is present for S-type measurements only. G is a polarizing grid which ensures that the radiation is effectively plane polarized. The block F is placed on the rotating table E and the radiation reflected into the detector D.

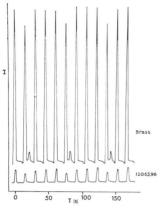

Fig. 2. Pen recording showing reflection from the block 12063,96 (type A) and a brass block. The intensity of the reflected beam is shown in the vertical direction, the horizontal axis showing time in seconds. The regular rotation of the sample gives a rotation of the reflected beam, and this intersects the detector giving the regular peaks shown in the figure. The lack in consistency in the reflection from the lunar block is attributed mainly to surface roughness.

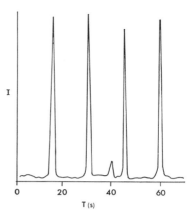

Fig. 3. Reflection from the block 12063,96 after polishing. In all other respects, the measurements are the same as those shown in Fig. 2. The very small peak between the second and third reflections is caused by a corner facet which could not have been removed except with an excessive fractional loss to the mass of the block.

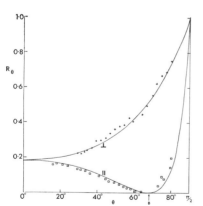

Fig. 4. Reflection measurements for the block 12063,96. The coefficient of reflection R_θ is shown as a function of the angle of incidence θ. Plane polarized 338 μm radiation was used, (a) when the E vector was in the plane of incidence indicated by squares, (b) when the E vector was perpendicular to the plane of incidence indicated by crosses. The continuous curves refer to the corresponding Fresnel expressions for reflection from a non-absorbing material of dielectric constant 6.25. B indicates Brewster's angle.

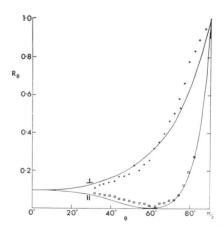

Fig. 5. Reflection measurements from the block 10065,30 of breccia. The symbols are used in the same context as in Fig. 4. The continuous curves are computed for the value of dielectric constant 3.7.

microcrystalline surfaces with respect to the incident beams are thought to be the cause of the deviations of the observations from the predicted angular variation of reflection for a smooth surface. It would be expected that such a situation would in general make the *s* and *p* reflection coefficients more nearly equal than in the smooth case whilst, in particular, at the Brewster angle, the *s* term would be non zero, and all of these effects are in fact seen in Fig. 5. The measurements of absorption to be described in the next section show that the loss tangent is sufficiently low at this wavelength for the simple Fresnel formula for reflection to be applied. The dielectric constant deduced from comparison with experiment is shown in Table 1 for a number of samples.

Absorption measurements in the wavelength range 200–5000 μm

The object of these measurements was to confirm with greater accuracy the results already reported (BASTIN *et al.*, 1970) for the far infrared absorption spectrum, and to extend the analysis both to other sample types and to a range of higher wavelengths. Figure 6 shows the results of measurements on lunar fines (sample 10084,111) using the 338 μm cyanogen laser source. The high intensity of the source enables the measurements to be extended to a sample thickness where the fractional transmission is very low. A simple analysis shows that, as the sample thickness is increased, the effects of reflections and inhomogeneities in the sample have progressively smaller effect on the calculated absorption coefficient. The cyanogen laser source is known to have other emission frequencies of lower intensities (STEFFEN *et al.*, 1966a and b), which might be thought to vitiate the measurements where the transmission is very low but a simple analysis of the situation shows that the effect on the calculated result can be ignored. For the Apollo 11 fines (sample 10084,111) a value of $2.5 \pm 0.2 \text{ kg}^{-1} \text{ m}^2$ has been deduced for the electromagnetic absorption coefficient at 338 μm. At the

Table 1. Dielectric constants of rock samples.

Rock	Dielectric Constant
10065,30 breccia Apollo 11	3.7 ± 0.2
10017,64 type A basalt Apollo 11	7.05 ± 0.3
12063,96 type A basalt Apollo 12	6.25 ± 0.2
terrestrial olivine basalt (New Zealand)	6.35 ± 0.3

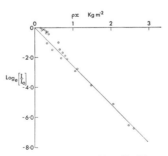

Fig. 6. Transmission intensities for lunar fines (Apollo 11 sample 10084,111) at 338 μm. A cyanogen laser source and Golay pneumatic cell detector were used. This arrangement makes possible a high sensitivity so that samples which reduce the initial intensity by several orders of magnitude can be used, thus reducing the effect of reflections on the calculated absorption. I and I_0 refer respectively to the intensities before and after transmission through the sample. ρ is the density and x the thickness of the sample.

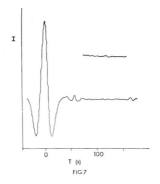

FIG 7

Fig. 7. Interferogram using polarizing interferometer (MARTIN and PUPLETT, 1969) and a 4 mm thick sample of Apollo 11 fines (10084,111). The figure shows the intensity I recorded as a function of the time T indicated here in seconds. The relatively broad zero order maxima and absence of structure in the interferogram indicate an absorption coefficient, which increases with frequency, as well as the absence of sharp spectral features. The inset shows the detector noise level with the interferometer plates kept at a constant level.

same wavelength, a number of samples of microbreccia and crystalline rock, kindly lent by Dr. S. O. Agrell, have been investigated and these are found to have mass absorption coefficients in the range 1.2–2.0 kg^{-1} m^2.

The polarising interferometer developed by MARTIN and PUPLETT (1969) has been employed to investigate the absorption spectrum in the range 200 μm–5000 μm. A typical interferogram is shown in Fig. 7. The corresponding transform gives the transmission intensities from which the absorption coefficient was calculated. The results are shown for the case of the lunar fines in Fig. 8. The spectra were obtained with a resolution of 0.5 cm^{-1} although, in an attempt to see structure in the spectrum, some measurements were made with a resolution of 0.1 cm^{-1}. In no case was any structure noticeable.

DISCUSSION

The infrared absorption properties of lunar fines previously reported (BASTIN et al. 1970) indicates a submillimetre absorption which increases very rapidly with decrease in wavelength. Below 100 μm the measurements show a much more constant mean value of the absorption coefficient but with obvious spectral structure presumably associated with vibrational absorption due to the crystalline lattice bonding. More recently this mid-infrared spectral structure has been examined at greater wavelength resolution by (ESTEP et al., 1971). Their measurements were made on separated fines and each mineral type showed considerable structure in the 8–40 μm region. A well marked transmission minima associated with a Si–O bond was found in all cases at 10 μm confirming the previously reported absorption maxima measured at this wavelength for unseparated fines (BASTIN et al., 1970). Although the measurements of ESTEP et al. (1971) are of high wavelength resolution and therefore of great value in determining structural bonding, it is unfortunate that as reported they give insufficient data to be used to determine absolute absorption coefficients in a wavelength range in which the most of the radiative component of thermal transfer in the fines takes place.

At wavelengths above 100 μm the measurements made with the polarising interferometer (MARTIN and PUPLETT, 1969) indicate no rapidly varying spectral features although the measurements have been made with high wavelength resolution. The extension of the measurements to wavelengths above 1000 μm is interesting because of the correlation between these measurements and the earth-based telescopic observations of the lunar brightness temperature as a function of lunar phase (see for example TROITSKI, 1967; BASTIN and GEAR, 1967). These telescopic observations interpreted in terms of a smooth homogeneous regolith with temperature independent thermal parameters indicate that the mass electromagnetic absorption coefficient κ_λ at wavelength λ is given by the expression

$$\kappa_\lambda = \frac{\omega^{1/2}}{2^{1/2}B\lambda}\left(\frac{c}{k\rho}\right)^{1/2} \tag{1}$$

where $(2\pi/\omega)$ is the lunar synodic period, c the specific heat, k the conductivity, and ρ the density of the top lunar surface. B is a constant determined from earth-based telescope observations in the microwavelength region from which measurements it is found that

$$B = 2.4 \pm 0.3 \times 10^2 \text{ m}^{-1}. \tag{2}$$

Naturally, the plane parallel temperature independent assumptions are open to considerable criticism although, if c and k refer to values corresponding to the mean lunar temperature, we would expect the deductions to be relatively exact (i.e., fractionally correct to about $0.1 \sim 0.3$). From equations (1) and (2) the quantity $(c/k\rho)^{1/2}$ for the lunar regolith may be determined. Since the thermal balance of the regolith is controlled largely by the surface fines, we will expect the deduced value of the parameter to relate to these fines provided that our laboratory measurement of κ_λ is also made on this material. Furthermore, the value deduced should relate to the material in situ on the lunar surface. Since ρ and c are known with relatively high accuracy, it is convenient to regard the comparison as essentially a determination of the thermal conductivity k, especially since this quantity may well change with sample handling, whereas c would remain unchanged.

Figure 8 immediately shows a problem with the comparison. It is clear from this figure that in no part of the range, except possibly above 1500 μm does the absorption coefficient vary inversely with wavelength. If the measured absorption coefficient is written over any short wavelength range in the form

$$\kappa_\lambda = A\lambda^n \tag{3}$$

where A is a constant, then n varies from -3.2 at 200 μm to -1.3 at 1500 μm. Although above 1000 μm there is considerable scatter in Fig. 8, careful inspection shows that measurements made with the thicker samples agree well with each other. In this wavelength range one ought properly to disregard measurements made with the thinner range of samples: these thinner samples were included only to give a measureable transmission at the lower wavelengths.

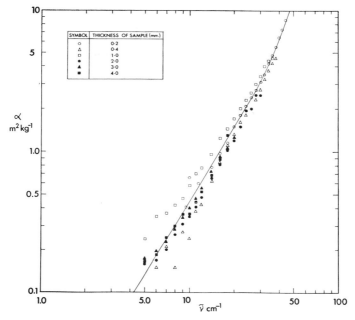

Fig. 8. Absorption spectrum in the range 200 μm–2000 μm for Apollo 11 fines (10084,111). α represents the absorption coefficient and $\tilde{\nu}$ the wave number. Measurements at a given wavelength with different sample thicknesses cover a large range of fractional transmissions (see the discussion). In particular above 1000 μm wavelength, only the thicker samples (solid symbols) can be expected to have appreciable significance.

Considering only the measurements on the thicker samples, it seems reasonable that at 2000 μm the wavelength variation of attenuation of the laboratory measurements is in agreement with equation (1) and, if we set the two values numerically equal at this wavelength, we obtain

$$\frac{k\rho}{c} = 1.8 \times 10^{-4} \, \text{s}^{-1} \, \text{m}^{-4} \, \text{kg}^2 \tag{4}$$

Using a value of $\rho = 1.3 \times 10^3$ kgm^{-3} and the mean specific heat 5.5×10^2 J kg^{-1} K^{-1} taken from the data of ROBIE et al. (1970), the conductivity becomes

$$k = 8 \times 10^{-5} \, \text{Jm}^{-1} \, \text{K}^{-1} \, \text{s}^{-1} \tag{5}$$

There is thus a large discrepancy between this value and the value determined by direct laboratory measurement (CREMERS et al., 1970). Previous determinations by the method described here (BASTIN et al., 1970) give lower values for the conductivity than that obtained by the direct laboratory method: the present measurements confirm the discrepancy but show it to be greater in magnitude so that the ratio of the two measured determinations is approximately 20. Although both the theory and experimental techniques used in the two cases would entail some error, it seems very unlikely that this very large factor could be accounted for in this way. In discussing reasons for this discrepancy, we note that the two determinations refer essentially to different situations: in one case the fines are packed into a container whilst, in the other case, the mean conductivity of the top few centimeters of the surface of the moon is effectively determined in situ. A lower degree of compaction of the fines on the lunar surface, and also the possibility of adsorbed gases at intergrain junctions in the laboratory determination, are both factors which could account for a discrepancy in the observed direction and we are at present considering both these possibilities. Finally, it should be noted that the value for the purely radiative term in the conduction process may be determined from near infrared absorption measurements.

The only reported values for the lunar dust absorption coefficient in this wavelength range have a mean of about 50 m² kg⁻¹ (BASTIN *et al.*, 1970). This figure, together with the theoretical expression given by CLEGG *et al.* (1966), suggests a radiative term almost as large as that given in equation (5). This very low value of the conductivity of the top surface layers of the lunar rock is thus still consistent with the near infrared data although a more thorough check of the values of absorption coefficients in the near and middle infrared ranges (3–100 μm) is clearly indicated. The very rapid increase of absorption coefficient with frequency which we have observed for submillimetre wavelengths has also important consequences for astronomy. Our results indicate that below 500 μm wavelength during a lunation the depth of penetration of the thermal wave into the lunar surface is very much greater than the corresponding penetration depth for electromagnetic radiation. Thus, at these wavelengths, the lunar brightness temperature is, apart from an albedo factor, simply a measure of the thermodynamic surface temperature. This result has two important contexts. Firstly, in sharp contrast to observations at mid infrared wavelengths (ALLEN and NYE, 1969), far infrared measurements below 500 μm can, by virtue of the exactness of the Rayleigh-Jeans approximation, give a true mean of the surface temperature of localised regions and not one which at lunar night is almost entirely controlled by a relatively small fraction of exposed rock. Secondly, the mean lunar brightness temperature at these wavelengths can now be calculated with greater accuracy as a function of lunar phase, thus providing a useful calibration for the development of astronomy at these wavelengths.

Measurements of the dielectric constant of various rock samples agree in general with an extrapolation to high frequencies of the measurements of CHUNG *et al.* (1971) and COLLETT and KATSUBE (1971). The lower values of dielectric constant for the breccia seems to us almost certainly a result of the relatively high void ratio for this material, a factor which we believe has considerable effect even in the type A samples.

Acknowledgments—We wish to thank Dr. P. E. Clegg for a great deal of advice with this work. We also are indebted in a number of ways to Dr. A. B. Poole and Dr. W. J. French of the Geology department of this College. The spectroscopic absorption measurements reported here were obtained with a Beckman—R.I.I.C. FS 720 Michelson interferometer converted to operate as a polarising interferometer.

REFERENCES

ALLEN D. A. and NYE E. P. (1969) Lunar thermal anomalies: Infrared observations. *Science* **164**, 419–421.

BASTIN J. A., CLEGG P. E., and FIELDER G. (1970) Infrared and thermal properties of lunar rock. *Geochim. Cosmochim. Acta* Suppl. 1, Vol. 3, 1987–1991. Pergamon.

BASTIN J. A., CLEGG P. E., and GOUGH D. O. (1971) Thermal and infrared properties of lunar rock. Second Lunar Science Conference (unpublished report). Heat transfer in lunar rock.

BASTIN J. A. and GEAR A. E. (1967) Observations in the wavelength range 1–3 mm. *Proc. Roy. Soc.* **A296,** 348–353.

CHUNG D. H., WESTPHAL W. B., and SIMMONS G. (1970) Dielectric properties of Apollo 11 samples and their comparison with earth materials. *J. Geophys. Res.* **75,** 6524–6531.

CHUNG D. H., WESTPHAL W. B., and SIMMONS G. (1971) Dielectric behaviour of lunar samples. Second Lunar Science Conference (unpublished report).

CLEGG P. E., BASTIN J. A., and GEAR A. E. (1966) Heat transfer in lunar rock. *Mon. Not. R. Astron. Soc.* **133,** 63.

COLLETT L. S. and KATSUBE T. J. (1971) Electrical properties of Apollo 11 and Apollo 12 lunar samples. Second Lunar Science Conference (unpublished report).

CREMERS C. J., BIRKEBAK R. C., and DAWSON J. P. (1970) Thermal conductivity of fines from Apollo 11. *Proc. Apollo 11 Lunar Sci. Conf., Geochim. Cosmochim. Acta* Suppl. 1, Vol. 3, pp. 2045–2050. Pergamon.

ESTEP P. A., KOVACH J. J., and KARR C. (1971) Infrared vibrational spectroscopic studies of minerals from Apollo 11 and 12 lunar samples (Jan. 1971, Lunar Science Conference unpublished report).

KROTIKOV, V. D. (1962) Some electrical characteristics of terrestrial rocks and their comparison with the characteristics of the Moon's surface layer. *Izv. vyssh. ucheb Zaved* (*Radiophysics*) **5,** 1057. Pergamon.

MARTIN D. H. and PUPLETT E. (1969) Polarized interferometric spectrometry for the millimetre and submillimetre spectrum. *Infrared Physics* **10,** 105–109.

ROBIE R. A., HEMMINGWAY B. S., and WILSON W. H. (1970) Specific heats of lunar surface materials from 90°K to 350°K. *Science* **167,** 749–750.

STEFFEN H., STEFFEN J. MOSER J. F., and KNEBÜHL F. K. (1966(a)) Stimulated emission up to 0.538 mm wavelength from cyanic compounds. *Phys. Lett.* **20,** 20–21.

STEFFEN H., STEFFEN J., MOSER J. F., and KNEBÜHL F. K. (1966(b)) Comments on a new laser emission at 0.774 mm wavelength from ICN. *Phys. Lett.* **21,** 425–426.

TROITSKI V. S. (1967) Investigation of the surfaces of the Moon and planets by means of thermal radiation. *Proc. Roy. Soc.* **A296,** 366–398.

Proceedings of the Second Lunar Science Conference, Vol. 3, pp. 2213–2221
The M.I.T. Press, 1971.

Physical characterization of lunar glasses and fines

W. B. WHITE, E. W. WHITE, H. GÖRZ, H. K. HENISCH,
G. W. FABEL, R. ROY, and J. N. WEBER
Materials Research Laboratory, The Pennsylvania State University
University Park, Pennsylvania 16802

(*Received* 23 *February* 1971; *accepted in revised form* 25 *March* 1971)

Abstract—Lunar fines and glasses from Apollo 11 and Apollo 12 samples have been characterized by diffuse reflectance, Raman spectroscopy, X-ray emission spectroscopy and thermally stimulated currents. Diffuse reflectance, in agreement with other workers, shows Apollo 12 fines to be more reflecting than Apollo 11. Comparison with spectra of synthetic anorthite glass shows that the glassy component of the fines may well contribute to the dark background absorption of the lunar glass. However, the absorption bands in the synthetic glass do not coincide with those in the spectra of the lunar glass. Thermally stimulated current and thermally stimulated depolarization measurements show the existence of low-temperature traps in the lunar fines and offers an alternative route to trap characterization in non-luminescent specimens. Individual fragments of lunar glass from Apollo 11 and Apollo 12 have been examined by X-ray emission and Raman spectroscopy. The glass fragments are very heterogeneous. Both methods indicate unusual structures in the glass, and many of the glass fragments show incipient crystallization as evidenced by sharp bands in the Raman spectrum.

INTRODUCTION

PHYSICAL PROPERTIES of lunar fines and lunar glass Apollo 11 and Apollo 12 localities have been measured. In the case of the unsorted lunar fines, the object was to obtain information about the electronic behavior of the bulk material, and in particular the contribution of the glass to the optical properties. In the case of the glass, the object was structural characterization by nondestructive methods. Glass structure is difficult to determine under any circumstances and is doubly so for the lunar glass because of small sample size and heterogeneity.

DIFFUSE REFLECTANCE SPECTRA

Studies of the reflectance spectra of lunar soil from the Apollo 11 material show a rather monotonously increasing absorption from the near infrared through the visible. The smooth curve is broken by two pronounced absorption bands, one at 940 nm, and a weaker one near 1900 nm (ADAMS and MCCORD, 1970; BIRKEBAK *et al.*, 1970). The absorption features were associated with the characteristic spectrum of Fe^{2+} in clinopyroxene.

In the present study, we examine the diffuse reflectance spectra of three specimens of lunar soil from Apollo 12 and compare the spectra with one specimen from Apollo 11. In particular, we are concerned with the role of glass in the lunar fines and have therefore also compared the spectra of the lunar materials with spectra of several synthetic glasses containing iron and titanium.

Diffuse reflectance spectra were obtained on a Beckman DK–2A spectro-photometer with an integrating sphere attachment. The sphere was coated with $BaSO_4$ paint, and $BaSO_4$ was used as a reference. Spectra of the synthetic glasses were obtained on a Cary model 14 spectrophotometer in the transmission mode using polished slabs of glass.

The lunar fines

Reflectance spectra obtained from pressed pellets of lunar fines are shown in Fig. 1. The absorption spectra from the Apollo 12 fines are all remarkably similar. The absorption of the material from the Apollo 11 site is distinctly higher (i.e., the reflectivity is lower), as has also been noticed by other workers concerned with the spectra of the lunar fines (ADAMS, 1971; BIRKEBAK and DAWSON, 1971; GARLICK *et al.*, 1971). There appears a distinct absorption feature at 940 nm which seems to be somewhat more intense in the Apollo 12 material than the equivalent feature in the Apollo 11 material. The expected spectral feature at 1900 nm appears as a broad and rather indistinct shoulder. The small dip that occurs in the spectra at this wavelength is an artifact caused by water absorption in the instrument. No additional features that can be specifically correlated with iron or titanium crystal field or charge transfer absorption can be distinguished.

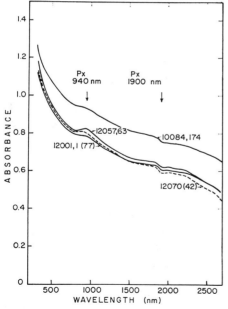

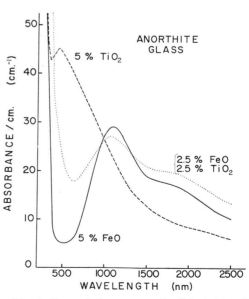

Fig. 1. Diffuse reflectance spectra of bulk samples of lunar fines. All spectra are adjusted to a base line of the BaSO₄ reference.

Fig. 2. Transmission spectra of polished slabs of synthetic anorthite glass. Concentrations of transition oxides in weight % are given on the curves.

Synthetic glasses

Spectra of three synthetic glasses are shown in Fig. 2. The base glass had the anorthite composition, $CaAl_2Si_2O_8$, to which was added the specified concentration of FeO and TiO_2. The glasses were melted in molybdenum crucibles at 1700°C for 30 minutes. Spectra were obtained from polished slabs of glass.

The characteristic band of 6-coordinated Fe^{2+} occurs at 1080 nm in anorthite glass. The shoulder at 1850 nm does not usually appear in glass apectra, and many indicate Fe^{2+} in some other coordination. When titanium is the only transition metal present, a Ti^{3+} band occurs at 430 nm in the visible. When both Fe^{2+} and Ti^{3+} are present in the glass, the 1080 band of Fe^{2+} still appears, but the titanium band is washed out and only a monotically rising absorption curve appears in the visible. The long tail of this absorption feature extends into the near infrared with the iron band superimposed on it. This continuous background absorption is due to charge transfer processes between titanium ions and between iron and titanium ions.

Origin of spectra

The reflectance spectra of the lunar fines may well contain a contribution from charge transfer processes between the various valence states of iron and titanium in the lunar glasses. The absorption bands that appear superimposed on the monotonic background appear to be due mainly to Fe^{2+} in the distorted M2 site of the pyroxenes. Low-calcium ortho- and clinopyroxene both have characteristic bands near 900 nm, while high-calcium pyroxenes, olivine, and the glass all have the octahedral iron band above 1000 nm (BURNS, 1970; LEWIS and WHITE, 1971). The coincidence of the rather broad absorption maxima, however, does not exclude the possibility that some contribution comes from the iron in the glassy phases.

Titanium does not appear to give a distinct band in the bulk spectra of the lunar fines. Such absorption if present is lost in the general charge transfer (or coloration from radiation damage centers) background. Only Ti^{3+} would give a crystal field band. Ti^{4+} does not have an absorption band of its own although it contributes very strongly to the charge transfer processes. The general increase in reflectance in the Apollo 12 lunar fines correlates rather well with the lower titanium concentrations. Also lower, of course, are all of the hosts that carry the coloring ions—opaques, pyroxenes, and glasses. The feldspars generally do not accept the coloring ions in solid solution.

THERMALLY STIMULATED CURRENTS AND THERMALLY STIMULATED DEPOLARIZATION

Several workers (DALRYMPLE and DOELL, 1970; HOYT et al., 1970; GEAKE et al., 1970) have demonstrated the presence of trapping levels in Apollo 11 samples by thermoluminescent measurements performed above room temperature. BLAIR and EDGINGTON (1970) and BLAIR et al. (1971) found a thermoluminescent glow peak below room temperature (near −135°C) in sorted fractions of Apollo 11 and Apollo 12 fines. They also reported the presence of a water-induced glow peak at −90°C which could be annealed out by heating above 200°C.

In the present investigation on unsorted fines, no thermoluminescence was detected in the temperature range −190°C to 100°C. The study of trapping levels in near-insulating materials of this type is severely limited by a shortage of applicable experimental procedures. One method which has been applied to such solids is the measurement of thermally stimulated currents (TSC). However, the interpretation of these measurements is hindered by a voltage dependent conductivity which arises from contact resistance and space-charge controlled processes in the bulk (HENISCH and FABEL, 1970). Alternative procedures, which are free from some of the difficulties associated with non-ohmic behavior, depend on electret processes (FABEL and HENISCH, 1970); one of these is thermally stimulated depolarization (TSD). They regard the material as a medium for polarized charge storage. The amount of charge that can be stored is a measure of the trap density and the temperature at which the overall dipole is discharged is (in the most general terms) a measure of the trap depth. Although a quantitative assessment of trapping levels by either of the above methods is not likely to be straightforward, the methods can be reliably used as sensitive detectors of the presence of trapping levels in near-insulating materials.

Measurements

A powder sample of lunar fines (12057,63) was pressed into a pellet 1 cm in diameter and 0.2 cm thick. To explore the possibility of trap characterization in this near-insulating specimen, the methods of thermally stimulated currents (TSC) and thermally stimulated depolarization (TSD) were employed. The sample was far from ideal for measurements of this kind, but since the specimen was found to be non-luminescent, a characterization of its trapping levels could be performed only by electrical methods.

The TSC measurements were performed in the usual manner. The pressed sample was cooled to −190°C and irradiated with UV (3650 Å) for a predetermined time interval ranging from 10 to 45 minutes. After irradiation, a field of 250 v/cm was applied across the pressed pellet, and the current was monitored while the temperature of the specimen was increased at a linear rate (30°C/min). For electrical contact, the specimen was gently clamped between a heated brass block and "Nesa" glass electrode. A current maximum at approximately −5°C was recorded for all UV exposure times tested (Fig. 3a). The magnitude of this peak varied within wide limits for a given irradiation dosage, presumably as a result of changes in the intergranular contact conditions. However, the temperature of the peak varied from measurement to measurement by only a few degrees. In some cases an additional peak was recorded near 40°C, but this could not be reproduced consistently.

For TSD measurements, the specimen was irradiated with UV at −190°C while in the presence of a large electric field (2500 v/cm). The UV and then the field were removed and the specimen stored for 15 minutes at −190°C before depolarization was measured. The specimen was then connected to the electrometer and the discharge was monitored while the temperature was increased at a uniform rate. The depolarization curve possessed one broad peak at approximately the same temperature as the TSC peak and one at 80°C (Fig. 3b). A smaller depolarization was recorded when the electret was formed without UV irradiation, i.e., by application of the field alone.

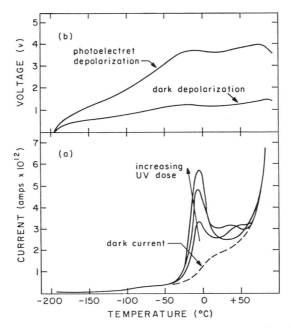

Fig. 3. Electrical measurements on sample 12057, 63, (a) thermally stimulated current after different UV exposures. UV exposure times: 10, 30, and 45 minutes; (b) photo-electret depolarization. Heating rate: 30°C/min. UV lamp output ~ 10 μ watt-sec/cm².

The peak at $-5°C$ in the TSC glow curve can be explained by carrier release from a single trapping level in the bulk or a closely bunched set of levels. A corresponding maximum in the TSD curve supports this hypothesis. The TSD method generally yields better resolution than TSC (SEMAK *et al.*, 1967). However, because of the nature of the space-charge polarization process, it favors the surface regions (versus bulk), and the broad response in Fig. 3b suggests that these are highly disturbed. Presumably this is due to the bombardment the material had received in the lunar environment.

For a sample of this type, a quantitative analysis of the above results would not be straightforward. The usual problems associated with electrical measurements on near-insulating materials as discussed above apply, of course, in this case. However, the fact that structure of the kind given in Fig. 3 can be revealed shows that the methods are intrinsically capable of useful application.

X-RAY EMISSION WAVELENGTH SHIFTS

Aluminum, silicon, and oxygen X-ray emission band spectra were recorded for a series of lunar glasses and a simulated lunar glass in two stages of densification.

The AlKβ and SiKβ bands were recorded as reported earlier (WHITE and GIBBS, 1967, 1969). The oxygen Kα emission band shifts were recorded as described in the work by GIGL *et al.* (1970) and measured with respect to an α-quartz standard. All spectra were obtained using an Applied Research Laboratories electron microprobe

Table 1. X-ray emission shift data for glasses (all shifts (Δ) recorded as Å $\times$ 10^{-4}).

Sample	ΔAlK_β	ΔSiK_β	ΔOK_α
10084,110			
2.8	18	43	190
2.9	48	35	210
3.0	39	23	190
3.1	35	36	290
10085,49	15	65	320
APOLLO 12 GLASS	22	39	160
High Ti, Fe, Al *synthetic*			
n = 1.727	30	15	190
n = 1.789	14	50	265

model EMX operated at 20 KeV and from 0.10 to 0.20 μA specimen current. A large spot diameter ($\sim$ 100 μm) was used throughout. Specimens were prepared as polished grain mounts using Lakeside 70 cement. Shift (Δ) measurements for AlKβ and SiKβ are precise to about ±5, whereas the $\Delta OK\alpha$ values are good to about ±15.

There are significant shifts in all three emissions bands among the lunar glass specimens, but no extensive interpretation can be made at this time for lack of descriptive data on the specimens. Interpretation is further complicated by the inhomogeneous nature of the glasses. Table 1 summarizes the results from the glasses.

The magnitude of the AlKβ shift (arbitrarily measured with respect to α-Al$_2$O$_3$ as used by WHITE and GIBBS (1969) is about the same as obtained from the synthetic glasses.

The average SiKβ shift (measured with respect to α-quartz) is on the order of 0.0040 Å, corresponding to a mean Si—O tetrahedral bond length of about 1.630 Å. The shift (Δ) of 65 for the glass from 10085,49 is the highest we have found for any silicate material. Most noteworthy of the SiKβ shift data is the large change with densification of the synthetic glass. If one can assume that this is due to changes in the Si—O bond lengths, then this represents a change from a value of about 1.615 Å for the less dense glass to about 1.635 Å for the densified material. If these tentative data are correct, then it means that the Si—O bonds are relatively more affected by changes in glass density than are the Al—O bonds. The large $\Delta SiK\beta$ and $\Delta OK\alpha$ values for the 10085,49 glass may possibly indicate an unusually large degree of covalency in the Si—O bonds.

RAMAN SPECTRA

Raman spectra were measured on individual spheres or shards of glass, using a Spex Ramalog double spectrometer with an ionized-argon laser source. The 488 or 514.5 nm lines of the laser can be focused to an area of less than 0.01 mm² so that measurements could be made on individual particles. The Raman scattered light was observed at 90° to the incident beam. The glass particles were placed at the focal point supported by a thin cast collodion tube which itself gave no spectrum under measurement conditions.

Glass spectra are extremely weak under all conditions and the dark lunar glasses gave extremely feeble spectra because of poor beam penetration. The Raman spectrum, however, appears to be very sensitive to traces of crystalline phases or to incipient

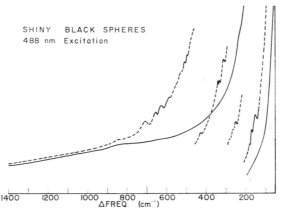

Fig. 4. Raman spectra of two specimens of black glassy spheres from sample 10084,110. Vertical axis is scattered intensity. Solid curve is a completely glassy specimen; dashed curve indicates recrystallization.

crystallization of the glass. The presence of crystals gives rise to numerous sharp bands superimposed on the low-frequency scattering curve. Fig. 4 shows spectra of two black glassy spheres from sample 10084. One appears to be completely glassy with only faint suggestions of spectral features (solid lines). The other (dashed lines) is partly recrystallized and shows a complex set of sharp bands in the low-frequency region. This variation from one particle to the next was observed in most of the glassy samples. Many particles of "glass" were shown by the Raman spectra to be partly crystalline.

Fig. 5 shows some representative spectra of several kinds of lunar glass. Apollo 11

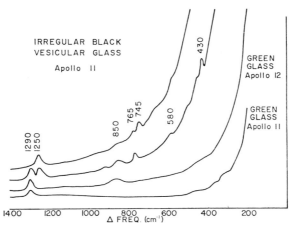

Fig. 5. Raman spectra of four typical lunar glasses. Both upper curves are for two different fragments (0.1 to 0.3 mm across) of vesicular green-black glass from 10084, 110. Apollo 11 and Apollo 12 green glass spectra are each from single particles, 0.1 to 0.3 mm in size, of transparent pale green equant fragments of glass. All spectra are very weak and instrument was operated near the limit of its sensitivity.

sample 10084 contained tiny irregular fragments of highly vesicular green-black glass. The significant features are the bands near 1200 to 1300 cm⁻¹ and the weak broad features at 850 and near 750 cm⁻¹. These represent stretching and bending vibrations respectively of the SiO_4 tetrahedra. Different fragments of the same glass from the same subsample gave quite different Raman spectra, and one showed some evidence of crystallization. One of these glasses yielded a stretching frequency at 1250 cm⁻¹, while the second fragment exhibited two such bands. The shards of clear green glass, one from 10084 of Apollo 11 and one from Apollo 12, gave rather similar spectra with identical stretching frequencies.

Both stretching frequencies are usually high. Raman data on silicate glasses are extremely sparse, but such data as exist indicate that the highest stretching frequency observed is at 1200 cm⁻¹ in fused silica. The addition of alkali to the silica generates a new band at 1100 cm⁻¹ and the 1200 cm⁻¹ band disappears (HASS, 1970).

These data are preliminary, but indicate (1) that reasonable-quality Raman spectra can be obtained from non-ideal samples such as the specks of lunar glass, (2) that there is an extreme degree of variability and heterogeneity among the particles of lunar glass from the same split of the same sample, and (3) that true glass is rare. Most specimens, even those that appear microscopically to be glass, show Raman evidence for some crystallinity.

Raman spectra measured on synthetic anorthite, albite, and silica glasses show only broad spectral features. In particular, the aluminum-containing feldspar glasses show very broad smeared-out scattering in the 900–1200 cm⁻¹ region, indicating that the stretching motions of SiO_4 and AlO_4 tetrahedra are mixed by the glassy structure. The sharp bands in this spectral region of the lunar glasses become even more puzzling and their final interpretation must await further work.

Acknowledgments—This work was supported under NASA Grant NGR-39-009-(152) and through the Space Sciences and Engineering Laboratory of The Pennsylvania State University, operated under NASA Grant NGL-39-009-015. We are indebted to Miss S. Kurtossy for preparing the synthetic glasses.

REFERENCES

ADAMS J. B. (1971) Second Lunar Science Conference (unpublished Proceedings).
ADAMS J. B. and MCCORD T. P. (1970) Remote sensing of lunar surface mineralogy: Implications from visible and near-infrared reflectivity of Apollo 11 samples. *Proc. Apollo 11 Lunar Sci. Conf.*, Geochim. Cosmochim. *Acta* Suppl. 1, Vol. 3, pp. 1937–1945. Pergamon.
BIRKEBAK R. C., CREMERS C. J., and DAWSON J. P. (1970) Directional spectra and total reflectance of lunar material. *Proc. Apollo 11 Lunar Sci. Conf.*, Geochim. Cosmochim, *Acta* Suppl. 1, Vol. 3, pp. 1993–2000. Pergamon.
BIRKEBAK R. C. and DAWSON J. P. (1971) Second Lunar Science Conference (unpublished proceedings).
BLAIR I. M. and EDGINGTON J. A. (1970) Luminescence and thermoluminescence under 159 MeV proton bombardment of the lunar material returned by Apollo 11. *Proc. Apollo 11 Lunar Sci. Conf.*, Geochim. Cosmochim. *Acta* Suppl. 1, Vol. 3, pp. 2001–2012. Pergamon.
BLAIR I. M., EDGINGTON J. A., and JAHN R. A. (1971) The luminescent properties of Apollo 11 and 12 material. Second Lunar Science Conference (unpublished proceedings).
BURNS R. G. (1970) *Mineralogical Applications of Crystal Field Theory*, Cambridge University Press.

DALRYMPLE G. B. and DOELL R. R. (1970) Thermoluminescence of lunar samples from Apollo 11. *Proc. Apollo* 11 *Lunar Sci. Conf., Geochim. Cocmochim. Acta* Suppl. 1, Vol. 3, pp. 2081–2092. Pergamon.

FABEL G. W. and HENISCH H. K. (1970) UV dosimetry by ZnS electret discharge. *Solid State Elec.* **13**, 1209.

GARLICK G. F. J., LAMB W., STEIGMANN G. A., and GEAKE J. E. (1971) Second Lunar Science Conference (unpublished proceedings).

GEAKE J. E., DOLLFUS A., GARLICK G. F. J., LAMB W., WALKER G., STEIGMANN G. A., and TITULAER, C. (1970) Luminescence, electron paramagnetic resonance and optical properties of lunar material from Apollo 11. *Proc. Apollo* 11 *Lunar Sci. Conf., Geochim. Cosmochim. Acta* Suppl. 1, Vol. 3, pp. 2127–2147. Pergamon.

GIGL P. D., SAVANICK G. A., and WHITE E. W. (1970) Characterization of corrosion layers on aluminum by shifts in the aluminum and oxygen X-ray emission bands. *J. Electrochem. Soc.* **117**, 15–17.

HASS M. (1970) Raman spectra of vitreous silica, germania, and sodium silicate glasses. *J. Phys. Chem. Solids* **31**, 415–422.

HENISCH H. K. and FABEL G. W. (1970) Characterization of non-luminescent, near-insulating solids. *Mater. Res. Lab. Spec. Pub. No.* 70–101, 56–60.

HOYT H. P., KARDOS J. L., MIYAJIMA M., SEITZ M. G., SUN S. S., WALKER R. M., and WITTELS M. C. (1970) Thermoluminescence, X-ray and stored energy measurements of Apollo 11 samples. *Proc. Apollo Lunar Sci. Conf., Geochim. Cosmochim. Acta* Suppl. 1, Vol. 3, pp. 2269–2287. Pergamon.

LEWIS J. F. and WHITE W. B. (1971) Electronic spectra of the pyroxenes. *J. Geophys. Res.* (in press).

SEMAK D. G., CHEPUR D. V., and ZOLOTAREV V. F. (1967) Thermally stimulated current method under photoelectret conditions. *Fiz. Tver. Tela* **9**, 1242.

WHITE E. W. and GIBBS G. V. (1967) Structural and chemical effects on the $SiK\beta$ X-ray emission line for silicates. *Amer. Mineral.* **52**, 985–993.

WHITE E. W. and GIBBS G. V. (1969) Structural and chemical effects on the $AlK\beta$ X-ray emission band among aluminum containing silicates and aluminum oxides. *Amer. Mineral.* **54**, 913–936.

Proceedings of the Second Lunar Science Conference, Vol. 3, pp. 2223–2233
The M.I.T. Press, 1971.

Luminescence of Apollo 11 and Apollo 12 lunar samples

Norman N. Greenman and H. Gerald Gross

Space Sciences Department, McDonnell Douglas Astronautics Company,
Huntington Beach, California 92647

(*Received* 22 *February* 1971; *accepted in revised form* 30 *March* 1971)

Abstract—Luminescence measurements have been made of Apollo 11 and Apollo 12 samples with far, middle, and near u.v., X-rays, and protons. Efficiencies were found to be low, of the order of 10^{-6} or less, for all irradiations. These cannot account for the astronomical observations of luminescence on the moon, at least not in the Apollo 11 and Apollo 12 landing areas and at least to the extent that our samples are representative (we do not have, however, any sialic rocks, as 12013, among our samples). In general, the efficiencies agree well with those for terrestrial rocks in which efficiency decreases with increasing basic character. We could find no evidence from measurements of both interior and exterior specimens that micrometeorite impact, space radiation, and other mechanisms operating at the lunar surface have affected the luminescence character of the rocks. Such effects are too small to be seen in our present series of measurements because of low luminescence efficiency and consequent weak signal.

Introduction

We have reported the first results of our luminescence studies of the Apollo lunar samples in previous papers (Greenman and Gross, 1970a, 1970b); in this paper we present additional results of measurements of the Apollo 11 samples together with the first results of measurements of the Apollo 12 samples. To recapitulate briefly, our objectives in these luminescence studies are (1) to understand how the luminescence behavior reflects the origin, history, and environment of the lunar rocks, (2) to discover luminescence characteristics of the lunar rocks that might aid in geologic mapping and other lunar exploration activities, and (3) to evaluate the reports of luminescence on the moon based on astronomical observations.

To achieve these objectives, we are measuring the luminescence spectra and efficiencies and comparing the results with those of similar measurements of terrestrial rocks and minerals. Our excitation sources are those of importance in the space environment—u.v. (1216 and 2000–4000 Å), X-rays (0.2–8 Å), protons (up to 150 keV), and electrons (up to 150 keV)—and our measurements of the luminescence spectra are from around 1216 Å (or from the exciting wavelength if it is longer than 1216 Å) to 6000 Å, except that in cases where luminescence is found near 6000 Å we extend the measurements to the cutoff wavelength of the detector, near 8000 Å.

Experimental Procedure

The experimental arrangements for the various irradiations are shown in Figs. 1–3. The system for middle and near u.v. irradiation was modified in some respects from that described in our previous papers in that a Bausch and Lomb grating monochromator was substituted for the Gaertner quartz prism monochromator for obtaining a single line or band, light-collecting and focusing optics were added at the entrance and exit slits of the Jarrell Ash output monochromator, and a photon-counting

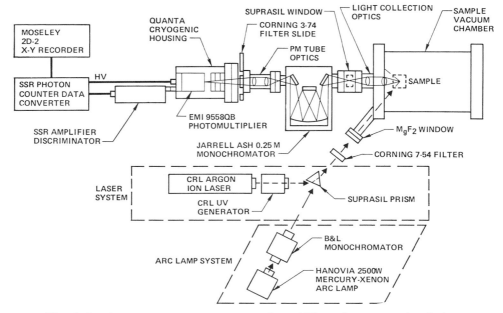

Fig. 1. Luminescence measurement system for middle and near u.v. irradiation.

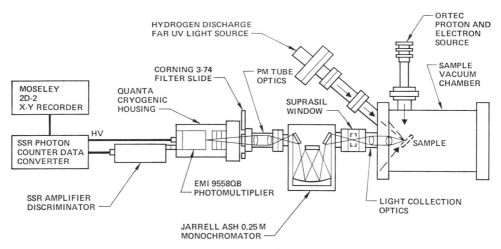

Fig. 2. Luminescence measurement system for far u.v. and low energy (up to 5 keV) proton and electron irradiation.

system of signal measurement was added. Also, an argon ion laser with a u.v. generator, which gave a line at 2573 Å, was used for one series of measurements.

The light-collecting and focusing optics were also used in the far u.v. measurements; otherwise this system was the same as previously described.

The samples used in the measurements are as follows: Apollo 11: 10044,38 (exterior) and 10044,53 (interior), coarse-grained igneous; 10022,55 (exterior) and 10057,45 (interior), fine-grained igneous;

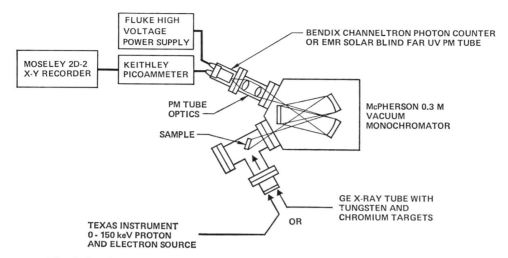

Fig. 3. Luminescence measurement system for soft X-ray and high energy (up to 150 keV) proton and electron irradiation.

10048,36 (interior) and 10048,37 (exterior), breccia; Apollo 12: 12002,99 (interior), 12002,106,107 (interior), and 12002,114 (exterior), medium-grained olivine dolerite; 12020,54 (interior) and 12020,55 (exterior), fine-grained olivine basalt; Terrestrial: granite (California), gabbro (California), willemite (New Jersey), andesine (Norway). The efficiencies given in the following section are total efficiencies; that is, they represent the ratio of energy in the luminescence band to the energy incident upon the sample.

DATA ANALYSIS AND RESULTS

Middle and near u.v. irradiation with xenon-mercury arc lamp

The results obtained from the Apollo 11 samples with 3000 Å irradiation were reported in our previous papers. The luminescence signals, if present at all, were weak and indicated upper limits of efficiency from 5×10^{-6} to 2×10^{-5}. In the present series of measurements, we irradiated the Apollo 12 samples with 3000 Å and both the Apollo 11 and Apollo 12 samples with 3650 Å and 2516 Å. In this last case, other lines were present in the incident beam, the 2516 Å line accounting for about half the flux and lines at 2800, 2900, 2970, 3000, 3140, 3340, and 3650 Å together accounting for the other half. Except for the willemite, none of the samples gave a detectable luminescence signal so that we have no basis for revising the efficiency upper limits already reported.

Middle u.v. irradiation with laser

The 2573 Å line from an argon ion laser was used in this series of measurements. Lunar sample 12020,54 showed a luminescence band in the region 6198–7765 Å above the $BaSO_4$ reference level, although the signal was only slightly above the noise (Fig. 4). Granite gave a detectable band at 6375–7820 Å, gabbro a marginally detectable band at 6280–7120 Å, and willemite a pronounced peak at about 5316 Å

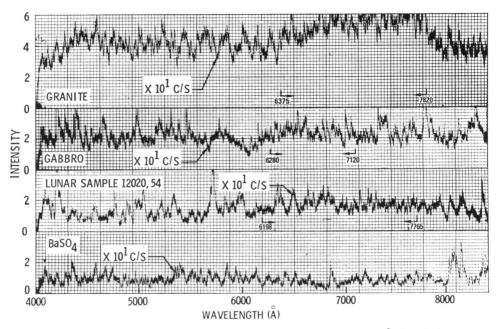

Fig. 4. Curves for terrestrial and lunar samples with u.v. laser (2573 Å) irradiation. Wavelength limits of luminescence band, shown by arrows, determined by comparison with BaSO₄ standard curve (Greenman and Gross, 1970b). C/S = counts per second.

Table 1. Luminescence data for samples irradiated with 2573 Å from an argon ion laser.

Sample	Peak wavelength (Å)	Bandwidth* (Å)	Efficiency upper limit at peak wavelength	Efficiency upper limit of total band
Willemite	5316	424	4×10^{-5}	1×10^{-2}
	7638	407	1×10^{-7}	3×10^{-5}
Granite	7169	1190	4×10^{-9}	4×10^{-6}
Gabbro	6700	655	2×10^{-9}	1×10^{-6}
12020,54	6826	915	2×10^{-9}	2×10^{-6}

* Full width at half maximum.

with a secondary peak at 7638 Å. Similar red luminescence bands had been found earlier in various silicates by Gross and Hyatt (1970) when they irradiated with the 4480 Å line from the argon laser, although the efficiencies were lower than with the 2573 Å line. The results of the 2573 Å irradiation, together with preliminary values for the upper limits of the efficiency, are summarized in Table 1.

Far u.v. through visible irradiation with hydrogen discharge

These measurements were made to observe the luminescence effect of hydrogen Lyman-alpha. Total band irradiation was used first before attempting measurements with the hydrogen Lyman-alpha line alone. The results are shown in Figs. 5–7.

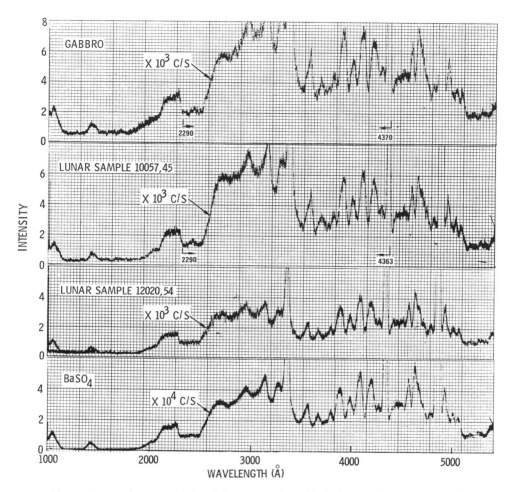

Fig. 5. Curves for terrestrial and lunar samples with hydrogen discharge (1120 Å through visible) irradiation. Wavelength limits of luminescence band, shown by arrows, determined by comparison with BaSO₄ standard curve (GREENMAN and GROSS, 1970b). Curve for sample 12020,54, which shows no luminescence in this part of the spectrum, is typical of curves for all lunar samples except 10057,45. C/S = counts per second.

Sample 10057,45 shows distinct luminescence in the band 2260–4363 Å with a peak around 3140 Å. Gabbro shows an almost identical band. In the remainder of the samples there are indications of a broad red band in the Apollo 11 and terrestrial samples and a narrower red band in the Apollo 12 samples. That these signals are real, though weak, is supported by the presence of red bands with the laser, X-ray, and proton irradiations. Calibration of the system is still in progress so that no efficiency values are yet available.

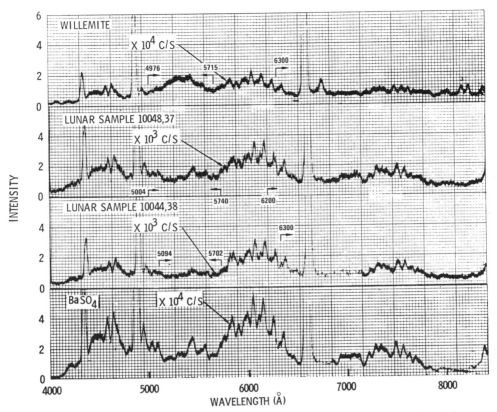

Fig. 6. Curves for terrestrial and Apollo 11 samples with hydrogen discharge (1120 Å through visible) irradiation. Wavelength limits of luminescence bands, shown by arrows, determined by comparison with BaSO₄ standard curve (GREENMAN and GROSS, 1970b). C/S = counts per second.

X-ray irradiation

The X-rays for these measurements were obtained from a tungsten target; the tube was operated at 70 kV and 45 mA. Under these conditions the irradiation band was from about 0.2 Å to the cutoff of the beryllium window at 8 Å. Distinct luminescence spectra were obtained from willemite, granite, and gabbro in the band from about 4000 Å to near 8000 Å. No detectable luminescence was found in the lunar samples in this band, and none was found in any of the samples in the band from 1000 Å to 4000 Å. The gabbro and lunar samples 10044,53 and 12020,55 were also measured with the source-to-sample distance reduced by about 30%; this resulted in an increase of about 50% in the intensity of the gabbro spectrum and no change in that of 12020,55. The curve for 10044,53, however, showed a barely discernible rise with a maximum in the 5300–5800 Å range. The granite and gabbro curves (Figs. 8 and 9) display prominent peaks at 5800 Å and at about 7350 Å and a third very faint peak at about 4500 Å in granite and about 4850 Å in gabbro. The curve for

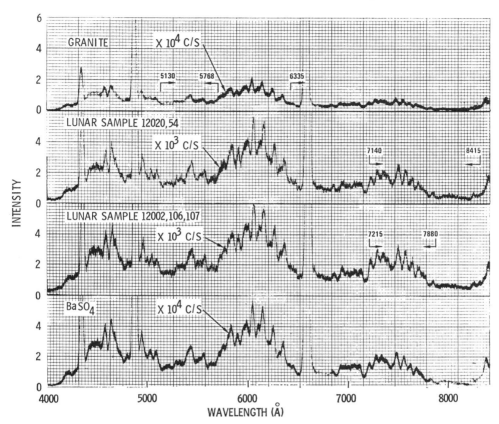

Fig. 7. Curves for terrestrial and Apollo 12 samples with hydrogen discharge (1120 Å through visible) irradiation. Wavelength limits of luminescence bands, shown by arrows, determined by comparison with BaSO₄ standard curve (GREENMAN and GROSS, 1970b). C/S = counts per second.

willemite was found to be virtually identical to the one obtained with u.v. irradiation— a single peak at 5350 Å and a band width of 550 Å at full width, half maximum. Preliminary efficiency calculations are shown in Table 2.

Proton irradiation

In this irradiation we used protons of 5 keV energy and a proton flux density at the sample of 2.0×10^{14} protons/cm² sec 5 keV, or an energy flux density of 1.6×10^6 ergs/cm² sec. As the protons were extracted from an RF-excited glow discharge of hydrogen, we ran a glow discharge reflectance curve for each sample before turning up the accelerating voltage to bombard with protons. With the accelerating voltage turned up, we also ran curves from an aluminum mirror, a non-luminescent material, to determine whether for any reason changes in the reflected light level occurred under conditions of bombardment by the proton beam. In all

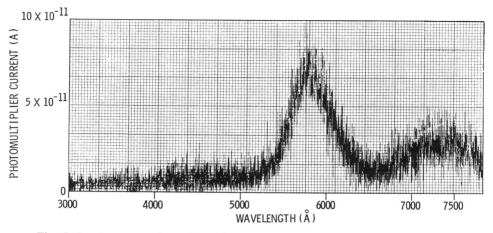

Fig. 8. Luminescence of granite with X-ray irradiation (tungsten target, 70 kV, 45 mA).

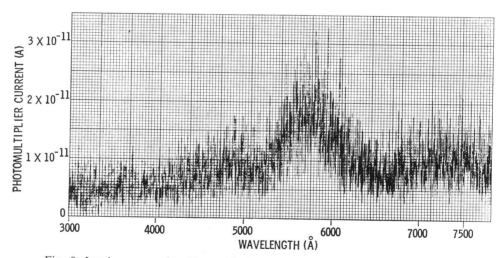

Fig. 9. Luminescence of gabbro with X-ray irradiation (tungsten target, 70 kV, 45 mA).

Table 2. Luminescence efficiencies of terrestrial and lunar samples with X-ray irradiation.

Sample	Efficiency	Remarks
Willemite	4×10^{-3}	
Granite	2×10^{-4}	Total for all three bands
Gabbro	6×10^{-5}	Total for all three bands
All lunar samples	$< 5 \times 10^{-6}$	Based on minimum detectable signal with source-to-sample distance used
10044,53	8×10^{-7}	With reduced source-to-sample distance

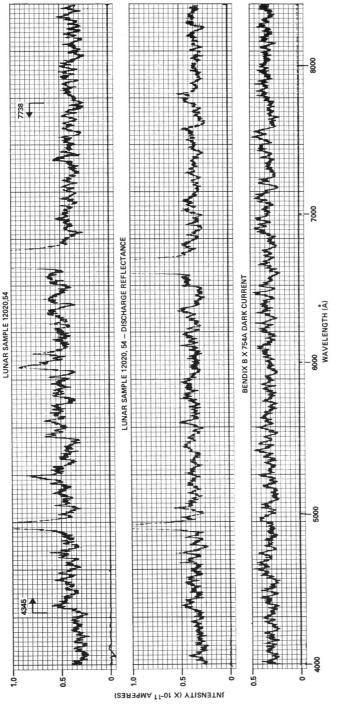

Fig. 10. Curves for lunar sample 12020,54 with proton (5 keV) irradiation. Other lunar samples show similar curves. Wavelength limits of luminescence band, shown by arrows, determined by comparison with curves for discharge reflectance and dark current.

Table 3. Luminescence efficiencies of terrestrial and lunar samples in the 1000–4000 Å band with 100 keV proton irradiation.

Sample	Efficiency
Granite	2×10^{-7}
Gabbro	5×10^{-8}
10022,55	9×10^{-9}
10044,53	1×10^{-8}
10048,36	1×10^{-8}
12002,114	8×10^{-9}
12020,55	1×10^{-8}

cases, the reflected light due to the glow discharge was below the dark current of the sensor except for the strong atomic hydrogen lines at 4861 Å and 6563 Å.

Comparison of the glow discharge reflectance curve with the proton irradiation curve for each sample showed that all samples displayed luminescence over a very broad band. A typical set of curves, with the wavelength limits so determined indicated by the numbered arrows, is shown in Fig. 10. In addition, most samples show a narrow line feature in the 5950–6000 Å interval. Willemite, 10044,38, and perhaps 10057,45 do not have this line; $BaSO_4$ has such a line at 5600 Å. The source of this feature has not yet been determined. Calibration is still in progress so that efficiency values are not yet available.

A second set of measurements was made with protons of 100 keV energy at 8×10^{-6} A, for a proton flux density of 10^{13} protons/cm² sec 100 keV, or an energy flux density of 1.6×10^6 ergs/cm² sec. The luminescence in this case was measured over the range 1000–4000 Å. In all cases, the flash effect described by Nash (1966), in which an initially high signal decays rapidly in a matter of minutes, was observed. Nash and Greer (1970) also observed this effect in most Apollo 11 samples. Repeated runs showed that the signal leveled off after about 6 to 9 minutes so that the efficiency values given below are for the last rather than the first runs.

Granite and gabbro show a low peak at 2220 Å and a prominent peak at 2760 Å, with a barely discernible peak at 3140 Å. The lunar samples are similar except that the 3140 Å peak is much more prominent. Also, in some lunar samples, the decay rates of the 2760 Å and 3140 Å peaks appear to be different, the former peak being most intense initially but declining more rapidly such that the latter is most intense in the stabilized luminescence curve. Efficiencies for the 100 keV proton irradiation in the 1000–4000 Å band are shown in Table 3.

Discussion

The luminescence efficiencies of the lunar samples, despite some variation with irradiation type, are uniformly low. They cannot, therefore, account for the astronomical observations of luminescence on the moon, at least not in the Apollo 11 and Apollo 12 landing areas, if our samples are representative and if our laboratory irradiation conditions do not differ radically from those at the surface of the moon (Greenman and Gross, 1970b; Nash and Greer, 1970). It should be observed, however, that rocks of more sialic character must exist on the moon, as evidenced by Apollo 12 rock 12013, but rocks of this type are not represented among our samples. Such types can be expected to show higher luminescence efficiencies if they behave at all like their terrestrial counterparts (GREENMAN *et al.*, 1965; GREENMAN and MILTON, 1968; NASH, 1966).

We could find no evidence from measurements of both interior and exterior specimens and of the various lunar lithologic types that meteorite and micro-meteorite impact, space radiation, and other processes operating at the lunar surface have affected the luminescence character of the rocks. SIPPEL and SPENCER (1970) report luminescence differences in the Apollo plagioclases both with respect to terrestrial plagioclases and with respect to degree of shock damage. They were working with individual grains under the microscope, however. We measure the whole rock, and under these circumstances the significant content of non-luminescent material greatly lowers the rock efficiency, as opposed to the mineral efficiency. This lowering is great enough to mask the effects noted by SIPPEL and SPENCER. The low efficiency and consequent weak signal have also masked the presence of degraded spectra, described by NASH and GREER (1970), in this series of measurements.

Acknowledgments—We wish to thank W. M. HANSEN, R. R. CARLEN, T. H. MILLS, and D. J. WILLIAMS for their valuable assistance. This work was supported by NASA Contract NAS9–7966.

REFERENCES

GREENMAN N. N., BURKIG V. W., GROSS H. G., and YOUNG J. F. (1965) Feasibility study of the ultraviolet spectral analysis of the lunar surface. Douglas Aircraft Co. Rep. SM–48529.

GREENMAN N. N. and GROSS H. G. (1970a) Luminescence of Apollo 11 lunar samples. *Science* **167**, 720–721.

GREENMAN N. N. and GROSS H. G. (1970b) Luminescence studies of Apollo 11 lunar samples. *Proc. Apollo 11 Lunar Sci. Conf., Geochim. Cosmochim. Acta* Suppl. 1, Vol. 3, pp. 2155–2161. Pergamon.

GREENMAN N. N. and MILTON W. B. (1968) Silicate luminescence and remote compositional mapping. Proceedings of the Sixth Annual Meeting of the Working Group on Extraterrestrial Resources, NASA SP–177, 55–63.

GROSS H. G. and HYATT H. A. (1970) Optical laser induced luminescence in natural materials: a tool for active remote sensing. *Trans. Amer. Geophys. Union* **51**, 770–771.

NASH D. B. (1966) Proton-excited silicate luminescence: experimental results and lunar implications. *J. Geophys. Res.* **71**, 2517–2534.

NASH D. B. and GREER R. T. (1970) Luminescence properties of Apollo 11 lunar samples and implications for solar-excited lunar luminescence. *Proc. Apollo 11 Lunar Sci. Conf., Geochim. Cosmochim. Acta* Suppl. 1, Vol. 3, pp. 2341–2350. Pergamon.

SIPPEL R. F. and SPENCER A. B. (1970) Luminescence petrography and properties of lunar crystalline rocks and breccias. *Proc. Apollo 11 Lunar Sci. Conf., Geochim. Cosmochim. Acta* Suppl. 1, Vol. 3, pp. 2413–2426. Pergamon.

Proceedings of the Second Lunar Science Conference, Vol. 3, pp. 2235–2244
The M.I.T. Press, 1971.

Luminescence and reflectance of Apollo 12 samples

D. B. NASH and J. E. CONEL

Jet Propulsion Laboratory, California Institute of Technology, Pasadena, California 91103

(*Received* 22 *February* 1971; *accepted* 31 *March* 1971)

Abstract—Luminescence, thermoluminescence, and spectral reflectance properties of samples from Oceanus Procellarum are qualitatively similar to those of Mare Tranquillitatis samples. Detailed differences are controlled by mineralogy; fines from Procellarum have higher plagioclase content relative to glass and opaques, and higher pyroxene content relative to plagioclase than fines from Tranquillity. Luminescence properties and reflectance do not vary systematically with depth in the core. The variations observed are attributable to differences in relative abundances of mineral and glass phases and are not indicative of variations in particle size, radiation damage, or surface coatings on individual grains.

INTRODUCTION

LUMINESCENCE (0.25–0.72 μm), thermoluminescence (23–450°C) and spectral reflectance (0.23–2.5 μm) were measured on lunar rock and soil samples collected from Oceanus Procellarum. The objectives of these laboratory measurements were to (1) compare the luminescence and reflectance properties of Apollo 12 surface and core-sample material with those of the surface material of Apollo 11, (2) obtain a better estimate of the average luminescence characteristics of the lunar surface under solar irradiation, and (3) obtain additional spectral reflectance data on lunar samples for more accurate interpretation of telescopic observations of the lunar surface. Experimental procedures and instrumentation used were the same as in our Apollo 11 study (NASH and GREER, 1970; CONEL and NASH, 1970). Luminescence was excited by 2- and 4-keV protons and 0.25–0.43 μm ultraviolet radiation. Hemispherical reflectance was measured relative to smoked MgO. Samples examined are listed in Table 1.

LUMINESCENCE PROPERTIES

Proton irradiation produces weak luminescence in all Apollo 12 material examined; spectral distribution, luminescence intensity, and intensity vs. dose are similar to those of Apollo 11 samples (NASH and GREER, 1970). Ultraviolet excitation at solar-equivalent intensity yields no detectable luminescence. Simultaneous proton and ultraviolet irradiation is no more effective in exciting luminescence than the proton irradiation alone.

Energy efficiencies of Apollo 12 core samples range from 1 to 3 × 10^{-6} (Table 1), slightly higher than Apollo 11 fines and breccia. Variations in core-sample luminescence efficiency with depth are plotted in Fig. 1; no systematic relation exists between efficiency and depth. However, there is a correlation between luminescence efficiency and average visible reflectance for the core fines as shown in Fig. 2; the luminescence is in general more efficient for core materials of higher reflectivity. This relationship is consistent with a higher concentration of plagioclase in the brighter core material

Table 1. Luminescence and reflectance data for Apollo 12 samples examined.

	LRL number	Our lab number	Core depth (cm)	Proton beam $\mu a/cm^2$	Lumin. effic. $\times 10^{-6}$ Start	Lumin. effic. $\times 10^{-6}$ 10^3 sec	Reflectance $\lambda = 0.55\,\mu$
Fines	12070,115	12–1	Surface	—	—	—	0.078
	12025,29	12–2	1.2–1.7	8.4	1.10	0.48	0.085
Core	12025,36	12–3	2.5–3.3	8	1.79	0.69	0.106
Fines	12025,53	12–4	7–8	7.2	1.14	0.45	0.081
	12028,94	12–5	19.7–20.8	7.3	3.11	0.94	0.137
	12028,125	12–6	32.2–33.2	7.3	1.16	0.53	0.095
Ground rock ($< 0.50\,\mu m$)	12065,65	12–7	—	8.4	2.59	0.99	0.236
Chip (interior, sawed)	12052,31	12–8	—	—	—	—	—
Chip (top)	12052,24	12–9	—	3.9	1.7	~0.85	—

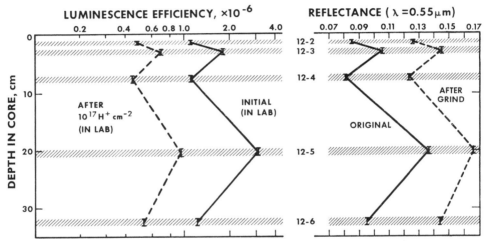

Fig. 1. Proton-excited luminescence efficiency and average reflectance of < 0.5 mm fines at various depths in Apollo 12 core tube. Each data point represents 0.25 g of core material from within the hachured depth zone. On left, solid line joins efficiency values at lab proton dose of $10^{14}H^+$ cm^{-2}; dashed line at $10^{17}H^+$ cm^{-2}. Proton energy was 4.0 kev. On right, solid line joins reflectance values of original material; dashed line after grinding to $< 50\,\mu m$. For brevity, core samples are numbered with out lab numbers; see Table 1 for equivalent LRL numbers.

(SELLERS et al., 1971), because plagioclase has been found to be the primary luminescing phase in lunar materials (NASH and GREER, 1970).

A possible explanation of the minimum in the upper curve of Fig. 2 is that samples numbered 2, 3, and 6 have somewhat higher solar-wind exposure ages than samples 4 and 5; i.e., the ability of the former samples to luminesce has been partially diminished by solar wind bombardment. This argument suggests that the core material at a depth of ~ 32 cm (our sample 6) has had longer solar wind exposure than any of the overlying material.

Two igneous rock samples were examined for comparison with the regolith fines. Rock 12065, an olivine-bearing pigeonite ilmenite basalt, contains 65–70% clinopyroxene, 20–25% plagioclase, 1–2% olivine, and 10–13% opaque material (LSPET,

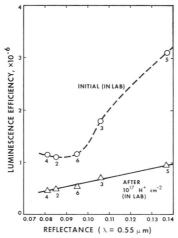

Fig. 2. Luminescence efficiency vs. reflectivity of Apollo 12 core material. Upper curve shows efficiencies at initial proton-beam exposure, lower curve after dose of $10^{17}H^+$ cm^{-2}.

1970). A fragment, powdered to about 50 μm, has a proton luminescence efficiency of 2.5×10^{-6}, which is slightly less than the efficiency of the most luminescent core fines from Apollo 12 and less than any of the Apollo 11 igneous rocks examined (Fig. 3). The relatively low luminescence efficiency of rock 12065 is due to its high pyroxene and low plagioclase content. A section of the topmost, zap-pitted surface of rock 12052 was examined for evidence of solar-wind decay of its luminescence. The laboratory proton luminescence shows a typical luminescence decay pattern of irradiated silicates, suggesting that this surface received low-energy proton bombardment on the moon's surface amounting to at least $10^{17}H^+$ cm^{-2}. Comparative spectra from a fresh interior fragment of this rock were not obtained because the sample provided was inadequate for analysis.

The luminescence properties of the lunar surface at Oceanus Procellarum as represented by Apollo 12 samples examined are essentially identical to luminescence properties of the surface at Mare Tranquillitatis as represented by Apollo 11 samples. The Apollo 12 sample suite includes a wider variety of igneous rock types than the Apollo 11 suite. Although not available for examination in this study, highly differentiated acidic rocks like 12013, which contain abundant K-feldspar and quartz, can be expected to have luminescence efficiency much higher than their counterpart basic rocks (NASH, 1968). But even if certain areas on the moon are composed of such acidic rocks, solar-excited luminescence would be insufficiently intense for detection from earth (NASH et al., 1970; NASH and GREER, 1970; NASH, 1968).

THERMOLUMINESCENCE OF CORE SAMPLES

The thermoluminescence (TL) of Apollo 12 core material was measured on 50-mg samples using equipment and procedures previously described (NASH and GREER, 1970). Natural TL in the range 23–120°C was undetectable for all samples, as was

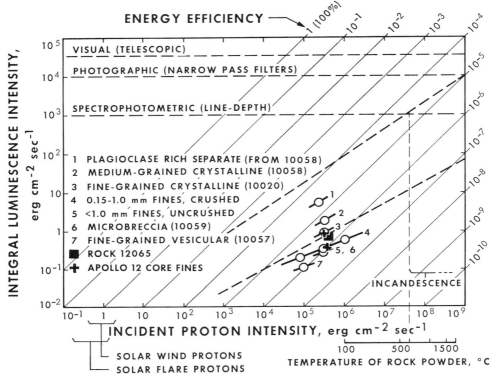

Fig. 3. Comparison of Apollo 11 and Apollo 12 sample luminescence, and summary of lunar luminescence detection problem. This is Fig. 7 of NASH and GREER (1970) with Apollo 12 data added. For detailed explanation of figure, see NASH and GREER (1970).

the case for all Apollo 11 materials. Natural TL above 120°C was observed in all Apollo 12 core samples. In general, the TL intensities are about a factor of 10 greater than that of the Apollo 11 fines. The natural TL intensity (N), integrated between 300–400°C varies with depth in the core as shown in Table 2. Note that the natural TL intensity varies with sample depth more or less in the same pattern as the proton luminescence and visible reflectance shown in Fig. 1.

HOYT *et al.* (1970) found that after heating the Apollo 11 core samples, a standard dose of laboratory radiation produced approximately equal TL intensity in all samples. In contrast, we found this not so in the Apollo 12 core samples; artificial TL intensity (A) (induced with a 100-min exposure to 40 keV Cu X-rays) varies with depth as shown in Column A in Table 2, and this variation pattern is similar to that of the natural TL intensity, proton luminescence, and average reflectance (albedo) of the sample material. We interpret this pattern as an indication of the variation in relative abundance of plagioclase in the core samples.

In addition, our ratios of natural to artificial (N/A) TL do not vary as systematically with depth as found in the upper 10 cm of the Apollo 11 core by DALRYMPLE and DOELL (1970) and HOYT *et al.* (1970) and in the Apollo 12 core by HOYT *et al.* (1971).

Table 2. Thermoluminescence data for Apollo 12 core samples examined.

Our lab* number	Core depth (cm)	TL intensity†		N/A
		Natural (N)	Induced (A)	
12–2	1.2–1.7	0.19	0.60	0.032
12–3	2.5–3.3	0.28	0.76	0.037
12–4	7–8	0.08	0.46	0.017
12–5	19.7–20.8	0.50	1.35	0.037
12–6	32.2–33.2	0.30	0.85	0.035

* See Table 1 for corresponding LRL numbers. † Area in arbitrary units under glow curve between 300 and 400°C.

Thus compositional effects appear to dominate over diurnal thermal draining and irradiation buildup of TL in the upper regolith stratigraphy as represented by our Apollo 12 core material. Since none of the deep core samples showed detectable TL between 23 and 120°C (which is the equatorial daytime temperature range), there is no evidence that sudden exposure of buried regolith material would produce any more TL than would occur at the dawn terminator. Thus, TL from either sudden daytime exposure or dawn heating would be undetectable from earth by present methods (NASH and GREER, 1970).

Unfortunately, these TL measurements were made approximately 14 months after the Apollo 12 core material was removed from the moon's surface. DALRYMPLE and DOELL 1970) give evidence suggesting that natural TL from lunar fines diminishes in intensity once the material is removed from its equilibrium radiation environment on the moon. They estimated the TL "half-lifes" of Apollo 11 samples to be on the order of one month, depending upon composition and texture. Thus, our results and conclusions may be confounded by this decay in TL prior to our analysis, especially if different samples decayed at different rates.

REFLECTANCE PROPERTIES

The spectral reflectance of rock sample 12065.65, ground to < 50 μm, and of contingency fines 12070 are shown in Fig. 4. The rock spectrum has a strong band centered at 0.96 μm and a broad minimum near 2.15 μm. The nearly linear variation in reflectance between 1.15 and 1.4 μm and the structure between 0.4–0.8 μm are also significant. Subtle structures appear near 0.5 μm. Below 0.4 μm the reflectance decreases more rapidly to 0.32 μm and is constant throughout the remainder of the ultraviolet studied. The infrared spectrum of contingency fines shows two absorption features characteristic of all samples of Apollo 12 fines studied: (1) a broad shallow feature located approximately at 0.95 μm, and (2) a broad structure near 2.0 μm. The visible and ultraviolet spectra of contingency fines are essentially structureless.

The reflectance spectra of original Apollo 12 core samples are shown in Fig. 5. The average visible reflectance does not vary systematically with depth in the core tube for the samples we examined (see Fig. 1). Crushing the samples to ≤ 50 μm increases the average reflectance by 30–50% (Fig. 1). These reflectance differences are on the order of differences determined for maria and highland sites using photographic and telescopic techniques (GOETZ et al., 1971). The spectral features in core

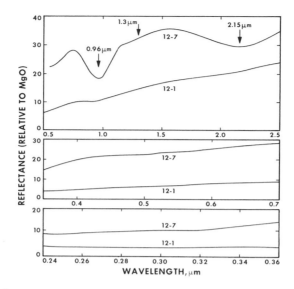

Fig. 4. Reflectance spectra of powdered chip of rock 12065 (curve 12-7) and of contingency fines (curve 12-1) from Apollo 12.

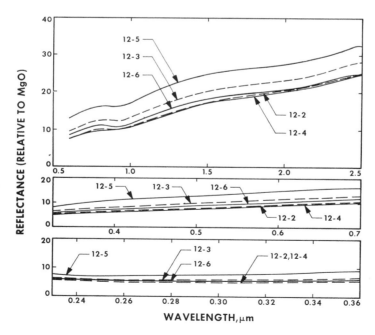

Fig. 5. Reflectance spectra of original fines from Apollo 12 double-drive core tube taken near Halo Crater. Spectra are numbered with out lab numbers; see Table 1 for equivalent LRL numbers.

fines for both original and crushed are in all cases similar to those described for
contingency fines; the visible spectral reflectance of core fines more closely approximates the spectral reflectance of powdered igneous rock as reflectance increases.

The relative strengths of the infrared structures at 0.95 and 2.0 μm in the spectra
of the core samples are functions of sample reflectance. This subtle relationship for
the 0.95 μm structure was quantified by estimating the area A in the band for both
original and crushed material and plotting it in Fig. 6 as a function of average visible
reflectance R computed for the wavelength interval 0.4–0.7 μm. Band strength tends
to increase with increasing sample brightness and, except for sample 12-6, is further
increased by grinding and reduction of particle size. Similar relationships are apparent
for absorption bands in terrestrial gabbro and basalt in which variations in reflectance
result from variations in particle size alone (ADAMS and FILICE, 1967).

The correlation shown in Fig. 6 suggests that differences in brightness between
samples of core fines result primarily from variations in abundances of mineral (and
possibly glass) phases that give rise to the band. In original core material effects of
particle size variation are minimal since SELLERS et al. (1971) show nearly uniform
particle size distributions for core samples near those studied here. Thus it seems
unnecessary to invoke more exotic causes for the observed reflectance variations
such as solar wind irradiation. No change in reflectance properties of any of the core
samples was observed as a result of bombardment by approximately 2×10^{17} H$^+$/cm^2
with energy of 4 KeV.

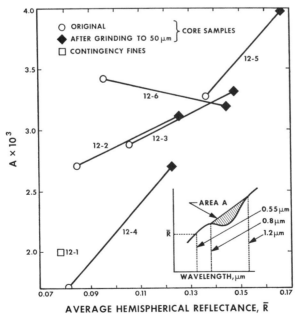

Fig. 6. Relationship between 0.95 μm band strength and average visible reflectance for
core samples before and after grinding and for contingency fines. Continuum reflectance
is arbitrarily taken as a linear function of wavelength between 0.8 and 1.2 μm. The
value of A for Apollo 11 fines is 1.09×10^{-3}.

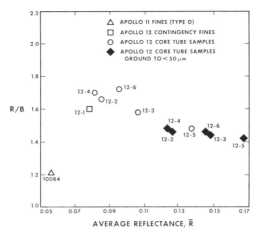

Fig. 7. Relationship between red/blue ratios (R/B) and average visible reflectance $\overline{R}$
in samples of Apollo fines. Solid points are for crushed Apollo 12 core fines.

The spectral reflectance from 0.23 to 2.5 μm of Apollo 12 core samples and contingency fines is higher than that of Apollo 11 fines (NASH et al., 1970; CONEL and
NASH, 1970). In addition, the 0.95 μm structure is stronger in the spectrum of Apollo
12 fines; a structure near 2.0 μm was not detected in our analysis of Apollo 11 fines.
Reflectance between 0.4–0.7 μm of all lunar fines examined is very nearly a linear
function of wavelength. This is not true for lunar rocks or synthetic glass (CONEL and
NASH, 1970).

Red/blue ratios (R/B) for original core material may be computed from the spectra
in Fig. 5. For wavelengths of 0.4 and 0.7 μm, the values of R/B plotted in Fig. 7
generally decrease with increasing values of average visible reflectance for Apollo 12
fines. This is contrary to relations observed for crushed terrestrial rocks that have
low reflectance because of large particle size (ADAMS and FILICE, 1967), and in which
reflectance variations result from particle size variations alone. However, for basic
rocks with particle size on the order of 75 μm or less, R/B decreases with increasing
reflectance. Thus the observed relations with Apollo 12 fines are consistent with
reflectance data for terrestrial basalts and gabbro of small particle size, if their reflectance is decreased to lunar values by addition of, say, a neutral density opaque component. The R/B value for Apollo 11 fines is lower than the values for Apollo 12 core
samples and contingency fines. The range of R/B exhibited by uncrushed Apollo 12
core samples is greater than the difference in R/B observed between Oceanus Procellarum and the Fra Mauro formation as determined by photographic and other
techniques (GOETZ et al. 1971). Values of R/B for crushed Apollo 12 core samples
are also shown in Fig. 7. In addition to increased average reflectance, crushed material
shows far less variation of R/B with $\overline{R}$ than do uncrushed samples. Comparing data
in Figs. 6 and 7 we see that decreasing values of R/B are associated with increasing
strength of the 1 μm band. Thus variations in R/B can at least be associated with
increased pyroxene or olivine abundance relative to opaque or optically absorbing
phases, or decreasing particle size with constant abundances of all phases.

The origin of some of the features in the Apollo 11 and Apollo 12 sample spectra can be established from the observed mineralogy and from earlier reflectance work on mineral separates from Apollo 11 rock 10058 (CONEL and NASH, 1970). The Apollo 11 rock spectra shown in CONEL and NASH (1970), Fig. 3 indicate that structures at 0.96 and 2.15 μm in the spectrum of rock 12065 (Fig. 4, this paper) arise principally from clinopyroxene, allowances being made for possible distortion of bands by other minerals, and for possible differences in composition of the pyroxenes. Because of the low abundance of olivine present, these bands may be attributed principally to Fe^{2+} in the pyroxene (ADAMS and FILICE, 1967; BURNS and FYFE, 1967). The linearity of the spectrum between 1.15–1.4 μm for rock 12065 is attributed principally to an absorption band near 1.3 μm arising from plagioclase, which, by analogy with work on other plagioclase feldspars, is due to Fe^{2+} (CONEL, 1970). The broad shallow structure between 0.4 and 0.8 μm in rock 12065 (Fig. 4) is essentially equivalent to that observed in the spectrum of Apollo 11 rock 10020 (CONEL and NASH, 1970; ADAMS and JONES, 1970). The structure has been attributed apparently on theoretical grounds to Ti^{+3} in the pyroxene (ADAMS and McCORD, 1970). Below 0.4 μm, reflectance may decrease because of an absorption band in the ultraviolet. The ionic species involved is undetermined, but a charge transfer band of Ti^{4+} in either pyroxene or plagioclase may be responsible (CONEL and NASH, 1970; ADAMS and McCORD, 1970; CONEL, 1970).

In the spectra of Apollo 12 fines, the bands at 0.95 μm and near 2.0 μm are distorted composite structures with contributions from clinopyroxene, and possibly orthopyroxene (WOOD et al., 1971) and glass. The relative contributions of each phase are indeterminate. A band from Fe^{2+} in olivine may also contribute to the 0.95 μm structure. We attribute differences in strength of the 0.95 μm band in the spectra of Apollo 12 core samples principally to differences in pyroxene abundance.

CONCLUSIONS

From our measurements of luminescence and reflectance properties of Apollo 11 and Apollo 12 samples, we make the following conclusions regarding lunar surface material: (1) Solar-excited luminescence at the Apollo 12 site is approximately the same as at the Apollo 11 site; the luminescence at either site would be undetectable from earth by present methods. (2) Apollo 12 fines contain more pyroxene relative to plagioclase, glass, and opaque phases than fines from the Apollo 11 site. (3) Spectra of Apollo 11 and Apollo 12 rocks and fines contain structures attributed principally to plagioclase and pyroxene. (4) The variation of visible reflectance for both Apollo 11 and 12 fines is nearly a linear function of wavelength. (5) Both luminescence and reflectance data suggest that reflectance variations in core samples arise from variations in mineral or glass content and not from radiation damage from solar wind bombardment. (6) The reflectance ratio R/B decreases with increasing average reflectance; R/B can be controlled among other variables either by particle size or pyroxene abundance. (7) Telescopic spectral reflectance measurements in the 1.1 to 1.5 μm may be useful in establishing the existence of feldspar-rich (i.e., anorthositic regions on the moon's surface).

Acknowledgments—We thank WARREN RACHWITZ, PATRICIA CONKLIN, and CHERYL CONEL for their assistance. This paper presents the results of one phase of research carried out at the Jet Propulsion Laboratory, California Institute of Technology, under NASA Contract NAS 1-700.

REFERENCES

ADAMS J. B. and FILICE A. L. (1967) Spectral reflectance 0.4 to 2.0 μ of silicate rock powders. *J. Geophys. Res.* **73**, 5705–5715.

ADAMS J. B. and JONES R. L. (1970) Spectral reflectivity of lunar samples. *Science* **167**, 737–739.

ADAMS J. B. and McCORD T. B. (1970) Remote sensing of lunar surface mineralogy: Implications from samples. *Proc. Apollo 11 Lunar Sci. Conf., Geochim. Cosmochim. Acta* Suppl. 1, Vol. 3, pp. 1937–1945. Pergamon.

BURNS R. G. and FYFE W. S. (1967) Crystal field theory and the geochemisty of transition elements. In *Researches in Geochemistry* (editor P. H. Abelson), Vol. 2, pp. 259–285. John Wiley.

CONEL J. E. (1970) Coloring of synthetic and natural lunar glass by titanium and iron. *Jet Propulsion Laboratory, SPS 37–62*, Vol. 3, pp. 26–31.

CONEL J. E. and NASH D. B. (1970) Spectral reflectance and albedo of Apollo 11 lunar samples; Effects of irradiation and vitrification and comparison with telescopic observations. *Proc. Apollo 11 Lunar Sci. Conf., Geochim. Cosmochim. Acta* Suppl. 1, Vol. 3, pp. 2013–2023. Pergamon.

DALRYMPLE G. B. and DOELL R. R. (1970) Thermoluminescence of lunar samples from Apollo 11. *Proc. Apollo 11 Lunar Sci. Conf., Geochim. Cosmochim. Acta* Suppl. 1, Vol. 3, pp. 2081–2092. Pergamon.

GOETZ A. F. H., BILLINGSLEY F. C., HEAD J., McCORD T., and YOST E. (1971) Apollo 12 lunar orbital multispectral photography experiment, Second Lunar Science Conference (unpublished proceedings).

HOYT H. P., KARDOS J. L., MIYAJIMA M., SEITZ M. G., SUN S. S., WALKER R. M., and WITTELS M. C. (1970) Thermoluminescence, X-ray and stored energy measurements of Apollo 11 samples. *Proc. Apollo 11 Lunar Sci. Conf., Geochim. Cosmochim. Acta* Suppl. 1, Vol. 3, pp. 2269–2287. Pergamon.

HOYT H., KARDOS J., MIYAJIMA M., WALKER R., and ZIMMERMAN D. (1971) Thermoluminescence and stored energy measurements in Apollo 12 samples. Second Lunar Science Conference (unpublished proceedings).

LSPET (Lunar Sample Preliminary Examination Team) (1970) Preliminary examination of lunar samples from Apollo 12. *Science* **167**, 1325–1339.

NASH D. B. (1968) Efficiency of proton-excited luminescence of silicate rock powders. In *Thermoluminescence of Geological Materials* (editor D. McDougall), p. 587. Academic Press.

NASH D. B., CONEL J. E., and GREER R. T. (1970) Luminescence and reflectance of Tranquillity samples: Effects of irradiation and vitrification. *Science* **167**, 731–724.

NASH D. B. and GREER R. T. (1970) Luminescence properties of Apollo 11 lunar samples and implications for solar-excited lunar luminescence. *Proc. Apollo 11 Lunar Sci. Conf., Geochim. Cosmochim. Acta* Suppl. 1, Vol. 3, pp. 2341–2350. Pergamon.

SELLERS G. A., WOOD C. C., BIRD M. L., and DUKE M. B. (1971) Descriptions of the composition and grain-size characteristics of fines from the Apollo 12 double-core tube. Second Lunar Science Conference (unpublished proceedings).

WOOD J. A., MARVIN URSULA, REID J. B., TAYLOR G. J., BOWER J. F., POWELL B. N., and DICKEY J. S. (1971). Relative proportions of rock types, and nature of the light-colored fragments in Apollo 12 soil samples, Second Lunar Science Conference (unpublished proceedings).

Proceedings of the Second Lunar Science Conference, Vol. 3, pp. 2245–2263
The M.I.T. Press, 1971.

Radiation dose rates and thermal gradients in the lunar regolith: Thermoluminescence and DTA of Apollo 12 samples

H. P. Hoyt, Jr., M. Miyajima,* R. M. Walker,
D. W. Zimmerman, and J. Zimmerman

Laboratory for Space Physics, Washington University, St. Louis, Missouri 63130

and

D. Britton and J. L. Kardos

Materials Research Laboratory, Washington University, St. Louis, Missouri 63130

(*Received* 23 *February* 1971; *accepted in revised form* 24 *March* 1971)

Abstract—The thermoluminescence (TL) of samples from core 12025/28 is in, or near, saturation above glow curve temperatures of $\sim$ 550°C. The TL is thermally drained at lower temperatures and below $\sim$ 225°C is in thermal equilibrium. This equilibrium value increases with depth to $\sim$ 15 cm because of the attenuation of the diurnal heat wave. Below 17 cm the equilibrium level decreases with depth. This decrease exceeds that expected from the attenuation of galactic cosmic ray dose rate and indicates a temperature increase of (2 ± 2)°K/m with depth, presumably due to outward heat flow from the moon. The TL emitted above 225°C by samples between 4 and 13 cm shows anomalies resulting from disturbances $\gtrsim 10^4$ years ago. In rock 12063, the normalized TL in the 1 mm top-surface sample is 4.5 times higher than the interior and bottom samples because of solar-flare irradiation. This confirms the recent orientation of the rock based on ^{22}Na data.

Differential thermal analysis (DTA) in air reveals three exotherms: (A) 250°–450°C, < 1 cal/g; (B) 350°–800°C, ≤ 15 cal/g; (C) 770°–870°C, ≤ 20 cal/g. Heating in nitrogen eliminates peak A, reduces peak B, and does not affect peak C. A large exotherm at 200°–400°C appears in nitrogen scans of the 20–26 cm levels of the core. A small endotherm at $\sim$ 540°C, along with the presence of the glass crystallization exotherm (peak C), may yield a measure of the sample glass content.

Introduction

THE PAPER IS DIVIDED into two parts, thermoluminescence and differential thermal analysis. Each part contains a description of the results and a discussion. The dominant effect previously observed in the thermoluminescence (TL) of lunar cores was the rapid increase in TL light output with depth throughout both the 10 and 12 cm Apollo 11 cores (Hoyt *et al.*, 1970; Dalrymple and Doell, 1970b). This was interpreted as arising from the decrease in thermal drainage of the TL with the decrease in amplitude of the diurnal heat wave with depth. We report here TL studies of the Apollo 12 double core 12025/12028 which is 40 cm deep (60 cm undisturbed). This core extends well beyond the point where the diurnal heat wave should have a measurable effect on the TL response and enables a study to be made of the above interpretation. As a further aid to the study the maximum glow curve temperature has been increased from 500° to 650°C. This extends the TL measurements to a region where there should be little if any thermal drainage. Because the amount of stored TL is strongly affected by the ambient temperature, measurements on the core below the depth at which the

---

* Present address: Waseda University, Tokyo, Japan.

effect of the diurnal heat wave has died out should reflect the presence of any positive temperature gradient due to heat flow outward from the moon. The TL should also be sensitive to dose-rate gradients at these depths and the TL data are interpreted in terms of these two effects. TL studies in rock 12063 are also reported. The much higher thermal conductivity in the rock reduces the effect of the thermal gradient and allows a study of the dose-rate gradients from solar flares near the surface of the rock.

We would like to emphasize that we are not reporting a TL phenomenon which is barely measurable but rather one which is quite marked. The natural TL light levels are very high in both of the following senses: (1) They are easily measurable above the background of photomultiplier (PM) tube noise and blackbody radiation, and (2) The statistical error in the number of photons detected is very small. We estimate a pulse rate at the PM tube anode of the order of 10^6 ct/sec at the glow peak maximum of a typical core sample. The natural light levels are approximately two orders of magnitude higher than those measured in TL dating of archeologic ceramics (AITKEN et al., 1968). It is not surprising that the light levels are high, considering the large radiation doses that the samples have received and the reasonable TL efficiency of the material.

Measurements on Apollo 11 samples revealed that energy stored from radiation damage was small (HOYT et al., 1970; KARDOS, 1970). The Apollo 12 samples are generally the same but show discrete releases of energy which are relatively large in some levels of the core sample. DTA may also provide a measure of the glass content in lunar samples.

THERMOLUMINESCENCE

Experimental technique

The TL apparatus is similar to that described previously for the Apollo 11 measurements (HOYT et al., 1970). A feedback system using a chromel-alumel thermocouple, spot welded onto a 1 in. by 2 in. nichrome heating plate, gives a linear heating rate of 6.5°C/sec. The glow oven is evacuated and then filled with argon ($<$ 1 ppm oxygen) before the sample is heated. The light is detected by a PM tube (EMI 9635QB) operated at 750 volts, and the current from it measured with a picoammeter.

To reach a region of thermal stability, the maximum glow curve temperature has been extended from 500°C (for the Apollo 11 measurements) to 650°C. To discriminate sufficiently against the blackbody radiation at the higher temperatures, four color glass filters (two 5-60 and one 7-59, Corning, Corning, N.Y.; and one HA3, Chance Pilkington, St. Asaph, Flintshire, G.B.) are placed between the sample and the PM tube. The system responds to wavelengths in the range 360 to 470 nm. To obtain more uniform core samples, only those grains passing through a 200 mesh sieve are now used for the TL measurements. The TL samples are placed directly onto the heating plate, and not embedded in silicone oil as was the case for the Apollo 11 samples. Figure 1 shows a comparison of the natural TL glow curves from a 7 cm core sample with and without these modifications.

Irradiations are made with a 1 Ci ^{90}Sr–^{90}Y β source whose dose rate (3.9×10^5 rads/hr) was calibrated with LiF TL dosimeters by R. S. Landauer, Jr. and Co., Glenwood, Ill.

Double core 12025/12028

Dependence of glow curves on radiation dose and depth in the core. To investigate the basic nature of the TL phenomenon in the core we first took samples from eight different depths and measured their light output as a function of radiation dose. For each position two samples were heated in the as-received condition to measure the

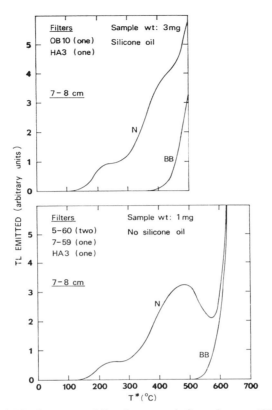

Fig. 1. Natural TL glow curve of 7 to 8 cm sample from the core 12025, as measured with and without the modifications described in the text.

amount of natural TL. Additional samples from the same position were then given increasing β doses prior to their initial heating.

Figure 2 shows the resulting glow curves for depths of 0.2 and 36 cm. (Unless otherwise indicated, the depths given will refer to the sample core depths and not the undisturbed depths.) The β-induced TL is quite similar in all of the core samples, indicating that the TL response of the samples from different depths is similar. The β-induced glow curves are also similar to those in the Apollo 11 core samples in shape, but are about 10 times more sensitive per unit dose.

The natural TL on the other hand varies considerably with depth. We consider first the behavior of the deeper core sample, which was stored at a low ambient temperature on the moon. At the higher glow-curve temperatures ($\geqslant 550°C$), there is no difference between the as-received samples and those which have received additional irradiation. We interpret this as simply indicating that all the available deep (high temperature) charge traps have already been filled by the intense radiation received on the moon.

In contrast, the TL output below 550°C is enhanced by additional β irradiation. Since the radiation dose received on the moon is much larger than the β doses, it is

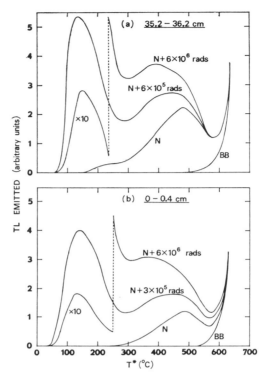

Fig. 2. TL glow curves of as-received samples (N) and of samples that have been given additional β doses before heating (N + indicated dose) for 0–0.4 cm and 35.2–36.2 cm core samples.

clear that draining of these shallower traps has taken place. In this paper we show that the assumption of *thermal* draining gives a consistent, semi-quantitative picture of the TL behavior. Alternatively, as suggested by Geake *et al.* (1970) and Garlick *et al.* (1971), the draining could be due in part to a *nonthermal* process. However the traps that have been suggested as being nonthermally drained have a decay constant of the order of one week. Since the time scale for establishing equilibrium on the moon in the traps that we measure is more than 10^4 times as long, such short-lived traps could have little influence on our results. Apollo 12 samples stored for 7 months at room temperature give the same TL, within the $\pm 15\%$ experimental error, as samples measured immediately upon reception.

Qualitative support of the thermal-draining model is obtained by comparing the results for the 0.2 cm core with those for the 36 cm core (see Fig. 2). The TL response of the shallower sample is enhanced by β irradiation at the highest temperatures, in contrast to the TL saturation of the deeper sample. We attribute this difference to the effect of the diurnal heat wave which causes thermal draining of even the deepest traps in core samples close to the lunar surface.

In Fig. 3 we show the normalized TL output at three different glow-curve temperatures as a function of core sample depth. (Also shown are the corresponding undisturbed depths determined by CARRIER *et al.*, 1971.) The normalization was done by dividing the natural TL output by the amplitude of the artificial TL at 250°C induced by a subsequent β dose of 1.5×10^5 rads (the irradiation being performed after the initial heating). Within experimental error the points from 10 to 40 cm could form a monotonically decreasing curve. In our preprint, a similar figure showed apparent structure. Subsequent measurements using the more homogeneous sieved material has not confirmed the existence of any such structure. The general features of the lunar TL are similar to those observed by DOELL and DALRYMPLE (1971) in their companion study of the double core. (We are deeply indebted to B. Dalrymple for his generous provision of the core samples previously studied by him which alternate in depth with ours.)

The initial rapid rise in TL as a function of depth, leveling off at about 13 cm, is qualitatively what one would expect from the attenuation of the diurnal heat wave. The relatively slow decrease at deep depths is attributed primarily to the attenuation of the galactic cosmic rays. In succeeding sections we show that the results are also in reasonable quantitative agreement with these views.

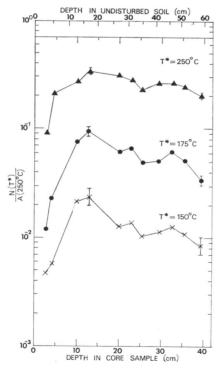

Fig. 3. Depth dependence of normalized natural TL emitted at 150°, 175°, and 250°C, for double core 12025,28. A(250°C) is the intensity at 250°C for a second heating of the same sample after a β dose of 1.5×10^5 rads. The lower horizontal axis gives the core sample depth, and the upper axis gives the estimated undisturbed depth. Error bars represent estimated error of $\pm 2\sigma$.

Penetration of the lunar diurnal heat wave. In order to predict the depth dependence of the TL quantitatively, we calculate first the penetration of the diurnal heat wave and second the dependence of the stored TL on ambient temperature.

The average lunar surface temperature is about 240°K but during the lunar day the temperature rises to a maximum of about 390°K and falls during the lunar night to a minimum of about 90°K (KOPAL, 1966). The amplitude, A, of the diurnal temperature fluctuation is attenuated with depth, d, below the lunar surface, as given by

$$A = A_0 \exp(-2\pi d/\lambda), \tag{1}$$

where λ is a constant called the thermal wavelength. This constant is defined as

$$\lambda = 2\pi \sqrt{\frac{2K}{\omega \rho C_p}},$$

where K is the thermal conductivity, $\omega = 2\pi/(\text{period of a lunation})$, ρ is the density, and C_p is the specific heat. Using the following values at 320°K

$$\left. \begin{array}{l} K = 1.95 \times 10^{-3} \text{ watt/m-deg} \\ \rho = 1.3 \text{ g/cm}^2 \end{array} \right\} \text{(CREMERS and BIRKEBACK, 1971)}$$

$$C_p = 0.1845 \text{ cal/g-deg} \qquad \text{(ROBIE } et al., 1970),$$

we get $\lambda = 24$ cm for Apollo 12 fines. Although K and C_p are temperature dependent, they vary with temperature in similar ways so that λ is effectively independent of temperature between 200° and 400°K.

Using this value for λ, we calculate the amplitude A from equation (1) above. When A is added to the average temperature of 240°K, it gives the maximum temperature, $T_{\max}$, whose depth dependence is shown in Fig. 4.

We next calculate the dependence of stored TL on ambient temperature using the following simple model: We first assume that the decay of TL is a first-order process; the rate of decay of filled traps, n, is then given by $-\beta(T)n$, where $\beta(T)$ is a temperature-dependent rate constant. The rate of filling of traps by incident radiation is taken to be proportional to the total number of unfilled traps present. Thus, the change of filled traps in time dt is

$$dn = \alpha R(N - n) \, dt - \beta(T)n \, dt, \tag{2}$$

where N is the total number of existing traps, R is the radiation dose rate, and α is the probability per unit dose of filling an unfilled trap. If the TL is in thermal equilibrium, then

$$\alpha R(N - n) = \beta n. \tag{3}$$

When $n \ll N$, i.e. well below the saturation level, this approximates to

$$\alpha R N = \beta n. \tag{4}$$

Assuming that α, R, and N are constants for the different samples studied, n (and hence the accumulated TL) is inversely proportional to β. We take β to be given by

$$\beta(T) = s \exp(-Q/kT), \tag{5}$$

where k is Boltzmann's constant, Q is the thermal activation energy of filled TL traps, and s is a constant called the frequency factor. Assuming that the frequency factor

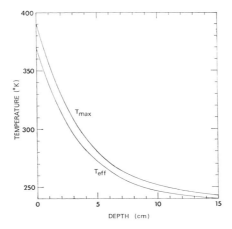

Fig. 4. Theoretical depth dependences of the maximum temperature, T_{max}, and the effective temperature, T_{eff}, in the lunar regolith. T_{eff} is the constant temperature which will give the same amount of thermal decay as the sinusoidal oscillation with amplitude T_{max}.

does not change between samples, the ratio of the TL at an effective ambient temperature of T_1 to that at an effective ambient temperature of T_0 is

$$\frac{L_1}{L_0} = \exp\left[-\frac{Q}{k}\left(\frac{1}{T_0} - \frac{1}{T_1}\right)\right], \tag{6}$$

where L_1 is the equilibrium value of the TL at an ambient temperature of T_1, and L_0 is the value at T_0.

Finally, to relate the stored TL to the diurnal heat wave, it is useful to define an effective temperature T_{eff} such that the amount of TL remaining in a sample after storage for one lunation period at an ambient temperature of T_{eff} is the same as for a sample whose temperature has been cycled about the average temperature of 240°K with an amplitude A. We assume that the lunar surface temperature varies sinusoidally about the average value, so that

$$T(t) = 240°K + A \sin(t), \tag{7}$$

where t is the phase angle. Then T_{eff} is given by

$$\exp(-Q/kT_{eff}) = \frac{1}{2\pi}\int_0^{2\pi} \exp(-Q/kT(t))\, dt. \tag{8}$$

To evaluate the integral numerically, the value of Q is needed. From measurements on as-received core samples, using the "initial rise" method of GARLICK and GIBSON (1948), Q lies between 0.85 and 1.4 eV. (This spread in the measured values is possibly caused by an overlap of low-temperature TL traps.) Since the calculated result is insensitive to changes in Q, we take Q to be 1 eV. Numerical integration of equation (8) gives T_{eff} as a function of A, and thus as a function of T_{max} through the equation

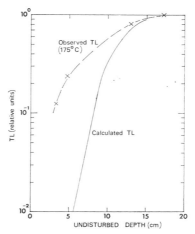

Fig. 5. Variations of observed and calculated TL with undisturbed depth, for double
core 12025,28.

$T_{\max} = 240°K + A$. We know the depth dependence of $T_{\max}$, and so can obtain the
depth dependence of T_{eff}, which is shown in Fig. 4. We now use equation (6) to calculate
the variation of equilibrium TL with effective temperature T_{eff}, taking T_1 to be T_{eff}, T_0
to be the mean subsurface temperature of 240°K, and Q to be 1 eV. This dependence
of TL on T_{eff} is then converted to a dependence of TL on depth using the values of
T_{eff} vs. depth shown in Fig. 4.

The calculated variation of the equilibrium TL with depth is shown in Fig. 5.
For comparison the experimental curve of 175°C TL as a function of undisturbed
depth is replotted from Fig. 3. The observed TL lies appreciably above the calculated
curve for shallow depths. However, as predicted, the decrease of stored TL becomes
experimentally detectable at undisturbed depths which are shallower than about
15 cm. This agreement gives strong support to the hypothesis that the observed TL
decrease is caused primarily by thermal drainage.

The disagreement which exists between the theoretical and experimental curves
at depths shallower than 10 cm probably arises from several different causes. In the
calculation we assumed that the dose rate was constant with depth whereas in fact it
increases by two or three orders of magnitude from depths of about 5 cm towards
the lunar surface (due to solar flares, see RYAN, 1964). Also we have found that the
natural TL does *not* decay in the simple exponential manner assumed in the above
analysis.

The sub-surface lunar temperature. The low temperature TL may be used to give
an estimate of the temperature of the moon just below the penetration depth of the
diurnal heat wave. We assume that the TL is in thermal equilibrium, so that we have
equation (3) above

$$\alpha R(N - n) = \beta n.$$

The values of α and n/N are estimated from the TL at 300°C (this temperature
being chosen in order to avoid complications caused by the high sensitivity of the

170°C peak to an additional artificial dose). For the 13 cm level, n/N is 0.12, and for the 36 cm level it is 0.14. For both of these levels, α is 1.3×10^{-6} rad^{-1}. We assume that the dose rate R is 10 rads/yr (after RYAN, 1964). Then, from equation (3), β averages 9×10^{-5} yr^{-1}.

Rearranging eq. (5), we can now calculate the subsurface lunar temperature as

$$T = \frac{Q}{k} \ln \frac{\beta}{s}. \qquad (9)$$

Assuming that the activation energy Q is 1 eV, and that the frequency factor s is about 10^{10} sec^{-1} (from preliminary thermal decay measurements, again assuming first-order kinetics), we find that the value of T corresponding to the observed TL results is 241°K. The agreement between this and the accepted value of 240°K is strong evidence for the validity of the assumed thermal equilibrium. (In view of the approximations used, the great similarity of the calculated and accepted values is obviously fortuitous).

The calculated value of T is fairly insensitive to errors in any parameters occurring inside the logarithmic term; for example, a change by an order of magnitude in such a term produces a change of only about 4% in T. However, T depends directly on Q, which may possibly be as low as 0.85 eV or as high as 1.4 eV. We hope to obtain a more precise value of Q from further studies, and use this and our improving knowledge of the kinetics involved to make a more reliable determination of the subsurface lunar temperature.

The dose-rate and temperature gradients at deep depths. The stored TL at low glow-curve temperatures decreases as the depth increases beyond an undisturbed depth of about 17 cm (see Fig. 3). At an undisturbed depth of 60 cm, the TL at 150° and 175°C is only $(36 \pm 6)\%$ of its value near 17 cm. (The TL above 200°C is not considered in this section because it may not yet have reached thermal equilibrium at these higher glow curve temperatures; this matter is considered in detail later.)

The most likely cause of the decreasing TL is the attenuation of the external radiation flux. Data of LALOU et al. (1970) indicate that the proton dose would be attenuated significantly between 17 and 60 cm. Using a normally incident proton beam it was found experimentally that the dose at a depth equivalent to 60 cm of lunar regolith would be about 67% of its value at 17 cm. Assuming an isotropic flux over a 2π solid angle, we calculate that the dose at 60 cm would be only 42% of its value at 17 cm.

This attenuation of an isotropic proton beam (to 42%) is very close to the experimentally-observed decrease of the TL to $(36 \pm 6)\%$. The reduction of radiation dose rate thus appears to be the primary cause of the decrease of TL.

It has been suggested by J. A. Bastin of Queen Mary College, London (private communication to RMW and HPH) that if the temperature increases towards the center of the moon then the stored low-temperature TL should decrease as depth increases. Comparison of the TL change observed between 17 and 60 cm with that predicted by dose attenuation indicates that the temperature variations produce an additional decrease of between 0% and 28%. Such a temperature-dependent decrease in the stored TL would be produced by a temperature rise of (0.8 ± 0.8)°K in these

43 cm. Therefore we infer that the temperature increases with depth in the regolith by $(2 \pm 2)°K/m$.

The temperature gradient of $(2 \pm 2)°K/m$ is consistent with the known thermal properties of the moon. If the regolith is thin (e.g., ~ 5 m as suggested by OBERBECK and QUAIDE, 1968) then the thermal gradient in the underlying bedrock will scale directly as the ratio of the thermal conductivity of the regolith to that of the bedrock. Since this ratio is $\sim 10^{-3}$ (BASTIN et al., 1970) the inferred gradient in the bedrock is $(2 \pm 2)°K/km$—in agreement with the value deduced by SONNETT et al. (1971) from electromagnetic observations. Such a thermal gradient in the upper bedrock permits a nonmolten inner core, as indicated by a number of other observations.

It is well-known that a direct measurement of the heat flow from the moon can give important information about the thermal state of the interior. This in turn can be used to set limits on the amount of radioactivity in the interior of the moon and hence (coupled with measurements made on lunar surface rocks) on the possible initial chemical composition of the material from which the whole moon was formed. We prefer to delay comments on these fundamental questions until we have had the opportunity to improve our results.

A direct measurement of the heat flow is planned for Apollo 15 (LANGSETH et al., 1970). A bore hole is to be drilled with a hollow bit and it is planned to recover the core material. If this experiment is successful it will provide an essential "ground truth" comparison with TL data from the same core. Once validated, the data can then be used to obtain heat-flow data from any of the longer cores in the Apollo missions and hence provide a larger area coverage of the moon than will be possible with the direct heat flow experiment itself.

It is noted that it would be most useful if it were possible to obtain cores at depths of two meters and beyond. This is because, at such depths, the external radiation dose rate will be attenuated to a value comparable with the internal radiation dose rate of about 0.1 to 0.5 rads/yr (estimated from the U concentration in lunar fines); the depth variations of the dose rate will be much less significant here than near the surface, thus making the thermal gradient determination much more accurate.

Thermal equilibrium and past disturbances. We have seen that the TL is well below saturation at glow-curve temperatures below about 550°C, because of thermal decay. A question of fundamental importance to this study is whether or not this TL is in thermal equilibrium (as was assumed in the analysis above). The time required to reach a given fraction of equilibrium is a function of the thermal-decay rate (and the dose rate); the faster the rate of decay, the shorter is this time. If the amount of TL is measured by the height of the glow curve at a given temperature T^*, then the rate of decay at a constant ambient temperature will decrease with increasing glow-curve temperature, and the time required to reach equilibrium will increase.

The total time of the samples on the moon would seem to be more than adequate to reach thermal equilibrium at all glow-curve temperatures. However, recent disturbances may upset the equilibrium level, as suggested by DALRYMPLE and DOELL (1970a). Evidence for this effect was found previously in the Apollo 11 cores in the form of "cross-overs" in the TL glow curves (DALRYMPLE and DOELL, 1970b; HOYT et al., 1970).

We have attempted to answer this question by the following analysis of the Apollo 12 core data: We use the model introduced earlier, which assumed that the TL is in equilibrium and the decay is a first-order process. This led to equation (3), which can be rearranged as

$$\frac{\beta}{R} = \alpha \left(\frac{N}{n} - 1 \right). \tag{10}$$

If we assume that we have a continuous distribution of trap depths, then β should be a monotonic function of the glow-curve temperature T^*. Figure 6 shows a plot of β/R vs. T^* for eight core depths. On the righthand axis is plotted the corresponding time to reach $1/e$ of the equilibrium value, assuming that R equals 10 rads/yr. This is a reasonable assumption for depths below about 5 cm, but for the shallower depths the dose rate is higher and the time to reach $1/e$ of equilibrium is shorter. The curves are displaced parallel to one another because of the variations in the dose rate R as a function of depth and, more important, the variations of the ambient temperature with depth.

Four of the levels show smoothly varying curves in agreement with the assumption of thermal equilibrium. However, three or four depths (4, 7, 13, and possibly 29 cm) show a sudden break in the curves, apparently from one line of equilibrium to another. A probable explanation of this shift is that these samples were exposed near the surface and then covered over recently. For each of these glow curves, the lower temperature

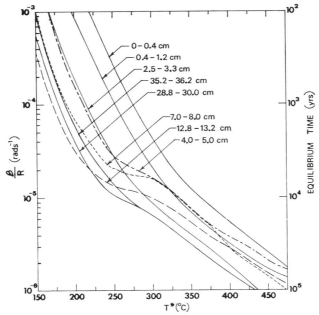

Fig. 6. Dependence upon glow-curve temperature, T^*, of TL decay rate divided by dose rate, for eight core depths of double core 12025,28; also shown on right-hand axis is time TL takes to reach $1/e$ of its equilibrium value.

part has changed to the equilibrium level for the present depth while the higher temperature part retains the equilibrium level characteristic of the former depth of the material.

In principle, the time of the past disturbances can be determined by the position at which the break occurs. Assuming a dose rate of 10 rads/yr gives an estimate of about 10^4 yr (see Fig. 6, right-hand axis). This seems low in terms of what is known about the rate of stirring of the soil (see, for example, RANCITELLI *et al.*, 1971). The basic source of error is probably that the experimentally measured values of α at a given T^* are too high because the freshly irradiated sample contains filled lower temperature traps which contribute to the TL at T^*. We hope to overcome this problem by proper annealing of the samples after artificial irradiation before the sensitivity is measured.

Rock 12063

Two bars were cut from a vertical section of rock 12063, one from the top and one from the bottom, each extending inward a distance of about 1 cm from the surface (see Fig. 7). These bars were diced with a diamond wire saw at 1 mm intervals for TL studies as a function of depth near each of the surfaces. Only three of these samples have been measured to date. Figure 8 shows the normalized TL glow curves of the

Fig. 7. Top end of vertical section from rock 12063. Outside surface is to the top of the photograph. The bar, cut off the right as shown, was diced at 1 mm intervals for TL studies as a function of depth. The left-hand piece was mounted for cosmic ray track studies (see CROZAZ *et al.*, 1971). Two divisions equal 1 mm.

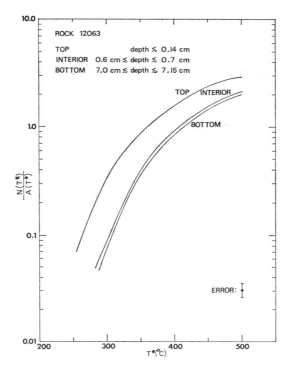

Fig. 8. Normalized TL emitted as a function of glow-curve temperature, T^*, for rock 12063. The natural TL intensity, $N(T^*)$, is divided by the artificial intensity, $A(T^*)$, for a second heating of the same sample after a β dose of 1.5×10^5 rads. Results are shown for samples taken from top and bottom surfaces, and from the interior. The error bar represents estimated $\pm 2\sigma$.

top-surface sample, the interior sample (~ 0.65 cm below the top) and the bottom-surface sample (~ 7 cm below the top). At a glow-curve temperature of 300°C, the TL in the top sample is 4.5 times that of the bottom sample; the difference is largest for lowest temperatures.

We attribute the higher top-surface TL to ionization produced by solar flares. This effect should be more pronounced in the rocks than in the core because the higher thermal conductivity of the rocks limits thermal gradients. (Compared to $\lambda = 24$ cm, calculated above for Apollo 12 fines, we find $\lambda = 425$ cm from the data of ROBIE *et al.* (1970) on crystalline rocks.) Thus we expect differences in ionization rate to be the predominant cause of the TL depth dependence.

Our interpretation of the high top-surface TL as due to solar flares is supported by radioactivity measurements on this rock (RANCITELLI *et al.*, 1971). These workers find the highest ^{22}Na ($T_{1/2} = 2.6$ yr) concentration on the top and the highest ^{26}Al ($T_{1/2} = 7.4 \times 10^5$ yr) on the bottom. They conclude that the surface which we identify as the top is indeed the most recent top, but that the rock has been turned over in the past few hundred thousand years. Since we estimate the TL equilibrium time for the top surface of a rock to be $\lesssim 10^2$ yr, our agreement with the ^{22}Na results is reasonable.

However, it is important that TL measurements be made on a rock that has been irradiated by solar flares on one side only, to confirm the interpretation of the TL results.

We consider the establishment of the solar-flare effect in rocks to be important because it should be possible to use this effect, when treated more quantitatively, to give an independent measurement of the average solar-flare energy spectrum over a time span inaccessible to radioactivity measurements.

Conclusions

With the extension of the natural TL glow curves to 650°C, a region in or near thermal stability, the qualitative features of the lunar TL have become clear. For the deeper core samples, which were stored at lower ambient temperatures, the TL is in saturation above about 550°C, the glow-curve temperature where thermal decay becomes negligible. At lower temperatures the thermal decay has been significant in all material, and reflects the existing thermal and dose-rate gradients. The decay is most pronounced in the samples shallower than about 15 cm (undisturbed), because of the penetration of the diurnal heat wave. The limit of 15 cm for a measurable effect of the diurnal heat wave is in agreement with a theoretical prediction based on knowledge of the thermal wavelength and the calculated change in TL response as a function of ambient temperature. The rapid increase of TL with depth over the first 12 cm is quantitatively very similar in the Apollo 11 and Apollo 12 cores, indicating that the measurements of thermal effects have not been seriously disturbed by local conditions.

Preliminary analysis shows that there is a positive temperature gradient of $(2 \pm 2)°K/m$, between 17 and 60 cm (undisturbed), representing heat flow outward. The subsurface (near 17 cm, undisturbed) lunar temperature is directly related to the absolute magnitude of the TL in thermal equilibrium and is calculated to be approximately 240°K. The TL at temperatures less than 225°C appears to be in thermal equilibrium for all samples. The TL above 225°C shows anomalies at some core depths which can be interpreted as resulting from past disturbances.

The TL in rock 12063 increases by a factor of 4.5 near the top surface, presumably because of solar flares. The establishment of this effect should give a method for determining the recent orientation of lunar rocks and for measuring the average solar-flare energy spectrum on a time scale (about $10^3–10^4$ yr) inaccessible by other methods.

DIFFERENTIAL THERMAL ANALYSIS

Experimental results

Differential thermal analysis (DTA) measurements were made on fines from the outer flank of the Surveyor Crater (12042,43), the trench (15 cm depth) on the northwest rim of Head Crater (12033,49), and from 14 levels of the double core tube (0.4 to 37.3 cm). The scans were made at 30°C/min from 25° to 900°C in both air and ultra-pure nitrogen atmospheres, with the exception of the core samples which were scanned only in nitrogen. The scans were made on ~ 12 mg samples using a DuPont model 900 unit equipped with a standard cell for the 25°–500°C range and a less sensitive, intermediate temperature cell covering the 150°–900°C range. Powdered

DTA SCANS OF LUNAR FINES

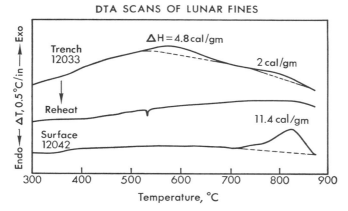

Fig. 9. Differential thermal analysis (DTA) scans at 30°C/min in nitrogen of lunar fines from Surveyor Crater (12042,43) and the trench (15 cm depth) on Head Crater (12033,49). The differences between the exotherm limits shown here and in Fig. 10 and the temperatures quoted in the text are due to thermocouple corrections. Note the small but sharp endotherm in the reheat scan of the Surveyor Crater sample.

Al_2O_3 was used as the reference material and the system was calibrated using the α-β transition ($\sim$ 580°C) in quartz (LAKODEY et al., 1956).

Generally, except for some notable exceptions in the core, energy is released in three broad temperature regions during heating in air: (A) 250°–450°C, $<$ 1 cal/g; (B) 350°–800°C, $\leq$ 15 cal/g; (C) 770°–870°C, $\leq$ 20 cal/g. For scans in air of the trench sample, peak A was not observed and both peaks B and C were much smaller than those in the Surveyor Crater sample; in the latter, peaks B and C were about the same in both position and intensity as found in Apollo 11 soil and breccia samples (HOYT et al., 1970; KARDOS, 1970). Generally, peak A was considerably smaller in Apollo 12 ($\leq$ 1 cal/g) than in Apollo 11 samples ($\leq$ 6 cal/g).

Heating in ultra-pure nitrogen after evacuation to less than 1 mm Hg eliminates peak A, reduces peak B considerably, and does not affect peak C. Typical nitrogen scans for the Trench and Surveyor Crater samples are shown in Fig. 9. Baselines for calculations of the energy releases were drawn based on the cooling curve and the slope of the heating curve before and after the exotherms.

In the core-tube samples, which were scanned only in ultra-pure nitrogen with the intermediate temperature cell, peak C was present at all levels. What is normally peak A could not be detected with the low-sensitivity cell, and peak B varied considerably in both its position and intensity as a function of depth. The energy release values and their temperature ranges are summarized in Table 1. Scans from three levels in the core are presented in Fig. 10.

The most striking feature of the core DTA scans is the large exotherm which appears in the 200°–400°C range at levels from 20–26 cm. It begins to build up at the 19.8–20.7 cm level (16 cal/g), increases to 47 cal/g at the 22.5–23.5 cm level (lowest scan of Fig. 10), rises to a maximum of 232 cal/g for the 25.4–26.1 cm level, and then disappears at greater depths. Although it occurs in the temperature range normally

Table 1. Energy releases in core-tube samples

Core depth (cm)	Peak B Range, °C	ΔH, cal/g	Peak C Range, °C	ΔH, cal/g
0.4–1.2	350–697	19.5	745–876	11.4
2.5–3.3	555–735	5.0	735–875	14.3
4.0–5.0	334–452	5.6	793–866	7.2
7.0–8.0	343–485	10.0	778–876	7.0
9.4–11.0	362–527	5.5	697–875	23.9
12.8–13.2	348–569	40.9	745–866	7.5
17.4–18.4	negligible	—	740–866	25.9
19.8–20.7	207–428*	16.0	769–817	5.8
22.5–23.5	198–338*	47.0	710–900	14.3
25.4–26.1	198–402*	232.0	800–900	< 2.0
28.8–30.0	494–744	18.3	830–920	4.7
32.2–33.2	negligible	< 1	770–910	12.2
35.2–36.2	negligible	< 1	750–900	8.4
36.7–37.3	negligible	< 1	730–910	16.4

* Range normally assigned to peak A.

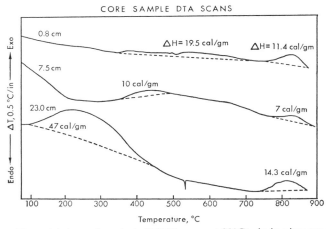

Fig. 10. Differential thermal analysis (DTA) scans at 30°C/min in nitrogen of samples from three levels of the core tube. Note the large low temperature exotherm in the scan of the 23 cm depth.

occupied by peak A, we have identified it as peak B in Table 1 because when it occurs, there are no detectable exotherms between it and the high temperature peak C, and because peak A has never before been detected during heating under nitrogen.

Reheating the samples in either air or nitrogen after initial heating to 900°C frequently produces a small but sharp endotherm at about 540°C (see Fig. 10). In the core samples, the endotherm appeared sporadically as a function of depth. Reheating caused the endotherm to intensify in scans where it was originally present and to appear in some but not all of the scans in which it did not appear initially.

Discussion

In general, the low temperature exotherm (peak A) seen during heating in air is most likely due to oxidation, since it was not detected during heating under nitrogen.

In the case of the core-tube samples, the situation is not as clear since air scans were not made due to limited sample quantities.

Of the three exotherms found, peak B is the most likely to include some radiation-induced stored energy. Although considerably reduced, it is not eliminated by heating under nitrogen. Furthermore, irradiated terrestrial hypersthene shows a large exotherm in this temperature region, which is not present in unirradiated samples (HOYT *et al.*, 1970; KARDOS, 1970). The anomalous behavior of peak B in the 20–26 cm level of the core is presently unexplained. If it is due to radiation-induced stored energy, those particular core depths must have received an unusually large amount of radiation damage. Although nothing unusual was found in the nuclear track densities in the larger crystals removed from this layer (CROZAZ *et al.*, 1971), it is possible that an unusually large number of heavily irradiated small grains reported by the Orsay and Berkeley groups (BORG *et al.*, 1970; BARBER *et al.*, 1971) are present. This possibility is currently being examined. What may be regarded as "normal" behavior for peak B, namely energy releases of less than 20 cal/g, is consistent with our current knowledge of the lunar radiation history and can be explained in terms of damage from high energy cosmic rays and solar flare particles, solar wind-induced damage, and recovery of defects produced by spallation reactions (HOYT *et al.*, 1970).

The high temperature exotherm (peak C) is very similar to that seen in Apollo 11 samples (HOYT *et al.*, 1970; KARDOS, 1970) and is believed to arise from crystallization of glass. Occasionally the peak took on a weak doublet appearance similar to that found in Apollo 11 fines and breccias. Evidence supporting crystallization of glass as the cause of peak C, in addition to that cited in the Apollo 11 study (HOYT *et al.*, 1970), comes from consideration of the origin and behavior of the small but sharp endotherm at about 540°C. Such a peak is characteristic of a solid-solid crystal transition such as the α-β transition in quartz. In cases when there is initially no endotherm but a prominant peak C, the endotherm always appears on reheating and is characteristic of a new crystalline solid presumably formed by crystallization of the glass. In a few instances, notably in the core samples, the endotherm appears in the initial heating and intensifies on reheating. In these cases, some of the crystalline material to which the glass crystallizes is probably already initially present and is enhanced by the high temperature crystallization. The magnitude of the endotherm is so small that a quantitative correlation with peak C is not feasible. Evidence for partially crystallized glass has been found by WHITE *et al.* (1971) in the Raman spectra of lunar glasses. The presence of such crystals as nucleating sites in the glass particles would greatly increase the ability of the glass to crystallize.

If our interpretation of peak C and the endotherm is correct, DTA would yield a measure of the glass present. There appears to be no obvious correlation between the intensity of peak C and the visible boundaries in the core profile. Final verification of our hypothesis that peak C and the endotherm are related to the sample glass content obviously rests on detailed quantitative measurements of the glass.

Acknowledgments—We thank G. BRENT DALRYMPLE for a critical review of the paper. This work was supported by NASA Contract NAS 9-8165. The support of one of us (D. B.) by the National Science Foundation under Grant GY 7501 is gratefully acknowledged.

Since this paper was submitted for publication, we have measured the natural TL for some depths intermediate to those reported above. The general features of the TL depth dependence have not changed. However, the low-temperature TL of samples at 14 and 15 cm is 50% lower than the TL from samples at the adjacent depths of about 13 and 17 cm. We have found that the natural TL may be reduced if the samples are illuminated with white light, but this optical-bleaching phenomenon is *not* responsible for the anomalously low TL at 14 and 15 cm. These results will be discussed in another paper.

REFERENCES

AITKEN M. J., ZIMMERMAN D. W., and FLEMING S. J. (1968) Thermoluminescent dating of ancient pottery. *Nature* **219**, 442–445.

BARBER D. J., HUTCHEON I., and PRICE P. B. (1971) Extralunar dust in Apollo cores? *Science* **171**, 372–374.

BASTIN J. A., CLEGG P. E., and FIELDER G. (1970) Infrared and thermal properties of lunar rock. *Proc. Apollo 11 Lunar Sci. Conf., Geochim. Cosmochim. Acta* Suppl. 1, Vol. 3, pp. 1987–1991. Pergamon.

BORG J., DRAN J. C., DURRIEU L., JOURET C., and MAURETTE M. (1970) High voltage electron microscope studies of fossil nuclear tracks in extraterrestrial matter. *Earth Planet. Sci. Lett.* **8**, 379–386.

CARRIER W. D., JOHNSON S. W., WERNER R. A., and SCHMIDT R. (1971) Disturbances in samples recovered with Apollo core tubes. Second Lunar Science Conference (unpublished proceedings).

CREMERS C. J. and BIRKEBAK R. C. (1971) Thermal conductivity of fines from Apollo 12. Second Lunar Science Conference (unpublished proceedings).

CROZAZ G., WALKER R., and WOOLUM D. (1971) Nuclear track studies of dynamic surface processes on the moon and the constancy of solar activity. Second Lunar Science Conference (unpublished proceedings).

DALRYMPLE G. B. and DOELL R. R. (1970a) Thermoluminescence of lunar samples. *Science* **167**, 713–715.

DALRYMPLE G. B. and DOELL R. R. (1970b) Thermoluminescence of lunar samples from Apollo 11. *Proc. Apollo 11 Lunar Sci. Conf., Geochim. Cosmochim. Acta* Suppl. 1, Vol. 3, pp. 2081–2092. Pergamon.

DOELL R. R. and DALRYMPLE G. B. (1971) Thermoluminescence of Apollo 12 lunar samples. *Earth Planet. Sci. Lett.* **10**, 357–360.

GARLICK G. F. J. and GIBSON A. F. (1948) The electron trap mechanism of luminescence in sulphide and silicate phosphors. *Proc. Phys. Soc.* (London) **60**, 574–590.

GARLICK G. F. J., LAMB W. E., STEIGMANN G. A., and GEAKE J. E. (1971) Thermoluminescence, EPR and diffuse reflection spectra of Apollo lunar samples. Second Lunar Science Conference (unpublished proceedings).

GEAKE J. E., DOLLFUS A., GARLICK G. F. J., LAMB W., WALKER G., STEIGMANN G. A., and TITULAER C. (1970) Luminescence, electron paramagnetic resonance and optical properties of lunar material from Apollo 11. *Proc. Apollo 11 Lunar Sci. Conf., Geochim. Cosmochim. Acta* Suppl. 1, Vol. 3, pp. 2127–2147. Pergamon.

HOYT H. P., KARDOS J. L., MIYAJIMA M., SEITZ M. G., SUN S. S., WALKER R. M., and WITTELS M. C. (1970) Thermoluminescence, X-ray and stored energy measurements of Apollo 11 samples. *Proc. Apollo 11 Lunar Sci. Conf., Geochim. Cosmochim. Acta* Suppl. 1, Vol. 3, pp. 2269–2287. Pergamon.

KARDOS J. L. (1970) Stored energy measurements in Apollo 11 lunar samples by differential thermal analysis. In *Analytical Calorimetry* (editors R. S. Porter and J. Johnson), Vol. 2, pp. 269–280. Plenum.

KOPAL Z. (1966) *An Introduction to the Study of the Moon*, pp. 339. Reidel.

LAKODEY P., EYRAUD C., and PRETTRE M. (1956) Étude énergétique des transformations "secondaires" de la silice. *Compt. Rend.* **242,** 3071–3074.

LALOU C., BRITO U., NORDEMANN D., and MARY M. (1970) Thermoluminescence induite dans des cibles épaisses par des protons de 3 GeV. *Compt. Rend.* **270,** Série B, 1706–1708.

LANGSETH M. G., WECHSLER A. E., DRAKE E. M., SIMMONS G., CLARK S. P., and CHUTE J. (1970) Apollo 13 lunar heat flow experiment. *Science* **168,** 211–217.

OBERBECK V. R. and QUAIDE W. L. (1968) Genetic implications of lunar regolith thickness variations. *Icarus* **9,** 446–465.

RANCITELLI L. A., PERKINS R. W., FELIX W. D., and WOGMAN N. A. (1971) Cosmogenic and primordial radionuclide measurements in Apollo 12 lunar samples by nondestructive analysis. Second Lunar Science Conference (unpublished proceedings).

ROBIE R. A., HEMINGWAY B. S., and WILSON W. H. (1970) Specific heats of lunar surface materials from 90° to 350°K, *Proc. Apollo 11 Lunar Sci. Conf., Geochim. Cosmochim. Acta* Suppl. 1, Vol. 3, pp. 2361–2367. Pergamon.

RYAN J. A. (1964) Corpuscular radiation produced crystalline damage at the lunar surface. In *The Lunar Surface Layer* (editors J. W. Salisbury and P. E. Glaser), pp. 265–312. Academic Press.

SONETT C. P., SMITH B. F., COLBURN D. S., SCHUBERT G., SCHWARTZ D., DYAL P., and PARKIN C. W. (1971) The lunar electrical conductivity profile: Mantle-core stratification, near surface thermal gradient, heat flux and composition. Second Lunar Science Conference (unpublished proceedings).

WHITE W. B., WHITE E. W., GORZ H., HENISCH H. K., FABEL G. W., and ROY R. (1971) Examination of lunar glass by optical, Raman, and X-ray emission spectroscopy, and by electrical meaurements. Second Lunar Science Conference (unpublished proceedings).

Proceedings of the Second Lunar Science Conference, Vol. 3, pp. 2265–2275
The M.I.T. Press, 1971.

Luminescence of Apollo lunar samples*

J. E. GEAKE and G. WALKER
U.M.I.S.T., Manchester, England

A. A. MILLS
University of Leicester, England

and

G. F. J. GARLICK
University of Hull, England

(*Received* 5 *March*, 1971; *accepted in revised form* 20 *April* 1971)

Abstract—Luminescence emission spectra in the visible and near IR regions are shown for lunar samples under 60 KeV proton excitation. The luminescence is mainly from the plagioclase present, whose emission spectrum consists mainly of a strong green peak at about 5600 Å, probably caused by Mn^{2+} as an activator; there is also a weaker blue peak at about 4500 Å which is common to most silicates and is probably caused by a lattice defect; and a weak IR peak at about 7700 Å whose cause is unknown. Comparison is made with meteoritic material, and also with terrestrial plagioclases most of which have the IR peak as the dominant one. Four different lunar fines samples from Apollo 11 and Apollo 12 all show similar spectra and differ only in their efficiencies, which range from 10^{-4} to 10^{-5} depending on their albedos, which in turn depend on their plagioclase content. Plagioclase separated from a lunar rock has an efficiency of about 10^{-3}. Luminescence photographs of lunar rocks and breccias under 6 KeV electron excitation show mainly an emission pattern of plagioclase crystals; one breccia only (10059) shows a variety of luminescence colours from constituents yet to be identified. Proton bombardment produces a slight darkening of two fines samples, but not of two others, and also an erosion effect involving the demolition of the initial complex grain assemblies.

INTRODUCTION

THIS PAPER and the two which follow describe the continuation of the work described in our earlier paper (GEAKE *et al.*, 1970), using the Apollo 11 material there listed, and its extension to Apollo 12 material received subsequently. From Apollo 12 we received a total of 3.04 g from three different samples of fines of grain size less than 1 mm, and 6 chips of total weight 7.45 g from two rocks, i.e., material from five different samples as follows:

12032 light-coloured fines

12033 light-coloured fines, from a trench 15 cm deep

12070 dark fines, from the contingency sample

12002 a grey-brown fine-medium grained crystalline igneous rock

12051 a grey-speckled brown medium-grained crystalline igneous rock.

These samples were used for the collaborative investigations described in this group of three papers (pp. 2265–2300), and divided as follows:

Paper 1, the present paper: Luminescence under proton and electron excitation, investigated at Manchester (U.M.I.S.T.) and at Leicester University.

---

* Paper 1 of three collaborative papers; papers 2 and 3 follow.

Paper 2, which follows (Garlick *et al.*, 1971: Thermoluminescence, investigated at Hull University.

Paper 3, which follows (Dollfus *et al.*, 1971): Optical polarisation and albedo, investigated at Paris Observatory, Meudon and at Manchester.

The present paper gives luminescence emission spectra for lunar materials in comparison with terrestrial and meteoritic materials, discusses the cause of this luminescence, and shows the distribution of the luminescent constituents within rock and breccia samples.

1. Luminescence Emission Spectra

All of our Apollo 12 dust and rock samples were inspected visually under UV lamps giving mainly 2537 Å and 3650 Å emission, but negligible luminescence was observed. However, when excited by protons or electrons all the samples showed some luminescence, mainly from their plagioclase parts. Emission spectra under 60 KeV proton excitation were scanned by means of a photoelectric grating spectrophotometer (Geake *et al.*, 1967 and 1970); an EMI 9558B trialkali photomultiplier was used for the visible region and a Mullard 150 CVP Ag–AgO–Cs photomultiplier, with its photocathode cooled to liquid nitrogen temperature, for the near IR region. A further photomultiplier monitored the total emission and corrected the output for time fluctuations and for efficiency fall-off caused by proton damage. Fig. 1 shows the emission spectra, corrected for the spectral response of the instrument, for all three of our Apollo 12 dust samples (12032,39; 12033,60; and 12070,113); our previously published spectrum for Apollo 11 dust sample 10084,6 is also shown for comparison (from Geake *et al.*, 1970). We have two types of Apollo 12 fines—12070,113 is very dark in appearance and resembles the Apollo 11 fines 10084,6, whereas 12032,39 and 12033,60 are much lighter because they contain more plagioclase and less ilmenite, and they are probably ray material (see Dollfus *et al.*, 1971). However, all four samples show very similar emission spectra, with a weak 4500 Å peak, a stronger peak at 5600 Å, and some emission in the near IR, each of which we ascribe to plagioclase. It is only the efficiencies that are different for the four samples, and these seem to be related to the plagioclase content. The efficiency increases faster than the plagioclase content, probably because there are two effects: as plagioclase is the most efficient luminescing component, more plagioclase means more emission, but it also means a lower content of absorbing material (e.g., ilmenite), as indicated by the albedo, so that the emitted light can get out more easily and can better survive scattering by surface roughnesses, giving an even greater apparent efficiency, as given in Table 1. The true efficiency of the actual luminescing material is probably about the same for all four samples.

The emission spectra of some terrestrial plagioclases are shown in Figs. 2, 3, and 4. Apart from the relative intensity of the IR emission, as discussed later, the main difference is that their overall efficiencies are 10–100 times that of the lunar fines. However, the plagioclase found in lunar *rocks* does have an efficiency comparable with that of terrestrial plagioclase; thus the separated plagioclase from lunar rock sample 10044,43 has about the same efficiency as labradorite-bytownite, as shown in Fig. 2. The low efficiency of the lunar fines is probably due to the admixture of the

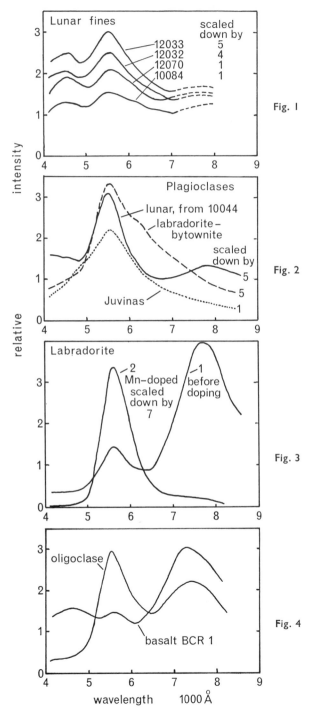

Figs. 1–4. Luminescence emission spectra with
60 KeV proton excitation.

Table 1. Albedo, efficiency, and plagioclase content of samples.

		Albedo* %	Efficiency† $\times 10^{-5}$	% Plagioclase
Apollo 12	12032	15	8	high (amount not given)
	12033	17	10	75‡
	12070	9	1.3	low (contingency sample, not analysed)
Apollo 11	10084	8	1	15‡
Basalt BCR-1		40	40	75
Labradorite			100	90
Lunar plagioclase from 10044			100	95

* From DOLLFUS *et al.* (1971); † approx. integrated energy conversion efficiency; ‡ LRL data.

plagioclase with opaque material, and also possibly to the coating of individual grains with a thin opaque layer, as suggested by HAPKE *et al.* (1970 and 1971) and others.

2. CAUSE OF PLAGIOCLASE EMISSION

Reasoning from crystal field theory suggested that the prominent green peak shown by both lunar and terrestrial plagioclase might be caused by the presence of Mn^{2+}. To test this, the proton-excited emission spectrum of a typical terrestrial plagioclase—labradorite—was scanned, as shown in Fig. 3, curve 1. It was then doped with about 0.1 % of Mn by heating it to 1050°C for 30 min in argon with $MnSO_4$. The result was a 16-fold increase in the height of the green peak, as shown by curve 2, accompanied by a halving of the blue and IR emission. In order to check that simply heating the plagioclase was not producing the change (perhaps by converting the low-temperature form of anorthite to the high-temperature form) a further sample was given the same heat treatment without any $MnSO_4$: its emission spectrum was found to be unchanged.

Plagioclases usually contain more than 100 ppm of Mn, as determined by X-ray fluorescence analysis. The plagioclase separated from lunar rock 10044 was found to contain about 200 ppm of Mn (about the same as an average terrestrial sample); its emission spectrum, shown in Fig. 2, has a strong green peak like that for the Mn-doped labradorite in Fig. 3, but rather more blue and IR emission as well. SIPPEL and SPENCER (1970) have pointed out that albites do not show a green peak, and they conclude that when it is present this peak is caused by a divalent activator substituting for Ca^{2+} in the lattice. Our results seem consistent with Mn^{2+} being the activator responsible. The situation may be compared with that of the Mn^{2+} emission peak which is found for α-$CaSiO_3$:Mn (α-wollastonite) and which also occurs at about 5600 Å (LANGE and KRESSIN, 1955). In view of the charge compensation and the ionic size it is to be expected that Mn^{2+} will replace Ca^{2+} in the lattice, as it also does in CaF_2:Mn, CaO:Mn, $CaCO_3$:Mn and in the manganese-activated calcium halophosphates. The colour of the Mn^{2+} emission in these phosphors ranges from green to orange-red.

It might be expected that if the green emission band for plagioclase is due to Mn^{2+}, then its intensity would be an indication of the Mn content. However, the emission spectra of the terrestrial plagioclases that we have examined do not show a simple correlation between the intensity of this band and the Mn content as measured by

X-ray fluorescence analysis. The reason may be partly that this method of analysis measures the total Mn content, and not the amount of Mn in metal cation sites, which is what really matters. Moreover, different plagioclases contain different amounts of "killers," such as iron, which probably affect the relative intensity of the Mn^{2+} emission.

The peak of the Mn^{2+} emission band varies a little in wavelength from one plagioclase to another, but it usually lies in the range 5550–5650 Å. The band has a half-width of about 1000 Å, but is not symmetrical—the intensity falls less steeply on the long wavelength side, as it also does for Zn_2SiO_4:Mn.

The cause of this green emission band will now be considered in more detail by examining the crystal field situation. In the low-temperature form of anorthite there are four possible calcium sites of slightly different symmetry but with similar metal–oxygen distances. All four are seven-fold co-ordinated if metal–oxygen distances up to 3.1 Å are counted, although one approximates to six-fold co-ordination. FLEET et al. (1966) have determined the structure of bytownite and have found that the calcium environment is changed very little by the introduction of a few sodium atoms into the structure. The environment of Ca^{2+} in the anorthite structure will therefore be considered as being applicable to calcium-rich plagioclases in general. Bond lengths are given by MEGAW et al. (1962), who describe the site symmetries as distorted cubes with one corner missing (or with two corners missing for the approximately six-fold co-ordinated site); four of the bonds approximate closely to cube-corner directions, although the sites are in general of low symmetry. The Ca–O bond lengths range from 2.28 to 3.09 Å, with an average of 2.54 Å for two of the sites and 2.50 Å for the other two. It is not clear whether Mn^{2+} will have a preference for any particular site, since it has zero "crystal field stabilisation energy" and all the sites are more than big enough to accommodate this ion. As the magnitude of the crystal field splitting parameter Δ has not been determined for any transition metal substituting in a calcium site, it is difficult to predict the wavelength of the emission band due to Mn^{2+} in these sites. Δ might be expected to be somewhat smaller than for 8-fold cubic co-ordination with a similar average metal–oxygen distance, and therefore possibly 20–30% smaller than for the corresponding octahedral case.

In considering the effect of the average metal–oxygen distance on the wavelength of the emission peak we will first consider the cases of CaO:Mn and $CaSiO_3$:Mn. In CaO:Mn the peak is at about 5900 Å and the crystal structure is of the NaCl type (simple face-centred cubic). The Ca^{2+} ion is therefore in octahedral co-ordination, and the Ca–O bond length is 2.40 Å (DAVEY, 1923). The emission for anorthite might therefore be expected to be at a wavelength considerably shorter than 5900 Å. In the pyroxenoid β-$CaSiO_3$:Mn (β-wollastonite) the wavelength of the emission peak is about 6200 Å, and the average metal–oxygen distance is 2.39 Å (PREWITT and PEACOR, 1964). There are three slightly different calcium sites of distorted octahedral symmetry (MAMEDOV and BELOV, 1956; TOLLIDAY, 1958). In α-$CaSiO_3$:Mn (pseudo- or α-wollastonite) the wavelength of the emission peak is about 5600 Å, which implies a much reduced crystal field, but unfortunately the detailed structure with bond lengths has not yet been determined. In β-wollastonite a splitting of the manganese $^4T_{1g}$ level, from which the emission is assumed to occur, would be expected because of the distortion of its octahedral symmetry, and this may be the reason why the Mn^{2+} emission occurs at a longer wavelength than in CaO:Mn, although differences in the Stokes shift may also occur. Moreover, although all the ligands are oxygen atoms they are not all equivalent, and this may also affect the value of Δ. As the possible Mn^{2+} sites in anorthite are of low symmetry, considerable splitting of the $^4T_{1g}$ level is again likely. Thus we have two factors tending to shift the emission peak wavelength in opposite directions—the large metal–oxygen distance with a crystal field weaker than for 8-fold cubic co-ordination imply a low value of Δ and a short wavelength, possibly in the green, whereas the $^4T_{1g}$ level splitting tends to shift the emission towards the red. In view of these factors it seems reasonable to expect the emission to occur somewhere in the region from 5500 to 6000 Å.

The other main emission bands of plagioclase are one in the blue and one in the near IR. All plagioclases show some blue emission, as do iron-free enstatites, forsterites and many other silicates (GEAKE et al., 1966). The emission band is broad, with a somewhat variable peak wavelength. It is likely that this blue emission band has a similar origin in all these relatively iron-free silicates, and that it is due to a particular type of lattice defect rather than to any impurity activator. SIPPEL and SPENCER (1970) have found that some heavily shocked plagioclases and silica mineral phases show an enhanced blue emission, which also implies that lattice defects are responsible.

Most terrestrial plagioclases show a prominent emission band in the near IR, as shown for labradorite in Fig. 3, curve 1, and for basalt BCR-1 in Fig. 4. However, sometimes this band is only of comparable intensity to the green band, as for oligoclase (Fig. 4), and occasionally it is virtually absent as for labradorite-bytownite (Fig. 2). We differ here from SIPPEL and SPENCER (1970), who suggest that all terrestrial plagioclases show a strong IR emission band. The lunar materials we have examined tend to show a weak emission band in the IR, as shown by the four samples of lunar fines in Fig. 1 and by the separated lunar plagioclase in Fig. 2. We also find that meteorites in which plagioclase is the main luminescent phase, e.g., hypersthene and bronzite chondrites and pyroxene-plagioclase achondrites, do not show an IR peak— see, for example, Juvinas (a pyroxene-plagioclase achondrite) shown in Fig. 2.

The wavelength of the IR emission peak is variable, ranging from about 7300 to 7700 Å, and the band is usually broader than the green band. It is often the dominant emission band for terrestrial plagioclase, as for the labradorite examined, and it does not appear to be associated with manganese since the addition of manganese was found to reduce its intensity, as shown in Fig. 3. If this emission is due to an impurity activator, then it should be possible by selective doping with suspected activators to identify the element responsible, which is probably present naturally in amounts of not less than 100 ppm. As most plagioclases contain up to 1% of iron, and as Fe^{2+} often gives rise to absorption bands in the near IR in silicates, this might be a possible candidate, although Fe^{2+} also acts as a luminescence "killer" in silicates. Preliminary tests in which iron was added to labradorite (by heating it to 1050°C in argon with 0.1% hydrated ferrous sulphate) did not show any increase in the (already high) IR emission—in fact there appeared to be a slight reduction. A program of selective doping and of comparative analysis of different plagioclases is now under way, in the hope of identifying the cause of the IR emission.

LUMINESCENCE DISTRIBUTION WITHIN ROCK AND BRECCIA SAMPLES

All of our Apollo 11 and Apollo 12 rock and breccia chips have been inspected under 6 KeV electron excitation, and in all cases where significant luminescence was observed colour photographs of the emission were taken. A typical exposure was 30 sec at $f/2$ with 1.85 magnification, on Kodak High-Speed Ektachrome (Daylight) reversal film; some microphotographs were taken with exposures of up to half an hour. The flatter parts of the rough chips were chosen, as the depth of focus obtainable was a major limitation. The current density used was about 8 μA/cm², giving a power absorption at the surface of the sample of about 50 mW/cm². The use of electrons in preference to protons permitted long exposures with little damage. The results cannot be reproduced here in colour, but some of them are shown in black and white in Figs. 5–8.

All our rock chips showed bluish-white luminescence from their plagioclase parts, and little else. For example, rock chip 10058,38, which is shown in Fig. 5, was observed visually and so arranged that one could "blink" from luminescence emission to direct illumination. It was evident that of its three main constituents, the white or clear plagioclase parts were luminescing bluish-white or blue, whereas the brown pyroxene and black ilmenite were completely nonluminescent. One small lath luminesced red.

Other rocks showed coarse or fine patterns of plagioclase emission, consistent with their stated grain sizes. Fig. 6 shows the luminescence emission from two Apollo 12 rocks: (a) is a medium-grained rock, 12051, showing fan-shaped groups of plagioclase laths; and (b) is of rock 12002 showing smaller and more irregular plagioclase crystals. Colour variations from bluish-white to blue across some plagioclase crystals, and from one crystal to another, probably indicate variations in the concentration of the manganese activator, an effect that we have also noticed with manganese-activated enstatite in meteorites (GEAKE and WALKER, 1967). We have observed that when plagioclase samples are inspected visually while being scanned in our luminescence spectrophotometer, then those whose luminescence appears nearly white give spectral profiles having a moderate green peak; when this peak is weak the emission appears blue, and when it is very strong, as for the Mn-doped labradorite shown in Fig. 3 and for the plagioclase-rich meteorite Juvinas shown in Fig. 2, the appearance is yellow-green. Several of the lunar rock chips also showed occasional small pink-luminescent grains. Some chips showed very bright blue-luminescent specks, but these we take to be contamination, probably by saw material.

Two breccias were available, both from Apollo 11, and their luminescence appearance was quite different from that of the rocks, showing many luminescent grains embedded in a non-luminescent matrix. Sample 10023,8 (Fig. 7) showed white-luminescent grains (probably plagioclase), and little else, but sample 10059,36 (Fig. 8) showed a surprising variety of emission colours including red, orange, yellow, green, blue, and a range of mauves and pinks which usually vary in hue across the grains concerned. Luminescent grains of all sizes from 1 mm or so down to a few μm are seen. This one breccia sample appears to contain so much luminescence information that it would be useful to identify the materials responsible, but this will only be practicable with a thin section of the same sample on which electron microprobe analysis

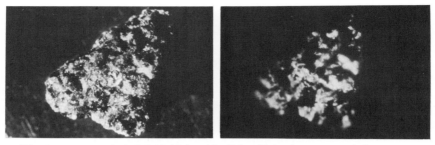

Fig. 5. Rock sample 10058,38 (a) in white light, (b) luminescence emission.

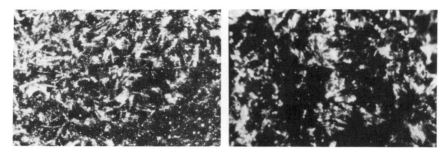

Fig. 6. Luminescence emission of two Apollo 12 rock samples: (a) 12051,16, (b) 12002,102.

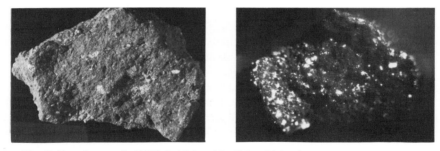

Fig. 7. Breccia sample 10023,8, (a) in white light, (b) luminescence emission.

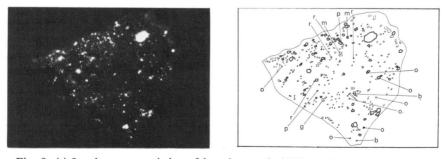

Fig. 8. (a) Luminescence emission of breccia sample 10059,36, (b) a map indicating some of the luminescence colours seen: r = red, p = pink, o = orange, g = green, b = blue, m = mauve. Most of the other emitting regions appear white or bluish white.

Figs. 5–8. The height of each print represents about 1 cm in Figs. 5, 7 and 8, and about 5 mm in Fig. 6. Luminescence excitation is by 6 KeV electrons.

can be carried out, using the luminescence photograph as a map. Luminescence may then provide a quick and almost non-destructive method of exploring the surface of a rock, and of identifying at least some of its constituents, without making sections. We have noticed that most of the rock chips show a few grains of some luminescence colour other than that of plagioclase.

It is already possible to say that the red luminescence may be due to apatite. We have shown by electron microprobe analysis that some luminescent regions in two hypersthene chondrite meteorites (APPLEY, BRIDGE, and MANGWENDI) contained Ca and P, but no Fe, indicating apatite. The emission varied from red to orange across the grains, probably due to a variation in the Cl/F ratio.

Two samples of lunar fines (10084,6 and 12032,39) have also been observed and photographed under electron excitation. They were much fainter than the luminescent parts of the rocks: they both appeared generally mauve, with brighter white-luminescent specks and larger grains, which were probably plagioclase. No other colours were seen. This low efficiency in comparison with similar constituents in rock samples is probably due to the surface damage and grain-coating effects discussed earlier.

DAMAGE AND EROSION BY PROTONS

The effect of proton bombardment has now been investigated for all four of our samples of lunar fines. We reported earlier (GEAKE et al., 1970) that the very dark Apollo 11 fines material 10084 was not further darkened by proton bombardment, even after several hours exposure at 60 KeV and 1 $\mu A/cm^2$. We now find that a similar dose slightly darkens the darker Apollo 12 fines (12070) and one of the lighter ones (12033)—but not the other light one (12032).

Microphotographs have been taken of all the fines samples before and after

Fig. 9. Lunar fines sample 12033,60 (a) before, and (b) after proton bombardment (4 hr at 1 $\mu A/cm^2$). The regions shown are not the same, but are typical. The widths shown represent 2 mm.

bombardment. There is a marked change in appearance in each case, amounting to a degradation of complex grain assemblies: before bombardment many complex "tree" and "cross" formations of grains are visible, as shown in Fig. 9a, whereas afterwards the appearance is different, as shown in Fig. 9b, and these complex structures are not to be found. This demolition effect is probably due to electrostatic charging by the ion beam. It is possible that similar effects may occur locally on the lunar surface. The effect of proton irradiation on the optical polarisation characteristics of the surface is discussed by Dollfus et al. (1971), later in this volume.

Conclusions

We conclude that for the Apollo 11 and Apollo 12 material we have examined the main luminescent constituent is plagioclase, that the main emission of lunar plagioclase is a green peak at about 5600 Å, and that the cause of this peak is about a hundred parts per million of manganese in the form of Mn^{2+} substituting for Ca^{2+} in the plagioclase lattice. The less efficient blue emission at about 4500 Å is probably caused by a lattice defect and is common to most silicates; the weak IR peak at about 7700 Å is probably caused by an impurity activator yet to be identified. Most terrestrial plagioclases show much stronger IR emission than the lunar material.

Electrons and protons in the KeV region are both effective in exciting luminescence in lunar material; UV is ineffective. The overall efficiencies of lunar rock and breccia samples depend largely on their content of plagioclase, which itself has an efficiency of about 10^{-3}. Samples of lunar fines also have efficiencies depending on their plagioclase content, but they are at least an order of magnitude less efficient than the rock samples, possibly because the grains have a thin surface layer which is either opaque or damaged. Two light-coloured Apollo 12 fines samples, thought to be ray material, are an order of magnitude more efficient than the dark fines material from Apollo 11 and Apollo 12, which has an efficiency of about 10^{-5}.

Lunar rock and breccia samples generally show blue-white luminescence distributed texturally as their plagioclase component, with a few grains of other luminescent materials one of which may be apatite. One breccia (10059) shows a wide variety of luminescence colours from components yet to be identified.

Proton bombardment seems to cause a very marginal darkening of some fines material and no darkening of others; the main effect was found to be demolition of the complex grain assemblies which were observed initially.

Acknowledgements—We are grateful to M. L. Gould for technical assistance; to F. Kirkman of U.M.I.S.T., Leicester University photographic section, Hull University photographic section, N. W. Scott, D. Rendell and Colour 061 Ltd. for photographic assistance; to J. Zussman, A. C. Dunham, and J. Esson of Manchester University Geology Department for advice and for Mn determination by X-ray fluorescence analysis; also for the loan of the sample of separated lunar plagioclase; to P. Suddaby of the Geology Department Imperial College, University of London, for help with the electron microprobe investigation of meteorites; to C. J. E. Kempster for advice on the structure of plagioclase; to The British Museum (Natural History) for the loan of meteorite samples; to The Science Research Council for financial support, and finally to NASA, and especially to all the Apollo 11 and Apollo 12 astronauts, for providing the lunar samples.

REFERENCES

DAVEY W. P. (1923) Precision measurements of the crystal structures of CaO, CaS and CaSe. *Phys. Rev.* **21**, 213.

DOLLFUS A., GEAKE J. E., and TITULAER C. (1971) Polarimetric properties of the lunar surface and its interpretation. Part IV: Apollo 11 and Apollo 12 lunar samples. Second Lunar Science Conference (unpublished proceedings).

FLEET S. G., CHANDRASEKHAR S., and MEGAW H. D. (1966) The structure of bytownite. *Acta Crystallogr.* **21**, 782–801.

GARLICK G. F. J., LAMB W. E., STEIGMANN G. A., and GEAKE J. E. (1971) Thermoluminescence of lunar samples and terrestrial plagioclases. Second Lunar Science Conference (unpublished proceedings).

GEAKE J. E. and WALKER G. (1966) The luminescence spectra of meteorites. *Geochim. Cosmochim. Acta* **30**, 927–937.

GEAKE J. E. and WALKER G. (1967) Laboratory investigations of meteorite luminescence. *Proc. Roy. Soc. London* **296**, 337–346.

GEAKE J. E., DOLLFUS A., GARLICK G. F. J., LAMB W., WALKER G., STEIGMANN G. A., and TITULAER C. (1970) Luminescence, electron paramagnetic resonance and optical properties of lunar material from Apollo 11. *Proc. Apollo 11 Lunar Sci. Conf., Geochim. Cosmochim. Acta* Suppl. 1, Vol. 3, pp. 2127–2147. Pergamon.

HAPKE B. W., COHEN A. J., CASSIDY W. A., and WELLS E. N. (1970) Solar radiation effects on the optical properties of Apollo 11 samples. *Proc. Apollo 11 Lunar Sci. Conf., Geochim. Cosmochim. Acta* Suppl. 1, Vol. 3, pp. 2199–2212. Pergamon.

HAPKE B. W., CASSIDY W. A., and WELLS E. N. (1971) Albedo of the Moon: evidence for vapor-phase deposition processes on the lunar surface. Second Lunar Science Conference (unpublished proceedings).

LANGE H. and KRESSIN G. (1955) Der Einfluss der Kristallstruktur aus die Lumineszenz des Calcium-silikates. *Z. Phys.* **142**, 380–386.

MAMEDOV K. S. and BELOV N. V. (1956) The structure of wollastonite. *Dokl. Akad. Nauk. SSSR* **107**, 463–487.

MEGAW H. D., KEMPSTER C. J. E., and RADOSLOVICH E. (1962) The structure of anorthite. *Acta Crystallogr.* **15**, 1017–1035.

PREWITT C. T. and PEACOR D. R. (1964) The crystal chemistry of the pyroxenes and pyroxenoids. *Amer. Mineral.* **49**, 1527–1542.

SIPPEL R. F. and SPENCER A. B. (1970) Luminescence petrography and properties of lunar crystalline rocks and breccias. *Proc. Apollo 11 Lunar Sci. Conf., Geochim. Cosmochim. Acta* Suppl. 1, Vol. 3, pp. 2413–2426. Pergamon.

TOLLIDAY J. (1958) The structure of parawollastonite. *Nature* **182**, 1013.

Proceedings of the Second Lunar Science Conference, Vol. 3, pp. 2277–2283
The M.I.T. Press, 1971.

Thermoluminescence of lunar samples and terrestrial plagioclases*

G. F. J. Garlick, W. E. Lamb, and G. A. Steigmann

Department of Physics, University of Hull, England

and

J. E. Geake

Physics Department, U.M.I.S.T., Manchester, England

(*Received* 5 *March* 1971; *accepted in revised form* 20 *April* 1971)

Abstract—Studies have been made of natural and X ray-excited thermoluminescence in Apollo 12 samples and compared with data on selected terrestrial plagioclases and on Apollo 11 samples. Similar thermoluminescence curves are obtained for all samples indicating similar trapping state distributions in each. Residual natural thermoluminescence in Apollo 12 samples is much greater than in Apollo 11 samples and is consistent with the larger plagioclase content and lower optical absorption of the Apollo 12 samples. An investigation has been made of the "drainage" of electrons from relatively deep trapping states even at ordinary temperatures. Analysis of data including those of other workers indicates that a nonthermal trap emptying process is present at such temperatures although thermal activation intervenes as the temperature is raised. Similar but more rapid "drainage" of trapped electrons is found in terrestrial plagioclases.

Introduction

In this paper we report that part of our joint research which has been carried out at Hull University and which has been mainly concerned with the investigation of the thermoluminescence characteristics, both natural and X ray-induced, of Apollo 12 samples and also of terrestrial plagioclases. In previous studies of Apollo 11 samples (Geake *et al.*, 1970) we also made measurements of diffuse reflexion spectra and electron paramagnetic resonance of specimens since defects in solids often have characteristic absorption bands, have unpaired electron spins, and also give rise to some of the electron traps responsible for thermoluminescence. However, in that work we did not find, by annealing experiments, that there was a significant association between trapping states and the systems giving optical absorption or spin resonance. This was found to be the case by other workers also (Hoyt *et al.*, 1970) with respect to a correlation between traps and radiation-induced defects. For this reason we concentrate in this paper on the thermoluminescence of lunar and terrestrial samples. It has been established by several workers (Dalrymple and Doell, 1970; Hoyt *et al.*, 1970; Nash and Greer, 1970; Geake *et al.*, 1970; Sippel and Spencer, 1970) that the luminescence effects in lunar samples are located in the plagioclase feldspar component and so we have selected several natural plagioclases with different albite-anorthite ratios and have examined their thermoluminescence characteristics.

There are two main areas in our investigations: (1) a comparison of the thermoluminescence characteristics of lunar and terrestrial samples after a standard excitation by X rays to see whether there are significant similarities, or differences, between the

---

* Paper 2 of three collaborative papers: paper 1 precedes and paper 3 follows.

electron trap distributions in lunar samples and those in terrestrial plagioclases. Such an investigation is intended to explore the possibility of using thermoluminescence as a method of distinguishing features in the past history or in the formation of the material. The present results refer only to studies at and above room temperature. Current work is extending measurements to much lower temperatures; and (2) a study of the "drainage" effects on electron traps in lunar samples which cause the isothermal decay of the stored electrons at temperatures well below those at which they become sufficiently thermally unstable to give rise to thermoluminescence at temperatures above 300°C. Terrestrial plagioclases have also received our attention in this respect. Analysis of the "drainage" effect has involved the use of data from other groups previously reported (DALRYMPLE and DOELL, 1970; HOYT *et al.*, 1970) together with the information from our own measurements.

THERMOLUMINESCENCE STUDIES

We made use of experimental facilities and conditions as described previously (GEAKE *et al.*, 1970). All samples were measured in vacuo (10^{-5} torr) and a warming rate of 1°K/sec was used in all experiments. It might be pointed out that better signal/noise ratios are obtained with higher warming rates but serious temperature differences and lags can occur in samples which lower the resolution and accuracy of thermoluminescence curves. For laboratory excitation of specimens a 30 kV X-ray system was used and all samples received a standard dose of 10^7 rad at room temperature. A delay time of 15 minutes was allowed between cessation of excitation and the start of warming the sample. This removed from the initial part of the thermoluminescence curves the contributions from relatively shallow traps and enabled the peak between 200 and 300°C to be accentuated for comparisons between lunar and terrestrial samples and estimates of the shape of the initial rise of emission with temperature from which thermal activation energies for trapped electrons could be estimated.

The natural thermoluminescence of Apollo 12 samples

As found previously for Apollo 11 samples (DALRYMPLE and DOELL, 1970; HOYT *et al.*, 1970) the amount of natural thermoluminescence obtained on warming is very much smaller than that to be expected from the radiation dose received on the moon during the cosmogenic and radiogenic life time of lunar material. In our measurements of natural thermoluminescence of Apollo 12 samples a similar observation was made but the total light sum was an order of magnitude greater than that from Apollo 11 samples and is consistent with the much higher particle excited luminescence for Apollo 12 material (GEAKE *et al.*, 1971). Our measurements of diffuse reflexion spectra show that part of this improvement in efficiency comes from the lower optical absorption in these samples, while a further contribution is undoubtedly due to the higher fraction of plagioclase feldspar present in Apollo 12 samples.

X ray-excited thermoluminescence of lunar and terrestrial samples

Samples of lunar and terrestrial plagioclases have been shown by others (GEAKE *et al.*, 1970; NASH and GREER, 1970; NASH 1966; SIPPEL and SPENCER, 1970) to have

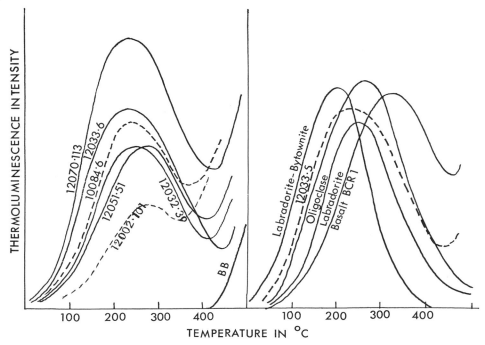

Fig. 1. Thermoluminescence curves of Apollo 12 samples compared with curves for terrestrial plagioclases and an Apollo 11 fines sample (10084,6). All samples excited at room temperature by 10^7 rad of 30 kV X rays and held in dark for 15 min before heating. The curve for the Apollo 11 sample has been multiplied by 10 to make the comparison. BB is the curve for black-body background emission.

similar luminescence spectra when excited by electrons or protons although the dominant infrared emission band of terrestrial samples is often missing or present to a much smaller degree in lunar samples (see also SIPPEL, 1971). There are also marked similarities in the thermoluminescence characteristics of lunar and terrestrial samples. A selection of our measurements is given in Fig. 1. All samples show a pronounced peak in the 250°C temperature region but its exact location is a little different for different samples. Apollo 11 and Apollo 12 samples are not substantially different in thermoluminescence characteristics and correlate with labradorite among the terrestrial samples. We had thought that the peak position might have shown a regular shift with albite-anorthite ratio, but the significant shifts found show no systematic change in position. We found no effect on the thermoluminescence curves of annealing samples in a stream of pure argon (5N) at temperatures just above 600°C, at which temperatures damage tracks in lunar specimens are reported to anneal out (CROZAZ et al., 1970). This confirms the previous observations of HOYT et al. (1970).

As part of an attempt to determine the emission centres responsible for luminescence in lunar and terrestrial feldspars and their possible relation to electron traps, we doped a sample of plagioclase (labradorite) with a small amount of manganese. This treatment enhanced the green emission peak in the particle excited sample as

reported by Geake *et al.* (1971). We found that the total intensity of thermoluminescence was also increased by more than five times due to the incorporation of manganese, but the shape of the thermoluminescence curve was not altered. Thus the traps
giving rise to thermoluminescence are not closely associated with the green emission
centres. However, the type of curve shown in Fig. 1 is characteristic of many different
types of luminescent silicates, as is the blue emission band in the plagioclases. It is
likely that the traps are due to a natural defect in the silicate chains. However, there is
a need of much more crucial experiment to ascertain the nature of trapping states in
these materials.

As part of out studies of X ray-induced thermoluminescence in plagioclases we
have looked at the thermoluminescence emitted in the infrared band of terrestrial
plagioclases and have also found such emission in Apollo 12 samples. Two examples
of these results are given in Fig. 2. The thermoluminescence is much more intense for
the terrestrial labradorite, due mainly to the much higher efficiency of the infrared
emission band and not necessarily to differences in the total numbers of trapped
electrons. The higher gain needed to measure the Apollo 12 sample emission also
means that the background, black-body emission intervened at a lower temperature.
However, curve A indicates that a different distribution of trapping states is associated
with the infra red emission. The samples were held for 15 min after X ray excitation
like all other cases but there appears to be a more stable group of trapped electrons
involved in the thermoluminescence below 200°C. For these experiments care was
taken to exclude contributions from visible emission bands. An infrared sensitive
Mullard CVP 150 type photomultiplier was used together with a Wratten 88A gelatin
filter to exclude radiation below 0.7 μm. The peak response of the photomultiplier
was at 0.8 μm with a limit of 1.1 μm at the long wavelength side of the response. The
photomultiplier was used in the cooled state (about -80°C).

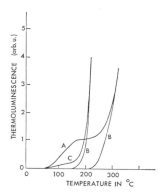

Fig. 2. Infrared (0.7–1.1 μm wavelengths) thermoluminescence curves of Apollo 12
fines and terrestrial labradorite. A. Labradorite, B. Black body background for each
sample, C. Lunar fines 12070,113 (gain of recording system raised by factor 16 compared
to curve A). Both samples excited at room temperature by 10^7 rad of 30 kV X rays and
held in dark for 15 min before heating.

Investigation of trapped electron instability in lunar and terrestrial samples

It is known that trapped electrons escape from quite deep electron traps in lunar samples, even at room temperatures (HOYT *et al.*, 1970; DALRYMPLE and DOELL, 1970). We have found this effect also in terrestrial plagioclases. In order to compare our data on lunar and terrestrial samples with those of the above workers we have made the composite diagram of Fig. 3. The diagram shows how the number of residual trapped electrons varies with the time for which the sample is held at a fixed temperature, shown in parenthesis on each curve, after laboratory excitation. The ordinate, representing the number of trapped electrons, is obtained by measuring the intensity, or height, of the thermoluminescence curve at selected temperatures, shown on each curve without parenthesis. Curves for the "drainage" effects at 22° and 100°C are from the data of DALRYMPLE and DOELL (1970), that for a temperature of 150°C is from the data of HOYT *et al.* (1970), while other curves are from our own data.

It is evident that at room temperature or thereabouts the decay of trapped electrons in lunar samples is very similar for all parts of the thermoluminescence curves selected, i.e., for temperatures up to 300°C. Those for labradorite are much faster decays but again have the same slope for the two different regions of the thermoluminescence curves. However, when the "holding" temperature is raised the shallower traps, represented by the 200°C curve at 100°C holding temperature, quickly lose their electrons, presumably by the occurrence of thermal activation. Our conclusion is that there are two different mechanisms possible for the release of electrons from traps. One is the thermal activation which can occur if the sample temperature is high enough but which is strongly temperature dependent (see, for example, Bonfiglioli, 1968). The other which can occur at lower temperatures is the non-thermal activation or loss of trapped electrons by return to their ground states in the crystal. The latter effect is found in zinc sulphide phosphors at very low temperature (RIEHL, 1970). A most

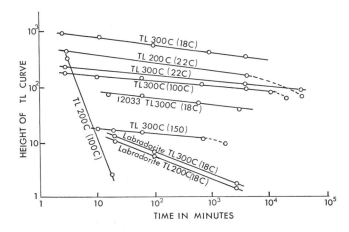

Fig. 3. Decay of thermoluminescence curves by holding samples at selected fixed temperatures for different times before warming. Point on thermoluminescence curve given by first temperature; holding temperature given in parenthesis. Curves for decay at 22 and 100°C are from DALRYMPLE and DOELL (1970); curve for decay at 150°C is from HOYT *et al.* (1970); both sets are for Apollo 11 samples.

probable cause is the finite overlap of wave functions for the electron in the trapping state and the ground state to which it might return. It is a process often ignored in the analysis of thermoluminescence but it is included in some theoretical treatments (e.g., DIEHL *et al.*, 1968).

In the case of lunar samples, or terrestrial plagioclases, a quantitative analysis of the effect is made difficult by the wide distribution of trapping states present. We do not at present wish to comment on the decay law shown by the curves of Fig. 3, i.e., a very low inverse power of time. For a uniform trap distribution the decay of phosphorescence at a fixed temperature follows an inverse time law (RANDALL and WILKINS, 1945) if first order kinetics are involved. This might be evident for lunar samples if the holding temperature is high enough to allow thermal activation since the trap distributions indicated by thermoluminescence curves are almost or at least quasi-uniform. HOYT *et al.* (1970) have in fact made such an experiment by measuring the decay of light output with time at 150°C over a restricted range (compared with times in Fig. 3) of some 160 sec. The range is too small to infer that thermal drainage will account for the decay at much later times.

DISCUSSION AND CONCLUSIONS

Comparison of lunar and terrestrial plagioclases

After considerable time spent in examining the luminescence effects in lunar samples it is important to ask what part such studies ought to play in future examination of lunar material from other lunar locations. In particular what can be gained from a study of thermoluminescence? As a means of exploring defects and their origins in crystals it suffers from the fact that only trapped electrons which give rise to luminescence are detected and these may be unrepresentative of the total distribution. Even when emission of luminescence is present thermoluminescence from one sample to another will depend for its overall intensity on luminescence centre concentrations and efficiencies and on the optical absorption of samples. For example, data for core samples (i.e. luminescence or thermoluminescence) must be taken together with optical characteristics such as diffuse reflexion spectra and relative luminescence efficiency (e.g., as shown by laboratory excitation by electrons or protons). Thermoluminescence is also limited at higher temperatures by onset of appreciable black-body radiation. In spite of these various problems luminescence phenomena offer another experimental parameter to be taken alongside other data. We are extending our studies to laboratory produced thermoluminescence at 77°K upwards since shallower traps may show correlation with such states of the samples as presence of damage tracks, metamorphism due to shock etc. Other components of the lunar dust and rocks, e.g., pyroxenes, may provide luminescence effects if the temperature is low enough. Hence we intend to pursue studies of terrestrial silicates and of lunar samples as they become available, over a much wider range of physical conditions.

Our studies of the isothermal drainage of trapped electrons by means of thermoluminescence experiments and comparison of data with those other workers referred to above have indicated there to be a considerable non-thermal effect present at the lower temperatures. However, it is now important to see whether this effect applies to the trapped electrons giving thermoluminescence above 400°C. It may be necessary to use the method of thermally stimulated currents to minimise the effects of black-body background encountered when measuring the luminescence. There are more crucial experiments to be done in order to obtain a more certain idea of the extent of non-thermal drainage. One such has been suggested by one of our referees which we here acknowledge. From the theoretical point of view we are planning a computer programme to extend the "single trap depth" models for thermal and non-thermal activation processes to cases of wide trap depth distributions.

Acknowledgements—We are grateful to NASA for the supply of lunar samples and to the Geology department of this university for plagioclase samples. We wish to thank our senior technician Mrs Irene Robinson for her invaluable assistance in experiments and the Science Research Council of London for a grant to carry out the research.

REFERENCES

BONFIGLIOLI, G. (1968) Thermoluminescence: What it can and cannot show. In *Thermoluminescence of Geological Materials* (editor D. J. McDougall) Chap. 2.1, pp. 15–24, Academic Press.

CROZAZ G., HAACK U., HAIR M., MAURETTE M., WALKER R., and WOOLUM D. (1970) Nuclear track studies of ancient solar radiations and dynamic lunar surface processes. *Proc. Apollo 11 Lunar Sci. Conf.*, *Geochim. Cosmochim. Acta* Suppl. 1, Vol. 3, pp. 2051–2080. Pergamon.

DALRYMPLE G. B. and DOELL R. R. (1970) Thermoluminescence of lunar samples from Apollo 11. *Proc. Apollo 11 Lunar Sci. Conf.*, *Geochim. Cosmochim. Acta* Suppl. 1, Vol. 3, pp. 2081–2092. Pergamon.

DIEHL H., GRASSER R., and SCHARMANN A. (1968) Discussion of a simplified model for the thermo-luminescence of inorganic photoconducting phosphors, In *Thermoluminescence of Geological Materials* (editor D. J. McDougall) Chap. 2.3, pp. 39–50, Academic Press.

GEAKE J. E., DOLLFUS A., GARLICK G. F. J., LAMB W., WALKER G., STEIGMANN G. A., and TITULAER C. (1970) Luminescence, electron paramagnetic resonance and optical properties of lunar material from Apollo 11. *Proc. Apollo 11 Lunar Sci. Conf.*, *Geochim. Cosmochim. Acta* Suppl. 1, Vol. 3, pp. 2127–2148. Pergamon.

GEAKE J. E., WALKER G., MILLS A. A., and GARLICK G. F. J. (1971) Luminescence of Apollo lunar samples. Second Lunar Science Conference (unpublished proceedings).

HOYT H. P., KARDOS J. L., MIYAJIMA M., SEITZ M. G., SUN S. S., WALKER R. M., and WITTELS M. C. (1970) Thermoluminescence, X-ray and stored energy measurements of Apollo 11 samples. *Proc. Apollo 11 Lunar Sci. Conf.*, *Geochim. Cosmochim. Acta* Suppl. 1, Vol. 3, pp. 2269–2288. Pergamon.

NASH D. B. (1966) Proton excited luminescence of silicates: Experimental results and lunar implications. *J. Geophys. Res.* **71**, 2517–2534.

NASH D. B. and GREER R. T. (1970) Luminescence properties of Apollo 11 lunar samples and im-plications for solar-excited lunar luminescence. *Proc. Apollo 11 Lunar Sci. Conf.*, *Geochim. Cosmochim. Acta* Suppl. 1, Vol. 3, pp. 2341–2350. Pergamon.

RANDALL J. T. and WILKINS M. H. F. (1945) Phosphorescence and electron traps, II. The interpre-tation of long-period phosphorescence. *Proc. Roy. Soc.* (London) **A184**, 390–407.

RIEHL N. (1970) Tunnel luminescence and infrared stimulation *J. Luminescence* **1**, 1–15.

SIPPEL R. F. and SPENCER A. B. (1970) Luminescence petrography and properties of lunar crystalline rocks and breccias. *Proc. Apollo 11 Lunar Sci. Conf.*, *Geochim. Cosmochim. Acta* Supp. 1, Vol. 3, pp. 2413–2426. Pergamon.

Proceedings of the Second Lunar Science Conference, Vol. 3, pp. 2285–2300
The M.I.T. Press, 1971.

Polarimetric properties of the lunar surface and its interpretation. Part 3: Apollo 11 and Apollo 12 lunar samples*

A. Dollfus

Observatoire de Paris, Meudon, France

J. E. Geake

U.M.I.S.T., Manchester, M60 1QD England

and

C. Titulaer†

Observatoire de Paris, Meudon, France

(*Received* 5 *March* 1971; *accepted in revised form* 20 *April* 1971)

Abstract—The polarimetric and photometric properties of seven Apollo 12 fines samples, three Apollo rocks, and one Apollo 11 breccia have been studied in addition to the measurements of Apollo 11 reported earlier by Geake *et al.* (1970). The full polarization curves and the normal albedoes are given for at least 5 color filters in the wavelength range 3540 to 6400 Å. The optical properties of the Apollo 12 samples correspond to different types of lunar terrain. Some of the samples can be identified as either continental or ray material. The wavelength dependence of the optical properties was investigated and some interpretations are given.

The effect of proton irradiation was studied for four samples; only two of them were darkened. It seems possible that the other samples were already saturated with solar wind protons. The lunar telescopic photometric determinations are compared with laboratory measurements on Apollo samples, and when reduced to a phase angle of 5°, a correction factor of 1.40 for the telescopic photometric scale is indicated. All these measurements on fines and rocks provide the astronomer with a catalogue to compare with telescopic observations of other solar system objects, especially asteroids.

Introduction

This is the third of three papers describing work carried out in collaboration between Manchester and Hull universities, England and Paris Observatory, France. It is also Part 4 of the series "Polarimetric Properties of the Lunar Surface and its Interpretation." The previous parts were published by Dollfus and Bowell (1970), Dollfus *et al.* (1971), and Dollfus and Titulaer (1971). This latter part deals with the analysis of the Apollo 11 and Apollo 12 returned lunar samples. The study was conducted as a joint program between the University of Manchester Institute of Science and Technology (U.M.I.S.T), England, The University of Hull, England, and the Meudon Observatory, France. Dr. J. E. Geake, Physics Department, U.M.I.S.T, was NASA Principal Investigator and Dr. A. Dollfus, Meudon Observatory, Laboratory "Physique du Système Solaire" was NASA Co-Investigator. The first results of this collaborative project for the Apollo 11 samples were published by Geake *et al.* (1970). The present report extends the investigations to Apollo 12 samples.

---

* Paper 3 of three collaborative papers: papers 1 and 2 precede.
† ESRO Research Fellow at Paris Observatory, Meudon, France.

In all, 8 Apollo fines samples from different sites and 6 rock samples and breccia were studied. The preliminary analysis of some Apollo 11 lunar fines samples (GEAKE *et al.*, 1970) showed that they reproduce identically the photometric and polarimetric properties measured by telescopic observations of the landing site region in Mare Tranquillitatis. Curves for lunar rock samples show properties appreciably different from the fines. Our new results for Apollo 12, which follow, make it clear that not all of this material was originally from the landing site in Oceanus Procellarum; some of it can be identified partly as ray material splashed near the landing site and possibly originating from Copernicus.

POLARIZATION OF APOLLO 12 SAMPLES

The polarization measurements were carried out with the photoelectric polarimeter at Meudon, and with a Lyot visual fringe polarimeter at Manchester. The samples investigated are listed in Table 1.

The polarization curves for all the samples of fines measured (except 10084 previously published in GEAKE *et al.*, 1970) are presented in Figs. 1–7. Each figure shows the measured polarization curves at different wavelengths, as indicated for each curve. The degree of linear polarization P is given in parts per thousand on the vertical scale. The horizontal scale indicates the phase angle V in degrees; in order to separate the different measurements, each curve has a translation of 40 deg. from the previous curve.

Fig. 1 shows the curves for sample 12070, a dark powder. The polarization of this sample has been measured visually for 4 wavelengths at Manchester, and for 3 more wavelengths in the blue and UV regions with the photoelectric polarimeter at Meudon. A separate study of the negative branch has been inserted in Fig. 19.

Sample 12032, a lighter powder, was also measured partly at Manchester and partly at Meudon, and the results are shown in Fig. 2. Sample 12033, still lighter, was measured at Manchester, and the results are shown in Fig. 3. All the measurements on the 12028 core tube samples (parts 55, 98, 155, and 203) have been carried out at Meudon. The results are presented in Figs. 4, 5, 6 and 7.

Table 1. Apollo samples measured.

Sample	Allocated to	Albedo A at 5800 Å (%)	P_{max} at 5800 Å (parts per thousand)	Remarks
12032,39	(Geake,			
	(Maurette	14.8	64	Fines
12033,60	Geake	16.5	55	Fines
12070,113	(Geake,			
	(Maurette	9.5	115	Fines
12028,55	Maurette	10.2	134	Core tube at 13 cm depth
12028,98	Maurette	15.0	92	Core tube at 20 cm depth
12028,155	Maurette	15.2	68	Core tube at 21 cm depth
12028,203	Maurette	12.3	98	Core tube at 41 cm depth
12002,102	Geake	13.8	373	Rock
12051,51	Geake	26.0	313	Rock
12020,42	Geake	10.2	540	Rock
10059,36	Geake	9.6	400	Breccia
10084,6	Geake	7.8	140	Fines ⎫
10057,54	Geake	—	340	Rock ⎬ see GEAKE *et al.*, 1970
10058,37	Geake	—	345	Rock ⎭

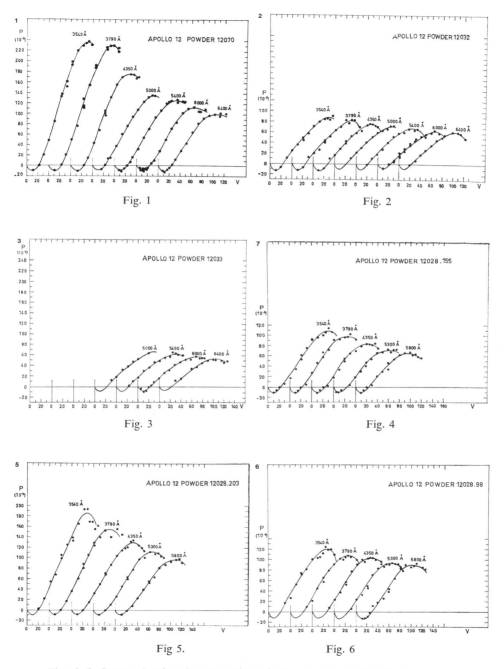

Figs. 1–7. Curves showing the proportion of plane polarised light (*P*, in parts per thousand) versus phase angle *V*, at several wavelengths, with the curves shifted horizontally to avoid overlap.

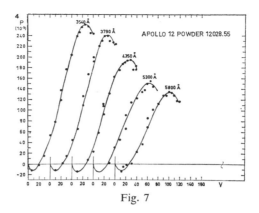

Fig. 7

PHOTOMETRY OF APOLLO LUNAR SAMPLES

The opposition effect

For phase angles V larger than 5 deg, the photometric function of the Moon is almost linear (Rougier's curve). Gehrels *et al.* (1964) discovered a discrepancy for small phase angles ($V < 5$ deg). This brightness excess was called the *opposition effect* (sometimes referred to also as the "halo" or "spike" effect).

Hapke (1963) pointed out that the direction of incident radiation is a preferred direction for the reflected light: "A bundle of light rays entering the surface will be attenuated on its way in, but light scattered from deep in the interior of the surface and reflected directly back towards the source can escape unattenuated .... The light reflected from that object which exactly retraces the path of the incident rays can do so without being blocked."

Hapke (1966) calculated a photometric function of the Moon for a model surface. In his theoretical model the opposition effect is explained well by the shadowing effect within the complex porous clumps of lunar rock powder. Van Diggelen (1965) measured radiances for phase angles smaller than 5 deg on plates taken during a lunar eclipse, showing the reality of the opposition effect. O'Leary and Rea (1968) studied the opposition effect for the planet Mars in six colors, and they report that the departure from the linear magnitude relation depends on the wavelength. Whitaker (1969) used Apollo 8 photographs, taken in orbit around the Moon and showing the shadow of the craft cast on the lunar surface and surrounded by the halo. He derived a photometric curve of the lunar surface down to very small phase angles (see Fig. 8).

To study the effect in still more detail, we measured the photometric function of lunar sample 120032 and of a terrestrial basalt GF/H2 b (Dollfus *et al.*, 1971) for phase angles $1° < V < 20°$. The measurements are corrected for the opposition effect of the MgO screen used for comparison, on the basis of determinations published by O'Leary and Rea (1968). The results are presented in Fig. 8; it will be noted that an excess in brightness starts at a phase angle of 10°. We have rescaled Whitaker's curve (given in relative units) to fit our value at 10°. In Fig. 8 the solid line represents the curve calculated from Whitaker, and the dots show our laboratory observations. Our measurements are in perfect agreement with Whitaker's results.

The halo effect is highly dependent on the microscopic packing of the grains, as shown by HAPKE and as also illustrated by the great difference between the lunar and basaltic curves in Fig. 8. The agreement between our Laboratory curve for the lunar sample, and the direct measurements of the lunar surface by WHITAKER, confirms that the fine texture of Apollo fines as used in our laboratory is representative of the natural structure on the surface of the Moon.

The wavelength dependence of the albedo

For the photometric study of lunar samples it is convenient to decide on one phase angle at which to do the measurements. In view of the fact that for phase angles $V > 10°$ the linear curve is due mainly to large-scale lunar cavities, not reproduced by the laboratory deposit, we should select a small phase angle; but for too small phase angles, the opposition effect, described above and depending on the microscopic texture, adds its effects. We therefore measured the albedoes at $V = 5°$. The residual opposition effect already present at this angle must be taken into account, with the help of Fig. 8, if our value at $V = 5°$ is to be compared with other measurements.

Figure 9 presents the results of the albedo measurements as a function of

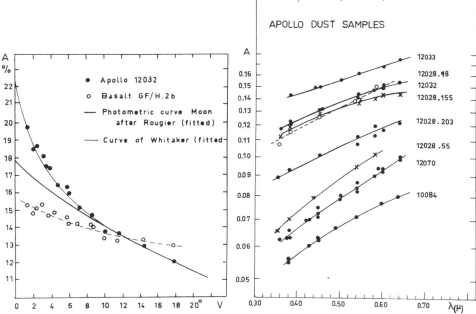

Fig. 8. Albedo versus phase angle, at small phase angles, for lunar fines (dots) and for a pulverised basalt GF/H.2 (circles). Heavy line: photometric curve for the Moon, after Rougier, fitted for $V = 10°$. Faint line: photometric curve for the lunar halo, after Whitaker, fitted for $V = 10°$.

Fig. 9. Albedo versus wavelength, for $V = 5°$, for eight samples of lunar fines. The albedo is plotted on a logarithmic scale.

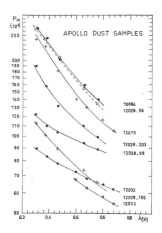

Fig. 10. Maximum degree of polarisation ($P_{\max}$, on a logarithmic scale) versus wavelength, for samples of lunar fines.

wavelengths for eight lunar dust samples. Four of these (12028,55, 98, 155, and 203) are core tube samples. Figure 10 presents, for the same samples, the values of maximum degree of polarization $P_{\max}$ as a function of wavelength. The curve for sample 12028,98 seems to have a relatively low slope, and 12028,155 to have a particularly steep slope.

The Relation Between Polarization and Albedo

In the three previously published parts of this series of papers, attention was focused on the fact that the maximum degree of polarization $P_{\max}$ and the albedo A are linearly related on a log A versus log P_m plot. Telescopic observations of the Moon occupied a fairly well-defined and almost linear domain in this diagram. Theoretical explanation of this linear relation, for powders with small dark absorbing grains, is given in Dollfus and Titulaer (1971).

In Fig. 11, we show as circles the Apollo fines results at 5800 Å; the dots show the telescopic observations of the Moon. Roughly, the top part corresponds to continental areas on the Moon and the lower left part to the maria. In Fig. 12 are presented, in the same log A versus log P_m diagram, the measurements at different wavelengths for each sample. We again find straight lines with almost the same slopes.

Interpretation of the Optical Properties

Several results can be derived from the study of the log A versus log P_m plots of Figs. 11 and 12.

Photometric calibration of the lunar telescopic observations

The Apollo lunar samples are systematically displaced on the diagram in Fig. 11 as compared to the telescopic values. The discrepancy must be explained by a

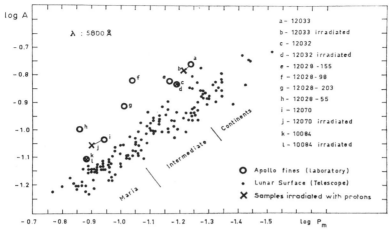

Fig. 11. Albedo versus P_{max}, on a log-log plot, for lunar samples (circles) and for the telescopic lunar measurements published in Dollfus and Bowell (1969 and 1970)(dots). The crosses relate to proton-irradiated samples and lines link the points before and after irradiation. All the measurements are at 5800 Å.

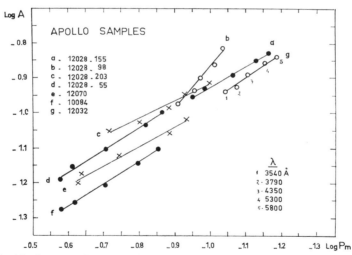

Fig. 12. Albedo versus P_{max}, on a log-log plot, for lunar samples. For each sample the measurements at 5 wavelengths lie roughly on a straight line.

difference either in photometric or in polarimetric results. As the calibration of the polarimetry is easy, we consider it most unlikely that the discrepancy can be explained by errors in the values of P_{max}.

There may, of course, be a difference between a small localized sample and a large area measured telescopically on the Moon. In DOLLFUS et al. (1971) and DOLLFUS and TITULAER (1971), however, we proved that physical processes generally displace a given point on our plot along a line almost parallel to the lunar domain. Thus, no such effect can explain the difference.

The explanation of the discrepancy is more likely to be given by photometry. In Fig. 11, the displacement on the logarithmic brightness scale is -0.15, corresponding to an error of the order of 1.4 in albedo. As the photometry in the Laboratory is more reliable than that at the telescope, we have to look into the problems of the astronomically observed albedoes. It is well known that absolute calibration at a telescope is extremely difficult, as discussed by Bowell (Dollfus and Bowell 1970). He showed that there are large differences between the photometric scales used by various observers. Furthermore, it was not possible at that time to enter into the details of the opposition effect, as both measurements and theoretical analysis were then in progress. It is important to compare only values reduced to the same angle.

The influence of the phase angle can be derived from Fig. 8. Apparently, the telescopic normalized lunar albedo scale, when reduced to phase angle 5°, has to be multiplied by a factor of 1.40.

The nature of the lunar terrain

The Apollo 11 and Apollo 12 samples show a remarkable variety in albedo and P_{max} and the results have stratigraphic implications. For the Apollo 11 sample 10084 the maximum degree of polarization is 140‰ in orange light, with an albedo of 7.8%, both these values being typical of lunar maria. Geake et al. (1970) concluded that the observed polarization curve of the sample was identical to the telescopic observations in this area of Mare Tranquillitatis. However at the landing site of Apollo 12, also in a mare (Oceanus Procellarum), we see from Fig. 11 that some samples behave optically as though they are of continental or intermediate origin, and only 12070 shows mare properties. This sample has a maximum degree of polarization slightly lower than the average value for Oceanus Procellarum (see Dollfus and Bowell, 1970, Figs. 8 and 11), corresponding to a somewhat lighter material.

The other surface samples, especially 12032 and 12033, are definitely not typical mare material. The suggestion that a ray of the crater Copernicus crosses the landing area was published in the Apollo 12 Preliminary Science Report (NASA, 1970). Our polarization measurements add new evidence to this suggestion: sample 12032 was taken at the surface, on the north rim of Bench Crater. Thus it seems likely that the high albedo for these samples can be explained largely by a major contribution of ray material to these samples. Sample 12033 was taken at a depth of 15 cm below the surface, near the LM, at the north rim of the crater in which Surveyor III landed. The astronauts noted the distinctly brighter material here, and this sample has the highest albedo of all the samples examined.

There is evidence of stratification at the Apollo 12 site, as shown by the four parts of core tube 12028 examined. These samples are from the bottom of a double core collected during the second EVA. Polarization curves for the four parts 55, 98, 155, and 203 are shown on Figs. 4–7. In Fig. 11, we see that sample 55 has mare characteristics, 155 is almost continental, and 98 and 203 are intermediate.

We can conclude that Apollo 11 gave an uncomplicated picture, resulting in confirmation of the telescopic polarization measurements, but that Apollo 12 has provided us with a larger variety of samples.

Wavelength dependence of the optical properties

In the log A versus log P_m plot of Fig. 12, each sample is characterized by a line fitting 5 measurements made at each of the following wavelengths: 3540, 3790, 4350, 5300, and 5800 Å. The length of the line indicates the range of the optical change in the spectral range between 3540 and 5800 Å; for an optically grey material, all five measurements will be close together, whereas for a vividly coloured sample they will be spread all along the diagonal of the figure. All the lunar samples have the measurements for the longest wavelength at the right, thus the albedo is high in the red and low in the ultraviolet.

The position of each line characterizes the absorption of the sample; upper right is for the lightest powders and bottom left for the darkest one. The slope of each line is related to the refractive index of the material and the effect of absorption within the grains. More detailed analysis of these processes will need further investigation. For the core tube sample 12028,98, the slope is somewhat more pronounced than average.

THE EFFECT OF PROTON BOMBARDMENT

In DOLLFUS and TITULAER (1971) we have already compared the results of Fig. 11 and the lines in Fig. 12 with terrestrial samples. It was demonstrated that, optically speaking, properties of the lunar samples in the spectral range 3540–5800 Å can be almost exactly matched by some specific types of terrestrial samples e.g., volcanic pulverized basalts from lava flows. Thus, it is not necessary to call upon a specific darkening process operating at the lunar surface to explain the low albedo of the Moon.

However, it has been suggested that some processes may be working on the Moon which may darken or lighten the surface materials (e.g., HAPKE, 1965). The implications of various processes have been discussed, with regard to their photometric and polarimetric properties, by DOLLFUS et al. (1970).

An important possible process is the solar wind. HAPKE (1965) studied the effects of hydrogen-ion bombardment on terrestrial samples and on a crushed lunar rock (1970). KENKNIGHT et al. (1967) and NASH (1967) contributed to the subject, and DOLLFUS and GEAKE (1965) discussed the differences in the optical properties of a powdered sample of the Khor-Temiki meteorite before and after irradiation with protons.

Experiments have now been carried out with the Manchester equipment, which was described by DERHAM and GEAKE (1963) and later improved by GEAKE and WALKER (1967) to eliminate any possibility of oil contamination. One Apollo 11 and three Apollo 12 dust samples were irradiated. Two of these samples, 12033 and 12070, have a lower albedo after a proton irradiation of 4 μA.cm^{-2}.hr, at 60 keV. The polarization characteristics are also changed. However the other samples, 10084 and 12032, showed no change either in albedo or in polarization.

For sample 12070, Fig. 13 shows four portions of the polarization curves near P_{max}, for different wavelengths, on the virgin sample before irradiation (dots) and after irradiation by 60 keV protons for about 7 hours (crosses and circled crosses). The curve of polarization is given in full for 6000 Å; irradiation processes have not modified the negative branch, for $V < 40°$. Near P_{max}, the irradiated sample gives a

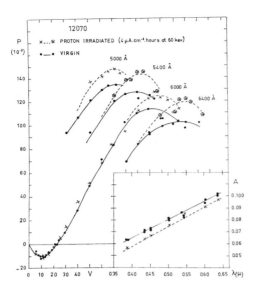

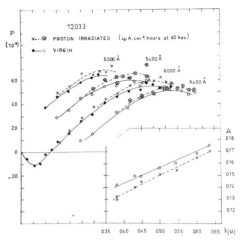

Fig. 13. The effect of proton irradiation on fines sample 12070,113, showing albedo versus wavelength before and after irradiation, and polarisation curves for four wavelengths before and after irradiation (only the curve for $\lambda = 6000$ Å is shown in full; the rest are shown around P_{max} only).

Fig. 14. The effect of proton irradiation on fines sample 12033,60, as Fig. 13.

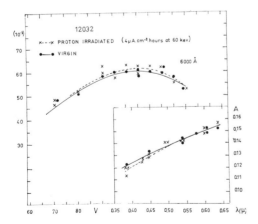

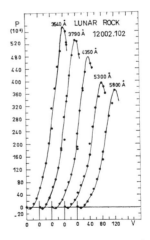

Fig. 15. The effect of proton irradiation on fines sample 12032,39, as Fig. 13, except that the polarisation curve is only shown around P_{max}, at larger scale, and for one wavelength only (6000 Å).

Fig. 16. Polarisation curves for rock 12002,102, for 5 wavelengths.

systematic increase at all wavelengths of about 10%. The albedo curve is also given in Fig. 13; the irradiation has caused a decrease in albedo of about 5% in the red and 12% in the ultra-violet. Sample 12070 is a dark powder, typical of mare material.

Figure 14 refers to sample 12033, a light powder characteristic of a continental surface. Again, the polarization P_{max} is increased by about 7% in the visible range, and the albedo is decreased by about 2% in the red and 7% in the ultraviolet.

Sample 12032 is also a light powder, somewhat less bright than 12033. No darkening or change in P_{max} were detected. In Fig. 15, a portion of the polarization curve near P_{max} (for $\lambda = 6000$ Å) is enlarged $\times 2$; dots (virgin sample) and crosses (irradiated sample) are intermixed within the accuracy of the measurements, and any change in P_{max} is less than 2%. The same is true for the albedo, except perhaps in the ultra-violet, where the small decrease may be real.

Sample 10084, already measured by GEAKE et al. (1970) also shows no detectable changes either in P_{max} or in A. It is a very dark powder.

Values of P_{max} and A, in orange light before and after irradiation, are marked with circles and crosses respectively in the log-log plot in Fig. 11. As pointed out by DOLLFUS et al. (1971) and DOLLFUS and TITULAER (1971), a change of the physical properties normally results in a displacement along a line roughly parallel to the observed lunar values. Figure 11 shows that laboratory proton darkening also has this effect.

It should be noted that 12032 and 10084 are, respectively, light and dark powders, and neither of them darkens; the two others, 12033 and 12070, are also of high and low albedo, respectively, but both of these did darken. It therefore seems that if proton irradiation has any effect at all on the lunar surface, then this effect may vary from place to place, and may be complicated by other effects, such as heating and UV bleaching, etc.

ROCK AND BRECCIA

In addition to the results for rocks 10057,54 and 10058,37 reported previously (Geake et al., 1970, Fig. 12) we have now investigated rock samples 10020,42,–12002, 102 and 12051,51 and breccia sample 10059,36. Their polarization characteristics are shown in Figs. 16–20; Figs. 21, 22, and 23 show scanning electron microscope photographs of the surfaces responsible. The photometric albedo for these samples, measured at 5 wavelengths, are shown in Fig. 24, on a log scale. Sample 12051,51 has the very high albedo of 22% in orange light; for breccia sample 10059,36, the albedo is so uniform with wavelength that it is almost neutral grey in colour. All these samples show high values of P_{max}; rock 10020,42 show the highest value we have recorded for a lunar sample. The samples differ in the negative branch of the polarization curve (see Fig. 20). Breccia sample 10059,36 shows a deep negative branch, comparable to that for dark fines material. This breccia consists of white regions, some of them plagioclase, set in a very dark matrix: it is this dark material that is probably responsible for the multiple scattering which results in the negative branch.

Figure 21(a) shows the texture of such a region, consisting of a mixture of clumps of very fine-grained material and coarser glassy pieces. Figure 21(b) shows some of each on a larger scale; although grains smaller than the wavelength are resolved on this

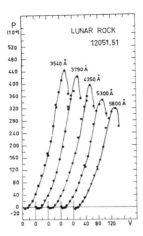

Fig. 17. Polarisation curves for rock 12051,51, for 5 wavelengths.

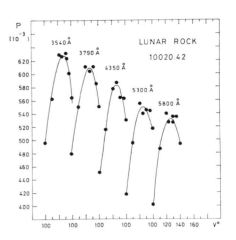

Fig. 18. Polarisation curves for rock 10020,42, near P_{max} only, for 5 wavelengths.

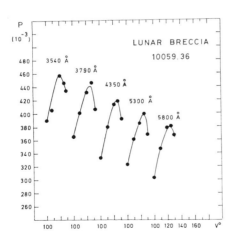

Fig. 19. Polarisation curves for breccia 10059,36, near P_{max} only, for 5 wavelengths.

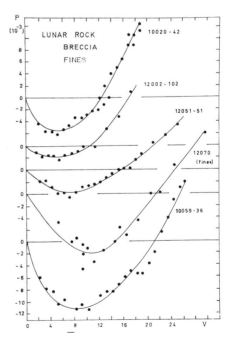

Fig. 20. Polarisation curves for small phase angles only, showing the negative branches, for three crystalline rocks (12002,102, 12051,51, and 10020,42), a breccia (10059,36), and a fines sample (12070,113), at a wavelength of 5800 Å.

Fig. 21. Scanning electron microscope photographs of breccia sample 10059,36. (a) shows region A on (d); (b) is central on (a), at about 4 times the scale. (c) shows region B on (d) at about 10 times the scale. The widths of the photographs represent (a) 160 μ, (b) 40 μ, (c) 420 μ, (d) 4.0 mm.

photograph, most of the grains present in the fine-grained component seem to be a few wavelengths across. This very fine-grained texture is similar in appearance to the fines material 10084,6 and also to the fine-grained rock 10057,54 shown previously (GEAKE *et al.*, 1970, Fig. 13(a) and (c)). The negative branch at small phase angles is evidently not affected by the coarser glassy pieces seen to be present in the dark matrix; these will contribute to the high value of P_{max}, as also will the glassy white component of the breccia, which together will dominate the reflected light. Figure 21(c) shows a typical part of one of these white areas. Figure 21(d) shows, on a smaller scale, the locations of the above regions.

In contrast, rock sample 12002,102 shows a negligible negative branch, as shown in Fig. 20. Its surface is mostly glassy in texture, as shown in Fig. 22(a), (b), and (c) on three different scales, with rounded polished features and very few small grains. There is thus an absence of the sort of material that would lead to the multiple scattering required to produce a strong negative branch, while the presence of many highly reflecting surfaces gives a very high value of P_{max}. This surface texture may be compared with that of sample 10058,37 (GEAKE *et al.*, 1970, Fig. 13(e) and (f)), which also gave a negligible negative branch and a high value of P_{max}.

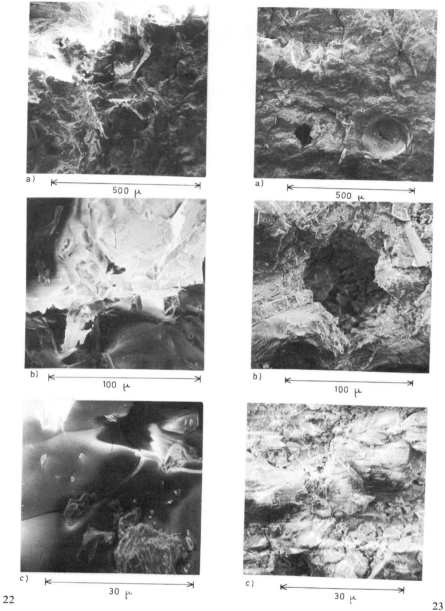

22   23

Fig. 22. Scanning electron microscope photographs of rock sample 12002,102.
(b) shows a region just above the centre of (a) at about 5 times the scale; (c) shows a piece
of glassy material on a scale about 200 times that of (a). The widths of the photographs
represent (a) 760 μ, (b) 150 μ, (c) 38 μ.

Fig. 23. Scanning electron microscope photographs of rock sample 12051,51. (b) shows
the dark hole, lower left on (a), at about 5 times the scale; (c) shows a typical region in
the middle of (a) at about 200 times the scale. The widths of the photographs represent
(a) 910 μ, (b) 180 μ, (c) 45 μ.

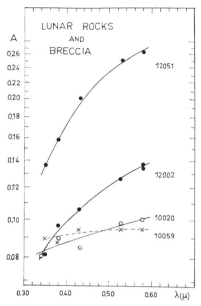

Fig. 24. Albedo versus wavelength for lunar rock and breccia samples. The albedo is plotted on a logarithmic scale.

Sample 12051,51 is intermediate in character. Figure 23(a), (b), and (c) show the surface to consist mostly of large crystals and fragments, and to be of medium roughness. There are not very many small grains, but there are more than for sample 12002,102, giving a lower P_{max} and a deeper negative branch, as shown in Figs. 17 and 20.

Lunar rocks and breccias thus show a wide variety of polarization characteristics, and in particular a full range of negative branch depths; terrestrial rocks, on the other hand, rarely show a deep negative branch.

One reason for investigating lunar rocks and breccias is their relevance to the study of asteroids and small satellites; it is possible that asteroids will resemble lunar materials as they have received similar impact treatment and exposure in space. The accumulated data from laboratory studies of the polarization characteristics of lunar rocks and breccias thus provides a catalogue to assist with the interpretation of telescopic observations of the polarization characteristics of asteroids and satellites. Furthermore, it is mainly the negative branch that is important in this connection, as asteroids can usually be observed only at small phase angles.

Acknowledgements—We are grateful to NASA for providing us with the lunar samples and to M. MAURETTE for his collaboration in lending us further samples; to G. WALKER of UMIST for his contributions to the proton irradiation of the samples; to the Textile Technology Department, UMIST, for the use of their scanning electron microscope and to B. LOMAS for operating it; to the Science Research Council for supporting the work in Manchester, and to the European Space Research Organisation, ESRO, for sponsoring the work of one of us at Meudon (C. T.).

References

Derham C. J. and Geake J. E. (1963) A Proton Source for Luminescence excitation. Techn. Note Nr. 2. Contract AF.61(052)379 UMIST.

Van Diggelen J. (1965) The radiance of lunar objects near opposition. *Planet. Space Sci.* **13**, 271–279.

Dollfus A. and Geake J. E. (1965) L'altération des propriétés polarimétriques du sol lunaire par l'action des protons du vent solaire. *Compt. Rend. Acad. Sci. Paris* **260**, 4921–4923.

Dollfus A. and Bowell E. (1970) Polarimetric properties of the lunar surface and its interpretation, Part I. Telescopic observations *Astron. Astrophys.* **10**, 29–53.

Dollfus A., Bowell E., and Titulaer C. (1971) Polarimetric properties of the lunar surface and its interpretation, Part II. Terrestrial samples in orange light. *Astron. Astrophys.* **10**, 450–466.

Dollfus A. and Titulaer C. (1971) Polarimetric properties of the lunar surface and its interpretation, Part III. Volcanic samples in several wavelengths. *Astron. Astrophys.* (in press).

Geake J. E. and Walker G. (1967) Laboratory Investigations of meteorite luminescence. *Proc. Royal Soc. London* **A296**, 337–346.

Geake J. E., Dollfus A., Garlick G., Lamb W., Walker G., Steigmann G., and Titulaer C. (1970) Luminescence, electron paramagnetic resonance and optical properties of lunar material from Apollo 11. *Proc. Apollo 11 Lunar Sci. Conf.*, Geochim. Cosmochim. Acta Suppl. 1, Vol. 3, pp. 2127–2147. Pergamon.

Gehrels T., Coffeen D., and Owings D. (1964) The Wavelength dependence of polarization, III. The Lunar surface. *Astronomical. J.* **69**, 826–852.

Hapke B. (1963) A Theoretical photometric function for the lunar surface. *J. Geophys. Res.* **68**, 4571–4586.

Hapke B., Cohen A. J., Cassidy W. A., and Wells E. N. (1970) Solar radiation effects on the optical properties of Apollo 11 Samples. *Proc. Apollo 11 Lunar Sci. Conf.*, Geochim. Cosmochim. Acta Suppl. 1. Vol. 3, pp. 2199–2212. Pergamon.

Kenknight C. E., Rosenberg D. L., and Wehner G. K. (1967) Parameters of the optical properties of the lunar surface powder in relation to solar-wind bombardment. *J. Geophys. Res.* **72**, 3105–3130.

NASA (1970) SP-235 Preliminary Science Report.

Nash D. (1967) Proton irradiation darkening of rock powders. *J. Geophys. Res.* **72**, 3089–3104.

O'Leary B. T. and Rea D. G. (1968) The Opposition Effect of Mars and Its Implications. *Icarus* **9**, 405–428.

Whitaker E. A. (1969) An Investigation of the Lunar Heiligenschein. NASA SP-201 38–39.

Proceedings of the Second Lunar Science Conference, Vol. 3, pp. 2301–2310
The M.I.T. Press, 1971.

Apollo 12 multispectral photography experiment

A. F. H. GOETZ and F. C. BILLINGSLEY

Jet Propulsion Laboratory, Pasadena, California 91103

J. W. HEAD

Bellcomm, Inc., Washington, D.C. 20024

T. B. McCORD

Department of Earth and Planetary Sciences, Massachusetts Institute of Technology
Cambridge, Massachusetts 02139

and

E. YOST

Science Engineering Research Group, Long Island University
Greenvale, New York 11548

(Received 24 February 1971; accepted in revised form 30 March 1971)

Abstract—Apollo 12 carried a 4-band camera system for orbital lunar surface photography. New image processing techniques were developed to delineate accurately subtle spectral reflectivity differences, independent of brightness differences within selected areas of the lunar surface. Ground-based photoelectric photometry was used to verify large area color differences.

In general the highlands areas covered are quite uniform in normalized spectral reflectivity on a 200 m scale. Differences were detected in the Descartes region, which can be attributed to exposed rock in the ejecta blanket of Dollond E. No color difference was detected across the mare-highland boundary at Fra Mauro. With few exceptions, the highlands areas studied are extremely uniform and the variation in spectral reflectivity in the wavelength region covered seen in any frame is less than that found in some Apollo 12 core samples.

INTRODUCTION

Lunar color has been the subject of intensive investigation during the last few years using ground based telescopes (McCORD, 1969; McCORD and JOHNSON, 1969, 1970; SODERBLOM, 1970). The lunar surface and the bulk fines returned from the Apollo 11 and Apollo 12 sites appear gray to the eye. However, in fact, the absolute spectral reflectivity curve in the visible wavelength region displays an increasing reflectivity with wavelength. The telescope data show good agreement with returned sample fines, as demonstrated by ADAMS and McCORD (1970) and shown in Fig. 1. Slight variations in the shape of the reflectivity curve can be mapped and are loosely called color differences. In this study all ratios of spectral reflectance are normalized to unity at 0.56 μm (green) so that direct comparison of the curve shapes can be made independent of brightness variations. In general an area that is "bluer" than another will have a normalized blue reflectance ratio greater than unity when compared with the standard area. If the spectral reflectivity curves are nearly linear over the visible wavelength region, then a "bluer" area will have a smaller red/blue ratio than the comparison area. The interpretation of the differences will be discussed in more detail below.

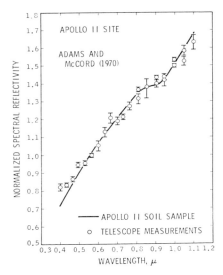

Fig. 1. Normalized spectral reflectivity of Apollo 11 sample fines and telescopic data from an 18 km diameter area surrounding the Apolo 11 site showing the good agreement between the two types of measurement (Adams and McCord, 1970).

The Lunar Multispectral Photography Experiment (S–158) was flown in lunar orbit aboard Apollo 12 (1) to record subtle color differences at two to four orders of magnitude higher areal resolution than achieved from earth, and (2) to obtain information on the scale of surface compositional and mineralogical heterogeneity. Once the mineralogical or compositional effects causing color differences are established, we will then be able to extrapolate information from sample areas to wide areas of the lunar surface. The advantage of photography over standard photoelectric photometry is that it produces an image, which is a more useful mapping tool than a discrete point measurement. However, the use of film limits the photometric accuracy as discussed below.

Camera System

S–158 consisted of four electric Hasselblad cameras equipped with 80 mm lenses and mounted in a ring attached to the Command Module hatch window. Each camera was fitted with a different Wratten type filter and black and white film. The combination film-filter band passes are shown at the base of Fig. 2. In the visible wavelength region a Plus-X type emulsion was used. A total of 142 frames in each of the three visible wavelengths was returned. A filter centered at 0.85 μm was also used; however, no usable IR frames were returned because of a focusing problem and lack of dynamic range in the emulsion.

The photographic ground track is shown in Fig. 3. All frames were taken with the camera axes oriented toward the nadir with the exception of special target photography of Theophilus, the proposed Descartes landing site, and the Apollo 14 Fra Mauro site. More than 80% of the area covered was in the highlands.

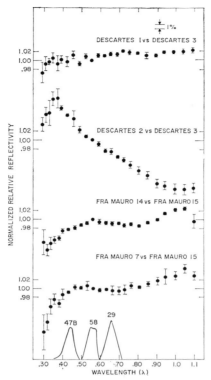

Fig. 2. Normalized ratios of the spectral reflectivity of 18 km areas within the Apollo 12 photography. Descartes 1 and 3 are points near the bright crater Dollond E but outside the bright ejecta blanket. Descartes 2 is centered on the ejecta blanket. Fra Mauro 14 is the Apollo 14 landing site and Fra Mauro 15 is in Oceanus Procellarum, due west of the landing site. Fra Mauro 7 is an area in the crater Fra Mauro. Each 0.02 reflectivity interval is approximately equivalent to one DN in a processed picture. The curves at the bottom show the relative film-filter responses for each of the three cameras used.

METHOD

Separation of spectral reflectivity variations from brightness changes due to albedo and lunar surface slope, necessitated the use of a ratio technique. McCORD (1969) has demonstrated the high sensitivity of this method for recording lunar color variations, and BILLINGSLEY *et al.* (1970) applied computer ratio techniques to display color differences in telescopic images of Mare Imbrium. The two color technique was applied to a frame of the Fra Mauro region shown in Fig. 4. In this study we developed new methods for combining ratios of three color filtered images by both computer image processing and photographic techniques.

The routine followed in image processing by computer was the following.

(1) Each set of frames was digitized by means of a video film converter (VFC) flying spot scanner system with a density interval of approximately 0.01. Before each run the calibration step wedge from the appropriate film strip was also scanned. The spot size used was 50 μm yielding approximately 10^6 picture elements per frame.

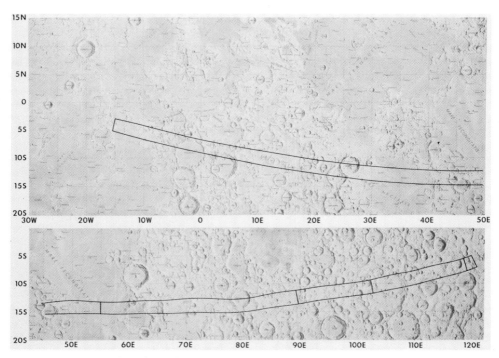

Fig. 3. Ground track on the lunar surface for Apollo 12 multispectral photography. Each frame covered approximately 75 × 75 km on the surface with 60% overlap between frames.

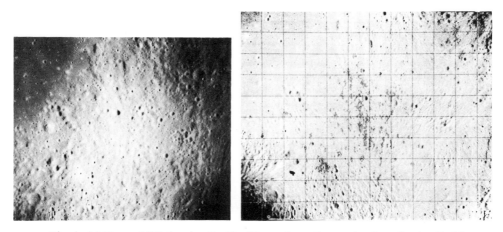

Fig. 4. (a) Frame 8438 showing the Fra Mauro formation centered on the Apollo 14 landing site. The upper left-hand corner contains the Fra Mauro formation—Oceanus Procellarum boundary. (b) Red/blue diffpic showing lack of color contrast at the mare-highland boundary. Shadows and improper registration yield false color differences. The dark areas at the center of the frame result from overexposure and concomitant loss of data in the very bright regions.

(2) The frames of the same area were registered by an automatic rubber sheet stretch program which compensated for differential geometric distortion between cameras, mainly due to focal length differences and film buckling.

(3) By means of the appropriate calibration data, densities were converted to log-exposure values (log E) so that frames of different colors were comparable and all film processing and duplication variables were removed.

(4) Each pair of frames was differenced point for point. The resulting values then represented

$$\log_{10} A_{ij} - \log_{10} B_{ij} + K_1 = \log_{10} \frac{A_{ij}}{B_{ij}} + K_2 \tag{1}$$

where A_{ij} corresponds to a picture element (pixel) exposure value in one color and B_{ij} is the same scene pixel exposure value in another color for pixel coordinates (i, j). K_1 and K_2 are arbitrary normalization constants which may be different for each frame set. In practice, for display, the constants were chosen so that the maximum in the distribution of difference picture (diffpic) values was set to neutral gray. The pixel values represent the normalized two color exposure ratios in which all effects of the shutter, lens and window transmission, photographic development and duplication, and scene brightness effects have been removed.

(4) A bivariate plot of red/green vs. blue/green ratios of the type used by Soderblom (1970) (Fig. 5) for displaying photometry data was formed, with each point in the plot assigned one of 256 hues. As a final output, a computer frame was produced with each pixel assigned its appropriate color. The sequence of operations is shown in Fig. 6. A very large (16 times) contrast stretch was used to maximize subtle color discrimination. At this high contrast, a number of noise effects are also seen. The details of the color display techniques are given by BILLINGSLEY (1971).

The advantage of the computer processed image is that the only errors introduced into the system are in the film and wedge scanning processes. All other operations on the images are essentially noiseless and error free. The resulting image is a quantitative, photometrically accurate display of a combination of the original data.

A parallel program was carried out to combine the frames by purely photographic techniques. In order to eliminate the brightness component, WHITAKER's (1965) two color sandwich printing technique, in which a blue negative plate and a red positive plate were superimposed and printed, was extended to three colors. From each blue, green and red negative a positive mask was made, having half the gamma, or density range, in the straightline portion of the $D \log E$ curve. Each negative was

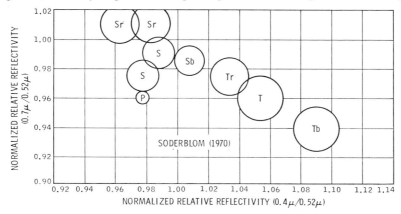

Fig. 5. Bivariate plot of red/green and blue/green ratios of areas on the lunar surface (SODERBLOM, 1970). Each circle corresponds to a spectral classification unit derived from photoelectric photometry. Each square corresponds to approximately one DN in the pictures processed in this study. Photographic methods are therefore capable of distinguishing among spectral types.

sandwiched together with the positive masks of the other two colors and printed. The three prints were placed in a color reconstruction projector and photographed.

The advantages of the photographic over the computer reduction method are (1) The ultimate resolution of the film is utilized. (2) Color composites can be made with relatively simple equipment.

On the other hand, extreme accuracy is required to match the slopes of the $D \log E$ curves in the first negative and subsequent positive masks. Perfect registration could not be achieved because of residual geometric distortions in the camera systems. The dynamic range of the composites is limited by the straight-line portion of the $D \log E$ curve of the original film material. Any areas outside of this brightness range appear falsely colored. Details of the photographic method are given by Yost et al. (1971). In this study, the exposure dynamic range of the computer method was approximately a factor of 4 greater than in the photographic method.

An independent calibration for both methods was provided by ground-based photoelectric photometry. Areas where large albedo variations were present, such as across mare-highland boundaries and the bright ejecta blanket surrounding Dollond E in the Descartes region, were measured to provide confidence in the differences observed on film. The results are shown in Fig. 2. Differences in the telescopic ratios at the filter center wavelengths can be directly related to the diffpic averages over the same areas.

Discussion of Errors

The sensitivity of the method is basically limited by the photometric resolution of the film emulsion which is governed by film grain noise. Each digital step or digial number (DN) corresponds approximately to a 2% change in exposure. For a uniform area in a difference picture the standard deviation is 1–2 DN or 2–4% in the ratio of the exposures.

A number of systematic errors were also defined. (1) Calibration wedge non-uniformities and light leaks. In some frames, higher densities were observed on the film than in the most dense wedge step. No conclusions were drawn from data in dense frames. (2) Drift in the VFC causing shift of the calibration at high densities. (3) Differential vignetting among cameras. Vignetting maps of the camera image planes were made and corrections applied where appropriate. (4) Nonuniform film processing. Continuous development methods were used and density control of 0.02 was maintained over the length of the 10 m film strip. Any error will result in a false color difference being registered. The fact that we have found the highlands to be nearly uniformly colored gives us confidence in the method.

Results

Forty frames have been analyzed, of which seventeen were studied in detail. Most frames are conspicuous by the absence of significant color differences. Preliminary data analysis indicates that no major color differences exist (1) over a large portion of the Central Highlands; (2) between some areas of upland basin fill, upland plateau volcanics, and apparently older homogeneous highland material; (3) between certain parts of mare regions and adjacent highlands; (4) within several maria themselves; and (5) within the portion of the Fra Mauro Formation in the vicinity of Apollo 14 landing site, and between Fra Mauro and nearby mare areas.

Several areas, however, show evidence of color differences of varying degree and are described below. A detailed investigation of the geology of these areas and its relation to the observed color differences will be reported elsewhere (Goetz and Head, 1971).

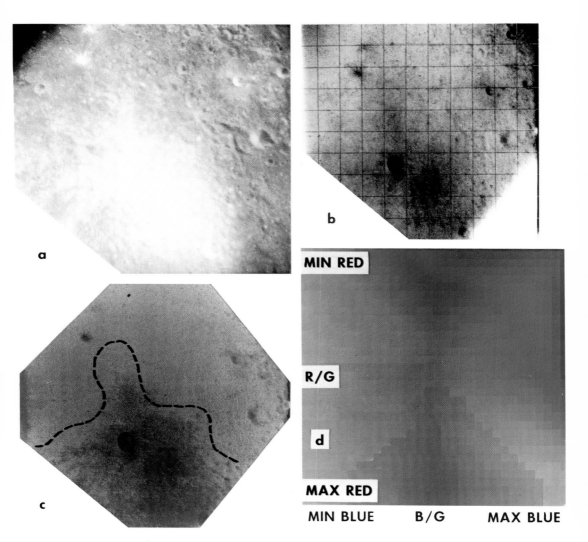

Fig. 6. (a) Frame 8436 taken with a red filter centered on Dollond E. The proposed Apollo 16 Descartes landing site is in the upper left-hand corner of the frame. (b) Black and white rendition of the ratio red/blue; darker areas are bluer. The corners are covered because the round spacecraft hatch window obscured a different corner in each frame. (c) Computer composite of red/green and green/blue frames. The area outlined corresponds to the ejecta pattern of Dollond E. The colors can be interpreted in terms of the ratios by reference to Fig. 6d. (d) Computer produced color display key for color composite frames. In this paper, each DN corresponds to 4 × 4 small squares in this color key. The center represents gray and in any one frame set, the maximum in the DN histogram is forced to the center of the plot allowing the full range of hues to be used.

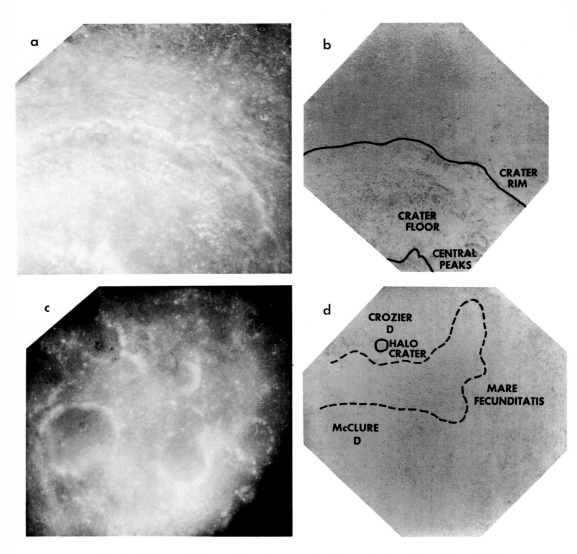

Fig. 7. (a) Frame 8434 an oblique photograph of the north rim of Theophilus. (b) Color ratio composite. Bright areas beyond the rim are redder than the bright floor. There is no direct correlation between relative color and brightness. (c) Frame 8328, a high sun angle view of the highland–Mare Fecunditatis boundary. (d) Color ratio composite showing the possible mare filling of McClure D and Crozier D.

Fra Mauro region (Fig. 4). Color differences are not seen across the western mare boundary of the Fra Mauro Formation in spite of a 50% change in normal albedo. This observation was substantiated by ground based photoelectric photometry (Fig. 2). In frame 8392, not shown here, rays originating from the bright highland crater Lalande are blue with respect to underlying Mare Nubium, in agreement with ground based measurements made on larger, analogous structures (McCORD *et al.*, 1971).

Descartes region (Fig. 6). The region to the north of the crater Descartes lies along the western margin of the Kant Plateau, an area of supposed highland volcanism of Imbrian age (MILTON, 1968). The surface morphology of the units comprising the Kant Plateau differs from the surrounding and underlying older highland terrain, and from the more level plains which are generally stratigraphically younger than materials of the Kant Plateau. The proposed Apollo 16 landing site is located on the northwest margin of the Kant Plateau. Major color differences are not visible at the margins of most of these units. The area surrounding the crater Dollond E, however, is, for the highlands, strongly colored and is some 6–8% bluer than the surrounding material of the Kant Plateau. Four frames from two different orbital passes contain the same area and the colors observed are internally consistent. These values were substantiated by telescope photometry; Descartes 2 vs. Descartes 3 in Fig. 2. The distribution of the color differences is more related to the high albedo area associated with the two Copernican age craters Dollond E and Descartes C than it is to mapped geologic units in the area. This association with brightness and young craters suggests that the color differences are due to freshly exposed material excavated from these two craters and lying in their ejecta blankets.

Theophilus (Fig. 7a,b). Theophilus is a major Copernican age crater (MILTON, 1968) located on the northwest edge of Mare Nectaris. Frame 8434 covers a segment of Theophilus from the central peaks north to the outer ejecta blanket. The central peaks, the floor, and part of the crater rim are considerably bluer than most of the surrounding terrain including the ejecta blanket. Large portions of the hummocky crater rim are also bluer than the surrounding area but the outer portions of the ejecta blanket (the radial facies) and some parts of the inner rim and crater wall appear redder in comparison. Color is not directly related to brightness in this frame.

McClure region (Fig. 7c,d). Along the southwestern boundary of Mare Fecunditatis, in the vicinity of the McClure and Crozier craters, a color difference is noted between the mare surface and parts of the adjoining highlands. Fecunditatis mare material borders the highlands in this region and appears to flood the floors of several adjacent craters such as McClure D and Crozier D. A regional color difference is noted between the mare filled areas (Fecunditatis and the filled craters) and the highland area, with the mare filled areas being bluer. A dark halo crater in Crozier D appears blue-green and more mare-like than its highland surroundings or the crater fill. However, nearby craters of similar appearance do not exhibit any color contrast.

Discussion

Differences in the ratio of spectral reflectivities between areas on the lunar surface can be attributed to several causes: (1) chemical composition; (2) roughness; (3)

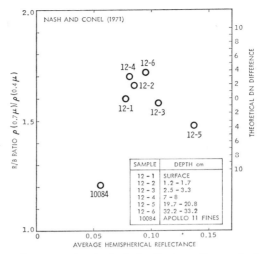

Fig. 8. Red/blue ratios versus hemispherical reflectance for a set of Apollo 12 core samples (Nash and Conel, 1971). At the right the theoretical DN difference is shown. If photographs were taken of the samples and the techniques under discussion applied, it should be possible to discriminate between 12–1 and 12–2, but probably not between 12–1 and 12–3.

mineralogy, including crystal field effects and the presence of glass; and (4) relative age in which areas are affected by contamination with foreign materials, roughness differences, and the formation of glass *in situ*.

Compositional differences in silicates produce color differences in the visible wavelength regions. Basic rocks are, in general, bluer than their fines (Adams and Filice, 1967) and this is particularly true for the Apollo 11 samples (Adams and McCord, 1970). Therefore the color of an area will be affected by the percentage of rocks on the surface. The presence of Fe and Ti rich glass will also contribute to the reddening in the fines (Conel and Nash, 1970; Adams and McCord, 1971). There is also an apparent correlation of color with age based on telescope studies (McCord *et al.*, 1971) which may be primarily a function of the glass content (Adams and McCord, 1971). In interpreting color differences in a given area, it is usually possible to rule out one or more causes.

Analysis of some Apollo 12 core samples shows a range of color and brightness (Nash and Conel, 1971). The spectral reflectivity curves are very nearly linear functions of wavelength in the 0.4–0.7 μm range. A plot of red/blue ratios reveals that brighter samples appear to be bluer (Fig. 8). However, two samples, 12–1 and 12–3, with a 40% difference in average reflectance have essentially the same red/blue ratios. Based on luminescence studies, the brighter core materials appear to contain greater quantities of plagioclase. Results of Sellers *et al.* (1971) using a 125–250 μm fraction in nearly the same portions of the core, show increases in feldspar and pyroxene which correlate with brightness in core samples of Nash and Conel (1971). Therefore, it appears possible, with the range of compositions found in the Apollo 12 core samples, to produce an albedo difference corresponding to the albedo difference between mare-highland boundary at the Fra Mauro formation which is consistent with the spectral reflectivity measurements. We do not imply that the mare-upland difference is found uniquely in the Apollo 12 core. On the other hand, the range of red-blue ratios in the Apollo 12 core samples is greater than we have found in any of our frames except for the Descartes region.

The area surrounding Dollond E is clearly covered with bright ejecta, most likely a higher percentage of rocks on the surface gives rise to the relative blue signature. The differences seen in the Theophilus region cannot be explained entirely by a differential roughness model. A more thorough study is necessary to explain the color differences unrelated to brightness. However, differences in relative proportions of crystalline and glassy materials is a simple possible explanation. The Mare Fecunditatis-highland boundary is diffuse. The color variations could be explained by the apparent age differences (ADAMS and McCORD, 1971) or by a chemical difference. Crozier D and McClure D appear to be flooded by mare material.

In other areas not shown here, we find that the highlands are very uniform in spectral reflectivity on a 200 m scale, the approximate limit of resolution in a final color composite; if inhomogeneities in physical or chemical properties exist, they are of smaller linear dimension than 200 m. Previously, similar results were obtained on a 10 km scale (McCORD, 1969; McCORD and JOHNSON, 1969). This homogeneity extends over areas which have been mapped from pre-Imbrian to Copernican in age, rims and floors of craters, upland basin fill, and crater chains.

Based on the fact that sharp color boundaries between areas of different ages in the maria and sharp albedo boundaries at the mare-highland intersections are seen (Whitaker, 1965), it is reasonable to assume that the majority of the fine surface material is generated in situ. The color and the albedo in the highlands is quite uniform along the Apollo 12 flight path. Therefore the average surface compositional variation is probably quite small.

REFERENCES

ADAMS J. B. and FILICE, A. L. (1967) Spectral reflectance 0.4 to 2.0 microns of silicate rock powders. *J. Geophys. Res.* **72**, 5705–5715.

ADAMS J. B. and McCORD, T. B. (1970) Remote sensing of lunar surface mineralogy: Implications from visible and near-infrared reflectivity of Apollo 11 samples. *Proc. Apollo 11 Lunar Sci. Conf.*, *Geochim. Cosmochim. Acta* Suppl. 1, Vol. 3, pp. 1937–1945. Pergamon.

ADAMS, J. B. and McCORD, T. B. (1971) Alteration of lunar optical properties: age and composition effects. *Science* **171**, 567–571.

BILLINGSLEY F. C., GOETZ A. F. H., and LINDSLEY J. N. (1970) Color differentiation by computer image processing. *Photo. Sci. and Eng.* **14**, 28–34.

BILLINGSLEY, F. C. (1971) 2 × 2 dimensional color display by computer. In preparation.

CONEL J. E. and NASH D. B. (1970) Spectral reflectance and albedo of Apollo 11 lunar samples: Effects of irradiation and vitrification and comparison with telescopic observations. *Proc. Apollo 11 Lunar Sci. Conf.*, *Geochim. Cosmochim. Acta* Suppl. 1, Vol. 3, pp. 2013–2023. Pergamon.

GOETZ A. F. H. and HEAD J. W. (1971) Lunar surface compositional variation from orbital multispectral photography. In preparation.

McCORD, T. B. (1969) Color differences on the lunar surface. *J. Geophys. Res.* **74**, 3131–3142.

McCORD T. B. and JOHNSON T. V. (1969) Relative spectral reflectivity 0.4–1 μ of selected areas of the lunar surface. *J. Geophys. Res.* **74**, 4395–4401.

McCord T. B. and Johnson T. V. (1970) Lunar spectral reflectivity 0.3–2.5 μ and implications for remote mineralogical analysis, *Science* **169,** 855–858.

McCord T. B., Charrette M., Johnson T. V., Lebofsky L. A., and Pieters C. (1971) Lunar spectral types. In preparation.

Milton D. J. (1968) Geologic map of the Theophilus quadrangle of the moon: USGS geologic atlas of the moon I–546.

Nash D. B. and Conel J. E. (1971) Luminescence and reflectance of Apollo 12 samples. Second Lunar Science Conference (unpublished proceedings).

Sellers G. A., Woo C. D., Bird M. L., and Duke M. B. (1971) Descriptions of the composition and grain-size characteristics of fines from the Apollo 12 double core tube. Second Lunar Science Conference (unpublished proceedings).

Soderblom L. A. (1970) The distribution and ages of regional lithologies in the lunar maria. Ph.D. thesis, California Institute of Technology.

Whitaker E. A. (1965) in Heacock *et al.*, Ranger VII, Part II: Experimenters analysis and interpretations. TR 32–700, JPL, California Institute of Technology.

Yost E., Goetz A. F. H., and Anderson R. (1971) A three color photographic method for detection of small spectral reflectivity differences, in preparation.

Proceedings of the Second Lunar Science Conference, Vol. 3, pp. 2311–2315
The M.I.T. Press, 1971.

Thermal conductivity of fines from Apollo 12

C. J. Cremers and R. C. Birkebak

Department of Mechanical Engineering, University of Kentucky,
Lexington, Kentucky 40506

(Received 9 February 1971; accepted 23 March 1971)

Abstract—The thermal conductivity of the fines returned by the Apollo 12 astronauts (sample 12001,19) was measured under vacuum conditions using the line heat-source technique. It was found to vary from about 0.12×10^{-2} w/m-°K at 160°K to about 0.35×10^{-2} w/m-°K at 428°K for a sample density of 1300 kg/m^3.

Introduction

THE SAMPLES WHICH were returned to earth by the Apollo 12 astronauts consist of basaltic igneous rocks, microbrecias, and lunar soil. As was the case at the Apollo 11 site, it was found that the Apollo 12 site was covered by a thick layer of the soil or fines. Consequently, calculations of lunar heat flow and heat transfer to lunar systems depend heavily on the properties of this material.

The fine nature of the soil eliminated some problems in the measurement of the thermal conductivity but added some others. A particle sizing study by GOLD et al. (1971) showed that the particles varied in diameter from 100 μm down to less than 1 μm with most of the particles being at the lower end of the range. Consequently, one would expect any effects of thermal contact resistance on the measurements to be negligible. On the other hand, because the soil is porous, measurements must be made under vacuum conditions to eliminate gas conduction effects.

A porous material at atmospheric pressure transfers heat through a complicated interaction of solid conduction through the particles and their contact points; gaseous conduction and convection in the voids; and scattered, emitted, and transmitted radiation in the voids and particles. Because of the low pressure the gaseous effect is absent on the moon. WECHSLER and GLASER (1964) have shown that for pressures below about 10^{-2} torr, heat transfer in a powdered basaltic rock is not affected by residual gases. Therefore, to simulate the pressure conditions on the moon, one must measure the conductivity at pressures below this level.

While pressure is not a factor under lunar conditions the temperature variation does affect the conductivity. At higher pressures gaseous contributions to the heat transfer effectively mask radiation effects. These radiation effects become important at low pressures. Previous studies, e.g., WATSON (1964) and WILDEY (1967), have indicated that heat transfer in powdered rocks under vacuum is made up of two components. A normal conductive component describes the usual solid conduction and a radiative component describes the radiative portion of the total heat transfer. Analyses show that if one wishes to represent heat transfer through the medium by Fourier's law, then the thermal conductivity can be expressed as a constant (conductive portion) plus a term proportional to the temperature cubed (radiative portion). Cast in this form, the thermal conductivity is only an effective one. Consequently, the thermal conductivity must be determined as a function of temperature.

The present experiment was set up, then, to measure the thermal conductivity of the finely powdered lunar sample under vacuum conditions and over a range of temperatures. Because of the scarcity of sample material, the experiment was set up to require only about 6 grams of material. The sample used was 12001,19 as cataloged by the Lunar Receiving Laboratory at the Manned Spacecraft Center, NASA, Houston.

The Experiment

The size of the sample available for testing dictated the size of the experimental apparatus and, to a large degree, the method to be used. It was decided to employ the line heat-source technique which has been used in the past for vacuum conductivity measurements of silicate materials. The method is well suited for measurements of the thermal conductivity of small samples and was used for measurements on the Apollo 11 fines by Cremers *et al.* (1970). The basic mathematical treatment of the working equation is given by Carslaw and Jaeger (1959) and the experimental errors incurred in deviating from the mathematical model have been considered by Blackwell (1959).

The application of this method requires that a long (length to diameter ratio greater than 30) line heat-source be imbedded in the material to be tested. For such a source in an infinite medium, it can be shown that after an initial period during which the probe heat-capacity is dominant, the temperature change at any point in the medium over a time period from t_1 to t_2 is given by

$$T_2 - T_1 = \frac{q}{4\pi k} \ln \frac{t_2}{t_1} \qquad (1)$$

Here q is the heat-source strength and k is the thermal conductivity of the medium. Note that k must be considered constant over the temperature range $(T_2 - T_1)$ under consideration.

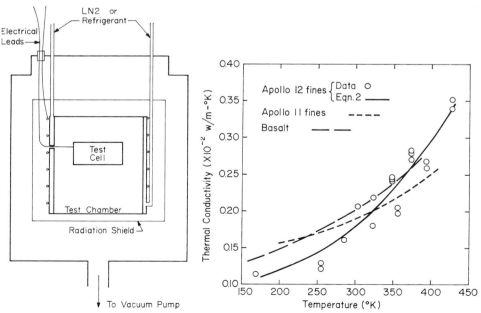

Fig. 1. Conductivity test chamber showing vacuum chamber, thermal environmental chamber, and cell placement.

Fig. 2. Thermal conductivity of Apollo 12 fines compared with Apollo 11 fines and basalt for a density of 1300 kg/m³.

Equation (1) is the usual working equation for the line heat-source method. It is apparent that if the conditions for the model are met in the experiment, a plot of temperature versus the logarithm of time will result in a straight line, the slope of which is $q/4\pi k$. The measurement of q and the slope of the curve then yields the thermal conductivity k.

The lunar samples which were made available for analysis were limited in volume so that it became imperative to measure the temperature as close to the source as possible to minimize deviations from the infinite medium assumption. For this reason the line source itself, a 32 AWG (0.203 mm dia.) Nichrome-V wire was calibrated as a resistance thermometer. Then with the wire in place in the sample the voltage change over about 22 mm of the wire was controlled by a constant-current power supply which supplied a current constant to within four significant figures. The voltage change during a run was 0.25% at the maximum and so the heat generation was constant to within this value as well. Axial heat conduction loss was minimized by providing an extra 10 mm of heating wire beyond the voltage taps.

The test cell was constructed of teflon and held 5.49 g of the lunar fines at a density of 1300 kg/m³. The size of the sample when in the cell was about 25 × 13 × 13 mm. The density of the fines when loosely poured was about 1300 kg/m³. To achieve greater densities the cell was vibrated with a Vibrotool to cause uniform settling and packing.

The test chamber is shown schematically in Fig. 1. The vacuum chamber is of stainless steel and is approximately 0.3 m diameter by 0.4 m high. The chamber was pumped with a Welsh Turbo-molecular pump to provide a pressure on the order of 10^{-6} torr. Electrical feed-throughs were provided for power and for temperature control and sensing.

An inner stainless steel chamber of double-walled construction was used for ambient temperature control. A heating tape wrapped about it provided higher than room temperatures and liquid nitrogen or expanding freon passed through the jacket provided low temperatures. A thermistor attached to the chamber actuated the heater for temperature control and a thermocouple immersed in the sample indicated the sample temperature.

RESULTS AND DISCUSSION

The thermal conductivity of the Apollo 12 fines for a density of 1300 kg/m³ is shown as a function of temperature in Fig. 2 and the data are tabulated in Table 1. For comparison, the vacuum thermal conductivities of the Apollo 11 fines of CREMERS et al. (1970) and powdered terrestrial basalt of FOUNTAIN and WEST (1970), both at a density of 1300 kg/m³, are also shown. Further vacuum data for basalt presented by WECHSLER and GLASER (1964) and also BERNETT et al. (1963) as a function of temperature are substantially in agreement with the FOUNTAIN and WEST data and are not shown here. Because of its similarity in chemical composition to the lunar fines, basalt is probably the best terrestrial material for comparison.

Table 1. Apollo 12 thermal conductivity data (1300 kg/m³).

Temperature (°K)	Thermal conductivity (w/m-°K)	Temperature (°K)	Thermal conductivity (w/m-°K)
169	1.14 × 10⁻³	349	2.46 × 10⁻³
256	1.21	356	1.97
256	1.29	356	2.05
286	1.61	374	2.70
286	1.61	374	2.78
304	2.07	374	2.83
324	1.80	393	2.59
325	2.19	394	2.68
349	2.41	429	3.39
349	2.43	429	3.52

Table 2. Coefficients of equation (2) for a density of
1300 kg/m³

Sample	A w/m-°K	B w/m-°K⁴
Apollo 12	0.922×10^{-3}	0.319×10^{-10}
Apollo 11	0.142×10^{-2}	0.173×10^{-10}
Basalt	0.124×10^{-2}	0.243×10^{-10}

The curves given in Fig. 2 are all of the form

$$k = A + BT^3 \tag{2}$$

as elementary theory suggests they should be. (A curve of the form $k = AT^{-1} + BT^3$ (suggested in review by Kanamori) reflecting an inverse temperature dependence for the lattice component of the conductivity was also tried. The fit was acceptable at high temperatures but not at the low end of the range. It was also tried without success on the more extensive data from Apollo 11 samples. This lack of agreement is probably due to the amorphous nature of the finely divided lunar samples.) The coefficients are given in Table 2. Consideration of the accuracy of representation of the data by an equation of the form of equation (2) suggests that a cubic dependence on temperature is probably the case. However, because of data scatter and a present lack of data at lower temperatures the evidence is far from conclusive.

The data given for the Apollo 11 samples are not yet sufficiently complete for a critical analysis of the temperature dependence of the thermal conductivity. On the other hand, the FOUNTAIN and WEST data for terrestrial basalt, taken at a number of densities ranging from 790 to 1500 kg/m³ are extensive. These were much larger samples and so the errors should be expected to be less significant. Analysis of these data shows that a cubic least-squares fit works well in some cases (one such case is the density of 1300 kg/m³) but not so well for other densities. This suggests that the elementary theory is close to correct but that some vital elements are still missing.

It is of interest to note that the magnitudes of the conductivities shown in Fig. 2 are roughly the same. What is of more importance, however, is the apparent difference in temperature dependence between the different sets of data shown. The slope of each curve, at a given value of the temperature is an indication of the magnitude of the radiative component which is expressed through the coefficient B. It appears that the radiative effects in the terrestrial basalt are intermediate in importance as compared with those in the two lunar samples. Most probably, the causes of these deviations are differences in particle size distribution and possibly shape as well as variations in the amounts of glass present in each of the lunar samples.

The data of FOUNTAIN and WEST (1970) are for particulate basalt 37–62 μm in diameter. That is a much narrower size range than was found for the Apollo 11 lunar fines. The abundance of micron-sized and smaller particles in the latter samples would tend to suppress the radiative transfer mode as the powder would more closely resemble a solid. The study by WATSON (1964) indicated that the radiative component in powdered media depends strongly on particle size being much more important for larger particles, on the order of 100 μm diameter than for particles on the order of 10 μm diameter. The Apollo 12 data indicate that there is perhaps a greater abundance of larger particles present resulting in stronger radiative effects. The sizing study by GOLD *et al.* (1971) suggests this possibility.

There is a possibility for a number of errors in the present study. A local variation in density may be caused if the line source expands as it heats, and bends, compressing the soil on one side and creating a void on the other side. The wire is spring loaded to prevent this, but such an action may occur. Some variation in the data is also possible in the plotting of the temperature—log time graphs and finding the slope of the straight portion graphically. The curve is S-shaped with a straight portion in the middle and an error can easily be made in measuring the slope of this linear part.

The most significant systematic error present is caused by using a mathematical model which is imperfectly matched to the physical situation. In the present experiment, sample size constraints were severe and although the method is well adapted for small samples, truncation of the series expression for the exponential integral leads to an error on the order of plus 10% at the most for this experiment. A second error, that due to end loss, is important at longer times. BLACKWELL (1956) recommends a length to diameter ratio of greater than 30. For the present experiment this ratio was 116. However, because of the extremely low conductivities of the lunar vacuum samples, this may not be enough. The error due to contact resistance at the line source surface is expected to be negligible for samples as finely divided as these.

Acknowledgments—We thank CARLA CREMERS and MICHAEL POLAK for their assistance in data reduction and acquisition. Support by NASA under contract NAS 9–8098 is also gratefully acknowledged.

REFERENCES

BERNETT E. C., WOOD H. L., JAFFE L. D., and MARTENS H. E. (1963) Thermal properties of a simulated lunar material in air and vacuum. *Amer. Inst. Aeronaut. Astronaut. J.* **1**, 1402–1407.

BLACKWELL J. H. (1956) The axial flow error in the thermal conductivity probe. *Can. J. Phys.* **34**, 412–419.

BLACKWELL J. H. (1959) A transient method for determining the thermal constants of insulating materials in bulk. *J. Appl. Phys.* **25**, 137–144.

CARSLAW H. S. and JAEGER J. C. (1959) *Conduction of Heat in Solids*, Oxford University Press, London, 334–345.

CREMERS C. J., BIRKEBAK R. C., and DAWSON J. P. (1970) Thermal conductivity of fines from Apollo 11. *Proc. Apollo 11 Lunar Sci. Conf., Geochim. Cosmochim. Acta* Suppl. 1, Vol. 3, pp. 2045–2050. Pergamon.

FOUNTAIN J. A. and WEST E. A. (1970) Thermal conductivity of particulate basalt as a function of density in simulated lunar and martian environments. *J. Geophys. Res.* **75**, 4063–4069.

GOLD T., O'LEARY B. T., and CAMPBELL M. J. (1971) Physical properties of the Apollo 12 lunar fines. Second Lunar Science Conference (unpublished proceedings).

WATSON K. (1964) Thermal conductivity measurements of selected silicate powders in vacuum from 150° to 300°K Part I of Ph.D. dissertation, California Institute of Technology, Pasadena.

WECHSLER A. E. and GLASER P. E. (1964) Thermal conductivity of nonmetallic materials summary report. Contract No. NAS 8–1567, NASA, Huntsville, Ala.

WILDEY R. L. (1967) On the treatment of radiative transfer in the lunar diurnal heat flow. *J. Geophys. Res.* **72**, 4765–4767.

Proceedings of the Second Lunar Science Conference, Vol. 3, pp. 2317–2321
The M.I.T. Press, 1971.

Thermal expansion of lunar rocks

W. Scott Baldridge and Gene Simmons*
Department of Earth and Planetary Sciences, Massachusetts Institute of
Technology, Cambridge, Massachusetts 02139

(*Received* 23 *February* 1971; *accepted in revised form* 23 *March* 1971)

Abstract—The thermal expansion of lunar samples 10020, 10046, 10057, and 12022,95 was measured over the temperature interval −100°C to +200°C using a Brinkmann dilatometer. Measured values of the volume coefficient of thermal expansion were 40% to 70% of those calculated from Turner's equation for the volume coefficient of an aggregate. This discrepancy is attributed to the presence of a large number of micro-cracks in the lunar rocks. A model for the variation of the thermal expansion coefficient with depth is presented.

Introduction

The variation of density with depth in the moon is a function of temperature and pressure for any given composition. To evaluate the effects of temperature on volume as part of a larger program to study the physical properties of returned lunar rocks, we measured the room pressure thermal expansion of four Apollo samples. Our measurements reflect the structural properties, especially microfractures, of the rock aggregate. We have then attempted to infer the behavior of thermal expansion with depth in the outer few kilometers of the moon.

Experimental Techniques

The thermal expansion of lunar samples 10020, 10046, 10057, and 12022,95 was measured over the temperature interval −100°C to +200°C using a Brinkmann model TD IX dilatometer. All the samples were rectangular prisms measuring approximately 1 by 1 by 2 cm. Ends were ground flat and parallel to 0.003 cm. The (linear) thermal expansion was measured along the longest dimension of the samples only, and anisotropy was not investigated. The dilatometer was calibrated with a 2 cm long by 1 cm diameter rod of monocrystalline quartz cut parallel to the c-axis. We used the combined data of Buffington and Latimer (1926) and Kozu and Takane (1929) as the correct values for quartz over our experimental range of temperature. Each data set was recalculated to the relative expansion $\Delta L/L_0$, fitted to a third degree polynomial, and adjusted to a reference temperature of 25°C (Fig. 1).

Results and Discussion

The relative expansion of the lunar samples is presented in Fig. 2. A second degree polynomial was fitted to the data by the method of least squares. The uncertainty in $\Delta L/L_0$ is approximately 1.5×10^{-4}. In Table 1 are tabulated values of the mean volume coefficient α_V of thermal expansion

$$\alpha_V = 3\,\Delta L/[L_0 \cdot \Delta T(°C)]$$

derived from the data in Fig. 2 as well as values for α_V calculated from Turner's

---

* Present address: NASA, Manned Spacecraft Center, Houston, Texas 77058.

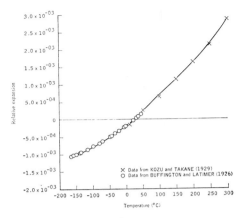

Fig. 1. Relative thermal expansion $\Delta L/L_0$ of quartz parallel to the c-axis.

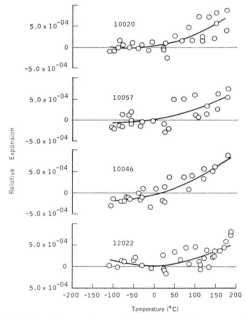

Fig. 2. Relative thermal expansion $\Delta L/L_0$ of lunar samples.

Table 1. Measured and calculated values of $\Delta V/[\Delta T \cdot V_0]$ for lunar rocks $[\times 10^{-6} \ (°C^{-1})]$.

	Measured −100 to 25°C	Measured 25 to 200°C	Calculated 25 to 200°C
10020	0	13	22
10046	−5	15	22
10057	−2	12	22
12022	0	9	$23\frac{1}{2}$

Table 2. Values of thermal expansion and bulk modulus and sources of data used in calculating thermal expansion of lunar rocks

Mineral	Thermal expansion (at 200°C) $\alpha_V \times 10^6$ (°C^{-1})	Source of data
Plagioclase	13.3	Skinner (1966)
Pyroxene	20.0	Skinner (1966)
Olivine	28.3	Skinner (1966)
Hematite	26.9	Skinner (1966)
Glass	29.1	Peters and Cragoe (1920)

Mineral	Bulk Modulus (at 1 atm) Bulk modulus (megabars)	Source of data
Plagioclase	0.68	Simmons and Wang (1971)
Pyroxene	0.95	Simmons and Wang (1971)
Olivine	1.30	Simmons and Wang (1971)
Hematite	2.07	Simmons and Wang (1971)
Glass	0.63	Birch (1966)

equation for the thermal expansion of an aggregate (KINGERY, 1960):

$$\alpha_r = \Sigma \alpha_i K_i V_i / \Sigma K_i V_i$$

where α_r and α_i are the volume coefficients of the aggregate and of the ith phase, respectively, and K_i and V_i are the bulk modulus and volume fraction of the ith phase. The modal analyses of rocks 10020 and 10046 were taken from HORAI et al. (1970), of 10057 from HAGGERTY et al. (1970), and of 12022 from NASA (1970). The values of thermal expansion and bulk modulus as well as the sources of this data are indicated in Table 2. Where minerals are not stoichiometric values of thermal expansion and bulk modulus of compositions appropriate to lunar rocks were chosen. Since the bulk modulus and thermal expansion coefficient of ilmenite have not been measured data for hematite was used in these calculations. Because of the incompleteness of existing data these calculated values can be considered only as a first approximation.

However, it is readily apparent that the measured values of the thermal expansion coefficient are significantly lower (40% to 70%) than the calculated values. This discrepancy is almost certainly due to the presence of a very large number of microcracks in the lunar rocks, resulting in part from the repeated thermal cycling which these rocks have undergone on the moon's surface. Low values for compressional and shear velocities (KANAMORI et al. 1970; ANDERSON et al. 1970; and WANG et al. 1971) and for thermal conductivity at confining pressures less than 100 bars (HORAI et al. 1970) as well as high values for compressibility (STEPHENS and LILLEY, 1971) also indicate that the lunar rocks are extremely cracked, probably more so than terrestrial rocks. Hence it is not unexpected that the coefficient of thermal expansion of lunar rocks should be somewhat lower than that of terrestrial rocks of similar composition. SKINNER (1966) gives a value of α_V of approximately 16×10^{-6} °C^{-1} for terrestrial basalts.

We believe our measurements made at atmospheric pressure are valid at reduced pressure. THIRUMALAI and DEMOU (1970) found that for a quartzite, a granodiorite,

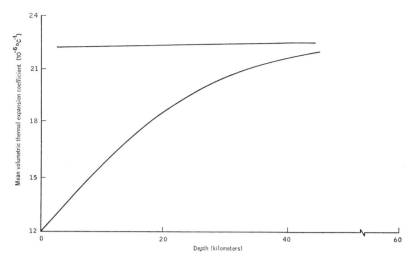

Fig. 3. Predicted dependence of α_V on depth of burial in moon (for rock 10057) based
on model of large number of microcracks closing with pressure.

and an obsidian thermal expansion behavior was independent of reduced environmental pressure down to 10^{-5} Torr.

The thermal expansion of the microbreccia 10046 is slightly larger than for the lunar igneous rocks. McKay *et al.* (1970) suggest that lunar breccia consists of lithic fragments and glass shards sintered together at contacts by microsized glass. If this model is correct the relatively large thermal expansion of glass ($\alpha_V = 29 \times 10^{-6}\,°C^{-1}$ for glass of SiO_2 content of approximately 44%; Peters and Cragoe, 1920) may dominate the thermal expansion of the breccia. Breccia 10046 may be taken as a model for some rocks in the regolith.

Our model for the variation of α_V with depth in the moon is presented in Fig. 3. Kanamori *et al.* (1970) observed that the values of V_p and V_s for lunar rocks reached their intrinsic values at a pressure of about 5 kb (corresponding to a depth of about 90 km), concluding that at this pressure all microcracks had closed. Travel times derived from the impact of the LEM (Apollo 12) and the Saturn IVB (Apollo 13) serve to confirm the low seismic velocities to a depth of at least 30 km (Latham *et al.* 1970, 1971). This behavior is in marked contrast to that of terrestrial rocks, where a pressure of 1 to 2 kb is sufficient to close cracks and hence bring V_p and V_s to their intrinsic values. We postulate a similar behavior for the variation of thermal expansion with pressure, i.e., that α_V should increase from its atmospheric pressure values as a smooth function of depth with closing of microcracks until at a depth of about 90 km α_V reaches its predicted (intrinsic) value of 21 to $23\frac{1}{2} \times 10^{-6}\,(°C^{-1})$.

Because the extreme variations in temperature characteristic of the lunar surface extend no more than a few meters into the moon, some mechanism of forming microcracks other than thermal stress must be sought. Perhaps repeated impacts are sufficient to cause extensive microfracturing to depths of tens of kilometers.

REFERENCES

ANDERSON O. L., SCHOLZ C., SOGA N., WARREN N., and SCHREIBER E. (1970) Elastic properties of a micro-breccia, igneous rock and lunar fines from Apollo 11 mission. *Proc. Apollo 11 Lunar Sci. Conf.*, *Geochim. Cosmochim. Acta* Suppl. 1, Vol. 3, pp. 1959–1973. Pergamon.

BIRCH F. (1966) Compressibility; elastic constants. In *Handbook of Physical Constants* (editor S. Clark), Section 7, pp. 97–173, Geol. Soc. Amer. Mem. 97.

BUFFINGTON R. M. and LATIMER W. M. (1926) The measurement of coefficients of expansion at low temperatures, some thermodynamic applications of expansion data. *J. Amer. Chem. Soc.* **48**, 2305–2319.

HAGGERTY S. E., BOYD F. R., BELL P. M., FINGER L. W., and BRYAN W. B. (1970). Opaque minerals and olivine in lavas and breccias from Mare Tranquillitatis. *Proc. Apollo 11 Lunar Sci. Conf.*, *Geochim. Cosmochim. Acta* Suppl. 1, Vol. 1, pp. 513–538. Pergamon.

HORAI K., SIMMONS G., KANAMORI H., and WONES D. (1970) Thermal diffusivity, conductivity and thermal inertia of Apollo 11 lunar material. *Proc. Apollo 11 Lunar Sci. Conf.*, *Geochim. Cosmochim. Acta* Suppl. 1, Vol. 3, pp. 2243–2249. Pergamon.

KANAMORI H., NUR A., CHUNG D. H., and SIMMONS G. (1970) Elastic wave velocities of lunar samples at high pressure and their geophysical implications. *Proc. Apollo 11 Lunar Sci. Conf.*, *Geochim. Cosmochim. Acta* Suppl. 1, Vol. 3, pp. 2289–2293. Pergamon.

KINGERY W. D. (1960) *Introduction to Ceramics*, p. 478, Wiley.

KOZU S. and TAKANE K. (1929) Influence of temperature on the axial ratio, the interfacial angle and the volume of quartz. *Sci. Rep. Tohoku Univ.* **3** (3rd series) 239–246.

LATHAM G., EWING M., DORMAN J., PRESS F., TOKSOZ N., SUTTON G., MEISSNER R., DUENNEBIER F., NAKAMURA Y., KOVACH R., and YATES M. (1970) Seismic data from man-made impacts on the moon. *Science* **170**, 620–626.

LATHAM G., EWING M., PRESS F., SUTTON G., DORMAN J., NAKEMURA Y., MEISSNER R., TOKSOZ N., DUENNEBIER F., KOVACH R., and LAMMLEIN D. (1971) Results from the Apollo 12 passive seismic experiment. Second Lunar Science Conference (unpublished proceedings).

MCKAY D. S., GREENWOOD W. R., and MORRISON D. A. (1970) Origin of small lunar particles and breccia from the Apollo 11 site. *Proc. Apollo 11 Lunar Sci. Conf.*, *Geochim. Cosmochim. Acta* Suppl. 1, Vol. 1, pp. 673–694. Pergamon.

NASA (1970) Lunar Sample Information Catalog, Apollo 12. Manned Spacecraft Center, Houston, Texas.

PETERS C. G., and CRAGOE C. H. (1920) Measurements on the thermal dilatation of glass at high temperatures. *J. Optical Soc. Amer.* **4**, 105–144.

SKINNER BRIAN J. (1966) Thermal expansion. In *Handbook of Physical Constants* (editor S. Clark), Section 6, pp. 75–96, Geol. Soc. Amer. Mem. 97.

SIMMONS G. and WANG H. (1971) *Single Crystal Elastic Constants and Calculated Aggregate Properties; A Handbook*, 2nd edition, 370 pp., MIT Press.

STEPHENS D. R. and LILLEY S. M. (1971) Pressure-volume properties of two Apollo 12 basalts. Second Lunar Science Conference (unpublished proceedings).

THIRUMALAI K. and DEMOU S. G. (1970) Effect of reduced pressure on thermal-expansion behavior of rocks and its significance to thermal fragmentation. *J. Appl. Phys.* **41**, 5147–5151.

WANG H., TODD T., WEIDNER D., and SIMMONS G. (1971) Elastic properties of Apollo 12 rocks. Second Lunar Science Conference (unpublished proceedings).

Proceedings of the Second Lunar Science Conference, Vol. 3, pp. 2323–2326
The M.I.T. Press, 1971.

Elastic wave velocities of Apollo 12 rocks at high pressures

H. Kanamori

Earthquake Research Institute, University of Tokyo, Tokyo, Japan

and

H. Mizutani and Y. Hamano

Geophysical Institute, University of Tokyo, Tokyo, Japan

(*Received* 20 *February* 1971; *accepted* 24 *March* 1971)

Abstract—New results of P- and S-wave velocity measurements on two Apollo 12 rocks, 12052 and 12065, under pressures up to 10 kbars are presented. These rocks are basalt-like crystalline rocks with a bulk density of about 3.26 g/cm³ and a mean atomic weight of 24.5. Like the Apollo 11 rocks, the velocities and the wave transmission efficiency are surprisingly low at low pressures despite their relatively tight texture; at pressures below 200 bars, Q is estimated to be less than 100. The velocities increase very rapidly with pressure and approach 7.0 km/sec (P wave) and 3.9 km/sec (S wave) towards 10 kbars. No evidence is found for an increase of Q at 1 MHz with a reduction of the ambient pressure to 3×10^{-3} torr.

THIS REPORT presents new results of P- and S-wave velocity measurements on two Apollo 12 crystalline rocks, 12052,35 and 12065,68 under pressures up to 10 kbars at room temperature. The chemical composition of these rocks has been given by LSPET (1970) and Kushiro and Haramura (1971). These two rocks closely resemble one another in composition and are, on the whole, of basaltic composition. The mean atomic weight of these rocks as calculated from the data given by Kushiro and Haramura is 24.5 and is significantly larger than that of ordinary terrestrial basalts.

The measurement method described by Mizutani *et al.* (1970) and the high-pressure system used by Kanamori and Mizutani (1965) are employed. Since the method of Mizutani *et al.* was originally devised for very small samples (several millimeters in dimension), it ensures a high accuracy when applied to samples the size of the Apollo 12 rocks; the approximate dimension of the samples is $1 \times 1 \times 2$ cm³. At pressures above 1 kbar, the accuracy of the present measurement is probably better than 0.7% for P waves, and 1.5% for S waves. At pressures below 200 bars, however, the wave transmission efficiency is so poor (low Q) that the onset of the signal becomes blunt and the accuracy drops considerably.

The results are summarized in Table 1 and Figs. 1 and 2; Figs. 1 and 2 give the original readings, and Table 1 lists the smoothed values. Because the samples have large compressibilities, the correction for the pressure shortening of the sample is estimated. This correction is made according to Cook (1957) but the difference between the isothermal and adiabatic bulk modulus is ignored. In such case the true P- and S-wave velocities $\alpha(P)$ and $\beta(P)$ at a pressure P can be obtained from the

Table 1. Bulk density and velocity (in km/sec) of samples

Sample	Wave	Pressure (kb)								
		0.0	0.2	0.5	1.0	2.0	3.0	5.0	7.0	10.0
12052	P	4.30	4.90	5.55	5.93	6.32	6.55	6.80	6.90	7.01
$\rho = 3.27$ g/cm³*	S	2.59	2.70	2.84	3.03	3.34	3.55	3.74	3.82	3.88
12065	P	3.27	4.44	5.21	5.80	6.24	6.47	6.74	6.86	6.96
$\rho = 3.26$ g/cm³*	S	2.14	2.42	2.73	3.04	3.38	3.54	3.72	3.82	3.86

* No correction is made for the porosity.

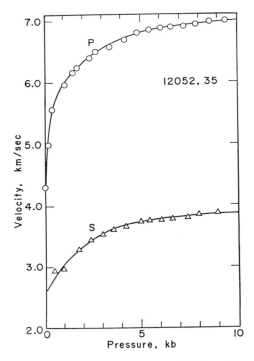

Fig. 1. The P- and S-wave velocities of sample 12052,35 as a function of pressure.

uncorrected P- and S-wave velocities $\alpha'(P)$ and $\beta'(P)$ by

$$\frac{\alpha(P)}{\alpha'(P)} = \frac{\beta(P)}{\beta'(P)} = \left[1 + \frac{1}{3\rho_0}\int_0^P \frac{dP}{(\alpha'(P)^2 - \frac{4}{3}\beta'(P)^2)}\right]^{-1}$$

where ρ_0 is the density at 0 pressure. Numerical integration of $\alpha'(P)$ and $\beta'(P)$ listed in Table 1 leads to a correction of only 0.4% at 10 kbars; this correction is therefore not meaningful in view of other experimental uncertainties. It may be argued that the static compressibility data are more appropriate for this correction than the ultrasonic data. The static compression data on the Apollo 12 rocks reported by Stephens and Lilley (1971) lead to a correction of about 0.8% (at 10 kbars) which is still

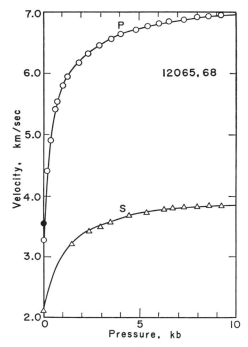

Fig. 2. The P- and S-wave velocities of sample 12065,68 as a function of pressure. The solid circle at 0 pressure indicates the velocity obtained after the pressure run.

insignificant. The densities are measured by the Archimedes method at 0 (atmospheric) pressure, and no correction is made for the porosity.

The overall elastic and anelastic behaviors of the Apollo 12 crystalline rocks are surprisingly similar to those of the Apollo 11 rocks reported by KANAMORI et al. (1970) and SCHREIBER et al. (1970). The rapid increase of the velocity for the initial 2 kbar pressure increase found for the Apollo 11 rocks is also typical of the Apollo 12 rocks. Although the velocity of the Apollo 12 rocks is slightly larger than that of the Apollo 11 rocks, it is still consistent with the travel times obtained by the Apollo 12 seismic experiments (LATHAM et al., 1970), if the vertical velocity gradient beneath the lunar surface is caused by compaction alone. Thus the conclusion that the shallow part (to a depth of about 20 km) of the mare region consists of relatively homogeneous basalt-like material (KANAMORI et al., 1970; LATHAM et al., 1970) seems to be substantiated.

The wave transmission efficiency, at low pressures, of the Apollo 12 samples is surprisingly poor; it is much poorer than would be expected from the apparently tight textures of these samples. The wave transmission efficiency is frequently specified by the quality factor Q, where $2\pi/Q$ is the fractional loss of energy per cycle of oscillation of a vibrating system. Although the value of Q could not be measured accurately, a crude comparison of the amplitude of ultra-sonic waves transmitted through these lunar rocks with those through ordinary terrestrial rocks suggests that the value of Q

cannot be larger than 100 at pressures below 200 bars. This value may be compared with the value $Q \sim 20$ obtained by Wang *et al.* (1971) at a frequency of a few Hz, and with the value on the order of 100 obtained by Warren *et al.* (1971) over a frequency range of 40 to 13 kHz. These values are much smaller than that required to explain the seismic ringing in terms of a diffusive and a dispersion process (Latham *et al.*, 1970). Pandit and Tozer (1970) suggested on an experimental basis that, when the ambient pressure is reduced to 10^{-2} torr, the value of Q in porous terrestrial rocks increases by a factor of 5 over the value measured at 1 atmosphere. In order to see whether the pressure effect on Q is significant or not, we bonded 1 MHz transducers directly on the sample 12065,68, suspended it by a thin wire in a vacuum chamber, and observed the change with pressure of the decay rate of the ultra-sonic reverberation. No significant change of the decay rate, however, was observed over the range from 1 atmosphere to 3×10^{-3} torr. Since this experiment was made on a sample which had been subjected to high confining pressures during the velocity measurements, it is possible that the lossless "welded" contact had been destroyed and that the frictional dissipation became significant. It is also possible that the scattering at the grain boundaries is so large at such a high frequency, 1 MHz, that any pressure effect on the attenuation is masked. In any case, the increase of Q with a reduction of ambient pressure could not be confirmed in our experiment.

References

Cook R. K. (1957) Variation of elastic and static strains with hydrostatic pressure; A method for calculation from ultrasonic measurements. *J. Acoust. Soc. Amer.* **29**, 445–449.

Kanamori H. and Mizutani H. (1965) Ultrasonic measurement of elastic constants of rocks under high pressures. *Bull. Earthquake Res. Inst. Tokyo Univ.* **43**, 173–194.

Kanamori H., Nur A., Chung D. H., Wones D., and Simmons G. (1970) Elastic wave velocities of lunar samples at high pressures and their geophysical implications. *Science* **167**, 726–728.

Kushiro I. and Haramura H. (1971) Major element variation and possible source materials of Apollo 12 crystalline rocks. *Science* **171**, 1235–1237.

Latham G., Ewing M., Dorman J., Press F., Toksoz N., Sutton G., Meissner R., Duennebier F., Nakamura Y., Kovach R., and Yates M. (1970) Seismic data from man-made impacts on the moon. *Science* **170**, 620–626.

LSPET (Lunar Sample Preliminary Examination Team) (1970) Preliminary examination of the lunar samples from Apollo 12. *Science* **167**, 1325–1339.

Mizutani H., Hamano Y., Ida Y., and Akimoto S. (1970) Compressional wave velocities of fayalite, Fe_2SiO_4 spinel, and coesite. *J. Geophys. Res.* **75**, 2741–2747.

Pandit B. I. and Tozer D. C. (1970) Anomalous propagation of elastic energy within the moon. *Nature* **226**, 335.

Schreiber E., Anderson O. L., Soga N., Warren N., and Scholz C. (1970) Sound velocity and compressibility for lunar rocks 17 and 46 and for glass spheres from the lunar soil. *Science* **167**, 732–734.

Stephens D. R. and Lilley E. M. (1971) Pressure-volume properties of two Apollo 12 basalts. Second Lunar Science Conference (unpublished proceedings).

Wang H., Todd T., Weidner D., and Simmons G. (1971) Elastic properties of Apollo 12 rocks. Second Lunar Science Conference (unpublished proceedings).

Warren N., Schreiber E., Scholz C., Morrison J., Kumazawa M., and Anderson O. L. (1971) Elastic and thermal properties of Apollo 12 and Apollo 11 rocks. Second Lunar Science Conference (unpublished proceedings).

Proceedings of the Second Lunar Science Conference, Vol. 3, pp. 2327–2336
The M.I.T. Press, 1971.

Elastic properties of Apollo 12 rocks

Herbert Wang, Terrence Todd, Donald Weidner, and
Gene Simmons*
Department of Earth and Planetary Sciences, Massachusetts Institute of
Technology, Cambridge, Massachusetts 02139

(*Received* 22 *February* 1971; *accepted* 29 *March* 1971)

Abstract—Compressional and shear velocities in samples 12002,54 and 12022,60 were measured in three orthogonal directions by the standard Birch pulse transmission method to 6.5 kb. Both P and S wave velocities doubled over this pressure range and both showed less than 3% anisotropy which is about the experimental error. Average *five kilobar* values are given below:

Sample	ρ (g/cc)	V_p (km/sec)	V_s (km/sec)
12002,54	3.30	7.6	4.0
12022,60	3.32	7.4	3.8

The values of Q determined ultrasonically and in a torsional pendulum assembly at room conditions are somewhat lower than in terrestrial basalts: Q (lunar sample) $\simeq 20$ vs. Q (Fairfax diabase) $\simeq 100$. These velocities and Q are similar to those obtained on Apollo 11 samples. The Q are far lower than those inferred from lunar seismograms ($Q \simeq 3000$). The velocities as a function of pressure give a travel time curve consistent with data obtained from the Apollo 13 SIV-B impact. Current petrologic models indicate that below depths corresponding to a pressure of 10 kb, compositional changes would occur so that the present velocity data are probably applicable to this depth only.

Introduction

THE LABORATORY MEASUREMENT of sound velocities of lunar samples under pressure provides basic data on the structure of the outer tens of kilometers of the moon when it is combined with seismic data. We have measured P and S wave velocities on samples 12002 and 12022 to 6.5 kb by the standard Birch (1960) pulse transmission technique. This pressure would occur at a depth of about 120 km. Maximum seismic ray penetration resulting from the Apollo 13 S-IVB rocket impact was about 40 km (Latham *et al.*, 1970).

We have also measured the Q of samples 12022,60 and 10057 in a torsional pendulum apparatus (Jackson, 1969) at a frequency of a few Hertz. The value obtained for Q is about 20 which is of the same order of magnitude as that obtained ultra-sonically by Kanamori *et al.* (1970) on Apollo 11 sample 10057. Such values of Q are somewhat lower than terrestrial igneous rocks (Q from 50 to 300) but they do not agree with the apparently high values ($Q = 3000$) obtained from the seismograms that resulted from the Apollo 12 LM and Apollo 13 S-IVB impacts. The difference between the subsurface lunar material in situ and in the laboratory must account for the discrepancy if our samples are representative of the subsurface layer.

---

* Now at NASA, Manned Spacecraft Center, Houston, Texas 77058.

Table 1. Modal analyses of samples in volume %. The modal counts for 12002 were made by D. WONES; those for 12022 were taken from the Lunar Sample Information Catalogue (p. 108).

12002		12022	
olivine	10.8	olivine	32.8
pyroxene	58.9	pyroxene	29.9
plagioclase	17.7	plagioclase	25.5
opaques	7.7	ilmenite	9.1
other	4.9	spinel	2.0
		iron	0.5
		troilite	0.2

SAMPLE DESCRIPTION

Velocity measurements were made on four samples—two each from 12002 and 12022. Though samples 12002 and 12022 have very nearly equal densities (3.30 g/cm^3 and 3.32 g/cm^3 as determined by hydrostatic weighings) their mineralogy is quite different (Table 1). In appearance sample 12002 was a lighter gray and more friable than sample 12022. Both looked like typical fine-grained terrestrial basalts.

Sample 12002,54 was irregularly shaped with two flat parallel sides 1.6 cm apart. The other samples were parallelipipeds with approximate dimensions either 1.5 × 1.5 × 2.0 cm or 1.0 × 1.0 × 2.0 cm. Velocities were measured in 3 orthogonal directions of samples 12002,58 and 12022,60 and in the longest direction of 12002,54 and 12022,95.

VELOCITY RESULTS

The method for making the velocity measurement is described by KANAMORI et al. (1970). All the samples were dried for about 4 hours at 80°C in a vacuum oven and jacketed with Sylgard, an electronic encapsulating material to keep the pressure medium, petroleum ether, from penetrating the rock. Coaxially plated, 1 MHz,

Table 2. Velocity (in km/sec) of samples.

Sample		0	100	250	500	750	1000	1500	2000	3000	5000	6500
						Pressure (bars)						
12002,58	P	4.50	5.10	5.65	6.10	6.40	6.55	6.85	7.05	7.30	7.60	7.70
A-direction	S	2.45	2.70	2.90	3.20	3.40	3.55	3.75	3.90	4.00	4.10	4.15
12002,58	P	4.05	4.90	5.50	6.00	6.30	6.55	6.85	7.05	7.30	7.60	7.80
B-direction	S	2.30	2.70	2.95	3.20	3.35	3.50	3.70	3.80	3.95	4.10	4.15
12002,58	P	3.85	4.50	5.05	5.65	6.00	6.30	6.65	6.85	7.10	7.40	7.55
C-direction	S	2.45	2.55	2.70	2.95	3.10	3.20	3.40	3.55	3.75	3.90	4.00
12002,54	P	3.60	4.45	5.05	5.65	6.05	6.25	6.60	6.90	7.20	7.55	7.70
	S	2.10	2.30	2.55	2.85	3.00	3.10	3.25	3.40	3.60	3.85	3.95
12022,60	P	3.55	3.95	4.50	5.25	5.80	6.15	6.60	6.90	7.10	7.50	7.60
A-direction	S	1.70	1.85	2.05	2.30	2.50	2.70	3.00	3.25	3.50	3.85	3.95
12022,60	P	3.60	3.85	4.25	4.80	5.20	5.50	6.15	6.55	7.05	7.35	7.40
B-direction	S	1.55	1.70	1.90	2.20	2.50	2.70	3.05	3.25	3.60	3.90	4.00
12022,60	P	3.55	3.80	4.10	4.65	5.10	5.55	6.20	6.55	6.90	7.30	7.40
C-direction	S	1.75	1.85	2.05	2.35	2.55	2.75	3.05	3.25	3.50	3.75	3.85
12022,95	P	3.65	4.05	4.60	5.20	5.70	6.00	6.40	6.70	7.05	7.35	7.40
	S	2.25	2.35	2.55	2.75	2.90	3.10	3.35	3.50	3.70	3.85	3.90
Fairfax diabase	P	6.10	6.30	6.60	6.90	7.00	7.10	7.15	7.20	7.25	7.35	7.45

Lunar samples have $\rho = 3.3$ g/cm^3. Fairfax has $\rho = 3.2$ g/cm^3.

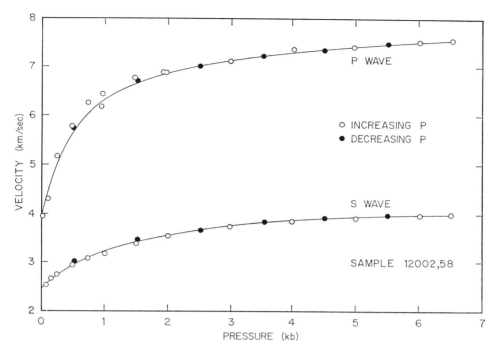

Fig. 1. *P* and *S* wave velocities of sample 12002 as a function of pressure.

barium titanate transducers were used for *P* waves while AC-cut quartz transducers were used for *S* waves (SIMMONS, 1964). The *S* wave arrival was free of *P* noise but its amplitude was small. The transducers were $\frac{3}{8}$ or $\frac{1}{2}$ inch in diameter and were bonded to the sample either with epoxy or Dow Resin 276-V9. The anisotropy in velocity of samples 12002,58 and 12022,60 was less than 3 to 5% which is about the error of measurement. Data are tabulated in Table 2 and typical velocity vs. pressure curves are given in Figs. 1 and 2. As was the case with the Apollo 11 rocks, the velocities nearly double in going from zero pressure to 2 kb. The velocities begin to show intrinsic pressure behavior at about 4 kb. Crack closings presumably are the reason for the initial rapid increase in velocity. This doubling of velocity over a 2 kb range is not typical of earth rocks. As an example we present in Fig. 2 the velocity data for a Fairfax diabase sample which was cut to the approximate dimensions of sample 12022,60 and measured in the same manner. Constant meteoritic bombardment could be responsible for introducing more cracks into lunar rocks.

From the chemical analyses of samples 12002 and 12022 (WILLIS *et al.*, 1971 and LSPET, 1970) we calculated mean atomic weights of $\bar{M} = 23.3$ and 23.8, respectively. However, from BIRCH's (1961) velocity-density plots we would obtain a value of $\bar{M} \simeq 22$ in both cases. This discrepancy is somewhat larger than typical deviations of 0.5 mean atomic weight units, and suggests limitations of estimating composition by this scheme.

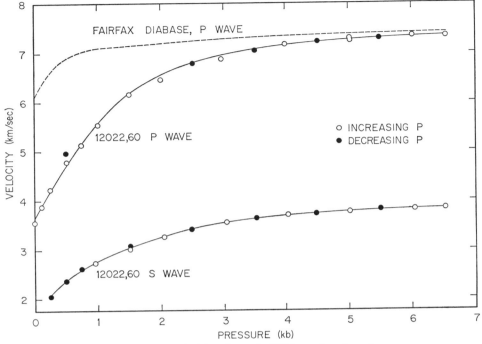

Fig. 2. *P* and *S* wave velocities of sample 12022 as a function of pressure.

From Table 1 we have calculated an estimate of the velocities (shown in Table 4) by a simple volume average of the constituent minerals. Table 3 gives the velocities we chose from our recent compilation (Simmons and Wang, 1971) as reasonable for the composition of olivine, pyroxene, plagioclase, and opaques in the lunar samples. These calculated averages agree fairly well with the observed velocities.

Comparison with Seismic Results

Since our velocity results are quite similar to those obtained by Kanamori *et al.* (1970) and Anderson *et al.* (1970) their discussions as well as that by Latham *et al.* (1970) apply to the present results. We note, however, that the applicability of our

Table 3. Velocities of individual minerals used in Table 4 for volume averages of lunar sample velocities.

Mineral	Mode	Velocity (km/sec)
olivine	P	8.0
	S	4.5
pyroxene	P	7.0
	S	4.1
plagioclase	P	6.8
	S	3.8
opaques	P	6.5
	S	3.5

Table 4. Calculated 5-kb velocities in lunar samples.

Sample	Mode	Velocity (km/sec)
12002	P	7.0
	S	4.0
12022	P	7.2
	S	4.1

pressure data on lunar surface samples to the interpretation of seismograms will probably be limited to rays shallower than about 200 km. Current petrologic models (eg., RINGWOOD, 1970) based upon phase diagrams for lunar basaltic composition show that at about this depth the lunar basalt undergoes a transformation to eclogite. Thus though our samples have densities about 3.30 g/cm³ vs. a mean lunar density of 3.34 g/cm³, the phase studies show that the entire moon cannot remain basaltic and satisfy the mean moment of inertia.

COMPARISON WITH STATIC MEASUREMENTS

STEPHENS and LILLEY (1971) have measured the compressibility by a static strain gauge method on the same two Apollo 12 rocks as we. To ten kilobars they have made the measurements under hydrostatic pressure. The volume compressibility can also

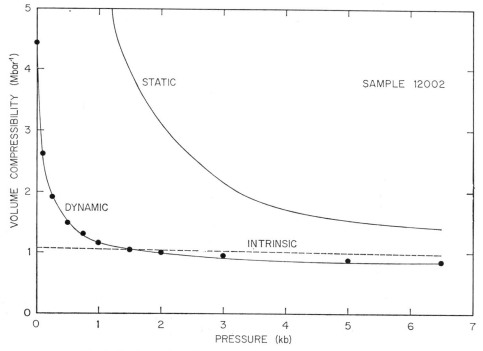

Fig. 3. Static vs. dynamic volume compressibility for sample 12002.

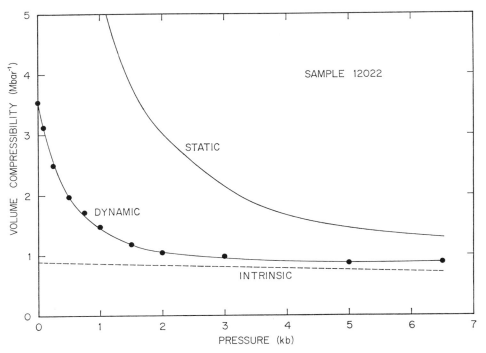

Fig. 4. Static vs. dynamic volume compressibility for sample 12022.

be calculated from our velocity data through the relation

$$\beta = [\rho(V_p{}^2 - (\tfrac{4}{3})V_s{}^2)]^{-1} \tag{1}$$

Figures 3 and 4 show the values obtained by the two methods. The "static" and "intrinsic" curves are from Stephens and Lilley while the "dynamic" curve is from the present velocity results. At a pressure of 5 kb the static compressibility exceeds the dynamic compressibility by 50% compared to the 8% larger value found in Frederick diabase by Simmons and Brace (1965). At lower pressures the difference between static and dynamic values are also much larger than in terrestrial rocks. We interpret this again as evidence for the presence of more microcracks and pores in the lunar samples. As is discussed in more detail by Simmons and Brace, the discrepancy is probably due to a 1-MHz elastic pulse being less affected by cracks than is the strain due to zero frequency static stress.

$$Q$$

In a homogeneous system without dispersion, the quality factor Q is defined by the attenuation of an harmonic wave in time or distance (see, for example, Knopoff, 1964). If the amplitude of a signal of frequency f is A_0 at time $t = 0$, the amplitude at time t is given by

$$A = A_0 \exp\left(-\pi f t / Q\right). \tag{2}$$

The value of Q may be estimated ultrasonically by comparing the amplitudes of the received signals through two different materials. By comparing sample 10057 with steel, KANAMORI *et al.* (1970) obtained a value of $Q = 10$ for the lunar sample at $P = 200$ bars. We have compared our Apollo 12 samples with Fairfax diabase (Fig. 5) which we consider to be an igneous terrestrial analogue in terms of texture, density, and 5-kb velocity. The amplitude ratio of 12022 to Fairfax diabase at $P = 0$ bars for the first half cycle received is about $\frac{1}{4}$. Since the travel times and geometries of the samples and the initial pulse are the same, we can use equation (2) to convert this amplitude ratio of $\frac{1}{4}$ to $Q_{12022} = 15$ by taking $Q_{\mathrm{Fairfax}} = 100$. Note that the length of the two wave trains are similar. However, many variables other than intrinsic attenuation, such as transducer bonding and electrical connections, affect the amplitude of the received signal.

We then measured the Q of two lunar samples and a number of terrestrial igneous rocks in a torsional pendulum apparatus specifically designed for attenuation measurements (JACKSON, 1969) in the Hertz frequency range (as compared to the MHz

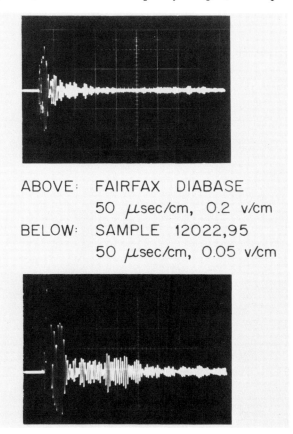

Fig. 5. Oscilloscope traces of the received elastic signal at 1 MHz for Fairfax diabase and sample 12022.

ultrasonic frequency region). A schematic of the apparatus is shown in Figs. 6 and 7. Each sample, in the shape of a cylinder or rectangular parallelipiped, was surface ground so that the long ends were parallel. These ends were then epoxied to the metal bars as shown in Fig. 6 and dried for at least two hours in a vacuum oven at 100°C. After cooling in vacuum the Q was determined at room temperature and pressure.

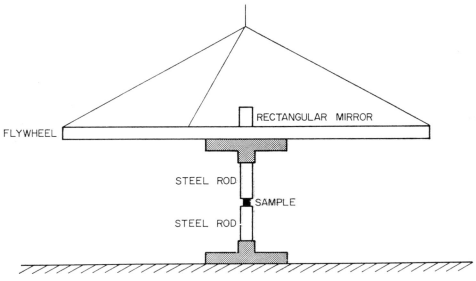

Fig. 6. Torsional pendulum assembly for determining Q.

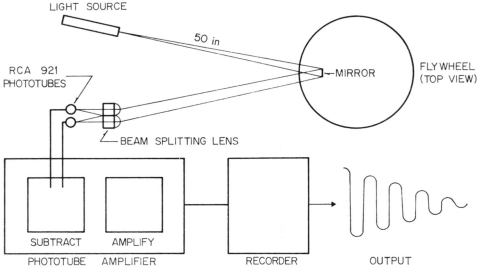

Fig. 7. Schematic of electro-optical system for measuring rotational displacement of flywheel.

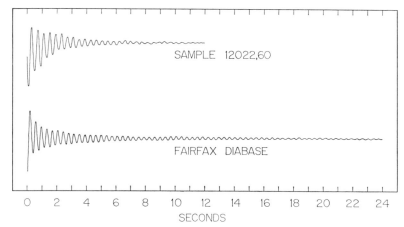

Fig. 8. Typical output records for Q determinations on Fairfax diabase and sample 12022.

By measuring the damping of angular oscillations initiated across the sample by a large fly wheel, the Q of the sample can be calculated from equation (2).

The different geometries of the samples gave no systematic change in Q values. WARREN et al. (1971) found no difference in Q determinations made in nitrogen at atmospheric pressure and in vacuum. They also show that a 100% increase in the humidity only decreases the Q by a factor of 2. BORN (1941) shows that a small amount of water in a sample is of little importance in the low frequency range at which we are working. These facts indicate that our results would be the same if we measured them in vacuum and on completely dry samples.

Typical records for sample 12022,60 and Fairfax diabase are shown in Fig. 8. The values $Q = 17$ for sample 12022,60, $Q = 35$ for sample 10057, and $Q = 100$ for the Fairfax diabase were obtained. The Q for other terrestrial samples were between 50 and 250. The low Q for the lunar samples implies that a given input signal is damped more in the lunar samples than in the terrestrial igneous rock samples. One possible source for this increase is frictional damping on a larger number of micro-cracks. Thus, these observations are consistent with our hypothesis that more micro-cracks are present in the lunar samples than in terrestrial rocks.

The Q of the lunar samples is not 3000 under laboratory conditions. Values of 3000 are typical of glasses and metals—not earth rocks nor lunar rocks in earth atmosphere. Our measurements indicate that the Q is between 10 and 50 for the lunar samples from ultrasonic amplitudes and torsional damping. If the seismically determined values of Q are correct, then we believe the subsurface lunar environment to be responsible for the difference.

Acknowledgments—We wish to thank PETER MCFARLIN for surface grinding the samples, DAVID WONES for determining the mineralogical composition of sample 12002, and DAVID JACKSON for suggesting that we measure the Q of the lunar samples on the torsional pendulum apparatus. This work was supported by NASA contract NAS 9-8102.

REFERENCES

ANDERSON O. L., SCHOLZ C., SOGA N., WARREN N., and SCHREIBER E. (1970) Elastic properties of a micro-breccia, igneous rock and lunar fines from Apollo 11 mission. *Proc. Apollo 11 Lunar Sci. Conf., Geochim. Cosmochim. Acta* Suppl. 1, Vol. 3, pp. 1959–1973. Pergamon.

BIRCH F. (1960) The velocity of compressional waves in rocks to 10 kilobars, Part 1. *J. Geophys. Res.* **65**, 1083–1102.

BIRCH F. (1961) The velocity of compressional waves in rocks to 10 kilobars, Part 2. *J. Geophys. Res.* **66**, 2199–2224.

BORN W. T. (1941) The attenuation constant of earth materials. *Geophysics* **6**, 132–148.

JACKSON D. D. (1969) Grain boundary relaxation and the attenuation of seismic waves. Sc.D. Thesis, Massachusetts Institute of Technology.

KANAMORI H., NUR A., CHUNG D., and SIMMONS G. (1970) Elastic wave velocities of lunar samples at high pressures and their geophysical implications. *Proc. Apollo 11 Lunar Sci. Conf., Geochim. Cosmochim. Acta* Suppl. 1, Vol. 3, pp. 2289–2293. Pergamon.

KNOPOFF L. (1964) *Q. Rev. Geophys.* **2**, 625–660.

LATHAM G., EWING M., DORMAN J., PRESS F., TOKSOZ N., SUTTON G., MEISSNER R., DUENNEBIER F., NAKAMURA Y., KOVACH R., and YATES M. (1970) Seismic data from man-made impacts on the moon. *Science* **170**, 620–626.

LSPET (Lunar Sample Preliminary Examination Team) (1970) Preliminary examination of the lunar samples from Apollo 12. *Science* **167**, 1325–1339.

NASA (1970) *Lunar Sample Information Catalogue, Apollo 12.* Manned Spacecraft Center, Houston.

RINGWOOD A. E. (1970) Petrogenesis of Apollo 11 basalts and implications for lunar origin. *J. Geophys. Res.* **75**, 6453–6479.

SIMMONS G. (1964) Velocity of shear waves in rocks to 10 kilobars, 1. *J. Geophys. Res.* **69**, 1123–1130.

SIMMONS G. and BRACE W. F. (1965) Comparison of static and dynamic measurements of compressibility of rocks. *J. Geophys. Res.* **70**, 5649–5656.

SIMMONS G. and WANG H. (1971) *Single Crystal Elastic Constants and Calculated Aggregate Properties: A Handbook*, 2nd edition, 370 pp. MIT Press.

STEPHENS D. R. and LILLEY E. M. (1971) Pressure-volume properties of two Apollo 12 basalts. Second Lunar Science Conference (unpublished proceedings).

WARREN N., SCHREIBER E., SCHOLZ C., MORRISON J., KUMAZAWA M., and ANDERSON O. L. (1971) Elastic and thermal properties of Apollo 12 and Apollo 11 rocks. Second Lunar Science Conference (unpublished proceedings).

WILLIS J. P., AHRENS L. H., DANCHIN R. V., ERLANK A. J., GURNEY J. J., HOFMEYR P. K., McCARTHY T. S., and ORREN M. J. (1971) Some inter-element relationships between lunar rocks, fines and stony meteorites. Second Lunar Science Conference (unpublished proceedings).

Proceedings of the Second Lunar Science Conference, Vol. 3, pp. 2337–2343
The M.I.T. Press, 1971.

Surface elastic wave propagation studies in lunar rocks

B. R. TITTMANN and R. M. HOUSLEY

North American Rockwell Science Center, Thousand Oaks, California 91360

(*Received* 22 *February* 1971; *accepted* 29 *March* 1971)

Abstract—Elastic surface wave measurements on rock 12038,47 are reported. The absolute surface wave velocity was measured for different directions on the rock surface and was found to vary from 0.97 to 1.45 × 10⁵ cm/sec. Substantial relative changes in surface wave amplitude and velocity were observed when the absolute pressure of air was changed from 1 atmosphere to 6 × 10⁻⁷ mm of Hg. The amplitude increased by a total of about 25%, the velocity increased by about 3% as the pressure was reduced, with most of the change occurring between 1 and 10 mm of Hg. A small but significant part of the changes also occurred reproducibly between 6 × 10⁻⁷ and 1 × 10⁻³ mm of Hg.

INTRODUCTION

ANALYSIS OF LUNAR SEISMIC data is expected to yield valuable information about the subsurface structure of the moon (LATHAM *et al.*, 1970a). So far this expectation has proved to be an elusive goal since the lunar results do not even resemble familiar terrestrial seismograms. Although a number of different types of events have been noted in the more than 200 natural lunar seismic events so far recorded (LATHAM *et al.*, 1971) the most effort has been devoted (LATHAM *et al.*, 1970b) to attempts to understand the simple L-type events, which are thought to be produced by meteoritic impacts. The slow buildup and decay of all components of displacement in this type of event in a largely uncorrelated way strongly indicates (1) that the seismic energy is transferred in diffusive manner due to multiple scattering and (2) that the *Q* of the lunar material must be very high (LATHAM *et al.*, 1970a).

The necessity to further postulate a grossly inhomogeneous structure to depths of several kilometers in order to provide the necessary scattering centers can be avoided if a mechanism can be found which will keep a sizeable fraction of the seismic energy concentrated near the surface. Three mechanisms have so far been suggested. The first and, in our opinion, most attractive possibility was suggested by KANAMORI *et al.* (1970) on the basis of their measurements of the elastic properties of two Apollo 11 lunar basalts. They found that the lunar rocks have unusually low sound velocities at atmospheric pressure which increase rapidly with applied pressure and approach the values expected for rocks of their composition at about 5 kilobars. They pointed out that if the rocks under the lunar mare surfaces behave in this way it could create an effective "wave guide," keeping seismic energy concentrated near the surface. The other Apollo 11 basalt (ANDERSON *et al.*, 1970) measured so far shows qualitatively similar behavior with low elastic wave velocities which increase rapidly with pressure.

STEG and KLEMENS (1970) showed that mass anomalies, shallow compared to the wavelength, scatter surface waves largely into other surface waves and hence suggested that the observed seismograms might be interpreted as resulting from multiply scattered surface waves. The amount of scattering predicted for any given distribution of mass anomalies depends on the fifth power of the frequency, and it appears that it would be orders of magnitude too small at the dominant frequency of about 1.5 Hz (LATHAM et al., 1970b) actually observed in the lunar seismograms.

GOLD and SOTER (1970) have analyzed a ray-optics model of seismic propagation in a medium in which elastic velocities increase linearly with depth. They showed that if waves refracted back to the surface are assumed to be reflected in a somewhat random manner by gentle undulations, the general character of the lunar L-type seismic events can be matched. GOLD and SOTER (1970) assumed the subsurface material to be dust in which the velocity change resulted from increasing compaction with depth. WARREN et al. (1971) have pointed out that the actual velocity gradient obtained from high pressure experiments on lunar dust is much smaller than the one required by GOLD and SOTER (1970) to get a semiquantitative fit to the lunar seismograms and is in fact smaller than the gradient determined for the solid rocks.

It is clear that an unambiguous interpretation of the lunar seismic data will depend on a good understanding of the unusual elastic properties of the lunar rocks. The low velocities at atmospheric pressure have been tentatively ascribed to the presence of an anomalous density of microcracks in the rocks which close with increasing pressure (KANAMORI et al., 1970; ANDERSON et al., 1970). Assuming this explanation is correct it is important to establish whether the high microcrack density is intrinsic to rocks of this composition solidifying under lunar vacuum conditions or whether it is the result of secondary damage caused for example by shock or thermal cycling suffered at the lunar surface. One way to shed light on this question is to study the elastic properties of a fairly large number of lunar rocks and look for correlations of the elastic properties with modal composition and texture.

There is another interesting problem surrounding the lunar rocks. The interpretation of the seismic data seems to require a medium with a Q value of 3000–5000. The Q values so far measured on lunar rocks on earth range from 10–100 (KANAMORI et al., 1971; WARREN et al., 1971). In an attempt to understand this difference, Q measurements of the lunar rocks under high vacuum conditions and at temperatures similar to the subsurface lunar temperature seem important. The results reported here represent preliminary data obtained in a program centered around the above problems.

EXPERIMENTAL RESULTS

The elastic surface wave measurements were performed on rock chip 12038,47 mounted on a sample holder with low melting point wax. This sample weighing about 0.9 g had a smooth saw-cut plane on one side, which we dry polished with SiC and Al_2O_3 papers, obtaining a working surface about 13 mm long by 10 mm wide with a thickness $h \approx 4$ mm over two-thirds of the specimen decreasing to a minimum of 2 mm in some spots. The impulse technique (TITTMANN et al., 1971) with 0.1 μsec wide transmitter pulses was used to measure the absolute surface wave velocity by translating the receiver with respect to the source and recording the change in arrival time. The size of the source and receiver transducers (1 mm in diameter) and the avoidance of proximity effects from the sample edges gave a useful measuring range of about 3 mm to 5 mm of travel, with source and receiver a minimum of 3.5 mm apart. The signals observed were comparable to those we saw in fine-grained terrestrial basalts.

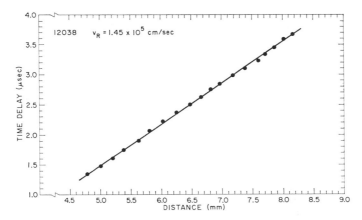

Fig. 1. Change in signal time delay as a function of the transducer separation for rock 12038,47 in the A direction (see Fig. 2). The reciprocal of the slope gives the elastic surface wave velocity v_R.

The data points were taken about 0.2 mm apart with good run-to-run reproducibility (see Fig. 1). Spectrum analysis of the received signal gave a maximum at a frequency of about 2 Mhz. This corresponds to a $h/\lambda \approx 5$, where λ is the wavelength. Thus any dispersion because of the wax and sample holder was negligible.

Absolute velocity data for different directions on rock surface

Figure 2 shows a mosaic of several micrographs taken of the sample working area. Also shown are solid lines indicating the direction and approximate location of the paths of transducer motion during various runs. The dashed line outlines a region where no meaningful measurements could be made because of the presence of large vugs. The Rayleigh wave velocity obtained for each path from data such as shown in Fig. 1 is shown at the side of the photograph. Data obtained from runs along parallel path *directions* but different locations on the rock gave roughly the same velocities. In contrast, a change in direction gave a noticeable change in velocity with a maximum in the B direction and a minimum in the roughly orthogonal C direction with intermediate values observed in intermediate directions. This observation suggests the possible presence of a systematic orientation dependence of the velocity and therefore an anisotropy in the elastic behavior of this rock.

Relative velocity and amplitude data down to 6×10^{-7} mm of Hg

Additional experiments with the sample and testing fixture in a high vacuum chamber were made down to 6×10^{-7} mm of Hg. The measurements were performed along the A direction (see Fig. 2) with a fixed transducer separation of about 6.5 mm. The amplitude and delay time of the received Rayleigh wave signal were monitored as the pressure around the sample was slowly changed from atmospheric to 6×10^{-7} mm of Hg at room temperature and back again. As seen in Fig. 3 the surface wave

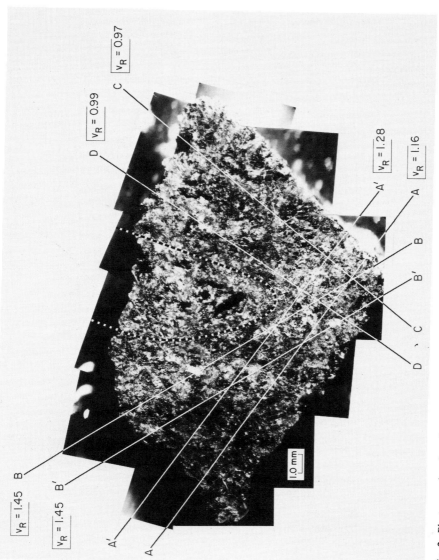

Fig. 2. Photomosaic of surface of rock 12038,47 consisting of several assembled micrographs. The solid lines are directions of transducer movement during measurements. The dashed line surrounds a region not useful for measurements because of the presence of vugs. At one end of each line is stated the measured elastic surface wave velocity v_R (10^5 cm/sec).

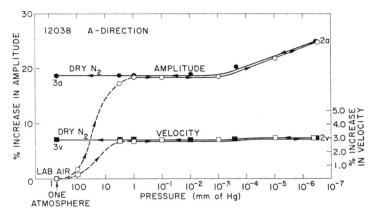

Fig. 3. Percent change in surface wave amplitude and velocity on different scales vs. absolute gas pressure of environment of rock 12038,47. The dashed lines indicate the part (about 60 minutes long) of the cycle in which the vacuum chamber is pumped down after approximately 100 minutes exposure to laboratory air at nominally 40 ± 5% relative humidity. Point 1 at atmospheric pressure of laboratory air marks start of typical test run with signal amplitude and time delay being observed. Points 2a and 2v give % changes in amplitude and velocity relative to the values at point 1 at 6 × 10⁻⁷ mm of Hg (after 8 hours). Points 3a and 3v give % changes of amplitude and velocity after gradual (about 1 hour) pressurization with dry nitrogen.

amplitude and therefore the quality factor Q as well as the surface wave velocity increased as the pressure was reduced. In tests where the sample was previously outgassed then pressurized with dry nitrogen gas, the predominant change in Q and v_R took place between 10^{-3} mm and 10^{-7} mm of Hg. The amplitude increased by 5% and the velocity by 0.2% as the pressure decreased in this range. On the other hand, when the cycling was performed such that the sample was exposed to laboratory air, an additional pronounced change in Q and v_R took place between atmospheric pressure and 1 mm of Hg. The amplitude increased 20%, and the velocity increased 2.8% as the pressure was reduced from 760 mm to 1 mm of Hg. This result is similar to the results obtained in bulk wave experiments by WARREN *et al.* (1970) who reported measurements from 760 to 10^{-2} mm of Hg and suggested that the pronounced change in Q on exposure to air, not present in dry N_2 cycling, may be due to adsorption of water. The relation between the attenuation of surface and bulk waves has been treated by PRESS and HEALY (1957) and VIKTOROV (1964), who show that for most materials the Rayleigh wave attenuation factor is usually within about 20% of the transverse wave attenuation factor.

DISCUSSION

Table 1 shows a compilation of calculated and measured Rayleigh wave velocities on various rocks. The calculations were made from reported bulk wave data under the dubious assumption that the rocks are isotropic and homogeneous. Immediately

Table 1. Calculated Poisson's ratio σ and Rayleigh wave velocity v_R (10^5 cm/sec) based on reported densities ρ (gm/cm^3) and compressional (P) and shear (S) wave velocities (10^5 cm/sec) compared with measured v_R on 12038.

Sample	Type	Obs. ρ	Obs. P	Obs. S	Calc. σ	v_R	Reference
10017	U	3.10	1.84	1.05	0.258	0.967	ANDERSON et al. (1970)
		3.10	1.85	1.03	0.275	0 952	
10020	PB	3.18	3.50	2.20	0.173	1.99	KANAMORI et al. (1970)
10057	IB	3.38	2.82	1.70	0.216	1.55	KANAMORI et al. (1970)
12002	PB	3.30	3.60	2.10	0.242	1.93	WANG et al. (1971)
12018	PB	3.20	1.79	1.07	0.236	0.981	WARREN et al. (1971)
		3.20	2.04	1.17	0.255	1.08	
		3.20	1.81	1.10	0.210	1.00	
12038	GB					1.45*	
						1.45*	
						1.28*	
						1.16*	
						0.99*	
						0.97*	
12052	PB	3.27	4.30	2.59	0.215	2.36	KANAMORI et al. (1971)
12063	GB	3.10	2.93	1.71	0.241	1.57	WARREN et al. (1971)
		3.10	3.32	1.80	0.261	1.64	
		3.10	2.67	1.45	0.291	1.34	
		3.10	2.51	1.37	0.212	1.27	
12065	PB	3.26	3.27	2.14	0.125	1.92	KANAMORI et al. (1971)

* Our present measured values for v_R. The other values calculated from bulk wave data under the dubious assumption that the medium is isotropic and homogeneous.

The identification for the various basaltic rocks is according to WARNER and ANDERSON (1971): PB—porphyritic, GB—granular and ophitic, IB—intersertal, and U—unclassified basalt. Where more than one set of data appear for the same rock, the direction and/or location was altered between measurements.

apparent is the wide range of calculated values for v_R from 0.95 to 2.36 $\times$ 10^5 cm/sec with the measured velocities grouped in the low range of values. Noticeable is also the spread in the bulk wave data (and therefore in the calculated v_R) for the *same* rock when direction and/or location was altered between measurements. The experimental results discussed in the previous section actually suggest an orientational dependence of the velocity. For definitive conclusions, further measurements on different surfaces of a large specimen are required. On such a sample, the anisotropy could be tested on many different locations in order to show conclusively whether real anisotropy or only inhomogeneity is being observed. If the anisotropy is indeed demonstrated as real, this could account in part for the spread in the elastic constants reported thus far in lunar rocks.

In addition to the substantial change in relative amplitude and velocity in the range from 1 mm to 760 mm of Hg, there is a noticeable change between 10^{-3} to 10^{-7} mm of Hg. This latter change is significant in that it represents an increase in Q under conditions approaching more nearly the lunar vacuum.

Acknowledgment—We are very grateful to G. V. LATHAM for suggesting the importance of measurements in high vacuum. We are also grateful to M. ABDEL GAWAD for the petrographic examinations and thank G. A. ALERS, M. BLANDER, E. A. KRAUT, and T. C. LIM for many helpful discussions and suggestions. This work was partially supported by NASA contract NAS 9-10208

REFERENCES

ANDERSON O. L., SCHOLZ C., SOGA N., WARREN N., and SCHREIBER E. (1970) Elastic properties of a micro-breccia, igneous rock and lunar fines from Apollo 11 mission. *Proc. Apollo 11 Lunar Sci. Conf., Geochim. Cosmochim. Acta* Suppl. 1, Vol. 3, pp. 1959–1973. Pergamon.

GOLD T. and SOTER S. (1970) Apollo 12 seismic signal: Indication of a deep layer of powder. Science **169**, 1071–1075.

KANOMORI H., NUR A., CHUNG D. H., and SIMMONS G. (1970) Elastic wave velocities of lunar samples at high pressures and their geophysical implications. *Proc. Apollo 11 Lunar Sci. Conf., Geochim. Cosmochim. Acta* Suppl. 1, Vol. 3, pp. 2289–2293. Pergamon.

KANAMORI H., MIZUTANI H., and HAMANO Y. (1971) Elastic wave velocities of Apollo 12 rocks at high pressure. Second Lunar Science Conference (unpublished proceedings).

LATHAM G. V., EWING M., PRESS F., SUTTON G., DORMAN J., NAKAMURA Y., TOKSOZ N., WIGGINS R., DERR J., and DUENNEBIER F. (1970a) Apollo 11 passive seismic experiment. *Proc. Apollo 11 Lunar Sci. Conf., Geochim. Cosmochim. Acta* Suppl. 1, Vol. 3, pp. 2309–2320. Pergamon.

LATHAM G. V., EWING M., DORMAN J., PRESS F., TOKSOZ N., SUTTON G., MEISSNER ROLF, DUENNEBIER F., NAKAMURA Y., KOVACH R., and YATES M. (1970). Seismic data from man-made impacts on the moon. *Science* **170**, 620–626.

PRESS F. and HEALY I. (1957) Absorption of Rayleigh waves in low-loss media. *J. Appl. Phys.* **28**(11), 1323–1325.

STEG R. C. and KLEMENS P. G. (1970) Scattering of Rayleigh waves by surface irregularities. *Phys. Rev. Lett.* **24**, 381–383.

TITTMANN B. R., KRAUT E. A., and LIM T. C. (1971) Direct measurement of surface wave velocity and time delay temperature variation using a pulse technique. *Proc. Letters IEEE* **59**, 86–88.

VIKTOROV I. A. (1964) Damping of surface and spatial ultrasonic waves. *Akust. Zh.* **10**(1), 116–118.

WANG H., TODD T., WEIDNER D., and SIMMONS G. (1971) Elastic properties of Apollo 12 rocks. Second Lunar Science Conference (unpublished proceedings).

WARNER J. L. and ANDERSON D. H. (1971) Lunar crystalline rocks: Petrology, geology and origin. Second Lunar Science Conference (unpublished proceedings).

WARREN N., SCHREIBER E., SCHOLZ C., MORRISON J., KUMAZAWA M., and ANDERSON O. L. (1971) Elastic and thermal properties of Apollo 12 and Apollo 11 rocks. Second Lunar Science Conference (unpublished proceedings).

Proceedings of the Second Lunar Science Conference, Vol. 3, pp. 2345–2360
The M.I.T. Press, 1971.

Elastic and thermal properties of Apollo 11 and Apollo 12 rocks*

N. Warren, E. Schreiber,† and C. Scholz

Lamont-Doherty Geological Observatory of Columbia University,
Palisades, New York 10964

J. A. Morrison and P. R. Norton‡

Institute for Material Research, McMaster University, Hamilton, Ontario, Canada

and

M. Kumazawa§ and O. L. Anderson

Lamont-Doherty Geological Observatory of Columbia University,
Palisades, New York 10964

(Received 22 February 1971; accepted in revised form 31 March 1971)

Abstract—Low temperature (2° to 5°K) specific heat and thermal conductivity measurements were made on igneous rock sample 10017 and microbreccia sample 10046. Measurements of $1/Q$ by resonance methods were made as a function of pressure and temperature on rocks 12063 and 12038. Velocities v_p and v_s were measured on 12063 and 12018 at room pressure and temperature. The main findings to date are that density and average elastic properties of these lunar rocks are similar to those of the Apollo 11 rocks. The bulk densities are about 3.1 to 3.2 gm/cm³ and compressional velocities range from 2 to 3.5 km/sec. Velocity v_s ranges from about 1.0 to 1.7 km/sec. Low temperature heat capacities of both 10017 and 10046 are anomalous. The values of C_v are over 100 times greater than predicted from the acoustic wave data on these rocks. Q was found to be on the order of 100, with a weak temperature dependence and a strong dependence on total pressure and partial pressure of water vapor from 25° to 125°C. A discussion of lunar Q and the problem of lunar seismic signal character is made based on these experimental and other data.

Introduction

In this paper, elastic, thermal, and Q measurements are presented which were made on Apollo 11 and Apollo 12 returned lunar rocks 10017, 10046, 12063, 12038, and 12018.

Elastic and thermal properties of returned lunar samples are anomalous with respect to values of these properties expected for terrestrial type materials. Unexpectedly low velocities (v_p and v_s) were observed for Apollo 11 rocks (Schreiber et al., 1970, Kanamori et al., 1970). At room P, T conditions velocity density data fall well below the Birch curve for terrestrial materials (Schreiber and Anderson, 1970). This same trend is observed in the recent measurements on Apollo 12 samples presented here.

The heat capacity measured at 2 to 5°K on two returned samples do not agree with simple Debye theory, i.e., with values predicted from acoustic velocity data.

---

* Lamont-Doherty Geological Observatory no. 1655.

† Queens College, C.U.N.Y. Flushing, New York 11367.

‡ Present address: Atomic Energy of Canada, Ltd., Chalk River, Ontario.

§ Present address: Nagoya University, Chikusa, Japan.

Furthermore, as is known, Q deduced from lunar seismic data seems to be anomalously high, a fact which seems inconsistent with the observed low velocities for the returned lunar samples. A discussion of this seismic problem is given.

Thermal Properties

Heat capacity measurements and estimates of thermal conductivity have been made on portions of lunar rocks 10017 and 10046 in the temperature region 2 to 5°K. A discussion of the calorimeter assembly used, and its calibration, is given in Morrison and Norton (1970). Measurements were made originally in hopes of providing a check on our estimates of the intrinsic velocities of the lunar rocks (Schreiber *et al.*, 1970).

Heat capacity

The derived specific heats for the two lunar rocks are given with their estimated accuracies in the middle column of Table 1. The results for the rock 10017 are more extensive and more accurate because the equilibrium time after heating was much shorter ($< \frac{1}{2}$ min compared with 10–15 min for rock 10046). The difference in the behavior of the two rocks is consistent with their bulk structures. From the Debye theory of solids (see, e.g., Blackman (1955)), heat capacity at low temperature and acoustic velocity are related by

$$C_v = \tfrac{16}{15} k \pi^5 V \left(\frac{k}{h}\right)^3 \left(\frac{2}{v_s{}^3} + \frac{1}{v_p{}^3}\right) \times T^3 \tag{1}$$

where V is volume and k and h are Boltzmann's and Planck's constants, respectively. Using (1) for both specimens, the heat capacity is over 100 times larger than is predicted from acoustic wave velocities measured in the same rocks.

In the last column of Table 1 we list the quantity C/T^3 that will be used in discussion and interpretation of the data. To estimate C/T^3 at $T \to 0°$K from wave velocities

Table 1. Measured specific heats.

T (°K)	$10^4 \times C$ (cal/g-deg)	$10^5 \times C/T^3$ (cal/g-deg⁴)	T (°K)	$10^4 \times C$ (cal/g-deg)	$10^5 \times C/T^3$ (cal/g-deg⁴)
Rock 10017 (weight of specimen—13.26 g)			*Rock 10046* (weight of specimen—8.63 g)		
2.344	4.62 ± 0.11	3.59 ± 0.09	3.08$_3$	2.8 ± 0.5	0.95 ± 0.17
2.393	4.66 ± 0.03	3.40 ± 0.02	3.26$_8$	4.2 ± 0.4	1.2 ± 0.1
2.472	4.59 ± 0.18	3.04 ± 0.12	3.32$_7$	$3.5\{^{+\,0.6}_{-\,0.5}$	0.95 ± 0.14
2.713	4.83 ± 0.04	2.42 ± 0.02			
2.876	4.86 ± 0.03	2.04 ± 0.01	3.54$_8$	$4.5\{^{+\,1.2}_{-\,0.8}$	$1.0\{^{+\,0.27}_{-\,0.20}$
3.028	4.98 ± 0.05	1.79 ± 0.02			
3.399	5.29 ± 0.05	1.35 ± 0.01	3.71$_7$	3.8 ± 0.5	0.74 ± 0.10
3.483	5.31 ± 0.04	1.26 ± 0.01	4.05$_0$	$2.4\{^{+\,0.5}_{-\,0.3}$	$0.95\{^{+\,0.20}_{-\,0.12}$
3.819	5.59 ± 0.06	1.00 ± 0.01			
4.036	5.72 ± 0.05	0.87 ± 0.01			
4.27$_0$	6.08 ± 0.15	0.78 ± 0.02			
4.43$_3$	6.1$_8$ ± 0.15	0.71 ± 0.02			
4.52$_5$	6.4$_9$ ± 0.07	0.70 ± 0.01			
4.53$_7$	6.5$_6$ ± 0.07	0.70 ± 0.01			
4.54$_6$	6.3$_3$ ± 0.15	0.67 ± 0.02			
4.70$_8$	5.9$_6$ ± 0.15	0.57 ± 0.02			
4.97$_5$	6.7$_1$ ± 0.15	0.54 ± 0.02			

we should, strictly speaking, allow for the temperature dependence of v_s and v_p. Since this is not likely to be more than a few % (ANDERSON 1963), however, we substitute the values obtained by SCHREIBER et al. (1970) for the glass spheres at room temperature (in particular, $v_p = 7$ km/sec and $v_s = 4$ km/sec) into equation (1) and so obtain $C/T^3 = 1.11 \times 10^{-7}$ cal/g deg^4. Inspection of the last column of Table 1 shows that at the lowest temperatures reached, C/T^3 from the calorimetric data is larger by 2 orders of magnitude at least. It is evident, therefore, that there must be a large extra contribution to the heat capacities of the lunar rocks at these low temperatures.

It is interesting to note that similar results to 10017 have recently been obtained by Morrison on Palisades diabase (to be published). A more detailed discussion of possible origins of these extra contributions is given in MORRISON and NORTON (1970). However, here we note that the most likely origin is that of extra vibrational modes at very low frequency. Other possible sources such as occluded helium (FRIEDMAN et al., 1970; FUNKHOUSER et al., 1970), do not provide sufficient contribution to the specific heat. The occluded helium could contribute no more than 0.4% to the specific heat of rock 10046 and much less to that of rock 10017.

A schottky-type heat capacity anomaly (FOWLER and GUGGENHEIM, 1949) is not expected since from Table 1, C_v of 10017 increases by about 50% for a two-fold change in temperature. This rate is considerably less than predicated for the simplest possible system of two energy levels.

Assuming additional frequency modes, however, the experimental results could be accounted for quantitatively by assuming that about 1.6% of the total vibrational modes occurred in peaks of frequency ≤ 40 cm^{-1}.

The data in the third column of Table 1 indicate that C/T^3 for rock 10017 is perhaps approaching a maximum at the lowest temperature reached, which corresponds to $T < \theta/230$. (θ is the characteristic temperature, which SCHREIBER et al. (1970) computed for the glass beads as 530° using the measured wave velocities. We suggest that this may just be a more exaggerated form of the effects that occur in the vibrational spectrum of glasses. In support of this view, we note that the large heat capacity of rock 10017 can be closely accounted for between 3° and 5°K by the introduction of a monochromatic frequency of 5 cm^{-1} containing 0.26% of the total number of modes. We will not speculate as to their origin.

Thermal conductivity

The equilibrium time for rock 10046 was sufficiently long for a numerical estimate of the thermal conductivity of the rock to be made. For the computation, it was assumed that the temperature gradient across the specimen after heating was linear. At about 4°K, the result was $K = 2.5(\pm 0.5) \times 10^{-6}$ cal/sec cm deg. This is to be compared with the value $K = 1.66 \times 10^{-3}$ cal/sec cm deg obtained for this type of rock in the region $-130°$ to $+150°$C by HORAI et al. (1970). It is important to note, however, that the latter measurements were made in air at atmospheric pressure, which would be expected to increase K considerably. Our present result is consistent with values obtained for lunar fines (e.g., 2.5×10^{-6} cal/sec cm deg, BASTIN et al.,

1970, 5×10^{-6} cal/sec cm deg, BIRKEBAK *et al.*, 1970) from which the microbreccia are formed.

Little can be said quantitatively about the thermal conductivity of rock 10017. Its behavior in the calorimeter assembly indicates that K for it is 10 to 100 times larger than that for rock 10046. The so-called thermal parameter γ calculated from our data agrees with the estimates of ROBIE *et al.* (1970) obtained from calorimetric data, and with the estimates of TROITSKY (1967) derived from earth-based microwave experiments: γ lies in the range 500 to 2000 cm² sec$^{1/2}$ deg cal^{-1} for type C rocks and lunar fines.

ACOUSTIC WAVE DATA

Acoustic wave velocities v_p and v_s were measured on rocks 12063 and 12018. The measurements were made by the pulse transmission method (BIRCH, 1960; MATTABONI and SCHREIBER, 1967). We used 1 MHz PZT compressional and shear transducers. Both samples were approximately 2 cm $\times$ 1.5 cm $\times$ 1.5 cm.

Table 2. Compressional and shear velocities (km/sec) for samples 12063 and 12018 at ambient conditions.

Direction	12063*		12018†	
	v_p	v_s	v_p	v_s
1–1′	2.93	1.71	1.79	1.07
2–2′	3.32§ 2.67‡	1.80§ 1.45‡	2.04	1.17
3–3′	2.51	1.37	1.81	1.10
	$\rho = 3.10 \pm 0.01$ g/cm³		$\rho = 3.20 \pm 0.05$ g/cm³	

v_p and v_s measured at 1 MHz; * accuracy $\pm 5\%$; † accuracy $\pm 2\%$; ‡ lowest velocity, measured along center axis; § greatest velocity, measured toward one end of specimen.

The results are given in Table 2. In the table, the long direction axis is denoted by 3–3′ and the two shorter axes by 2–2′ and 1–1′. It is noticed that v_p in the 3–3′ direction and along the (2–2′) axis is almost a km/sec slower than the fastest velocities measured in the rock, even though the distances between measurement locations were only about a centimeter. The data (in an average sense) for rock 12063 are very similar to those reported for 10020 (KANAMORI *et al.*, 1970).

Rock sample 12018 with a mean density of 3.20 gm/cm³ exhibits velocity and density characteristics very similar to those of 10017 reported by ANDERSON *et al.* (1970) (see Table 2).

Discussion

As noted, the values obtained for v_p and v_s on the lunar igneous rock at ambient conditions is a factor of 2 to 3 below the values expected from their density and from Birch's relation and indeed agrees with in situ lunar velocities. Strong velocity dependence of the rocks on pressure was observed by SCHREIBER *et al.* (1970) and by KANAMORI *et al.* (1970), and the common suggestion has been that the lunar rocks are much more extensively microfractured than terrestrial rocks of equivalent density. This problem of low velocities pertains to the igneous rocks only, since WARREN (1971) and ANDERSON *et al.* (1970) have shown that the velocity-density relationship for microbreccias 10046 and 10065 and the in situ lunar soil (LATHAM *et al.*, 1970b)

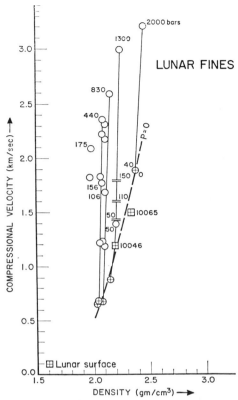

Fig. 1. Compressional velocity of lunar fines as a function of uniaxial pressure and density. Velocity data on microbreccias 10046 from ANDERSON *et al.* (1970); microbreccia 10065 from KANAMORI *et al.* (1970); lunar surface from LATHAM *et al.* (1970b). Circles indicate lunar fine data. Double horizontal bars indicate data for 10046 on first pressure run.

are quantitatively accounted for by simple (cold) compaction of the lunar soil (Fig. 1).

Recently CHAO *et al.* (1971) characterized porous microbreccias ($\rho \simeq 2.2$ to 2.4 gm/cm^3) as being formed without shock welding, suggesting that they represent consolidated mixtures of fallout ejecta produced at small (meter) depths by series of small impact events. Our elastic velocity data suggest formation predominantly by pressure rather than by thermal sintering. Effective pressures of 1 to 2 kb are required to compact loose fines to the densities of these microbreccias.

In the case of the igneous rocks, however, the case is not so simple. WARREN (1971) shows that, for a firmly sintered material of porosity ω, upper bound equations for the elastic moduli may be derived. The equations are in terms of an integral of the tangential stress concentrations on the surface of a finite pore of chosen geometry imbedded in an infinite elastic medium. Different upper bound equations are generated for different pore shapes.

If, for the igneous rock, high pressure (closed pore) velocities are assumed to be about $v_p = 6$ km/sec, $v_s = 4$ km/sec with $\rho_0 = 3.3$ to 3.5 gm/cm^3, and at ambient pressures values are $v_p = 3$ km/sec and $v_s = 1.7$ km/sec with $\rho = 3.1$ gm/cm^3, then the ratio of the effective bulk modulus K of a typical rock to its high pressure value K_0 is about

$$\frac{K}{K_0} = \frac{[v_p{}^2 - (\tfrac{4}{3})v_s{}^2]\rho}{[v_p{}^2 - (\tfrac{4}{3})v_s{}^2]_0\rho_0} \approx 0.36.$$

If porosity is about 5%, and if pores are considered approximately spherical (at grain boundary junctions), then the highest upper bound equations (MacKenzie, 1950) predicts $K/K_0 \approx 0.95$. From Warren (1971), if the stress integral around the pore is increased, say, by a factor of 10 because of the extreme curvature of the pore, then at $\omega = 0.05$, $K/K_0 \simeq 0.5$.

Although pore geometry effects may account for a considerable portion of the low bulk modulus, it is not yet clear whether the total decrease of bulk modulus can be accounted for simply by open microcracks in an otherwise continuously welded rock unless the distribution of cracks is such that at least a large portion of the individual rock grains are point welded more in the fashion of a compacted soil. Point welding allows a free deformation of grains in certain directions at low strains, an effect that can lower elastic moduli and velocity considerably, independent of porosity. Such a structure would predict a high Q only under the conditions of low strain and dry atmosphere or vacuum, as discussed in the next section.

The textural assemblage of Apollo 11 rocks indicate near surface rapid crystallization from low viscosity basaltic magmas under low oxygen pressure (Brown, 1970). A rapid cooling (and probably outgassing) history may be conducive to the general rock weakening. Very low level shock wave passage through essentially unshocked lunar igneous rocks can contribute to this effect. Radcliffe et al. (1970a) made high voltage transmission electron microscopic study of a type B igneous rock. The rock was relatively free of shock damage (Radcliffe et al., 1970b). From the transmission study, however, there was found to be a general system of irregular and cleavage cracks crossing all of the minerals, more marked than in terrestrial rock sections. This again suggests the possibility of high stress concentrations.

Dissipation Factor $1/Q$

Introduction

The specific dissipation function $1/Q$ was measured for the Apollo 12 specimens by a resonance method as a function of humidity, temperature, and gas environment. The importance of such an experiment lies in the problem of understanding the unexpectedly long duration of the seismic vibrations observed on the surface of the moon (Latham et al., 1970a). Values of Q suggested from Apollo seismic results, based on the duration of the signal dominant frequency are as high as 3000 (dissipation, $1/Q = 3 \times 10^{-4}$).

Laboratory measurement of Q on lunar rocks at low pressure by Kanamori et al. (1970) showed anomalously low values, $Q \approx 10$. However, these data are not comparable with the observed in situ Q for two reasons: (1) As pointed out by Kanamori *et al.*, the method of measurement by the use of ultrasonic pulse involves some ambiguity; and (2) the wave length used is so short that the scattering of waves due to the microscopic structure of the specimens might have affected the results.

In this study, the samples were placed in a bell jar and dissipation was measured by resonance methods as a function of pressure, temperature, and gas environment (N_2, H_2O). Unfortunately the irregular shape of these samples, as well as the limited nature of the vacuum facility, prevented complete interpretation of specimen resonance, so this study must be considered to be preliminary.

Experimental techniques

Measurements were made in a vacuum bell jar system in which pressure, gas environment, and temperature could be varied independently. The resonance technique used for Q determination is the same as that of sphere resonance (FRASER and LeCRAW, 1964; SOGA and ANDERSON, 1967) and cube resonance (DEMAREST, 1971).

It was found that the most satisfactory results were obtained by the arrangement of cube resonance. The sample is simply mounted from two diagonally opposite corners with 1 MHz PZT transducers in such a manner that mechanical loading of the specimen was minimized. Resonant frequencies were measured in a frequency range of 40 to 130 kHz, a temperature range of 25°C to 125°C, and in pressures as low as about 10^{-2} torr.

A saw-tooth voltage function was used to drive the sweep of an oscilloscope and to simultaneously control the frequency output of a frequency synthesizer. The scope displayed a steady curve of resonance amplitude versus frequency. Photographs were taken for data reduction. The speed of the frequency scanning was 0.5 to 1.0 kHz. When the scanning speed was less than 10 kHz, the resonance curve did not show any distortion. Photographs of typical oscilloscope displays are shown in Fig. 2. The specific dissipation of the specimen is calculated directly from the observed resonances, and $1/Q$ is given by

$$1/Q = (1/\sqrt{3})(\Delta f/f_0)$$

where f_0 is the resonance frequency, and Δf is the half amplitude frequency range. Accuracy in $1/Q$ is estimated as $\pm15\%$, based on Q measurements by this technique on single crystal KCl.

Experimental results

Original prototype measurements of $1/Q$ versus humidity were carried out on a sample of welded tuff cut to the dimensions of the lunar samples. The rock is very heterogeneous and porous with a bulk density of 1.92 g/cm³, immersion porosity of 14.8%, and anisotropic acoustic wave velocities of $v_p = 3.6$ to 4.1 km/sec and $v_s = 2.2$ to 2.4 km/sec.

As shown in Fig. 3, increasing humidity definitely increases the mechanical dissipation as anticipated from the results of KATAOKA and OGURI (1959a, b) and also by PANDIT and TOZER (1970). However, the change in $1/Q$ observed here is only a factor of 2. It is noted that there was no detectable difference in the value of $1/Q$ measured in vacuum and dry nitrogen. Therefore, both the data at 1 atm dry nitrogen and vacuum are plotted at zero humidity. The same measurement was made on lunar sample 12063. Again it was noted that there was no difference in $1/Q$ in vacuum and in dry nitrogen at $P = 1$ atm.

Figure 4 shows the temperature variation of the mechanical dissipation in lunar sample 12063 in vacuum. The data at 26°C, 48°C, and 76°C are those for the cooling

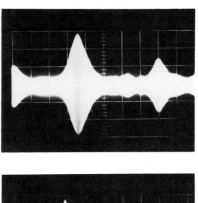

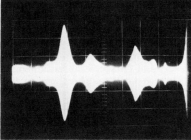

Fig. 2. Typical oscilloscope photographs showing specimen resonance.

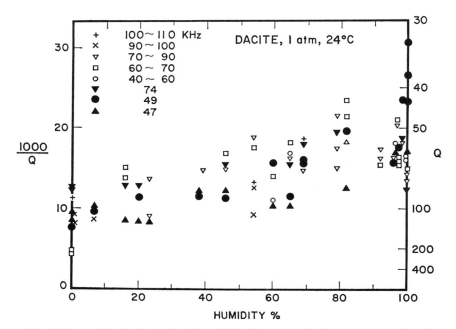

Fig. 3. Internal dissipation in dacitic welded tuff at atmospheric pressure as a function of relative humidity of the surrounding gas.

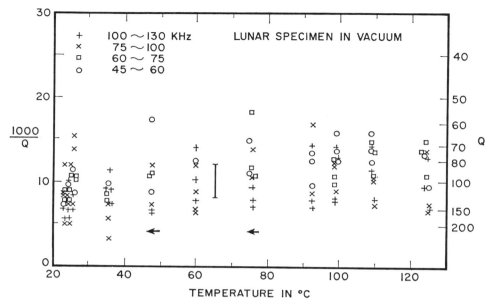

Fig. 4. Internal dissipation in lunar specimen 12063.97 in vacuum as a function of temperature. Data from the cooling run are indicated with arrows.

run. No appreciable hysteresis in dissipation is observed. Figure 5 shows the change of the resonance curve during this temperature run. Temperature dependence of the dissipation is very small but may show a small maximum at about 110°C. In almost all the cases, the frequency dependence of mechanical dissipation is small and within the experimental error in the frequency range of the present experiments. The mechanical dissipation of 12063 is $1/Q = 1/(130 \pm 50)$ in vacuum of 10^{-2} torr at room temperature.

Measurements on sample 12038 also show fairly low dissipation, $1/Q = 1/(300 \pm 50)$ in vacuum and dry nitrogen at room temperature. These values are a factor of 10 greater than $1/Q = 1/3500$ observed from lunar seismology.

DISCUSSION

The resonance experiment indicates an average Q for the returned lunar rocks that is 10 times greater than values obtained by acoustic transmission methods, but a factor of 10 less than values observed from lunar seismology. Values of Q for the earth's lower lithosphere have been determined to be on the order of 1000 or higher (OLIVER and ISACKS, 1967; MOLNAR and OLIVER, 1969). On earth these Q values are determined from attenuation of high frequency over distance, while lunar Q is determined from essentially the half power decay time of a signal with a dominant frequency which characterizes the signal over the entire record.

At this point, it is important to note that the problem of lunar in situ Q cannot really be separated from that of the character of lunar seismic signal envelope, phases, and spectral content. The signal envelopes of the lunar seismograms can be

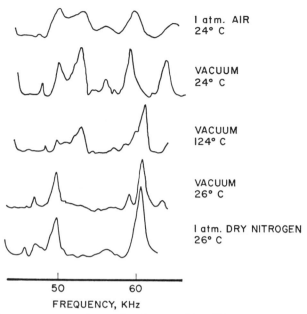

Fig. 5. Change of resonance curve due to the changes of humidity, temperature, and introduction of dry nitrogen.

modeled by a diffusion or scattering, random walk propagation function acting on an impulse source function (Latham et al., 1970a, b, Gold and Soter, 1970). Earth seismic records usually show definite phases with relative strong onset and short duration in marked contrast to the character of lunar seismic signals. (However, an important exception will be discussed later in the section.)

From the above it is not clear that high Q on earth can easily be related to high Q on the moon. For an extensively scattered signal, extreme attenuation is intuitively expected. Therefore, even though lunar Q values are about the same magnitude as lithospheric Q (measured by the two different methods), the difference in signal character makes the lunar Q seem anomalously high. Experimental Q (or the dissipation $1/Q$) was determined in our experiments by a process analogous to the lunar Q determination.

We first discuss the difference of experimental Q and lunar in situ Q. Two reasons for the difference may exist. First, the lunar rocks may have been mechanically degraded during their terrestrial history. Second, lunar mare Q may be higher than obtainable for small rock samples. In the first case mechanical degrading may have been predominantly caused by effects of exposure to water vapor. Adsorbed water cannot be baked out at our experimental temperatures. Mechanical dissipation is a structurally sensitive property, and a small amount of adsorbed phase may affect the dissipation greatly. A second major possible contributor to the lower Q is cold working, such as may result from machining the sample in the process of cutting the rock for distribution. Cold working and sample size effects on bulk strength and dissipation have been discussed by Kataoka and Oguri (1959a).

On the other hand, seismic in situ Q may be higher than measurable on an isolated rock or cutting of rock due to (1) relative frequency ranges of measurements (and wavelength range), (2) in situ temperature, and (3) in situ overburden pressures with zero pore pressures and humidities. Although the effect of temperature should not be the major cause of the order of magnitude difference of Q between the laboratory and in situ Q, this effect should be reserved as one of the possible interpretations.

If large-scale heterogeneity does not exist below maria surfaces, then simple hydrostatic pressure may increase Q by an order of magnitude, as observed in high pressure rock experiments. However, the assumption of closure of structure with pressure may not be comparable with the observed character of the lunar seismic signals. Furthermore, it is not clear that near surface fissures close, or indeed whether pressures are essentially hydrostatic in the outer portions of maria.

There are several reasons to believe that at least the outer portions of the moon are supported by rock strength withstanding gravitational compaction. Lunar rock chemistry indicates an apparent lack of in situ water (including magmatic water). The strength of silicates is decreased by the presence of water (SCHOLZ, 1970); and, therefore, on the moon gravitational compaction due to rock failure, creep, and adjustment of stress to a more hydrostatic distribution may not have occurred. Portions of the moon may be supported by the strength of the rocks to great depths. This is further evidenced by the existence of Mascons, by the reasonable fit of an heterogeneous igneous model as an upper bound to Apollo seismic data (Fig. 6 model IIb; see SCHREIBER et al., 1970; ANDERSON et al., 1970 for more complete model description), by the high porosities of lunar microbreccias and igneous rocks under pressure up to 40 kb (500-km depth) (STEPHENS and LILLEY, 1970), and by the complicated nature of the lunar seismic signals felt to be a result of propagation path (LATHAM et al., 1970a, b).

The possibility of seismic (long wavelength) Q being greater than rock sample Q for a heterogeneous mare cannot be ruled out. KUMAZAWA and ANDERSON (1971) point out one possible interpretation: that regions beneath the lunar craters are occupied by masses of high Q glassy material. However, it is not known if sufficient glassy material exists beneath the craters. Evidence of simultaneously heterogeneous and high Q terrestrial regions may have already been presented in seismology literature (MITRONOVAS et al., 1969).

Figure 7 shows the character of earthquakes observed in the Tonga Island arc. Any one earthquake is observed by stations in both the volcanic and nonvolcanic ridges. Seismic paths are high Q ($Q > 1000$) in both cases. In the volcanic ridge the character of the signal is changed from impulsive to a more diffusive signal. Normally sharp phases, such as P and S, are smeared out and the signal length extended in duration by about a factor of 2. The integrated energy in the diffusion like records is apparently not less than the integrated energy in the impulsive records. Unfortunately, frequency content and dispersion for these records cannot be well determined. It must also be noted that the total time duration of the "diffusion-like" Tonga records is still on the order of minutes, rather than hours as observed for the lunar signals.

Two points, therefore, may be raised: (a) that the body wave Q in the earth's lithosphere is as high as expected for the lunar Q; and (b) that seismic data suggest

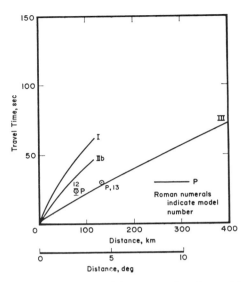

Fig. 6. Travel times derived from models compared to observed travel times for Apollo 12 LEM (P-12) and Apollo 13 Saturn SIVB (P-13) impacts. (After LATHAM *et al.*, Passive Seismic Experiment, Apollo 12 Preliminary Science Report to NASA.)

that introducing heterogeneity changes the character of the seismic signal in the sense observed in lunar seismograms without introducing significant energy losses. This may imply for scattered signals that (a) the same energy passes the station more than once, or (b) the gathering area for scattered energy is much greater than for directed ray energy.

The question may not be how long the record is, but rather does scattering (which does lengthen the record) introduce extreme attenuation also. If not, the lunar seismic problem may be more closely related to terrestrial seismic cases than previously suspected. The Tonga records have implications for the problem of the heterogeneity scale used in lunar models. In a volcanic zone, large-scale heterogeneity is expected. LATHAM *et al.* (1970b) state that the major character (time-amplitude behavior and frequency character) of the LEM and SVIB impact data can be accounted for by assuming body wave scattering due to a scattering structure of a few hundred meters to 1 km in size. On the other hand, GOLD and SOTER (1970) suggests multiple scattering of body waves from the lunar surface and use a linear velocity gradient of 1.35 km/sec/km in a self-compacting lunar dust model with a thickness on the order of kilometers. The large-scale heterogeneity expected under volcanic ridges supports the idea of large scale heterogeneity in lunar maria.

WARREN (1971) made theoretical and experimental studies in porous systems with complex structures. For wavelengths on the order of 4 to 10 times the longest continuously distributed structural unit, something akin to a "filtering" effect seems to occur. That is, the predominant frequency which carries most of the signal energy seems to be governed by the wavelength, which is structurally determined, and by a velocity which depends on the bulk density of the material and the structural elastic

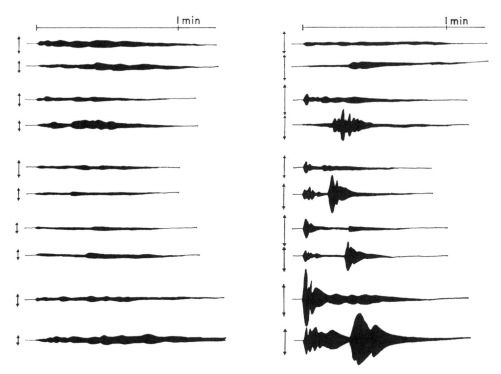

Fig. 7. Comparison of amplitude envelopes from the two ridges in Tonga for some shallow earthquakes arranged in increasing hypocentral depths. Arrows indicate equal ground motions. Volcanic ridge on left, nonvolcanic ridge on right (after MITRONOVAS et al., 1969).

moduli. Low pass filtering to very narrow band filtering seems to occur, so that only a single frequency may characterize the first arrival signal. This frequency increases proportionally with velocity and compaction pressure. The observation of a dominant frequency may be analogous to the frequency "peak" observed in the lunar seismic records.

The dimension of lunar structural units predicted from the experimental results is close to that predicted by diffusion theory, i.e., hundreds of meters to about 1 km and velocities are equivalent to those observed in situ in the moon.

On the other hand, given a simple fine fragmentary or "dust" model, a problem arises regarding reasonable velocity gradients for dry compaction. In Table 3, data from ANDERSON et al. (1970) on lunar soil are given. Column a is for an initial compaction cycle, and column b is for an elastic compaction cycle after precompaction to 1.3 kb.

Velocity gradients decrease in value along initial compaction curves for pressures higher than tens of bars. Indeed, if compaction effects alone predominate for pressures up to 1 kb, (lunar depths of about 20 km for densities of about 3 gm/cm^3) soil velocity

Table 3. Velocity (km/sec), density (gm/cm³), and pressure (bars) for Apollo 11 lunar soil compaction experiment, from Anderson *et al.* (1970).

	a Initial Compaction			b Pre-Compacted to 1300 bars		
v_p	ρ	P		v_p	ρ	P
0.1	1.6	1		1.1	1.2	1
2.1	2.0	175		1.8		150
2.4	2.1	440				
2.6	2.16	830				
3.0	2.24	1300		3.0	2.24	1300

gradients become lower than for "solid" (micro-fractured) rock (Models 2 and 3, Anderson *et al.*, 1970).

Estimating from the start of the compaction data on the lunar soil, a gradient of $\Delta v_p / \Delta Z = 0.75$ km/sec/km seems reasonable. In comparison, Gold and Soter (1970) assumed a linear velocity gradient of 1.35 km/sec/km. However, experimental velocity gradients of 1 km/sec/km were observed only at low pressure. At high pressure (or depth) experimental velocity gradients appear to be perhaps 20 times lower. These values are presently being confirmed. Since large velocity gradients will not extend to great depth, it is suggested that a low velocity channel cannot be satisfactorily generated by simple compaction of a fine-grained medium layer with a thickness on the order of kilometers.

If, in the lunar case, low velocities reflect large scale (greater than 10^2 meters) structural effects only, then small igneous rock samples (such as the returned lunar samples) if at most only weakly shocked, should more typically show higher velocities than in situ velocities. As discussed in the section on acoustic measurements, probably the majority of the sample velocities agree very well with in situ velocities and have been used to generate reasonable velocity profiles. However, samples 10020 (Kanamori *et al.*, 1970) and 12018 (this report) do have higher velocities at ambient conditions and so lead to a prediction of faster in situ travel times than observed. The number of lunar samples measured on earth is too small to resolve the problem.

Acknowledgment—The authors wish to thank B. Isacks, P. Molnar, W. Mitronovas, and J. Dorman for useful and important discussions on aspects of this paper. This work was supported by Contract NAS 9–7890.

References

Anderson O. L. (1963) A simplified method for calculating the Debye temperature from elastic constants. *J. Phys. Chem. Solids* **24**, 909–917.

Anderson O. L., Scholz C., Soga N., Warren N., and Schreiber E. (1970) Elastic properties of a microbreccia, igneous rock and lunar fines from Apollo 11 mission. *Proc. Apollo 11 Lunar Sci. Conf., Geochim. Cosmochim. Acta*, Suppl. 1, Vol. 3, pp. 1959–1973. Pergamon.

Bastin J. A., Clegg P. E., and Fielder G. (1970) Infrared and thermal properties of lunar rock. *Science* **167**, 728–730.

Birch F. (1960) The velocity of compressional waves in rocks to 10 kilobars. *J. Geophys. Res.* **65**, 1083–1102.

Birkebak R. D., Cremers C. J., and Dawson J. P. (1970) Thermal radiation properties and thermal conductivity of lunar material. *Science* **167**, 724–726.

Blackman M. (1955) The specific heat of solids. *Handbuch der Physik* **7** (part 1), 325–382.

BROWN G. M. (1970) Petrology, mineralogy, and genesis of lunar crystalline igneous rocks. *J. Geophys. Res.* **75**, 6480–6496.

CHAO E. C. T., BOREMAN J. A., and MINKIN J. A. (1971) Unshocked and shocked Apollo 11 and Apollo 12 microbreccias: Characteristics and some geologic implications. Second Lunar Science Conference (unpublished proceedings).

DEMAREST H. H., JR. (1971) Cube resonance method to determine the elastic constants of solids. *J. Acoust. Soc. Amer.* **49**, 768–775.

FOWLER R. H. and GUGGENHEIM E. A. (1949) *Statistical Thermodynamics*, pp. 102–106, Cambridge University Press, London.

FRASER D. B. and LeCRAW R. C. (1964) Novel method of measuring elastic and anelastic properties of solids. *Rev. Sci. Instr.* **35**, 1113–1115.

FRIEDMAN I., O'NEILL J. R., ADAMI L. H., GLEASON J. D., and HARDCASTLE K. (1970) Water, hydrogen, deuterium, carbon, carbon-13 and oxygen 18 content of selected lunar material. *Science* **167**, 538–540.

FUNKHOUSER J. G., SCHAEFFER O. A., BOGARD D. D., and ZAHRINGER J. (1970) Gas analysis of the lunar surface. *Science* **167**, 561–563.

GOLD T. and SOTER S. (1970) Apollo 12 seismic signal: Indication of a deep layer of powder. *Science* **169**, 1071–1075.

HORAI K., SIMMONS G., KANAMORI H., and WONES D. (1970) Thermal diffusivity and conductivity of lunar material. *Science* **167**, 730–731.

KANAMORI H., NUR A., CHUNG D., WONES D., and SIMMONS G. (1970) Elastic wave velocities of lunar samples at high pressures and their geophysical implications. *Science* **167**, 726–728.

KATAOKA A. and OGURI M. (1959a) Effects of humidity and temperature on dynamic properties of rocks. *ZISIN* **12**, 83–85.

KATAOKA A. and OGURI M. (1959b) Some dynamical properties of rocks at room temperatures. *ZISIN* **12**, 91–100.

KUMAZAWA M. and ANDERSON O. L. (1971) Preliminary determination of specific internal dissipation function in lunar rock specimens. In preparation.

LATHAM G. V., EWING M., PRESS F., SUTTON G., DORMAN J., NAKAMURA Y., TOKSOZ N., WIGGINS R., DERR J., and DUENNEBIER F. (1970a) Passive seismic experiment. *Science* **167**, 455–457.

LATHAM G. V., EWING M., PRESS F., SUTTON G., DORMAN J., NAKAMURA Y., TOKSOZ N., MEISSNER R., DUENNEBIER F., KOVACH R., and YATES M. (1970b) First seismic data from manmade impacts on the moon. *Science* **170**, 620–626.

LEADBETTER A. J. (1968) The thermal properties of glasses at low temperatures. *Phys. Chem. Glasses* **9**, 1–13.

MACKENZIE J. K. (1950) The elastic constants of a solid containing spherical holes. *Proc. Phys. Soc. London, B*, **63**, 2–11.

MATTABONI P. and SCHREIBER E. (1967) Method of pulse transmission measurements for determining sound velocities. *J. Geophys. Res.* **72**, 5160–5163.

MITRONOVAS W., ISACKS B., and SEEBER L. (1969) Earthquake locations and seismic wave propagation in the upper 250 km of the Tonga Island arc. *Bull. Seismol. Soc. Amer.* **59**, 1115–1135.

MOLNAR P. and OLIVER J. (1969) Lateral variations of attenuation in the upper mantle and discontinuities in the lithosphere. *J. Geophys. Res.* **74**, 2648–2682.

MORRISON J. A. and NORTON P. R. (1970) The heat capacity and thermal conductivity of Apollo 11 rocks 10017 and 10046 at liquid helium temperatures. *J. Geophys. Res.* **75**, 6553–6557.

OLIVER J. and ISACKS B. (1967) Deep earthquake zones, anomalous structures in the upper mantle, and the lithosphere. *J. Geophys. Res.* **72**, 4259–4275.

PANDIT B. I. and TOZER D. C. (1970) Anomalous propagation of elastic energy within the moon. *Nature* **226**, 335.

RADCLIFFE S. V., MEUER A. H., FISHER R. M., CHRISTIE J. M., and GRIGGS D. T. (1970a) High-voltage transmission electron microscopy study of lunar surface material. *Science* **167**, 638–640.

RADCLIFFE W. V., MEUER A. H., CHRISTIE J. M., and GRIGGS D. T. (1970b) High voltage (800 kv) electron petrography of type B rock from Apollo 11. *Proc. Apollo 11 Lunar Sci. Conf. Geochem. Cosmochim. Acta*, Suppl. 1, Vol. 1, pp. 731–748. Pergamon.

ROBIE R. A., HEMINGWAY B. S., and WILSON W. H. (1970) Specific heats of lunar surface materials from 90 to 350 degrees Kelvin. *Science* **167**, 749–750.

SCHOLZ, C. H. (1970) Static fatigue of quartz (abstract), *Trans. Amer. Geophys. Union* **51**, 827.

SCHREIBER E., and ANDERSON O. L. (1970) Properties and composition of lunar materials: Earth analogies. *Science* **168**, 1579–1580.

SCHREIBER E., ANDERSON O. L., SOGA N., WARREN N., and SCHOLZ C. (1970) Sound velocity and compressibility for lunar rocks 17 and 46 and for glass spheres from the lunar soil. *Science* **167**, 732–734.

SOGA N. and ANDERSON O. L. (1967) Elastic properties of tektites measured by resonant sphere technique. *J. Geophys. Res.* **72**, 1733–1739.

STEPHENS D. R. and LILLEY E. M. (1970) Compressibilities of lunar crystalline rock microbreccia, and fines to 40 kilobars. *Science* **167**, 731–732.

TROITSKY V. S. (1967) Investigation of the surfaces of the moon and planets by means of thermal radiation. *Proc. Roy. Soc.* **A296**, 366–398.

WARREN N. (1971) Elastic properties of bulk complex porous materials. In preparation.

Proceedings of the Second Lunar Science Conference, Vol. 3, pp. 2361–2365
The M.I.T. Press, 1971.

Specific heats of the lunar breccia (10021) and olivine dolerite (12018) between 90° and 350° Kelvin*

RICHARD A. ROBIE and BRUCE S. HEMINGWAY

U.S. Geological Survey, Silver Spring, Maryland 20910

(*Received* 22 *February* 1971; *accepted* 30 *March* 1971)

Abstract—The specific heat of the Apollo 11 breccia (10021,41) and the Apollo 12 olivine dolerite (12018,84) have been measured over the temperature range 95° to 340°K. The specific heat of 12018 increases monotonically from 0.050 at 90°K to 0.201 cal/gram-deg at 350°K. The specific heat of the breccia (10021) increases from 0.054 at 90°K to 0.197 cal/gram-deg at 350°K. The temperature variation of the specific heat of the breccia is quite similar to that of the regolith (10084) from the Sea of Tranquillity.

As PART of a continuing study of the thermal properties of the lunar surface materials we have measured the specific heats of the breccia from Tranquillity Base (10021) and of the olivine dolerite from the Ocean of Storms (12018). The magnitudes of the specific heats of these two samples are similar to the Apollo 11 regolith and vesicular basalt which we have studied previously (ROBIE *et al.*, 1970).

For the measurements on sample 10021 the calorimeter was filled with 35.097 grams of breccia. The total weight of the aluminum sample holder including the thermometer and heater was 24.7 grams. For the measurements on sample 12018 the calorimeter contained 29.931 grams.

The heat capacity of the empty sample container was measured in a separate set of experiments at 10 degree intervals over the temperature range 90 to 350°K. Our experimental specific heat data for samples 10021 and 12018 are listed in Tables 1 and 2, and are shown graphically in Figs. 1 and 2. The smoothed values at integral temperatures are listed in Table 3.

In an earlier paper, ROBIE *et al.* (1970) calculated the temperature dependence of the thermal parameter, $\gamma = (k\rho C)^{-1/2}$, where k is the thermal conductivity, ρ the density, and C the specific heat. They used their specific heat data, the measured densities for the Apollo 11 basalt (10057) and regolith (10084) (LSPET, 1969) and an assumed temperature-independent value for the conductivity of the basalt of 0.004 cal/cm sec deg and for the regolith (in vacuum) of 0.000004 cal/cm sec deg. The pressure on the lunar surface is less than 10^{-7} torr (JOHNSON *et al.*, 1970). Their calculations showed that γ for the basalt varied from 34.3 at 100°K to 19.4 at 350°K and the γ for the regolith decreased from 1543 to 898 cm² sec$^{1/2}$ deg cal^{-1} over the same temperature range.

HORAI *et al.* (1970) measured the thermal diffusivity of sample 10057 (basalt) using a modified Angstrom method between 149 and 436°K. From their diffusivity data and an *estimated* temperature dependent specific heat for sample 10057 they

---

* Publication authorized by the Director, U.S. Geological Survey.

Table 1. Experimental specific heat measurements for lunar sample 10021,41 (breccia) from Tranquillity Base.

Temperature °K	Specific heat cal/gram-deg	Temperature °K	Specific heat cal/gram-deg	Temperature °K	Specific heat cal/gram-deg
98.52	0.0618	170.23	0.1187	274.64	0.1692
106.64	0.0686	180.38	0.1238	284.90	0.1734
115.66	0.0761	190.50	0.1292	294.84	0.1777
124.71	0.0834	200.17	0.1351	302.83	0.1801
134.06	0.0904	215.92	0.1435	311.81	0.1831
144.60	0.0983	224.86	0.1473	320.58	0.1862
149.77	0.1012	233.44	0.1509	329.14	0.1894
155.34	0.1051	242.60	0.1555	337.33	0.1926
161.04	0.1105	253.10	0.1600		
166.52	0.1130	264.05	0.1652		

Table 2. Experimental specific heat measurements for lunar sample 12018,84 olivine dolerite from the Sea of Storms.

Temperature °K	Specific heat cal/gram-deg	Temperature °K	Specific heat cal/gram-deg	Temperature °K	Specific heat cal/gram-deg
96.05	0.0557	167.66	0.1116	249.38	0.1593
104.40	0.0628	176.30	0.1172	258.64	0.1640
112.88	0.0701	184.58	0.1224	267.60	0.1683
121.69	0.0773	193.14	0.1278	276.60	0.1721
127.19	0.0826	201.98	0.1329	285.60	0.1767
134.76	0.0883	210.50	0.1381	293.03	0.1784
141.92	0.0937	218.75	0.1429	301.98	0.1830
148.73	0.0987	222.65	0.1453	310.70	0.1869
149.06	0.0989	231.20	0.1499	319.59	0.1904
158.61	0.1056	240.23	0.1550	328.64	0.1933

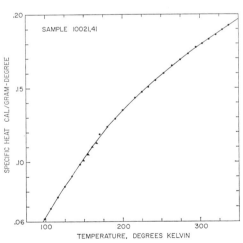

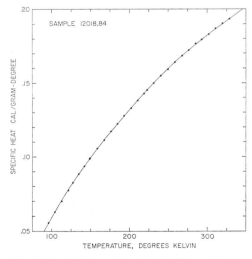

Fig. 1. Specific heat of Apollo 11 lunar breccia sample 10021,41. Filled circles indicate experimental observations for which the temperature rise was approximately 10°K. Filled triangles are for a separate set of measurements made with temperature changes of about 5°K. Full line is the least squares fit to the data.

Fig. 2. Specific heat of Apollo 12 olivine dolerite 12018,84. Filled circles indicate the experimental observations. Full line is the least squares fit to the data.

Table 3. Specific heats (smoothed values) for samples 10021,41 and 12018,84.

Temperature °K	Specific heat cal/gram-deg	
	10021,41	12018,84
90	(0.0539)*	(0.0498)
100	0.0631	0.0591
120	0.0798	0.0766
140	0.0947	0.0921
160	0.1158	0.1064
180	0.1235	0.1193
200	0.1348	0.1317
220	0.1450	0.1438
240	0.1545	0.1549
260	0.1632	0.1647
280	0.1714	0.1738
300	0.1778	0.1823
320	0.1864	0.1903
340	0.1934	(0.1977)
350	(0.1969)	(0.2014)

* Extrapolated values are indicated by parentheses.

calculated thermal conductivities which showed a large temperature variation. We have recalculated the data of HORAI et al. (1970) using our measured values of the specific heat of 10057 (ROBIE et al. 1970) and the measured density 3.4 gm cm^{-3} (LSPET, 1969). The calculations are listed in Table 4 and are shown graphically in Fig. 3. As can be seen by comparison with the curve of HORAI et al. (1970), the *observed* data incorporating the measured specific heats, support our previous assumption that the conductivity of the dense basalt is essentially independent of temperature between 90° and 350° Kelvin and has a value near 0.004 cal/sec cm deg.

The temperature of the lunar surface varies periodically between approximately 390° and 90°K over a period of 29 days, 12 hours, 44.05 minutes (synodic period). This is shown schematically in Fig. 4.

From the calculations of WESSELINK (1948), LINSKY (1966), and SINTON (1962) it can be shown that the numerical value of the thermal parameter used to calculate the lunar surface temperature over the period of a lunation will affect the calculated temperature-time curve of those areas covered by the regolith only at times greater

Table 4. Calculated thermal conductivity of lunar basalt (10057) between 150° and 436°K.

Temperature °K	Measured diffusivity[a] 10^{-3} cm^2/sec	Measured specific heat[b] cal/gm-deg	Calculated thermal conductivity cal/cm sec deg
149	11.94	0.0991	0.00402
166	11.03	0.1118	0.00419
208	9.06	0.1386	0.00427
304	6.44	0.1799	0.00394
313	6.86	0.1830	0.00427
344	6.06	0.1930	0.00398
345	5.91	0.1934	0.00389
435	5.37	0.2230	0.00407
436	6.55	0.2233	0.00497

[a] HORAI et al. (1970); [b] ROBIE et al. (1970).

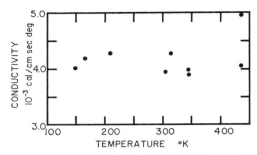

Fig. 3. Calculated thermal conductivity of lunar basalt 10057 based on the diffusivity
data of Horai *et al.* (1970), and specific heat results of Robie *et al.* (1970).

than 0.25 of the total period, that is 7 days after full moon (between sunset and sunrise),
and when the temperature is less than 150°K. From the specific heat data of Robie
et al. (1970) and our present results, and using the thermal conductivity of sample
10084 at 200°K, 0.0000034 cal/cm sec°K in vacuum as measured by Cremers *et al.*
(1970), (also see Morrison and Norton, 1970), the thermal parameter for the
regolith during this dark phase of a lunation will be in the range 1300–1500 cm²
sec$^{1/2}$ deg cal^{-1}.

The depth, *l*, beneath the surface at which this 300 degree temperature variation
will be attenuated to $e^{-2\pi}$ of its initial variation is (Wesselink, 1948)

$$l = 2\left(\pi \frac{k}{C\rho} P\right)^{1/2}$$

where *P* is the synodic period (2,551,443 seconds). From our specific heat data and
the observed conductivities of Cremers *et al.* (1970) we obtain $l = 23.1$ centimeters
at 350°K and 33.2 centimeters at 90°K. At these depths the synodic temperature
variation will be reduced to about 6°K and the maximum will lag one-eighth of a
period behind the maximum temperature at the surface (Wesselink, 1948). These
complications must be kept in mind when the lunar heat flow experiment (Langseth
et al., 1970) is eventually undertaken.

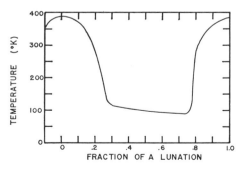

Fig. 4. Temperature variation of the lunar surface over a period of 1.1 lunation
(29.57 days).

Acknowledgements—We wish to thank our U.S. Geological Survey colleagues W. H. WILSON, R. W. WERRE, R. TRULI, and E. SEGINAK for help with the reduction of the measurements and construction of the apparatus. H. R. SHAW and E. C. ROBERTSON offered many helpful suggestions for improving the manuscript. We are particularly grateful to NASA for support of this research under contract T–75405.

REFERENCES

CREMERS C. J., BIRKEBAK R. C., and DAWSON J. P. (1970) Thermal conductivity of fines from Apollo 11. *Proc. Apollo 11 Lunar Sci. Conf., Geochim. Cosmochim. Acta* Suppl. 1, Vol. 3, pp. 2045–2050. Pergamon.

HORAI K.-I, SIMMONS G., KANAMORI H., and WONES D. (1970) Thermal diffusivity, conductivity and thermal inertia of Apollo 11 lunar material. *Proc. Apollo 11 Lunar Sci. Conf., Geochim. Cosmochim. Acta* Suppl. 1, Vol. 3, pp. 2243–2249. Pergamon.

JOHNSON F. S., EVANS D. E., and CARROLL J. M. (1970) Cold cathode gage (Lunar atmosphere detector). Apollo 12 preliminary science report, NASA SP–235, pp. 93–97.

LANGSETH M. G., Jr., WECHSLER A. E., DRAKE E. M., SIMMONS G., CLARK S. P., Jr., and CHUTE J., Jr. (1970) Apollo 13 lunar heat flow experiment. *Science* **167,** 211–217.

LINSKY J. L. (1966) Models of the lunar surface including temperature-dependent thermal properties. *Icarus* **5,** 606–634.

LSPET (Lunar Sample Preliminary Examination Team) (1969) Preliminary examination of lunar samples from Apollo 11. *Science* **165,** 1211–1227.

MORRISON J. A. and NORTON P. R. (1970) The heat capacity and thermal conductivity of Apollo 11 lunar rocks 10017 and 10046 at liquid helium temperatures *J. Geophys. Res.* **75,** 6553–6557.

ROBIE R. A., HEMINGWAY B. S., and WILSON W. H. (1970) Specific heats of lunar surface materials from 90 to 350°K. *Proc. Apollo 11 Lunar Sci. Conf., Geochim. Cosmochim. Acta* Suppl. 1, Vol. 3, pp. 2361–2367. Pergamon.

SINTON W. M. (1962) Temperatures on the lunar surface. In *Physics and Astronomy of the Moon* (editor Z. Kopal), Chapter 11, pp. 407–428, Academic Press.

WESSELINK A. F. (1948) Heat conductivity and nature of the lunar surface material. *Bull. Astron. Inst. Netherlands* **X,** 351–363.

Proceedings of the Second Lunar Science Conference, Vol. 3, pp. 2367–2379
The M.I.T. Press, 1971.

Electrical properties of Apollo 11 and Apollo 12 lunar samples

T. J. Katsube and L. S. Collett

Geological Survey of Canada, Ottawa, Ontario, Canada

(*Received* 22 *February* 1971; *accepted in revised form* 2 *April* 1971)

Abstract—Three samples (10017,30; 10065,22; 10084,83) from Apollo 11 and three (12002,84; 12002,85; and 12070,107) from Apollo 12 were measured for dielectric conductivity, relative permittivity, and loss tangent in the frequency range of 10^3 to 10^7 Hz. Except for one sample (10017,30) usual contact electrode method of the capacitance type was used on the samples in a dry nitrogen atmosphere at room temperature. In the absence of moisture, the samples behave as dielectric material.

In order to cover the frequency range of measurement, two systems are used. A high precision capacitance bridge is the main part of the system for 10^3 to 10^5 KHz, and a modified Q-meter method for the system for 10^5 to 10^7 Hz. Problems in measurement caused mainly by the irregular shapes of certain lunar samples has necessitated application and development of various measuring methods and techniques.

The investigations revealed the following general characteristics: (1) the dielectric conductivity increases from below 10^{-7} mhos/m for all samples at 10^3 Hz to almost 10^{-4} mhos/m for the solid rocks and to 10^{-5} mhos/m for the fines at 10^7 Hz; (2) the relative permittivity is $8.0 \pm 10\%$ for the solid rocks and is 3.0 to 3.8 for the fines; (3) the loss tangent ranges from 0.14 to 0.01 for the solid rocks, decreasing most rapidly above 10^6 Hz, and from 0.13 to 0.005 for the fines, decreasing gradually over the frequency range.

The results of this investigation suggest that the general dielectric behaviour of the lunar samples do not seem to differ from those of the dry terrestrial rocks, and that the dielectric properties seem to be mainly affected by the mineral content and the void space of the samples.

Introduction

THIS STUDY AND SET of measurements has been carried out to determine the electric properties of the lunar samples and to study the relation between these properties and the petrology. The dielectric characteristics of lunar samples are of interest because lunar rocks were supposedly formed in a water-free environment and for planning future electromagnetic soundings of the moon.

In this paper, a report is made on the measured results for permittivity, loss tangent, and dielectric conductivity over the frequencies from 10^3 Hz to 10^7 Hz, for six lunar samples. And in addition, the measuring techniques and equipment which are used in this investigation are described, besides discussing the relation between the results and the petrology of the lunar samples. The six samples used in this measurement include three igneous rocks (one of type A and two of type B), one breccia (type C) and two fines (type D). Their mineral composition, density, and other descriptions are compiled in Tables 1 and 2. It should be noted that samples 10017,30 (type A) 12002,85 (type B), and 12002,84 (type B) are of irregular and partially irregular shapes which are not ideal for electrical measurements.

2367

Table 1. Type, weight, density, and description of the lunar samples.

Sample	Type Code	Weight (gr)	Bulk Density (gm/cm³)	Description
10017,30	A	13.85	3.11[1]	solid, irregular
10065,22	C	4.52	2.452[2]	solid, rectangular
10084,83	D	5.05	1.944[4]	fines
12070,107	D	2.50	1.744[4]	fines
12002,84	B	11.14	3.103[3]	solid, cube (partially irregular)
12002,85	B	2.43	3.043[3]	solid, slab (partially irregular)

[1] ANDERSON et al. (1970); [2] SCHMITT et al. (1970); [3] Calculated; [4] Packing density at time of measurement.

DEFINITION OF PARAMETERS

The definition of the parameters used in this measurement is based on VON HIPPEL (1954) and the ASTM (D 150–68).

The loss tangent (tan δ) is the ratio of the loss current (I_1) to the dielectric charging current (I_c) which are considered to flow across the sample parallel to one another, that is,

$$\tan \delta = \frac{I_1}{I_c} . \tag{1}$$

I_1 is inphase with the applied voltage, and I_c is 90 degrees out of phase with it. This loss current I_1 does not necessarily consist only of an ohmic current, but can contain an inphase equivalent current due to any other mechanism. This situation is expressed by the introduction of the complex permittivity:

$$\varepsilon^* = \varepsilon' - j\varepsilon'', \tag{2}$$

where ε^*, ε', and ε'' are the complex permittivity, real permittivity, and the loss index, respectively. The relation with these parameters to I_1 and I_c are

$$\begin{cases} I_c = \omega\varepsilon' \dfrac{C_0}{\varepsilon_0} V \\[2mm] I_1 = \omega\varepsilon'' \dfrac{C_0}{\varepsilon_0} V \end{cases} \tag{3}$$

where ω is the angular frequency, C_0 is the vacuum capacitance of the sample, V is the applied voltage, and ε_0 is the vacuum permittivity. Therefore, from equation (1),

$$\tan \delta = \frac{\varepsilon''}{\varepsilon'} . \tag{4}$$

The vector of the total current is

$$\vec{I} = jI_c + I_1 = (j\omega\varepsilon' + \omega\varepsilon'') \frac{C_0}{\varepsilon_0} V. \tag{5}$$

The dielectric constant or the real relative permittivity (K') is defined as follows:

$$K' = \frac{\varepsilon'}{\varepsilon_0} \tag{6}$$

Table 2. General rock classification and mineral composition of two lunar rocks.

Sample	10017	12002
General rock classification	Very fine-grained olivine-basalt	Fine to medium-grained holocrystalline basalt
Type code	A	B
Major minerals		
pyroxene	45–50%*	50%†
plagioclase	30–35%	30%
ilmenite	15–20%	3–4%
olivine	<1%	15%

* SCHMITT *et al.* (1970); † LSIC, Apollo 12, p. 61.

The dielectric conductance is equivalent to I_1 over V, and the dielectric conductivity (σ') is defined as follows:

$$\sigma' = \frac{I_1}{V}\frac{d}{A}$$

where d and A are thickness and area of the sample, respectively. And by substituting equation (3) and equation (4) in this equation, we obtain

$$\sigma' = \omega\varepsilon' \tan \delta \qquad (7)$$

since

$$C_0 = \varepsilon_0 \frac{A}{d}. \qquad (8)$$

Results of these measurements, both for solids and fines, do not show any indication significant enough to suggest the existence of a relaxation process. Therefore, the Debye-relaxation process, considered by SAINT-AMANT and STRANGWAY (1970), has not been taken into consideration in this case.

GENERAL MEASUREMENT PROCEDURE

For the first step in the procedure, the impedance of the sample is measured (KATSUBE and COLLETT, 1971) over the frequency range from 10^3 to 10^7 Hz. The dielectric constant of the sample is derived from the impedance. This is possible because the lunar samples show a very low value for loss tangent.

In order to determine the loss tangent and the dielectric conductivity, a high-resolution and high-precision capacitance bridge assembly (GR 1620-A) is used for frequencies 10^3 to 10^5 Hz, and a modified Q-meter method (using standard coils of Boonton Radio Corp., type 103-A) for frequencies 10^5 to 10^7 Hz. The sample holder is basically of the capacitance type, which consists of two parallel electrode plates. The rocks are placed directly between the electrode plates while the fines are placed in a container of a similar type which fits into the same assembly.

Various techniques have been applied in order to eliminate electrode fringe effect, stray capacitances, and the electrode contact effect. The use of the variable air-gap method (KATSUBE and COLLETT, 1971) which is basically a technique to eliminate the electrode contact effect makes it possible to measure the dielectric properties of the irregular shaped rocks.

The assembly containing the sample is placed for measurement in a dry nitrogen atmosphere in a vacuum chamber in order to simulate the moisture free environment of the moon. The measurements were carried out at room temperature.

In order to measure the loss tangent (tan δ) and the dielectric conductivity (σ'), the samples are inserted between the electrode plates with a good contact. No other methods for electrode-sample contact, such as the standard ones stated in ASTM (D 150–68) are used, in order to prevent contamination of samples.

The measurement accuracy of these systems have not yet been fully evaluated. From experience, it has been seen that the repeatability of the measurement falls within the order of $\pm 3\%$, the instrumental error is in the order of $\pm 3\%$, and error caused by applied techniques varies according to the shape of the sample. Therefore, the total accuracy is estimated to be in the order of $\pm 10\%$ (average) for permittivity, and somewhat inferior to that for loss tangent measurements. The accuracy of the conductivity depends on both that of permittivity and loss tangent.

THEORY OF MEASURING TECHNIQUES

The real relative permittivity or the dielectric constant of the sample is derived from the impedance. The impedance (Z) of the sample is determined by measuring the total current (I) that passes through the sample and the applied voltage (V), as shown in Fig. 1, and is shown by the equation

$$Z = \frac{V}{I} = \frac{V}{v} R_V \tag{9}$$

Since the loss tangent of these samples is small, the capacitance can be calculated from

$$C = \frac{1}{\omega Z} \tag{10}$$

where C is the capacitance related to the sample and ω is the angular frequency. Equation (10) is derived from equations (5), (1), and (6):

$$Z = \frac{V}{I} = \frac{1}{\sqrt{(\omega\varepsilon')^2 + (\omega\varepsilon'')^2}} \frac{\varepsilon_0}{C_0}$$

$$= \frac{1}{\sqrt{1 + \tan^2 \delta}} \frac{1}{C_0 K' \omega}.$$

Since for these samples tan $\delta < 0.3$, then

$$Z \simeq \frac{1}{\omega C_0 K'} = \frac{1}{\omega C}.$$

In order to eliminate the electrode fringe effect and other stray capacitances, the capacitance of the sample is measured with the sample in and removed (KATSUBE and COLLETT, 1971). To avoid the electrode contact effect, the variable air-gap method is used (KATSUBE and COLLETT, 1971). As a result of these measurement procedures

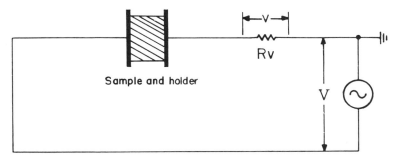

Fig. 1. Circuit for measurement of impedance from 10^3 to 10^7 Hz. V, output voltage of generator; v, voltage across the known resistor (Rv); I, current ($I = v/Rv$).

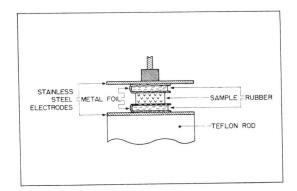

Fig. 2. Sample-electrode contacting method for measurement of loss tangent and dielectric conductivity.

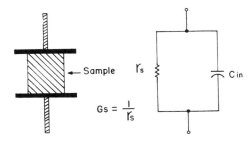

Fig. 3. Equivalent circuit of sample holder when sample is clamped between the electrode plates.

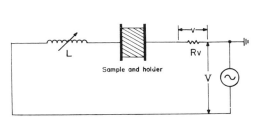

Fig. 4. Circuit of modified Q-meter system. V and v, same as those in Fig. 1; L, inductance.

the permittivity of the sample can be calculated from

$$\varepsilon' = \frac{d}{A} [\Delta C]_{x=0} + \varepsilon_0 \tag{11}$$

where $x =$ thickness of the air-gap between the sample and the electrode plates;
$\Delta C =$ difference of capacitance obtained by measuring with the sample in the holder and removed from it.

$[\Delta C]_{x=0}$ is the value of ΔC when $x = 0$, which is extrapolated from measuring ΔC for various values of x.

For the measurement of loss tangent or the dielectric conductivity, an attempt is made to minimize the sample-electrode contact effects by using metal foil and rubber layers (Fig. 2). The equivalent circuit of the sample holder, with the sample in it, is shown in Fig. 3. The first step in this measurement is to determine the conductance of the sample, G_s. By employing this procedure, the stray capacitance effects are eliminated. The relation between dielectric conductivity and G_s are

$$\sigma' = \frac{d}{A} G_s . \tag{12}$$

Then the loss tangent of the sample can be calculated from the following equation, which is derived from equations (7) and (12):

$$\tan \delta = \frac{\sigma'}{\omega \varepsilon'} = \frac{d}{A} \frac{G_s}{\omega \varepsilon'} . \tag{13}$$

For measurements in the 10^3 to 10^5 Hz range, a high-precision capacitance bridge assembly (GR 1620-A) is used, and $\tan \delta'$ is measured directly from the bridge. This $\tan \delta'$ is based on the equivalent circuit shown in Fig. 3 which contains some stray capacitance, shown by

$$\tan \delta' = \frac{G_s}{\omega C_{in}} . \tag{14}$$

where C_{in} is the value of the capacitance measured on the bridge. C_{in} also contains stray capacitance including the sample capacitance. Therefore the true conductance of the sample is

$$G_s = \omega C_{in} \tan \delta'. \tag{15}$$

For the loss tangent measurement in the frequency range of 10^5 to 10^7 Hz, first the series resistance, R_s, of the sample is measured by the modified Q-meter method (Fig. 4) and then G_s is derived from

$$G_s = \frac{R_s}{Z^2} \tag{16}$$

where Z is the impedance of the sample holder containing the sample.

Results of Measurement

The results of the impedance measurement indicate that these lunar samples can be treated as good dielectric materials which do not show significant relaxations.

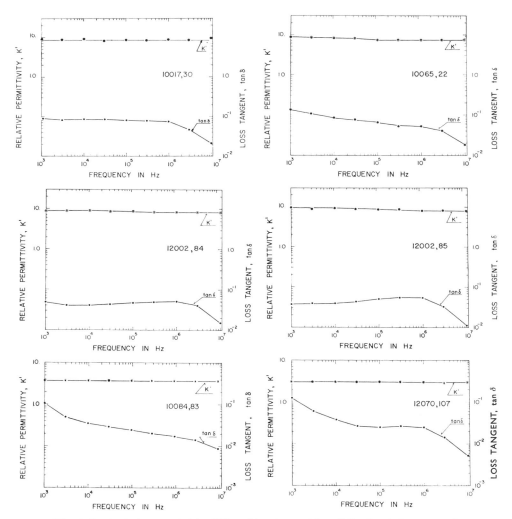

Fig. 5. Real relative permittivity (K') and loss tangent (tan δ) for the six lunar samples.

Results of the measurement for real relative permittivity and loss tangent for the 6 samples are shown in Fig. 5 and for dielectric conductivity in Fig. 6 (see Table 3). The various trends observed in these results are compiled below.

Results of permittivity measurements for samples 10017,30; 10084,83; and 12070,107 indicate no variation with frequency. Results for samples 12002,84; 12002,85; and 10065,22 can be considered constant, but they do show a slight rise in permittivity with decrease in frequency below 3×10^5 Hz.

There is a certain order of permittivity values for samples at frequencies above 3×10^5 Hz, but below this frequency this trend is fairly distorted. Fluctuations of the measured values (Table 3) for sample 10017,30 and the difference between the

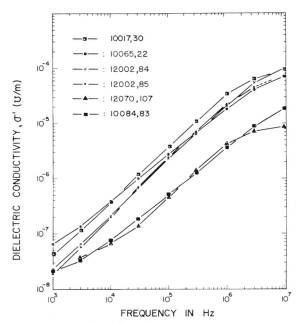

Fig. 6. Dielectric conductivity of the lunar samples.

values for 12002,85 and 12002,84 are probably due to measurement errors caused mainly by the irregular and partially irregular shapes of these samples. However, a comparison of the mean values of relative permittivity for the four sample types in the frequency range from 3×10^5 to 10^7 Hz show a decrease from type A to type D, as seen in Fig. 7.

The loss tangent, in general, decreases with increase in frequency as displayed in Figs. 5 and 8. In detail, the curve shape can be classified into two types. One type exhibits a continuous decrease with a slight relaxation between 10^4 to 10^6 Hz as shown by the fines (type D) and the breccia (type C) samples. The other type show a rather large relaxation or a partial increase between 3×10^3 to 10^6 Hz as displayed by the igneous samples (types A and B). These trends are also seen in terrestrial rocks (SAINT-AMANT and STRANGWAY, 1970; PARKHOMENKO, 1967). Between 10^6 and 10^7 Hz, the rate of decrease of loss tangent is very pronounced for most samples.

The conductivity of lunar samples generally increases with frequency, which is usual for all terrestrial rocks when in a dry state (PARKHOMENKO, 1967; KELLER and FRISCHKNECHT, 1966; and FULLER and WARD, 1970). The solid samples show larger conductivity than the fines as shown in Fig. 6. And among the solid samples, sample 10017,30 generally shows a higher conductivity compared to the others.

Comparing these results with other dielectric measurement results on lunar samples (CHUNG et al., 1970), the general trends are similar, except for the low frequency dispersion seen in their results. Our results suggest constant permittivity with frequency at room temperature which is similar to theirs for measurements of lunar samples at 77°K and for terrestrial basalt in vacuum, but differs from theirs for

Table 3. Summary of dielectric properties of Apollo 11 and Apollo 12 samples*

Electrical parameter	Frequency	10017,30† Type A	10065,22 Type C	10084,83 Type D	12070,107 Type D	12002,84 Type B	12002,85 Type B
K'; real relative	10^3 Hz	9.1	8.8	3.8	3.1	8.8	9.4
permittivity	3×10^3	8.7	8.4	3.8	3.1	9.0	8.5
(dielectric	10^4	9.0	8.0	3.8	3.1	9.0	9.0
constant)	3×10^4	8.1	8.0	3.8	3.1	8.6	8.5
	10^5	8.5	7.3	3.8	3.1	8.8	8.3
	3×10^5	8.3	7.3	3.8	3.1	8.3	8.5
	10^6	8.8	7.3	3.8	3.0	8.3	7.8
	3×10^6	8.5	7.3	3.8	2.9	8.3	7.9
	10^7	9.3(?)	7.3	3.8	3.0	8.3	8.8
tan δ; loss	10^3 Hz	0.090	0.14	0.108	0.128	0.049	0.038
tangent	3×10^3	0.080	1.11	0.051	0.061	0.040	0.039
	10^4	0.085	0.087	0.036	0.038	0.040	0.039
	3×10^4	0.085	0.078	0.030	0.027	0.044	0.044
	10^5	0.080	0.067	0.025	0.025	0.047	0.051
	3×10^5	0.079	0.053	0.021	0.027	0.049	0.056
	10^6	0.075	0.053	0.0175	0.025	0.051	0.056
	3×10^6	0.047	0.040	0.0143	0.0145	0.040	0.034
	10^7	0.021	0.019	0.0089	0.0053	0.0158	0.0114
σ'; dielectric	10^3 Hz	4.2×10^{-8}	6.7×10^{-8}	2.2×10^{-8}	2.2×10^{-8}	2.4×10^{-8}	1.89×10^{-8}
conductivity	3×10^3	1.15×10^{-7}	1.4×10^{-7}	3.2×10^{-8}	3.6×10^{-8}	6.6×10^{-8}	5.7×10^{-8}
in mhos/m	10^4	3.9×10^{-7}	3.9×10^{-7}	7.7×10^{-8}	6.1×10^{-8}	2.1×10^{-7}	1.97×10^{-7}
	3×10^4	1.20×10^{-6}	1.0×10^{-6}	1.90×10^{-7}	1.40×10^{-7}	6.7×10^{-7}	6.8×10^{-7}
	10^5	3.8×10^{-6}	2.8×10^{-6}	5.2×10^{-7}	4.6×10^{-7}	2.3×10^{-6}	2.4×10^{-6}
	3×10^5	1.11×10^{-5}	6.7×10^{-6}	1.31×10^{-6}	1.41×10^{-6}	6.6×10^{-6}	7.1×10^{-6}
	10^6	3.5×10^{-5}	1.9×10^{-5}	3.7×10^{-6}	4.3×10^{-6}	2.1×10^{-5}	2.2×10^{-5}
	3×10^6	6.7×10^{-5}	4.2×10^{-5}	9.1×10^{-6}	7.1×10^{-6}	5.5×10^{-5}	4.4×10^{-5}
	10^7	9.7×10^{-5}	7.1×10^{-5}	1.88×10^{-5}	8.7×10^{-6}	1.0×10^{-4}	—

* Samples measured at room temperature and in a dry atmosphere.
† Measurement precision ($\pm 15\%$ for real relative permittivity, K') for this sample is lower than that of the others, due to its irregular shape.

measurement of lunar samples at room and higher temperatures where their results show low frequency dispersion. Similarly in connection with loss tangent measurements, there is a dispersion seen at the lower frequencies for igneous rocks in their results, whereas our results for the same type of rock do not indicate much dispersion at those frequencies.

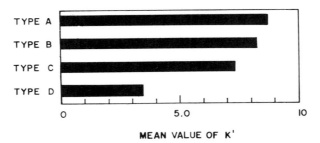

Fig. 7. Comparison of mean values of the real relative permittivity for the four types of lunar material. Values are averaged over the frequency range from 3×10^5 to 10^7 Hz.

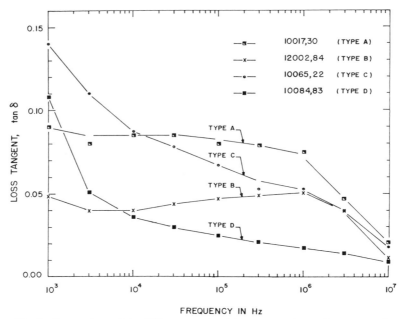

FREQUENCY IN Hz

Fig. 8. Comparison of the different trends of loss tangent for various types of samples.
(Note: the ordinate is linear.)

Discussion

The intrinsic real permittivity of these lunar samples at room temperature appear to be constant with frequency within the range from 10^3 to 10^7 Hz. Presumably this is due to the lack of moisture within the grain boundaries of the samples. All samples used for these measurements have been transported, stored, and measured in a dry nitrogen atmosphere. But samples 12002,84; 12002,85; and 10065,22, which show a slight rise in permittivity at the low frequencies, have been temporarily exposed to room atmosphere in the winter and early spring of 1970 when the room humidity was generally below 55% for an accumulated time of about 2 to 10 hours, to make some primary dielectric tests. This temporary exposure may be the cause of the slight rise in permittivity seen at the lower frequencies.

The values of the intrinsic real permittivity for the lunar samples, perhaps are expressed best by the mean values of the higher frequency measurements, shown in Fig. 7. The decreasing order of these values are from types A, B, C to D. This sequence in permittivity values is perhaps due to two factors: one being the mineral and chemical composition, and the other being the amount of the void volume (vesicles, vugs, grain-boundaries, fractures, microfractures). The small permittivity values of the fines is apparently due to the large void volume between the grains. The difference of permittivity between the igneous rocks (types A and B) and the breccia (type C) is perhaps mainly due to the difference of void volume, in which the latter is presumably larger than the former. The lunar breccia is made up of fragments of lunar igneous rocks of types A and B which contain vesicles and vugs (LSIC, Apollo 11,

1969, pp. 44–46). In addition the breccia contains a large amount of voids and micro-fractures due to shock or impact metamorphism (LSIC, Apollo 11, 1969, pp. 47–49). These facts suggest that the total void space of the lunar breccia could be larger than that of the igneous rocks. The densities of the igneous rocks (see Table 1), being larger than that of the breccia sample, may be an indication of this difference of the total void space. The ilmenite content of sample 10017 (type A) is 15–20% compared to 3–4% (Table 2) in sample 12002 (type B), and is presumably the main reason for the permittivity of 10017 to be larger than that of 12002.

The reason that the fines show lower values for loss tangent and dielectric conduc-tivity at the higher frequencies as compared to the rocks is due to the same reason as that given for the difference in permittivity of rocks and fines, that is, due to large void volume between the grains. Sample 10017 (type A) shows a larger loss tangent and dielectric conductivity at the higher frequencies, compared to the other rocks. This is due, perhaps to the higher content of ilmenite and presumably to a certain extent to the low void space compared to the others. No basic data has been located that could be used to explain, sufficiently, the cause of the rapid decrease of loss tangent at the frequencies between 10^6 and 10^7 Hz. Preliminary studies on this subject suggest that this rapid decrease may indicate the ending of grain boundary effect which, perhaps, started at the lower frequencies. Or it might indicate the existence of a slight Debye relaxation effect, which is not pronounced enough to show up on the real permittivity curves. It is not thought to be due to measurement error, because of extensive caution taken for measurements at those frequencies.

The dielectric measurements of CHUNG et al. (1970) at room temperature show a low frequency dispersion which is not exhibited by our measurements. This low frequency dispersion is perhaps due to moisture as suggested by them and indicated above. The dielectric measurements on ice by SMYTH and HITCHCOCK (see KELLER 1966, p. 570) suggest that the reason for the lower frequency dispersion to disappear at low temperature in their work may be due to the freezing of the moisture within the grain boundaries. The loss tangent of the igneous rocks in our results do not show a low frequency dispersion as they do in those of CHUNG et al. (1970). This may be because of the moisture free condition under which our samples were measured.

The abundance of ilmenite in the lunar rocks is perhaps the most outstanding difference from terrestrial basalts, in the aspect of major mineral composition and in relation to factors which may affect the dielectric properties. Since ilmenite is a semiconductive material (KELLER and FRISCHKNECHT, 1966, p. 6) its abundance can be considered a factor which increases the bulk permittivity of the rock in which it is contained. However, actually the permittivity values of terrestrial basalts vary over a wide range. CAMPBELL and ULRICHS (1969) report relative permittivity values of basalt in the range from 5 to 10; SAINT-AMANT and STRANGWAY (1970) 10; CHUNG et al. (1970) 6 and 7; VOLOROVICH, TARASOV and BONDARENKO (PARKHOMENKO 1967, p. 40) 10 and 16, and our measured values on terrestrial basalt are 14 to 15. Our basalt is a trap-amygdaloid from Nova Scotia, having few vesicles and a density of 2.9 gm/cc. Therefore the permittivity value of this basalt is perhaps nearer to the true value of the void-free material, compared particularly to the vesicular type basalts. This indicates that the relative permittivity of the void-free material of basalt can be quite high, and that the relative permittivity of the bulk rock can vary from any value near that of air to that of the void-free material, depending on the environment in which it was formed. The same statement can be made for lunar rocks as well, that is, the true permittivity value of the void-free material is presumably larger than the values reported on so far, and it can vary according to the amount of the void space. A type A sample has a void space which is in the order of 15% (see LSIC Apollo 11, 1969, p. 43). Therefore, it may be possible that the permittivity of the void-free material of the lunar rocks is larger than that of terrestrial basalts due to the ilmenite, as stated by CHUNG et al. (1970). But as yet no evidence has been obtained in order to make a general statement as such.

CONCLUSIONS

(1) The intrinsic real permittivity or dielectric constant of these lunar samples appear to be constant with frequency, over the frequency range from 10^3 to 10^7 Hz, due to the lack of moisture within the grain boundaries.

(2) Comparing values of the intrinsic real permittivity for the four sample types, a decrease in the order from types A ($K' = 8.7$), B (8.2), C (7.3) to D (3.4) can be seen. This is perhaps due to two factors: one being the mineral composition and the other being the amount of void space within the sample.

(3) Evidence is not sufficient, to date, to state whether the permittivity of the lunar samples is larger than that of the terrestrial basalts, or vice versa. Permittivity of terrestrial basalts vary over a wide range, and there is a lack of studies and data in this field which prevents a sufficient comparison. However, the abundance of ilmenite may cause the void-free material of the lunar rocks to exhibit a larger permittivity than that of the terrestrial basalts.

(4) The following trends are seen for loss tangent and dielectric conductivity at the higher frequencies: type A shows larger values than that for other rocks, and the rocks generally show larger values than those for the fines. This is presumably due to mineral composition, void space, and rock fabrics.

(5) The low frequency dispersion for permittivity of all rocks and for the loss tangent for igneous rocks, which have been reported by other scientists, is perhaps the result of the Maxwell-Wagner effect; in particular it may be due to moisture within the grain boundaries. It is therefore not an intrinsic characteristic of the moisture-free lunar rocks.

(6) The dielectric studies on dry rocks, up to the present, do not seem to indicate outstanding differences between the general dielectric properties of lunar rocks and terrestrial basalts.

Acknowledgments—We are very grateful to Dr. A. BECKER for discussion and guidance throughout the project. We thank R. H. AHRENS, P.Eng., for helpful discussions in connection with the electronic measuring techniques, and to J. FRECHETTE for carrying out part of the measurements, and others who are connected with our laboratory for their assistance. We also thank H. R. STEACY, curator of the GSC Systematic Reference Series, National Mineral Collection, for supplying terrestrial rock specimens.

REFERENCES

ANDERSON O. L., SCHOLZ C., SOGA N., WARREN N., and SCHREIBER E. (1970) Elastic properties of microbreccia, igneous rock and lunar fines from the Apollo 11 mission. *Proc. Apollo 11 Lunar Sci. Conf., Geochim. Cosmochim. Acta.* Suppl. 1, Vol. 3, pp. 1959–1973. Pergamon.

ASTM (American Society for Testing Materials) (1968) Standard methods of testing for ac loss characteristics and dielectric constant of solid electrical insulating materials, ASTM D 150–68, 29–54.

CAMPBELL M. J. and ULRICHS J. (1969) Electrical properties of rocks and their significance for lunar radar observations. *J. Geophys. Res.* **74,** 5867–5881.

CHUNG D. H., WESTPHAL W. B., and SIMMONS G. (1970) Dielectric properties of Apollo 11 lunar samples and their comparison with earth materials. *J. Geophys. Res.* **75,** 6524–6531.

FULLER B. D. and WARD S. H. (1970) Linear system description of the electrical parameters of rocks, IEEE Transactions on Geoscience Electronics, GE-8, No. 1.

KATSUBE T. J. and COLLETT L. S. (1971) in preparation.

KELLER G. V. and FRISCHKNECHT F. C. (1966) *Electrical Methods in Geophysical Prospecting.* Pergamon.

KELLER G. V. (1966) Electrical properties of rocks and minerals; in *Handbook of Physical Constants,* Geol. Soc. Amer. Mem. 97, 553–571.

LSIC (Lunar Sample Information Catalog) Apollo 11 (1969) Lunar Receiving Laboratory, Science and Applications Directorate.

LSIC (Lunar Sample Information Catalog) Apollo 12 (1970) Lunar Receiving Laboratory, Science and Applications Directorate.

PARKHOMENKO E. I. (1967) *Electrical Properties of Rocks.* Plenum Press.

SAINT-AMANT M. and STRANGWAY D. W. (1970) Dielectric properties of dry geologic materials *Geophys.* **35,** 624–645.

SCHMITT H. H., LOTGREN G., SWANN G. A., and SIMMONS G. (1970) Apollo 11 samples: Introduction. *Proc. Apollo 11 Lunar Sci. Conf., Geochim. Cosmochim. Acta* Suppl. 1, Vol. 1, pp. 1–55. Pergamon.

VON HIPPEL A. R. (1954) *Dielectric Materials and Applications,* MIT Press.

Proceedings of the Second Lunar Science Conference, Vol. 3, pp. 2381–2390
The M.I.T. Press, 1971.

Dielectric behavior of lunar samples: Electromagnetic probing of the lunar interior

D. H. CHUNG,* W. B. WESTPHAL,† and GENE SIMMONS*‡
Massachusetts Institute of Technology, Cambridge, Massachusetts 02139

(*Received* 22 *February* 1971; *accepted in revised form* 17 *March* 1971)

Abstract—Presented first in this paper are measurements of apparent dielectric constant and dissipation factor of Apollo 12 lunar samples 12002,58, 12022,60, and 12022,95, over a range of frequencies from 100 Hz to 10 MHz and temperatures from −196°C to +200°C. The dielectric properties determined on earth basalts and simulated lunar materials with the composition of Surveyor V analyses, along with the corresponding properties measured on Apollo 11 lunar samples, are used to characterize the dielectric properties of these lunar samples. These data are used as the basis of two major models (wet and dry) of lunar depth variation in the complex dielectric constant to investigate the EM probing of the moon (at a few kHz frequency) on a horizontal scale of about 100 km. Our conclusions follow.

(1) The lunar samples have higher values of dielectric constant than earth materials of similar chemical compositions, an effect probably due to high conductivity materials present in the lunar samples. The high frequency dielectric constant of the lunar igneous samples is about 7 to 14 at room temperature, and increases slightly with increasing temperature.

(2) The lunar samples show greater dielectric losses than terrestrial basalts. The high-frequency loss tangents for the lunar samples range from 0.03 to 0.2 at room temperature, and they are temperature dependent.

(3) The lunar samples exhibit a temperature-dependent low-frequency dispersion, a characteristic of excessive impurity charges present in these lunar samples.

(4) If significant moisture is present in the moon, then a transverse magnetic surface wave mode will be possible and its characteristics will be sensitive to the depth of the ice-water transition (expected at about 1 km depth) and to the dielectric properties of the material above the transition depth. If the moon is dry, as it seems most likely at present, then a thermally activated region of a conductive basement is expected at a depth of about 100 km, and the surface wave will be sensitive to dielectric properties of this large region.

INTRODUCTION

MEASUREMENT OF THE dielectric properties of lunar samples has two objectives: (1) to study properties of rocks which were formed under extraterrestrial environments; and (2) to provide the basic data necessary for planning future exploration of the moon. The two principal electrical properties of interest are the dielectric constant and the conductivity. The dielectric loss is also of interest, since this gives a measure of the relative amounts of charge transferred in conduction and stored in polarization. Presented first in this paper are measurements of apparent dielectric constant and dissipation factor of Apollo 12 lunar samples 12002,58, 12022,60, and 12022,95, over a range of frequencies from 100 Hz to 10 MHz and temperatures from −196°C to +200°C. The dielectric properties determined on terrestrial basalts and simulated

---

* Department of Earth and Planetary Sciences; † Laboratory for Insulation Research; ‡ Manned Spacecraft Center, NASA, Code TA, Houston, Texas 77058. On leave from MIT.

Table 1. Mineralogical composition of lunar samples.

Mineral	10020	10057	10046	12002	12022
Plagioclase	25	20	20	18	26
Pyroxene	53	51	57	59	30
Olivine	4	—	—	11	33
Ilmenite	15	16	9	8	9
Troilite	0.6	—	—		0.2
"SiO₂"	1	0.1	—	5*	—
"Glass"	—	10	14		—
Unidentified	1.4	2.9	—		2

* Includes cristobalite, glass vugs, and possibly zircon (D. R. Wones, personal communications, 1970).

lunar materials with the composition of Surveyor V analyses, along with the corresponding properties measured on Apollo 11 lunar samples, are used to characterize the dielectric properties of lunar samples. Next, with these data applied to two major models (wet and dry) of lunar depth variation in the complex dielectric constant, the electromagnetic probing of the moon (at a few kHz frequency) on a horizontal scale of about 100 km is investigated. We assume stratified media that consist of N electromagnetically linear, isotropic, homogeneous layers.

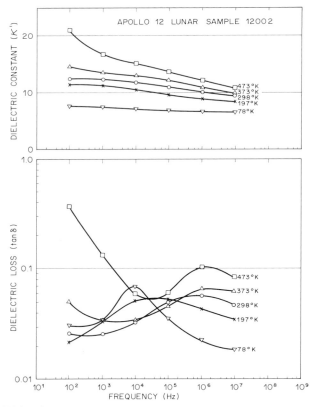

Fig. 1. Dielectric constant and loss tangent of sample 12002,58 as a function of frequency and temperature.

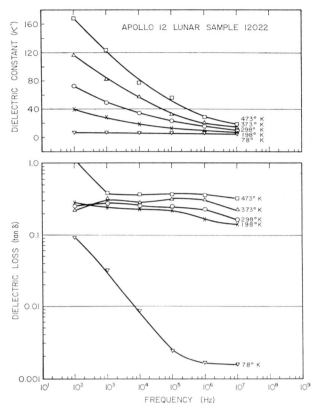

Fig. 2. Dielectric constant and loss tangent of samples 12022,60 and 12022,95 as a function of frequency and temperature.

Experimental Methods

Three samples in the form of a rectangular prism of about $2 \times 2 \times 1$ cm were provided by NASA Manned Spacecraft Center. These samples were coded as 12002,58, 12022,60, and 12022,95; a description of them may be found in *Apollo 12 Lunar Samples Information Catalog* (NASA/MSC, Houston, Texas). Samples were dense crystalline rocks with bulk densities of about 3.30 (for 12002) and 3.32 gm/cm³ (for 12022 series). The sample 12002,58 was a speckled brownish-gray rock with some dust-like particles adhering to the surface, and was a fine-to-medium-grained holocrystalline basalt. The samples of the 12022 series were dark green, medium-grained, crystalline rocks. Our 12022 series samples were slightly porphyritic, rich in olivine (about 33% by volume). Details of mineralogy of these lunar samples, together with some other materials, are given in Table 1.

The two-terminal capacitance substitution method described in detail by Von Hippel (1954) was used throughout this study. The method had been used in our earlier study of Apollo 11 lunar samples (Chung *et al.*, 1970). Methods of transportation, handling, heating, and drying of lunar samples, along with other experimental procedure, utilized in the present work were exactly the same as ones described in that report, and this same description will not be repeated here.

Dielectric Properties of Lunar Samples

Values of the dielectric constant and loss tangent of lunar sample 12002,58 are presented in Fig. 1 as a function of frequency and temperature. The values of these properties obtained for 12022,95 are shown in Fig. 2. Data obtained on sample

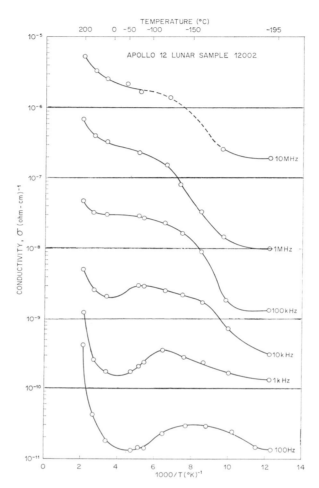

Fig. 3. Electrical conductivity of sample 12002,58 as a function of frequency and temperature.

12002,58 represent the dielectric properties of *dry* lunar igneous rocks. (Sample 12002,58 had been baked at 300°C in vacuum for 10 hours before the dielectric measurements. Our experience with terrestrial rocks indicates that this temperature and time required to remove moisture from a rock sample is sufficient to warrant our sample 12002,58 is free from water contamination. Moisture, once trapped in cracks in rocks, is very difficult to remove by ordinary baking procedure. The baking of rock samples must be carried out in vacuum at elevated temperatures.) The low frequency values of the dielectric constant and loss tangent measured on lunar samples 12022,60 and 12022,95 represent the dielectric properties of water-contaminated lunar samples, and they compare well with our previous measurements on Apollo 11 lunar samples.

The electrical conductivity of lunar sample 12002,58 as a function of frequency and temperature is presented in Fig. 3. The conductivity of samples 12022,60 and 12022,95 is shown in Fig. 4.

All the lunar samples show a distinct dispersion at low frequency. The dispersion characteristics are larger for those lunar samples contaminated with water (see Chung *et al.*, 1970). The extent of low-frequency dispersion observed for sample

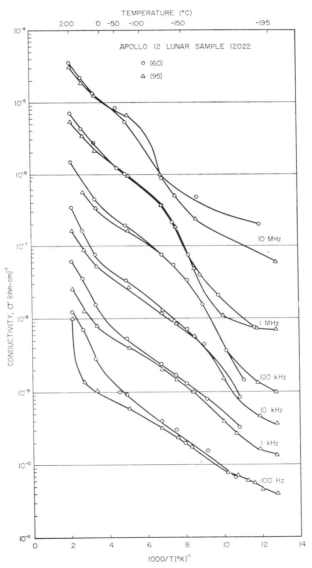

Fig. 4. Electrical conductivity of samples 12022,60 and 12022,95 as a function of frequency and temperature.

Table 2. Summary of the dielectric properties of Apollo 11 and Apollo 12 lunar samples and comparison with earth materials.

Material	Bulk density (gm/cm³)	High freq. dielectric constant	Dielectric loss	Low-temperature electrical conductivity (ohm-cm)⁻¹
Lunar samples				
10020	3.18	10–15	0.09–0.2	10^{-9}–10^{-11}
10057	2.88	9–13	0.09–0.2	10^{-9}–10^{-11}
10046	2.21	6–9	0.05–0.09	10^{-10}–10^{-11}
12002	3.30	8–10	0.02–0.09	10^{-10}–10^{-12}
12022	3.32	7–14	0.002–0.2	10^{-9}–10^{-11}
Earth basalts				
Hawaiian Oahu 2.68	2.68	7	0.03	10^{-10}–10^{-12}
Cape Neddick	2.60	6	0.05	10^{-10}–10^{-12}
Simulated sample				
LMT-J	3.01	6	0.002	10^{-10}–10^{-12}

12002,58 is at least one order of magnitude smaller than that observed for water-contaminated lunar samples. As stated before, however, the general behavior of the dielectric characteristics of all these lunar samples is somewhat like that associated with the Maxwell-Wagner effect. The presence of grains or particles of materials of various dielectric properties and conductivities probably causes the low-frequency dispersion. A discussion on the nature of this low-frequency dispersion observed for terrestrial rocks has been presented by Parkhomenko (1967) and Strangway (1969) and also by Saint-Amant and Strangway (1970); the low-frequency dispersion observed by us in lunar samples probably is also associated with the same origin as that of terrestrial rocks. Many different phenomena occur in rocks that contribute to electrical polarization. These electrical polarizations are more intense at low frequencies in general, and they are strongly temperature-dependent.

Table 2 summarizes the dielectric properties of lunar samples, together with some typical data measured on terrestrial materials. In the Second Lunar Science Conference, Collett and Katsube (1971) reported the electrical properties of samples 12002,84 and 12002,85; results of these authors compare very well with our data on 12002,58.

Dielectric Properties of the Lunar Interior

Strangway (1969) presented a discussion of electrical properties of lunar regolith (in connection with a lunar probing at megahertz frequencies) by calculating a depth penetration of radio frequency waves. Our measurement of the dielectric properties of lunar samples supports Strangway's estimate for the penetration depth of megahertz frequency waves in the first few kilometers of lunar interior. In the present paper, we are concerned with electrical properties of lunar interior down to about 150 kilometers depth. We consider a possibility of lunar probing with use of kilohertz frequency waves.

The electromagnetic properties of the moon on a horizontal scale of about 100 km can be approximated by a horizontally stratified medium consisting of N electromagnetically linear, isotropic, homogeneous layers. For oscillating fields the electromagnetic properties of each layer are contained formally in dimensionless parameters like K^* and K^c, which are, respectively, the relative permeability and the complex relative dielectric constant. The electromagnetic propagation characteristics of the solar wind plasma at the earth's orbit can be described adequately at frequencies above a few hertz by the tensor form of K^c, provided effects of collision and ion motion are negligible (see, for example, Rossi and Olbert, 1970). At frequencies much smaller than the electron cyclotron frequency, the influence of the solar wind magnetic field becomes negligibly small and K^c is isotropic. The solar wind plasma on the sunlit portion of the moon impinges directly on the lunar surface, as was noted also by Johnson and Midgley (1968). At frequencies above 1 kHz, a satisfactory model is that of an isotropic halfspace overlying a stratified moon. A formal presentation of mathematical theory will be published separately at a later date. In this paper we present only the results of some computation.

Two major models of lunar depth variation in K^c are (1) a wet moon and (2) a dry moon. In the wet model, we assume there is a transition of ice to water at lunar depth of about 1 km or so (England et al., 1968). In the dry model we assume a thermally activated conductive basement at about 100 km. Based on our values of the dielectric properties of lunar samples given by these two lunar models, Reisz (1970) calculated the relative phase velocity given by ($\omega/k_\rho c$) and associated attenuation coefficient ρ (i.e., the imaginary part of k_ρ) as a function of depth and frequency. The parameter ω is the angular frequency, k_ρ is the wave propagation vector, and c is the velocity of light.

Figure 5 is a graphical representation of the effect of the variation of depth of

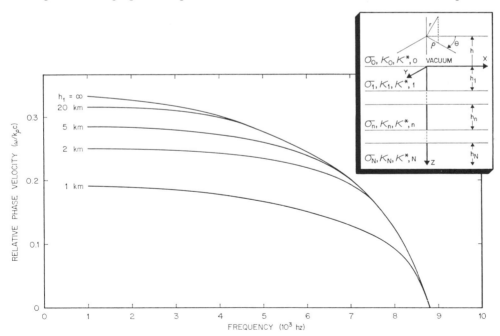

Fig. 5. Coordinate system for a horizontally layered lunar model and a depth variation of the relative phase velocity with frequency. The plasma frequency assumed is 27 kHz and $\kappa_1 = 9$.

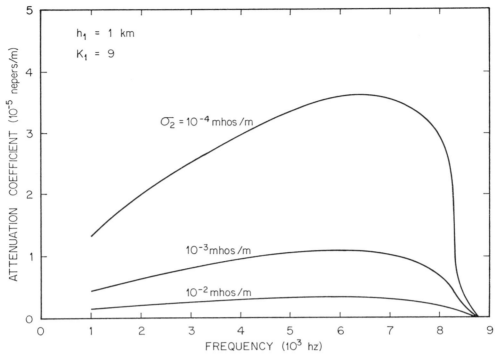

Fig. 6. Depth variation of the attenuation coefficient with frequency for different values of the basement conductivity.

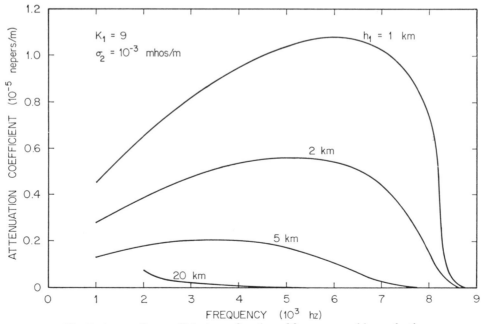

Fig. 7. Attenuation coefficient as a function of frequency and lunar depth.

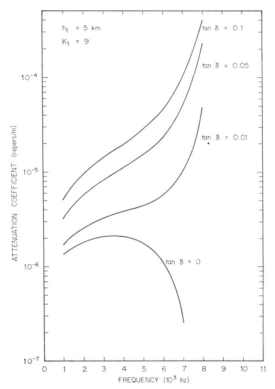

Fig. 8. Frequency variation of the attenuation coefficient for different values of loss tangent.

the upper layer on the relative phase velocity as a function of frequency. Presented in Fig. 6 is the associated attenuation coefficient. Clearly evident from these two figures is the strong possibility of EM probing of the moon at frequencies below 8 kHz. The attenuation coefficient for the wet model is shown in Fig. 7 as a function of frequency and the basement conductivity. The attenuation coefficient for the dry model as a function of frequency is presented in Fig. 8 for several different representative values of loss tangent; it illustrates several points of interest to electromagnetic probing. The first is that megahertz frequencies cannot provide information about the lunar interior at a depth greater than a few kilometers. Use of 8 kHz frequency would probably run into trouble due to high attenuation, as is shown in Fig. 8. The best frequency range for use in passive probing would therefore be 4–7 kHz. Since one nepers/meter corresponds to about 4.346 dB/meter, the attenuation expected is in the order of 10^{-2} dB/km over this range of frequency.

Acknowledgments—We are grateful to Professor A. R. von Hippel for his encouragement and interest in our work, and to Mr. A. C. Reisz for his computation. We express our appreciation to Dr. D. R. Wones for sample analysis. Our sincere thanks are also expressed to Dr. S. H. Ward and Dr. D. W. Strangway for their review of this paper and for helpful suggestions. Financial support was provided by the National Aeronautics and Space Administration under Contract NAS 9-8102 and Grant NGR-22-009-597.

REFERENCES

COLLETT L. S. and KATSUBE T. J. (1971) Electrical properties of Apollo 11 and Apollo 12 rock samples. Second Lunar Science Conference (unpublished proceedings).

CHUNG D. H., WESTPHAL W. B., and SIMMONS G. (1970) Dielectric properties of Apollo 11 lunar samples and their comparison with earth materials. *J. Geophys. Res.* **75,** 6524–6531.

ENGLAND A. W., SIMMONS G., and STRANGWAY D. W. (1968) Electrical conductivity of the moon. *J. Geophys. Res.* **73,** 3219–3226.

JOHNSON F. S. and MIDGLEY J. E. (1968) Notes on the lunar magnetosphere. *J. Geophys. Res.* **73,** 1523–1532.

REISZ A. C. (1970) Electromagnetic probing of the lunar interior in the frequency region from 1 to 10 kHz. M.Sc. thesis, Massachusetts Institute of Technology, Cambridge, Mass.

PARKHOMENKO E. I. (1967) *Electrical Properties of Rocks.* Plenum Press.

ROSSI B. and OLBERT S. (1970) *Introduction to the Physics of Space.* McGraw-Hill.

SAINT-AMANT M. and STRANGWAY D. W. (1970) Dielectric properties of dry, geological materials. *Geophysics* **35,** 624–645.

STRANGWAY D. W. (1969) Moon: Electrical properties of the uppermost layers. *Science* **165,** 1012–1013.

VON HIPPEL A. R. (1954) *Dielectric Materials and Applications.* MIT Press.

Proceedings of the Second Lunar Science Conference, Vol. 3, pp. 2391–2413
The M.I.T. Press, 1971.

The Apollo 12 magnetometer experiment: Internal lunar properties from transient and steady magnetic field measurements

PALMER DYAL and CURTIS W. PARKIN

NASA-Ames Research Center, Moffett Field, California 94035

(*Received* 22 *February* 1971; *accepted in revised form* 31 *March* 1971)

Abstract—The electrical conductivity of the lunar interior has been determined from analysis of magnetic field step-transient events measured simultaneously on the lunar surface and in circumlunar orbit. The data fit a spherically symmetric three-layer lunar model having a thin outer crust of very low electrical conductivity. The intermediate layer of radial thickness $R_1 - R_2$, where $0.95 R_{moon} \leq R_1 < R_{moon}$ and $R_2 \sim 0.6 R_{moon}$, has electrical conductivity $\sigma_1 \sim 10^{-4}$ mhos/meter; the inner core has radius $R_2 \sim 0.6 R_{moon}$ and conductivity $\sigma_2 \sim 10^{-2}$ mhos/meter. For the example of an olivine moon, the temperatures of the layers are as follows: crust, $\leqslant 440°$K; intermediate layer, $\sim 810°$K; core, $\sim 1240°$K. The bulk relative magnetic permeability of the moon has been calculated to be $\mu/\mu_0 = 1.03 \pm 0.13$. A steady field of 38 ± 3 gammas has been measured at the Apollo 12 site. Restrictions on size and location of the source are calculated and discussed. Three-hour averages of the magnetic field during the first lunation indicate a compression of the steady field by the solar wind.

INTRODUCTION

THE APOLLO 12 lunar surface magnetometer (LSM) was deployed on the moon in order to study the intrinsic lunar magnetic field and the global response of the moon to large-scale solar and terrestrial magnetic fields. The whole-moon inductive response to magnetic step transients (e.g., tangential discontinuities) in the solar wind is studied by simultaneously measuring the magnetic field on the lunar surface and in circumlunar orbit. The Apollo 12 magnetometer measures the vector sum of the lunar response field and the external driving field at the surface, while the Explorer 35 magnetometer, when positioned in the free-streaming solar wind, measures the driving field alone.

In this paper we study the surface magnetic field in the different regions of the lunar orbit and investigate electrical properties of the lunar interior. Three-hour magnetic field averages during one lunation show contrasts in field amplitude and time characteristics as the moon orbits the earth. Measurements in a quiet region of the free-streaming solar wind allow an upper limit to be placed on the unipolar induction field and the near-surface electrical conductivity.

The time-dependent characteristics of induced eddy-current fields are analyzed for magnetometer data obtained during lunar nighttime at the Apollo 12 site. These time-dependent decay characteristics are a function of the interior electrical conductivity, which in turn is a strong function of temperature. We calculate a conductivity profile of the lunar interior and determine a corresponding temperature profile for assumed material compositions of the moon. Analysis of data taken while the moon was immersed in the quiet geomagnetic tail yields a value of the lunar magnetic permeability and the local steady field at the Apollo 12 site.

Fig. 1. The Apollo 12 lunar surface magnetometer deployed on the moon in Oceanus Procellarum. Sensors are at the top ends of the booms and approximately 75 cm above the lunar surface.

EXPERIMENTAL TECHNIQUE

The magnetometer developed for the Apollo 12 mission is shown in Fig. 1, and the critical instrument properties are discussed in this section. A detailed description of the instrument is reported by DYAL *et al.* (1970a).

1. *Instrument geometry*

The three orthogonal vector components of the magnetic field are measured by three fluxgate sensors (GEYGER, 1964; GORDON *et al.*, 1965) located at the ends of three 100-cm-long orthogonal booms. The sensors are separated from each other by 150 cm and are 75 cm above the ground. Prelaunch test results show that magnetic fields due to artificial sources are less than 0.2 gamma at the sensor locations.

The Apollo 12 magnetometer is deployed at 23.35° west longitude, 2.97° south latitude in selenographic coordinates on the lunar surface. The instrument geometry as seen in Fig. 1 is such that each sensor is directed approximately 35° above the horizontal with the Z-sensor pointed toward the east, the X-sensor toward the northwest, and the Y-sensor completing a right-hand orthogonal system. Orientation measurements with respect to lunar coordinates are made with two devices. A shadow graph is used by the astronaut to align and measure the azimuthal orientation with respect to the moon-to-sun line to an accuracy of 0.5°. Gravity-level sensors measure the tilt angle to an accuracy of 0.2° every 4.8 seconds.

Table 1. Apollo 12 magnetometer characteristics.

Parameter	Value
Range	0 to ± 400 gammas
	0 to ± 200 gammas
	0 to ± 100 gammas
Resolution	0.2 gamma
Frequency response	dc to 3 Hz
Sensor geometry	Three orthogonal sensors at the end of 100-cm booms. Orientation determination to within 1° in lunar coordinates
Analog zero determination	180° flip of sensor
Internal calibration	0, ± 25, ± 50, and $\pm 75 \%$ of full scale.
Power	3.4 W average in daytime
Weight	8.9 kg
Size	25 × 28 × 63 cm

The gradient of the magnetic field as well as the vector components can be measured by sending commands to operate three motors in the instrument which rotate the sensors such that all simultaneously align parallel first to one of the other two boom axes, then to each of the other two boom axes in turn. This rotating alignment permits the vector gradient to be calculated in the plane of the sensors and also permits an independent measurement of the magnetic field vector at each sensor position.

2. *Internal calibration and data processing*

Long-term stability is attained by extensive use of digital circuitry and by internally calibrating the instrument every 24 hours and mechanically rotating each sensor by 180° in order to determine the sensor zero offset. Internal data processing of the magnetic-field measurements requires the major portion of the magnetometer electronics. The analog output of the sensor electronics is internally processed by a low-pass digital filter and a telemetry encoder; the output is transmitted to earth via the central-station *S*-band transmitter.

The magnetometer has two data samplers: the analog-to-digital converter (26.5 samples/second) and the central-station telemetry encoder (3.3 samples/second). The prealias filter following the sensor electronics has attenuations of 3 db at 1.7 Hz, 64 db at 26.5 Hz, and 58 db at the Nyquist frequency (13.2 Hz), with an attenuation rate of 22 db/octave. The four-pole Bessel digital filter limits the alias error to less than 0.05% and has less than 1% overshoot for a step-function response. This filter has an attenuation of 3 db at 0.3 Hz and 48 db at the telemetry-sampling Nyquist frequency (1.6 Hz) and has phase response that is linear with frequency. The response of the entire magnetometer measurement system to a step function input is shown in Fig. 2. The digital filter can be bypassed by ground command in order to pass higher frequency information.

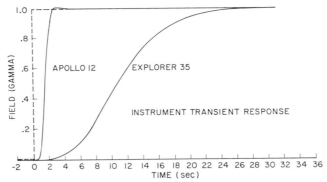

Fig. 2. Laboratory measurements comparing responses of the Apollo 12 and Explorer 35 magnetometers to a 1.0-gamma magnetic field step input.

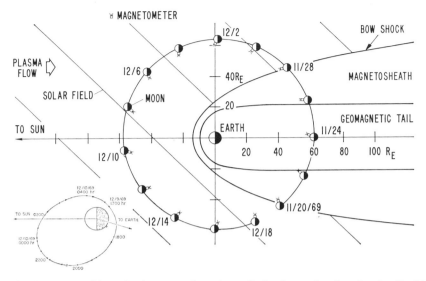

Fig. 3. Lunar orbit projection onto the solar ecliptic plane, showing the Apollo 12 magnetometer during the first post-deployment lunation, 1969. During a complete revolution around the earth, the magnetometer passes through the earth's bow shock, the magnetosheath, the geomagnetic tail, and the interplanetary region dominated by solar plasma fields. The insert shows the Explorer 35 orbit around the moon, projected onto the solar ecliptic plane. The period of revolution is 11.5 hours. This particular Explorer 35 orbit shows the nighttime positions of the Apollo 12 and Explorer 35 magnetometers at the time of the transient event shown in Fig. 10.

3. *Explorer 35 magnetometer*

The ambient steady-state and time-dependent magnetic fields in the lunar environment are measured by the Explorer 35 satellite magnetometer. The satellite has an orbital period of 11.5 hours, aposelene of 9390 km, and periselene of 2570 km (see Fig. 3 insert). The Explorer 35 magnetometer measures three magnetic-field vector components every 6.14 seconds and has an alias filter with 18-db attenuation at the Nyquist frequency (0.08 Hz) of the spacecraft data-sampling system. This instrument has a phase shift linear with frequency, and its step-function response is slower than that of the Apollo 12 instrument (see Fig. 2). For further information on the Explorer 35 magnetometer, see SONETT *et al.* (1967).

MAGNETIC FIELD AVERAGES DURING ONE LUNATION

The electromagnetic environment at the moon varies strongly with the lunar spatial position during each orbit around the earth (see Fig. 3). The electromagnetic properties of the geomagnetic tail, magnetosheath, and free-streaming solar wind are distinct and cause different responses by the moon.

The total magnetic field $\vec{B}_A$ measured at the lunar surface by Apollo 12 will be the vector sum of the following fields:

$$\vec{B}_A = \vec{B}_E + \vec{B}_\mu + \vec{B}_S + \vec{B}_T + \vec{B}_P + \vec{B}_D + \vec{B}_F \qquad (1)$$

where $\vec{B}_E$ is the total external (solar or terrestrial) driving magnetic field, measured by Explorer 35 while outside the lunar cavity; $\vec{B}_\mu$ is the field induced in lunar permeable material; B_S is the steady remanent field at the site; B_T is the toroidal field corresponding to unipolar currents driven through the moon by the $\vec{v} \times \vec{B}_E$ electric field; $\vec{B}_P$ is the poloidal field due to eddy currents induced in the lunar interior by changing external fields; $\vec{B}_F$ is the field associated with the hydromagnetic solar wind flow past the moon; and $\vec{B}_D$ is the field due to the diamagnetic lunar cavity. Each of these vector fields can be investigated by choosing times in the lunar orbit when one field mode predominates.

1. Apollo 12 magnetometer averages

A plot of the total surface magnetic field measured at the Apollo 12 site during the first lunation is shown in Fig. 4. All data are expressed in the ALSEP surface coordinate system $(\hat{x}, \hat{y}, \hat{z})$ which has its origin at the Apollo 12 magnetometer site; $\hat{x}$ is directed radially outward from the surface, while $\hat{y}$ and $\hat{z}$ are tangent to the surface, directed eastward and northward, respectively. The plots represent vector averages over a three-hour period and the main features characterizing the different

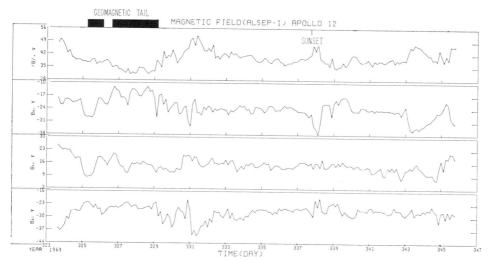

Fig. 4. Three-hour averages of the total surface magnetic field measured by the Apollo 12 magnetometer during the first postdeployment lunation (19 November–12 December, 1969). The lower three graphs are the field components expressed in the ALSEP coordinate system which has its origin on the lunar surface at the Apollo 12 site. The x-axis (corresponding to B_{Ax}) is directed radially outward from the lunar surface; the y-axis (B_{Ay}) and z-axis (B_{Az}) are tangential to the surface, directed eastward and northward, respectively. The top graph is the field magnitude

$$\vec{B}_A = (B_{Ax}^2 + B_{Ay}^2 + B_{Az}^2)^{1/2}.$$

Refer to the orbit plot (Fig. 3) to locate the position of the moon relative to the earth-sun line.

field regions of the lunar orbit are shown in Fig. 3. The external driving field, interaction fields, and internal induced fields given in equation (1) are superimposed upon the intrinsic steady field $\vec{B}_S$ at the surface site to give the resultant measured field $\vec{B}_A$.

2. Magnetic field difference averages

Magnetic field data $\vec{B}_E$ (obtained from the Explorer 35 magnetometer during times when the satellite was in sunlight) and the intrinsic lunar steady field $\vec{B}_S$ are vectorially subtracted from the Apollo 12 field $\vec{B}_A$ to give $\overrightarrow{\Delta B} = \vec{B}_A - (\vec{B}_E + \vec{B}_S)$. These three-hour average differences are shown in Fig. 5 along with the Kp index, which shows that large-amplitude effects (e.g., on days 326, 331, and 340) generally correlate with solar activity.

Two primary effects are obvious in the component plots of Fig. 5. There is an increase in the magnitude $|\overrightarrow{\Delta B}|$ measured during lunar nighttime at the Apollo 12 site, which is consistent with the effects due to plasma diamagnetism reported by

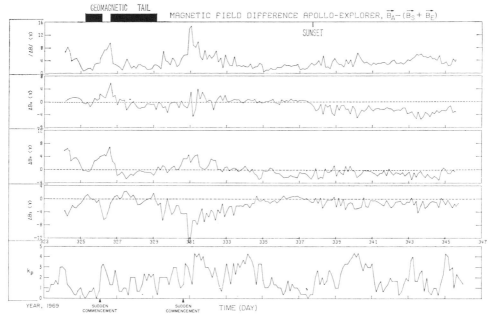

Fig. 5. Three-hour averages of the total surface magnetic field B_A (measured by the Apollo 12 magnetometer) minus the external driving field $\vec{B}_E$ (measured by the lunar orbiting Explorer 35 magnetometer) and the local steady field $\vec{B}_S$ at the Apollo 12 site. The middle three graphs, expressed in the ALSEP coordinate system, represent $\Delta B_i = B_{Ai} - (B_{Ei} + B_{Si})$, $i = x, y, z$; the top graph is the magnitude $|\overrightarrow{\Delta B}| = (\sum_i \Delta B_i^2)^{1/2}$.

Reference to equation (1) shows that these components represent summations of lunar induced fields and moon-plasma interaction fields. The large deviations from zero on days 331 and 338 are related to high solar activity, as indicated by the geomagnetic activity index Kp.

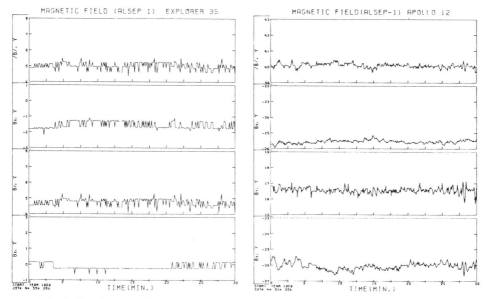

Fig. 6. Simultaneous time-series data taken when the moon was immersed in the free-streaming solar wind during a period of very low solar activity ($Kp = 0+$). Day 337 corresponds to 3 December 1969.

COLBURN et al. (1967) and NESS et al. (1967). A field effect suspected to correlate with the hydromagnetic solar wind flow past the moon is also seen, notably on the z and y axes on day 331. This surface amplification field is directly proportional to the solar wind plasma energy density as measured by SNYDER (1971) and EGIDI et al. (1970). The flow field $(\vec{B}_F)$ increase is prominent in both the horizontal components ΔB_y and ΔB_z and is largest in the z direction. It is noted that these field increases with high plasma energy are directly proportional to the Apollo 12 steady field components $B_{Sy} = +13.0\gamma$ and $B_{Sz} = -25.6\gamma$. This correlation suggests that the field increase is due to a compression of the local steady field $\vec{B}_S$ by the solar wind.

3. Toroidal field calculations

The Kp plot in Fig. 5 shows that solar activity is especially low on day 337. An estimated upper limit for the induced toroidal field has been determined from data obtained during this very quiet time ($Kp = 0+$) in the solar wind (see Fig. 6). In this period the time-dependent poloidal field B_P is assumed negligible for frequencies 10^{-4} Hz (SONETT et al., 1971); the magnetometer is on the sunward side and so $\vec{B}_D \rightarrow 0$; $\vec{B}_\mu \rightarrow 0$, as will be shown in a later section on steady field measurements, and the solar wind flow field $\vec{B}_F$ is assumed to approach zero due to the low plasma energy density of day 337. The toroidal field is then calculated from equation (1) to be $\vec{B}_T = \vec{B}_A - \vec{B}_S - \vec{B}_E$. Substituting values from Fig. 6 and values for the steady field $\vec{B}_S$ components, to be discussed later in this paper, the toroidal field vector in the ALSEP system is calculated to be $\vec{B}_T = (0.3, -1.2, -0.3)$ in gammas. This is within the error of the experimental measurements of $\vec{B}_S$ so that it represents an upper limit for $\vec{B}_T$ (in fact $\vec{B}_T$ may be zero).

The electrical conductivity upper limit of the outer crust is estimated for this calculated $\vec{B}_T$ by the method of SCHUBERT and SCHWARTZ (1969) to be $\sim 10^{-9}$ mhos/meter for an outer layer 5 km ($0.003R_{\text{moon}}$) thick, using the solar wind velocity of day 337 reported by EGIDI *et al.* (1970). This value is consistent with the electrical conductivity estimated by SIMMONS (1970) for the lunar crystal material.

Qualitative properties of magnetic field three-hour averages during one lunation show the number and complexity of magnetic interactions as the moon orbits the earth, posing an especially complicated situation on the sunlit side. For our first attempt to analyze the eddy current inductive response of the moon, therefore, we have chosen nighttime conditions where plasma interaction effects are minimized. More complex analysis of lunar magnetic properties will be facilitated by a future magnetometer network on the moon.

TRANSIENT MAGNETIC-FIELD MEASUREMENTS

For our analysis magnetic step-transient measurements were selected from time periods when the moon was in the free-streaming solar wind and the surface magnetometer was on the lunar dark side (see Fig. 3). During these times the sunlit hemisphere of the moon obstructs the solar-wind flow, forming an elongated cavity which extends downstream for several lunar radii (SERBU, 1969; SNYDER *et al.*, 1970).

In the lunar darkside cavity several terms in equation (1) can be neglected to a first approximation. For data taken on the lunar dark side, the surface magnetometer is isolated from the plasma flow; hydromagnetic effects such as surface-field compression will not occur and $\vec{B}_F$ can be disregarded. Subsequent analysis in this paper will show that the bulk lunar magnetic permeability is close to that of free space; thus the induced magnetization field $\vec{B}_\mu$ is neglected. For time-series step-transient analysis $\vec{B}_T$ and $\vec{B}_D$ are separated from poloidal ($\vec{B}_P$) response since they are time-independent for constant applied driving field $\vec{B}_E$. $\vec{B}_T$ is negligible compared to $\vec{B}_P$ as shown in the previous section. $\vec{B}_D$ can also be neglected for reasons that will be pointed out later in the analysis. In this case, therefore, equation (1) reduces to

$$\vec{B}_A = \vec{B}_P + \vec{B}_E + \vec{B}_S.$$

Over one hundred step-function transients which have penetrated the moon have been measured and analyzed using simultaneous data from the Apollo 12 lunar surface magnetometer and the lunar orbiting Explorer 35 magnetometer. These are the first measurements which clearly show a whole-moon spherical response to magnetic-field transients in interplanetary space (DYAL *et al.*, 1970b).

A first overall scan of the Apollo 12 data revealed that the dark-side transient measurements show a remarkable similarity to the eddy current response of a conducting sphere in a vacuum: measured surface magnetic-field radial components show a characteristic damped response to solar-wind field step transients, while components tangent to the lunar sphere show rapid response and overshoot initially, followed by decay to a steady-state value. The characteristics of these transient measurements suggest the use of a simple two-layer model of the moon (i.e., a homogeneous core surrounded by a nonconducting outer shell) as a first approximation.

1. Two-layer model and theory

This two-layer model has the following assumed properties: the spherically symmetric model has a homogeneous inner core of scalar electrical conductivity σ_i and radius R_i surrounded by a nonconducting outer shell of outer radius R_m (the lunar radius). The sphere is in a vacuum and permeability is everywhere that of free space ($\mu = \mu_0$). Conduction currents dominate displacement currents within the sphere, and dimensions of external field transients are large compared to the diameter of the sphere. A transient event is assumed to affect all parts of the lunar sphere instantaneously since the solar wind transports a step discontinuity across the entire moon in less than 9 seconds. This time period has been found to be much less than the decay time of lunar induced eddy currents.

Total magnetic fields measured at the surface of the sphere are noted by the subscript A (corresponding to Apollo 12 lunar surface magnetometer measurements: $\vec{B}_A$) and external applied magnetic fields are noted by the subscript E (corresponding to lunar orbiting Explorer 35 magnetometer measurements: $\vec{B}_E$). The external field step-change vector ($\overrightarrow{\Delta B_E}$) and the radius vector to the Apollo 12 site lie in the same plane. All data are expressed in the ALSEP surface coordinate system ($\hat{x}, \hat{y}, \hat{z}$) which has its origin at the Apollo 12 magnetometer site; $\hat{x}$ is directed radially outward from the surface, while $\hat{y}$ and $\hat{z}$ are tangent to the surface, directed eastward and northward, respectively.

Using the above assumptions, the magnetic field solutions of Maxwell's equations for the eddy current response of a spherical two-layer model are derived following the methods of SMYTHE (1950) and WAIT (1951) and a detailed derivation is given by DYAL et al. (1970b). The vector components of the magnetic field at the lunar surface are listed below for the case of an external magnetic-field step transient of magnitude $\overrightarrow{\Delta B_E} = \vec{B}_{Ef} - \vec{B}_{Eo}$ applied to the lunar sphere at time $t = 0$. $\vec{B}_{Eo}$ are $\vec{B}_{Ef}$ are Explorer 35 initial and final external fields, respectively. The solutions for the components of the vector field measured on the lunar surface ($\vec{B}_A = \vec{B}_P + \vec{B}_E + \vec{B}_S$) can be expressed as follows:

$$B_{Ax} = -3 \left(\frac{R_1}{R_m}\right)^3 (\Delta B_{Ex})F(t) + B_{Ex} + B_{Sx} \tag{2}$$

$$B_{Ay,z} = \frac{3}{2} \left(\frac{R_1}{R_m}\right)^3 (\Delta B_{Ey,z})F(t) + B_{Ey,z} + B_{Sy,z} \tag{3}$$

Here $\Delta B_{Ei} = B_{Eif} - B_{Eio}$, $i = x, y, z$; R_1 and R_m are radii of the conducting core and the moon, respectively. B_{Eio} and B_{Eif} are the initial and final external applied field components, respectively, and are both measured by Explorer 35. B_{Ai} are total surface fields measured by the Apollo 12 magnetometer, and the time-dependence of the magnetic field is expressed as

$$F(t) = \frac{2}{\pi^2} \sum_{s=1}^{\infty} \frac{1}{s^2} \exp\left(\frac{-s^2\pi^2 t}{\mu_0\sigma_1 R_1^2}\right) \tag{4}$$

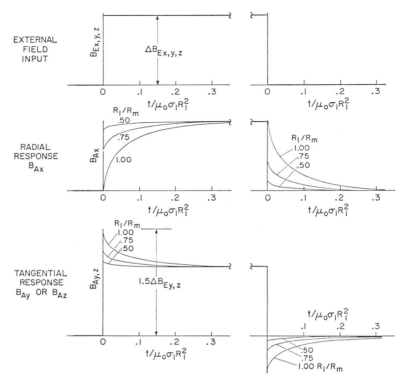

Fig. 7. Theoretical solutions for the poloidal magnetic-field response of a homogeneous conducting lunar core of radius R_1 to a step-function transient in the driving solar wind magnetic field. For a step-function change $\overrightarrow{\Delta B_E}$ in the external driving field (measured by Explorer 35), the total magnetic field at the surface of the moon $\vec{B}_A$ (measured by the Apollo 12 magnetometer) will be damped in the radial (B_{Ax}) component and will overshoot in the tangential (B_{Ay} and B_{Az}) components. The initial overshoot magnitude is limited to the maximum value $\Delta B_{Ey}/2$ or $\Delta B_{Ez}/2$ for the case $R_1 \to R_m$. A family of curves is shown for different values of the parameter R_1/R_m.

Solutions (2) and (3) for radial and tangential transients are shown graphically in Fig. 7.

2. Comparison of the data to the two-layer model

The general appearance of the one hundred analyzed transients was similar to the theoretical curves shown in Fig. 7. For all data the Apollo 12 and Explorer 35 scales differ since the 38 ± 3 gamma steady field at the surface site has been retained in the Apollo 12 data.

Typical step-function transient involving all three vector components are shown in Figs. 8 and 9. One can easily see the characteristic damped decay in the Apollo radial data (B_{Ax}) and overshoot in the tangential data (B_{Ay} and B_{Az}). One can also see from the external field magnitude $|\vec{B}_E|$ curve that this transient is a discontinuity involving primarily a field direction change. This external field direction change does, however, produce a change in the Apollo 12 magnitude $|\vec{B}_A|$ curve since it vectorally adds to the local steady field at the Apollo 12 site.

All transient measurements exhibit characteristic damping in the surface field radial component and overshoot in the tangential component. Eight step events occurred while the Explorer 35 and Apollo 12 magnetometers were simultaneously in the cavity and within one lunar radius apart; these events exhibited the same characteristics as those transients obtained when Explorer 35 was on the sunward side of the moon. The difference between radial and tangential measurements clearly shows symmetry with respect to the lunar sphere; this strongly indicates that the transient response is a whole-moon effect and discounts the possibility that the measured induction fields are due to a smaller conducting body set off to one side of the Apollo 12 magnetometer.

The theoretical $F(t)$ function given in equation (4) has been fitted by the least-squares method to the post-transient $(t > 0)$ Apollo 12 measurements over four-minute intervals. A theory-to-data least-squares fit for one particular transient is shown in Fig. 10. For this transient, values of $\sigma_1 = 1.5 \times 10^{-4}$ mhos/meter and $0.95 R_m \leq R_1 < R_m$ best fit the data. Dashed curves in Fig. 10 show the strong dependence of $F(t)$ on the electrical conductivity σ_1 for values of σ_1 varying from 1.5×10^{-4} by factors of 0.1, 0.5, 2, and 10. The average value of the electrical

MAGNETIC FIELD

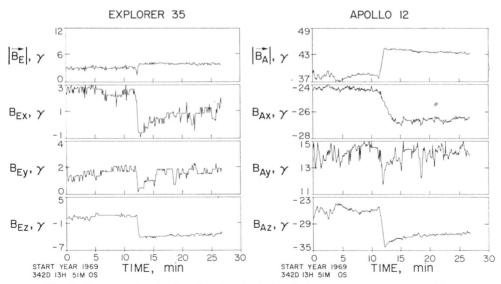

Fig. 8. Response to step-function transients in all three vector components. Note damping on the radial x-axis and overshoot on tangential y and z-axes. Apollo 12 and Explorer 35 field magnitudes and component data scales differ due to the existence of a 38 ± 3 gamma steady field at the Apollo 12 site.

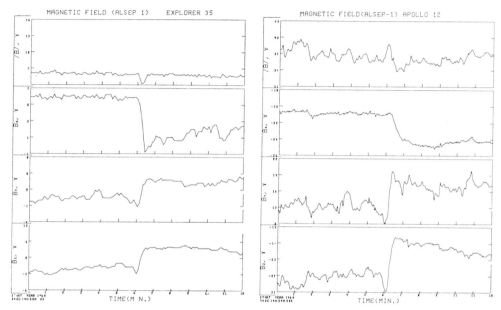

Fig. 9. Another transient event showing response to step transients in all three vector components. Again note damping on the radial x-axis and overshoot on the tangential y- and z-axes. All components are expressed in the ALSEP coordinate system and ordinate labels correspond exactly to those of Fig. 8.

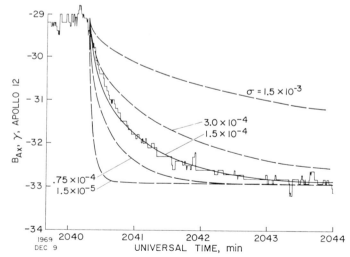

Fig. 10. Comparison between Apollo 12 magnetometer data and two-layer lunar model theory for an x-axis (radial) step transient. Best fit is for a homogeneous core of conductivity $\sigma = 1.5 \times 10^{-4}$ mhos/meter and radius R in the range of $0.95 R_m \leq R < R_m$. Superimposed theoretical decay curves for other core conductivities illustrate the sensitive dependence of decay characteristics on the conductivity. This transient event occurred at a time when the Apollo 12 magnetometer was on the lunar dark side and Explorer 35 was in the free-streaming solar wind (see Fig. 3 insert).

conductivity for ten transient events, using the two-layer model field solution, was calculated to be $1.7 \pm 0.4 \times 10^{-4}$ mhos/meter.

A determination of the radius of the conducting core is made by comparing the features at the start of the rise or fall portion of the response curves (as shown in Fig. 7) with the data. The transients in the radial component indicate that the instrument is relatively close to the conducting core. By comparison with theoretical curves, the conducting-core radius is calculated to be in the range $0.95R_m \le R_1 < R_m$. For these average calculated values of σ_1 and R_1, the theoretical time-dependent function $F(t)$ of equation (4) will decay to less than 2% of its initial value within four minutes.

3. Deviations from the two-layer model and introduction of a three-layer model

Examination of the time series step response for periods longer than four minutes reveals that there is a consistent deviation of the data from the two-layer model. In particular, the total surface field measured by Apollo 12 does not decay exactly to the Explorer 35 final field value within this four-minute time period for either radial or tangential components; however, it does decay for periods of the order of 60 minutes.

The Apollo 12 field radial components generally fall short of the Explorer 35 radial components by about 20% after four minutes (see Fig. 11). For tangential components, on the other hand, the Apollo 12 data fails to decay back to the final Explorer 35 asymptote by about 35% at four minutes after the initial overshoot. These field differences do not vary as a function of lunar orbit position for times when the magnetometer is greater than 400 km inside the optical shadow. This implies that the deviation from the two-layer model is not a function of position within this

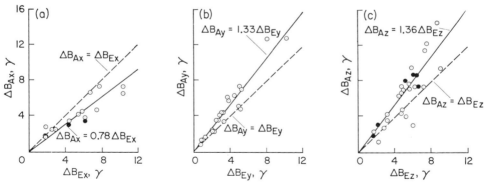

Fig. 11. Differences between field component values measured at $t = 0$ and at $t = +4$ minutes in the Apollo 12 data (ΔB_{Ai}), plotted versus corresponding differences in the Explorer 35 data (ΔB_{Ei}). Each point represents a separate transient event occurring when the Apollo 12 magnetometer was on the lunar dark side. A reference line of slope one is shown on each plot, indicating the functional dependence of ΔB_{Ai} on ΔB_{Ei} for a two-layer model which decays to an asymptotic value within four minutes. Deviations of the data from the reference line indicate that currents are still flowing in a deep lunar core at times $t > 4$ minutes. (a) Radial x-axis. (b) Tangential y-axis. (c) Tangential z-axis.

region of the lunar cavity, and therefore effects of solar-wind hydromagnetic flow past the moon can be neglected. These differences are all independent of the Explorer 35 orbital position. Eight step-transient events occurring when the Explorer 35 and Apollo 12 magnetometers were simultaneously inside the cavity have been analyzed. The solid data points in Fig. 11 represent field differences for these eight events, supporting our earlier contention that the diamagnetic field $\vec{B}_D$ effects can be neglected.

In order to investigate the departure from two-layer theory, three possible conditions that were excluded in the assumptions of the two-layer model are qualitatively re-examined here: (1) the existence of a shell of permeability several times greater than μ_0 in the moon, (2) a toroidal induction mode large enough to be measured, and (3) a deep core of conductivity much higher than 10^{-4} mhos/meter.

If a permeable ($\mu > \mu_0$) shell exists in the moon, an induced magnetization field would produce an external radial and tangential field of exactly the opposite sign as that required to explain the deviations from the two-layer model. The toroidal field induction mode does not account for the deviation from the two-layer model since it would produce no radial component at the lunar surface and furthermore, for a statistical number of transient events, the tangential components would either add to or subtract from the poloidal tangential components.

The third alternative, that of a highly conducting core deep in the lunar interior, would cause the deviation of the data from the two-layer model. For this case, the eddy-current decay time for the deep core would be much longer than that in the outer regions, causing a poloidal field to persist long after the currents in the outer regions have decayed. This deep core eddy-current field, which would persist much longer than four minutes, would cause the Apollo 12 radial component to fall short of the Explorer 35 post-transient value and would cause the Apollo 12 tangential component, after the initial overshoot, to fall short of the corresponding Explorer 35 value (see Fig. 9). The deep core, therefore, would indeed produce the characteristics of the measured deviations from the two-layer model. A time-series solution of Maxwell's equations for an eddy current response of a multilayer conducting sphere to a step transient has not been completed at this time. In lieu of this solution, the properties of the inner core will be studied by a psuedo three-layer model, which is used to place bounds on conductivity and dimensions of the deep core.

4. *Properties of the three-layer moon*

Figure 12 schematically shows the three-layer conductivity model of the moon and depicts a time sequence of field-line configurations for a magnetic step transient. Transient events which persist long enough to permit study of this long-term decay of inner-core currents are rare, but four good cases have been found. Limits on the conductivity and radius of the inner core are found by fitting the theoretical function $F(t)$ of equation (4) to the data for times greater than four minutes after transient arrival. The parametric fit yields $\sigma_2 \sim 10^{-2}$ mhos/meter and $R_2 \sim 0.6R_m$. The conductivity σ_1 calculated for the two-layer model should approximate that of the intermediate shell of the three-layer model since the intermediate shell has a volume

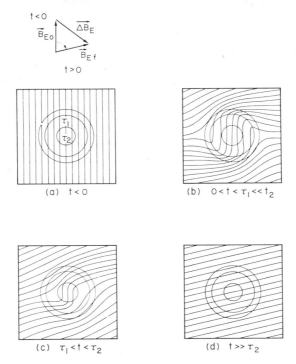

Fig. 12. A graphical representation of the total-field poloidal response of a three-layer sphere to a step directional change in the external driving field. The external field changes at $t = 0$ from an initial value $\vec{B}_{Eo}$ to a final $\vec{B}_{Ef}$ by the vector quantity $\overrightarrow{\Delta B}_E = \vec{B}_{Ef} - \vec{B}_{Eo}$. For the outer crust $\sigma_0 \sim 0$ so its decay time constant $\tau_0 \sim 0$, and for the inner layers $\sigma_1 \ll \sigma_2$ so $\tau_1 \ll \tau_2$. (a) Initial condition at $t < 0$. (b) External field has changed direction; the field has instantaneously penetrated the outer nonconducting shell and is diffusing into the middle shell. (c) Poloidal currents have mostly decayed in the middle shell, but they are still freely flowing in the highly conducting core. (d) Final state after a time long enough for eddy currents to have decayed throughout the sphere, allowing the external field to penetrate the entire sphere.

$\geqslant 4$ times that of the inner core. However, a highly conducting deep core will inhibit the decay during early times after transient arrival. Thus the earlier two-layer calculations of σ_1, which neglected contributions of an inner core, attributed a longer decay time to an overall conductivity rather than partially to a deep core, resulting in an overestimate of the actual outer conductivity at $\sigma_1 = 1.7 \times 10^{-4}$ mhos/meter. Within the restrictions of the psuedo three-layer model, therefore, a best estimate of the outer shell conductivity is $\sigma_1 \sim 10^{-4}$, with 2×10^{-4} as an upper limit and 5×10^{-5} mhos/meter as a lower limit. These values are obtained by fitting the initial post-transient values at $t = 0$ with an asymptotic approach to the deep-core decay curve. More accurate values for the conductivities and radii await an analytical three-layer time-series solution.

It is concluded, therefore, that the simplest model which qualitatively explains the general aspects of the dark-side transient-response data is a three-layer model having a thin outer crust of very low conductivity (i.e., conductivity so low that field changes diffuse through the crust in times too short to be detected by the Apollo 12 magnetometer), an intermediate layer of conductivity $\sigma_1 \sim 10^{-4}$ mhos/meter, and a deep conducting core of conductivity $\sigma_2 \sim 10^{-2}$ mhos/meter.

5. *Internal temperature calculations*

The electrical conductivity values obtained by measuring the time-response characteristics of the vector magnetic field transient can be used to calculate the internal temperature distribution if the material composition is known. For materials that have been used to model the lunar interior (see, e.g., UREY and MACDONALD, 1970; KOPAL, 1969), the electrical conductivities are expressed as functions of temperature by

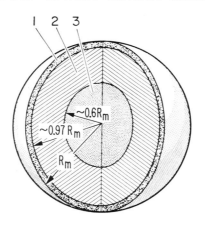

THREE-LAYER LUNAR MODEL

REGION	ELECTRICAL CONDUCTIVITY, σ, mhos/meter	TEMPERATURE, °K		
		OLIVINE	PERIDOTITE	APOLLO 11 SURFACE SAMPLE
1	$< 10^{-9}$	<440	<430	< 300
2	$\sim 10^{-4}$	~810	~890	~530
3	$\sim 10^{-2}$	~1240	~1270	~740

Fig. 13. Conductivity (σ) and temperature (T) contours for a three-layer moon. Temperature calculations are based on σ vs. T relationships for pure olivine and peridotite (ENGLAND *et al.*, 1968) and an Apollo 11 surface sample (NAGATA *et al.*, 1970).

NAGATA *et al.* (1970):

$$\sigma_{\text{olivine}} = 55 \exp\left(-0.92/kT\right) + 4 \times 10^7 \exp\left(-2.7/kT\right) (\text{ohm-m})^{-1}$$
$$\sigma_{\text{peridotite}} = 3.8 \exp\left(-0.81/kT\right) + 10^7 \exp\left(-2.3/kT\right) (\text{ohm-m})^{-1}$$
$$\sigma_{\substack{\text{Apollo 11} \\ \text{sample}}} = 7.9 \exp\left(-0.51/kT\right) + 3.1 \times 10^6 \exp\left(-1.25/kT\right) (\text{ohm-m})^{-1}$$

These conductivities are assumed to be independent of pressure below 50 kilobars (ENGLAND *et al.*, 1968) and independent of frequency below 10 Hz (KELLER and FRISCHKNECHT, 1966). A lunar cross section showing a temperature-conductivity profile of a three-layer lunar interior is given in Fig. 13.

STEADY MAGNETIC FIELD MEASUREMENTS

During times when the moon is immersed in steady field regions (i.e., fluctuations are of frequency $< 10^{-4}$ Hz) of the earth's magnetotail, the unipolar field $\vec{B}_T$, the hydromagnetic flow field $\vec{B}_F$, and diamagnetic cavity field $\vec{B}_D$ in equation (1) are negligible since the solar plasma is excluded from this region. The poloidal eddy-current induced field $\vec{B}_P$ can also be neglected for frequencies $< 10^{-4}$ Hz (SONETT *et al.*, 1971) as seen from the transient response analysis. Therefore in the steady magnetotail, equation (1) reduces to $\vec{B}_A = \vec{B}_E + \vec{B}_\mu + \vec{B}_S$. The magnetization field $\vec{B}_\mu$ is dipolar and proportional to the steady external field $\vec{B}_E$ (JACKSON, 1962). In the ALSEP surface coordinate system $(\hat{x}, \hat{y}, \hat{z})$ described earlier, the radial and tangential components for a spherically symmetric permeable shell become

$$B_{Ax} = (1 + 2F)B_{Ex} + B_{Sx} \tag{5}$$
$$B_{Ay,z} = (1 - F)B_{Ey,z} + B_{Sy,z}. \tag{6}$$

$$\text{Where } F = \frac{(2k_m + 1)(k_m - 1)\left(1 - \dfrac{R}{R_m}\right)^3}{(2k_m + 1)(k_m + 2) - 2\left(\dfrac{R}{R_m}\right)^3 (k_m - 1)^2} \tag{7}$$

Here k_m is the relative permeability μ/μ_0; R_m is the lunar radius; R is the radius of the spherical boundary which encloses lunar material with temperature above the Curie point. Equations (5) and (6) are linear and graphically intercept the ordinates at values of steady field components. The slopes of the equations are proportional to the permeability and radial dimensions of lunar permeable material.

1. *Lunar relative permeability calculations*

Apollo 12 magnetometer measurements B_{Ax}, are plotted versus corresponding geomagnetic tail field components B_{Ex} (measured by Explorer 35) in Fig. 14(a). These measurements were made during the first four post-deployment lunations (November 1969 to March 1970). A least-squares linear fit to the data in Fig. 14(a) yields a slope which corresponds to a value $F = 1.2 \pm 0.5 \times 10^{-2}$ obtained from

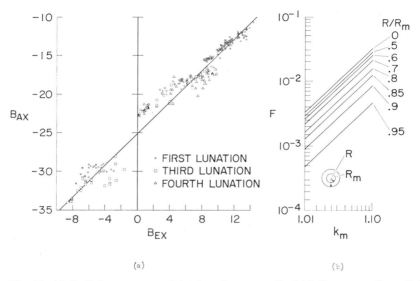

(a)   (b)

Fig. 14. (a) Radial component of total surface magnetic field B_{Ax} versus the radial component of external driving field B_{Ex}. Data points consist of measurements in quiet regions of the geomagnetic tail taken during the first four post-deployment lunations. The figure is a graphical representation of equation (5). The B_{Ax} intercept of the least-squares best-fit solid line gives the radial component of the Apollo 12 permanent field; the best-fit slope corresponds to a value of 1.03 ± 0.03 for the bulk relative permeability μ/μ_0 of the moon. (b) A graphical representation of equation (7), which relates the function F to relative magnetic permeability $k_m = \mu/\mu_0$ for various values of R/R_m. R and R_m are internal and external radii, respectively, of a global permeable shell.

equation (5). Then by referring to Fig. 14(b), this value of F is used to determine the relative permeability of lunar material for different values of the ratio R/R_m. For the limiting case of a moon of homogeneous permeability throughout, the bulk relative permeability is $\mu/\mu_0 = 1.03 \pm 0.13$.

2. The steady field at the Apollo 12 site

The steady field at the Apollo 12 site is determined by plotting the geomagnetic tail data for all the vector components B_{Ai} vs. B_{Ei}. Ordinate intercepts of linear least-squares fits to the data yield the steady field components $B_{Sx} = -24.4 \, \gamma$ (directed down into the lunar surface), $B_{Sy} = +13.0 \, \gamma$ (directed east), and $B_{Sz} = -25.6 \, \gamma$ (south). With all errors considered, the field has a magnitude 38 ± 3 and is directed down into the lunar surface, at an angle $40°$ down from the horizontal plane and azimuthally $63°$ clockwise from due east.

Two additional measurements place bounds on the size of the source producing this steady field (DYAL et al., 1970c). First, Explorer 35 data indicate that the global surface field is $\leq 4\,\gamma$ (BEHANNON, 1968); secondly, Apollo 12 magnetometer data specify that the local surface gradient must be $\leq 0.133\,\gamma/\text{meter}$. The first restriction implies that the $38\,\gamma$ source cannot be global and maximizes its size; the second limit minimizes the source size.

If the source of the steady field is assumed to be a single dipole lying on or near the surface, then the Apollo 12 gradient and Explorer 35 measurements require the source to lie outside the surface contour shown in Fig. 15 and inside a circular contour of radius 200 km. The minimum distance contour of Fig. 15 shows that all Apollo 12 mission artifacts are eliminated as sources. Only Surveyor III lies outside the contour, and it can be discounted as a possible source since it is small and contains

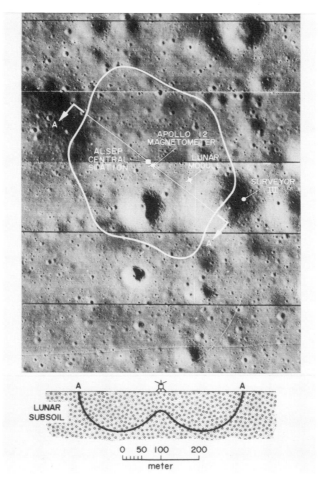

Fig. 15. Lunar-orbiter photograph of the Apollo 12 site. Field-gradient measurements by the surface magnetometer require that the source of the 38 gamma steady field, if it can be approximated as a point-dipole source, must lie outside the superimposed surface and sub-surface contours.

mostly tubular aluminum in its structure. Surveyor III is 265 meters from the Apollo 12 instrument and would have to produce a minimum field of 5500 *gauss* at one meter distance away in order to produce 38 γ at 265 meters. A minimum-distance contour calculated from the gradient measurements for a subsurface point dipole is shown in Fig. 15. This source must be greater than 50 meters deep if it is directly below the magnetometer. The minimum size of the source can be calculated by assuming the source is a sphere of uniform remanent magnetization 10^{-2} emu/cm³ (the upper limit of Apollo 12 samples). If lying near the inner contour 200 meters away from the surface magnetometer, the source would have a 50-meter diameter; at 200 km, a 50-km diameter. If it lay 50 meters below the magnetometer, it would have a diameter of about 13 meters.

Other experimental evidence concerning the dimensions of the steady field source is obtained by measuring the compression effects on the field by the solar wind (BARNES *et al.*, 1971). This effect is seen in Fig. 5 for days 330 through 334 and indicates that the steady field source has a scale size greater than 10 km.

It is possible, of course, that the Apollo 12 steady field could be the superposition of fields of several dipole sources or the nondipolar field of an extended near-surface source which has lost uniform magnetization by a process such as meteorite impact fracturing. In such a case the properties of the source could not be accurately calculated from the single-point field measurement.

Though it is possible that the source is a magnetized meteorite fragment or breccia shock-magnetized by meteorite impact, it is more likely that the source, if a *single* source, is a large deposit of native lunar material which was magnetized by a large ambient field sometime in the past (DYAL *et al.*, 1970b). This supposition is consistent with the unexpectedly high natural remanent magnetism found in Apollo 11 and Apollo 12 lunar samples; the extent of this magnetization would have required an ambient field $\geqslant 10^3 \gamma$, much higher than exists at present (see, e.g., RUNCORN *et al.*, 1970; PEARCE *et al.*, 1971; HELSLEY, 1971).

Our preliminary results of the Apollo 14 lunar portable magnetometer experiment show that steady fields of ~ 100 gammas and ~ 40 gammas exist at two Fra Mauro sites separated by 1.12 km. It is uncertain at present whether these two field magnitudes are due to the same source as the field at the Apollo 12 site 180 km away in Oceanus Procellarum. The three field measurements are not in the same direction, although they are all directed down into the lunar surface. It is certainly unlikely, however, that such high local surface fields are unique to the lunar region containing the Apollo 12 and Apollo 14 landing sites (BARNES *et al.*, 1971); indeed, many of the samples from the distant Apollo 11 site (e.g., STRANGWAY *et al.*, 1970; RUNCORN *et al.*, 1970) are strongly magnetized, and lunar orbiting Explorer 35 magnetometer has measured evidence of numerous local sections of the lunar surface magnetized to varying degrees (MIHALOV *et al.*, 1971). The cause for such a high degree of magnetization of much of the lunar surface remains one of the more interesting enigmas of lunar history. Possibilities include a much stronger solar field at some time in the past; a lunar orbit formerly much closer to the earth ($\sim 3R_E$); and an ancient lunar field due to dynamo action, thermoelectric currents, or unipolar induction currents.

CONCLUSIONS

1. *Electrical conductivity and internal lunar temperature*

The electrical conductivity of the lunar interior has been determined from magnetic field step transient measurements made on the lunar dark side. The data fit remarkably close to the classical theory of a conducting sphere in a magnetic field. Radial and tangential magnetic field component measurements indicate a global rather than a local response to these step transients. The simplest model which qualitatively explains the dark side transient-response data is a spherically symmetric three-layer model having a thin outer crust of very low conductivity. The intermediate layer, of radial thickness $R_1 - R_2$ where $0.95R_m \leq R_1 < R_m$ and $R_2 \sim 0.6R_m$, has an electrical conductivity $\sigma_1 \sim 10^{-4}$ mhos/meter; the inner core has radius $R_2 \sim 0.6R_m$ and conductivity $\sigma_2 \sim 10^{-2}$ mhos/meter. Electrical conductivities calculated for the three-layer model are related to the temperatures in the three regions. For the example of an olivine moon, the temperatures of the layers are as follows: crust, $\leqslant 440°$K; intermediate layer, $\sim 810°$K; core, $\sim 1240°$K.

2. *Relative magnetic permeability*

The whole-moon relative permeability has been calculated from the measurements to be $\mu/\mu_0 = 1.03 \pm 0.13$. If the inner core of radius $R \sim 0.6R_m$ has a Curie temperature $> 800°$C, then an upper bound of $\mu/\mu_0 = 1.05 \pm 0.14$ is determined for the outer layer.

3. *Steady field measurements*

A steady field of 38 $\pm$ 3 gamma has been measured at the Apollo 12 site in Oceanus Procellarum. Simultaneous measurements by the lunar orbiting Explorer 35 satellite have shown that the source of this steady field cannot be global in extent and must be confined to a region within 200 km of the Apollo 12 site. Gradient measurements and remanent magnetic measurements of returned samples place limits of 13 meters to 50 km for an assumed spherical diameter of a dipole source. One reasonable model for the magnetic source is a layer of material which was magnetized in the past by a large ($\geqslant 10^3 \gamma$) ambient field and was subsequently fractured by meteorite impacts.

4. *Magnetic field averages during one lunation*

Three-hour averages of the vector magnetic field during the first lunation showed an increase in the horizontal field components which was proportional to the steady field component magnitudes and the plasma energy density in the solar wind. This is interpreted to be a solar wind compressional effect on the steady field. The three-hour average plots also showed an increase in magnetic field over that in interplanetary space during the time that the Apollo instrument was in the diamagnetic cavity. During a quiet time in the free-streaming solar wind, an induced toroidal field component was measured to be $|\vec{B}_T| = 0 \pm 1.3$ gammas. This field was then used to calculate an approximate upper limit conductivity of $\sim 10^{-9}$ mhos/meter for an outer shell with thickness $0.03R_m$.

Acknowledgments—Drs. C. P. Sonett and D. S. Colburn are especially thanked for their help and stimulation during this entire experiment. We are also grateful to the Apollo Program Office, Astronauts Charles Conrad, Alan Bean, and Richard Gordon, and the personnel at the Manned Spacecraft Center for their direct participation in carrying out the experiment. We thank Philco-Ford and Bendix Aerospace Corporations for the experiment fabrication, testing and integration into the Apollo 12 system. Special thanks go to Dr. Thomas Mucha, Kenneth Lewis, Marion Legg and their team members for help in computer programming and data reduction.

References

Barnes A., Cassen P., Mihalov J. D., and Eviatar A. (1971) Permanent lunar surface magnetism and its deflection of the solar wind. *Science* **171**, in press.

Behannon K. W. (1968) Intrinsic magnetic properties of the lunar body. *J. Geophys. Res.* **73**, 7257.

Colburn D. S., Currie R. G., Mihalov J. D., and Sonett C. P. (1967) Diamagnetic solar-wind cavity discovered behind moon. *Science* **158**, 1040.

Dyal P., Parkin C. W., and Sonett C. P. (1970a) Lunar surface magnetometer. *IEEE Trans. on Geoscience Electronics* **GE-8 (4)**, 203–215.

Dyal P., Parkin C. W., Sonett C. P., and Colburn D. S. (1970b) Electrical conductivity and temperature of the lunar interior from magnetic transient response measurements. NASA TM X-62,012. Submitted to *J. Geophys. Res.*

Dyal P., Parkin C. W., and Sonett C. P. (1970c) Apollo 12 magnetometer: measurement of a steady magnetic field on the surface of the moon. *Science* **196**, 762.

Egidi A., Formisano V., Palmiotto F., and Saraceno P. (1970) Solar wind and location of shock front and magnetopause at the 1969 solar maximum. *J. Geophys. Res.* **75**, 6999–7006.

England A. W., Simmons G. and Strangway D. (1968) Electrical conductivity of the moon. *J. Geophys. Res.* **73**, 3219.

Geyger W. A. (1964) *Nonlinear-Magnetic Control Devices.* McGraw-Hill.

Gordon, D. I., Lundsten, R. H., and Chiarodo, R. A. (1965) Factors affecting the sensitivity of gamma-level ring-core magnetometers. *IEEE Trans. on Magnetics, MAG–1*, (4), 330.

Helsley C. E. (1971) Evidence for an ancient lunar magnetic field. Second Lunar Science Conference (unpublished proceedings).

Jackson J. D. (1962) *Classical Electrodynamics.* John Wiley.

Keller G. V. and Frischknecht F. C. (1966) *Electrical Methods in Geophysical Prospecting.* Pergamon.

Kopal Z. (1969) *The Moon.* D. Reidel.

Mihalov J. D., Sonett C. P., Binsack J. H., and Moutsoulas M. D. (1971) Crustal inhomogeneity inferred from lunar satellite magnetometer. *Science* **171**, 892–895.

Nagata T., Rikitake T., and Kono M. (1970) Electrical Conductivity and age of the moon. Proc. of COSPAR Assembly Committee on Space Research. *Space Science* **10**.

Ness N. F., Behannon K. W., Scearce C. S., and Cantarano S. C. (1967) Early results from the magnetic field experiment on Explorer 35. *J. Geophys. Res.* **72**, 5769.

Pearce G. W., Strangway D. W., and Larson E. E. (1971) Magnetism of two Apollo 12 igneous rocks. Second Lunar Science Conference (unpublished proceedings).

Runcorn S. K., Collinson D. W., O'Reilly W., Battey M. H., Stephenson A. A., Jones J. M., Manson A. J., and Readman P. W. (1970) Magnetic properties of Apollo 11 lunar samples. *Proc. Apollo 11 Lunar Sci. Conf., Geochim. Cosmochim. Acta* Suppl. 1, Vol. 3, pp. 2369–2387. Pergamon.

Schubert G. and Schwartz K. (1969) A theory for the interpretation of lunar surface magnetometer data. *The Moon* **1**, 106–117.

Serbu G. P. (1969) Explorer 35 measurements of low-energy plasma in lunar orbit. *J. Geophys. Res.* **74**, 372.

SIMMONS G. (1970) Electrical conductivity of the lunar interior. In *Electromagnetic Exploration of the Moon* (editor W. I. Linlor). Mono Book Corp.

SMYTHE W. R. (1950) *Static and Dynamic Electricity*. McGraw-Hill.

SNYDER C. W., CLAY D. R., and NEUGEBAUER M. (1970) The solar-wind spectrometer experiment. In Apollo 12 Preliminary Science Rept. NASA SP-235, pp. 55–81.

SONETT C. P., COLBURN D. S., CURRIE R. G., and MIHALOV J. D. (1967) The geomagnetic tail; topology, reconnection and interaction with the moon. In *Physics of the Magnetosphere* (editors R. L. Carovillano, J. F. McClay, and H. R. Radoski), D. Reidel.

SONETT C. P., SMITH B. F., COLBURN D. S., SCHUBERT G., SCHWARTZ K., DYAL P., and PARKIN C. W. (1971) The lunar electrical conductivity profile: mantle-core stratification, near-surface thermal gradient, heat flux and composition. Second Lunar Science Conference (unpublished proceedings).

STRANGWAY D. W., LARSON E. E., and PEARCE G. W. (1970) Magnetic studies of lunar samples— breccia and fines. *Proc. Apollo 11 Lunar Sci. Conf., Geochim. Cosmochim. Acta* Suppl. 1, Vol. 3, pp. 2435–2451.

UREY H. C. and MacDONALD G. J. F. (1970) Origin and history of the moon. In *Physics and Astronomy of the Moon* (editor Z. Kopal), Academic Press.

WAIT J. R. (1951) A conducting sphere in a time varying magnetic field. *Geophysics* **16,** 666.

Proceedings of the Second Lunar Science Conference, Vol. 3, pp. 2415–2431
The M.I.T. Press, 1971.

Lunar electrical conductivity from Apollo 12 magnetometer measurements: Compositional and thermal inferences

C. P. Sonett, G. Schubert,* B. F. Smith,* K. Schwartz,†
and D. S. Colburn

NASA Ames Research Center, Moffett Field, California 94035

(*Received* 19 *February* 1971; *accepted in revised form* 30 *March* 1971)

Abstract—The lunar electrical conductivity profile has been obtained from an iterative fit of theoretical electromagnetic transfer functions to empirical ones based on joint power spectral density analyses of data from the Apollo 12 Lunar Surface Magnetometer and the Ames Explorer 35 magnetometer. Seven selected two hour swaths and seven selected one hour swaths of data were used. The spectrum analyzed ranged in frequency from 0.00083 to 0.04 Hz. The amplification of the interplanetary magnetic field at the lunar surface showed a distinct increase from ~ 1 at 0.001 Hz to ~ 3 at 0.005 Hz. At higher frequencies the slope of the amplification vs. frequency curve decreases until the amplification levels off at about 4 at 0.025 Hz. An amplification curve with such a distinct bend is best fit by a lunar conductivity model with a sharp maximum around 1500 km radius, so that high frequency variations in the magnetic field are compressed into the outer shell while low frequency variations can penetrate to the interior. For a monotonic temperature profile, the electrical conductivity model requires a change in composition around 1450 km radius. A thermal and compositional model which appears to fit the data is a basalt-like outer layer, an olivine-like core and a temperature of 450°C at the conductivity peak, increasing to 800°C in the deep core.

Introduction

This paper extends our preliminary report (Sonett *et al.*, 1971a) of the bulk electrical conductivity profile of the Moon determined by analysis of data from the Apollo 12 Lunar Surface Magnetomer (LSM) and the Ames Explorer 35 lunar orbiter magnetometer. The account given here is still exploratory but more details are supplied on the mechanism of induction, the analysis, and the implications of our findings.

It has been recognized in recent years that electromagnetic excitation of the Moon by signals arising in the solar wind produce electrical currents in the deep lunar interior. The magnetic fields associated with these currents are detectable at the surface of the Moon (Sonett *et al.*, 1971a, b, c; Dyal and Parkin, 1971) and, in the case of a strong inductive response may even modify the plasma flow near the Moon. The establishment of Explorer 35 in lunar orbit permitted study of the plasma environment close to the Moon. However, the detection of electromagnetic events in the Moon appears to be feasible only from either the surface or a very low altitude orbiter. Using the LSM, lunar electromagnetic induction can be assured. We believe the corresponding theory is sufficiently well understood that lunar magnetometry can be used to determine the electrical conductivity of the interior and thence to make inferences of temperature and composition. Although a detailed analysis of spatial resolution has

---

* Professional affiliation: University of California, Los Angeles, California 90024.
† Professional affiliation: American Nucleonics Corp., Woodland Hills, California 91364.

so far not been carried out, our preliminary assessment is that volume elements of the order of 3% are resolvable in the region of the steep conductivity rise with depth.

A fundamental difference of the Moon from the earth arises from the solar wind. When the angle between the Moon-Sun line and the local vertical at the LSM is appreciably less than 90 deg, the dynamic pressure of the solar wind is known to compress the induced magnetic field into the less conducting subsurface layers of the Moon (SONETT *et al.*, 1971a), appreciably amplifying the induction over that in a pure vacuum or a nonconducting atmosphere. This distinction together with the presence of the diamagnetic cavity results in a day-night induction asymmetry.

MECHANISM OF THE INDUCTION

Electromagnetic induction in the Moon is dependent upon a variety of discontinuities and waves in the solar wind. The forcing function which drives the induction is composed of the steady interplanetary magnetic field upon which is superimposed a hydromagnetic radiation continuum due to (1) plasma waves arising presumably in the solar atmosphere and subsequently convected and propagated outwards, (2) waves due to local instabilities in the solar wind, and (3) discrete large amplitude events such as collision-free shock waves, tangential discontinuities, and Alfvén waves (cf. COLBURN and SONETT, 1966; (although a large body of experimental work has subsequently been published, this paper provides the essential background for large events).

The electromagnetic interaction includes toroidal and poloidal magnetic fields. The principal excitations for these are (a) the interplanetary electric field (in a reference frame comoving with the Moon) given by $\underline{E}_m = \underline{v} \times \underline{B}$, where $\underline{v}$ is the velocity of the Moon relative to the solar wind and $\underline{B}$ the instantaneous interplanetary magnetic field, and (b) the time rate of change of the interplanetary field, $\dot{B}$. The two modes correspond respectively to transverse magnetic (TM) and transverse electric (TE) excitation. Both modes display strong frequency (f) dependence. The TE transfer function is zero at $f = 0$ and increases with increasing f. The TE mode currents which close wholly in the lunar interior, tend to be concentrated where variations in the magnetic field are damped substantially. With increasing frequency, the poloidal fields become compressed into shells of decreasing thickness, thus increasing the amplification of the magnetic field as observed at the Moon's surface. Examination of the interior of the Moon to depths around 800 km is facilitated by this mode, but for greater depths is limited by the decreased response.

The TM mode attains peak value for $f = 0$; it remains approximately constant with increasing f until a combination of core and lithospheric conductivities forces the currents to pass wholly through the crust whereupon it decreases with further increase in f beyond about 0.03 Hz. This mode is responsible for steady state bow wave phenomena. Attempts to detect a bow wave in the solar wind upstream from the Moon have not met with success. The TM current system, which is required to pass through the lunar crust and close in the solar wind, is consequently limited to less than 10^5 amperes. However, as shown later, the total absence of the TM mode is not assured and evidence for this mode is present.

Theoretical treatments of the lunar interaction with the solar wind generally include the effect of the solar wind dynamic pressure in confining the induced field lines (SONETT and COLBURN, 1968; JOHNSON and MIDGLEY, 1968; BLANK and SILL, 1969; SCHUBERT and SCHWARTZ, 1969). This effect is provided in the model by a field confining surface current layer in the solar wind just ahead of the lunar surface. Preliminary examination of the lunar response using the LSM data shows a strong amplification of incident tangential discontinuities whose free stream properties are monitored by Explorer 35. The amplification occurs only for the vector components tangential to the surface; the normal component tends to follow the interplanetary value. Thus the existence of a confining current layer appears verified for the sunward side of the Moon. The very strong amplification implies that the lines of force are confined within the Moon to a crustal layer having an electrical conductivity substantially less than that of the deeper layers (SONETT et al., 1971b).

THEORY OF THE INDUCTION

A complete treatment of the electromagnetic interaction would require appropriate matching of the interior fields to those in the plasma surrounding the Moon. The exterior flow field shows no marked perturbations aside from the diamagnetic cavity. An analysis based upon an inhomogeneous Moon immersed within a perfectly conducting space is used for the sunward hemisphere. This simplification is justified since cavity current effects are much smaller on the sunward hemisphere.

We consider the interaction of a solar wind forcing magnetic field oscillation

$$H = \hat{\eta}\, H_0 \exp \left\{ 2\pi i \left(\frac{\zeta}{\lambda} - ft \right) \right\}, \tag{1}$$

with a radially inhomogeneous Moon. The cartesian coordinate system (ξ, η, ζ) with unit vectors $\hat{\xi}, \hat{\eta}, \hat{\zeta}$ is fixed relative to the Moon and has its origin at the Moon's center. The Moon moves with speed v in the negative ζ direction. The quantities H_0 and $\lambda = v/f$ are the amplitude and wavelength, respectively, of the magnetic field oscillation. The magnetic field forcing function alone drives the TE lunar response. A derivation for the TM mode response is not included here because it must be relatively unimportant (see the discussion in the following section). In the lunar interior the solution of Maxwell's equations for the TE mode can be represented by the potential Ω which satisfies

$$\nabla^2 \Omega + k^2 \Omega = 0, \tag{2}$$

where

$$k^2 = \omega^2 \mu \varepsilon + i \sigma \mu \omega, \qquad \omega = 2\pi f, \tag{3}$$

σ is the electrical conductivity, and μ and ε are, respectively, the magnetic permeability and permittivity of the Moon. Although some possibility exists that local effects are significant in increasing the value of μ over the free space value, there is presently no evidence for this. An upper bound for the global permeability of $1.18\,\mu_0$ is given by DYAL and PARKIN (1971). In the following discussion the values assumed for the global permeability and permittivity of the Moon are the free space values of these quantities. Possible departures of the permittivity from the free space value are insignificant in the subsequent application of this theory. The equations for determining $\underline{H}$ from the potential Ω are presented in the Appendix.

The boundary condition applied to the TE mode is the continuity of the normal component of the magnetic field at the lunar surface. This condition results from the confinement of the induced field, treated as a current sheet at the Moon-plasma interface. The potential Ω is given by

$$\Omega = \mu v H_0 \frac{a}{r} \sin \varphi \sum_{l=1}^{\infty} \beta_l G_l(r) P_l^1(\cos \theta), \tag{4}$$

where (r, θ, φ) are spherical polar coordinates with ζ the polar axis, a is the lunar radius, β_l is $i^l(2l+1)/l(l+1)$, and P_l^1 are associated Legendre polynomials. The functions G_l are solutions of

$$\frac{d^2 G_l}{dr^2} + \left\{ k^2 - \frac{l(l+1)}{r^2} \right\} G_l = 0, \tag{5}$$

with the boundary conditions

$$G_l(r = a) = j_l\left(\frac{2\pi a}{\lambda}\right), \tag{6}$$

where j_l are the spherical Bessel functions. The equation for the potential and the form of the boundary conditions follow, in part, from the character of the spherical harmonic expansion of the solar wind forcing field, given in the Appendix. LAHIRI and PRICE (1939) were the first to obtain the radial differential equation for the TE mode in connection with their investigation of the geomagnetic induction problem.

For the interpretation of LSM data we introduce the modal transfer function for the tangential components of the TE mode magnetic field at the lunar surface

$$\frac{a \dfrac{dG_l}{dr}}{\dfrac{d}{da}\left\{ aj_l\left(\dfrac{2\pi a}{\lambda}\right) \right\}} \qquad \text{(see Appendix).}$$

INSTRUMENTATION AND DATA

General properties of both the LSM and the Ames magnetometer on Explorer 35 are given elsewhere (DYAL and PARKIN, 1971; MIHALOV *et al.*, 1968). Provision has been made in the Explorer magnetometer for the suppression of spin tone modulation of the data spectrum by utilizing a pair of synchronous demodulators which operate in quadrature upon the two spin tone modulated signals. This suppression is done on the spacecraft and can be shown to lead to a time series from which spin modulation is ideally eliminated. The signals returned to Earth are formally identical to those obtained from a nonspinning inertially stable spacecraft (SONETT, 1966).

A sample and hold system samples the three components of the vector at a uniform rate. In order to avoid the risk of unacceptable alias the magnetometer outputs are passed through a low pass filter prior to sampling, according to the Nyquist criterion.

The effect of filter rolloff is significant for frequencies above 0.025 Hz, which is the 3 db point. *The power spectral densities have been corrected for this effect at all frequencies.*

Spectra have been obtained both for cases where the Moon is in the magnetosheath and also in the free stream solar wind. The forcing field defined by Explorer 35 measurements is transformed into the local LSM coordinate system (x is along the normal outwards from the surface, y is easterly, and z is northerly at the site of ALSEP and the LSM). The LSM data, which define the response field, are edited to eliminate spurious data and to insure time continuity. Time gaps and noise in the LSM data are aperiodic and are attributed to telemetry transmission and data processing. For the cases studied the amount of data missing or deleted because of gaps and spurious noise is small ($< 5\%$). The data are then numerically filtered and decimated in order to analyze segments of the record with approximately the same sampling interval and upper frequency limit as that from Explorer 35, the latter having a Nyquist limit approximately one order less than LSM. The Explorer data are similarly edited to insure time continuity and to remove obviously spurious signals. For the Explorer data the time series are continuous for the intervals studied. Spurious signals at random intervals, are removed and again constitute a small amount of the time interval ($< 5\%$).

The power spectral densities for the forcing and response fields are computed using standard techniques for determining the autocorrelation function, smoothing, and taking the Fourier transform (BENDAT and PIERSOL, 1966; JENKINS and WATTS, 1968). The spectra are analyzed using 20 degrees of freedom, and the autocorrelation function is smoothed with a Parzen weighting function (JENKINS and WATTS, 1968). This gives an approximate error of 25 per cent in each power spectral density estimate. We have applied these techniques to seven selected 2-hour swaths and seven selected 1-hour swaths of data. Longer swaths, which would have reduced the horizontal error bars in Fig. 2, were not used because of time gaps, on the order of 10–30 minutes, in either the LSM or Explorer data. These are caused by telemetry shadowing, calibration interruptions, etc.

A representative set of spectra at both Explorer and LSM for the y and z magnetic field components (tangent to the surface) is shown in Fig. 1. The forcing spectra are seen to display the expected f^{-2} dependence characteristic of interplanetary fluctuations, while the LSM shows an f^{-1} dependence. Thus significant frequency dependent power amplification is apparent in the record.

The empirical transfer function $A_i(f)$ is defined by

$$h_{2i}(f) + h_{1i}(f) = A_i(f)h_{1i}(f), \tag{7}$$

where $h_{2i}(f)$ and $h_{1i}(f)$ are the Fourier transformed time series of the magnetic field induced in the Moon and measured on the lunar surface, and the free stream interplanetary magnetic field, respectively, and the subscript i is x, y, or z. The LSM measures the sum $h_{2i}(f) + h_{1i}(f)$ while the free stream magnetic field, $h_{1i}(f)$, is measured by the Ames magnetometer on the lunar orbiting satellite, Explorer 35.

Figure 2 shows the average ratio of power spectral densities $A_i(f)$. The mean values are the arithmetic averages using the 14 swaths of data. The error bars are the one

C. P. SONETT *et al.*

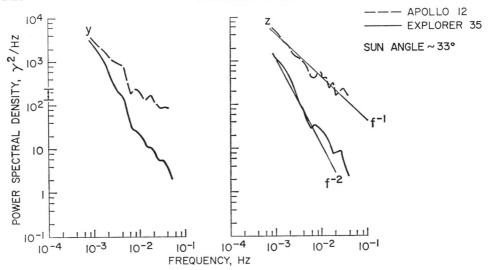

Fig. 1. Power spectral density determinations for simultaneous magnetic field observations on the lunar surface (Apollo 12) and in the solar wind near the Moon (Explorer 35). The *y* component is east and the *z* component north at the Apollo 12 site. Significant amplification is observed, increasing with frequency. Length of the time series is 2 hours.

standard deviation limits of the mean calculated from the transfer functions for the 14 time swaths. The bars do not include the error estimates in the calculation of the individual spectra. They are, however, consistent with the propagation of errors in the individual spectra. The values of A_x lie near unity at all frequencies indicating that the interior conducting region is relatively far from the magnetometer compared to the distance to the plasma current sheath on the sunward side of the Moon. Characteristically the amplification rises from near unity for both A_y and A_z at 0.001 Hz until values near 4 are attained above about 0.02 Hz. Values of A_y are less than those of A_z over the entire frequency range. This departure increases fractionally as frequency is decreased. It is plausible that the lunar TM response is frequency dependent at frequencies above about 0.01 Hz. If so the response for the TM mode should decrease with frequency in accordance with the observation. This is the strongest evidence for contributions from both the TE and TM modes. An alternate possibility that the basic response is purely TE and that the Moon is asymmetrically excited seems less convincing but cannot be ruled out at present. We have not yet found a completely satisfactory explanation for why the northward component of the excitation is systematically higher than the eastward component. The important facts are that they both have the same slope and change in slope around 0.01 Hz. We conclude that the lunar transfer function is dominated by the TE mode with possible contributions, especially at the lower frequencies, from the TM mode.

A significant scatter appears in the amplifications A_i. This scatter is substantially reduced by taking averages over many A_i, but is exact only if the errors have zero mean, which is approximately the case as shown in Fig. 3. A combination of all the

data points $\bar{A}(f) = \{\frac{1}{2}(A_y^2(f) + A_z^2(f))\}^{1/2}$ for all frequencies up to 0.035 Hz has
been made in order to investigate the distribution of the ratios of the power spectral
densities. The distribution shown in Fig. 3 is a histogram of the number of cases vs.
the difference of the measured value $\bar{A}$ from the mean at each frequency, normalized
by the standard deviation at that frequency and weighted by the expected error
$(\bar{A}(f_0)/\bar{A}(f))$. The figure shows that the distribution is somewhat more sharply
peaked than a normal one and also has a small bias.

The sources of the variations in the A_i may originate in any of several possible
ways: (1) the presence of contamination especially in the Explorer data from spurious
tones in the spacecraft; (2) the contribution to induction from higher order modes;
(3) fluctuations in cavity currents in turn due to changing diamagnetism in the solar
wind; (4) the possible contribution of TM modes to the predominantly TE spectra;
(5) the possibility that some field line leakage takes place where the component of
solar wind plasma pressure normal to the surface falls below the pressure of the induced
magnetic field. These factors may also contribute to the droop of the average A_i at
the higher frequencies. It should be noted that there is no correlation between the A_i
and sun angle, implying that field line confinement is effective for the data used here
(sun angles less than 65 deg).

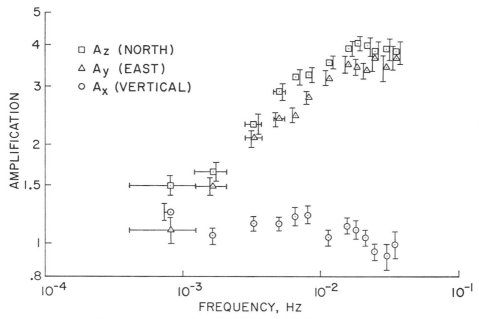

Fig. 2. Amplitude of the transfer function (ratio of lunar surface to free stream
magnetic measurements) vs. frequency. The error bars for frequency are the windows
defined by the lags in the autocorrelation calculation. The error bars in amplification
are the one standard deviation limits determined from the means of 14 data spectra.
Strong amplification is observed in the tangential (north and east) components of the
field, while for the normal component the amplification follows the expected value of
unity.

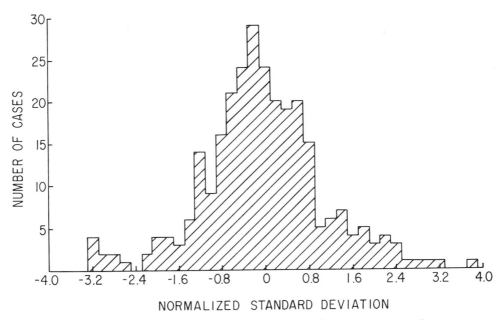

Fig. 3. Distribution of the values of the differences of $\bar{A}(f)$ from the mean $\bar{A}$ at each frequency for all frequencies up to 0.035 Hz in units of the normalized standard deviation.

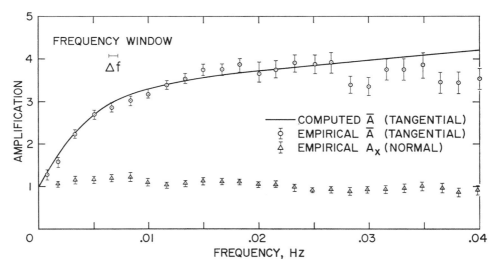

Fig. 4. The rms tangential lunar transfer function, $\bar{A} = [0.5(A_y^2 + A_z^2)]^{1/2}$ and the normal transfer function A_x as a function of frequency. The error bars, frequency windows, and directions are as shown in Fig. 2. The solid line is the value of amplification calculated from the conductivity profile whose corresponding amplification is fitted to experimental values at frequencies of 0.83, 1.7, 5, 12, 17, 22, 25, and 35 millihz.

LUNAR ELECTRICAL CONDUCTIVITY PROFILE

The theory of lunar induction, discussed earlier, has been used to derive a profile of electrical conductivity from the empirical transfer function. The theoretical amplification for the lowest TE mode, computed by numerically integrating equation (5) with a radially dependent conductivity, is matched to the empirical transfer function using a Newton-Raphson iterative scheme. This scheme readjusts the entire conductivity profile to yield amplifications which best fit the data in a least-squares sense over the entire frequency range. This has been carried out using frequency values of 0.83, 1.7, 5, 12, 17, 22, 25, and 35 milliHz. The conductivity profile is characterized by its values at the radial locations $r = 800, 1200, 1400, 1450, 1490, 1510, 1550$, and 1740 km. For $r < 800$ km the conductivity σ is set to the value at $r = 800$ km; elsewhere a linear interpolation of log σ is used.

The computer calculation is started with a continuous conductivity profile defined by 8 parameters as described above. Numerical integrations are carried out to obtain values of amplification at the eight frequencies. A comparison of these amplifications with the data provides the Newton-Raphson scheme with the input required to adjust the 8 conductivity parameters to yield a profile whose corresponding theoretical amplification curve is a better least-squares fit to the empirical transfer function. The iteration is continued until an adequate fit is obtained. The result of such a fit to the empirical amplification curve $\bar{A} = \{\frac{1}{2}(A_y{}^2 + A_z{}^2)\}^{1/2}$ is shown in Fig. 4. The introduction of $\bar{A}$ permits the use of both A_y and A_z data to improve the statistics. The differences between the theoretical amplification and the empirical $\bar{A}$ are, in our assessment, partially attributable to the various complications of the excitation process enumerated at the end of the previous section (not accounted for by the theory), as well as considerations of computer running time which limits the number of frequencies and conductivity parameters (spatial resolution) employed.

A conductivity profile is associated with the theoretical amplification curve of Fig. 4. In addition we have obtained conductivity profiles by fitting not only $\bar{A}$ but also A_y, A_z and their one standard deviation limits. The conductivity profiles derived from the iterative least squares inversion for A_y, A_z, and $\bar{A}$ are shown in Fig. 5. The prominent spike in each conductivity profile is an invariant characteristic of the inversions. It is centered at about $r = 1500$ km where the conductivity is nearly 10^{-2} mhos/m. The inner minimum lies at about $r = 1400$ km and the conductivity appears to rise at greater depth. Gross bounds on the conductivity profile are seen in the insert; these are determined from the one standard deviation limits of the various A, but do not themselves represent one standard deviation limits on the conductivity profile.

The computer calculations for Figs. 4 and 5 started with a constant conductivity of 10^{-4} mhos/m. However, a number of computations have been carried out using different values for uniform starting conductivities (e.g. 10^{-3} and 10^{-5} mhos/m) and different radial locations. In every case tested, initial convergence was rapid and the final conductivity profile invariably displayed the prominent spike near $r = 1500$ km. For the profiles reported here, several values of r were chosen in the neighborhood of $r = 1500$ km to better define the conductivity spike.

Whereas the large spike in conductivity is a persistent feature of the inversions, the character of the conductivity profile at greater depth, where the conductivity appears to rise, is not so certain and our results for the conductivity at these depths must remain tentative. The surface amplification is a rather insensitive function of core conductivity because of the relatively small core volume and the distance to the surface.

Figure 6 shows the manner in which the amplitude and phase of the relative field strength vary with depth in the Moon, indicating different degrees of field compression at various frequencies. The conductivity profile used in this calculation was the one associated with the theoretical fit of $\bar{A}$ (see Fig. 5). In the range $r = 1500$ to 1740 km, the field is essentially constant and phase shift is small, consistent with the low conductivity (large skin depth) in this region. The large rise in conductivity at $r = 1500$ km produces a penetration barrier (small skin depth) for the higher frequencies. Below this barrier the high frequency variations in the field are low in amplitude and have large phase shifts. Over the high frequency range the barrier essentially confines field lines to the outer region, producing a nearly constant amplification with frequency, consistent with the data. For the lower frequencies the barrier is not effective and the

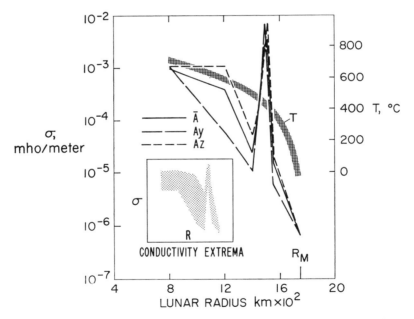

Fig. 5. Lunar bulk electrical conductivity profiles determined from the individual transfer functions A_y, A_z, and $\bar{A}$. The prominent rise of σ from the surface inwards to $r = 1500$ km is apparent for all three cases as well as the subsequent decrease inwards to $r = 1400$ km followed by a more gradual rise. A tentative version of a lunar thermal profile is shown as the grey overlay with temperatures indicated on the right hand margin. This profile is a fit of conductivities to a Nagata basalt in the mantle, an England olivine in the core, and the known subsurface temperature of $-30°C$. The insert is shown to suggest extreme values of the σ using the one standard deviation limits of the A to calculate conductivities.

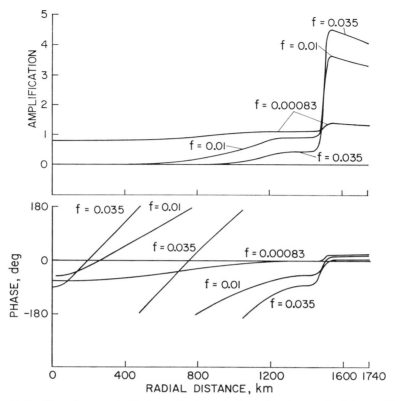

Fig. 6. Profiles of magnetic field amplitude and phase for the conductivity profile labeled $\bar{A}$ in Fig. 5. The values are normalized to the transfer function at the lunar surface. The region of high conductivity centered at $r = 1500$ km (see Fig. 5) is responsible for strong attenuation and large phase shift at the higher frequencies. There is a correspondence between the selective attenuation shown here and the amplification vs. frequency curve of Fig. 4, as discussed in the text. Negative phase angle in this figure represents a time delay with respect to the driving function.

relatively low conductivity in the range $r < 1500$ km allows the field to penetrate to the deep interior, yielding a smaller amplification. Where phase shifts are greater than 90 deg the field lines connected with the driving field have been excluded; the resulting compression leads to the amplification.

COMPARISON WITH OTHER CONDUCTIVITY PROFILES

A physical understanding of the appearance of the large conductivity spike can be obtained as follows. In a two layer model with an infinitely conducting core and a nonconducting shell the amplification of the tangential magnetic field components is given by

$$1 + \frac{3 \text{ (core volume)}}{2 \text{ (shell volume)}}$$

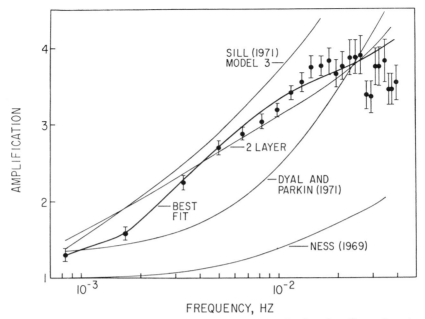

Fig. 7. The experimental r.m.s. tangential lunar transfer function $\bar{A}$ as a function of frequency. The error bars are the same as those in Figs. 2 and 4. The curve labeled "best fit" are the values of the amplification for the conductivity profile labeled $\bar{A}$ in Fig. 5. The "2 layer" curve was calculated for a model with a core of radius 1560 km and a constant conductivity of 7.6×10^{-4} mhos/m, and a shell of zero conductivity. This value of the core conductivity provides a best fit to $\bar{A}$ for the given core radius. The Dyal and Parkin (1971) amplification curve was calculated for a model with an inner core of radius 1044 km and a conductivity of 10^{-2} mhos/meter, a middle layer extending to 1653 km with a conductivity of 1.7×10^{-4} mhos/m and an outer layer of zero conductivity. Other amplification curves are based on conductivity models proposed by Ness (1969) and Sill (1971). Note that the "2 layer" curve does not duplicate the "S" nature of the experimental data.

This simple result requires that the high frequency skin depth barrier be near $r = 1500$ km for an amplification ≈ 4. At frequencies above 0.02 Hz essentially no wave penetration takes place through this barrier. Thus the volume available in the core for field lines is insignificant compared to that in the nonconducting shell. This means that the amplification becomes independent of frequency, i.e. $dA/df \rightarrow 0$ at high frequency. Concurrent with the high frequency limitation is the requirement that A drop to near unity at the lower frequencies. If it were assumed that the electrical conductivity were monotonic, so that the interior conductivity were uniformly high, then the low frequency amplification would be in excess of the observed values.

The inadequacies of monotonic conductivity profiles are shown in Fig. 7. The experimental r.m.s. amplification data $\bar{A}$ (circled points with associated error bars) can be compared with the amplification curve for the $\bar{A}$ conductivity profile of Fig. 5 (labeled "best fit") and amplification curves of several monotonic conductivity profiles. The "2 layer" amplification curve was calculated for a model with a core of radius of 1560 km and a constant conductivity of 7.6×10^{-4} mhos/m, and a shell of zero conductivity. This value of the core conductivity provides a best fit to the experimental $\bar{A}$ for the given core radius. Other best fit two layer models with different core radii have been investigated; the one used in Fig. 7 yields an amplification curve which matches the data most closely. The amplification curve labelled DYAL and PARKIN (1971) was computed from a 3 layer conductivity model proposed by these authors. The parameters of their model are $\sigma = 10^{-2}$ mhos/m for $0 < r < 1044$ km, $\sigma = 1.7 \times 10^{-4}$ mhos/m for 1044 km $< r < 1653$ km and $\sigma = 0$ in the outer shell. In other 3 layer conductivity models consistent with the data analysis of DYAL and PARKIN (1971), the outer boundary of the region of intermediate conductivity can be located anywhere between $r = 1653$ and 1740 km. We have computed the amplification curves for a number of these additional 3 layer models; none provide a better fit to the observational curve. Also shown in Fig. 7 are amplification curves from conductivity profiles proposed by SILL (1971) (model 3 in that paper) and NESS (1969). Other conductivity models investigated by SILL (1971) fit the experimental data no better than his model 3. The conductivity model of NESS (1969) consists of a core of radius 1426 km with $\sigma = 8 \times 10^{-5}$ mhos/m, and a nonconducting shell. Amplification curves from conductivity models of WARD (1969) were also considered but these were rejected upon comparison with the observational data.

The "2 layer" amplification curve shown in Fig. 7 is, compared to the experimental data, high at $f < 0.0035$ Hz and $f > 0.03$ Hz, and low in the range 0.01 Hz $< f < 0.02$ Hz. The data show a flat response at frequencies above 0.02 Hz, whereas the slope of the "2 layer" curve is high at these frequencies. Furthermore as the frequency increases the curvature of the data changes from positive to negative at about 0.003 Hz. The "2 layer" amplification curve is everywhere concave up. *This s-shaped character of the experimental data is faithfully reproduced only by our "best fit" conductivity profile.* Further work is in progress to improve the high frequency slope of our "best fit" model. This will probably lead both to a higher value of the maximum conductivity and a steeper slope of the conductivity profile at the outer edge, effects which will tend to sharpen the conductivity spike in order to allow penetration of low frequency magnetic field fluctuations.

COMPOSITIONAL AND THERMAL MODEL

The electrical conductivity profile cannot be explained by a uniform material and a plausible thermal profile. In the region from the surface to $r = 1500$ km where the conductivity attains its maximum value, the rise of conductivity with depth is a reasonable consequence of the accompanying increase of temperature in a material of uniform composition. Below $r = 1400$ km the apparent rise in conductivity is again explainable by an increase in temperature. On the other hand the precipitous decrease of electrical conductivity by 2 to 3 orders between $r = 1500$ and 1400 km cannot be explained as due to temperature. Either a compositional change, phase change, or a combination of the two is required. Thus a reasonable model for stratification of the Moon, limited by the present poor spatial resolution of the analysis indicates a core out to $r \approx 1400$ km overlain by a mantle of higher conductivity material, plus possibly a transition layer at $r = 1400 - 1500$ km.

2428   C. P. SONETT *et al.*

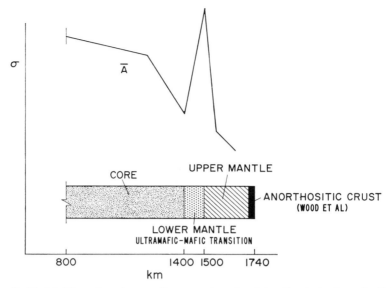

Fig. 8. Model Moon based upon the magnetometer data, the derived conductivity
profiles, and the consistent fit of σ to rocky material. The anorthositic crust is taken
from WOOD *et al.* (1970); it is not seen in the present analysis.

In order to infer a temperature profile from the conductivity profile it is necessary
to use conductivity-temperature functions of known rock materials. The Apollo
basalts (SCHWERER *et al.*, 1970; NAGATA *et al.*, 1970) are representative of the most
conducting rocky matter known. On the other hand olivine (dunite) or olivine-
peridotite (ENGLAND *et al.*, 1968) is representative of poorly conducting geological
material. The peak conductivity value found in the lunar mantle corresponds to a
temperature of about 450°C for lunar basalt or 950°C for olivine. Since the temperature
just under the lunar surface is −30°C, the corresponding thermal gradients in the
outer mantle are about 2°K/km and 4°K/km.

The precipitous decrease in electrical conductivity of 10^2–10^3 as depth increases
from 250 km to 350 km takes place in a distance where the temperature rises at least
100 degrees. While the absolute conductivities of the materials are unsure because of
impurities, etc., it seems clear that the change in conductivity around 300 km depth
requires a change in composition. Thus one model fitting the data has an olivine-like
core, possibly a lower mantle composed of a material transitional between the
substrate and the upper mantle, and an upper mantle having the conductivity of
Apollo basalt (see Fig. 8). Since the present frequency resolution is insufficient to
estimate the conductivity close to the surface, these results do not preclude the presence
of an anorthositic crust (WOOD *et al.*, 1970). The volume of the mantle comprises about
one half the total volume of the Moon. To fill it completely with a basaltic extract is
in conflict with geochemical constraints which imply at most a 10–15% contribution
by volume. The upper mantle may have to be composed only partly of a basalt-like
substance to account for the apparent dependence of the upper mantle conductivity

upon temperature. The actual basalt-like thickness cannot be assayed yet because of limited spatial resolution. An alternative model which avoids the difficulty of finding a source of so much basalt has a layer of iron concentration at $r = 1400$ to 1500 km, such as could occur from a differentiation of the outer 340 km (WOOD *et al.*, 1970; cf. UREY *et al.*, 1971).

For our compositional model the temperature at $r = 1500$ km is 450°C and the average near surface thermal gradient is 2°K/km. The temperature at $r = 800$ km, computed from the conductivity profiles shown in Fig. 5 together with an olivine conductivity function is in the range 750–800°C, decreasing to approximately 550°C at 1400 km where the local minimum in electrical conductivity is observed. A calculation based upon more conducting matter would depress the computed temperatures. R. REYNOLDS (private communication) has shown that after 4.5 billion years a non-convecting Moon of 25% ordinary chondritic radioactive composition and an initial uniform temperature of 0°C attains a near surface thermal gradient of approximately 2°K/km and a uniform core temperature of about 800°C. Hence a thermal model consistent with the conductivity requires appreciable depletion of the heat sources in the Moon relative to ordinary chondrites.

An approximate value for the lunar heat flux can be calculated from the near surface gradient determined above. For silicate material the thermal conductivity is 0.008 cal/cm sec deg though it is known that it will decrease with increasing temperature (HORAI *et al.*, 1970; MURASE and MCBIRNEY, 1970). This effect will not introduce a substantial error in the estimate of the heat flux which we find to be 1.6×10^{-7} cal/cm² sec using the "basalt" composition for the mantle; olivine would double this estimate. The former value is approximately $\frac{1}{8}$ that of the Pre-Cambrian regions of Earth (KAULA, 1968). For a Moon scaled Earth, the flux would be twice that actually measured for a "basalt" model and approximately equal for an olivine model.

Acknowledgments—We wish to thank our colleagues C. W. SNYDER, M. NEUGEBAUER, J. H. WOLFE, and the Vela group at Los Alamos Scientific Laboratory for supplying plasma data for testing the field line confinement problem. Constructive comments by W. M. KAULA have substantially improved this paper.

REFERENCES

BENDAT J. S. and PIERSOL A. G. (1966) *Measurement and Analysis of Random Data*. John Wiley.

BLANK J. L. and SILL W. R. (1969) Response of the Moon to the time-varying interplanetary magnetic field. *J. Geophys. Res.* **74**, 736–743.

COLBURN D. S. and SONETT C. P. (1966) Discontinuities in the solar wind. *Space Science Rev.* **5**, 439–506.

DYAL P. and PARKIN C. W. (1971) Electrical conductivity and temperature of the lunar interior from magnetic transient-response measurements. Second Lunar Science Conference (unpublished proceedings).

ENGLAND A. W., SIMMONS G., and STRANGWAY D. (1968) Electrical conductivity of the Moon. *J. Geophys. Res.* **73**, 3219–3226.

HORAI K., SIMMONS G., KANAMORI H., and WONES D. (1970) Thermal diffusivity, conductivity and thermal inertia of Apollo 11 lunar material. *Proc. Apollo 11 Lunar Science Conf.*, Geochim. Cosmochim. *Acta* Suppl. 1, Vol. 3, pp. 2243–2249. Pergamon.

JENKINS G. W. and WATTS D. G. (1968) *Spectral Analysis and Its Applications*. Holden-Day.

JOHNSON F. S. and MIDGLEY J. E. (1968) Notes on the lunar magnetosphere. *J. Geophys. Res.* **73**, 1523–1532.

KAULA W. M. (1968) *An Introduction to Planetary Physics.* John Wiley.

LAHIRI B. N. and PRICE A. T. (1939) Electromagnetic induction in non-uniform conductors, and the determination of the conductivity of the Earth from terrestrial magnetic variations. *Phil. Trans. R. Soc.* A**237**, 509–540.

MIHALOV J. D., COLBURN D. S., CURRIE R. G., and SONETT C. P. (1968) Configuration and reconnection of the geomagnetic tail. *J. Geophys. Res.* **73**, 943–959.

MURASE T. and MCBIRNEY A. R. (1970) Thermal conductivity of lunar and terrestrial igneous rocks in their melting range. *Science* **170**, 165–167.

NAGATA T., RIKITAKE T., and KONO M. (1970) Electrical conductivity and the age of the Moon. Presented at the Thirteenth Plenary Meeting, COSPAR, Leningrad, USSR.

NESS N. F. (1969) The electrical conductivity and internal temperature of the Moon. *Goddard Space Flight Center Report X-616-69-191* (also presented as paper K11 at 12th COSPAR, Prague, May, 1969).

SCHUBERT G. and SCHWARTZ K. (1969) A theory for the interpretation of lunar surface magnetometer data. *The Moon* **1**, 106–117.

SCHWERER F. C., NAGATA T., and FISHER R. M. (1971) Electrical conductivity of lunar rock and chondritic meteorites. Submitted to *The Moon.*

SILL W. R. (1971) Electrical conductivity and temperature of the lunar interior. *J. Geophys. Res.* **7,6** 251–256.

SONETT C. P. (1966) Modulation and sampling of hydromagnetic radiation. In *Space Research VI* (editor R. L. Smith-Rose), 280–322. Spartan Books.

SONETT C. P. and COLBURN D. S. (1968) The principle of solar wind induced planetary dynamos. *Phys. Earth and Plan. Int.* **1**, 326–346.

SONETT C. P., COLBURN D. S., DYAL P., PARKIN C. W., SMITH B. F., SCHUBERT G., and SCHWARTZ K. (1971a) The lunar electrical conductivity profile. *Nature* **230**, 359–362.

SONETT C. P., DYAL P., PARKIN C. W., COLBURN D. S., MIHALOV J. D., and SMITH B. F. (1971b) Whole body response of the Moon to electromagnetic induction by the solar wind, *Science* **172**, 256–258.

SONETT C. P., DYAL P., COLBURN D. S., SMITH B. F., SCHUBERT G., SCHWARTZ K., MIHALOV, J. D., and PARKIN C. W. (1971c) Induced and permanent magnetism on the Moon: structural and evolutionary implications. Transactions of the *IAU*, General Assembly, Commission 17, D. Reidel (in press).

UREY H. C., MARTI K., HAWKINS J. W., and LIU M. K. (1971) Model history of the lunar surface (preprint).

WARD S. H. (1969) Gross estimates of the conductivity, dielectric constant, and magnetic permeability distributions in the Moon. *Radio Sci.* **4**, 117–137.

WOOD J. A., DICKEY J. S., MARVIN U. B. and POWELL B. J. (1970) Lunar anorthosites and a geophysical model of the Moon. *Proc. Apollo 11 Lunar Science Conf., Geochim Cosmochim. Acta* Suppl. 1, Vol. 1, pp. 965–988. Pergamon.

APPENDIX

The spherical harmonic expansions for the potential and magnetic field of the transverse electric part of the solar wind excitation are

$$\Omega = \mu v H_0 \frac{\sin \varphi}{r} \sum_{l=1}^{\infty} \beta_l r j_l \left(\frac{2\pi r}{\lambda}\right) P_l^1(\cos \theta),$$

$$\begin{Bmatrix} H_\theta \\ H_\varphi \end{Bmatrix} = H_0 \begin{Bmatrix} \sin \varphi \\ \cos \varphi \end{Bmatrix} \sum_{l=1}^{\infty} \beta_l \frac{\lambda}{2\pi i r} \frac{d}{dr}\left(r j_l\left(\frac{2\pi r}{\lambda}\right)\right) \begin{Bmatrix} \dfrac{dP_l^1(\cos \theta)}{d\theta} \\ \dfrac{P_l^1(\cos \theta)}{\sin \theta} \end{Bmatrix}$$

$$H_r = H_0 \sin \varphi \sum_{l=1}^{\infty} \beta_l \frac{\lambda}{2\pi i r} l(l+1) j_l\left(\frac{2\pi r}{\lambda}\right) P_l^1(\cos \theta).$$

The TE part of the magnetic field is related to the TE potential by

$$H_r = -\frac{i}{\mu\omega}\left(\frac{\partial^2}{\partial r^2} + k^2\right)(r\Omega), \qquad H_\theta = -\frac{i}{\mu\omega r}\frac{\partial}{\partial r}\left(r\frac{\partial\Omega}{\partial\theta}\right),$$

$$H_\varphi = \frac{-i}{\mu\omega r \sin\theta}\frac{\partial}{\partial r}\left(r\frac{\partial\Omega}{\partial\varphi}\right)$$

Thus the TE magnetic field in the lunar interior is

$$\begin{Bmatrix} H_\theta \\ H_\varphi \end{Bmatrix} = H_0\begin{Bmatrix} \sin\varphi \\ \cos\varphi \end{Bmatrix}\sum_{l=1}^{\infty}\beta_l\frac{\lambda a}{2\pi i r}\frac{dG_l}{dr}\begin{Bmatrix} \dfrac{dP_l^1(\cos\theta)}{d\theta} \\ \dfrac{P_l^1(\cos\theta)}{\sin\theta} \end{Bmatrix},$$

$$H_r = H_0\sin\varphi\sum_{l=1}^{\infty}\beta_l l(l+1)\frac{\lambda a}{2\pi i r}\left(\frac{G_l}{r}\right)P_l^1(\cos\theta).$$

Proceedings of the Second Lunar Science Conference, Vol. 3, pp. 2433–2449
The M.I.T. Press, 1971.

Magnetic properties of individual glass spherules, Apollo 11 and Apollo 12 lunar samples*

S. Sullivan and A. N. Thorpe†
Howard University, Washington, D.C. 20001

and

C. C. Alexander, F. E. Senftle, and E. Dwornik
U.S. Geological Survey, Washington, D.C. 20242

(*Received* 24 *February* 1971; *accepted in revised form* 25 *March* 1971)

Abstract—The magnetic properties of eight glass spherules (0.03–0.24 mg) from the Apollo 12 lunar fines, one fragment (44 mg) from glass spatter collected during the Apollo 12 mission, and eleven glass spherules from the Apollo 11 fines have been determined. As in the case of the Apollo 11 specimens, previously studied, the specimens showed a strong paramagnetism, and an easily and difficultly magnetized ferromagnetic component. An intermediate ferromagnetic component was found which was small and contributed little to the total susceptibility. Subsequent remeasurements of the spherules from both the Apollo 11 and Apollo 12 samples show gradual changes in the magnetic properties with time in many of the specimens. Selected specimens have been heat treated in controlled atmospheres to determine the effect of oxidation which was found to be primarily a surface effect. The metallic iron varied from about 0.01 to 1% and the total iron calculated from the magnetic measurements compared well with electron probe analysis. The data indicate that the titanium is essentially all in the Ti^{4+} state.

Several specimens were also studied at temperatures as low as 4.2°K in order to determine if the Curie-Weiss law held at low temperatures. The data follow a Curie-Weiss law with a Weiss temperature of about 3°K.

Introduction

In a prior investigation magnetic studies have been made on individual glass spherules from the bulk lunar fines of the Apollo 11 sample (Thorpe *et al.*, 1970). All the specimens showed strong paramagnetism due to Fe^{2+} dissolved in the glass. Most of the specimens also had an easily and also a difficultly magnetized ferromagnetic component, but no major superparamagnetic component, such as found in the crystalline rock specimens of the lunar sample (Nagata *et al.*, 1970). The source of the difficultly magnetized ferromagnetic component was found to be metallic iron spheres, whereas the easily magnetized component was ascribed to odd-shaped fragmental iron or ferromagnetic minerals. To compare the magnetic properties of glass spherules from the bulk lunar fines of the Apollo 11 sample with similar specimens from the fines of Apollo 12 sample, glass spherules from the bulk fines of both the Apollo 12 and Apollo 11 missions were selected for further detailed magnetic measurements. The results are generally similar but some differences are evident. In addition, a bubble from glass spatter on an Apollo 12 rock was also studied.

---

* Publication authorized by the Director, U.S. Geological Survey.

† Also at U.S. Geological Survey, Washington, D.C.

Experimental Measurements

Specimens

Eight glass spherules from the bulk fines of Apollo 12, sample 12070,90; six glass spherules from Apollo 11, sample 10084,22; and five glass spherules from Apollo 11, sample 10085,17 were selected for the basic studies. One of the rock chips of sample 12057, coated on one side with glass spatter, had a large hollow scoriaceous glass bubble (designated 12057,10) which was removed. The magnetic properties of this hollow hemispherical glass bubble were compared with those of the glass spherules.

Physical measurements

All the specimens were weighed either on a Mettler balance or on a quartz helical spring balance and weights are good to 10 μg. The diameters were measured with a precision traveling microscope. Magnetic susceptibility and magnetization measurements were made on a quartz helical spring balance and also with a semiautomated Cahn suspension balance, depending on the size of the sample and the experiment being performed. The balance and technique are described elsewhere (SENFTLE *et al.*, 1958; THORPE and SENFTLE, 1959; SENFTLE and THORPE, 1963). Magnetic susceptibility (χ_a) measurements were made on all the specimens (1) as a function of the magnetic field (H_a) at room temperature and (2) as a function of temperature (T) down to 77°K at constant magnetic field. As described previously for similar specimens (THORPE *et al.*, 1970), the magnetic susceptibility extrapolated to infinite field (χ_0) and the low field (< 0.8 kOe) saturation magnetization (σ) were determined from the measurements in (1). The Curie constant (C) and the temperature independent susceptibility were determined from the measurements in (2). In addition, the saturation magnetization, $I_s = \chi H_s$, was determined from a plot of the magnetization, I, versus the magnetic field H_a up to 12 kOe. The susceptibility of all specimens was measured as a function of temperature down to 77°K, and a few specimens were studied down to liquid helium temperatures. Transmission Mössbauer measurements were made at room temperature on the hemispherical bubble using a spectrometer in the constant acceleration mode. Pure iron was used as a standard, and the centroid of the iron spectrum was used to standardize the isomer shift.

Upon completion of the magnetic studies, the spherules were prepared for electron probe analysis by embedding in epoxy, grinding and polishing to a diameter a little less than the maximum diameter to prevent loss of the spherule. Two analyzed hornblendes and synthetic Ca–Fe silicate glasses, somewhat similar in composition to the glass spherules, were used as standards. Corrections were made for background only, and the reported values are good to ±10% of the amount present; SiO_2 to ±5%.

Results and Discussion

Petrographic examination and electronprobe analyses

Petrographic examination of the polished surfaces of the spherules from the Apollo 12 sample revealed some unusual properties worthy of note. Microphotographs of some of the spherules from the Apollo 12 sample are shown in Fig. 1B. Of the eight samples only three—301, 304, and 305—proved compact, and free of voids. Number 307, a light green clear spherule, contained euhedral crystals of the order of 2 μm. These are similar to the metallic iron cubes observed by GLASS (1971) in greenish glass spherules recovered from sample 12057, and identified by him as octahedral Ni–Fe crystals. Spherules 302, 303, and 306 are an olive gray color, and contain numerous voids, mineral inclusions, and metallic spheres ranging in size from 2 μm down to ~ 0.2 μm. A qualitative probe analysis of the metal spheres showed a high iron content and the intensity of nickel radiation was higher than that of the matrix suggesting metallic iron with a small amount of nickel. Mineral inclusions of olivine

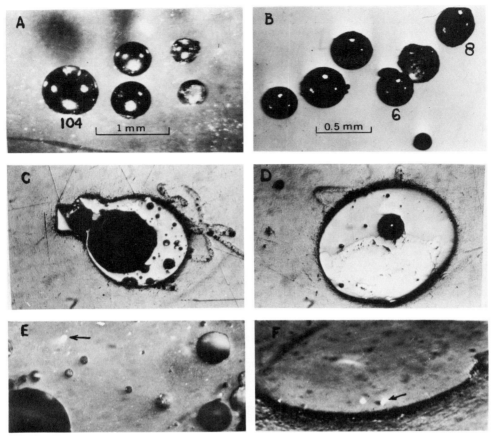

Fig. 1. (A) Spherules from soil of Apollo 11, sample 10085,17. Largest spherule (104) is 720 μm in diameter. (B) Spherules from soil of Apollo 12, sample 12070,90. Spherule number 6 is 300 μm in diameter. (C) Photomicrograph of spherule 306 (6 in photo B) showing interior and polished surface prepared for electron probe analysis. Note the large void in the interior and the crystalline appendage. (D) Photomicrograph of polished section of spherule 308 (8 in photo B) showing glassy upper half and devitrified or crystalline lower half of the specimen. (E) Photomicrograph of spherule 302 showing polished surface of typical spherical metallic inclusion (arrow). Size of largest inclusion is $\sim$ 2 μm. (F) Photomicrograph of spherule 307 (light green color) showing unusual octahedron (probably metallic iron). Octahedron is $\sim$ 2 μm in size.

and pyroxene up to 40 μm in size were also present. Spherule number 308 was almost half crystalline or devitrified. The electron probe analysis of the glass spherules is given in Table 1.

Of the Apollo 11 specimens only five were selected for petrographic and electron probe analysis (Fig. 1A). Spherules 101, 104, and 204 were dense, compact glass with a few spherical high iron inclusions, of which the largest was $\sim$ 6 μm in diameter and the smallest was less than 1 μm. Spherule 106 was similar but contained some gas

Table 1. Electron probe analysis of glass spherules from lunar soil 12070,90 and glass bubble 12057,10.

Spherule	SiO$_2$	TiO$_2$	Al$_2$O$_3$	FeO	MgO	CaO	K$_2$O	Remarks
301	40.1	3.1	13.5	15.7	11.0	11.5	< 0.1	Dark brown, dense
302	42.6	2.6	10.1	15.6	12.7	11.0	0.2	Olive gray vesicular, with mineral and metallic inc.
303	43.2	3.0	11.5	17.0	9.3	10.0	0.2	Same
304	41.9	2.8	12.3	16.7	8.5	10.9	< 0.1	Dark brown, dense
305	46.1	2.6	9.8	18.2	9.3	10.5	< 0.1	Dark brown, dense
306	44.9	2.6	11.5	15.5	10.4	10.4	0.5	Olive gray, virtually hollow, mineral inclusions
307	39.2	0.2	26.9	3.8	9.3	15.9	< 0.1	Clear, green with octa-hedral Ni-Fe crystals
308 A	45.9	1.8	8.3	17.3	13.8	8.5	< 0.1	Glassy half
308 B	48.9	0.6	0.8	17.1	19.0	4.8	< 0.1	Crystalline or devitrified half
GB–1	44.2	2.7	11.6	15.5	8.0	10.8	—	Hemispherical glass bubble

Na$_2$O content of these samples was of the order of 0.1%; GB–1 was 0.5%.
NiO content also of the order of 0.1% with a value of 0.2% obtained for metallic inclusions in spherules 2 and 3.

Table 2. Partial analyses, Apollo 11 spherules from soil 10085,17.

Spherule	FeO	TiO$_2$	Remarks
101	15.0	6.1	Dark amber, smooth, glossy, oblate
104	7.4	0.6	Light green, clear, slightly oblate with grayish material fused to about $\frac{1}{3}$ of surface
106	18.6	7.8	Dark amber, spherical, dull
204	4.1	0.9	Milky white, spherical glazed surface
205	3.9	0.3	Light green, clear, spherical

bubbles. Specimen 205 was milky white and showed an unusual banded texture. Electron probe analysis of iron and titanium is shown in Table 2.

Mass and apparent densities

The apparent density (as shown in Table 3) was calculated from the mass and mean radius assuming a perfect sphere and should only be considered as approximate. Many of the specimens were not spherical and some had deformed surfaces or attached fragments; those which appeared to be near perfect spheres are marked with an asterisk. The deviation from true sphericity in most cases was not large and therefore one can attach some significance to the density values. The specimens from Apollo 12 in general show a lower apparent density than those of Apollo 11. To some extent this may be due to their slight oblate shape and in some cases surface protuberances or blisters which would increase the volume. However, as confirmed by the photomicrographs (see Fig. 1), the glass in the Apollo 12 specimens is more vesicular and contains more crystalline inclusions than those of Apollo 11.

Table 3. Room temperature and low temperature magnetic measurements.

(1)	(2)	(3)	(4)	(5)	(6)	(7)	(8)	(9)	(10)	(11)
		Room temperature (300°K)						Low temperature (300°K–77°K)		
Spherule	Mass (mg)	Diameter (μm)	Calc. density (g/cm^3)	χ_0 ($\times 10^{-6}$ emu/g)	σ ($\times 10^{-4}$ emu/g)	I_s† (emu/g)	H_s (kOe)	Curie constant ($\times 10^{-3}$)	χ_I ($\times 10^{-6}$ emu/g)	$f(H_a)/H_a$ ($\times 10^{-6}$ emu/g)
		Individual glass spherules, Apollo 12 (12070,90)								
301	0.24	610	2.0	28.3	28.3	0.08	9.5	6.21	8.0	1
302	0.19	600	1.7	201.0	228.0	1.12	7.5	7.44	177.0	2
303	0.17	525	2.2	191.0	213.0	1.06	7.5	5.71	172.0	3
304	0.05	370	2.0	35.9	263.0	0.13	9.5	6.27	15.4	3
305	0.04	340	2.1	61.0	491.0	0.33	9.5	7.12	37.7	3
306	0.03	300	1.8	442.0	384.0	2.10	8.0	6.44	421.0	4
307	0.03	340	1.6	35.3	654.0	n.d.	n.d.	1.94	29.2	−1
308	0.04	340	1.9	68.4	194.0	0.26	6.5	8.69	39.8	5
		Glass bubble, Apollo 12 (12057,10)								
GB–1	44.4	~ 6000	n.d.	174.0	~ 0	n.d.	n.d.	5.26	157.0	1
		Individual glass spherules, Apollo 11 (10084,22)								
101	0.28	620	2.2	52.7	138.0	0.30	10.0	6.19	32.5	0
102	0.18	490	2.9	100.0	89.3	0.76	8.5	5.27	82.8	0
103	0.16	540	1.9	32.2	115.0	0.08	9.5	6.34	11.5	2
104	0.74	820	2.6	30.0	430.0	0.08	7.0	3.91	17.4	1
105	0.44	700	2.4	12.9	30.2	0.04	7.4	1.93	6.9	0
106	0.30	565	3.2*	126.0	82.6	0.74	10.0	6.67	104.0	0
		Individual glass spherules, Apollo 11 (10085,17)								
201	0.08	360	3.3*	29.3	53.8	0.06	9.0	5.98	9.8	2
202	0.09	410	2.5*	181.0	125.0	1.16	7.0	3.54	170.0	0
203	0.15	450	3.1*	60.4	19.6	0.52	8.5	3.09	50.5	2
204	0.16	440	3.6*	35.0	117.0	0.20	7.5	1.76	29.5	0
205	0.14	410	3.9*	23.0	490.0	0.02	7.0	1.80	17.4	2

n.d.—not determined; † Corrected for σ; * Near perfect spheres.

Room and low temperature results (300°K to 77°K)

In a previous paper (THORPE et al., 1970) it was shown that the magnetization of the glass spherules can generally be described by assuming two components of magnetization, an easily magnetized component σ (saturated at less than 800 Oe), and a more difficultly magnetized component I_s (saturated at fields in excess of 6000–7000 Oe). The easily saturated component was ascribed to iron, or iron minerals with a shape associated with a low demagnetizing factor, e.g., needle-like shapes along the major axis. The difficultly magnetized component was ascribed to near-perfect spheres of iron or nickel–iron which have been shown to have a field and temperature independent susceptibility ($<$ 6000 Oe and between 77°K and 300°K; see SENFTLE et al., 1964). This temperature independent susceptibility, X_I, is 0.03 emu/g for pure iron spheres and amounts to a significant fraction of the total susceptibility extrapolated to an infinite magnetic field.

The fact that the magnetization of the spherules of the Apollo 11 sample could be explained in terms of two components lends credence to the two component model. However, from the petrographic evidence it would appear that the metallic iron on the surface or within the glass may have shapes other than perfect spheres. For instance, MCKAY et al. (1970) and CARTER and MACGREGOR (1970) have shown that the surface of many of the glass spherules from the fines of the Apollo 11 sample are covered with hundreds of small nickel–iron droplets which have many odd shapes and which sometimes coalesce into a thin coating covering a small area on the glass surface. Nonspherical iron particles will have demagnetization factors different from $4\pi/3$ (i.e., that of a sphere). Thus, to more accurately describe the magnetic susceptibility a third component which reaches saturation in fields between 800 and 6000 Oe should be assumed to exist, although it may be relatively small. It is not clear how the magnetization will change with field and therefore the susceptibility will simply be designated as $f(H_a)/H_a$.

Therefore, in the glass spherules of the lunar sample the magnetic susceptibility χ_a (at 800 Oe $< H <$ 4000 Oe) can be written as

$$\chi_a = C/T + \chi_I + \chi_g + \sigma/H_a + f(H_a)/H_a, \tag{1}$$

where the first term is the paramagnetism of the Fe^{2+} ions in the glass and mineral inclusions at a temperature T, χ_I is the temperature independent paramagnetism of the metallic spheres (difficultly magnetized ferromagnetic component), σ/H_a is the susceptibility of the easily magnetized ferromagnetic component of iron or nickel–iron metal, $f(H_a)/H_a$ is the intermediate ferromagnetic component as described above, and χ_g is the basic diamagnetism of the glass which we have assumed to be about -0.4×10^{-6} emu/g based on previous measurements (THORPE et al., 1970). On a χ_a vs. $1/H_a$ plot for low magnetic fields (800 Oe $< H <$ 4000 Oe), the y-intercept, χ_0, is the susceptibility extrapolated to infinite magnetic field. From equation (1) this will be

$$\chi_0 = C/T + \chi_I + \chi_g. \tag{2}$$

Similarly, the intercept, χ_t, on the χ_a vs. $1/T$ plot, i.e., the susceptibility extrapolated to infinite temperature, will be

$$\chi_t = \chi_I + \chi_g + \sigma/H_a + f(H_a)/H_a. \tag{3}$$

The experimental and calculated magnetic parameters for temperatures between 77°K and 300°K are summarized in Table 3. As previously reported for the Apollo 11 sample (THORPE et al., 1970), the glass spherules obtained from the Apollo 12 mission and the additional specimens from the Apollo 11 sample all have similar magnetic properties. Most of the iron is in the form of ferrous iron in the glass or in mineral inclusions. The ferrous iron gives rise to temperature dependent paramagnetism the amount of which is reflected by the size of the Curie constant, C (column 9). The saturation magnetization of metallic iron forming the easily magnetized component is σ (column 6). The susceptibilities of iron or nickel–iron spherules, χ_I, calculated from equation (2) are given in column 10. The susceptibilities of the third component, $f(H_a)/H_a$, shown in column 11 were calculated from equation (3) using the calculated value of χ_I and the experimental values of the other terms. The contribution of the third component to the total susceptibility is almost negligible, and thus explains why the two component model (THORPE et al., 1970) adequately explains the experimental data. The surface coating of nickel–iron droplets reported by McKAY et al. (1970) and CARTER and MacGREGOR (1970) evidently saturates at a very low magnetic field, and comprises all or most of the easily magnetized ferromagnetic component. The negative value for spherule 307 is not clear, but may be associated with the presence of octahedral iron crystals in addition to metal spheres in this particular sample.

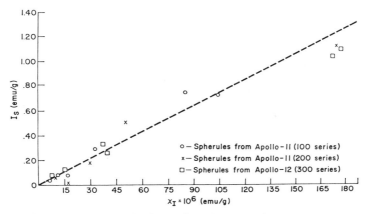

Fig. 2. The saturation magnetization I_s of the glass spherules plotted as a function of the susceptibility of the difficultly magnetized component χ_I. Dashed line is theoretical curve for pure iron spheres.

To show that the metallic iron spherules are the source of the difficultly magnetized component, the magnetization I_s was plotted as a function of χ_I, the susceptibility of the spheres (see Fig. 2). For pure iron spheres, $I_s = 218C_{Fe}$ and $\chi_I = 0.03C_{Fe}$ and, therefore, the slope of the line in Fig. 2 should be $I_s/\chi_I = 218/0.03 = 7266$. Most of the points fall close to this slope but a few are higher. The high points suggest that some spherules may have iron spheres containing a few percent nickel. Iron alloys containing less than 10% nickel have saturation moments slightly higher than that of pure iron, whereas larger amounts of nickel lead to a lower magnetic moment (DORFMAN, 1955). Also the higher density of nickel will tend to reduce the susceptibility of the metal spheres in accordance with equation (5). These two effects tend to increase the ratio of I_s/χ_I for metallic spheres containing a few % nickel.

Two of the spherules, spherule 306 ($I_s/\chi_I = 5050$; not shown) and spherule 104 ($I_s/\chi_I = 4770$), are further below the line than can be expected from the known experimental error of the measurements (error $\sim 2\%$). The presence of other than spherical iron particles in the glass spherules suggests that large deviations from the theoretical slope may be caused by a variation of the demagnetization factor from that of true spheres. Orthogonal susceptibility measurements of the glass spherules failed to show any anisotropy. Therefore, we have assumed that the major axis of the variously shaped metallic iron or nickel–iron particles are randomly distributed throughout the glass. Thus the susceptibility χ_a of an aggregate of individual metal particles is given by ($H < 4000$ oe)

$$\chi_a = \tfrac{1}{3}\chi_{||} + \tfrac{2}{3}\chi_\perp \tag{4}$$

where $\chi_{||}$ and $\chi_\perp$ are the sum of the susceptibilities parallel and perpendicular to the major axes, respectively. If the applied field, H_a, is less than the saturation field, H_s, then there is essentially zero field within a metal particle and the susceptibility of the particle will be

$$\chi = 1/D\rho \tag{5}$$

where D is the demagnetization factor and ρ is the density. Equation (4) can therefore be rewritten for an aggregate of similar-shaped particles

$$\chi_a = \frac{1}{3\rho}\,(1/D_1 + 2/D_2),\tag{6}$$

where D_1 and D_2 are the demagnetization factors along the low and high symmetry directions, respectively. Substitution of numerical values for D into this equation shows that iron spheroids or ellipsoids must be badly deformed (i.e., spheroidal axes $a/b < 0.5$, or ellipsoidal axes $a/b > 3$) before the susceptibility deviates significantly from 0.03 emu/g. Slight spheroidal or ellipsoidal deformation leads to a susceptibility just a little over 0.03 emu/g, which will slightly reduce the ratio of I_s/χ_I. However, as the demagnetization factors of platelets, needles, and cubical shapes are small, the susceptibility of such odd-shaped iron will be large, and thus, will also reduce the value of I_s/χ_I. Odd shaped iron fragments such as the Ni–Fe coating reported by MCKAY *et al.* (1970) and CARTER and MACGREGOR (1970) or Ni–Fe crystals (GLASS, 1970) could therefore account for the low value of I_s/χ_I found for spherules 104 and 306, and to a lesser extent for some of the other spherules with low values as shown in Fig. 2.

It appears possible to divide the glass spherules into two general groups: (1) those having compact glass and few iron spheres, and (2) those having highly vesicular glass and a larger number of metal spheres and mineral inclusions. In group (1) (e.g., spherules 301, 304, 305, 101, 104, and 204) the value of χ_I (column 10, Table 3) is not large, i.e., there are relatively few iron spheres. In contrast, χ_I is large for the spherules in group (2) (e.g., spherules 302, 303, 306, 106, 202).

Significance of Ni–Fe *surface coating*

The irregular coating of Ni–Fe droplets described by MCKAY *et al.* (1970) and CARTER and MACGREGOR (1970) not only is a source of irregular metal particles in the glass spherules, but being on the surface they can easily oxidize and thus may change the magnetic properties drastically. During the initial measurements it was observed that the magnetic susceptibility measurements could not always be repeated. Figure 3 shows changes in the magnetic susceptibility with reciprocal field for spherule 305, measured on three different days at room temperatures. The changes in σ (slope of the curves), while not large, were also not systematic. Repeat measurements, after removing the sample from the balance (thus exposing it to air) and storing it in nitrogen sometimes resulted in an increase and sometimes a decrease in σ depending on the past exposure history. It is possible that oxidation and reduction of iron on the surface of the glass spherules in air and nitrogen could be responsible for these changes in magnetic susceptibility. HANEMAN and MILLER (1971) have ascribed oxygen adsorption to broken bonds on metal silicate surfaces of freshly fractured lunar rocks. Oxygen is easily adsorbed on paramagnetic sites but is also easily removed by simple evacuation. Adsorption of oxygen could significantly alter the susceptibility if a similar process occurs on the surface of the spherules.

When the same sample (Fig. 3) was heated for four days in air at moderate temperature (473°K), the data were not significantly different from the original measurements

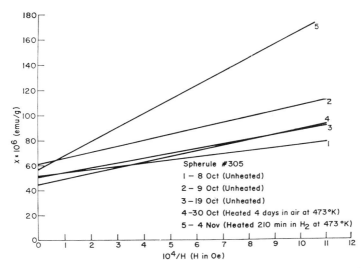

Fig. 3. Magnetic susceptibility, χ, as a function of the reciprocal magnetic field showing changes in the unheated specimen measured on three different days, and the effects of heating in air and hydrogen.

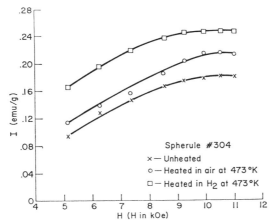

Fig. 4. The magnetization as a function of the applied magnetic field of the unheated spherule 304, and after heating in air and then in hydrogen for 3 hours and 18 hours, respectively.

on the unheated spherule indicating prior saturation of adsorption sites by oxygen atoms. However, when the spherule was heated at the same temperature in hydrogen there was a significant increase in the magnetization. Figure 4 shows how the magnetization changes with magnetic field before and after heating in air and then in hydrogen. If the observed effects are not entirely surface phenomenon, i.e., if the hydrogen were diffusing into the glass, the Fe^{2+} would be reduced to metallic iron within the glass and the Curie constant should decrease. A redetermination of the Curie constant after

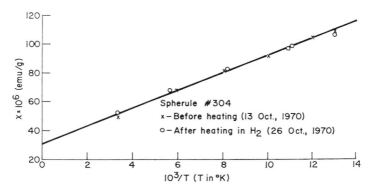

Fig. 5. Magnetic susceptibility, χ, as a function of reciprocal temperature after heating in air and hydrogen at 473°K.

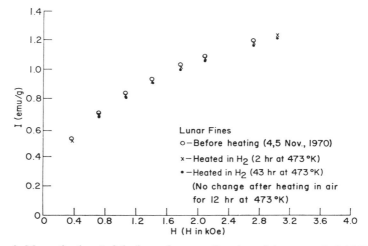

Fig. 6. Magnetization I of the lunar fines as a function of the magnetic field H before and after heating in hydrogen.

heating in hydrogen showed no change in the Curie constant (Fig. 5). Thus, these experiments suggest that the hydrogen does not permeate the glass and that the observed oxidation and reduction of iron takes place primarily on the surface of the spherule. The increase in the magnetization when the spherule is heated in hydrogen must be due to reduction of partially oxidized metallic iron on the glass surface, i.e., iron oxidized since the sample was first exposed to the atmosphere.

Consider a 250 μm diameter glass spherule with its surface partially coated with iron droplets and with an average thickness as given by MCKAY *et al.* (1971) of 1 μm. Assume the droplets cover 0.1 % of the total surface, and that this iron because of its nonspherical shape was magnetically saturated at 800 Oe. If there is no iron in the interior of the spherule, then it can be shown that $\sigma = 140 \times 10^{-4}$ emu/g for the glass sphere. Now assume an irregular iron particle with the same surface area as the glass spherule, and let its surface be oxidized to a depth of 1 nm (the depth necessary to oxidize the same amount of iron as assumed present on the surface of the glass spherule). Before oxidation the σ of the iron particle will be 218 emu/g, and after oxidation it can be shown that its σ will be diminished by only $10^{-3}\%$. Extending this argument to the bulk lunar fines, which contain metallic iron fragments, it would appear that if the fines were heated in oxygen, no significant changes in magnetization should be observed as most of the magnetization is due to iron which is not readily exposed to the atmosphere. Figure 6 shows how the magnetization of the bulk fines of Apollo 11 changes with magnetic field after exposure to air for over a month, after heating in hydrogen at 473°K for 43 hours, and then heating in oxygen at the same temperature for 12 hours. No change in the magnetization was evident in the fines, again indicating that the previously observed changes are due to a surface phenomenon peculiar to the glass spherules.

Comparison with glass in bubble

Magnetic measurements were made on the previously mentioned glass bubble shown in Fig. 7. It was hemispherical in shape and very vesicular. The electron probe analysis of the glass (Table 1) is similar to that of the glass spherules implying both glasses are from the same source. The magnetic properties (Table 3) of the glass are not unlike the glass spherules except for the magnetization of the easily magnetized ferromagnetic component σ. The value of σ was essentially zero indicating relatively few angular or nonspherical ferromagnetic inclusions in the glass or droplets on the surface. The outside surface of the bubble formed after the splash and represents a fresh surface. Thus, few droplets of metallic vapor would be expected on the surface in contrast to the glass spherules which cooled while passing through a vapor (McKay et al., 1970).

The mass of the glass bubble (44.4 mg) was large enough to obtain a Mössbauer spectrum, but with some difficulty, as it was desired not to crush the specimen. The

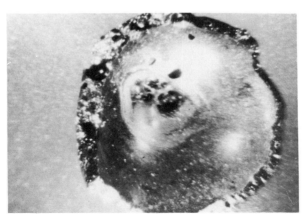

Fig. 7. Microphotograph of glass blister bubble (12057,10) removed from glass splash on specimen 12057. Diameter is 6 mm.

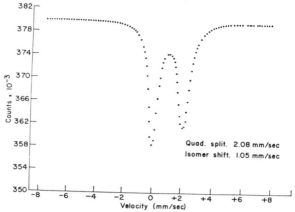

Fig. 8. Computer fit of Mössbauer spectrum of glass showing a typical Fe²⁺ pattern. Isomer shift given is with respect to the centroid of the metallic iron spectrum at room temperature.

spectrum shown in Fig. 8, computer processed and smoothed, is that of Fe²⁺ and shows a quadrapole splitting of 2.08 mm/sec and an isomer shift of 1.05 mm/sec. The isomer shift is somewhat less than that found for Fe²⁺ in the iron-bearing glasses of Apollo 11 (HERZENBERG and RILEY, 1970). There is no indication of the six-line spectrum of metallic iron or other inclusions such as ilmenite. The absence of the six-line spectrum of iron further confirms the small amount of metallic iron compared to the total iron shown by the magnetic measurements in this specimen.

Iron in glass and metallic inclusions

From the known analyses of lunar glass, iron is the major magnetic element which is present (1) as dissolved Fe²⁺ ions in the glass, (2) as metallic inclusions, or (3) as Fe²⁺ ions in iron bearing crystalline mineral inclusions. From the magnetic parameters one can obtain an approximation of the phase distribution of iron in the specimen. Assuming that Fe²⁺ is the only magnetic ion present in the glass the Curie constant, C, can be written as

$$C = N\beta^2 M\mu^2/3kA \qquad (5)$$

where N is Avogadro's number, β is the unit Bohr magneton, M is the Fe²⁺ concentration, μ is the magnetic moment of the Fe²⁺ ion, k is Boltzmann's constant, and A is the atomic weight of iron. In general the magnetic moment is given by

$$\mu = \sqrt{4S(S+1) + L(L+1)} \qquad (6)$$

where S and L are the spin and orbital quantum numbers, respectively. In our previous work on tektites (THORPE *et al.*, 1963; THORPE and SENFTLE, 1964), it was shown that the magnetic moment of Fe²⁺ was 4.91 Bohr magnetons, i.e., the orbital contribution to the magnetic moment is quenched, i.e., $L = 0$. Because of the highly reducing conditions under which the lunar glass was formed it is reasonable to assume as a first

Table 4. Comparison of the calculated iron and iron determined by electron microprobe analysis.

Spherule	(1) Fe^{2+} (%)*	(2) Fe^0 (%)‡	(3) Fe^0 (%)§	(4) Sum of columns 1, 2	(5) Total iron elec. probe
	Individual glass spherules, Apollo 12 (12070,90)				
301	11.6	0.03	0.04	11.6	12.0
302	13.8	0.59	0.51	14.4	12.1†
303	10.6	0.57	0.49	11.2	11.3†
304	11.7	0.05	0.06	11.8	13.0
305	13.3	0.13	0.15	13.4	14.1
306	11.9	1.40	0.96	13.3	12.1†
307	3.6	0.10	n.d.	3.7	2.9
308	16.2	0.13	0.12	16.3	13.2
	Glass bubble, Apollo 12 (12057,10)				
GB–1	9.8	0.52	n.d.	10.3	12.1
	Individual glass spherules, Apollo 11 (10084,22)				
101	11.5	0.11	0.14	11.6	11.7
102	9.8	0.28	0.35	10.1	n.d.
103	11.8	0.04	0.04	11.8	n.d.
104	7.3	0.06	0.04	7.4	5.8
105	3.6	0.02	0.02	3.6	n.d.
106	12.4	0.35	0.34	12.8	14.5
	Individual glass spherules, Apollo 11 (10085,17)				
201	11.1	0.03	0.03	11.1	n.d.
202	6.6	0.57	0.53	7.2	n.d.
203	5.7	0.17	0.24	5.9	n.d.
204	3.3	0.10	0.09	3.4	3.2
205	3.3	0.06	0.01	3.4	3.0

n.d.: not determined.
  † Value probably low. High iron inclusions in glass; * Calculated from the Curie constant; ‡ Calculated from equation (7); § Calculated from equation (8).

approximation that $\mu = 4.91$. Using this value of the magnetic moment in equation (5), the % Fe^{2+} was calculated and the data are shown in Table 4. In making this calculation it is also tacitly assumed that the Fe^{2+} in the mineral inclusions within the glass also has a spin-only moment of 4.91 Bohr magnetons, and that the contributions of other paramagnetic elements, primarily titanium and nickel, was not large enough to significantly affect the Curie constant. In the Apollo 11 sample it was shown (THORPE et al., 1970) that the titanium did not contribute to the Curie constant, i.e., it was in the Ti^{4+} state. A 0.5% nickel concentration (over twice that found in these specimens) would cause a change in the Curie constant equivalent to only 0.16% Fe^{2+}.

The amount of metallic iron in the form of spheres was calculated from the low field magnetic susceptibility of iron spheres, χ_I, in a manner similar to that reported for spheres in tektites (THORPE and SENFTLE, 1964). Thus,

$$C_{Fe^0} = (\chi_I/3 \times 10^{-2}) \times 10^2, \tag{7}$$

where C_{Fe^0} is the concentration of metallic iron in percent and χ_I is the field and temperature independent susceptibility of the metallic spheres (column 10, Table 3). From the data in Table 4 it is clear that the metallic iron is less than 1% in nearly all the specimens. As the metallic iron spherules are the source of the difficultly magnetized ferromagnetic component, it is also possible to obtain the concentration of metallic iron from the intensity of the high field saturation magnetization. The saturation magnetization of pure iron at room temperature is 218 emu/g. Thus, the concentration of metallic iron in percent is given by

$$C_{Fe^0} = (I_s/218) \times 10^2, \qquad (8)$$

where C_{Fe^0} is concentration of spherically shaped metallic iron, and I_s is the intensity of the saturation magnetization (column 7, Table 3). The sum of the ferrous iron and iron in the metal spheres is shown in column 4 of Table 4. The comparison with the total iron by electron probe is reasonably good. The calculated values do not include iron in inclusions which have an intermediate intensity of magnetization (column 12, Table 3) and this is a negligibly small contribution. The ferrous iron was calculated on the assumption that the magnetic moment was completely quenched as is generally found in highly reduced glass. The magnetic moment of the iron in the inclusions may not be entirely quenched, and a higher moment would yield a lower concentration of Fe^{2+}. The presence of small amounts of other paramagnetic ions in the glass would cause the calculated Fe^{2+} concentration to be high. Undoubtedly this latter effect will tend to compensate for neglecting the presence of iron inclusions having an intermediate intensity of magnetization or iron in the inclusions having only a partially quenched orbital magnetic moment. The fact that the comparison of columns 4 and 5 of Table 4 is reasonably close indicates that these latter effects are only minor. It also confirms that titanium is present in the Ti^{4+} state.

Liquid helium measurements

To see if the Curie law, which appears to hold well down to 77°K, was still valid at very low temperatures, magnetic susceptibility measurements were also made from room temperature down to 4°K on several glass spherules from both the Apollo 12

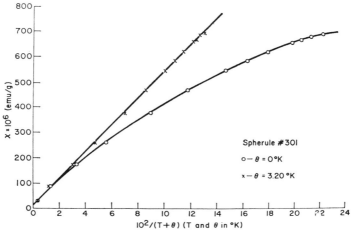

Fig. 9. Magnetic susceptibility, χ, as a function of reciprocal temperature for a typical lunar glass spherule. ○ are experimental points; × are the same points corrected to a Curie plot using a Weiss temperature, θ, as indicated.

Table 5. Comparison of the Weiss temperatures of lunar and tektites glasses.

Spherule	Sample	Weiss temperature, θ (°K)
301	12070,90	3.2
302	12070,90	3.0
303	12070,90	3.5
304	12070,90	3.3
307	12070,90	2.4
GB–1	12075,10	3.1
7	10084,86,2	3.2
104	10084,22	3.3
B–6	Bediasite	1.4
ED–1	Philippinite	1.4
Js–11	Javanite	1.5
Id–1	Indochinite	1.3

and the Apollo 11 missions. A typical plot of the susceptibility as a function of the reciprocal temperature is shown in Fig. 9. The experimental data clearly follows a Curie-Weiss relationship, i.e., $\chi = C/(T + \theta)$, where θ is the so-called Weiss temperature. A computer fit was made to determine the straight line shown and a value of θ determined. The values of θ, shown in Table 5, vary from 3.0 to 3.5 with the exception of spherule 307. It is interesting to note that this specimen was bottle-green with an exceptionally low iron concentration ($\sim 3\%$ total iron). The fact that the lunar glasses appeared to follow the Curie law down to 77°K is due to the small value of θ which is not easily observed except at very low temperatures.

Similar measurements on four tektites showed that they also obeyed the Curie-Weiss rather than the Curie law. The values of θ, however, were smaller as shown in the table. The Weiss temperature for each type of glass appears to fall within a narrow range characteristic of the type of glass. We plan to investigate this further.

SUMMARY AND CONCLUSIONS

Magnetic studies of individual glass spherules from the Apollo 11 and Apollo 12 samples show that the total susceptibility is caused by a paramagnetic component due to Fe^{2+}, a diamagnetic component due to the basic glass composition, and an easily, moderately difficult, and difficultly magnetized ferromagnetic components due to metallic iron or nickel-iron inclusions of various shapes in the glass. The ferromagnetic component which is moderately difficult to magnetize (800 Oe $< H <$ 6000 Oe) is small compared to the easily magnetized ($H < 800$ Oe) and difficultly magnetized ($H > 6000$ Oe) components. The latter component is due to iron spheres in the glass, whereas the easily magnetized component is thought to be due to irregularily shaped droplets of iron or nickel-iron on the surface of the glass spherules.

As a result of the metallic coating on most of the glass spherules there is a considerable change in magnetization after exposure to the atmosphere due to oxidation. Experiments are described which show that the change in the magnetization caused by atmospheric oxidation is a property easily observed for glass spherules but not for the bulk lunar fines. The ionic iron in the glass is in the ferrous state and the concentration can be calculated by assuming a magnetic moment of $\mu = 4.91$. Thus, the orbital contribution of the magnetic moment of iron in lunar glass is quenched. There is no evidence of Ti^{3+}. The metallic iron varies from 0.01 to 1 %. The total calculated iron compares well with the electron probe analysis within experimental error.

A large fragment of a glass bubble, part of a glass splash on the side of a rock from the Apollo 12 sample, showed no soft ferromagnetic component, i.e., was not coated with metallic droplets as was the glass spherules. Mössbauer analysis of this glass showed the pattern of Fe^{2+} only. Magnetic studies at very low temperatures (liquid helium) show that the lunar glass does not obey the Curie law but a Curie-Weiss law with a Weiss temperature of about 3°K. The Weiss temperature of some tektites were measured for comparison and averaged about 1.5°K, i.e., about half that of the lunar glass.

Acknowledgments—We are grateful to our colleagues, FRANK CUTTITTA of the U.S. Geological Survey and BRIAN MASON, KURT FREDRICKSON, and JOSEPH NELAN of the Smithsonian Institution, for making some of their samples available to us. Thanks are also due to B. FINKELMAN, who made the electron probe analysis of the glass bubble, FREDRICK OLIVER of Howard University for helping with the Mössbauer measurements, and to JOHN O'KEEFE of Goddard Space Flight Center for his helpful suggestions.

The lunar samples were made available for this investigation by the National Aeronautics and Space Administration, which also supported part of this work under contract NGL-09-11-006.

REFERENCES

CARTER J. L. and MACGREGOR J. D. (1970) Mineralogy, petrology, and surface features of some Apollo 11 samples. *Proc. Apollo 11 Lunar Sci. Conf., Geochim. Cosmochim. Acta* Suppl. 1, Vol. 1, pp. 247–265. Pergamon.

DORFMAN Y. G. (1955) *The Magnetic Properties and Structure of Mattter.* Pg. 308, State Publishing House of Technical-Theoretical Literature, Moscow, USSR.

GLASS B. P. (1971) Investigation of glass recovered from Apollo 12 sample No. 12057. Second Lunar Science Conference (unpublished proceedings).

HANEMAN D. and MILLER D. J. (1971) Surface properties of lunar material by electron paramagnetic resonance. Second Lunar Science Conference (unpublished proceedings).

HERZENBERG C. L. and RILEY D. L. (1970) Analysis of first returned lunar samples by Mössbauer spectrometry. *Proc. Apollo 11 Lunar Science Conf., Geochim. Cosmochim. Acta* Suppl. 1, Vol. 3, pp. 2221–2241. Pergamon.

MCKAY D., GREENWOOD W., and MORRISON D. (1970) Origin of small lunar particles and breccia from the Apollo 11 site. *Proc. Apollo 11 Lunar Sci. Conf., Geochim. Cosmochim. Acta* Suppl. 1, Vol. 1, pp. 673–694. Pergamon.

NAGATA T., ISHIKAWA H., KINOSHITA M., KONO M., and SYONO Y. Magnetic properties and natural remanent magnetization of lunar materials. *Proc. Apollo 11 Lunar Sci. Conf., Geochim. Cosmochim. Acta* Suppl. 1, Vol. 3, pp. 2325–2340. Pergamon.

SENFTLE F., LEE M., MONKEWICZ A., MAYO J., and PANKEY T. (1958) Quartz-helical magnetic susceptibility balance using the Curie-Chenevan principle. *Rev. Sci. Instrum.* **29**, 429–432.

SENFTLE F. E. and THORPE A. N. (1963) Technique and interpretation of magnetic susceptibility measurements of water in normal and tumor tissue. *Instr. Soc. Amer. Trans.* **2**, 117–120.

SENFTLE F. E., THORPE A. N., and LEWIS R. R. (1964) Magnetic properties of nickel-iron spherules in tektites from Isabela, Philippine Islands. *J. Geophys. Res.* **69**, 317–324.

THORPE A. N. and SENFTLE F. E. (1959) Absolute method of measuring magnetic susceptibility. *Rev. Sci. Instrum.* **30**, 1006–1008.

THORPE A. N. and SENFTLE F. E. (1964) Submicroscopic spherules and color of tektites. *Geochim. Cosmochim. Acta* **28,** 981–994.

THORPE A. N., SENFTLE F. E., and CUTTITTA F. (1963) Magnetic and chemical investigations of iron in tektites. *Nature* **197,** 836–840.

THORPE A. N., SENFTLE F. E., SULLIVAN S., and ALEXANDER C. C. (1970) Magnetic studies of individual glass spherules from the lunar sample 10084,86,2, Apollo 11. *Proc. Apollo 11 Lunar Sci. Conf., Geochim. Cosmochim. Acta* Suppl. 1, Vol. 3, pp. 2453–2462. Pergamon.

Proceedings of the Second Lunar Science Conference, Vol. 3, pp. 2451–2460
The M.I.T. Press, 1971.

Magnetism of two Apollo 12 igneous rocks

G. W. PEARCE* and D. W. STRANGWAY†
Department of Physics, University of Toronto, Toronto, Canada

and

E. E. LARSON
Department of Geology, University of Colorado, Boulder, Colorado

(*Received* 22 *February* 1971; *accepted in revised form* 30 *March* 1971)

Abstract—The magnetic properties of two igneous samples from the Apollo 12 landing site have been studied in some detail. A weak, natural remanent magnetization has been found that is believed to be of lunar origin. When added to previous evidence from the Apollo 11 samples this suggests that the moon had a field 3.3 and 3.6 billion years ago. Evidence is presented to suggest that this ancient lunar field was less than $\frac{1}{10}$ of the present earth's field. Data from the Apollo 12 magnetometer tend to confirm that the moon did have a magnetic field since it is difficult to account for the local steady field except by the presence of rocks which have a remanent magnetism. The remanence is carried in part by iron but also in part by material with a lower blocking temperature near 500°C. This may be troilite or very fine-grained iron. In the lunar soil from the Apollo 12 site the main magnetic constituent is nearly pure iron.

INTRODUCTION

One of the surprising discoveries upon the return of Apollo 11 samples to earth was the presence of a significant natural remanent magnetism (NRM). This remanence was found to be quite strong in some of the breccias and fairly weak, but measurable in the igneous samples (DOELL and GROMMÉ, 1970; STRANGWAY *et al.* 1970; LARO-CHELLE and SCHWARZ, 1970; DOELL *et al.* 1970; HELSLEY, 1970; NAGATA *et al.* 1970; RUNCORN *et al.* 1970). At that time there was some concern that this magnetism could have been acquired by exposure to large, local fields associated with laboratory equipment in the Lunar Receiving Laboratory. A preliminary survey has revealed that these fields are not very large, generally about the same as the earth's field, except in the immediate vicinity of the large vac-ion pumps. Even here the field reduces to essentially the same as the earth's field within a few feet. Moreover, laboratory tests which subjected the samples to A.F. demagnetization suggest that the remanence is fairly stable up to several hundred oersteds. A set of experiments done by applying known fields to a breccia sample from Apollo 11 at room temperature give the samples a remanence (IRM) which on demagnetization in alternating fields does not behave like the NRM (STRANGWAY *et al.* 1970). This leads one to believe that indeed the samples do have a remanence acquired on the lunar surface. Investigations by some workers (e.g. HELSLEY, 1970) suggest that there was a field of a few thousand gammas present on the lunar surface at the time the lunar basalts were formed about 3.5 b.y. ago. This was a tentative conclusion requiring a further understanding of the way in

---

* Present address: Lunar Science Institute, Houston, Texas 77058.
† Present address: NASA Manned Spacecraft Center, Houston, Texas 77058.

which the lunar samples become magnetized and requiring studies of many rocks from different areas on the moon.

As a result of studies of magnetic field data using the lunar orbiting satellite, Explorer 35, and the Apollo 12 station magnetometer, several lines of evidence point to the presence of local magnetic fields. Dyal et al. (1970) and Parkin et al. (1970) have shown that a steady field of about 38 ± 3 gammas is present at the Apollo 12 site. Binsack et al. (1970) have shown that some of the Explorer 35 data suggests the presence of local regions on the lunar surface which have magnetic fields. Finally, a study of charged particles arriving at the Solar Wind Spectrometer suggests that a local magnetic field is present north of the Apollo 12 site (Snyder et al. 1971). Since we know from the Explorer 35 that the moon has a total magnetic moment less than 10^{20} gauss cm³ and a resulting overall field of less than 7 gammas at the surface, we are forced to conclude that the magnetic effects observed are local and due to the presence of rocks carrying a remanent magnetism. This in itself is strong evidence for the presence of an ancient field since this magnetism can have been acquired only in the presence of such a field. It is not possible at this time to say much about the nature of the rocks carrying this magnetism except to point out that it cannot be due to the presence of uniform lava flows of very large horizontal extent since it is well known (Grant and West, 1965; Strangway, 1961) that these do not give rise to an external magnetic field except near the edges. The fields observed must be due to inhomogeneities in such flows (small horizontal extent or topographic features or nonuniform magnetic properties) or to bodies with dimensions more like those of intrusions.

It seems therefore that several lines of evidence point to the presence of an ancient lunar field, although the orbiting magnetometer in Explorer 35 has not suggested that there is a significant dipole moment. It is interesting to take the typical magnetizations of about 10^{-5} emu/gm found in the lunar samples and consider what dipole moment this would give. Using a density of about 3 gm/cc, a uniformly magnetized shell about 100 km thick would be consistent with the upper limit of the dipole moment. This suggests that the whole moon is not uniformly magnetized, suggesting either a temperature increase with depth to above the Curie point or a composition change to a material with less remanence in the interior. Several other lines of evidence suggest that a composition change is indeed present.

Opaque Mineralogy

The opaque minerals of the Apollo 11 samples have been studied by many investigators, and the same fact holds true for ongoing studies of the Apollo 12 samples. We wish to discuss here only those observations which pertain to the magnetic properties and will briefly discuss the results of microscopic observations of polished sections of rocks 12063,5 and 12021,134, which are slides from the two igneous rocks we studied in detail. *Sample 12021* is relatively coarse-grained and contains the following phases of interest to us.

(1) *Ilmenite* is the dominant opaque oxide with many grains, some up to 1000 μ in length. (2) *Troilite* grains up to 110 μ in length are present and generally irregular in outline. A few small grains 3 μ and smaller are found scattered in silicate grains. They

do not usually contain blebs of iron as found in the Apollo 11 samples. (3) *Iron* occurs as separate grains in and between silicates and in troilite in a few cases. The largest grain seen was about 22 μ and the smallest about 0.3 μ, the limit of resolution. Fifty-nine grains were counted giving a mean diameter of 3.7 μ, but many small ones were present that were hard to see. BRETT *et al.* (1971) and WALTER *et al.* (1971) showed using microprobe studies that this rock has very little nickel in the iron (< 0.5 wt.%). (4) *Spinel phases* occur in minor amounts. They are similar to those described by REID *et al.* (1970) and are probably chromian ulvospinels with varying compositions from chromite to ulvospinel.

Sample 12063 is relatively fine-grained and contains the following phases of interest.

(1) *Ilmenite* is much like that found in 12021 except that the maximum length is 250 μ. (2) *Troilite* is much like that seen in 12021. (3) *Iron* occurs in and between silicate grains and occasionally in troilite. The grain size distribution of the iron is almost like that in 12021. The largest grain seen was 75 μ and sizes ranged down to 0.3 μ, the limit of resolution. Ninety-four grains were measured and had a mean diameter of 4.5 μ, essentially the same as that in sample 12021. This is important to the magnetic studies because it means the grains of most interest have the same size distribution in spite of a wide range in the overall grain size. Nickel content in the iron is variable (3–4%) as is cobalt content (1–1.7%) (TAYLOR *et al.*, 1971). (4) *Spinel phases* are quite common in this sample. Again much of this is believed to be chromian ulvospinel and fits the description given by REID *et al.* (1970) exceedingly well, with distinct color gradations from rouge brown on the outside to blue gray on the inside.

MAGNETIC MINERALOGY

We have conducted three sets of experiments to help elucidate the nature of the magnetic minerals. These are high field magnetization vs. temperature curves above room temperature; high field magnetization vs. temperature curves at low temperatures; and hysteresis loops, the general results of which are described below:

Lunar soil

Lunar soil sample 12070,104 was found to be particularly simple magnetically. The Curie temperature of this material was measured and compared with a sample of pure iron. The result showed that the Curie temperature was slightly less than that of pure iron, indicating the presence of less than 5% of impurities in the iron such as nickel or silicon. In the Apollo 11 soil a distinct α–γ phase change was found by us (STRANGWAY *et al.*, 1970) suggesting the presence of an iron-nickel component of meteoritic origin. This same result was not found in our sample of Apollo 12 soil in spite of several attempts to detect it, and we can only assume that our sample of 12070 has less meteoritic material in it. Based on a value of 217.8 emu/gm for pure iron the soil contained about 0.31% metallic iron by weight as determined from the saturation magnetization of 0.68 emu/gm. Hysteresis loops of the soils as well as of the rocks show that above a few thousand oersteds the ferromagnetism is saturated and the B-H curve is linear due to paramagnetic materials. The ferromagnetic portion is determined by subtracting this out. A larger percentage of the iron in this soil as opposed to the Apollo 11 soil is derived from the lunar rocks rather than from meteorites.

Igneous rocks

Two small chips 12021,106 and 12063,55 were available to perform these tests including heating tests. Unlike reports given in the Apollo 11 volume (DOELL *et al.*, 1970; NAGATA *et al.*, 1970; RUNCORN *et al.*, 1970; SCHWARZ, 1970) where no investigators were able to heat lunar samples without oxidizing them, we have now been able to do this for grains removed from these chips. This has been achieved by a continuous pumping system in which the vacuum is kept below 0.1 μ at all times for the Js-T curves (Fig. 1A). The measurement of Curie temperature by high field magnetization curves suggests once again that the dominant ferromagnetic mineral present is nearly pure iron (Fig. 1A). (Note that heating curves were run with many fields but above a few thousand oersteds the ferromagnetism is saturated.) When the Curie temperature curves are compared with those for pure iron, it is found that the Curie temperature is slightly reduced in both cases indicating the presence of less than 5 % of impurities such as nickel in the samples. This is consistent with the general findings of REID *et al.* (1970), who did electron microprobe studies. BRETT *et al.* (1971) and WALTER *et al.* (1971) showed that 12021 had a nickel content in the iron of 0–0.5 %. TAYLOR *et al.* (1971) showed that 12063 had 3–4 % nickel and 1–1.7 % cobalt. RAMDOHR *et al.* (1971) obtained similar results for 12063, although they found one or two grains with up to 22 % nickel. Since cobalt tends to raise the Curie temperatures (BOZORTH, 1951) of

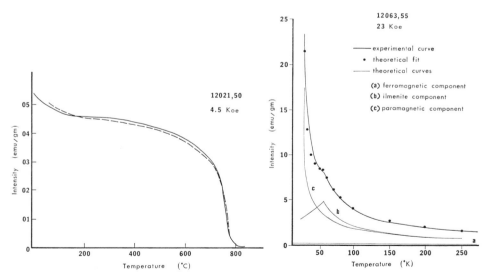

Fig. 1. High field magnetization versus temperature curves for (A) igneous sample 12021,50 between 0°C and 850°C showing iron Curie temperature, and (B) igneous sample 12063,55 showing ilmenite Néel temperature and theoretical fit to experimental curve based on 10 vol. % ilmenite, .12 vol. % free iron, and 35 vol. % FeSiO₃.

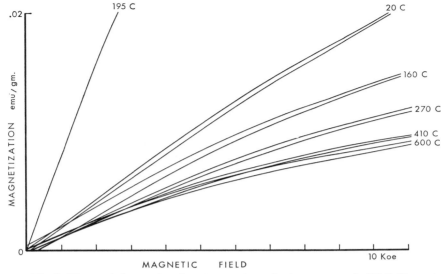

Fig. 2. Hysteresis loops at various temperatures for igneous sample 12063,55.

iron it is likely that the effects of the nickel and the cobalt offset each other giving a Curie temperature for 12063 near that of pure iron.

A series of hysteresis loops were performed on a sample of 12063 at 6 temperatures up to 600°C (Fig. 2). The magnetization can be seen to consist of a ferromagnetic component saturating at 4–5 Koe and producing a small hysteresis, and of a paramagnetic component which does not saturate in the maximum fields used. By decomposing the curves of Fig. 2 into paramagnetic and ferromagnetic components, we have deduced the ferromagnetic iron content assuming that all the ferromagnetism is due to iron. This turns out to be 0.06% by weight in 12021 and 0.035% by weight in 12063. These numbers are in good agreement with mode counts (BRETT, personal communication). The paramagnetic susceptibility at room temperature is 3.0×10^{-5}. This includes contribution from Fe^{2+}, ilmenite, and possibly superparamagnetic iron.

In Fig. 1a, attention should be drawn to the portion of the Js-T curve below about 150°C. This part of the curve clearly reflects the paramagnetic and antiferromagnetic portions of the sample and will be referred to in more detail in discussing the low temperature data.

Although troilite is known to be present there is no magnetic indication of its presence. It is antiferromagnetic if close to stoichiometric, and presumably in the presence of metallic iron and a considerable paramagnetic effect due to iron silicates, it cannot be detected. Since it appears not to be ferromagnetic we can support other observations that it is essentially stoichiometric FeS.

LOW TEMPERATURE

Several runs have been made to measure the saturation magnetization at. low temperatures in an attempt to identify various phases. In particular, ilmenite is antiferromagnetic with a Néel temperature of 55°K, ulvospinel is antiferromagnetic with a Néel temperature of 120°K, and chromite is ferromagnetic with a Curie point of 88°K. Very few studies have been made of iron silicates, but Shenoy et al. (1969) report that $FeSiO_3$ orthopyroxene is antiferromagnetic with a Néel temperature of 38°K (note that basaltic lunar samples contain primarily clinopyroxene). No indication of a Néel point associated with ulvospinel or a Curie temperature associated with chromite was found in spite of repeated attempts on samples 12021 and 12063 on cooling to the liquid nitrogen temperature (77°K). This indicates that only very small amounts of these phases were present.

Cooling a sample of 12063 down to 4.2°K revealed a distinct flexure at a Néel point of about 55°K corresponding to the presence of large quantities of ilmenite (Fig. 1B). Above 55°K the data fit a composite curve including a paramagnetic $1/T$ phase, an ilmenite $1/(T + \theta)$ where θ is -17°K (Nagata, 1961), and a ferromagnetic phase. The fit suggests the sample contains about 10 volume % ilmenite in agreement with Warner (1970); and about 0.05 weight % native iron. The paramagnetic phase represents a susceptibility at room temperature of 2.7×10^{-5} which could be accounted for by about 35 volume % $FeSiO_3$ and Fe_2SiO_4 combined, an amount typical for this rock (Warner, 1970). Superparamagnetic iron may therefore have a minor effect on the bulk susceptibility of this sample. There is no separate indication of a pyroxene antiferromagnetic effect but these results are otherwise similar to those reported by Nagata et al. (1970).

Natural Remanent Magnetization

All samples studied, 12063,98 (11.18 gm), 12063,55 (0.64 gm), 12021,50 (10.18 gm), and 12021,106 (0.96 gm), showed a definite natural remanence. This remanence was 0.94×10^{-5}, 1.5×10^{-5}, 1.15×10^{-5}, and 1.33×10^{-5} emu/gm, respectively, which is weak relative to earth basalts, but which nevertheless is readily measurable. On demagnetizing in alternating fields, results like those shown in Fig. 3 were obtained for all samples and moreover there was no significant change in the direction of magnetization. This is in contrast to the results of Hargraves and Dorety (1971), who found that the magnetization of chip 12063 became too weak to measure after demagnetizing in fields of 100 oe or so.

The natural remanence data seem to suggest the presence of a less stable component which can be demagnetized in alternating fields of 50 oe or less and a more stable component which cannot be demagnetized except in much larger fields. However since the direction of both components is the same they must have a common origin.

Only for sample 12063,98 could the lunar orientation be partially reconstructed since the documentation for 12021,50 was inadequate to permit orientation. The inclination of the NRM is $-31°$, but no information concerning orientation in the horizontal plane is available so we do not know the declination.

Artificial Remanent Magnetism

We have made several attempts to reproduce these A.F. demagnetization curves in the laboratory on the premise that we will not know how the samples became

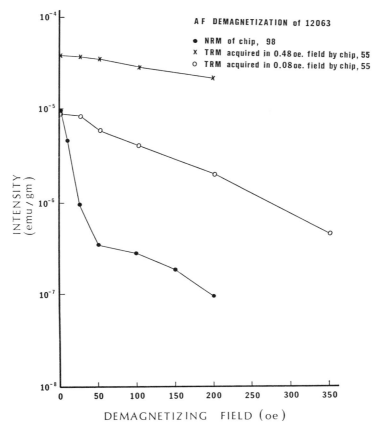

Fig. 3. Demagnetization of natural remanent magnetization and thermoremanent magnetization of sample 12063,98.

magnetized until we can do this. We know that their magnetization is not IRM since IRM does not behave this way when subjected to A.F. demagnetization as shown in Fig. 4 for sample 12063,98, and we suggest that a probable mechanism for the acquisition of an NRM by igneous rocks is thermoremanence. To determine whether the NRM of the lunar rock is thermoremanence we must be able to heat the sample to at least 780°C without altering its magnetic properties. We are not certain yet that we can do this but we have heated samples in a vacuum furnace up to 1000°C, which is continuously pumped and in which the vacuum never exceeds 0.001 μ. By this process we can heat samples and achieve identical hysteresis loops before and after heating, so we believe that we are not oxidizing the samples. This does not ensure that we have made no changes but it is at least a first test. The results of heating the sample in steps to 800°C and then cooling it in the earth's field are shown in Fig. 5. In this figure it can be seen that the remanence is mainly acquired at temperatures over 500°C, a higher temperature than shown by the results of HELSLEY (1970) when he thermally demagnetized igneous samples from Apollo 11. He found that much of the natural

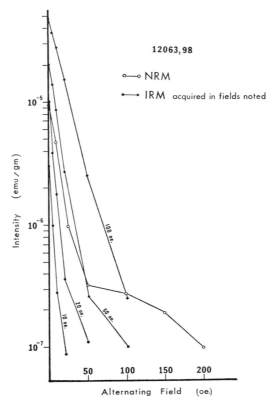

Fig. 4. Demagnetization of isothermal remanent magnetization and natural remanent magnetization of sample 12063,98.

remanence disappeared on heating above 400 or 500°C and cooling in a field-free space.

We have heated sample 12063,55 to over 800°C in a vacuum of 0.001 μ and then cooled it successively in fields of 0.48 and 0.08 oersteds. The samples were demagnetized in steps and the data shown in Fig. 3. It can be seen that the thermoremanence acquired in the earth's field (0.48 oe) is considerably greater than the NRM and much more stable to alternating fields. When cooled in the weaker field, the TRM acquired is less by a factor of 4, indicating that the acquisition of TRM is not linear with applied field. Moreover the A.F. demagnetization curve shown in Fig. 3 suggests that the magnetization is softer than that acquired in the higher field. The curve still does not correspond to the NRM curve but due to the nonlinearity it may be that similar results could be obtained at even weaker fields, perhaps even 1000 γ (0.01 oe) or less. We suggest that the stable component may be associated with imbalanced spins in the antiferromagnetic state of troilite acquired at its Néel temperature (Strangway et al., 1967), which is not well known, or by extremely small particles of iron with a low blocking temperature. The Néel temperature of troilite is usually quoted as 320°C as indicated by Thiel and Van Den Berg (1968), although phase changes make this a

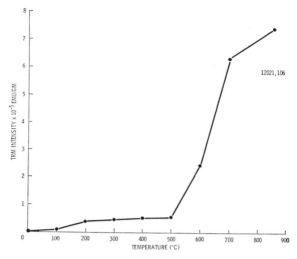

Fig. 5. Acquisition of thermoremanent magnetization in the earth's field by cooling from temperature shown.

complex system which has not been well studied. It may also be carried by very small particles of iron with high stability and low blocking temperatures. The less stable component we consider to be carried by iron acquired in the same cooling process in the same magnetic field.

The strength of the ancient field indicated in this way is less than $\frac{1}{10}$ of the present earth's field using the most stable data. The less stable data cannot be used for the intensity determination until more experiments have been done, since it acquires a TRM that is not linear with field in the range of interest.

CONCLUSIONS

We conclude the Apollo 12 igneous rocks have a remanent magnetism of lunar origin. When added to earlier evidence from the Apollo 11 rocks and to the independent geophysical evidence for local lunar fields the case for an ancient lunar field becomes strong. The samples show that this field was present 3.3 and 3.6 b.y. ago, the ages of the two sites, and suggests that the source field was not transitory in nature and that it existed long after the original formation of the moon 4.5 or 4.6 b.y. ago. The cause of the field is still not known, but the presence of an ancient dynamo is a strong contender in view of the long existence of the field (0.3 b.y.).

Acknowledgments—We have received a great deal of assistance in the laboratory from D. REDMAN and W. MOTTLEY. The laboratory part of the work was supported by a National Research Council grant to the University of Toronto.

REFERENCES

BINSACK J. H., MIHALOV J. D., SONETT C. P., and MOUTSOULAS M. D. (1970) Possible lunar surface fossil magnetism. EOS, *Trans. Amer. Geophys. Union*, **51** (abs.).
BOZORTH R. M. (1951) *Ferromagnetism*, Van Nostrand.

Brett R., Butler P., Jr., Meyer C., Jr., Reid A., Takeda H., and Williams R. (1971) Apollo 12 igneous rocks 12004, 12008, 12009, and 12022: A mineralogical and petrological study. Second Lunar Science Conference (unpublished proceedings).

Doell R. R. and Gromme C. S. (1970) Survey of magnetic properties of Apollo 11 samples at the Lunar Receiving Laboratory. *Proc. Apollo 11 Lunar Sci. Conf., Geochim. Cosmochim. Acta* Suppl. 1, Vol. 3, pp. 2093–2096. Pergamon.

Doell R. R., Gromme C. S., Thorpe A. N., and Senftle F. E. (1970) Magnetic studies of Apollo 11 lunar samples. *Proc. Apollo 11 Lunar Sci. Conf., Geochim. Cosmochim. Acta* Suppl. 1, Vol. 3, pp. 2097–2102. Pergamon.

Dyal P., Parkin C. W., and Sonett C. P. (1970) Apollo 12 magnetometer: Measurement of a steady field on the surface of the moon. *Science* **169**, 762–764.

Grant F. S. and West G. F. (1965) *Interpretation Theory in Applied Geophysics*, McGraw-Hill.

Hargraves R. B. and Dorety N. (1971) Magnetic properties of some lunar crystalline rocks returned by Apollo 11 and Apollo 12. Second Lunar Science Conference (unpublished proceedings).

Helsley C. E. (1970) Magnetic properties of lunar samples 10022, 10069, 10084, and 10085. *Proc. Apollo 11 Lunar Sci. Conf., Geochim. Cosmochim. Acta* Suppl. 1, Vol. 3, pp. 2213–2219. Pergamon.

Larochelle A. and Schwarz E. J. (1970) Magnetic properties of Apollo 11 lunar sample 10048,22. *Proc. Apollo 11 Lunar Sci. Conf., Geochim. Cosmochim. Acta* Suppl. 1, Vol. 3, pp. 2305–2308. Pergamon.

Nagata T. (1961) *Rock Magnetism*, rev. ed., Maruzen, Tokyo.

Nagata T., Ishikawa Y., Kinoshita H., Kono M., Syono Y., and Fisher R. M. (1970) Magnetic properties and natural remanent magnetization of lunar samples. *Proc. Apollo 11 Lunar Sci. Conf., Geochim. Cosmochim. Acta* Suppl. 1, Vol. 3, pp. 2325–2340. Pergamon.

Parkin C. W., Dyal P., Sonett C. P., and Colburn D. S. (1970) Steady magnetic field measurements on the surface of the Moon. EOS, *Trans. Amer. Geophys. Union* **51** (abs.).

Ramdohr P., El Goresy A., and Taylor L. A. (1971) The opaque minerals in the lunar rocks from Oceanus Procellarum. Second Lunar Science Conference (unpublished proceedings).

Reid A. M., Meyer C., Harmon R. S., and Brett R. (1970) Metal grains in Apollo 12 igneous rocks. *Earth Planet. Sci. Lett.*, **9**, 1–5.

Runcorn S. K., Collinson D. W., O'Reilly W., Battey M. H., Stephenson A., Jones J. M., Manson A. J., and Readman P. W. (1970) Magnetic properties of Apollo 11 lunar samples. *Proc. Apollo 11 Lunar Sci. Conf., Geochim. Cosmochim. Acta* Suppl. 1, Vol. 3, pp. 2369–2387. Pergamon.

Schwarz E. J. (1970) Thermomagnetics of lunar dust sample 10084,88. *Proc. Apollo 11 Lunar Sci. Conf., Geochim. Cosmochim. Acta* Suppl. 1, Vol. 3, pp. 2389–2397. Pergamon.

Shenoy G. K., Kalvius G. M., and Hafner S. S. (1969) Magnetic behavior of the $FeSiO_3$–$MgSiO_3$ orthopyroxene system from NRG in ^{57}Fe. *J. Appl. Phys.* **40**, 1314–1316.

Skinner G. J. (1970) High crystallization temperatures indicated for igneous rocks from Tranquility Base. *Proc. Apollo 11 Lunar Sci. Conf., Geochim. Cosmochim. Acta* Suppl. 1, Vol. 1, pp. 891–895. Pergamon.

Snyder D. W., Clay D. R., and Neugebauer M. (1971) An impact-generated plasma cloud on the moon. Second Lunar Science Conference (unpublished proceedings).

Strangway D. W. (1961) Magnetic properties of diabase dikes. *J. Geophys. Res.* **66**, 3021–3032.

Strangway D. W., McMahon B. E., and Honea R. M. (1967) Stable magnetic remanence in antiferromagnetic goethite. *Science* **158**, 785–787.

Strangway D. W., Larson E. E., and Pearce G. W. (1970) Magnetic studies of lunar samples— breccia and fines. *Proc. Apollo 11 Lunar Sci. Conf., Geochim. Cosmochim. Acta* Suppl. 1, Vol. 3, pp. 2435–2451. Pergamon.

Taylor L., Kullerud G., and Bryan W. (1971) Minerology of two Apollo 12 samples. Second Lunar Science Conference (unpublished proceedings).

Thiel R. C. and van den Berg C. B. (1968) Temperature dependence of hyperfine interactions in near stoichiometric FeS. *Phys. Stat. Sol.* **29**, 837–846.

Walter L. S., French B. M., Ghose S., Heinrich K., Spikerman J., Lowman P., Doan A., Jr. and Adler S. (1971) Mineralogical studies of Apollo 12 samples. Second Lunar Science Conference (unpublished proceedings).

Warner J. (1970) Apollo 12 lunar-sample information. NASA Tech. Rep. No. NASA TR R-353.

Proceedings of the Second Lunar Science Conference, Vol. 3, pp. 2461–2476
The M.I.T. Press, 1971.

Magnetic properties and remanent magnetization of Apollo 12 lunar materials and Apollo 11 lunar microbreccia

T. Nagata

Geophysical Institute, University of Tokyo, Japan

R. M. Fisher and F. C. Schwerer

U.S. Steel Corporation Research Center, Monroeville, Pennsylvania 15146

and

M. D. Fuller and J. R. Dunn

Earth and Planetary Sciences, University of Pittsburgh, Pittsburgh, Pennsylvania 15213

(*Received* 22 *February* 1971; *accepted in revised form* 30 *March* 1971)

Abstract—The natural remanent magnetization (NRM) of three Apollo 12 crystalline rocks range from 2 to 8 $\times$ 10^{-6} emu/gm in intensity, and in the case of rock 12038, is northward, eastward, and upward in direction. The magnetic properties of Apollo 12 crystalline rock 12053 and fines 12070 are not essentially different from those of Apollo 11 lunar materials, being mostly due to ferromagnetism of metallic iron, paramagnetism of pyroxenes, and antiferromagnetism of ilmenite. However, there is much less antiferromagnetic ilmenite present than in Apollo 11 material.

The NRM and other magnetic properties of weakly, moderately, and strongly impacted microbreccias were specifically examined. More strongly impacted microbreccias have stronger and stabler remanent magnetization. The observed magnetic characteristics particularly the thermoremanent, piezoremanent, and shock remanent magnetization of Apollo 12 crystalline rocks suggest that the magnetic field on the lunar surface when the rock was formed was between 500 and 3000 gamma.

Possible sources of this ancient lunar field considered are (1) classical dynamoaction of an early liquid lunar core, (2) an ancient solar wind field of the necessary intensity, and (3) enhancement of the field produced by a moderate pristine solar wind when hot conducting lava results in lunar unipolar generator action. The high electrical conductivity required by this latter model has been confirmed by measurements of lunar materials at elevated temperatures so that it seems the most reasonable in the light of presently available information.

Introduction

Preliminary studies of the magnetic properties of several samples of Apollo 11 lunar rocks and fines were reported in the Proceedings of the First Lunar Science Conference. This work revealed that the observed magnetic properties are mainly due to paramagnetic pyroxenes, antiferromagnetic ilmenite and ferromagnetic native iron. Qualitative estimates of the amounts of various minerals present were confirmed by Mössbauer analysis. In this paper results of more comprehensive measurements are reported. These include specific susceptibility, saturation magnetization, coercive force, natural and viscous remanent magnetization, AF demagnetization, and remanence acquisition characteristics of a wider range of lunar materials. In particular the acquisition of thermoremanent, piezoremanent, shock remanent, and inverted thermoremanent magnetization were investigated. Electrical conductivity measurements were made up to 800°C.

The samples used were lunar fines, types A and B lunar crystalline rocks from both Apollo 11 and Apollo 12 missions and three different Apollo 11 lunar microbreccias. The modal mineral composition of the samples was determined by optical petrography and confirmed qualitatively by electron probe and Mössbauer analysis.

The NASA identification and the composition of the crystalline rocks is as follows:

10024 60% pyroxene 25 plagioclase 15 ilmenite
12038 55% pyroxene 30 plagioclase 10 ilmenite 5 cristobalite
12053 80% pyroxene 10 plagioclase 10 ilmenite

The lunar fines were from samples 12070 and 10084, and the abundances of metallic iron in both was approximately 0.60% in contrast to 0.07% found in the crystalline rocks.

The microbreccias include

10021,32—very weakly impacted breccia
10048,55—moderately impacted breccia
10085,13—strongly impacted breccia.

The degree of shock in two of these samples was determined by I. Kushiro, who found evidence of strong shock metamorphism in 10085,13 as some plagioclase areas were remelted and others contained a large number of shock fissures and deformation markings. No trace of shock metamorphism was observed in sample 10021,32. According to J. M. Christie, pyroxene and feldspar grains in sample 10048,55 show evidence of moderate shock so that the three available breccias cover a broad range of shock history.

Natural Remanent Magnetization of Apollo 12 Crystalline Rocks

The intensity and direction of the natural remanent magnetization (NRM) of three Apollo 12 crystalline rocks (12053,47; 12038,29; and 12038,32) were measured. The approximate orientation of sample 12038 on the lunar surface was recorded by the astronauts, so that the direction of NRM of these samples in the lunar surface coordinates can be determined. Unfortunately, sample 12053 seems to have been rotated and therefore its original orientation is uncertain.

Standard storage tests of these three samples in nonmagnetic space (less than 200γ) for one full day revealed no observable change. The acquisition and decay experiments of the viscous remanent magnetization (VRM) of these samples also show that the viscosity coefficient S', in the expression of acquisition of VRM as I (VRM) = $S' \log t$ where t is expressed in unit of second, is extremely small, as shown in Table 1. (Because samples 12038,29 and 12038,32 are so small, exact values of S' could not be determined for them, so that only possible upper limits are indicated.) This result suggests that the observed NRM of these samples cannot be attributed to VRM acquired in the geomagnetic field during the period since they were returned to the earth.

Results of measurements of NRM and the AF demagnetizations of these three samples are shown in Figs. 1, 2, and 3 and summarized in Table 1, where the effective AF demagnetization field (H_o) is defined as the intensity of demagnetization field which reduced the intensity of NRM to $(1/e)$ of its initial value. The critical AF

Table 1. Intensity and direction of NRM of Apollo 12 crystalline rocks and their AC-demagnetization characteristics.

		12053,47	12038,29	12038,32	Unit
Mass of samples		9.147	0.566	0.772	gm
Intensity of					
NRM	In	$(2.28 \pm 0.02) \times 10^{-6}$	$(8.13 \pm 0.41) \times 10^{-6}$	$(6.25 \pm 0.20) \times 10^{-6}$	emu/gm
Direction NRM	D	—	$+70$	$+70$	degree
on the lunar surface	I	—	-38	-35	degree
Effective AC-demagnetization					
field $\tilde{H}_o$		8	55	12	oersteds
Critical AC-demagnetization					
field $\tilde{H}_*$		7	40	5	oersteds
IRM coefficient b		2.7×10^{-8}	—	2.7×10^{-8}	emu/gm/(Oe)^2
VRM visosity coefficient for 0.55 oe field at					
room temperature		3.2×10^{-8}	$<2 \times 10^{-7}$	$<1 \times 10^{-7}$	emu/gm

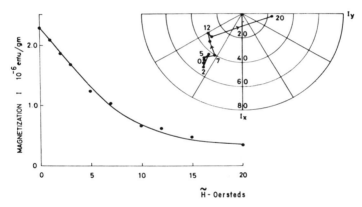

Fig. 1. AF demagnetization of NRM of Apollo 12 crystalline rock 12053,47.

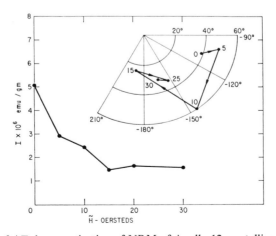

Fig. 2. Results of AF demagnetization of NRM of Apollo 12 crystalline rock 12038,29.

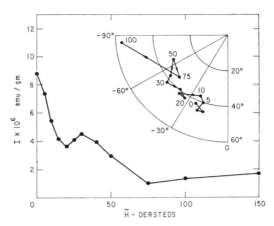

Fig. 3. Results of AF demagnetization of NRM of Apollo 12 crystalline rock 12038,32.

demagnetization field (H_*) is defined, on the other hand, as the maximum demagnetization field below which the NRM is approximately invariant (within $\pm 10°$ in most cases).

The intensity of NRM of three Apollo 12 crystalline rocks is rather weak, being 10^{-6} emu/g in the order of magnitude, and these NRM are not appreciably stable against the AF demagnetization, in comparison with thermoremanent magnetization or chemical remanent magnetization of terrestrial rocks. However, the coefficient b of the isothermal remanent magnetization TRM, in the expression IRM $= bH^2$ for weak magnetic fields H, commonly amounts to 2.7×10^{-8} emu/g/oe². If we assume that the observed NRM are due to IRM, the necessary magnetic field should be 9.2, 18.4, and 15.2 oe, respectively, for these three samples. Such intense magnetic fields are not likely in the neighborhood of the moon's surface.

By referring to the description of the orientation of sample 12038 on the moon's surface, the direction of NRM of 12038,29 and 12038,32 can be given in the moon's geographic coordinates. Defining the declination D and the inclination I of a direction at the Apollo 12 landing site just as they would be defined on the earth's surface (namely, $D =$ the eastward deviation angle from the north and $I =$ the downward inclination from the horizon), the direction of NRM of the two specimens are described as given in Table 1. The agreement between the two NRM's directions is reasonably good.

MAGNETIC PROPERTIES OF APOLLO 12 CRYSTALLINE ROCK
(12053,47) AND FINES (12070,102)

The initial magnetic susceptibility (χ_0), the paramagnetic susceptibility (χ_a), the saturation magnetization (I_s), the saturation remanent magnetization (I_R), the coercive force (H_c) and the remanence coercive force (H_{RC}) were determined at room temperature for Apollo 12 crystalline rock 12053,47 and fines 12070,102. Observed hysteresis curves of the two samples are illustrated in Figs. 4 and 5. The magnetic parameters

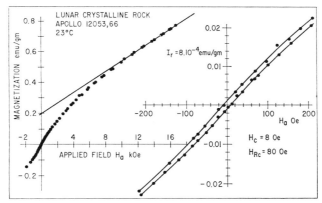

Fig. 4. Magnetic hysteresis curve of Apollo 12 crystalline rock 12053,47 at room temperature.

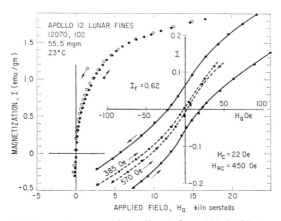

Fig. 5. Magnetic hysteresis curve of Apollo 12 fines 12070,102 at room temperature.

thus determined are summarized in Table 2, where the same magnetic parameters for Apollo 11 crystalline rock and fines (NAGATA *et al.*, 1970) also given for comparison.

Generally speaking, the basic magnetic properties of Apollo 12 crystalline rock and fines are not appreciably different from those of Apollo 11 materials. As in the case of Apollo 11 lunar materials (NAGATA *et al.*, 1970), the ferromagnetism of Apollo 12 rock and fines also is mostly due to metallic iron whose Curie point is about 780°C. The abundance of metallic iron estimated from the observed values of I_s is 0.071 and 0.59 weight %, respectively, for rock 12053,47 and fines 12070, which are comparable with rocks and fines, respectively. These values agree well with results of Mössbauer analysis by G. P. HUFFMAN.

The magnetization versus temperature curves for the low temperature range from 4°K to 300°K of the two Apollo 12 materials do not show a distinct antiferromagnetic peak of ilmenite, as shown in Figs. 6 and 7. This is a remarkable contrast between Apollo 12 and Apollo 11 materials, since the latter have a distinct antiferromagnetic

T. NAGATA *et al.*

Table 2. Magnetic properties of Apollo 12 crystalline rock and fines in comparison with these of Apollo 11 materials.

Magnetic parameters	Crystalline rock		Fines		Unit
	12053,47	10024,22	12070,102	10084,89	
χ_0	2.6×10^{-4}	2.6×10^{-4}	7.2×10^{-3}	8.8×10^{-3}	emu/gm
χ_a	3.2×10^{-5}	3.4×10^{-5}	2.5×10^{-5}	3.5×10^{-5}	emu/gm
I_s	0.195	0.155	1.28	1.17	emu/gm
I_R	8×10^{-4}	1.5×10^{-3}	6.2×10^{-2}	8.4×10^{-2}	emu/gm
H_c	8	45	22	36	oersted
H_{RC}	76	160	450	460	oersted

χ_0 = specific intensity of initial magnetic susceptibility; χ_a = specific intensity of paramagnetic susceptibility; I_s = specific saturation magnetization; I_R = specific saturation remanent magnetization; H_c = coercive force; H_{RC} = remanence coercive force.

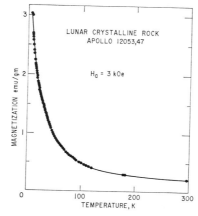

Fig. 6. Change of magnetization of Apollo 12 crystalline rock 12053,47 with temperature in a low temperature range.

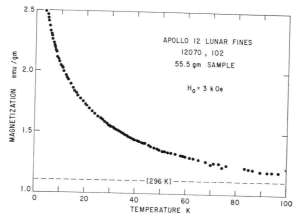

Fig. 7. Change of magnetization of Apollo 12 fines 12070,102 with temperature in a low temperature range.

peak of ilmenite at 57°K (NAGATA *et al.*, 1970). This result may indicate that the abundance of ilmenite is much less in Apollo 12 materials than in Apollo 11. However, the curves of magnetization versus temperature of the Apollo 12 materials cannot be represented by a simple sum of the ferromagnetic phase of metallic iron and the paramagnetic phase of pyroxenes and others. They seem to require the presence of at least an antiferromagnetic component, which may be ilmenite amounting to several weight %.

An interesting feature of the magnetic properties of the lunar materials is that the remanence coercive force (H_{RC}) is much larger than the coercive force (H_c), as shown in Table 2. This fact may imply that the ferromagnetism of these materials consist of a soft component having a small coercivity and a hard component having a large coercivity.

THERMOREMANENT MAGNETIZATION, PIEZO-REMANENT MAGNETIZATION, SHOCK REMANENT MAGNETIZATION, AND INVERTED THERMOREMANENT MAGNETIZATION OF APOLLO 12 CRYSTALLINE ROCK

Thermoremanent magnetization

An experiment of successive acquisitions of thermoremanent magnetization (TRM) for sample 12053,47 was carried out in a vacuum of $(2 - 10) \times 10^{-7}$ Torr. The temperature range was from 20°C to 850°C and the applied magnetic field 0.200 oe. As shown in Fig. 8, the TRM blocking temperature is about 800°C and the total TRM per unit magnetic field amounts to 9.0×10^{-5} emu/g/oe. These TRM characteristics are similar to those of Apollo 11 crystalline rock 10024,22, which has a TRM blocking temperature of about 800°C, and whose total TRM per unit magnetic field amounts to 2.90×10^{-4} emu/g/oe (NAGATA *et al.*, 1970).

Piezo-remanent magnetization

The piezo-remanent magnetization (PRM) of sample 12053,47 was examined as a function of applied magnetic field (H) and static compression. PRM acquired by terrestrial rocks (NAGATA and CARLETON, 1968, 1969) in a weak magnetic field and by moderate uniaxial stress is approximately expressed as

$$J_R(H_+P_+P_oH_o) \approx b(3HH_p - H^2) \qquad \text{for} \quad H \leq \tfrac{1}{2}H_p, \tag{1}$$

$$J_R(H_+P_+P_oH_o) \approx b(H^2 + HH_p + 2H_p{}^2) \quad \text{for} \quad H \geq \tfrac{1}{2}H_p, \tag{2}$$

where

$$H_p \equiv KP$$

and K is a material constant depending on the saturation magnetization and the magnetostriction coefficient. In the present case of the Apollo 12 sample, $K \simeq 5.2 \times 10^{-2}$ oe/bar.

Figure 9 illustrates at the left an observed linear relationship between $J_R(H_+ P_+P_oH_o)$ and H for a constant compression $P = 300$ bars, while at the right it shows

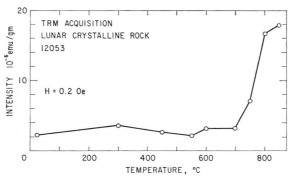

Fig. 8. Acquistion of thermoremanent magnetization of Apollo 12 crystalline rock 12053,47 in a magnetic field of 0.20 oersted.

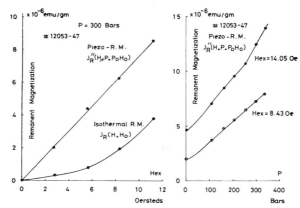

Fig. 9. Piezo-remanent magnetization characteristics of Apollo 12 crystalline rock 12053,47, (a) dependence of piezo-remanent magnetization, $J_R''(H_+P_+P_oH_o)$, on magnetic field H_{ex} where P = constant. (b) dependence of piezo-remanent magnetization, $J_R''(H_+P_+P_oH_o)$, on uniaxial compression P, where H_{ex} = constant.

the dependence of $J_R(H_+P_+P_oH_o)$ on P for two constant values of H, i.e., 8.43 and 14.05 oe. Figure 9 left corresponds to the case expressed by equation (1), while Fig. 9 right corresponds to equation (2). It must be noted that PRM for an extremely weak magnetic field and a large compression is given approximately by $3bKHP$, which becomes very large compared with the ordinary IRM expressed by bH^2. In the present case (PRM)/(IRM) = $(3KP)/H$ = $0.15P/H$, where P and H are expressed in bar and oe, respectively.

Shock remanent magnetization

When an impulse of mechanical compression is given to a magnetic specimen in the presence of a magnetic field, the sample acquires a kind of dynamical PRM (SHAPIRO and IVANOV, 1967). Since characteristics of this type of remanent magnetization are appreciably different from those of PRM which is caused by a static mechanical

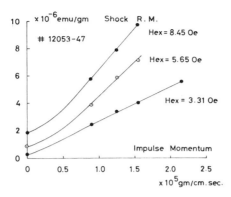

Fig. 10. Shock remanent magnetization characteristics of Apollo 12 crystalline rock 12053,47; dependence of shock remanent magnetization $J_H''(H + SH_o)$ on impulse momentum of shocks in three different constant magnetic fields.

stress, the former will be specifically called the shock remanent magnetization (SRM). SRM caused by a weak magnetic field (H) and moderate magnitude of the impulse momentum (M) of applied mechanical shock is approximately proportional to both H and M, as shown in Fig. 10. Here the applied impulse has a reasonably simple wave form consisting of a main pulse of about 0.4 millisecond in duration and a small secondary pulse. As in the case of the static PRM, SRM produced in a weak magnetic field is approximately represented by

$$J(H_+SH_o) \simeq bCHM.$$

For the Apollo 12 sample, $C = 3.2 \times 10^{-4}$ oe/(g/cm sec). Thus, for a very weak field, $(SRM)/(IRM) = (CM)/H = 3.2 \times 10^{-4} M/H$, where M and H are expressed in g/cm sec and oe, respectively.

Inverted thermoremanent magnetization

It has been reported (RUNCORN *et al.*, 1971) that some Apollo 12 lunar rocks (12020,23 and 12018,47) exhibit a low temperature transition phenomenon at about $-150°C$ which might indicate the presence of magnetite in these samples (NAGATA *et al.*, 1964). In view of this surprising evidence for the occurrence of magnetite in lunar rocks, the low temperature magnetic characteristics of available samples were investigated.

The acquisition of inverted thermoremanent magnetization due to low temperature transitions of the magnetic materials (NAGATA *et al.*, 1963) in samples 12053,47 and 10024,22 were examined in a magnetic field of 12 oe throughout the temperature range from $-190°C$ to $20°C$. As shown in Fig. 11, no definite phase transition of magnetite was detected in these samples, but an appreciable change in remanent magnetization was detected for a comparatively broad temperature range below $0°C$. A second attempt to detect the magnetite transition was made by observing the warming curves of saturated IRM given at liquid nitrogen temperature. Samples 10024,22 and 12053,47 revealed no discrete transition of remanence but rather a progressive decrease in

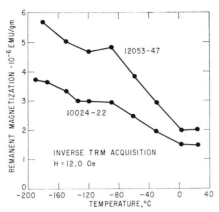

Fig. 11. Acquisition of inverted type thermoremanent magnetization of Apollo 11 crystalline rock 10024,22, and Apollo 12 crystalline rock 12053,47, in a magnetic field of 12 oersteds.

remanent magnetization when warmed from liquid nitrogen temperature to room temperature possibly related to the presence of fine metallic particles. It may be significant that sample 12018 is reported to be higher in olivine and have a wider range of nickel content in the metallic particles than either of the samples we have studied.

REMANENT MAGNETIZATION AND MAGNETIC PROPERTIES OF APOLLO 11 MICROBRECCIA

Three specimens of Apollo 11 lunar microbreccias representing a wide range of shock history as described in the Introduction were studied in detail. The magnetic parameters of the three samples are summarized in Table 3. As can be seen from the magnitude of I_s and the ratio χ_a/I_s in the table, the abundance of metallic iron and its ratio to Fe^{2+} in silicate minerals vary to some extent in different samples. There is however a remarkable correlation such that the coercive force (H_c) is larger for more strongly impacted breccias, and that this increase of H_c is accompanied by an increase

Table 3. Magnetic properties of Apollo 11 lunar microbreccia

Magnetic parameters	10021,32	10048,55	10085,13	Unit
χ_0	8.6×10^{-3}	9.6×10^{-3}	4.3×10^{-3}	emu/gm
χ_a	—	4.3×10^{-5}	4.4×10^{-5}	emu/gm
I_s	0.74	1.8	0.44	emu/gm
I_R	5.0×10^{-2}	1.3×10^{-1}	6.7×10^{-2}	emu/gm
H_i	19	50	125	oersted
H_{RC}	—	520	670	oersted
I_n	1.5×10^{-5}	5.6×10^{-5}	1.53×10^{-4}	emu/gm
(I_n/I_s)	(2.0×10^{-5})	(3.1×10^{-5})	(34.8×10^{-5})	
$\tilde{H}_o$	35	~ 400	~ 1400	oersted
$\tilde{H}_*$	40	~ 100	> 500	oersted
$\Delta I_v/I_n$	8.4	3.3	0.23	

ΔI_v = Viscous remanent magnetization acquired in the geomagnetic field (0.55 oe) which is saturated with respect to the short term time scale (i.e., $< 10^7$ sec)

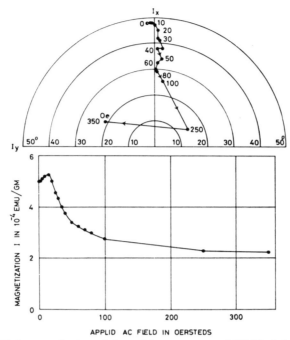

Fig. 12. AF demagnetization of the stable component of NRM of Apollo 11 micro-breccia 10048,55.

of I_n/I_s, H_o, and H_*, and by a decrease of $\Delta I_v/I_n$. In particular, the rate of increase of the stability of NRM against the AF demagnetization, represented by H_o and H_*, and the corresponding increase in H_c is markedly large. As an example of the behavior of these samples, the AF demagnetization curve of sample 10048,55 is illustrated in Fig. 12. It seems thus that the effects of impact metamorphism result in a larger magnetic coercivity as well as a stronger and stabler remanent magnetization of the lunar microbreccias.

The lunar microbreccias contain an extremely large number of fine grains. Such fine grains of ferromagnetic metallic iron behave more or less superparamagnetically, giving rise, in a magnetic field, to viscous remanent magnetization (VRM) having a variety of time constants at room temperature. Results of laboratory demonstrations of acquisition and decay of VRM are shown in Figs. 13, 14, and 15 for the three samples. In the very weakly impacted breccia 10021,32, such fine grains including metallic irons are the major constituents. Studies on VRM of this sample have indicated that the mean diameter of the very fine grains of metallic iron ranges between 170 and 185 Å. The moderately impacted breccia 10048,55 consists of a large number of very fine grains and a number of tiny chips of plagioclase, pyroxene, feldspar, etc. It seems that the very fine grains are responsible for the appreciably large VRM of this sample such as observed previously (LaRochelle and Schwarz, 1970). Results of analysis of VRM of this sample have shown that the mean diameter of metallic iron grains ranges between 160 and 180 Å.

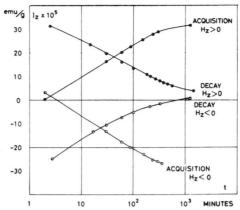

Fig. 13. Acquisition of VRM in the geomagnetic field and its decay in nonmagnetic
space of a weakly impacted lunar microbreccia 10021,32.

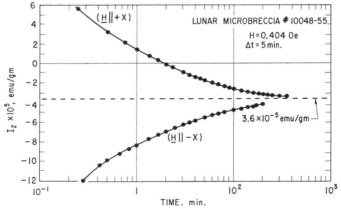

Fig. 14. Decay in nonmagnetic space of VRM acquired in the geomagnetic field of a
moderately impacted lunar microbreccia 10048,55.

Although the strongly impacted breccia 10085,13 also contains very fine grains of
metallic iron of about 180 Å in mean diameter, which results in a small portion of
VRM, the major parts of remanent magnetization are due to much stabler magnetic
phases which have extremely larger relaxation times. It may thus be presumed that
fine grains of metallic iron 160–180 Å in mean diameter would be transformed from
their free state which they had when they were constituents of the lunar fines to a
certain magnetically hard state by the impact metamorphism.

An experimental demonstration of TRM acquisition and the thermal demagnetization of TRM was carried out for microbreccia 10048,55. The blocking temperature of TRM of this sample also is about 800°C, and the intensity of total TRM
per unit magnetic field amounts to 4.7×10^{-3} emu/g/oe. If therefore the observed
stable component of NRM of this sample is assumed to be due to its TRM, the
magnetic field for the TRM acquisition is estimated to be about 1200 γ.

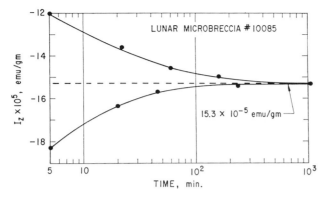

Fig. 15. Decay in nonmagnetic space of VRM acquired in the geomagnetic field of a strongly impacted lunar microbreccia 10085,13.

Electrical Conductivity of Lunar Crystalline Rocks and Microbreccia

The d.c. electrical conductivity of lunar crystalline rocks 10024,22 and 12053,47, and a lunar microbreccia, 10048,55, was measured as a function of temperature from 20°C to 800°C. Even in a helium-2% hydrogen atmosphere, the electrical conductivity of Apollo 12 lunar materials shows a large irreversibility with respect to temperature. Fig. 16 illustrates the dependence of electrical conductivity (σ) on temperature of these samples in their initial heating process. The electrical conductivity of all the lunar materials is much larger than that of typical terrestrial minerals (ENGLAND *et al.*, 1968) by two or three orders of magnitude throughout the whole temperature range. The large electrical conductivity of lunar materials may be interpreted as due to relatively easy activation of electrons and holes at the abundant Fe^{2+} ion sites in pyroxene and ilmenite. The hysteresis on heating may reflect a redistribution of the ions in different lattice sites, including resorption of fine metallic particles.

Hypothetical Acquisition Mechanisms of Natural Remanent Magnetization of Lunar Materials

Lunar crystalline rocks have a natural remanent magnetization (NRM) of $(2–8) \times 10^{-6}$ emu/g, while the NRM of the lunar microbreccias amounts to $(1.5–15.0) \times 10^{-5}$ emu/g. It has already been pointed out that these NRM do not seem to be attributable to the IRM of the lunar materials. If the NRM of lunar crystalline rocks is assumed to be due to their TRM, the magnetic field present when the rocks cooled can be estimated from a comparison of NRM with TRM to be about 2500 γ. A comparison of NRM data with TRM ones for a microbreccia, 10048,55, has suggested that the estimated lunar magnetic field is about 1200 γ. If the NRM of the crystalline rock 12053,47 is assumed to be due to PRM caused by a simple compression of 50 kb in magnitude, the required magnetic field is about 1000 γ. If alternatively it is assumed to be due to SRM caused by the impact of a 5 kg meteorite with cross section of order 10^2 cm^2 and having an impact speed of 5 km/sec., the necessary magnetic field amounts to 1000 γ also.

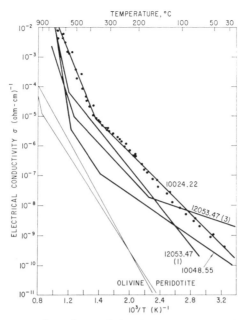

Fig. 16. Temperature dependence of electrical conductivity of Apollo 11 and Apollo 12 lunar materials in comparison with typical terrestrial minerals. Data for two samples from 12053, labeled (1) and (3), are shown.

All these estimates are based on the crude assumption that experimental results obtained in the laboratory can simply be extended to a case of very weak magnetic field, and in addition to a very strong impact for cases of PRM and SRM. Even if we dare assume so, a magnetic field of about $10^3 \gamma$ in order of magnitude is required for producing the observed NRM of lunar crystalline rocks and microbreccias.

At present, the average intensity of magnetic field in the solar wind is 10γ or less, and the magnetic field intensity observed by the Lunar Surface Magnetometer at Apollo 12 landing site is $36 \pm 5 \gamma$ (DYAL et al., 1970). However it is reasonable to speculate that the magnetic field on the lunar surface was as high as $10^3 \gamma$ at some time in the past. There may be three possibilities for the origin of such a magnetic field.

One is the common hypothesis that the moon had a small metallic core whose dynamoaction produced a small magnetic field. The second hypothesis is the direct assumption that the solar wind magnetic field itself was very strong, being $10^3 \gamma$ in order of magnitude. A third, more plausible, hypothesis is that the intensity of the solar wind was somewhat greater in the geological past than at present, and that the magnetic field of the solar wind was amplified significantly by its interacting with molten lava pools on the lunar surface. This hypothesis is just a special form of the lunar unipolar generator which has already been discussed (SONETT and COLBURN, 1967). In the present case, it may be assumed that only a part of the lunar surface layer is molten and therefore electrically conductive. Let us consider simply a lava pool of approximately a square shape of D in thickness and L in side length. If $D \ll L$, the unipolar magnetic field B can be approximately expressed as

$$B \simeq 2\pi(1 - K)\bar{\sigma}E_m\text{D} \qquad \text{(in cgs e.m.u. unit)}$$

where

$K =$ the loss factor defined by SONETT and COLBURN,

$\bar{\sigma} =$ mean electrical conductivity of the lava pool,

$\vec{E}_m = \vec{v} \times \vec{B}_o$,

with

$\vec{v} =$ relative velocity between the moon and the solar wind,

$\vec{B}_o =$ ambient magnetic field in the solar wind.

The interior of the cooling lava would remain very hot at temperature of 1000°C or more. Thus the experimental results represented by Fig. 16 suggest that σ of the inside could be $1(\Omega \text{ cm})^{-1}$ in order of magnitude. However, the side edges will be cooler so that the average value of σ parallel to the electric current flow within the pool would be considerably less. The thickness of the low conductivity edges may be several meters at most, according to experience of terrestrial lava pools and lava flows. Hence we may consider that $\bar{\sigma} \geqslant 10^{-1}(\Omega \text{ cm})^{-1}$, provided $L \geqslant 10$ km. Thus we may be able to expect a sufficiently high electrical conductivity for the hypothetical lava pool.

The loss factor K increases with an increase in B, as discussed by SONETT and COLBURN, the maximum limit of B thus being subjected mostly to the dynamic pressure of the solar wind. Given the present condition of the solar wind, the maximum limit of B is about 60 γ, whose magnetostatic pressure ($B^2/8\pi$) is identical to the solar wind pressure. The solar wind pressure is expressed approximately by nmv^2 where n and m denote, respectively, the number density of proton in the solar wind and mass of proton. The average values of these solar wind parameters are represented by $n = 5$ and $v = 400$ km/sec. Even at present the solar wind pressure of a severe magnetic storm frequently becomes more than 10 times as large as its average value. If n and v of the young sun several billion years ago were sufficiently large, then B on the lunar surface could readily attain the order of $10^3 \gamma$. For example, the maximum value of B can be about 1300 γ for $n = 10^2$ and $v = 2000$ km/sec.

At the present stage of our knowledge of lunar science, none of the three speculative hypotheses can clearly be ruled out of consideration. However, the solar wind seems the most reasonable source of lunar NRM in the light of presently available information.

Acknowledgments—The authors' thanks are due several members of the University of Pittsburgh faculty involved in planetary studies for helpful suggestions and for discussions of the results reported in this paper. They are also indebted to G. P. Huffman and J. W. Conroy of the U.S. Steel Fundamental Research Laboratory for Mössbauer analysis and for assistance with electrical and magnetic measurements, respectively. The intensity of shock effects in microbreccia samples was evaluated by Professor J. M. Christie of the University of California at Los Angeles and Professor I. Kushiro of the University of Tokyo. The authors also acknowledge the cooperation of Professor S. V. Radcliffe of Case-Western Reserve University for allowing us to make remanence measurements on samples 12038,29 and 12038,32, and similarly the cooperation of Professor I. Kushiro for loan of samples 10021,31 and 10085,16. Finally, the helpful assistance of the Curator of the Manned Spacecraft Center at Houston is much appreciated.

2476 T. NAGATA *et al.*

REFERENCES

DYAL P., PARKIN C. W., and SONETT C. P. (1970) Lunar surface magnetometer experiment. Apollo 12 Preliminary Science Report, NASA SP-235, pp. 55–73.
ENGLAND A. W., SIMMONS G., and STRANGWAY D. (1968) Electrical conductivity of the moon. *J. Geophys. Res.* **73**, 3219–3226.
LAROCHELLE A. and SCHWARZ E. J. (1970) Magnetic properties of lunar sample 10048,22. *Proc. Apollo 11 Lunar Sci. Conf.*, Geochim. Cosmochim. Acta Suppl. 1, vol. 3, pp. 2305–2308. Pergamon.
NAGATA T. and CARLETON B. J. (1968) Notes on piezo-remanent magnetization of igneous rocks I. *J. Geomag. Geoelect.* **20**, 115–127.
NAGATA T. and CARLETON B. J. (1969) Notes on piezo-remanent magnetization of igneous rocks II. *J. Geomag. Geoelect.* **21**, 427–445.
NAGATA T. and CARLETON B. J. (1969) Notes on piezo-remanent magnetization of igneous rocks III: Theoretical interpretation of experimental results. *J. Geomag. Geoelect.* **21**, 623–645.
NAGATA T., ISHIKAWA Y., KINOSHITA H., KONO M., SYONO Y., and FISHER R. M. (1970) Magnetic properties and natural remanent magnetization of lunar materials. *Proc. Apollo 11 Lunar Sci. Conf.*, Geochim. Cosmochim. Acta Suppl. 1, Vol. 3, pp. 2325–2340. Pergamon.
NAGATA T., KOBAYASHI K., and FULLER M. (1964) Identification of magnetite and hematite in rocks by magnetic observation at low temperature. *J. Geophys. Res.* **69**, 2111–2120.
NAGATA T., OZIMA M., and YAMA-AI M. (1963) Demonstration of the production of a new type of remanent magnetization: Inverted type of thermoremanent magnetization. *Nature* **197**, 444–445.
RUNCORN S. K., COLLINSON D. W., O'REILLY W., STEPHENSON A., GREENWOOD N. N., and BATTEY M. H. (1971) Magnetic properties of Apollo 12 lunar samples. Second Lunar Science Conference (unpublished proceedings).
SHAPIRO V. A. and IVANOV N. A. (1967) Dynamic remanence and the effect of shock on the remanence of strongly magnetic rocks. *Doklady Akad. Nauk USSR* **173**, 1065–1068.
SONETT C. P. and COLBURN D. S. (1967) Establishment of a lunar unipolar generator and associated shock and wake by the solar wind. *Nature* **216**, 340–343.

Proceedings of the Second Lunar Science Conference, Vol. 3, pp. 2477–2483
The M.I.T. Press, 1971.

Magnetic properties of some lunar crystalline rocks returned by Apollo 11 and Apollo 12

R. B. HARGRAVES and N. DORETY

Department of Geological and Geophysical Sciences, Princeton University,
Princeton, New Jersey 08540

(Received 22 February 1971; accepted in revised form 23 March 1971)

Abstract—Measurement before and after alternating field demagnetization of three Apollo 11 samples (10047, 0.76 gm; 10058, 0.87 gm; 10062, 0.60 gm) reveals the presence of a weak N.R.M. ($\simeq 1 \times 10^{-5}$ emu/gm) which is relatively stable with coercivity in excess of 100 oe (3 to 5 $\times 10^{-6}$ emu/gm after 100 oe). The same measurements on three Apollo 12 samples (12053, 2.74 gm; 12063, 2.86 gm; 12065, 3.18 gm) indicate an average N.R.M. of similar intensity ($\gtrsim 1 \times 10^{-5}$ emu/gm) but suggest that if any component of stable remanence is present, it is very much weaker (2 to 5 $\times 10^{-7}$ emu/gm after 100 oe). Isothermal R.M. induced in 8000 oe is highly stable, of intensity about 3 $\times 10^{-4}$ emu/gm, and essentially the same in both Apollo 11 and Apollo 12 samples.

INTRODUCTION

THE PRESENCE of a stable remanent magnetism has been demonstrated in Apollo 11 breccias (DOELL *et al.*, 1970; NAGATA *et al.*, 1970; STRANGWAY *et al.*, 1970) but the origin of this magnetization is uncertain. A thermo remanent magnetization (T.R.M.) following heating associated with lithification, and cooling in a magnetic field is one possibility; a piezomagnetization associated with impact lithification, or I.R.M. induced by lightning strikes associated with the impact, have also been contemplated.

In view of the uncertainties associated with the origin of remanent magnetism in the type C breccias, the unshocked type A and B crystalline rocks can offer relatively less ambiguous clues as to whether or not the lunar lavas cooled in the presence of a steady magnetic field. The results of study of magnetic properties of only four Apollo 11 crystalline rocks have been reported to date, three type A (HELSLEY, 1970, samples 10022, 10069; RUNCORN *et al.*, 1970, sample 10017), and one type B (NAGATA *et al.*, 1970; 10024.22) HELSLEY (1970) and RUNCORN *et al.* (1970) concluded that a stable remanence component was present. Helsley in fact, inferred an ambient field in excess of 1500 gammas, if the stable magnetization be a T.R.M.

Authority was obtained in January 1970 for magnetic measurements to be made on the coarse crystalline type B rocks provided to us for the investigation of lunar pyroxenes (HARGRAVES *et al.*, 1970; HOLLISTER and HARGRAVES, 1970; HOLLISTER *et al.*, 1971). The data we have obtained on three Apollo 11, and three Apollo 12 samples is provided to supplement the studies of other investigators.

PETROLOGY

All three Apollo 11 rocks (10047, 0.76 gm; 10058, 0.87 gm; 10062, 0.60 gm, are microgabbros, and consist principally of monoclinic pyroxene (50%), plagioclase

(35%) and ilmenite (15%). Sample 10062 is much finer grained (average 0.1 mm) in comparison with samples 10047 and 10058, and contains accessory olivine in place of cristobalite. Trace amounts of troilite with included globules of native iron were observed in all three samples. In addition, ulvospinel was observed in sample 10058, and chromite and rutile in sample 10062.

The Apollo 12 rocks (12053, 2.74 gm; 12063, 2.86 gm; and 12065, 3.18 gm) are melabasalts, and contain as essential minerals significantly more ferromagnesian minerals (65% to 69% pyroxene plus olivine), with less ilmenite (9%), and 20 to 25% plagioclase. The pyroxene is zoned from pigeonite toward augite; an early crystallizing chrome spinel, is rimmed by ulvospinel, and (in 12063) in turn by ilmenite. Metallic iron occurs in two distinct generations: As an early crystallizing phase intimately associated with (or being replaced by) chrome spinel, and also as the familiar globules associated with accessory troilite. Collectively, native metal is estimated to be more abundant in the Apollo 12 rocks than in the Apollo 11's, although it is at best a trace constituent in each. Additional petrographic and mineralogical detail for these rocks is provided in Hargraves *et al.* (1970); Hollister and Hargraves (1970) for Apollo 11, and Hollister *et al.* (1971) for Apollo 12.

High-Field Magnetic Property Measurements

Isothermal J/H curves measured on small aliquots of sample 10058 (63 mg) and 12063 (60 mg) are illustrated in Fig. 1. These curves for both are dominated by the paramagnetic susceptibility of the Fe ion in the pyroxene and ilmenite. The value obtained for 10058 is 4.5×10^{-5} emu/gm oe; that for 12063 is somewhat lower, 3.3×10^{-5} emu/gm oe commensurate with its lower ilmenite content. These values are similar to that reported by Nagata *et al.* for type B sample 10024: 3.4×10^{-5} emu/gm oe. The ferromagnetic component in 10058 is barely discernible in the data, and had at most a saturation moment of 0.03 emu/gm. This is equivalent to about 0.01% Fe if that is assumed to be the only ferromagnetic phase present. In keeping with the microscopic evidence of greater metal content, the ferromagnetic constituent in 12063 is more conspicuous, with a saturation moment of around 0.10 emu/gm, equivalent to about 0.04% iron.

The temperature dependence of induced magnetization (3.65 kilogauss) was measured in a vacuum ($< 10^{-3}$ torr), in three cycles: 25° to 400°, 25° to 600°, 25° to 820°C. This technique was employed to determine the limiting temperature to which heating could be carried without oxidation. The results for the 600°C and 820°C run are illustrated in Fig. 2. (The 400° run, in which no change occurred, is omitted for clarity.) Through 600° there was slight change in the heating and cooling curves, with, surprisingly, a small increase in the induced magnetization at room temperature. Heating to 820°C revealed a Curie temperature at approximately 785°C, attributed to the native iron. The inflection at this temperature is far less pronounced on cooling, suggesting that some oxidation has occurred. In both heating and cooling curves there is an inflection between 500° and 600° which suggests the presence of a second ferromagnetic phase, possibly magnetite. The cooling curve crosses the heating curve at about 500°C and remains above it until room temperature. Whereas the amount of native Fe has decreased the amount of this second phase appears to have increased.

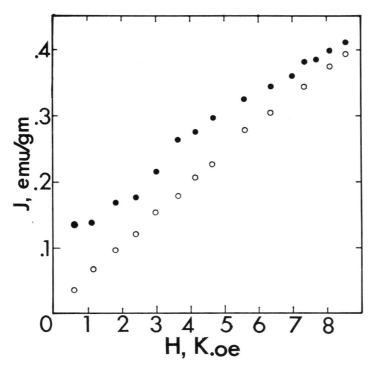

Fig. 1. J/H curves for samples 10058 (open circles), and 12063 (solid circles).

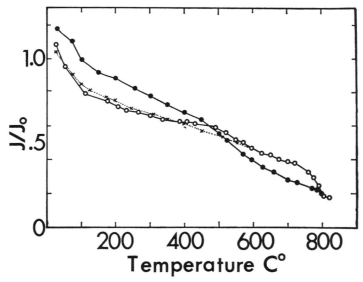

Fig. 2. Normalized J_s/T (3650 oersted) curve for sample 12063. Dotted line joining crosses represents heating to 600°C, cooling curve negligibly different: Open (closed) circles represent heating (cooling) to 820°C.

With our relatively poor vacuum ($< 10^{-3}$ torr) oxidation of native Fe to magnetite as suggested by SCHWARZ (1970) has doubtless occurred. In addition, however, sample 12063 contains approximated 2% fayalite (Fe_2SiO_4) as well as FeO bearing silicates. A partial pressure of oxygen of less than 10^{-15} bars (not attained with our apparatus) is required to prevent the oxidation of fayalite to magnetite plus quartz at 800° $3Fe_2SiO_4 + O_2 \rightarrow 2Fe_3O_4 + 3SiO_2$. Growth of magnetite by this oxidation reaction evidently more than compensates for the decrease in saturation moment which results from the oxidation of native iron.

All detailed investigations of the sulfide phase in Apollo 11 rocks have indicated that it is stoichiometric troilite (SKINNER, 1970), and hence nonmagnetic. Native iron (with some Ni and Co) appears to be the only phase present in these rocks which is ferromagnetic at room temperature, and it is presumed to be the carrier of the remanent magnetism.

REMANENT MAGNETIC PROPERTIES

The migration of the natural remanent vectors (arbitrary initial orientation) measured on Princeton Applied Research SM1 spinner magnetometer with increasing A.F. demagnetization (MCELHINNY, 1963) are shown in Fig. 3. The change in intensity on demagnetization of both N.R.M. and I.R.M (induced in 8000 oe), is shown in Fig. 4.

N.R.M. of these samples is weak, but measurable (10^{-5} to 10^{-6} emu/gm). Changes in direction and intensity after A.F. demagnetization in fields less than 100 oe suggest

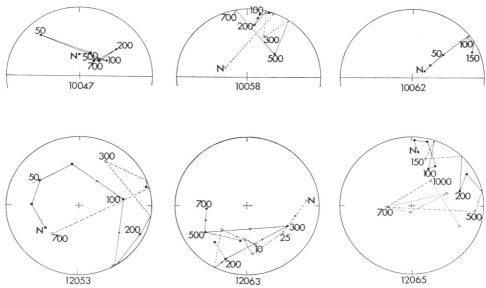

Fig. 3. Equal area projections showing migration of NRM vector with A.F. demagnetization for Apollo 11 samples 10047, 10058, 10062 (upper row) and Apollo 12 samples 12053, 12063, and 12065 (lower row). Initial orientation arbitrary; solid (open) circles on lower (upper) hemisphere.

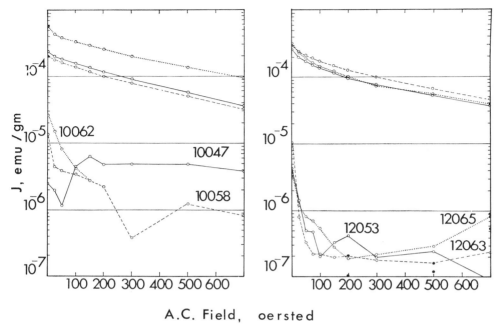

Fig. 4. Variation of specific remanent intensity with AF demagnetization, Apollo 11 samples on left, Apollo 12 on right. Lower set of curves in both diagrams represents NRM, upper set represents I.R.M. induced in 8000 oe.

the presence of a substantial viscous magnetization of unknown origin. In all likelihood this was induced during sample preparation, which for the Apollo 11 samples in our laboratory included cutting of the samples on a diamond saw for polished thin section preparation. (Note, 10062 was not cut). The Apollo 12 samples were cut at the Lunar Receiving Laboratory.

Between 100 and 200 oe the directions in the Apollo 11 samples appear to stabilize (Fig. 3; note that sample 10062 was only measured to 150 oe), becoming more scattered in higher fields. Sample 10047, in particular, appears to contain a remanent component ($J = 5 \times 10^{-6}$ emu/gm), which is stable with respect both to intensity and direction, up to 700 oe A.F.

The decline in intensity of the Apollo 12 samples is much more pronounced (2 to 5×10^{-7} emu/gm, after 100 oe A.F., Fig. 4). The migration of the vector above 100 oe A.F. is concomitantly much more erratic (Fig. 3), and there is no evidence of the stabilization between 100 and 200 oe that was recorded in the Apollo 11 samples.

In all samples, the presence of what is presumed to be a superparamagnetic component which apparently is anhysteretically magnetized during A.F. demagnetization, and decays during subsequent spinning makes measurements above 200 oe particularly tedious and uncertain. Nevertheless repeat measurements of 12065 at 700 oe, and 1000 oe gave moderately consistent directions at a signal level higher than that recorded after lower A.C. fields. Whether this vector represents a real natural remanent magnetism or is an artifact of A.C. demagnetization in high fields, is not certain.

Discussion

The IRM$_{8000\ oe}$ intensities of both Apollo 11 and Apollo 12 samples are essentially the same, indicating a comparable potential for acquiring stable remanent magnetism (Fig. 4). The base-level intensity of the demagnetized NRM in Apollo 11 is an order of magnitude greater than that in Apollo 12. The signals in the latter, in fact are so weak ($\sim 2 \times 10^{-7}$), that there is considerable uncertainty as to whether any primary permanent magnetism remains. The significance of the vector directions recorded in 12065 and 12063 after high field A.C. demagnetization is uncertain.

The Apollo 11 and Apollo 12 samples were measured at different times, but on the same apparatus, and following identical procedures. To the best of our knowledge and ability, the sensitivity of the instruments, and the degree of field cancellation in the spinner and A.F. demagnetizer field coils was the same. The Apollo 11 samples measured were smaller (0.6 to 0.9 gms) than the Apollo 12 (2.8–3.1 gms), but nevertheless their total moment after partial demagnetization was substantially higher.

We conclude that our data indicate the presence of remanent magnetism of at least intermediate stability (100 to 200 oe) and intensity ($J = 2$ to 5×10^{-6} emu/gm after 100 oe A.C. demagnetization), in Apollo 11 samples; if a similar stable (coercivity > 100 oe) component is present in Apollo 12 samples, it is at least an order of magnitude weaker.

Our results for Apollo 11 rocks are in qualitative agreement with those of Helsley (1970), Nagata et al. (1970) and Runcorn et al. (1970), but the treatment procedures employed to study the remanent magnetic properties were different in each case. Helsley undertook thermal demagnetization studies after less than 40 oe A.F. demagnetization; Nagata et al. (1970, p. 2336) ascribes a limiting stability of 30 oe to his sample, as the initial remanent intensity of 8.9×10^{-6} emu/gm was reduced to "practically zero" in 60 oe. In the sample of Runcorn et al. (1970, p. 2383) an NRM of intermediate hardness was reported in which an initial intensity of 5.6×10^{-6} emu/gm was reduced to 2.0×10^{-7} emu/gm in a field of 500 oe.

The results obtained on our Apollo 12 samples are likewise in good agreement with those reported by other workers at the Second Lunar Science Conference (generally N.R.M. of 5 to 10×10^{-6} emu/gm, demagnetizing sharply to 2 to 3×10^{-7} emu/gm after 100 to 200 oe).

Conclusions

Although N.R.M. of similar intensity (5×10^{-6} to 5×10^{-5} emu/gm) appear to be present in both Apollo 11 and Apollo 12 crystalline rocks, our data suggest that whereas a relatively stable remanence component is present in the Apollo 11 samples, this is absent or considerably weaker in Apollo 12. This contrast, if true, is important with regard to speculation as to the origin of the magnetizing field. It is hoped that additional data on the remanent magnetic properties of Apollo 11 crystalline rocks may yet be obtained by nondestructive measurements on those samples not yet examined.

REFERENCES

DOELL R. R., GROMME C. S., THORPE A. N., and SENFTLE F. E. (1970) Magnetic studies of Apollo 11 lunar samples. *Proc. Apollo 11 Lunar Sci. Conf., Geochim. Cosmochim. Acta* Suppl. 1, Vol. 3, pp. 2093–2096. Pergamon.

HARGRAVES R. B., HOLLISTER L. S., and OTALORA G. (1970) Compositional zoning and its significance in pyroxenes from three coarse-grained lunar samples. *Science* **167**, 631–633.

HELSLEY C. E. (1970) Magnetic properties of lunar 10022, 10069, 10084 and 10085 samples. *Proc. Apollo 11 Lunar Sci. Conf., Geochim. Cosmochim. Acta* Suppl. 1, Vol. 3, pp. 2213–2220. Pergamon.

HOLLISTER L. S. and HARGRAVES R. B. (1970) Compositional zoning and its significance in pyroxenes from two coarse grained Apollo 11 samples. *Proc. Apollo 11 Lunar Sci. Conf., Geochim. Cosmochim. Acta* Suppl. 1, Vol. 1, pp. 541–550. Pergamon.

HOLLISTER L. S., TRZCIENSKI W., HARGRAVES R. B., and KULICK C. (1971) Crystallization histories of two Apollo 12 basalts. In H. H. Hess Memorial Volume (Editor R. Shagam) *Geol. Soc. Amer. Mem. 132.*

MCELHINNY M. W. (1963) Theory and operating procedures; Rock magnetism instruments at Princeton. Princeton University, Department of Geological Engineering Report, 63–1.

NAGATA T., ISHIKAWA Y., KINOSHITA H., KONO M., SYONO Y., and FISHER R. M. (1970) Magnetic properties and natural remanent magnetization of lunar materials. *Proc. Apollo 11 Lunar Sci. Conf., Geochim. Cosmochim. Acta* Suppl. 1, Vol. 3, pp. 2325–2340. Pergamon.

RUNCORN S. K., COLLINSON D. W., O'REILLY W., BATTEY M. H., STEPHENSON A., JONES J. M., MANSON A. J., and READMAN P. W. (1970) Magnetic properties of Apollo 11 Lunar samples. *Proc. Apollo 11 Lunar Sci. Conf., Geochim. Cosmochim. Acta* Suppl. 1, Vol. 3, pp. 2369–2388. Pergamon.

SCHWARZ E. J. (1970) Thermomagnetics of lunar dust sample 10084.88. *Proc. Apollo 11 Lunar Sci. Conf., Geochim. Cosmochim. Acta* Suppl. 1, Vol. 3, pp. 2389–2398. Pergamon.

SKINNER B. J. (1970) High crystallization temperatures indicated for igneous rocks from Tranquillity Base. *Science* **167**, 652–654.

STRANGWAY D. W., LARSEN E. E., and PEARCE G. W. (1970) Magnetic studies of lunar samples, breccias and fines. *Proc. Apollo 11 Lunar Sci. Conf., Geochim. Cosmochim. Acta* Suppl. 1, Vol. 3, pp. 2435–2452. Pergamon.

Proceedings of the Second Lunar Science Conference, Vol. 3, pp. 2485–2490
The M.I.T. Press, 1971.

Evidence for an ancient lunar magnetic field*

Charles E. Helsley

Geosciences Division, The University of Texas at Dallas, Dallas, Texas 75222

(*Received* 22 *February* 1971; *accepted in revised form* 31 *March* 1971)

Abstract—Magnetic studies of samples 12002 and 12022 have shown that these samples contain a weak remanent magnetic moment. Progressive thermal demagnetization experiments suggest that these moments are stable and were not acquired as a result of exposure to the earth's magnetic field. The primary carrier of remanence is native iron. In addition, two spinel phases (chromite and ülvo-spinel) and troilite are also present in these samples. Progressive thermal demagnetization experiments, in which the samples were cooled in the presence of a known field, suggest that the remanent magnetization of these rocks was acquired on the moon in the presence of an applied field of up to 5000 gammas, provided that the initial magnetization was acquired by an ordinary partial thermo-remanent process. This would imply that the moon had a weak magnetic field 3.5 to 4.0 aeons ago and suggests that small conducting fluid core may have been present at that time.

Alternatively, the remanence could have been acquired in a much lower field as a result of the introduction of crystalline defects during thermal cycling or radiation exposure, and thus the pro-gressive thermal demagnetization experiments would imply an erroneously high estimate of the intensity of the paleofield.

Introduction

THE MAGNETIC STUDIES on Apollo 11 samples showed that lunar materials are capable of possessing an apparently stable remanent magnetization and that this magnetization was acquired by the samples in the lunar environment. The magnetic studies on the Apollo 11 material also indicate that the magnetization was acquired in a field of in excess of 1500 γ (HELSLEY, 1970). Similar studies on Apollo 12 materials are reported below.

The two rocks studied in this investigation are 12002 and 12022. These rocks are olivine microgabbros, and rock 12002 is conspicuously vesicular. Petrographic observations of these samples suggest a near surface, probably extrusive, origin. The potentially magnetic phases present in each sample (in order of decreasing abundance) are a zoned spinel (ülvospinel cores and chromite rims), troilite, and native iron. Of these only the native iron and the troilite are potentially magnetic above room temperature.

Thermomagnetic Analysis

Thermomagnetic analyses were made on thirty to eighty milligram fragments. The samples were crushed and sealed in evacuated quartz capillaries (pressure < 0.01 mm of Hg) and heated in a magnetic force balance to temperatures of up to 850°C in a field of about 3 koe. The heating and cooling rates varied from 5° to 20°C per minute. The results of these experiments are shown in Fig. 1. Both rocks show a progressive decrease in intensity that can be attributed to the changing susceptibility of the iron-bearing paramagnetic minerals in the sample. At temperatures in excess of 500°C the

---

* Contribution No. 174, Geosciences Division, The University of Texas at Dallas.

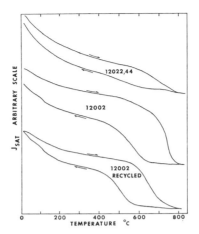

Fig. 1. Thermomagnetic analysis curves of samples 12002 and 12022. The temperatures recorded for 12002 are approximately 50°C too high as a result of unfavorable
sample geometry.

magnetization begins to drop steadily towards a Curie point near 770°C, indicating
the presence of native iron as the dominant magnetic phase. Upon cooling, the curve
is not reproduced and Curie points near 600°C and 120°C are indicated. This behavior
is reproduced upon recycling as is shown in the lower part of Fig. 1. These curves are
remarkably similar to those observed in iron-bearing meteorites (STACEY *et al.*, 1961;
LOVERING and PARRY, 1962). The Curie points observed suggest that the iron alloy
(Kamacite) has more than 7% nickel (LOVERING and PARRY, 1962, Table 3) and that a
small amount of taenite may also be present (Curie point near 120°C). No other Curie
points are present in the initial run. The curve observed on reheating of sample 12002
suggests that some chemical change has taken place since the heating curve upon
recycling begins to decrease rapidly near 600°C rather than near 700°C as in the initial
run.

The thermomagnetic observations support the polished thin section observations
in which iron, troilite, chromite, ülvospinel, and ilmenite were observed, and suggest
that the dominant carrier of the remanence observed in these rocks should be the iron
alloy. X-ray microprobe analysis of the opaque oxides in sample 12002 indicate that
nickel is present in the iron in quantities up to at least 20%. Iron grains internal to the
ülvospinel and chromites are the highest in nickel while those in contact with silicate
minerals, i.e., those formed later, have nickel contents near 5%.

NATURAL REMANENT MAGNETIZATION

Initial measurements of the natural remanent magnetization (NRM) of samples
12002 and 12022 yielded specific intensities of 0.153×10^{-5} and 0.208×10^{-4} emu
per gram, respectively. Each sample initially was stored in a field-free region for several
days with repeat observations being made each day. The observed intensity and direction changed progressively during these tests, suggesting that the samples had acquired
a small moment since return to the earth. Alternating field demagnetization was not

done since the mineralogy observed in polished thin sections was similar to that for Apollo 11 samples which were shown to have remanent coercivities near 50 oersted, and it was deemed advisable to concentrate on making a paleointensity determination rather than determine a coercivity spectrum.

Both samples were subjected to progressive thermal demagnetization experiments in 50°C steps. The furnace used was a conventional noninductively wound furnace operated in a region where the ambient field was maintained at 0 ± 10 gammas. Sample 12002 was heated in a sealed evacuated tube (pressure < 0.01 mm of Hg) while sample 12022 was heated in a vacuum furnace in which the pressure was maintained at < 0.01 mm of Hg during the entire run. In the sealed tube technique the thermocouple was outside the sample container while the vacuum furnace had an internal platinum thermocouple mounted about 2 mm from the sample. The results of these demagnetization experiments are shown in Fig. 2. After demagnetization at each step above 100°, each sample was reheated to the same temperature and allowed to cool for 50° or 100°C in the presence of a known field of 1000 to 10,000 γ depending upon the particular run. Thus the sample was allowed to acquire a partial thermal remanence so that paleointensity determinations by the double heating method of THELLIER and THELLIER (1959) could be calculated. For sample 12002 these measurements were carried out in the temperature range 20° to 150°C and the results are

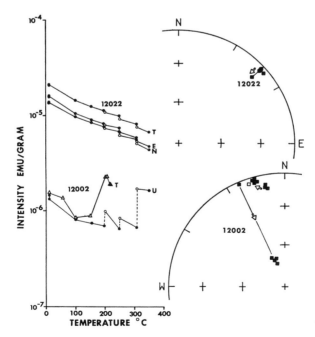

Fig. 2. Thermal stability of NRM of 12002 and 12022. Closed dots indicate observations before PTRM experiment (see text); open dots indicate observations after PTRM experiment; *T, N, E, U* indicate total moment, north component, east component, and upward component, respectively.

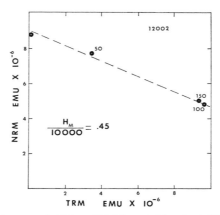

Fig. 3. Relation between field lost (NRM) to field gained (TRM) for sample 12002.

shown in Fig. 3. At temperatures above 200°C the sample did not demagnetize systematically, and thus these observations cannot be used for paleointensity determinations. Between 20°C and 150°C the ratio of the field lost to that gained in 10,000 gammas is 0.45. As is described below, the lunar surface temperature cycle is in this same range; consequently, the significance of the 4500 gamma paleointensity determination from this sample is in doubt.

Sample 12022 demagnetized systematically; i.e., the intensity decreased smoothly and the direction of magnetization was constant, up to 308°C. However, after the partial thermal demagnetization run at 308°C, the magnetization acquired during the PTRM experiment apparently was not completely removed. Upon heating to 350°C, the PTRM acquired at 308°C apparently remained; i.e., the component in the direction of the applied field remained unchanged. Since it is unlikely that the magnetization acquired in cooling from 308°C to 200°C should remain unchanged, even after heating to 350°C, it must be assumed that the magnetization of the sample has more than one component. In this particular case, one would surmise that the sample has a magnetic moment acquired below 300°C that is dominant at temperatures below 300°C and that a weaker component acquired above 300°C is also present. Studies on the high temperature portion of the magnetization have as yet not been completed. The average of the eight independent paleointensity determinations for 12022 was 4800 γ with the total range being 2800 to 10,060 γ.

PALEOINTENSITIES AND THE ORIGIN OF THE MAGNETIZATION

If one assumes that the observations made at temperatures below 300°C are valid indications of the field present at the time of origin of the rock, then one is led to the estimates of field intensity given in Table 1. Of the four rocks studied so far, 10022, 10069, 12002, and 12022, an estimate between 1000 and 5000 gammas seems to be indicated. This compares with the present lunar field of 38 gammas (DYAL et al., 1970) and suggests that the lunar field may have had a much larger value early in the moon's history (HELSLEY, 1970). This field could be of internal lunar origin or could have been acquired while the moon was in close proximity to the earth. The fact that fields of

Table 1. Intensity estimates for ancient lunar magnetic field

Sample	Estimated field (gammas)
10022	1500
10069	1000 to 170,000
12002	4500
12022	4800
Present (A-12)	30 to 50
Earth	24,000 to 70,000

reasonable size ($> 1000 \gamma$) are indicated for rocks of two distinctly different ages suggest that the observed paleofield was of internal origin since the moon would not be expected to be able to remain close to the earth (inside its magnetic field) for a period of 300 to 400 million years as is indicated by the age differences (CLIFF et al., 1971).

During the course of the experiments made on 12,022, corrections were made for the time the sample was kept at elevated temperatures according to the formula given by STACEY (1963):

$$\frac{T_1}{T_2} = \frac{\ln (3C\tau_2)}{\ln (3C\tau_1)},$$

where T is temperature in °K, τ is the time in question, and C is a constant with a value of 10^{10}. Thus an experimental run of 10 minutes duration at 300°C is equivalent to two weeks at a temperature of 120°C. Although the theory pertains strictly to a single magnetic phase with a single relaxation time, these calculations agreed qualitatively with the observations in that repeat runs at the same temperature produced a continued slow decrease in observed intensity. However, this theory, when applied to lunar rocks, suggests that there should be no change in the intensity below 300°C since all components with this stability range should have been removed during the 120+°C temperatures of the lunar day. Moreover, it would suggest that only the magnetization remaining above 525°C has a certain meaning in terms of paleointensity for it would be the only magnetization that could survive a billion years at 120°C. However, this conclusion involves a very large extrapolation that may not be valid since, as mentioned above, the theory upon which it is based pertains only to single domain, single relaxation time material. Moreover, the demagnetizing effects produced by repeated heating to a low temperature (120°C) may not be comparable to a long single heating to the same temperature.

The significant changes of magnetization observed at temperatures below 300°C, particularly for sample 12002, require the conclusion that STACEY's (1963) calculations are not entirely applicable to lunar material since no changes would have been observed if the theory were entirely correct. Since most of the iron is multidomain (the critical radius for single domain behavior is very small for iron: 20 Å CHIKAZUMI, 1964; 70 Å KITTEL, 1949; 160 Å NÉEL, 1947), the magnetization observed at temperatures below 300° could be due to domain wall pinning on dislocations resulting from stresses induced during rapid heating and cooling in the lunar surface environment. This pinning could enhance, or make more stable, any isothermal remanence acquired during an isolated magnetic event that occurred at some time since the original cooling of the rock. Thus, the samples could perhaps acquire their remanence as the result of thermal cycling in a weak field provided that a momentary stronger field had once been present. Such a field may have been caused by a meteorite impact or as a result of a strong solar flare (Sonnet *et al.*, 1970).

In all the samples studied so far the anomalous behavior observed at temperatures below 450°C could be attributed to the removal of the defects produced during thermal cycling on the moon, by laboratory heating to temperatures higher than those experienced on the lunar surface. Butler and Cox (1971) have recently confirmed that radiation induced damage can affect the magnetic properties of iron. In their experiment, changes in coercivity were produced as the result of irradiation of pure iron samples in a high neutron flux. Thus they propose that the stable remanence observed in lunar samples is induced by the defects produced during cosmic ray bombardment in a weak field.

In view of the fact that most of the stable remanence in lunar rock is only stable at low temperatures and that potential mechanisms exist for producing apparently stable remanence at low temperatures in the presence of a weak field or momentary strong field, the estimates of the paleointensity given in Table 1 should be used with caution until observations of stable remanence at higher temperatures are made to confirm, or refute, their validity.

Acknowledgments—Discussion and critical comments on the manuscript by Drs. A. L. Hales and J. L. Carter are greatly appreciated. X-ray microprobe studies were done with the assistance of J. B. Toney. This work was supported by NASA Contract NAS 9-8767.

References

Butler R. F. and Cox A. V. (1971) A mechanism for the production of stable magnetic remanence in chondritic meteorites and lunar samples by cosmic ray exposure. Submitted to *Science*.
Chikazumi Soshin (1964) *Physics of Magnetism*. John Wiley.
Cliff R. A., Lee-Hu C., and Wetherill G. W. (1971) Rb-Sr and U, Th-Pb measurements on Apollo 12 material; Second Lunar Science Conference (unpublished proceedings).
Dyal P., Parkin C. W., and Sonnett C. P. (1970a) Apollo 12 magnetometer: Measurement of a steady field on the surface of the moon. *Science* **169**, 762–764.
Helsley C. E. (1970) Magnetic properties of lunar 10022, 10069, 10084, and 10085 samples. *Proc. Apollo 11 Lunar Sci. Conf.*, Geochim. Cosmochim. Acta Suppl. 1, Vol. 3, pp. 2213–2219. Pergamon.
Kittel Charles (1949) Physical theory of ferromagnetic domains. *Rev. Mod. Phys.* **21**, 541–583.
Lovering J. F. and Parry L. G. (1962) Thermomagnetic analysis of co-existing nickel-iron metal phases in iron meteorites and the thermal histories of the meteorites. *Geochim. Cosmochim. Acta* **26**, 36–382.
Néel L. (1947a) Propriétés d'un ferromagnétique cubique en grains fins. *Comptes rendus* **224**, 1488–1490.
Sonnet C., Colburn D., Schwartz K., and Keil K. (1970) The melting of asteroidal-sized bodies by unipolar dynamo induction from a primordial T tauri sun. *Astrophys. Space Sci.* **7**, 446.
Stacey F. D. (1963) The physical theory of rock magnetism. *Advan. Phys.* **12**, 45–133.
Thellier E. and Thellier O. (1959) Sur l'intensite du champ magnetique terrestre dans le passe historique et geologique. *Ann. Geophys.* **15**, 285–376.

Proceedings of the Second Lunar Science Conference, Vol. 3, pp. 2491–2499
The M.I.T. Press, 1971.

Magnetic properties of Apollo 12 lunar samples 12052 and 12065*

C. Sherman Grommé and Richard R. Doell

U.S. Geological Survey, Menlo Park, California 94025

(Received 22 February 1971; accepted in revised form 25 March 1971)

Abstract—Magnetization versus field and magnetization versus temperature data indicate that the ferromagnetic material in Apollo 12 samples 12052 and 12065 is spheroidal grains of iron having Curie temperatures of 782° and 788°C respectively. The samples carry natural remanent magnetization that is stable with respect to the earth's field, of the order of 6 to 20×10^{-6} emu/gm in intensity. This magnetization consists of several components, all of which are destroyed by heating in vacuum (5×10^{-6} torr) to temperatures below 600°C. The ability of the rocks to acquire thermoremanent magnetization below 200°C is also destroyed by heating in vacuum to 600°C. The NRM in these rocks is concluded to have resulted from several magnetizing events that occurred at moderate temperatures some time later than the last cooling of the rocks.

Introduction

Natural remanent magnetization (NRM) has been reported in several specimens of crystalline rock returned by the Apollo 11 expedition to the moon (Helsley, 1970; Nagata et al., 1970; Runcorn et al., 1970). These NRM's ranged from 2×10^{-6} to 2×10^{-5} emu/gm in intensity, and exhibited low to moderate stability against alternating-field and thermal demagnetization. If these NRM's were acquired by the rocks while on the moon, as seems likely, the presence of a lunar magnetic field, either steady or transitory, is implied; a brief discussion of this implication was given by Runcorn et al. (1970). The magnetic experiments reported here were performed on two Apollo 12 crystalline rocks (12052 and 12065) and were designed with the following objectives: (1) to determine the amount, stability, and carrier of NRM, (2) to ascertain whether any observed NRM could be thermoremanent magnetization (TRM) acquired by the rocks during their initial cooling, and (3) to estimate the magnitude of lunar magnetic field necessary to produce such magnetization. The available samples were in the form of two sawn cubes weighing approximately 10 gm and smaller fragments weighing approximately 0.2 gm. Heating experiments according to the method of Thellier and Thellier (1959) were done in fields of a few thousand gammas on the larger specimens to investigate the thermal stability of NRM and the distribution of partial thermoremanent magnetization (PTRM). The smaller fragments were used for strong-field magnetic experiments to determine some of the intrinsic magnetic characteristics of the rocks. Owing to the limited amount of material, no alternating-field demagnetization was done.

High-Field Magnetization

Magnetization curves for specimen 12052,48 at four temperatures are shown in Fig. 1. The magnetization is seen to be strongly temperature dependent at low

---

* Publication authorized by the Director, U.S. Geological Survey.

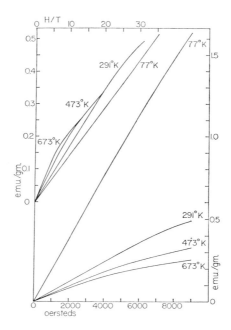

Fig. 1. Lower right: Magnetization versus magnetic field, crystalline rock 12052,48.
Upper left: The same magnetization data as a function of reduced field H/T.

temperatures, indicating a relatively large paramagnetic and/or antiferromagnetic contribution to the total susceptibility. To evaluate the possible contribution of the superparamagnetism that was found to be characteristic of the Apollo 11 fines and breccia and, to a lesser extent, crystalline rock (Nagata et al., 1970), the same magnetization data are shown in Fig. 1 as a function of reduced field H/T. The presence of a ferromagnetic component is clearly evident from the fact that the curves do not coincide.

A notable feature of the magnetization curves in Fig. 1 is that saturation of the ferromagnetic component is not approached until fields of about 5000 oe are exceeded. This may be interpreted to mean that the ferromagnetic mineral is iron, or iron alloy of high permeability, in the form of multidomain spherical particles. Because of their large demagnetizing factor and high permeability, the self-demagnetizing field of such particles is so great that in moderate external fields their net magnetization is virtually zero. Therefore at temperatures well below the Curie point and in low fields the apparent susceptibility is temperature-independent and is given by $\chi_a = 3/(4\pi\rho)$, and assuming a density $\rho = 7.8$ for the spheres, $\chi_a = 0.03$ gm^{-1} (Senftle et al., 1964; Runcorn et al., 1970). By plotting the observed initial susceptibilities as a function of reciprocal temperature, the temperature-independent part of the susceptibility of specimen 12052,48 is estimated to be between 1.5 and 2.0 $\times$ 10^{-5} gm^{-1}. This value is only 25% to 30% of the total initial susceptibility, 6.0 $\times$ 10^{-5} gm^{-1}, measured at 291°K; the remainder is due to paramagnetic and/or antiferromagnetic minerals.

The amount of metallic iron in the rock may be estimated from the susceptibility data. Using the values given above, the percentage of iron in rock 12052,48 is 0.05 to 0.07 by weight. This is in good agreement with percentages found using other magnetic methods for rock 12065 (TSAY et al., 1971), rocks 12021 and 12063 (PEARCE et al., 1971), and rock 12053 (NAGATA et al., 1971).

Magnetization versus temperature in the range $-180°C$ to $850°C$ was measured using the recording magnetic balance described by DOELL and COX (1967). The temperature calibration was checked with electrolytic iron (measured Curie temperature $775°C$) and pure natural magnetite (measured Curie temperature $583°C$). Curie temperatures were determined using a graphical method (GROMME et al., 1969) that tends to give a slightly high value; ARAJS and COLVIN (1964) have reported an accurately determined ferromagnetic Curie temperature of $771° \pm 2°C$ for pure iron. The relative precision of our Curie temperature measurements is about $\pm2°C$.

Thermomagnetic curves for specimens 12052,48 and 12065,37 are shown in Fig. 2. The Curie temperatures of these specimens are $788°C$ and $782°C$ respectively, significantly higher than that of pure iron. The only alloying elements of any importance that can raise the Curie temperature of iron are cobalt and vanadium (BOZORTH, 1951). GIBB et al. (1970) report the presence of 1.5% Co, 1.9–2.3% Ni, and 0.1–0.2% Cu in metallic iron in rock 12052, whereas CAMERON (1971) reports only 1.0 to 1.7% Ni in iron in rock 12065. To raise the Curie temperature of iron $13°C$ would require only 1.2% cobalt (FORRER, 1930), but as the effect of nickel is opposite, the exact Curie temperature of this quaternary alloy is difficult to predict.

The thermomagnetic curves in Fig. 2 were obtained in a vacuum of about 5×10^{-5} torr. The cooling curves nearly retrace the heating curves, indicating that at this pressure the rate of inward diffusion of oxygen was sufficiently reduced to minimize oxidation of the iron within the specimens. To evaluate further the effects of oxidation, these experiments were repeated using the same specimens, first in nitrogen and then in air. The resulting thermomagnetic curves are shown in Fig. 3, in which the effect of oxidation is clear. All the curves have inflections at the Curie temperature of magnetite and the magnetizations above this point are significantly reduced.

The form of the magnetization curve at room temperature has turned out to be a sensitive indicator of oxidation in these rocks, as illustrated in Fig. 4. Because the magnetite that results from oxidation has a much smaller self-demagnetizing field than do the iron spherules, it saturates in a much lower external field. This gives rise to convexity in the formerly straight parts of the curves below about 5000 oe. That a minor amount of oxidation occurred during heating in vacuum is evident from the curves marked a in Fig. 4. It is nevertheless clear from these experiments that heating to temperatures of the order of $800°C$ at pressures less than about 10^{-6} torr will produce no gross alteration of the phase responsible for the bulk magnetization curves in these rocks.

TIME-STABILITY OF NRM

All measurements of NRM were made using a spinner magnetometer in a field of a few hundred gammas to minimize possible effects of viscous magnetization. As

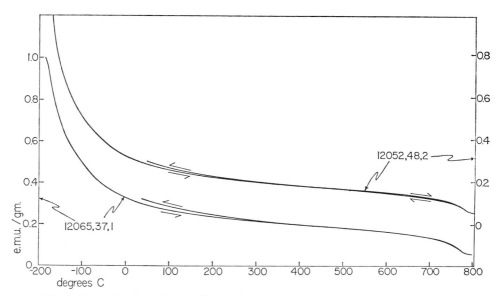

Fig. 2. Magnetizations of crystalline rocks 12052,48 and 12065,37 versus temperature in a field of 5500 oe. These are the first heating experiments on these specimens, and were done in a vacuum of 5×10^{-5} torr. Heating and cooling curves are indicated by arrows; the rate was 10°C per minute.

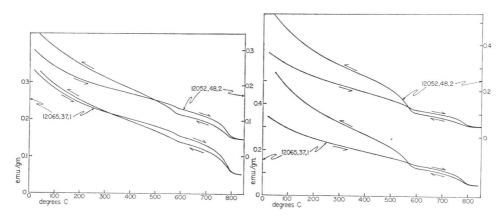

Fig. 3. Magnetizations of crystalline rocks 12052,48 and 12065,37 versus temperature in a field of 5500 oe. The second heating experiments on these specimens (left) were done in nitrogen, and the third heatings (right) were done in air. Heating and cooling curves are indicated by arrows.

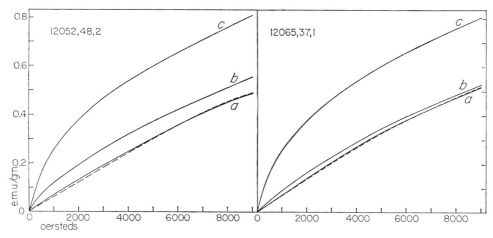

Fig. 4. Magnetizations versus magnetic field at room temperature, of crystalline rocks after successive heating experiments illustrated in Figs. 2 and 3. *a*, after heating in vacuum (5×10^{-5} torr); *b*, after heating in nitrogen; *c* after heating in air. For comparison, magnetization curves for the same specimens before heating are shown dashed.

received, specimens 12052,32 and 12065,67 had NRM amounting to 2.1×10^{-5} and 6.0×10^{-6} emu/gm, respectively. The specimens were remeasured after storage for 5 hours in the earth's magnetic field and again after 4 months. No significant changes in NRM direction or intensity were observed, and we conclude that viscous remanent magnetization in these rocks is negligible.

NRM-TRM EXPERIMENTS

The thermal stability of NRM and the production of artificial TRM in specimens 12052,32 and 12065,67 were investigated using the double heating method of THELLIER and THELLIER (1959). The specimens were heated and cooled in a vacuum of 5×10^{-6} torr in a nonmagnetic oven within a space that is field-free to within $\pm 10 \gamma$. Results of HELSLEY (1970) and NAGATA *et al.* (1970) on Apollo 11 crystalline rocks indicated that fields of the order of a few thousand gammas might be expected to produce TRM comparable in magnitude to the NRM; accordingly, fields of 5000 γ and 2000 γ were used.

The directions of magnetization after the successive heatings are shown in Fig. 5 and the corresponding intensities of NRM and PTRM are given in Table 1. After the first double heating, to 195°C in 5000 γ, the NRM in both specimens had diminished by about one-half to two-thirds and had changed directions significantly and significant PTRM's were produced. As a consistency test and because of the direction change, the specimens were then heated to 195°C and cooled in null field. The NRM intensities agreed well with those calculated from the double heating, and the NRM direction in specimen 12052 remained constant whereas that in specimen 12065 continued to change (Fig. 5). A second double heating to 195°C was done in a field of 2000 γ, which showed that the PTRM from 195° to 20°C was linear with applied field (Table 1). The third double heating was to 395°C in 2000 γ; as expected, the NRM was further diminished. The PTRM acquired in 2000 γ was less after heating to 395° than it had been after heating to 195°C, indicating that the carrier of the PTRM was being progressively destroyed by the heating. After heating to 590°C and cooling in 2000 γ, both NRM and PTRM were below the sensitivity limit of the magnetometer (approximately 3×10^{-7} emu/gm). Finally, to confirm that an irreversible change in the carrier of the NRM and PTRM had occurred, both specimens were heated to 195°C and cooled in 5000 γ; again no remanent magnetization could be detected. During the heating experiments, the NRM direction in specimen 12065,67 changed continuously and unpredictably up to 395°C, while the direction

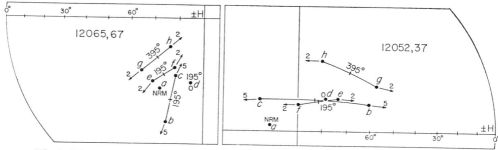

Fig. 5. Directions of remanent magnetization in crystalline rocks 12052,32 and 12065,67 after heating in vacuum (5×10^{-6} torr) in various applied fields. Directions are plotted on upper hemisphere of equal-area projection. Letters a through h indicate sequence of experiments. Direction of applied field is indicated by $\pm H$ and sign of applied field by short arrows; figures by arrows designate intensity of field in thousands of gammas. Point marked NRM is the natural remanent magnetization measured before heating. Maximum temperatures attained in successive heatings are shown in degrees C. Magnetization directions after paired heatings in opposing fields are joined by lines; midpoints of these lines are marked to indicate the calculated NRM directions. Note that heating d was done in zero applied field.

Table 1. Intensities of natural remanent magnetization and partial thermoremanent magnetization in specimens after heating and cooling in vacuum in a controlled magnetic field.

			Specimen			
			12052,32		12065,67	
Experiment	T	H	NRM	PTRM	NEM	PTRM
a	20	—	2.04	—	0.56	—
b	195	+5⎫				
c	195	−5⎭	(0.60)	0.23	(0.28)	0.11
d	195	0	0.61	—	0.31	—
e	195	+2⎫				
f	195	−2⎭	(0.62)	0.09	(0.29)	0.06
g	395	+2⎫				
h	395	−2⎭	(0.14)	0.04	(0.13)	0.02
i	590	+2	< 0.03	< 0.03	< 0.04	< 0.04
j	195	+5	< 0.03	< 0.03	< 0.03	< 0.03

T, maximum temperature in degrees Celsius; H, ambient magnetic field in units of 10^{-2} oe; NRM and PTRM in units of 10^{-5} emu/gm; NRM values in parenthesis calculated, values not in parenthesis directly measured.

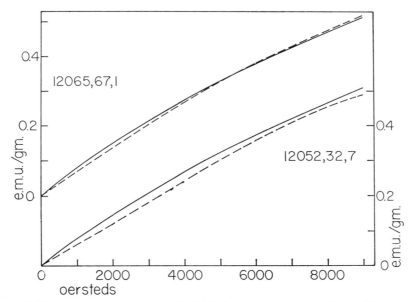

Fig. 6. Magnetizations versus magnetic field at room temperature, of crystalline rocks after successive heating experiments described in Table 1 and Fig. 5. For comparison, magnetization curves for specimens 12052,48 and 12065,37 before heating are shown dashed.

in specimen 12052,32 changed during the first heating to 195°C but remained constant until the maximum temperature was raised to 395°C.

In order to test whether appreciable oxidation had occurred during these heating experiments, magnetization curves were determined on fragments from both specimens. The results are shown in Fig. 6 together with the magnetization curves of unheated samples for comparison, and it appears that a small amount of iron was oxidized to magnetite.

DISCUSSION

From the experiments described above it is not possible to identify the carrier of the NRM and PTRM. One possibility is that it consists of single-domain grains of iron so small as to have very low blocking temperatures, i.e., not much greater than 160 Å (NÉEL, 1949), which were preferentially oxidized during the heating experiments. A difficulty with this hypothesis is that if an appreciable fraction of such small grains exists in the rocks, many of them should have sizes less than 160 Å and hence be superparamagnetic, but the absence of significant viscous magnetism and the reduced magnetization curves in Fig. 1 are evidence against this.

Another possibility is that over a long period of time after the rocks cooled, a structural state developed in the iron grains such that they were subdivided in part into extremely small effective grain sizes with correspondingly low blocking temperatures. A third and related possibility is exsolution from the original homogeneous iron alloy of very small grains of some other composition again having low blocking temperatures. In either case heating would result in homogenization of the iron grains and disappearance of NRM and PTRM. That exsolution of some kind did in fact occur is suggested by comparing the thermomagnetic curves in Figs. 2 and 3. In both specimens, the first reheating produced a small but unmistakable increase in Curie temperature, amounting to 8°C in specimen 12052,48 and 18°C in specimen 12065,37. The simplest explanation of these increases is resolution of exsolved cobalt in the iron. For homogenization of the iron grains to account for destruction of most of the NRM carrier at moderate temperatures would require appreciable resolution (and change in Curie temperature) to have occurred during the first heating of these specimens; the thermomagnetic curves in Fig. 2 do not preclude this.

Other more subtle hypotheses can be constructed to account for the changes in the magnetic properties of these specimens. The important point is that the heating experiments described in the previous section clearly show that the NRM in specimens 12052 and 12065 cannot be ordinary TRM that would have been acquired as the rocks initially cooled below their Curie temperatures. Furthermore, the NRM in specimen 12052,32 consists of at least two components and that in specimen 12065,67 apparently consists of three components. For rock 12065 the multicomponent nature and moderate degree of stability of the NRM has also been clearly shown by the alternating-field demagnetization results of Hargraves and Dorety (1971). The NRM in rocks 12052 and 12065 must have resulted from several magnetizing events that occurred at temperatures not much greater than those prevailing on the lunar surface and at some time later than the initial formation or last cooling of the rocks.

Acknowledgments—We thank Edward Mankinen for assistance with the magnetic measurements.

References

Arajs S. and Colvin R. V. (1964) Ferromagnetic-paramagnetic transition in iron. *J. Appl. Phys.* 35, 2424–2426.

Bozorth R. M. (1951) *Ferromagnetism.* Van Nostrand.

Cameron E. N. (1971) Opaque minerals in certain lunar rocks from Apollo 12. Second Lunar Science Conference (unpublished proceedings).

Doell R. R. and Cox A. (1967) Recording magnetic balance. In *Methods of Paleomagnetism* (eds. D. W. Collinson, K. M. Creer, and S. K. Runcorn) pp. 440–444. Elsevier.

Forrer R. (1930) Le problème des deux points de Curie. *J. Phys. Rad.* ser. 7, 1, 49–64.

Gibb F. G. F., Stumpfl E. F., and Zussman J. (1970) Opaque minerals in an Apollo 12 rock. *Earth Planet. Sci. Lett.* 9, 217–224.

Grommé C. S., Wright T. L., and Peck D. L. (1969) Magnetic properties and oxidation of iron-titanium oxide minerals in Alae and Makaopuhi Lava Lakes, Hawaii. *J. Geophys. Res.* 74, 5277–5293.

Hargraves R. B. and Dorety N (1971) Magnetic properties of some lunar crystalline rocks returned by Apollo 11 and Apollo 12. Second Lunar Science Conference (unpublished proceedings).

Helsley C. E. (1970) Magnetic properties of lunar 10022, 10069, 10084, and 10085 samples. *Proc. Apollo 11 Lunar Sci. Conf.*, Geochim. Cosmochim. Acta Suppl. 1, Vol. 3 pp. 2213–2219. Pergamon.

Nagata T., Ishikawa Y., Kinoshita H., Kono M., and Syono Y. (1970) Magnetic properties and natural remanent magnetism of lunar materials. *Proc. Apollo 11 Lunar Sci. Conf.*, Geochim. Cosmochim. Acta Suppl. 1, Vol. 2, pp. 2325–2340. Pergamon.

Nagata T., Fisher R. M., Schwerer F. C., Fuller M. D., and Dunn J. R. (1971) Magnetic properties and remanent magnetization of Apollo 12 lunar materials and Apollo 11 microbreccia. Second Lunar Science Conference (unpublished proceedings).

NÉEL L. (1949) Theorie du trainage magnétique des ferromagnetiques au grains fin avec applications aux terres cuites. *Ann. Geophys.* **5**, 99–136.

PEARCE G. W., STRANGWAY D. W., and LARSON E. E. (1971) Magnetism of two Apollo 12 igneous rocks. Second Lunar Science Conference (unpublished proceedings).

RUNCORN S. K., COLLINSON D. W., O'REILLY W., BATTY M. H., STEPHENSON A., JONES J. M., MANSON A. J., and READMAN P. W. (1970) Magnetic properties of Apollo 11 lunar samples. *Proc. Apollo 11 Lunar Sci. Conf., Geochim. Cosmochim. Acta* Suppl. 1, Vol. 3, pp. 2369–2387. Pergamon.

SENFTLE F. E., THORPE A. N., and LEWIS R. R. (1964) Magnetic properties of nickel-iron spherules in tektites from Isabela, Philippine Islands. *J. Geophys. Res.* **69**, 317–324.

THELLIER E. and THELLIER O. (1959) Sur l'intensité du champ magnétique terrestre dans le passé historique et géologique. *Ann. Geophys.* **15**, 285–376.

TSAY F-D., CHAN S. I., and MANATT S. L. (1971) Magnetic resonance studies of Apollo 11 and Apollo 12 samples. Second Lunar Science Conference (unpublished proceedings).

Proceedings of the Second Lunar Science Conference, Vol. 3, pp. 2501–2514
The M.I.T. Press, 1971.

Magnetic resonance properties of lunar samples: Mostly Apollo 12

J. L. KOLOPUS, D. KLINE,* A. CHATELAIN,† and R. A. WEEKS‡
Solid State Division, Oak Ridge National Laboratory
Oak Ridge, Tennessee 37830

(*Received* 22 *February* 1971; *accepted in revised form* 5 *April* 1971)

Abstract—The intense microwave absorption observed in samples of fines 12001,15; 12035,16; 12024,48; 12033,50; and 12070,125 has been verified as a ferromagnetic type of resonance. Little difference was found between our fines samples from Apollo 11 or Apollo 12. The spectra of the crystalline rocks 12021,55 and 12075,19 were attributed principally to several magnetic states of Fe^{3+}. Mn^{2+} was found predominantly in the plagioclase fraction of these rocks. Irradiation experiments showed that ionizing radiation produced valence changes of Fe^{3+} while 2 MeV electrons can also produce oxygen-vacancy type defects in the crystalline rocks. Annealing the fines produced changes in the intensity of the spectrum which were different depending on whether the sample was heated in air or in a vacuum. Nuclear magnetic resonance spectra of ^{27}Al, ^{23}Na, and ^{29}Si have been observed in several of the above samples as well as the Apollo 11 samples 10046,58; 10047,50; 10057,70; and 10062,21. Computer simulation of lineshapes has resulted in the determination of the asymmetry parameters and quadrupole coupling constants for ^{27}Al and ^{23}Na.

INTRODUCTION

ELECTRON PARAMAGNETIC RESONANCE (EPR) measurements have been made on samples of fines (12001,15,16; 12030,16; 12024,48; 12033,50; and 12070,125), and two crystalline rocks (12021,55 and 12075,19) at 9 and 35 GHz and at temperatures between 10 and 300 K. As was the case for our Apollo 11 samples, the spectra observed in the fines material was much more intense and apparently of a different origin than the spectra observed in whole crystalline rocks or mineral separates obtained from these rocks. We have made temperature, frequency, and thermal annealing dependence studies of the intense resonance in the Apollo 12 fines, in order to further verify its origin as a ferromagnetic resonance and to contrast it to the spectra observed in the crystalline rocks.

Nuclear magnetic resonance (NMR) measurements were made at room temperature and frequencies between 12 and 16 MHz on samples 12001,16, 12021,55; 10046,58; 10047,50; 10057,70; and 10062,21. Resonance spectra of ^{27}Al, ^{23}Na, and ^{29}Si have all been observed and with the help of computer simulated lineshapes, the asymmetry parameters and quadrupole coupling constants for ^{27}Al and ^{23}Na have been determined.

The experimental apparatus and procedures have been described elsewhere (WEEKS *et al.*, 1970a). The only modification for these investigations was the addition

---

* Present address: State University of New York at Albany, Albany, New York.

† Present address: Department of Physics, Federal Institute of Technology, Lausanne, Switzerland.

‡ Present address: The American University in Cairo, Cairo, Egypt, U.A.R., (on leave of absence from ORNL).

of a "cryotip" liquid helium cold finger apparatus to allow the 9 GHz EPR measurements to be made at temperatures as low as 10 K.

EXPERIMENTAL RESULTS

The resonance signal of fines

One specimen of fines (12001,16) was encapsulated in the F201 vacuum chamber at the Lunar Receiving Laboratory in a Weeks type container (WEEKS *et al.*, 1970a) with a vacuum pump attached. After removal it was maintained at a pressure of 10^{-8} torr and the spectrum was observed at 9 GHz at 300 K and 120 K and at 35 GHz at 300 K. After exposure first to dry nitrogen gas and finally to air, the spectrum was not observed to change—a result consistent with our previous measurements on similarly encapsulated Apollo 11 fines. The purpose of using vacuum encapsulated samples was to observe, if present, any paramagnetic surface states which might have been formed either by irradiation or by the fracturing of the material in the lunar vacuum (HANEMAN and MILLER, 1971). The negative results obtained thus far are probably due to the relatively poor vacuum conditions in the sample return container (ALSRC), or in the F201 vacuum chamber. It may not be possible, in fact, to return lunar material to earth at a pressure sufficiently low to preserve such paramagnetic centers with the techniques currently in use.

The intense resonance with a g value of about 2.1($g = (h\nu)/(\beta H)$ where h is Planck's constant, ν is the frequency of the spectrometer, β is the Bohr magneton, and H is the magnetic field) in the fines is characterized by an asymmetric lineshape and a linewidth which is both frequency and temperature dependent. At 9 GHz, the linewidth is about 800 gauss while at 35 GHz the line is about 950 gauss wide, with some variation between samples. In all cases, however, the linewidth is larger at 35 GHz than at 9 GHz for the same sample. The temperature dependence of the linewidth at 9 GHz was measured continuously between 10 K and room temperature. The results are shown in Fig. 1. As the temperature decreases, the linewidth, ΔH, increases while at the same time the peak-to-peak height of the derivative, A_{pp}, decreases. The intensity of a single symmetric resonance line should be proportional to $(\Delta H)^2 A_{pp}$ and while the observed spectrum may be composed of more than a single line, we assume that the one resonance dominates and apply the above criterion with the results also plotted in Fig. 1. Notice that within the experimental error of the measurement, (shown by bars on the curves) and taking into account the asymmetry of the line, the value of $(\Delta H)^2 A_{pp}$ could be considered almost constant over this temperature range. The continuity of the curves indicates that no phase changes or other anomalies are indicated by the resonance results over this temperature range.

Finally, up to the highest power of our klystron (~ 500 mw at 9 GHz) the intense resonance could not be saturated even at the lowest temperatures. This behavior along with the g value, the lineshape, and the data on ΔH as a function of temperature are consistent with the hypothesis that this resonance is predominantly a ferromagnetic (or possibly superparamagnetic) resonance absorption of particles whose diameter must be $< 10^{-5}$ centimeters (TUROV, 1964; NAGATA *et al.*, 1970; RUNCORN *et al.*, 1970). A similar conclusion was reached for the intense resonance observed in Apollo 11 fines (MANATT *et al.*, 1970).

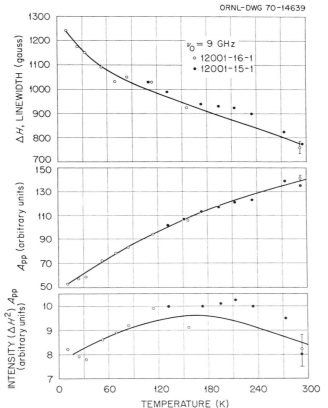

Fig. 1. Temperature dependence of the linewidth, ΔH, the peak to peak amplitude, A_{pp}, and the intensity $(\Delta H)^2 A_{pp}$ of the "characteristic" resonance of fines material. The weight of the samples was ~ 0.01 gm.

In most of our fines samples, (12001,16; 12030,16; and 12070,125) besides the strong characteristic resonance, another weaker resonance was observed at 35 GHz at lower magnetic fields with a g value of 4.2. In sample 12033,50,2, however, such a line was not resolved and the absorption at lower magnetic fields was much more intense than it was in the other three fines samples. At 9 GHz this latter sample showed an intense zero field absorption which was not typical of most of the samples of fines we have measured. Additional weak resonance peaks were observed in the fines at still higher g values. These lines are believed to arise from magnetic complexes which are different from that which is responsible for the ferromagnetic resonance.

Crystalline rock samples

Samples of the two crystalline rocks which were allocated for our measurements, 12021,55 and 12075,19, were prepared in such a way that their composition was

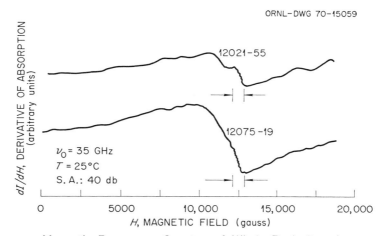

Fig. 2. The EPR spectra for two whole crystalline rocks are shown for $\nu_0 = 35$ GHz. The extent of the spectrum of Mn^{2+} is indicated by the arrows beneath the curves.

representative of the average rock composition. Their spectra, shown in Fig. 2, are composed principally of a broad absorption in the region of $g = 2.1$. It is very much weaker for a given sample size than the strong resonance in the fines. Superimposed on this absorption is the spectrum of Mn^{2+} which is identified on the curve by the brackets. Mineral separates of these rocks which were selected by hand under a $30\times$ microscope showed that the Mn^{2+} in these samples occurred predominantly in the plagioclase fraction. In fact, no Mn^{2+} at all was detected in the olivine fraction of rock 12075,19. This is unusual in that some Mn^{2+} is almost always detected in terrestrial and meteoritic olivines (Chatelain and Weeks, 1970; Chatelain *et al.*, 1970).

A weak absorption at a g-value near 4.2 is also observed in most of the lunar crystalline rocks for $\nu = 9$ GHz. Its intensity is significantly greater at 77 K than at room temperature and is also sensitive to ionizing radiation (see below). This line is due to isolated Fe^{3+} ions located in crystalline electric fields whose symmetry is lower than axial symmetry, (Dowsing and Gibson, 1969; Weeks *et al.*, 1970a) and is observed in both the plagioclase and pyroxene fractions of the rocks.

Annealing and irradiation experiments

Thermal annealing studies were made on two specimens of 12001,16. About 0.01 gm of particles whose diameter was less than 0.1 mm were loaded into quartz tubes. One was sealed off under 10^{-4} torr while the other was left open. The volume of the sealed tube (approximately 0.1 cm³) was such that in a vacuum of 10^{-4} torr, the oxidation of any iron present in the fines as FeO, from the remaining oxygen in the tube, would be negligible. Outgassing from the tube could produce some oxidation but again, because of the surface areas involved (approximately 100 cm² for the fines) not more than 10–15 % of the surface of the sample would have been contaminated. This would be a negligible fraction of the bulk volume of the sample. Isochronal

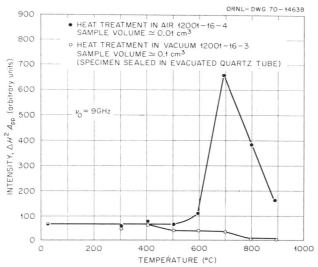

Intensity of the "Characteristic" Resonance Signal of 12001 – 16 vs Isochronal
Anneals of 30 – min Duration.

Fig. 3. The isochronal annealing behavior of Apollo 12 fines is shown for a sample
heated in air compared to a sample heated in a vacuum. The samples were measured at
$\pm 150°C$ after each anneal.

anneals of thirty minutes duration were made and the spectrum was recorded at
120 K after each anneal. The effect on the intensity of the ferromagnetic resonance is
shown in Fig. 3. The behavior is similar to that which we observed for our Apollo 11
fines (WEEKS *et al.*, 1970a).

The spectrum of isolated Fe^{3+} ions was not detected in either specimen before or
after the annealing procedure. The intensity of the resonance line of the vacuum
sample decreased by 99% after annealing to 800°C and a large zero field absorption
(at 9 GHz) appeared. This decrease may be due to a resolution of superparamagnetic
particles if they are, in fact, responsible for the initial absorption. The sample heated
in air increased its intensity by a factor of about 8 after the 700°C anneal with an
increase in width by about 50%. With increasing temperatures, the intensity dropped,
and after the 900°C anneal it was nearly the same as the initial intensity. After the
anneals the sample open to the air was reddish-brown in color, whereas, no color
change was observed for the vacuum sample. The increase in the intensity of the air
sample could be due to the oxidation of Fe^{2+} and metallic iron to Fe^{3+}. One difficulty
with this interpretation however, is that the final resonance which is experimentally
observed is not like that seen in Fe_2O_3 (WEEKS *et al.*, 1970a).

The effects of ionizing radiation on samples of rock 10047,49 and 12021,55 were
investigated at room temperature and liquid nitrogen temperature. A sample of the
plagioclase fraction of 10047,49 was irradiated with ^{137}Cs gamma rays at 77 K and
measured at 9 GHz without warming the sample above 150 K. The largest change
observed was the introduction of a very broad resonance ($\Delta H \sim 2000$ gauss) centered

approximately at $g = 2.9$. This absorption slowly annealed out at room temperature and could no longer be detected after three hours at 300 K. A group of absorption peaks in the region of $1.92 < g < 2.04$ were also generated. These had relatively narrow line widths ($\Delta H \sim 50$ gauss) and also annealed out when the sample was warmed to room temperature. One of the peaks with $g = 2.001 \pm 0.001$ annealed much more slowly than the other. However, because of its linewidth ($\Delta H = 12$ gauss) and the fact that the line could not be saturated with the power of our spectrometer, this line is not believed to be an oxygen vacancy center (E' center) (WEEKS and NELSON, 1960; CHATELAIN *et al.*, 1970) which in most silicate minerals has a g value very close to $g = 2.001$.

With the exception of the Fe^{3+} resonance at $g = 4.28$, the room temperature ^{60}Co gamma irradiations of the sample from 12021,55 had little effect. This Fe^{3+} resonance, however, behaved differently under gamma irradiation at room temperature than at 77 K. At low temperature the signal grew with increasing dose and then annealed out upon warming to room temperature. The room temperature irradiation, however, decreased the intensity of this line almost to zero after a dose of 5×10^7 R. These effects can be explained by the reactions

$$\text{at 77 K} \qquad Fe^{2+} + \gamma \rightarrow Fe^{3+} + e^-$$
$$\text{but at 300 K} \qquad Fe^{3+} + \gamma \rightarrow Fe^{2+} - e^-.$$

Those sites which act as the donors or acceptors of electrons are probably other impurities in the minerals which also change valence, but which are not paramagnetic and hence cannot be observed during these experiments. Finally, the plagioclase fraction of the rocks is observed to darken visibly with irradiation at 77 K whereas no color changes are observed after the 300 K irradiation.

Specimens from sample 12021,55 were also irradiated at room temperature with 2 MeV electrons from a Van de Graaff accelerator. The total dose was 2×10^{16} electrons incident on each sample. After the irradiation the oxygen vacancy E' centers were observed, presumably as a result of knock-on collisions. In the fraction of the sample consisting mostly of pigeonite the final concentration of E' centers was 7×10^{14} centers/cm^3, while the concentration in the augite fraction was 3.5×10^{15} centers/cm^3. Since both minerals are silicates, this factor of five difference in the production of E' centers is unexpected.

DISCUSSION

The first conclusion to which one is led by the EPR results is that the spectra in the fines arises from a different source than those observed in the crystalline rocks. Furthermore, none of the treatments we have carried out on crystalline samples or fines has succeeded in changing the spectra of one type of sample into that observed for the other. In this respect then, the major portion of the fines cannot be considered as pulverized crystalline rocks. All the EPR data so far (WEEKS *et al.*, 1970a; MANATT *et al.*, 1970; GEAKE *et al.*, 1970; TSAY *et al.*, 1971; HANEMAN and MILLER, 1971; DUCHESNE *et al.*, 1971) indicate that the spectra of the fines is a ferromagnetic (or superparamagnetic) resonance due to very small metallic particles. The as-returned crystalline rocks show spectra which are probably due to paramagnetic impurities and among which have been identified Fe^{3+}, Mn^{2+}, and possibly Ti^{3+} (WEEKS *et al.*, 1970b).

The annealing experiments show that when the fines are heated in air up to 700°C the resonance absorption increases. The difficulty in attributing this increase to an increase in the Fe^{3+} content of the fines is that the final resonance would be expected to look like that of Fe_2O_3 which has its principle absorption centered at $g = 4.31$ (WEEKS *et al.*, 1970a) whereas the resonance in the fines does not significantly change its lineshape or position. The decreased intensity of the vacuum annealed sample is most easily explained if the small metallic particles responsible for the resonance agglomerate to form larger particles. This would have the net effect of reducing the effective sample size contributing to the absorption with a corresponding intensity decrease.

The irradiation experiments showed that ionizing radiation was effective only in changing the valence states of various impurities. The 2 MeV electron irradiation, on the other hand, was able to produce damage in the silicates in the form of oxygen vacancies. The relative concentration differences in the pigeonite and augite samples are a bit difficult to explain. It was suggested (HAFNER, 1971) that this might have been due to a difference in oxidation conditions when the respective minerals crystallized. This may be possible, but one argument against it is that apparently knock-on collision processes were necessary to produce the oxygen vacancies. If these had been present in the as-returned minerals then one would expect to see the E' center resonance after an ionizing radiation exposure and this was not the case. A determination of the energy threshold and irradiation intensity dependence of the E' center production on future samples may help clarify these results.

NMR: EXPERIMENTAL RESULTS AND DISCUSSION

NMR spectra of ^{27}Al ($I = \frac{5}{2}$) and ^{23}Na ($I = \frac{3}{2}$) from samples 12001,16; 12021,55,2; 10046,58; 10047,50; and 10062,21 were recorded in the dispersion mode over a range of frequencies from 16 MHz to 12 MHz. These spectra, shown in Figs. 4 and 5, are proportional to the absorption curve itself rather than the dispersion first derivative (O'REILLY, 1958). Both the ^{27}Al and ^{23}Na spectra arise from second-order quadrupole broadened "central" transitions with slightly differing distributions of quadrupole interactions, indicative of atomic disorder and the presence of mineral components which contain strains. This disorder and strain, which may be associated with the rate of quenching and the crystallization history of the sample, appears to be qualitatively very similar in the Apollo 11 and Apollo 12 material (WEEKS *et al.*, 1970a).

The sharp narrow line appearing in Figs. 4 and 5 is an absorption derivative of ^{27}Al and ^{23}Na from aqueous solutions of $AlCl_3$ and $NaCl$, respectively, and is used as a marker to locate the position of the unperturbed resonant field, H_0. It should be made clear that the positions of liquid line markers used to locate H_0 were recorded following the accumulation of the lunar spectra. Because of the very weak ^{23}Na NMR signals and a drifting base line, thought to be due in part to a contribution from the ^{27}Al NMR probe line, a precise estimate of the range of ^{23}Na quadrupole parameters is not possible. However, on the basis of computer simulation studies (WEEKS *et al.*, 1970a) and also, from a comparison with the ^{23}Na dispersion mode spectrum from a

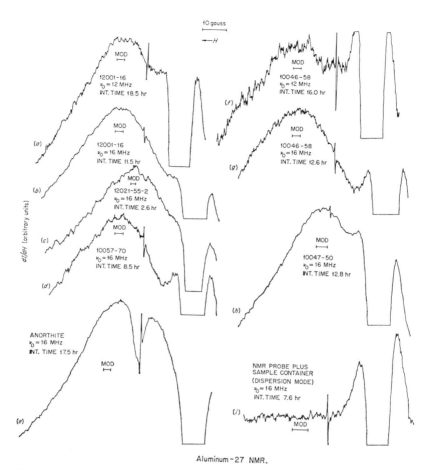

Aluminum-27 NMR.

Fig. 4. Comparison of NMR dispersion mode spectra of ^{27}Al from (*a*) 12001,16 at 12 MHz, (*b*) 12001,16 at 16 MHz, (*c*) 12021,55,2 at 16 MHz, (*d*) 10057,70 at 16 MHz, (*e*) powdered anorthite at 16 MHz, (*f*) 10046,58 at 12 MHz, (*g*) 10046,58 at 16 MHz, (*h*) 10047,50 at 16 MHz, and (*i*) NMR probe plus lunar sample container at 16 MHz.

powdered sample of albite (U.S.N.M. C-5390), Fig. 5*i*, $\eta = 0.25$, $e^2qQ/h = 2.62$ MHz, (HAFNER and HARTMANN, 1964) it is roughly estimated that the quadrupole coupling constants lie mainly in the range 2–4 MHz and the asymmetry parameters lie largely in the range 0.2 to 0.8 for the spectra of Fig. 5. Notice that the ^{23}Na spectrum from BCR-1 basalt, Fig. 5*h*, is quite similar qualitatively to those of the lunar samples. No background ^{23}Na spectrum was detected at 16 MHz from the NMR probe plus lunar sample container for an integration time of 16.5 hours, Fig. 5*e*. All the ^{23}Na spectra of Fig. 5 were recorded using identical spectrometer parameters. Differences in the noise levels of Fig. 5 are due only to differences in reading out the accumulated spectra from the CAT to the recorder.

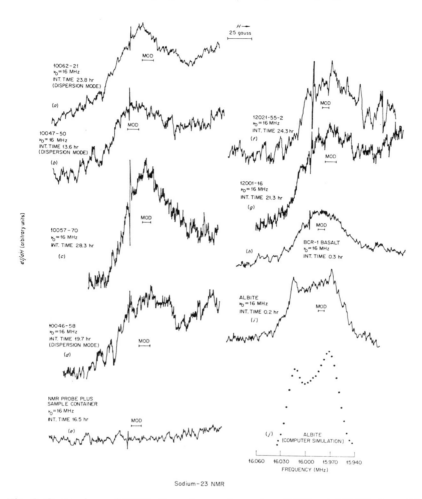

Sodium–23 NMR

Fig. 5. Comparison of NMR dispersion mode spectra of ^{23}Na at 16 MHz from (a) 10062,21, (b) 10047,50, (c) 10057,70, (d) 10046,58, (e) NMR probe plus lunar sample container, (f) 12021,55,2, (g) 12001,16, (h) BCR,1 basalt, (i) powdered albite. Shown in (j) is a computer simulated powder pattern based on the ^{23}Na quadrupole parameters of albite after convolution with a Gaussian curve of width $2\sigma = 6.4$ KHz for the case of the "central" transition with $I = \frac{3}{2}$.

Despite interference from the extraneous Knight-shifted aluminum response arising from the NMR probe (WEEKS *et al.*, 1970a), the shapes of the ^{27}Al high-field peaks, Fig. 4, and their splittings from H_0, the unperturbed resonance field, are easily resolved and indicate that the aluminum sites in each of the lunar samples can be characterized largely by non-zero asymmetry parameters and by quadrupole coupling constants of several MHz. The splitting of the high-field peak from H_0 varies from less than 5 gauss for samples 12021,55,2 and 10047,50 to more than 15 gauss in sample

10046,58, indicating a substantial variation from sample to sample. While the high-field portion of the ^{27}Al spectra in each of the lunar samples shows a slight broadening and noticeable reduction in intensity at 12 MHz relative to that at 16 MHz, indicating the presence of second-order quadrupole effects, this is not accompanied by a corresponding increase in the splitting of the high-field peak from H_0 as the magnetic field is decreased (compare Figs. 4a and 4b, and Figs. 4f and 4g). This behavior, and a corresponding behavior for the ^{23}Na spectra, may arise from small internal magnetic fields (5–10 gauss) at the aluminum and sodium sites which oppose the laboratory field.

While being very weak in intensity and relatively broad with signal-to-noise no greater than 2:1, the ^{29}Si ($I = \frac{1}{2}$) NMR dispersion mode spectrum from 12021,55,2,

ORNL-DWG 70-14654

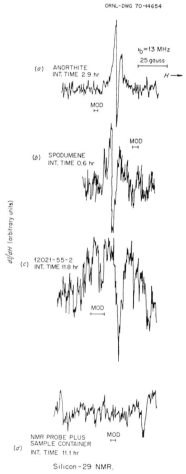

Silicon-29 NMR.

Fig. 6. Comparison of nuclear magnetic resonance dispersion mode spectra of ^{29}Si at 13 MHz from (a) powdered anorthite, (b) powdered spodumene, (c) 12021,55,2, and (d) NMR probe plus lunar sample container.

Fig. 6c, also appears to be affected by an internal field of a similar magnitude and sign as seen by a comparison with the positions of the ^{29}Si spectra from powdered samples of anorthite (U.S.N.M. 80524), Fig. 6a, and spodumene (U.S.N.M. 119434), Fig. 6b. It should be noted that in Fig. 6c, the modulation amplitude is larger than the apparent width of the ^{29}Si resonance, indicating that the sharp "spike" is caused in large part by the "noise." The actual width of the resonance is comparable to or greater than the modulation amplitude. No ^{29}Si background spectrum was detected at 13 MHz for an integration time of 11.1 hours. The absence of a ^{29}Si background signal is assumed to be related to saturation of the ^{29}Si nuclei in the probe and sample container.

Although it is not possible to discuss the ^{27}Al and ^{23}Na shifts in more than qualitative terms, due to the difficulty of separating electric quadrupole effects and internal magnetic field effects for the case of weak signals, it has been observed that, in general, a sample with a relatively large apparent shift for ^{27}Al, such as 10046,58, also has a relatively large apparent shift for ^{23}Na while a sample with a relatively small apparent shift for ^{27}Al, such as 10047,50, also has a relatively small apparent shift for ^{23}Na. This suggests that the internal magnetic fields, whose origins remain unclear, may extend over a large fraction of the sample rather than being restricted to some region immediate to a particular nucleus.

Finally, a comparison of the ^{27}Al spectra from lunar samples with that of a powdered sample of terrestrial anorthite (U.S.N.M. 50524), Fig. 4e, showed strong qualitative similarities. Computer simulations (WEEKS et al., 1970a) confirmed that with internal magnetic fields of slightly differing magnitudes the upper and lower bounds of the distributions of quadrupole coupling constants and asymmetry parameters in the lunar samples correspond, in large part, to those of the eight sites of anorthite (BRICKMANN and STAEHLI, 1968). It is appropriate that anorthite be included in any attempt to fit the ^{27}Al lunar spectra in view of the fact that the plagioclase fractions have been reported to be generally rich in anorthite (see, for example, AGRELL et al., 1970). Computer simulations made with other, somewhat arbitrary distributions of quadrupole interactions, were unable to produce a better fit to the ^{27}Al lunar spectra than that of anorthite. The strong qualitative similarities between the ^{27}Al spectra of anorthite and those of the lunar samples are noted, not for the purpose of excluding other equally reasonable distributions, but simply for the purpose of illustrating how rough quantitative estimates may be determined for the upper and lower bounds of the observed distributions within a reasonably realistic framework. As an illustration of attempts to fit ^{27}Al lunar spectra the computer simulated powder patterns of the "central" transition shown in Fig. 7 were obtained by combining the quadrupole coupling constants ($e^2qQ/h = 8.42$, 7.25, 6.81, 6.30, 5.54, 4.90, 4.30, and 2.66 MHz) and asymmetry parameters ($\eta = 0.66$, 0.76, 0.65, 0.88, 0.42, 0.42, 0.53, 0.66) of anorthite with isotropic magnetic field shifts of 0%, Figs. 7a and 7d, 0.05%, Figs. 7b and 7e, 0.1%, Figs. 7c and 7f. In the computed spectra shown in Fig. 7, the broadening due to dipole-dipole interactions is taken into account by a numerical convolution of the powder pattern with a Gaussian curve of appropriate width 2σ. Based on a comparison of the relative splitting of the low-frequency (high-field) peak from H_0 in the spectra of Fig. 4 and Fig. 7 as a function of frequency, a shift of approximately 0.05% appears to give a relatively better fit

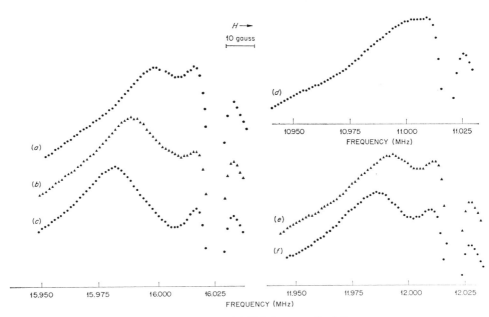

Aluminum—27 NMR Computer Simulations.

Fig. 7. Computer simulated powder patterns of the "central" transition for the case $I = \frac{5}{2}$ obtained by combining the ^{27}Al quadrupole interaction parameters of anorthite (see text) with isotropic magnetic field shifts of 0%, (a) and (d), 0.05%, (b) and (e), and 0.1%, (c) and (f) after convolution with a Gaussian curve of width $2\sigma = 8.4$ KHz. The computer simulations of Fig. 4 also include a representation of the 0.16% Knight-shifted ^{27}Al NMR probe line obtained by convolution with a Lorentzian curve of width $2\sigma = 18$ KHz and arbitrary amplitude.

to the experimental spectra than either of the other two values chosen. Although the precise amount of shift may vary from sample to sample, the presence of quadrupole effects accompanied by small internal magnetic fields is seen to provide one reasonable and consistent picture for understanding the ^{27}Al lunar spectra.

One feature of the ^{27}Al dispersion mode spectrum from anorthite, a sharp dip at H_0, is not seen in any of the lunar spectra, possibly reflecting shorter relaxation times in the lunar samples than in the anorthite sample (O'Reilly, 1958). This causes a discrepancy between the computed ^{27}Al powder pattern, Fig. 7b, and the experimental dispersion mode spectrum, Fig. 4e, for anorthite which indicates the need for modified line shapes for proper simulation of the experimental spectrum (Solomon and Ezratty, 1962; Pifer, 1968).

None of the lunar samples gave reproducible ^{1}H spectra at 16 MHz in either the absorption or dispersion modes for integration times greater than 24 hours, indicating that the proton concentration in these samples is less than 10^{21}/gm for resonance lines as wide as 40 gauss. The estimate given above is based on a study of ^{1}H NMR from several simulated lunar rocks of known proton content (Fogelson, 1968). It should be noted that internal magnetism present in all of the lunar samples to varying degrees appears to inhibit and distort the recording of meaningful NMR spectra much more seriously at low fields, e.g., for ^{1}H at $H_0 < 3800$ gauss, than at higher fields, e.g., for ^{23}Na and ^{27}Al for $H_0 > 10,000$ gauss.

Acknowledgments—We thank H. S. STORY (State University of New York at Albany) for his aid with the computer simulations and for his comments on the NMR results. M. M. ABRAHAM and J. T. LEWIS (Oak Ridge National Laboratory) helped with some of the measurements. We also thank A. BISHAY and the American University in Cairo for support of part of this research and S. S. HAFNER for his ideas on the E' center production. This research was performed under NASA Reg. no. 10-9729.

REFERENCES

AGRELL S. O., SCOON J. H., MUIR I. D., LONG V. P., MCCONNELL J. D. C., and PECKETT A. (1970) Observations on the chemistry, mineralogy and petrology of some Apollo 11 lunar samples. *Proc. Apollo 11 Lunar Sci. Conf., Geochim. Cosmochim. Acta* Suppl. 1, Vol. 1, 93–128. Pergamon.

BRINKMANN D. and STAEHLI J. L. (1968) Magnetische Kernresonanz von ^{27}Al im Anorthit, $CaAl_2Si_2O_8$. *Helv. Phys. Acta* **41**, 274–281.

CHATELAIN A., KLINE D., KOLOPUS J. L., and WEEKS R. A. (1970) Electron and nuclear magnetic resonance of three chondritic meteorites. *J. Geophys. Res.* **75**, 5681–5692.

CHATELAIN A. and WEEKS R. A. (1970) Electron paramagnetic resonance study of ordered Mn^{2+} in Mg_2SiO_4. *J. Chem. Phys.* **52**, 5682–5687.

DOWSING R. D. and GIBSON J. F. (1969) Electron spin resonance of high spin d^5 systems. *J. Chem. Phys.* **50**, 294–303.

DUCHESNE J., DEPIRCUX J., GERARD A., GRANDJEAN F., and READ M. (1971) A study by electronic paramagnetic resonance and Mössbauer spectroscopy of some lunar samples collected by Apollo 12. Second Lunar Science Conference (unpublished proceedings).

FRONDEL C., KLEIN C., ITO J., and DRAKE J. C. (1970) Mineralogical and chemical studies of Apollo 11 lunar fines and selected rocks. *Proc. Apollo 11 Lunar Sci. Conf., Geochim. Cosmochim. Acta* Suppl. 1, Vol. 1, pp. 445–474. Pergamon.

FOGELSON D. E. (1968) Simulated lunar rocks, U.S. Department of Interior, Twin Cities Mining Research Center.

GEAKE J. E., DOLLFUS A., GARLICK G. F. J., LAMB W., WALKER G., STEIGMANN G. A., and TITULAER C. (1970) Luminescence, electron paramagnetic resonance and optical properties of lunar material from Apollo 11. *Proc. Apollo 11 Lunar Sci. Conf., Geochim. Cosmochim. Acta* Suppl. 1, Vol. 3, pp. 2127–2147. Pergamon.

HAFNER S. and HARTMANN P. (1964) Electrische Feldgradienten und Sauerstoff-polarisierbarkeit in Alkali-Feldspaten ($NaAlSi_3O_8$ und $KAlSi_3O_8$). *Helv. Phys. Acta* **37**, 348–360.

HAFNER, S. S. (1971) Private communication.

HANEMAN D. and MILLER D. J. (1971) Surface properties of lunar material by electron paramagnetic resonance. Second Lunar Science Conference (unpublished proceedings).

MANATT S. L., ELLEMAN D. D., VAUGHAN R. W., CHAU S. I., TSAY F. D., and HUNTRESS W. T. (1970) Magnetic resonance studies of lunar samples. *Proc. Apollo 11 Lunar Sci. Conf., Geochim. Cosmochim. Acta* Suppl. 1, Vol. 3, pp. 2321–2323. Pergamon.

NAGATA T., ISHIKAWA Y., KINOSHITA H., KONO M., and SYONO, Y. (1970) Magnetic properties and natural remanent magnetization of lunar materials. *Proc. Apollo 11 Lunar Sci. Conf., Geochim. Cosmochim. Acta* Suppl. 1, Vol. 3, pp. 2325–2340. Pergamon.

O'REILLY D. W. (1958) Quadrupolar broadened nuclear magnetic resonance of polycrystalline solids. *J. Chem. Phys.* **28**, 1262–1264.

PIFER J. (1968) Modulation and cross-relaxation effects on the line shape of strongly saturated nuclear magnetic resonance in solids. *Phys. Rev.* **166**, 540–553.

RUNCORN S. K., COLLINSON D. W., O'REILLY W., BATTEY M. H., STEPHENSON A., JONES J. M., MANSON A. J., and READMAN P. W. (1970) Magnetic properties of Apollo 11 lunar samples. *Proc. Apollo 11 Lunar Sci. Conf., Geochim. Cosmochim. Acta* Suppl. 1, Vol. 3, pp. 2369–2387. Pergamon.

SOLOMON I. and EZRATTY J. (1962) Magnetic resonance with strong radio-frequency fields in solids. *Phys. Rev.* **127**, 78–87.

TSAY F. D., CHAU S. I., and MANATT S. L. (1971) Magnetic resonance studies of Apollo 11 and Apollo 12 samples. Second Lunar Science Conference (unpublished proceedings).

TUROV E. A. (1964) Specific features of ferromagnetic resonance in metals. In *Ferromagnetic Resonance* (editor S. V. Vonsovskii), Chap. 5, pp. 107–135. Translated from Russian by the Israel Program for Scientific Translations.

WEEKS R. A., CHATELAIN A., KOLOPUS J. L., KLINE D., and CASTLE J. G. (1970b) Magnetic properties of some lunar material. *Science* **167**, 704–707.

WEEKS R. A., KOLOPUS J. L., KLINE D., and CHATELAIN A. (1970a) Apollo 11 lunar material: Nuclear magnetic resonance of ^{27}Al and electron resonance of Fe and Mn. *Proc. Apollo 11 Lunar Sci. Conf., Geochim. Cosmochim. Acta* Suppl. 1, Vol. 3, pp. 2467–2490. Pergamon.

WEEKS R. A. and NELSON C. M. (1960) Trapped electrons in irradiated quartz and silica: II. *J. Amer. Ceram. Soc.* **43**, 399–404.

Proceedings of the Second Lunar Science Conference, Vol. 3, pp. 2515–2528
The M.I.T. Press, 1971.

Magnetic resonance studies of Apollo 11 and Apollo 12 samples*

Fun-Dow Tsay and Sunney I. Chan

Arthur Amos Noyes Laboratory of Chemical Physics, California Institute of
Technology, Pasadena, California 91109

and

Stanley L. Manatt

Space Sciences Division, Jet Propulsion Laboratory, California Institute of
Technology, Pasadena, California 91103

(Received 22 February 1971; accepted in revised form 19 April 1971)

Abstract—Electron spin resonance (ESR) studies at both X-band (9.5 GHz) and K-band (34.8 GHz) frequencies have been carried out on a selection of Apollo 11 and Apollo 12 lunar samples (10087,10,11; 10086; 10062,26,27; 10046,29,30; 10017,35,36; 12070,11; 12065,103; 12041,15; 12032,41; 12021,42; 12003,24). On the basis that no significant temperature dependence of the absorption intensity is noted together with the result that the signal intensity is at least three orders of magnitude greater than that expected for possible paramagnetism, we have assigned the broad, asymmetric resonance signal centered at $g = 2.09 \pm 0.03$ to ferromagnetic centers. A model study to simulate the polycrystalline spectra using computer simulation techniques has been carried out, from which it was possible to ascertain with some degree of certainty the size and shape of the ferromagnetic centers, and to ascertain that the ferromagnetism is due to metallic Fe particles. Metallic Fe contents have been measured for all the above lunar samples, and these range in concentration from 0.001 to 0.50 wt.%. These results suggest an apparent correlation between the metallic Fe contents and the geological and surface exposure ages of the samples.

Weak paramagnetic resonances attributable to octahedrally coordinated Mn^{2+} ions have also been observed in some samples (10086; 10062,26,27; 10017,35,36; 12065,103; 12021,42) in which the metallic Fe content is either low or partially removed. From the g value (2.002 ± 0.002) and the nuclear hyperfine coupling constant (-95.0 ± 2.0 gauss), it appears that these resonances originate from $Mn(H_2O)_6^{2+}$. No ESR signals ascribable to free electrons and/or holes, which would be indicative of radiation damage, or Ti^{3+} have been detected in these samples

Introduction

In our previous report (Manatt *et al.*, 1970) on the electron spin resonance (ESR) studies of the Apollo 11 lunar samples (10062,27; 10087,10), we suggested that the broad, asymmetric signals observed for these samples originate from ferromagnetic centers consisting primarily of metallic Fe particles. Our initial interpretations were based on the evidence that no temperature dependences of the integrated absorption intensities were observed and that the observed signal intensities were at least three orders of magnitude greater than those expected for possible paramagnetic species.

In this paper we summarize the ESR results which we have obtained for a larger selection of Apollo 11 and Apollo 12 lunar samples. In particular, we present metallic Fe contents which we have been able to extract from a detailed lineshape analysis of

* Contribution No. 4211, Division of Chemistry and Chemical Engineering, California Institute of Technology, Pasadena, California 91109.

the ferromagnetic spectra observed. In addition, we present here ESR evidence for the presence of octahedrally coordinated paramagnetic Mn^{2+} ions, possibly $Mn(H_2O)_6^{2+}$, in samples containing low metallic Fe contents.

EXPERIMENTAL TECHNIQUES

Electron spin resonance spectra were obtained at both X-band (9.5 GHz) and K-band (34.8 GHz) frequencies on a Varian 4500 spectrometer using 100 KHz field modulation. Measurements were made on bulk samples (6–60 mg) sealed in cylindrical quartz tubes. Samples were not handled in a dry box but every effort was made to ensure minimal atmospheric exposure during transfer or grinding. The coarser-grained rock and chip samples (12065,103, 12021,42) were ground to fine powder prior to being sealed in quartz tubes. One fine sample (10086) was extensively treated with organic solvents and acid (methanol, benzonitrile, naphthalene, and methane sulfonic acid) to remove part of the metallic Fe content (RHO et al., 1970), and was then dried under a nitrogen atmosphere. A detailed description of g-value measurements and methods used to obtain absorption intensities from the observed signals as well as a detailed account of the model and the method which we have developed to simultate the ferromagnetic resonance spectrum of a polycrystalline sample has been given elsewhere (TSAY et al., 1971).

RESULTS AND DISCUSSION

Ferromagnetic resonance—metallic Fe

Typical X-band and K-band spectra of the lunar samples recorded at room temperature are presented in Fig. 1. The ESR parameters obtained from computer simulation of the observed ferromagnetic signals are summarized in Table 1. For comparison, we have also listed in Table 2 literature values of these parameters for

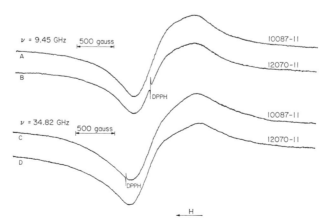

Fig. 1. Typical ferromagnetic resonance spectra of metallic Fe particles observed for the returned lunar samples. Spectra are recorded at 298°K and at X-band and K-band frequencies.

Table 1. Ferromagnetic resonance data for the lunar fines (10087,11).

	X-band frequency (9.501 GHz)		K-band frequency (34.825 GHz)	
	298°K	80°K	298°K	90°K
g value	2.08 ± 0.03	2.09 ± 0.03	2.08 ± 0.03	2.09 ± 0.03
Crystalline energy $(2K_1/M_z)$	$+500 \pm 50$ G	$+640 \pm 50$ G	$+500 \pm 50$ G	$+610 \pm 50$ G

Table 2. Ferromagnetic resonance parameters for metallic Fe, Co, and Ni (KITTEL, 1967).

	Fe	Co	Ni
g value	2.10	2.18	2.21
$2K_1/M_z$	$+500$ G	7000 G	-220 G
Structure	bcc	Hex.	fcc

pure metallic Fe, Co, and Ni in their ferromagnetic state. In our previous work (MANATT et al., 1970), we reported that no significant temperature dependence of the absorption intensity was observed for two Apollo 11 samples (10086,10, 10062,27). Careful measurements of the ESR absorption intensities for another fine sample (10087,11) over the temperature range of 80°–298°K again demonstrated that the absorption intensity remains constant within the experimental error of 5%. The lack of any noticeable temperature variation in the ESR intensities, the experimental g value of 2.09 and the first-order crystalline anisotropy energy $(2K_1/M_z)$ of the order of $+500$ gauss (at 298°K) and $+640$ gauss (at 80°K), strongly suggest that the ferromagnetic centers in the lunar sample consist mainly of metallic Fe. An effective g value of 2.09, which is close to the free-electron value, suggests that these ferromagnetic centers are essentially spherical in shape (KITTEL, 1948). Since no eddy current effects were detected in our ESR experiments at both X-band and K-band, we have set an upper limit of 1 μm average diameter for the size of the metallic Fe particles.

In a ferromagnetic system, the electron spins within a ferromagnetic domain are strongly exchange-coupled to each other. This results in a total magnetic moment which tends to align itself along the direction of an external polarizing magnetic field. The ferromagnetic resonance absorption intensity is thus expected to be proportional to the total number of electron spins only, and to be independent of the temperature at temperatures sufficiently below the Curie point and at sufficiently high magnetic field strengths. On the contrary, the absorption intensity for paramagnetic resonance, which is known to be proportional to $Nh\nu/2kT$ where N is the total number of electron spins, is expected to be both temperature and frequency dependent, increasing with decreasing temperature and increasing frequency. Since $h\nu/2kT \simeq 0.75 \times 10^{-3}$ for an ESR experiment at room temperature and X-band frequency, the ferromagnetic resonance absorption intensity of a ferromagnetic material is roughly three orders of magnitude greater than the ESR absorption intensity of a paramagnetic system for the same number of electron spins. Thus if the ESR signals observed in the present work were interpreted as arising from paramagnetic Fe^{3+} as has been suggested (WEEKS et al., 1970a, 1970b; GEAKE et al., 1970a, 1970b; GARLICK et al., 1971), our intensity measurement would indicate that the lunar fines (10087,11, 12003,24) contain at least 270 wt. % Fe^{3+}, which is impossible. On the other hand, if these same results were interpreted in terms of ferromagnetic centers of metallic Fe, our calculations would show that there is about 0.50 wt. % metallic Fe in the Apollo 11 fines (10087,11) and 0.41–0.46 wt. % in the Apollo 12 fines (12070, 12041, 12003), which is about the same order of magnitude as that determined by several other methods (Nagata et al., 1970, 1971; Strangway et al., 1970; Pearce et al., 1971; Runcorn et al., 1970; Herzenberg and Riley, 1970, 1971).

The metallic Fe contents determined in this work for a collection of lunar samples are summarized in Table 3. Our results show that metallic Fe content is much higher in the lunar fines and breccia than in the rock chips. Also some of our data suggest that the metallic Fe contents are higher in surface samples than in interior rock samples. This latter variation in metallic Fe content is thought to be due to inhomogeneous mixing of meteoritic Fe into these samples and/or thermal conversion of Fe^{2+} in these samples by impact heating or other processes (Keil et al., 1970; Simpson and Bowies, 1970). Since reduction of Fe^{2+} into metallic Fe requires a source of free electrons, it is possible that some of the electrons generated by ionizing radiation from solar wind, indigenous radioactivity and cosmic rays are trapped at Fe^{2+} sites to yield metallic Fe.

In the case of the lunar fines, where the metallic Fe content is high, the values we have determined are in general agreement with those reported by other investigators using other methods. For rock samples, however, our metallic Fe contents are lower than those reported by other investigators using either Mössbauer or magnetic susceptibility measurements. For comparison we have listed in Table 4 a few metallic Fe contents previously reported for some of the rock samples using these other methods. It appears that a significant variation with the analytical method exists for the metallic Fe contents determined by other workers. For example, the metallic Fe contents determined by the Mössbauer technique are all found to be of the order of 0.2 wt. % while metallic Fe contents derived from magnetic susceptibility measurements are of the order of 0.05 wt. % for similar rock types. Although we detect variations in metallic Fe content in different rock samples in our ESR measurements, it

Table 3. Metallic Fe contents for some Apollo 11 and Apollo 12 samples.

Sample	Metallic Fe (wt. %)	Sample type and other ESR results
10087,11	0.50 ± 0.05	Fines
10086	0.05 ± 0.01	Fines; after extensive treatment with organic solvents and acid, Mn^{2+} 70 ppm, H_2O 140 ppm.
10062,26	0.007–0.015*	Chips (external), Mn^{2+} 40 ppm, H_2O 80 ppm.
10062,27	∼0.003	Chips (internal), Mn^{2+} 20 ppm, H_2O 40 ppm.
10046,29	0.36 ± 0.05	Fines (external)
10046,30	0.31 ± 0.05	Fines (internal)
10017,35	0.001–0.013*	Rock (external), Mn^{2+} 20 ppm, H_2O 40 ppm.
10017,36	0.001–0.011*	Rock (internal), Mn^{2+} 10 ppm, H_2O 20 ppm.
12070,11	0.41 ± 0.05	Fines
12065,103	∼0.002	Rock; Mn^{2+} 30 ppm, H_2O 60 ppm.
12041,15	0.46 ± 0.05	Fines
12032,41	0.085 ± 0.015	Breccia
12021,42	∼0.001	Chips; Mn^{2+} 40 ppm, H_2O 80 ppm.
12003,24	0.45 ± 0.05	Fines

* Observed variations among the fine and coarse grained rock chips.

Table 4. Metallic Fe contents and ages used in plots.

Sample	Wt. % Metallic Fe	Method*	$Pb^{207}–Pb^{206}$ Age	Rb–Ar Age
10087,11	0.50 ± 0.05	ESR (TSAY *et al.*, 1971)	4.6 (SILVER, 1970)	4.5 (ALBEE *et al.*, 1970)
(fines)	0.45 ± 0.05	MS (TSAY *et al.*, 1971)		
12070,11	0.41 ± 0.05	ESR (this work)	4.5 (SILVER, 1971)	4.4 (ALBEE *et al.*, 1971)
(fines)	0.59	MS (NAGATA *et al.*, 1971)		
	0.31	MS (PEARCE *et al.*, 1971)		
12032,41	0.085 ± 0.015	ESR (this work)	4.40 (CLIFF *et al.*, 1971)	—
(breccia)	0.32	MOSS (HERZENBERG *et al.*, 1971)		
10017,35,36 (rock)	0.012 ± 0.005 (average)	ESR (this work)	4.13 (SILVER, 1970)	3.59 ± 0.08 (ALBEE *et al.*, 1970)
	0.22	MOSS (HOUSLEY *et al.*, 1970)		
12021,42	0.0013 ± 0.0010	ESR (this work)	3.94 (TATSUMOTO and KNIGHT, 1971)	3.28 ± 0.11 (CLIFF *et al.*, 1971)
(rock)	0.06	MS (PEARCE *et al.*, 1971)		
10024	0.071	MS (NAGATA *et al.*, 1971)	—	3.61 ± 0.07 (ALBEE *et al.*, 1971)
12053	0.093	MS (NAGATA *et al.*, 1971)	3.95 ± 0.02 (SILVER, 1971)	—
12063	0.035	MS (PEARCE *et al.*, 1971)	3.99 (TATSUMOTO and KNIGHT, 1971)	3.34 ± 0.10 (MURTHY *et al.*, 1971)
	0.046	MS (HARGRAVES and DOUTY, 1971)		
	0.25	MOSS (HERZENBERG *et al.*, 1971)		
10058	0.014	MS (HARGRAVES and DOUTY, 1971	—	3.63 ± 0.11 (MURTY *et al.*, 1971)
	0.15	MOSS (HOUSLEY *et al.*, 1970)		

* ESR = electron spin resonance; MS = magnetic susceptibility; MOSS = Mössbauer studies.

seems unlikely that less metallic Fe should be present in all the rock samples we have studied. We suggest that the discrepancies in metallic Fe content observed by different workers probably reflect problems in absolute calibration involved in these other methods.

The determination of metallic Fe contents by Mössbauer spectroscopy generally depends upon having prior or subsequent analytical data regarding the *total* iron content of the sample studied. The strong Fe^{2+} resonances which overlap the weak metallic Fe resonances can introduce large uncertainties in the metallic Fe content determination in rock samples with low metallic Fe contents. In the magnetic susceptibility measurements, the metallic Fe contents are estimated by comparing the saturation magnetization of the lunar sample with that of pure iron. The sensitivity of the method falls off considerably in the case of rock samples, because the metallic

Fe content is low in this case and the magnetization due to the paramagnetism of $FeTiO_3$, $FeSiO_3$, and free Fe^{2+} ions (Nagata et al., 1970) becomes much more significant as compared to the component from ferromagnetism of metallic Fe. In the ESR measurements, there is no interference from Fe^{2+} ions since the spin-lattice relaxation rates for these ions are rapid at both room and liquid nitrogen temperatures. Although the question of absolute calibration also arises in our ESR measurements, where intensities are measured relative to a DPPH/KCl standard in situ, the fact that our metallic Fe contents are in good agreement with those determined by others using other methods for samples containing high Fe contents would tend to argue in favor of the reliability of the ESR method. Any deterioration of the DPPH standard and/or failure on our part to take into account the contribution of other constituents to the ESR intensities would yield apparent metallic Fe contents which are high rather than low. Since our values are more likely upper limits, we find it difficult to understand why our determined Fe contents are consistently lower than those indicated by the Mössbauer and magnetic susceptibility measurements in the case of the samples with low metallic contents. Thus, the ESR technique appears to be unique for studying the metallic Fe phases of the lunar samples and can provide reliable metallic Fe contents.

Correlation of metallic Fe *content with rock age*

Perhaps one of the most unique features of the lunar samples is the ubiquitous, reasonably uniform distribution of metallic Fe phases in all the Apollo samples which have been examined in detail. Two different components of the metallic Fe phases have been discussed (Simpson and Bowie, 1970; Strangway et al., 1970). One component is associated with high Ni contents and the other with low Ni contents and some Co. The former probably represents the meteoritic component. Estimates of the meteoritic component to the total iron phases in the Apollo 11 fines samples have ranged from 25% to 75% (Goldstein et al., 1970; Keil et al., 1970). However, in the case of the Apollo 12 site the meteoritic component in the fines is believed to be much less than that at the Apollo 11 site based on detailed analyses of individual metal fragments in the fines (Goldstein and Yakowitz, 1971).

An important factor determining the meteoritic metallic Fe content obviously should be the surface exposure age. Thus in the case of the fines, less meteoritic component in the Apollo 12 material implies it has been on the surface a shorter period of time. In the three Apollo 12 fines samples we have investigated (12003,24, 12041,15, 12070,11) we find less metallic Fe than in our Apollo 11 sample (10087,11). If all the difference is due to meteoritic iron, then the Apollo 12 samples may have received 10–20% less exposure time to meteoritic influx. Most probably the formation of the metallic Fe in the lunar samples is a much more complicated process involving, in addition, contributions from indigenous radioactive decay, solar wind particles and cosmic particles. But the concept of a relation between exposure age and metallic Fe content prompted us to see if any trends were evident between established geological ages and metallic Fe contents. We have attempted to do this in Figs. 2 and 3 with the available age data on samples in which we have determined metallic Fe contents. It is evident that a trend does exist. Thus younger lunar rocks have less metallic Fe content than older rocks and fine soil. The potentially useful correlations demonstrated in

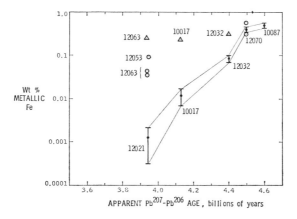

Fig. 2. Pb²⁰⁷–Pb²⁰⁶ age vs. metallic Fe content. Data from Table 4. ☉—magnetic susceptibility metallic Fe contents; △—Mössbauer metallic Fe contents, Φ —ESR metallic Fe contents. Sample numbers shown.

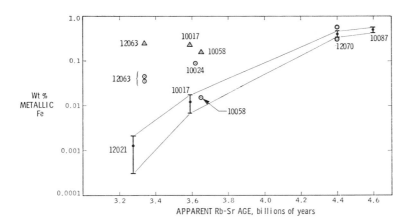

Fig. 3. Rb–Sr age vs. metallic Fe content. Data from Table 4. ☉—magnetic susceptibility metallic Fe contents, △—Mössbauer metallic Fe contents, Φ —ESR metallic Fe contents. Sample numbers shown.

Figs. 2 and 3 certainly warrant further testing. If indeed this type of relation does hold up under exhaustive testing, then the metallic Fe phases may become an even more important parameter of a bulk lunar sample.

In Figs. 2 and 3 are also plotted points for metallic Fe contents determined by the Mössbauer and magnetic susceptibility techniques. It can be seen that significant deviations exist between our data and that of others (except in one case). As discussed above these other techniques tend to overestimate considerably the metallic Fe contents for concentrations below about 0.3–0.4 wt. %. Points are not shown for Apollo 11 fines where general agreement between all three techniques exists.

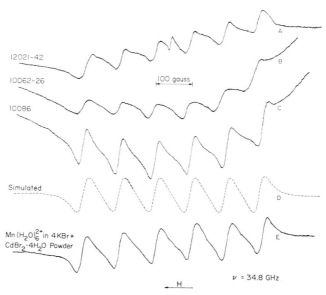

12021-42

10062-26

10086

100 gauss

A

B

C

Simulated

D

Mn $(H_2O)_6^{2+}$ in 4 KBr +
CdBr$_2$·4H$_2$O Powder

E

ν = 34.8 GHz

H

Fig. 4. Observed paramagnetic resonance spectra of Mn^{2+} ions for the lunar samples (spectra *A*, *B*, *C*) and for a powder sample of 4KBr + CdBr$_2$ · 4H$_2$O. A computer-simulated spectrum for Mn(H$_2$O)$_6$$^{2+}$ is also shown (spectrum *D*). Observed spectra are recorded at 298°K and at K-band frequency. The sharp peak observed at the center of spectrum *A* is due to free radicals in one of the quartz sample tubes used.

Paramagnetic resonance—Mn^{2+} ions

In addition to ferromagnetic resonance of metallic Fe, we have also observed in samples (10086; 10062,26, 27; 10017,35, 36; 12065,103; 12021,42), where the metallic Fe content is either low or partially removed (10086), weak paramagnetic signals with six resolved nuclear hyperfine components attributable to Mn^{2+} ions. The paramagnetic resonance spectra recorded for samples (10086, 10062,26, 12021,42) at K-band frequency and room temperature are presented in Fig. 4. A spectrum of Mn(H$_2$O)$_6$$^{2+}$ dissolved in a powder sample precipitated from an aqueous solution of KBr and CdBr$_2$·4H$_2$O (mole ratio of 4:1) is also shown in Fig. 4 for comparison, together with a theoretical spectrum expected for Mn(H$_2$O)$_6$$^{2+}$ ions. The X-band spectra recorded for the samples 10086 and 12021,42 at both room and liquid nitrogen temperatures are presented in Fig. 5. These X-band spectra were recorded on a Varian E-12 spectrometer using 100 KHz field modulation. As expected, the Mn^{2+} spectrum is considerably weaker at X-band field, even at liquid nitrogen temperatures, but its presence is definite and unequivocal. The Mn^{2+} spectrum observed for the lunar samples (see Fig. 4) is similar to that observed for Mn(H$_2$O)$_6$$^{2+}$ doped in the KBr/CdBr$_2$·4H$_2$O powder except that the presence of the ferromagnetic resonance of the metallic Fe displaces the base line to various extents depending on the lunar sample examined. Our sample 10086 had been treated with organic solvents and organic acid (Rho *et al.*, 1970); therefore, contamination with terrestrial water was expected. Since our other samples had been exposed to a limited extent to the atmosphere and thus to atmospheric moisture, it is possible that all or most of the water associated with the Mn^{2+} is of terrestrial origin as well.

To substantiate our claim that the Mn^{2+} signals have origin in $Mn(H_2O)_6{}^{2+}$ ions, we have heated sample 12021,42 at 150°C (roughly the maximum lunar daytime temperature) in vacuum (10^{-4} mm Hg) for 24 hours. The intensity of the $Mn(H_2O)_6{}^{2+}$ signal decreased roughly 10% after this treatment. Further heating (12 hours) at 200°C reduced the intensity of the $Mn(H_2O)_6{}^{2+}$ signal to about 70% of its original intensity. These crude experiments demonstrate that the water is fairly tightly bound at the Mn^{2+} sites and suggest the possibility that any indigenous lunar water trapped at Mn^{2+} sites could not have survived the lunar daytime temperatures for a reasonable fraction of the moon's history.

Although similar K-band Mn^{2+} signals were first reported by WEEKS et al. (1970a, 1970b), thus far no Mn^{2+} signals have been reported for the returned lunar samples at X-band frequency. WEEKS et al. (1970a, 1970b) suggested that their K-band signals were due to Mn^{2+} ions occurring in plagioclase and pyroxene minerals, but presented no information concerning the site symmetry or the coordination number of these

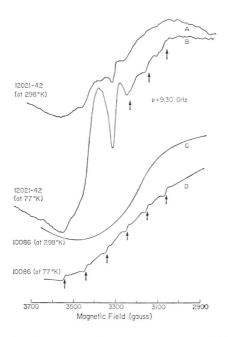

Fig. 5. Observed X-band paramagnetic resonance spectra of Mn^{2+} ions for several lunar samples at 298°K (spectra A, C) and 77°K (spectra B, D). Spectra were recorded on a Varian E-12 spectrometer using 100 kHz field modulation. The origin of the narrow line (width 20 gauss) centered at 3300 gauss in the spectra A and B has not been identified. The Mn^{2+} ($I = \frac{5}{2}$) hyperfine components which are discernible (spectrum B) or well-resolved (spectrum D) are indicated by labelling arrows. The sloping baseline is due to the ferromagnetic resonance absorption of metallic Fe.

Mn^{2+} ions. On the basis of the known crystal structures for the plagioclase (triclinic) and pyroxene (orthorhombic and monoclinic) minerals (MASON and BERRY, 1968), we would expect that the site symmetry (or symmetries) for the Mn^{2+} ions in these minerals would be low. As a result, large crystal field splittings due to the axial and/or rhombic crystal field terms (D and E) are expected to contribute significantly to the overall lineshape of the ESR spectrum of a polycrystalline sample. For Mn^{2+} ions ($S = \frac{5}{2}$, $I = \frac{5}{2}$) the crystal field contribution to the resonance field for the $M_s = \frac{1}{2} \leftrightarrow -\frac{1}{2}$ transition is known to be proportional to $D^2/g^2\beta^2H$ (g is the g value, β is the Bohr magneton, and H is the applied external magnetic field), whereas for the $M_s = \pm\frac{5}{2} \leftrightarrow \pm\frac{3}{2}$ and $M_s = \pm\frac{3}{2} \leftrightarrow \pm\frac{1}{2}$ transitions, it is proportional to $D/g\beta$ (DE WIJN and VAN BALDEREN, 1967). Therefore, if D and E are not sufficiently small so that the various ESR transitions are practically coincident, extra signals are expected from the $M_s = \pm\frac{5}{2} \leftrightarrow \pm\frac{3}{2}$ and $M_s = \pm\frac{3}{2} \leftrightarrow \pm\frac{1}{2}$ transitions, in addition to the six resolved nuclear hyperfine components for the $M_s = \frac{1}{2} \leftrightarrow -\frac{1}{2}$, $\Delta M_i = 0$ transitions centered at $g = 2.00$, which we are most certainly observing. Moreover, for sufficiently large D, the second-order effects arising from crystal field terms (proportional to $D^2/g^2\beta^2H$) can also become important for the $M_s = +\frac{1}{2} \leftrightarrow -\frac{1}{2}$ transition, and further splitting of these hyperfine components, corresponding to the parallel and perpendicular orientations in a polycrystalline sample, might also be observed. When $D \gg g\beta H$ and/or $E \gg g\beta H$, the $M_s = \pm\frac{5}{2} \leftrightarrow \pm\frac{3}{2}$ and $M_s = \pm\frac{3}{2} \leftrightarrow \pm\frac{1}{2}$ transitions may fall outside the range of observation of the spectrometer, but then extra peaks should be observed at $g = 6$ and/or $g = 4$, as has been demonstrated for Fe^{3+} in glass by CASTNER et al. (1960). The fact that no other resonance signals for Mn^{2+} ions were observed at both X-band and K-band frequencies, except for the one centered at $g = 2.002$, indicates the site symmetry for the Mn^{2+} ions under observation in the ESR spectra of the returned lunar samples must be high.

Computer simulation of the Mn^{2+} spectrum assuming axial symmetry with a crystal field splitting D of the order of 10 gauss, a g value of 2.002 ± 0.002, a nuclear hyperfine coupling constant of -95.0 ± 2.0 gauss, and hyperfine linewidth of 25.0 ± 3.0 gauss (spectrum D, Fig. 4) yielded results which can be seen to be in good agreement with all the K-band spectra observed for the Mn^{2+} ion in the returned lunar samples. For the fine sample 10086 which had been treated extensively with organic solvents, the $Mn(H_2O)_6^{2+}$ ions in this sample may well have a well-defined site symmetry with an axial crystal field splitting D of the order of 10 gauss. Comparison of the K-band (spectrum C, Fig. 4) and the X-band spectra observed at $77°K$ (spectrum D, Fig. 5) substantiates this conclusion. In both spectra, the hyperfine components exhibit isotropic behavior and undistorted line-shape with a width of 25 gauss, indicating that the crystal field splitting D is indeed smaller than the intrinsic linewidth of the hyperfine components and that the second-order effects ($D^2/g^2\beta^2H$) which can cause the hyperfine components to broaden by a factor of four at the X-band frequency is insignificant for the fine sample 10086. However, in the case of the rock sample 12021,42, the symmetry of $Mn(H_2O)_6^{2+}$ may vary from site to site, resulting in a distribution of axial crystal field splittings. If the average D is of the order of 100 gauss, contributions from the second-order effects and, possibly also, forbidden transitions ($\Delta M_s = \pm1$, $\Delta M_i = \pm1$) (DE WIJN and VAN BALDEREN, 1967) may result in the broader hyperfine components observed for the $Mn(H_2O)_6^{2+}$ ions in this sample at X-band frequency.

The broadening of the X-band hyperfine components due to variations in weak axial crystal field splittings ($E = 0$, and $D \simeq 100$ gauss) has been discussed in detail by GRISCOM and GRISCOM (1967) and by DE WIJN and VAN BALDEREN (1967). However, as pointed out by GRISCOM and GRISCOM (1967) in their analyses of the ESR spectra of Mn^{2+} in lithium borate glasses, even a wide distribution in the axial crystal field splitting D of the order of 600 gauss can still give rise to six well-resolved, though distorted, sharp hyperfine components with a linewidth of 20 gauss at X-band frequency, if there exists a significant rhombic distortion ($E/D = \frac{1}{3}$). (It is to be noted that when D is of the order of 600 gauss, the intensity of the extra peak at $g = 4.3$ is now quite large.) This is because the second-order effects lead to different lineshapes for the hyperfine components in powder samples in the presence and absence of rhombic distortions (GRISCOM and GRISCOM, 1967). The fact that the Mn^{2+} hyperfine components observed for the rock sample 12021,42 were wider and less resolved at X-band than at K-band frequency suggests that the site symmetry for the Mn^{2+} ions under observation cannot be of lower than axial symmetry.

On the basis of our ESR results, in particular, a g value of 2.002 ± 0.002 and a nuclear hyperfine coupling constant of -95.0 ± 2.0 gauss, and the ESR spin Hamiltonian parameters previously reported by CHAN et al. (1967) for a series of octahedral and tetrahedral Mn^{2+} complexes, we concluded that these Mn^{2+} ions are octahedrally coordinated and probably exist as $Mn(H_2O)_6^{2+}$. Our ESR intensity measurements indicate that the concentration of manganese existing in this environment is in the range of 10–70 ppm (depending upon the samples), indicating that only 3.5% or less of the total manganese content present in these returned lunar samples (average Mn concentration of 2000 ppm reported for several samples by LSPET, 1970) is in the form of $Mn(H_2O)_6^{2+}$ and gives rise to detectable ESR signals. Thus the amount of water tightly coordinated with the manganese is about 20–140 ppm. This estimate is close to the concentrations that have been measured by geochemical techniques. For example, FRIEDMAN et al. (1970) and EPSTEIN and TAYLOR (1970) liberated from breccia and lunar fines 150–455 ppm and 92–181 ppm water, respectively. EPSTEIN and TAYLOR (1970) believe that the origin of this water may be primarily terrestrial in origin but feel that the possibility that the lunar rocks contain indigenous water cannot be entirely eliminated.

Radiation damage and other paramagnetic ions

We have detected no signals characteristic of radiation damage. Although extensive radiation damage is expected in the lunar samples, it appears that radiation damage manifested in the form of free electrons and holes trapped in lattice defects is quite insignificant. In fact, on the basis of the known sensitivity of our ESR spectrometer, we estimated that the total radiation dose accumulated in these samples will not exceed 10^5 rad. The lack of extensive radiation damage noted here is in total agreement with the conclusions of the thermoluminescence and stored energy measurements of CROZAZ et al. (1970) and NASH and GREEN (1970), suggesting that some effective thermal annealing or other processes on the lunar surface empty the electrons or holes trapped in the lattice defects. As suggested above, some of the electrons generated by ionizing radiation may be trapped at Fe^{2+} sites to yield metallic Fe.

We have not been successful in detecting any ESR signals ascribable to Ti^{3+}. In addition, we have obtained no evidence for high spin Fe^{3+} as others have previously reported (Weeks et al., 1970a, 1970b; Geake et al., 1970a, 1970b; Garlick et al., 1971). High spin Fe^{3+}, like high spin Mn^{2+}, are both $^6S_{5/2}$ ions, and should exhibit relatively sharp ESR resonances if present at concentration levels in excess of several ppm. For low-spin Fe^{3+} and for paramagnetic Fe^{2+} as well, no room temperature spectra are expected in view of the rapid spin-lattice relaxation rates for these systems.

Conclusions

We have concluded that the ESR spectra centered at $g = 2.09$ for the returned Apollo 11 and Apollo 12 samples are ferromagnetic in origin. Our conclusions are based on the evidence that the ESR absorption intensities observed for these samples are at least three orders of magnitude greater than those expected for paramagnetic Fe^{3+} ions, and that no significant temperature dependence was observed for the ESR spectral intensities over a wide range of temperatures. On the basis of the g value and the crystalline anisotropy energy obtained from the detailed analysis of the poly-crystalline spectra, we further concluded that the ferromagnetic centers are prin-cipally metallic Fe particles and are present in the range of 0.001–0.50 wt.%. An apparent correlation has been noted between the metallic Fe content of the lunar samples and their apparent geological ages.

In addition, we have detected weak paramagnetic resonances attributable to octahedrally coordinated Mn^{2+} ions in some of our samples, and have presented evidence to indicate that the Mn^{2+} probably exists as $Mn(H_2O)_6^{2+}$ in these samples. If the bulk of the water in the lunar samples is coordinated with Mn^{2+}, our ESR measurements suggest water contents of 20–140 ppm in the returned lunar sample, which is in agreement with the concentration levels reported by other workers using geochemical techniques. Thus far, we have detected no ESR signals which would be indicative of radiation damage or the presence of Ti^{3+} and Fe^{3+} in the returned Apollo 11 and 12 lunar samples.

Acknowledgments—This work was supported by NASA under Contract No. NAS 7–100 to the Jet Propulsion Laboratory, California Institute of Technology. We wish to thank Professor Samuel Epstein for several discussions on the question of water contamination in the returned lunar samples, and Professor Robert Beaudet of the Department of Chemistry, University of Southern California, for the use of his Varian E-12 spectrometer.

References

Albee A. L , Burnett D. S., Chodos A. A., Eugster O. J., Huneke J. C., Papanastassious D. A., Podosek F. A., Price Russ G., II, Sanz H. G., Tera F., and Wasserburg G. J. (1970) Ages, irradiation history, and chemical composition of lunar rocks from the Sea of Tranquility. *Science* **167**, 463–466.

Albee A. L., Burnett D. S., Chodos A. A., Haines E. L., Huneke J. C., Podosek F. A., Papanastassious D. A., Price G., Tera F., and Wasserburg G. J. (1971) Rb–Sr ages, chemical abundance patterns and history of lunar rocks. Second Lunar Science Conference (unpublished proceedings).

CASTNER T., JR., NEWELL G. S., HOLTON W. C., and SLICHTER C. P. (1960) Note on the paramagnetic resonance of iron in glass. *J. Chem. Phys.* **32**, 668–673.

CHAN S. I. FUNG B. M., and LÜTJE H. (1967) Electron Paramagnetic Resonance of Mn(II) Complexes in Acetonitrile. *J. Chem. Phys.* **47**, 2121–2130.

CLIFF R. A., LEE-HU C., and WETHERILL G. W. (1971) Rb-Sr and U, Th–Pb measurements on Apollo 12 material. Second Lunar Science Conference (unpublished proceedings).

COMPSTON W., BERRY H., VERNON M. J., CHAPPELL B. W., and KAYE M. J. (1971) Rubidium-strontium chronology and chemistry of lunar material from the ocean of storms. Second Lunar Science Conference (unpublished proceedings).

CROZAZ G., HAACK U., HAIR M., MAURETTE M., WALKER R., and WOOLUM D. (1970) Nuclear track studies of ancient solar radiations and dynamic lunar surfaces processes. *Proc. Apollo 11 Lunar Sci. Conf., Geochim. Cosmochim. Acta* Suppl. 1, Vol. 3, pp. 2051–2080. Pergamon.

DE WIJN H. W. and VAN BALDEREN R. F. (1967) Electron spin resonance of manganese in borate glasses. *J. Chem. Phys.* **46**, 1381–1387.

EPSTEIN S. and TAYLOR H. P., JR. (1970) The concentration and isotopic composition of hydrogen, carbon and silicon in Apollo 11 lunar rocks and minerals. *Proc. Apollo 11 Lunar Sci. Conf., Geochim. Cosmochim. Acta* Suppl. 1, Vol. 2, pp. 1085–1096. Pergamon.

FRIEDMAN I., GLEASON J. D., and HARDCASTLE K. G. (1970) Water, hydrogen, deuterium, carbon and C^{13} content of selected lunar material. *Proc. Apollo 11 Lunar Sci. Conf., Geochim. Cosmochim. Acta* Suppl. 1, Vol. 2, pp. 1103–1109. Pergamon.

GARLICK G. F. J., LAMB W., STEIGMANN G. A., and GEAKE J. E. (1971) Thermoluminescence, EPR, and diffuse reflection spectra of Apollo lunar samples. Second Lunar Science Conference (unpublished proceedings).

GEAKE J. E., DOLLFUS A., GARLICK G. F. J., LAMB W., WALTER C., STEIGMANN G. A., and TITULAER G. (1970a) Luminescence, electron paramagnetic resonance, and optical properties of lunar material. *Science* **167**, 717–720.

GEAKE J. E., DOLLFUS A., GARLICK, G. F. J., LAMB W., WALTER C., STEIGMANN G. A., and TITULAER G. (1970b) Luminescence, electron paramagnetic resonance and optical properties of lunar material from Apollo 11. *Proc. Apollo 11 Lunar Sci. Conf., Geochim. Cosmochim. Acta* Suppl. 1, Vol. 3, pp. 2127–2147. Pergamon.

GOLDSTEIN J. I., HANDERSON E. P., and YAKOWITZ H. (1970) Investigation of lunar metal particles. *Proc. Apollo 11 Lunar Sci. Conf., Geochim. Cosmochim. Acta* Suppl. 1, Vol. 1, pp. 499–512. Pergamon.

GOLDSTEIN J. I. and YAKOWITZ H. (1971) Metallic inclusions and metal particles in the Apollo 12 lunar soil. Second Lunar Science Conference (unpublished proceedings).

GRISCOM D. L. and GRISCOM R. E. (1967) Paramagnetic resonance of Mn^{2+} in glasses and compounds of the lithium borate system. *J. Chem. Phys.* **47**, 2711–2722.

HARGRAVES R. B. and DORETY N. (1971) Magnetic properties of some lunar crystalline rocks returned by Apollo 11 and Apollo 12. Second Lunar Science Conference (unpublished proceedings).

HERZENBERG C. L. and RILEY D. L. (1970) Mössbauer spectrometry of lunar samples. *Science* **167**, 683–686.

HERZENBERG C. L., MOLER R. B., and RILEY D. L. (1971) Preliminary results from Mössbauer instrumental analysis of Apollo 12 lunar rock and soil samples. Second Lunar Science Conference (unpublished proceedings).

HOUSLEY R. M., BLANDER M., ABDEL-GAWAD M., GRAND R. W., and MUIR A. H., JR. (1970) Mössbauer spectroscopy of Apollo 11 samples. *Proc. Apollo 11 Lunar Sci. Conf., Geochim. Cosmochim. Acta* Suppl. 1, Vol. 3, pp. 2251–2268. Pergamon.

KEIL K., BUNCH T. E., and PRINZ M. (1970) Mineralogy and composition of Apollo 11 lunar samples. *Proc. Apollo 11 Lunar Sci. Conf., Geochim. Cosmochim. Acta* Suppl. 1, Vol. 1, pp. 561–598. Pergamon.

KITTEL C. (1948) On the theory of ferromagnetic resonance absorption. *Phys Rev.* **73**, 155–161

KITTEL C. (1967) *Introduction to Solid State Physics.* 3rd edition, pp. 491, 525, John Wiley.

LSPET (Lunar Sample Preliminary Examination Team) (1969) Preliminary examination of lunar samples from Apollo 12. *Science* **167**, 1325–1339.

MANATT S. L., ELLEMAN D. D., VAUGHAN R. W., CHAN S. I., TSAY F. D., and HUNTRESS W. T., JR. (1970) Magnetic resonance studies of some lunar samples. *Science* **167**, 709–711.

MASON B. and BERRY L. G. (1968) *Elements of Mineralogy*, pp. 467, 528, Freeman.

MURTHY R. V., EVENSEN N. M., JAHN B.-M., and COSCIO M. R., JR. (1971) Rb–Sr isotopic relations and elemental abundances of K, Rb, Sr, and Ba in Apollo 11 and Apollo 12 samples. Second Lunar Science Conference (unpublished proceedings).

NAGATA T., ISHIKAWA Y., HINOSHITA H., KONO M., SYONO Y., and FISHER R. M. (1970) Magnetic properties and natural remanent magnetization of lunar materials. *Proc. Apollo 11 Lunar Sci. Conf., Geochim. Cosmochim. Acta* Suppl. 1, Vol. 3, pp. 2325–2340. Pergamon.

NAGATA T., FISHER R. M., SCHWERER F. C., and FULLER M. D. (1971) Magnetic properties and remanent magnetization of Apollo 12 lunar materials. Second Lunar Science Conference (unpublished proceedings).

NASH D. B. and GREEN R. T. (1970) Luminescence properties of Apollo 11 lunar samples and implication for solar-excited lunar luminescence. *Proc. Apollo 11 Lunar Sci. Conf., Geochim. Cosmochim. Acta* Suppl. 1, Vol. 3, pp. 2341–2350. Pergamon.

PEARCE G. W., STRANGWAY D. W., and LARSON E. E. (1971) Magnetism of two Apollo 12 igneous rocks. Second Lunar Science Conference (unpublished proceedings).

RHO J. H., BAUMAN A. J., YEN T. F., and BONNER J. (1970) Fluorometric examination of the returned lunar fines from Apollo 11. *Proc. Apollo 11 Lunar Sci. Conf., Geochim. Cosmochim. Acta* Suppl. 1, Vol. 2, pp. 1929–1932. Pergamon.

RUNCORN S. K., COLLINSON D. W., O'REILLY W., BATTEY M. H., STEPHENSON A., JONES J. M., MANSON A. J., and READMAN P. W. (1970) Magnetic properties of Apollo 11 lunar samples. *Proc. Apollo 11 Lunar Sci. Conf., Geochim. Cosmochim. Acta* Suppl. 1, Vol. 3, pp. 2369–2387. Pergamon.

SILVER L. T. (1970) Uranium-thorium-lead isotopes in some Tranquillity Base samples and their implications for lunar history. *Proc. Apollo 11 Lunar Sci. Conf., Geochim. Cosmochim. Acta* Suppl. 1, Vol. 1, pp. 1533–1574. Pergamon.

SILVER L. T. (1971) U–Th–Pb isotope relations in Apollo 11 and Apollo 12 lunar samples. Second Lunar Science Conference (unpublished proceedings).

SIMPSON P. R. and BOWIE S. H. V. (1970) Qualitative optical and electron-probe studies of opaque phases in Apollo 11 samples. *Proc. Apollo 11 Lunar Sci. Conf., Geochim. Cosmochim. Acta* Suppl. 1, Vol. 1, pp. 873–890. Pergamon.

STRANGWAY D. W., LARSON E. E., and PEARCE G. E. (1970) Magnetic studies of lunar samples— breccia and fines. *Proc. Apollo 11 Lunar Sci. Conf., Geochim. Cosmochim. Acta* Suppl. 1, Vol. 3, pp. 2435–2451. Pergamon.

TATSUMOTO M., KNIGHT R. J., and DOE B. R. (1971) U–Th–Pb systematics of Apollo 12 lunar samples. Second Lunar Science Conference (unpublished proceedings).

TSAY F. D., CHAN S. I., and MANATT S. L. (1971) Ferromagnetic resonance of lunar samples. *Geochim. Cosmochim. Acta* (in press).

WEEKS R. A., CHATELIAN A., KOLOPUS J. L., KLINE D., and CASTLE J. G. (1970a) Magnetic resonance properties of lunar material. *Science* **167**, 704–707.

WEEKS R. A., KOLOPUS J. L., KLINE D., and CHATELIAN A. (1970b) Apollo 11 lunar material: nuclear magnetic resonance of ^{27}Al and electron resonance of Fe and Mn. *Proc. Apollo 11 Lunar Sci. Conf., Geochim. Cosmochim. Acta* Suppl. 1, Vol. 3, pp. 2467–2490. Pergamon.

Proceedings of the Second Lunar Science Conference, Vol. 3, pp. 2529–2541
The M.I.T. Press, 1971.

Clean lunar rock surfaces; unpaired electron density and adsorptive capacity for oxygen

D. Haneman and D. J. Miller

School of Physics, University of New South Wales, Sydney, Australia, 2033

(*Received* 23 *February* 1971; *accepted in revised form* 31 *March* 1971)

Abstract—A search was made for the presence of dangling bonds, or unpaired electrons, on surfaces of lunar material. The adsorptive capacity for oxygen was also investigated. Since the surfaces of returned material were contaminated by exposure to air, new clean surfaces were prepared in ultra high vacuum by crushing. Lunar fines (12070) and lunar rock (12021) surfaces were investigated in vacuo by electron paramagnetic resonance (X band). The strong pre existing ferromagnetic resonance from the fines prevented detection of any new small resonances after crushing. In the case of the rock however there were only relatively small resonances before crushing, and effects could be observed. After crushing, a new resonance appeared near $g = 2.006$. This resonance was due to surface centres, since it was altered (enhanced) by exposure to gases (oxygen). The density of surface spins was about 1 per 10^4 surface atoms. The oxygen adsorption that took place was reversible, since the enhancement in the resonance disappeared when the oxygen was pumped away. Evidence was also obtained by separate techniques that adsorption took place on non paramagnetic sites. This adsorption was not reversible. The rate of uptake decreased markedly at about one monolayer coverage. It is suggested from these results that the surfaces of soil and rocks on the moon may well be contaminated with adsorbed gases, due to the long period of exposures.

Introduction

THE PURPOSES of this work were (a) to search for the presence of unpaired electrons, or dangling (broken) bonds on clean lunar surfaces by electron paramagnetic resonance (e.p.r.) (b) to measure the adsorptive capacity for oxygen of clean surfaces. This would give information about the reactivity and nature of the surfaces of lunar material. The e.p.r. technique is well suited to detecting unpaired electrons in broken bonds at surfaces, and changes in the e.p.r. signal as a function of gas adsorption indicate whether the paramagnetic sites are affected by gas. Total adsorption was measured by volumetric techniques.

The material returned from the moon was contaminated by exposure to terrestrial gases. Therefore to study clean surfaces it was necessary to create new surfaces by crushing the material in ultra high vacuum, and study them in situ.

The methods were similar to those previously used for other specimens. (CHUNG and HANEMAN, 1966; HANEMAN, 1968; MILLER and HANEMAN, 1970.)

Processing took place in a mainly glass, bakeable ultra high vacuum system with background below 10^{-9} torr. The crushing device consisted of a glass cup into which a glass slug containing a sealed piece of iron was caused to reciprocate by external magnetic control. The ambient composition was monitored throughout by a quadrupole residual gas analyser. After crushing, the cup was moved by magnetic control over the mouth of a tube terminating in a quartz tube suitable for insertion in an electron paramagnetic resonance (e.p.r.) spectrometer cavity. By inverting the cup

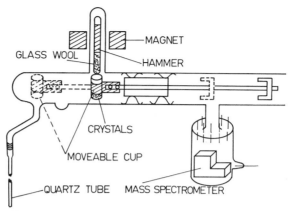

Fig. 1. Schematic diagram of portion of vacuum system and crushing apparatus. Part marked mass spectrometer has been replaced by quadrupole residual gas analyser.

(magnetic controls) the crushed sample fell to the bottom of the quartz tube. The relevant portion of the high vacuum system is shown in Fig. 1. The whole vacuum system was transposed by horizontal and vertical controls so as to locate the sample tube in the centre of the microwave cavity of the e.p.r. spectrometer. Bakeout was performed at only 120°C, to ensure that the lunar samples were not subjected before experimentation to higher temperatures than they might have experienced on the moon.

The measuring system used was a Varian 9.4 GHz, 9″ rotatable magnet system with 100 KHz modulation and cylindrical cavity. In order to attain the highest possible measurement sensitivity, accumulation techniques were used to improve the signal to noise ratio (by the square root of the number of accumulations). Signals were stored in a Northern Scientific digital memory oscilloscope, and the output was processed by a PDP8L computer. In this way it was possible to observe small effects that would otherwise have escaped detection.

Lunar Fines

Experiments were carried out on lunar fines and on lunar rock samples. In the case of the fines (12070, 117), crushing caused copious evolution of rare gases and hydrocarbons. A representative gas spectrum (concentration versus mass number) is shown in Fig. 2. Note that this was taken under a crushing stroke amplitude of about 1.0 cm and frequency of about 2 per second, which is quite mild. Furthermore, an 8 l/s ion pump was on during the mass scan.

The lunar fines displayed a large e.p.r. signal prior to crushing, as shown in Fig. 3. The absorption intensity was approximately unchanged on cooling to liquid nitrogen temperatures. From this evidence, and comparison with the results of others (Weeks *et al.*, 1970; Manatt *et al.*, 1970) it was identified as a ferromagnetic resonance, due principally to iron. After crushing, the signal remained, and made it difficult to detect any small additional resonances that could have been induced by the crushing process,

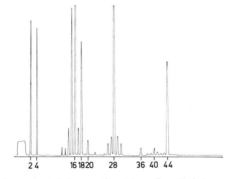

Gas analysis during crushing of lunar fines. Emission
current 230μA

Fig. 2. Note helium (mass 4), neon (20, 22) and argon (36, 40) emission. The helium
peak is never seen with other materials under crushing.

(such as by breaking atomic bonds). Likewise any signals or changes induced by exposure to oxygen, at room and liquid nitrogen temperatures, could not be detected above the strong background resonance. Attempts to separate out the ferromagnetic components were not successful, since all the fine material seemed to be attracted by a magnet.

To examine the effectiveness of the crushing process, the fines were studied by scanning electron microscopy before and after crushing. Due to their small particle size initially, the crushing was not as effective in breaking them up as for larger samples. Representative scans over the samples before and after crushing did not show markedly smaller particle sizes in the crushed samples. A degree of crushing had clearly taken place however, since a steady background of helium, neon and argon was produced during several hours of crushing, indicating constant fracturing. Nevertheless the ratio of new clean surfaces to the previous contaminated surfaces was not as large, according to the microscopy evidence, as desired. Due to their initial fine particle size, and high iron content, the fines are not suitable for this kind of experiment.

LUNAR ROCK

A sample of rock (12021, C-4, 104) was subjected to the same crushing, gas analysis and e.p.r. measurement technique as the fines. This sample was a porphyritic,

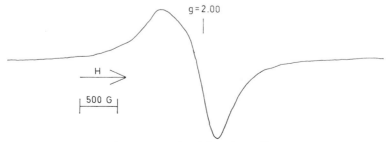

Fig. 3. E.p.r. signal from lunar fines.

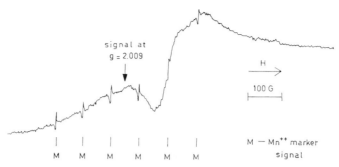

signal at
g = 2.009

H

100 G

M — Mn⁺⁺ marker
signal

| | | | | |
M   M   M   M   M   M

Fig. 4. E.p.r. signal from lunar rock 12021 before crushing.

very coarse grained rock, whose appearance was similar to the chips from 12021 described in the Apollo 12 Lunar Sample Information Catalog. The mineral composition range obtained there was plagioclase 40–60% (mainly calcic), zoned pyroxene 30–60% (probably mainly pigeonite), olivine 2–7%, brown mineral 2%, opaques up to 5% (mostly black metallic plates—ilmenite). The sample crumbled very easily into small (about 1 mm dimensions) crystal fragments, of greenish and honey yellow pigeonite, light green olivine, grey plagioclase and other fragments. After about 1 hour of the mild crushing action described above for the fines, the resultant powder consisted of distinct small green, yellow and dark brown crystals, of generally smaller size than those of the rock, plus a mass of fine grey powder. This showed that in addition to grains being separated by the crushing, they were also fractured.

The e.p.r. signals before crushing were much smaller than in the case of the fines, so the e.p.r. technique was fully applicable. The resonances are shown in Fig. 4, consisting mainly of a broad, powder type signal with g parallel $= 1.88$, g perpendicular $= 1.94$, width about 120 gauss; and a narrower line at $g = 2.009$, width about 20 gauss. The narrow line is barely visible on the scale of Fig. 4. Rare gases were not detected during crushing, indicating a content at least 300 times less than for the above sample of fines. It is mentioned that gas analysis was only used as a monitor of crushing, and not for the main investigation. (Otherwise much more vigorous crushing, and less ion pumping, would have been used, to bring up the evolved gas concentrations.)

The e.p.r. signal from the crushed sample differed from that before crushing. The change occurred in the region about $g = 2$, and was made clearly evident by displaying the output in the form of a teletype print out from the PDP8 computer, taken via an analogue to digital converter from the spectrum accumulator (digital memory oscilloscope). This facilitates accurate detection of changes by setting the manganese marker signals exactly coincident using the computer. The traces are shown in Fig. 5. The original resonance at $g = 2.009$ was replaced by a new resonance at $g = 2.006$, with width about 20 gauss.

The new signal is due to paramagnetic entities induced by the crushing. These entities are usually unpaired electrons at bonds broken either at surfaces created by rupture of bonds, or at internal damage centers. A distinction can be drawn if the signals are affected by exposure to gases, which principally affect the surfaces.

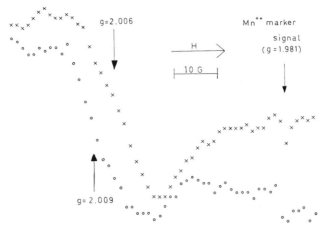

Fig. 5. Teletype output of e.p.r. signals before (○) and after (×) crushing. Manganese marker signals set coincident.

Experiments were performed with oxygen, and also nitrogen, hydrogen and water vapour.

GAS EXPOSURE

Oxygen was admitted to the system by diffusion through a heated silver tube, ensuring purity at least as high as mass spectrometer grade gas. Careful monitoring of the e.p.r. signal at $g = 2.006$, using accumulation to enhance sensitivity, revealed an effect due to oxygen at exposures of 10^{-2} torr min and higher. The signals are shown in Fig. 6. In the presence of oxygen, a small shoulder at about the center of the $g = 2.006$ line was enhanced. On pumping out the oxygen, the shoulder reduced

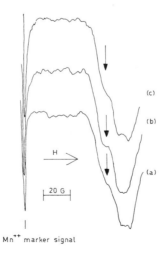

Fig. 6. E.p.r. signals (a) before (b) after exposure to oxygen (room pressure), and (c) after pumping out the oxygen.

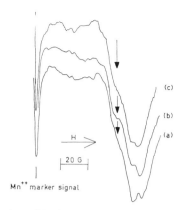

Fig. 7. E.p.r. signals (a) before, (b) after exposure to dry air (room pressure), and (c) after pumping out the air.

again. The experiment was repeated a number of times to establish that the change, though small, was reproducible. The size of the shoulder appeared to be proportional (not linearly) to oxygen pressure.

In subsequent measurements the samples were exposed to wet and dry air, wet and dry nitrogen and dry hydrogen. Effects were only observed with oxygen, indicating that nitrogen, hydrogen and water vapour did not induce changes detectable by e.p.r. The e.p.r. signals before and after exposure to air are shown in Fig. 7. Note that the change in shoulder is more pronounced, due to the higher pressure. Many repeats of this experiment were carried out to check that the only consistent reproducible change in the resonance region studied, was the change in shoulder on the line at 2.006.

Experiments were also carried out at liquid nitrogen temperatures. The broad signal at $g = 1.9$ increased by a factor of 3.9 which is expected according to Boltzmann statistics. However the signal at $g = 2.006$ only increased by a factor of 1.5. This is consistent with its being partly of surface origin, since some other surface signals (CHUNG and HANEMAN, 1966; MILLER and HANEMAN, 1970) have also been found to increase by less than the Boltzmann factor on cooling to liquid nitrogen temperatures. Admission of oxygen at low temperatures caused less changes than at room temperature.

The fact that the resonance changed in the presence of gas, indicates that the affected portion is located at the surface. To explain the results, it is thought that the resonance at $g = 2.006$ which is observed after the crushing process is probably a composite of the original line at $g = 2.009$ plus one (or two) new lines in the vicinity. One of these new lines is due to resonance centers at the surface and is affected by oxygen. This explains why only a portion of the resonance at $g = 2.009$ is affected (the "shoulder" near the center). The fact that the line is composite can not be directly deduced from the line shape, since, being a powder resonance, the details are to some extent averaged and in addition the signal to noise ratio, even after accumulation, is insufficient for detailed shape analysis.

An estimate of the density of surface spins was made by estimating the number of spins in the surface resonance, and subsequently measuring the surface area of the powder by the BET (Brunauer Emmett Teller) krypton adsorption technique. The figure obtained was 1 spin per 10^4 surface atoms, to an accuracy of only about a factor of 10, due to the uncertainty in the width and height of the surface resonance line.

TOTAL OXYGEN ADSORPTION

The above work has described the effects of oxygen on paramagnetic sites. However these form only a small proportion (1 in 10^4) of the surface atom sites. It is desirable to study the adsorption of oxygen on all the surface sites. Such measurements were made in two separate systems.

In the first experiments the same vacuum chamber as used for the e.p.r. measurements was employed. Oxygen uptake was studied by admitting the gas at a known rate in to the chamber, and monitoring the change in pressure both without and in the presence of, the crushed rock surfaces. The best method is to admit a known quantity of gas from a reservoir. However due to lack of suitable valving arrangements on this system the following method was used.

First volumetric technique

The oxygen diffusion leak (silver tube) was heated to a given temperature (monitored by a thermocouple, and checked by the power input to the heating coil). The 8 l/s ion pump was left on during this process of bringing the leak to temperature, thus keeping the base pressure at about 5×10^{-8} torr. When the correct operating temperature was attained and was stable, corresponding to the required rate of admission of oxygen, the ion pump was turned off. The oxygen pressure then built up rapidly till a predetermined pressure was reached, and at this point the leak was turned off. Thus a requisite amount of oxygen was in the system, and the subsequent changes in pressure gave information on adsorptive processes in the chamber.

In all cases the pressure began to fall. It is believed that this was due, at least in part, to pumping by the sputtered titanium film left in the ion pump body. This experiment was repeated several times to establish the reproducibility of the technique. A dummy crushing procedure was used, in which the hammer was caused to reciprocate without actually crushing any rock. A set of curves is shown in Fig. 8. Pressure measurements were made by switching on an ionization gauge with a thoria coated filament, 50 microamps emission, momentarily as required. This minimized filament reactions with the oxygen.

When the experiment was repeated after crushing a rock sample, two effects were noted. The rise of oxygen pressure after switching off the ion pump was slower. In addition the rate of fall of oxygen pressure after switching off the leak was markedly faster. This is shown in the bottom curve of Fig. 8. Although the method used in this experiment was not optimum, there is a clear demonstration of oxygen uptake by the rock. This was confirmed and made more quantitative in a separate set of experiments in a different vacuum system.

The tendency for the adsorbed oxygen to desorb on heating, was tested by transferring the crushed powder into the attached quartz e.p.r. tube, and immersing this

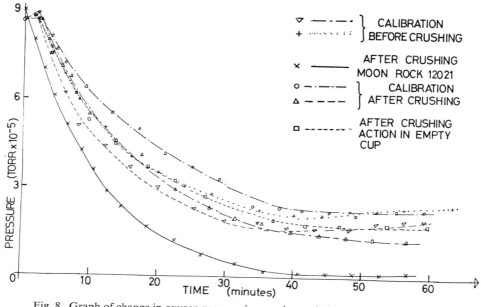

Fig. 8. Graph of change in oxygen pressure in experimental chamber, versus time after oxygen diffusion leak is switched off, in first volumetric technique. Note lowest curve, for case where freshly crushed lunar rock is present.

in a coil furnace. Twenty minutes was allowed at each temperature for the heat to spread throughout the powder. It was found that the usual background gases H_2O, CO, and CO_2 increased in the normal way on heating. The evolution of oxygen by comparison was slight, and comparable with that obtained when non oxygen covered materials were heated. The relative amounts of gas evolved are shown in Fig. 9. Note that the mass 32 peak (oxygen) has been increased by a factor of 50 for display on the graph. The desorption of oxygen was negligible. This indicates that the heat energy of adsorption is high, probably several electron volts.

Second volumetric technique

In the second technique, a separate ultra high vacuum system was used, designed specifically for straightforward volumetric type adsorption experiments. The main system was of pyrex glass, separated by bakeable metal valves from the liquid nitrogen trapped oil diffusion pump, and from the 50 l/s metal ion pump. The latter valve prevented oxygen uptake by remanent titanium films. The crushing arrangements were similar to those in Fig. 1 except that two glass cups were provided, one containing the rock sample, and one being empty, into which a dummy crushing action could be carried out. An oxygen reservoir was used, consisting of a 1 litre flask which was filled from a heated silver diffusion leak to a pressure of about 0.1 torr. This flask was separated from the experimental chamber by two glass stopcocks. By opening and closing these in sequence, a slug of gas comprising the small volume between the stopcocks, could be admitted into the chamber. The slug volume was only 0.3 % of the storage volume, so that several slugs could be produced and admitted in sequence at approximately the same pressure.

As before, the pressure in the chamber was monitored after a slug of oxygen was admitted. Dummy runs were made in which a "crushing" process was first carried out in the empty cup. The chamber was pumped out again, valved off, and the pressure behaviour after a slug of oxygen was admitted, was monitored. Such curves are shown in the top portion of Fig. 10. The lunar rock was then crushed, with the chamber pumped to below 10^{-9} torr and valved off. When a slug of oxygen was admitted, the pressure in the chamber was immediately much lower than at this stage in the preceding trials, and remained so. This is shown by the bottom curve in Fig. 10. Subsequent slugs of oxygen were admitted, but the adsorptive action of the crushed rock continued over a large number of slugs.

The sample area was subsequently measured by the BET (Brunauer Emmett Teller) technique as 600 cm². From this data, and knowing the oxygen pressure drop and chamber volume, it was possible to calculate the surface coverage assuming 7×10^{14} atoms per cm² of surface. Figure 11 shows the increase in coverage as a

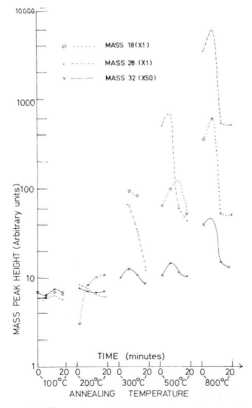

Fig. 9. Relative amounts of gas released during heating of crushed lunar rock. Quartz container tube was held for 20 minutes in furnace at each temperature.

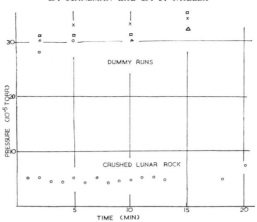

Fig. 10. Second volumetric experiment. Pressure in experimental chamber as function of
time after admission of standard slug of oxygen. Top graphs refer to dummy experiments,
involving crushing action in an empty cup, pumping, and then admission of slug at time
zero. Lowest curve shows greatly reduced resultant pressure when slug is admitted in
presence of crushed lunar rock.

function of the number of slugs of oxygen admitted, plotted on a logarithmic scale.
After the point corresponding to 25 slugs, the system including the samples was
heated, keeping the ion pump on, up to 170°C, at a maximum pressure due to background desorption, of 10^{-6} torr. Subsequently a large measured quantity of oxygen
was admitted, but the corresponding point on the graph (equivalent to 110 slugs)
remained on the straight line. This indicated that the activity of the surface had not
been changed by the heat treatment as already suggested by the negligible desorption
shown in Fig. 9. This was followed by still larger exposures, obtained by opening
the storage volume directly to the experimental tube. From the pressure measurements
the equivalent number of slugs could be calculated, and the two points are plotted

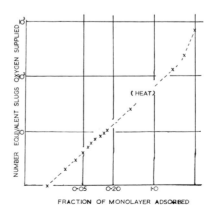

Fig. 11. Plot of total amount of gas adsorbed on lunar rock, in fractions of monolayer,
as standard slugs of oxygen are admitted. From powder area measurement, and assuming
7×10^{14} atoms per cm² of surface, monolayer coverage could be calculated.

in Fig. 11. They indicate that after these exposures there is a significantly reduced rate of adsorption. The coverage at which this change occurs is between 1 and 2 monolayers, but the results are consistent with a value of 1 monolayer since the coverage calculation, being based on the area determination, is only accurate to a factor of 2.

DISCUSSION

Paramagnetic sites

The fact that exposure to oxygen caused an enhanced resonance shows that the centres responsible are at the surface, since the gas must interact intimately with them. Furthermore this provides positive evidence that oxygen adsorbs on the paramagnetic sites produced by the fracture process. However the adsorption is weak since the change in e.p.r. signal is removed when the oxygen is pumped out, showing that the oxygen desorbs when vacuum is restored. In summary,

(a) surfaces of lunar rock exposed by crushing display an e.p.r. signal due to newly created unpaired electrons

(b) oxygen adsorbs on the paramagnetic sites at atmospheric pressures

(c) oxygen desorbs from these sites under high vacuum.

Non paramagnetic sites

Apparently a monolayer and more of oxygen can form on the crushed rock surfaces. This suggests that a large proportion of the surface sites adsorb oxygen. Since negligible desorption occurs even up to 800°C, the interaction is quite strong. The lunar rock studied in this work was from the moon's surface and therefore had a long history of cosmic ray and solar wind bombardment. The extent to which these factors affected the adsorption is not certain. Proper control experiments using very similar terrestrial rocks, or lunar ones far removed from the surface of the moon, would throw light on this aspect.

The degree to which the surfaces of the rock sample produced in our experiments by crushing, correspond to original lunar surfaces must be considered. It is clear that lunar material is subjected to meteorite bombardment. Outer surfaces show numerous effects of impacts of micrometeorites. All stages of impact metamorphism are apparent, from stress twinning, to crater pits, up to complete melting (RAMDOHR and EL GORESEY, 1970). Thus over the millions of years, fresh surfaces due to impact fracture have occurred to a considerable degree on material at the lunar surface. Many of these should be similar to the fracture surfaces created by the crushing operation. The conclusions for this should therefore apply to much of the original lunar surface.

The density of surface paramagnetic centres on the crushed lunar rock is about 1 per 10^4 surface atoms. This low figure makes it difficult to identify the precise nature of the surface paramagnetic entities, due to the inhomogeneous nature of the lunar samples. They could be associated with broken bonds on a constituent of very low concentration, such as ZrO_2 or NiO. Alternatively they could be associated with a low density of topographical inhomogeneities, such as steps, on fracture surfaces of a more abundant constituent. This may be more likely. However the signal does not correspond to that known to be obtained from crushing any of the allotropic modifications of SiO_2 (HOCHSTRASSER et al., 1969).

Extrapolation of these results (1 paramagnetic site per 10^4 surface atoms) suggests that the number of unpaired electrons on surfaces of lunar rock may be appreciable. They provide chemically active sites and also make a contribution to the surface charge. Surface charge effects have been found to be marked on vacuum cleaved lunar rocks (GROSSMAN et al., 1970). However some anomalous phenomena were noted on the lunar surface, such as the almost complete absence of dust on the surfaces of rocks. Deposits would have been expected due to meteoritic impacts on the neighbouring powdery ground. Surface creep processes as a possible cleansing agent were suggested (SIMMONS and CALIO, 1970). Such processes would be markedly affected by the state of charge and number of broken chemical bonds on the rock surfaces. These factors must be taken into account in more detailed investigations of the dust coverage phenomena.

Finally, one can make inferences about the pristine state of the surfaces of lunar material from the adsorption measurements. At pressures of 10^{-4} torr, adsorption of oxygen took place readily, exceeding a monolayer when sufficient gas was supplied. The partial pressure of oxygen on the moon is very low, certainly less than 10^{-9} torr. However even at extremely low pressures, the long periods to which the material was exposed, result in substantial values for the total exposure (pressure × time). Thus even at 10^{-15} torr, the exposure after 10^9 years is of order 1 torr minute, which is order of magnitude greater than that required to produce a monolayer coverage. There is thus good reason to expect that lunar material near the moon's surface is not clean, but is covered with layers of gas. This may not apply to material located far beneath the moon's surface.

Acknowledgments—Mr. L. G. VAN LOON performed the oxygen adsorption measurements in the second volumetric experiment. The surface area of the sample used there was measured by Mr. J. HIGINBOTHAM. The computer program used was written by Mr. D. HERON. Equipment used in these investigations was supplied by the Australian Research Grants Committee, and by the U.S. Army Research and Development Command (Far East).

REFERENCES

CHUNG M. F. and HANEMAN D. (1966) Properties of clean silicon surfaces by paramagnetic resonance. *J. Appl. Phys.* **37**, 1870–1889.

GROSSMAN J. J., RYAN J. A., MUKHERJEE N. R., and WEGNER M. M. (1970) Surface properties of lunar samples. *Science* **167**, 743–745.

HANEMAN D. (1968) Electron paramagnetic resonance from single crystal cleavage surfaces of silicon. *Phys. Rev.* **170**, 705–718.

HOCHSTRASSER G., ANTONINI J. F., and PEYCHES I. (1969) M.S. and e.s.r. studies of dangling bonds and adsorbed ions on the pristine surface of silica. In Structure and Chemistry of Solid Surfaces (editor G. Somorjai) pp. 36–1 to 36–11. John Wiley.

MANATT S. L., ELLEMAN D. D., VAUGHAN R. W., CHAN S. I., TSAY F. D., and HUNTRESS W. T., JR. (1970) Magnetic resonance studies of some lunar samples. *Science* **167**, 709–711.

MILLER D. J. and HANEMAN D. (1970) Evidence for carbon contamination on vacuum heated surfaces by e.p.r. surface. *Science* **19**, 45–52.

RAMDOHR P. and EL GORESEY A. (1970) Opaque minerals of the lunar rocks and dust from mare tranquillitatis. *Science* **167,** 615–618.

SIMMONS G. and CALIO A. J. (1970) Summary of scientific results. Apollo 12 Preliminary Science Reports, NASA, 1–6.

WEEKS R. A., CHATELAIN A., KOLOPUS J. L., KLINE D., and CASTLE J. G. (1970) Magnetic resonance properties of some lunar materials. *Science* **167,** 704–707.

Proceedings of the Second Lunar Science Conference, Vol. 3, pp. 2543–2558
The M.I.T. Press, 1971.

Nuclear track studies of dynamic surface processes on the moon and the constancy of solar activity

G. Crozaz

Service de Géologie et Géochimie Nucleaires Université Libre de Bruxelles, Belgium
and Laboratory for Space Physics, Washington University, St. Louis, Missouri 63130

and

R. Walker and D. Woolum

Laboratory for Space Physics, Washington University, St. Louis, Missouri 63130

(*Received* 19 *February* 1971; *accepted in revised form* 30 *March* 1971)

Abstract—Most of the samples of the core, as well as those from 12042 and 12044, show nuclear track densities $> 10^8/cm^2$ indicating that the regolith is well mixed down to a depth of at least 70 cm. However, the coarse grained layer at ~ 13 cm in the core has low densities. This layer was exposed on the surface for less than 15×10^6 years before being covered over. The light trench sample 12033 also gives low track densities and was deposited $\leqslant 40 \times 10^6$ years ago. Two rocks, 12064 and 12063, are particularly interesting—each has at least one surface that was exposed to the sun for an integrated time $\leqslant 1.5 \times 10^6$ years. From 12063, a lower limit for the rate of impact pitting of 2 pits/cm² (> 1 mm) per 10^6 years is inferred. These data support the view of a reasonably active lunar surface. Solar flare maps have been made in polished sections of rocks 10057 and 12063. Starting at points of highest track density (typically $\sim 10^9/cm^2$ at 10 microns) and proceeding inwards, the track density, ρ, varies as $R^{-\alpha}$ with $\alpha \sim 1$. The difference between this result, which is typical for a wide range of lunar samples, and the $R^{-2.5}$ spectrum observed in a sample of Surveyor glass is attributed mainly to the effects of lunar erosion. The short exposure age of 12063, coupled with the fact that it is apparently in erosion equilibrium leads to an estimated erosion rate $\geqslant 7 \times 10^{-8}$ cm/yr. Independent estimates in this and other rocks lead to values between 3×10^{-8} and 10^{-7} cm/yr. Comparison of four different rocks gives a maximum variation of 3.5 in the ratio of the solar flare activity to the rate of rock erosion at different times in the past.

Introduction

TRACK RESULTS in the Apollo 12 samples differ in important respects from those of Apollo 11 (Crozaz *et al.*, 1970b; Fleischer *et al.*, 1970; Lal *et al.*, 1970; and Price and O'Sullivan, 1970). This paper is a brief summary of our work on both Apollo 11 and Apollo 12 samples. In it we give values for surface exposure ages, erosion rates, solar flare spectra, and surface stirring. We also give a preliminary estimate of the absolute rate of impact pitting based on studies of rock 12063.

Almost all our measurements were made on polished feldspar crystals etched between 1 and 15 min in a constant boiling solution of NaOH (6 g NaOH to 8 g H_2O). Track densities were measured using either a scanning electron microscope (SEM) or an optical microscope. In the SEM tracks appear as deep, geometrically shaped holes. Comparison of optical and SEM data in low track density crystals taken from rocks show that SEM track densities are typically higher than those measured in the optical microscope where a characteristic "comet" appearance is taken as a criterion for track identification. If only pits whose diameter is ≥ 0.5 the maximum diameter are counted in the SEM, we obtain agreement between the optical and electron microscopes within $\pm 5\%$. If all deep pits are counted, the SEM values are greater by a factor of 2 ± 0.2. As previously noted in Apollo 11 samples, crystals removed from the same rock show a highly variable background of very shallow pits. Although the density of these shallow pits is sometimes so high that it is impossible to measure a true density of long tracks, the deep pits generally stand out well enough to be counted reliably.

Track Results in Cores and Fines:
Stirring of the Regolith

About 15 crystals were studied from each of twenty locations in the double-core tube 12025 and 12028 (near Halo Crater), and from the samples 12030 (between LM and Head Crater), 12033 (trench sample from Head Crater), 12042 (~ 20 meter northwest of Halo Crater), and 12044 (south rim of Surveyor Crater). The crystals were selected from either the > 100 mesh or > 200 mesh fraction and ranged in size from $\sim 100\,\mu$ to $300\,\mu$. Most of the samples of the core as well as those from 12042 and 12044 showed track densities $> 10^8/\text{cm}^2$ (see Fig. 1) characteristic of those found in the cores and bulk fines of Apollo 11. There is apparently some disagreement between track groups on the proportion of grains with densities $> 10^8$ t/cm^2. In

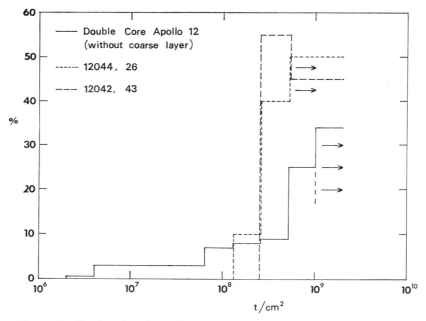

Fig. 1. Distribution of track densities in feldspar crystals from different samples of fine material. The densities are the *total* pit densities as seen in the scanning electron microscope. The core tube results represent the average over 11 positions with a total of ~ 200 crystals. Samples from the coarse-grained layer at ~ 13 cm are not included in the average. Where noticeable gradients were found the track densities were measured at the center.

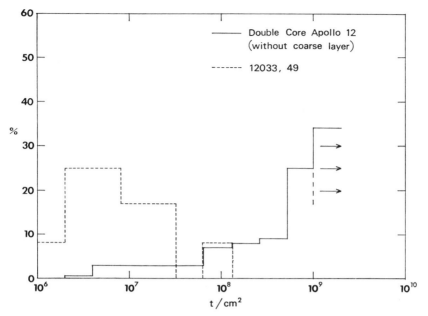

Fig. 2. Track density distribution in the light trench sample 12033,49 compared to the average of the double core.

agreement with our observations, BARBER *et al.* (1971a) find that $\geqslant 70\%$ of all grains $\geqslant 50\ \mu$ in size have densities $> 10^8$ t/cm^2. In contrast, FLEISCHER *et al.* (1970) report only 36% for the core sample 10005 from Apollo 11. COMSTOCK *et al.* (1971) and BHANDARI *et al.* (1971) also report $< 50\%$ for the soil samples of Apollo 12. It is not clear whether these differences arise from different methods of sampling or from differences in methods of observation. Grain-size bias is probably particularly important since it has been shown that very small grains ($\leqslant 1\ \mu$) have track densities $\geqslant 10^{11}$ t/cm^2 (BORG *et al.*, 1970; BARBER *et al.*, 1971a).

Also, track gradients near the edges were frequently found ($\sim$ one out of five crystals) indicating that the high track densities were induced by solar flare particles when the crystals were on the very surface of the moon. All crystals containing easily visible gradients ($\geqslant$ a factor of two variation from edge to center) had densities $> 4 \times 10^8$ t/cm^2. Only one crystal was found with a uniform irradiated border characteristic of crystals removed from gas-rich meteorites (LAL and RAJAN, 1969; and PELLAS *et al.*, 1969). No significant difference was found between the uppermost and bottom-most core positions showing that the different layers (with one exception) have had a similar radiation history down to a depth of at least 70 cm.

In striking contrast to Apollo 11, however, certain of the fines samples are clearly different and must have been added to the regolith comparatively recently. As can be seen from Figs. 2 and 3 both sample 12033 and samples from the coarse layer of the double core tube (12028,61, 12028,67, and 12028,69) have much lower track densities than the other fines samples.

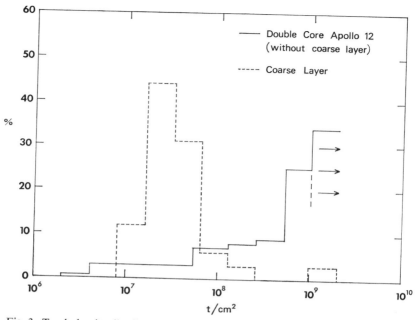

Fig. 3. Track density distribution in samples removed from the coarse-grained layer in double core compared to the average for all other core positions.

The low track density crystals in sample 12033 cannot have remained at the position where they were found for a very long time. At the sampling depth of 15 cm we estimate that the rate of production of tracks from galactic cosmic rays is 3.2×10^4 $t/cm^2 \times 10^6$ yrs. (In calculating exposure ages we have used the same spectra and assumptions previously described (Crozaz et al., 1970b). It should be noted that, in common with other track groups, our absolute numbers are subject to future revision, probably not exceeding 25%, as more detailed information of the track registration behavior of lunar minerals becomes available.) Taking the lowest density observed in the optical microscope ($1.3 \pm 0.1 \times 10^6$ t/cm^2) we obtain a maximum residence time of $(40 \pm 5) \times 10^6$ years. If it is assumed that this crystal formed part of the original ejecta blanket of Head Crater and has remained essentially undisturbed, then this would correspond to the age of formation of Head Crater. Various authors have suggested a remote origin for the light material of 12033; in any event it is clearly a relatively new addition to the regolith that has not been extensively stirred and irradiated.

Crystals from sample 12030 also show anomalously low track densities, similar, but somewhat higher and more variable than 12033. This is therefore also relatively fresh regolith material. Since 12030 was also removed from near Head Crater, this supports the association of this fresh material with this feature. The anomalous core tube samples come from a coarse-grained layer that is evident to the naked eye. The track results are consistent with a model in which the coarse layer was first laid down on the surface as a single unit and irradiated with both solar flare and galactic cosmic rays. To explain the range of track densities found in the bottom-most region of this layer, it is necessary to assume that the layer was stirred somewhat during this period. Following the surface irradiation the layer must have been covered over by well-mixed regolith material thrown out by a nearby impact. The maximum period of surface exposure can be obtained from an argument similar to that used above for 12033. Taking the lowest density crystal and assuming that it was irradiated at the bottom of the coarse layer, we obtain a maximum surface exposure age of $(15 \pm 1) \times 10^6$ years. It is interesting to note that this age is typical of surface exposure times found for lunar rocks. Contrary to the views of a number of investigators expressed at the Second Lunar Science Conference, we thus believe that the existence of a layered core is perfectly compatible with previous estimates of lunar gardening rates.

GALACTIC TRACKS IN ROCKS: EVIDENCE FOR FRESHLY EXPOSED MATERIAL AND LIMITS ON ABSOLUTE RATE OF IMPACT PITTING

Track densities in rocks are dominated by solar flares in the first few mm and by galactic cosmic rays at deeper depths. The galactic track densities are not substantially affected by the low erosion rates found from lunar rocks (CROZAZ et al., 1970b); meteorite studies have also shown that galactic cosmic rays are constant in time (ARNOLD et al., 1961; and MAURETTE et al., 1969). For these reasons, galactic tracks measured at $\geqslant 1$ cm from the surface can be used to measure surface exposure ages.

Using this measure we have found that at least two of the Apollo 12 rocks have been on the surface for much shorter periods of time then any of the Apollo 11 rocks so far measured. Sample 12064,13 is a 2 g chip removed from the surface labeled "top" in the Lunar Receiving Laboratory cutting diagram. The external surface is not readily identifiable since no impact pits are visible. However, samples removed from different edges show a forty-fold variation in track density, typical of what would be expected from the variation of solar flare track densities. This proves that the nominal "top" was indeed exposed to the sun at one time on the lunar surface.

The lowest track density measured was $1.2 \pm 0.2 \times 10^6/\text{cm}^2$ at a point that was at a maximum of 1 ± 0.1 cm from the external surface. This is the *lowest* track density by far that we have found in any lunar rock. With the maximizing assumption that all the tracks were produced by galactic cosmic rays during the time that the "top" surface was exposed, we calculate a maximum surface exposure of $\leqslant 1.5 \pm 0.2 \times 10^6$ years. Detailed investigation of the track profiles in well-documented samples from the rock may lower this age; it cannot increase.

Much more detailed work has been done by us on rock 12063 where we had a vertical section running between the nominal "top" and "bottom" of the rock from which to select samples. This section was in the form of a thin slab 1.5 cm wide and 7.6 cm long. Two saw cuts were made parallel to the surface to give samples extending from the surface to a depth of ~ 1 cm. These samples were mounted in epoxy, polished, and then studied in the SEM. Individual grain mounts were also made from carefully located samples along the remaining interior section. An additional vertical bar was cut from the end samples to provide material for thermoluminescence measurements described in a companion paper (HOYT et al., 1971).

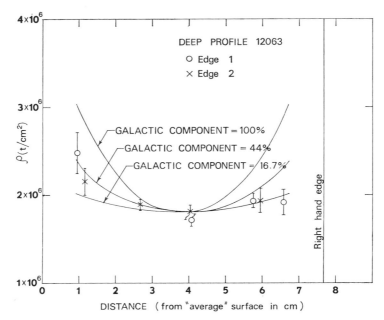

Fig. 4. Track density vs. depth in interior region of rock 12063. The solid curve labeled "100% galactic component" shows the expected variation of galactic cosmic-ray tracks with depth if all the galactic tracks had been accumulated while the rock was on the surface. The other solid curves show the expected variation if 44% and 16.7% respectively, of the total track density was accumulated during such a surface exposure, the remainder being added while the rock was buried in the surface at shallow depths.

As shown in Fig. 4 the track density drops to $\sim 2 \times 10^6/\text{cm}^2$ at a depth of ~ 1 cm and then stays quite constant from 1 cm to the center. The density mounts as either surface is approached, and it is clear that *both* sides of the rock have been exposed to the sun. This is also evident from the impact pits which are found on *all* surfaces of the rock. The track density of $2 \times 10^6/\text{cm}^2$ at 1 cm shows that neither surface could have been exposed more than 3×10^6 years.

In fact, the surface exposures must have been considerably shorter. If the tracks found at 1 cm depth were produced while the rock was on the surface, we would expect to find a reduction of a factor of 0.57 in going to the center. No such reduction is found. We must either assume that the cosmic-ray energy spectrum was very different in the past or that many of the tracks were accumulated while the rock was buried at shallow depths, and there was essentially no "inside" or "outside" to the rock. As shown in Fig. 4, the data can be fit if we assume that only between 16% and 44% of the tracks at the rock center were accumulated during a time in which the rock was fully exposed at the surface.

The rock does not appear to have been irradiated symmetrically, and we estimate that the surface that we have labeled A was exposed $\leq 1.5 \pm 0.5 \times 10^6$ years and the opposite surface D for $< 0.7^{+0.3}_{-0.7} \times 10^6$ years. Recently received documentation photographs show that A is the surface labeled "top" and "D" the surface labeled "bottom" in the Lunar Receiving Laboratory cutting diagram. Radioactivity data (RANCITELLI et al., 1971) show a complex history for rock 12063. Surface A has the highest Na^{22} content and must have been exposed most recently. This is also consistent with the thermoluminescence data from our laboratory (HOYT et al., 1971). Surface D on the other hand contains the largest proportion of Al^{26} and must have been uppermost during the largest proportion of time during the last $\sim 10^6$ years. Both surfaces show impact pits and solar flare tracks.

We wish to emphasize the importance of rocks with short exposure ages. Simply counting the impact pits on surfaces so young that equilibrium has not been achieved measures the absolute rate of arrival of interplanetary dust particles. In 12064 no pits were found by us in an area of ~ 1 cm^2. However this rock is very fragile and arrived in an obviously deteriorated condition. The absence of pits is therefore not significant. Rock 12063, which is much sturdier, is a more likely candidate for this work. The total surface area in our sample, however, was small (0.3 cm^2) and no pits were observed. However, close-up photographs of the rock taken at the LRL show pits on all sides. Working from these photographs we count $\geqslant 3$ pits/cm^2 on one surface and $\geqslant 1$ pit/cm^2 on the other in the vicinity of our samples. This suggests that the surfaces may not be in equilibrium (~ 10 pits (> 1 mm)/cm^2) (HORZ et al., 1970). However, the photographs are unsatisfactory, and it is quite possible that the surfaces are saturated. In any event a lower limit of pit formation would appear to be ~ 2 pits/cm$^2 \times 10^6$ years.

Galactic track data for all the Apollo 12 rocks that we studied are shown in Table 1.

Table 1. Summary of Apollo 12 rock results

Sample	Track density (t/cm^2)	Depth	Surface dwell time	Comments
12013,10,44 (internal chip)	$1.2 \pm 0.2 \times 10^7$	10^{+2}_{-1} mm	$14 \pm 3 \times 10^6$ yrs	similar to Apollo 11 values
12013,10,14 (external chip)	$6 \pm 1 \times 10^8$	external surface	—	—
12034,25 (external chip)	$3 \pm 0.3 \times 10^7$	0 to 4 mm	—	probably lithic fragment in breccia
12040,10 (internal chip)	$5.4 \pm 0.5 \times 10^6$	≈ 7 mm ± 3 mm	$\geqslant 6 \pm 1 \times 10^6$ yrs	assumes max depth also 2π irradiation
12040,9 (external chip)	$3.6 \pm 0.3 \times 10^6$	0 to 8 mm	—	surface labeled B may be buried surface
12040,11 (external chip)	$2.1 \pm 0.15 \times 10^8$	0 to 8 mm	—	high density shows surface labeled A and top was exposed to sun
12063,32 (vertical section)	$1.77 \pm 0.06 \times 10^6$	40 mm	surface $1 \leqslant 1.5 \times 10^6$ yrs	see text for a detailed discussion of this rock
	$1.9 \pm 0.2 \times 10^8$	10^{-2} mm	surface $2 \leqslant 7 \times 10^5$ yrs	
12064,13 (external chip)	$1.2 \pm 0.2 \times 10^6$	10 ± 1 mm from presumed surface	$1.5 \pm 0.2 \times 10^6$ yrs	age assumes 2π irradiation
	$4 \pm 0.4 \times 10^7$	0 to 1 mm from presumed surface		

Solar Flare Maps: Lunar Erosion Rates

Solar flare tracks completely dominate galactic cosmic-ray tracks in the first few mm of any rock that has been exposed for an appreciable time to the sun. (This statement describes the present state of affairs in the solar system. It is possible, though unlikely, that low-energy galactic particles were more important in the past due either to reduced solar modulation or to increase in the number of low energy galactic particles.) We have made detailed studies of these tracks in polished sections of rocks 12063 and 10057. The lengths of exposed surface were 1.3 cm and 1 cm respectively and the maximum depths were 1 cm and 4 mm respectively. A picture of the most studied section of rock 12063 is shown in a companion paper by Hoyt *et al.* (1971). After polishing, the surface edge was photographed at 500 × in the SEM and a mosaic map of the edge was constructed (see Fig. 5). Regions that appeared to be definitely on the surface of the rock (as seen in three dimensions in a low-power optical stereo microscope) were then selected for detailed study. Following etching, the selected regions were examined to find crystals with a maximum track density such as the one shown in Fig. 6. The track densities were then measured as a function of depth in the rock. This method is far superior to that used previously for Apollo 11 samples (Crozaz *et al.*, 1970b) where crystals were hand-picked and separately

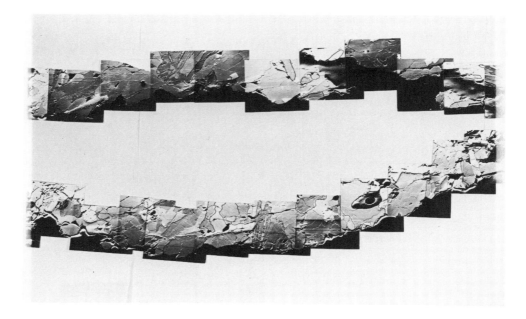

Fig. 5. Mosaic SEM photograph of edge of section A of vertical section of rock 12063. The bottom set of photographs are simply a continuation along the same edge as shown in the top set. Certain promising areas containing feldspar crystals (identifiable by their relatively dark appearance) have been outlined in ink on the original photographs. The original maps are made at a magnification of 500×. The total length shown here is ∼ 7 mm.

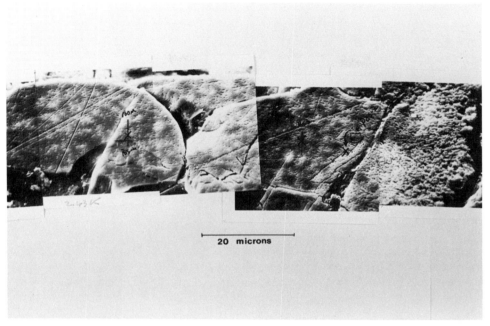

Fig. 6. Close-up of a moderately high track density region in section A following etching. The rapid increase of solar flare tracks towards the outside surface (right hand edge) is clearly visible.

mounted. Figure 7 shows a track profile obtained this way for rock 12063 and Fig. 8 a comparison with similar data for rock 10057.

In both rock 10057 and rock 12063 we find that the density, ρ, vs. depth, R, can be represented by the equation $\rho = CR^{-\alpha}$ where α is typically ~ 1. In Table 2 we show that gradients measured in other lunar samples and in gas-rich meteorites also have values of $\alpha \sim 1$. The data in Table 2 have not been corrected by subtracting a galactic contribution. This correction varies from sample to sample but does not increase α in any case by more than ~ 0.3.

These data in lunar rocks differ considerably from those found in a glass filter removed from Surveyor III (CROZAZ and WALKER, 1971; FLEISCHER et al., 1971; and BARBER et al., 1971b) where an effective slope of $\alpha \sim 2.5$ was found between equivalent depths of 10 μ to 500 μ. We now wish to discuss this difference and show how measurement of track profiles in lunar rocks can be used to measure lunar erosion rates and to set limits on the constancy of the sun's activity.

It is first important to realize that *in the absence of erosion* the parameter α is a rough measure of the spectral index γ of the energy spectrum of the bombarding particles represented as $dN/dE = CE^{-\gamma}$. The general equation for the calculation of track density vs depth has the following form (FLEISCHER et al., 1967):

$$\rho(R) = \left(\frac{dN}{dE}\right)\left(\frac{dE}{dR}\right)_R \exp\left(-R/L\right) \Delta R_c, \qquad (1)$$

where ΔR_c is a constant and L is the mean free path for nuclear interaction. In the near

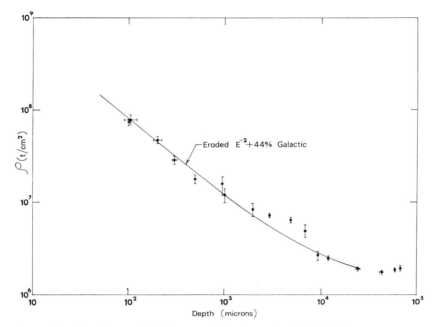

Fig. 7. Track density vs. depth in a vertical section of rock 12063. The measurements were taken starting in a region of maximum track density. The solid curve represents the sum of a galactic component of 44% (see Fig. 4) and a solar flare energy spectrum given by $dN/dE = CE^{-2}$. An erosion rate of 10^{-7} cm/yr is assumed. The break in the experimental curve at ~ 2 mm probably arises from the presence of a local valley in the surface topography of the rock in the direction perpendicular to the plane of the section.

surface regions, R/L is ~ 0 and the exponential term is ~ 1. If we make the approximation that $R = kE^{\beta}$, then it is easy to show that $\rho(R) = k'R^{[(-\gamma-\beta+1)/\beta]}$. At low energies, $\beta \sim 1$ (the actual value is ~ 1.3) and $\rho(R) \approx R^{-\gamma}$.

In the case of the Surveyor glass, erosion is negligible, and the measured value of the slope implies an energy spectrum for the bombarding particles of $dN/dE = CE^{-\gamma}$ with γ between 2 and 3. Detailed computer fits to the Surveyor data confirm this simple interpretation (Crozaz and Walker, 1971; and Fleischer et al., 1971).

Erosion of the rocks modifies the density vs. depth profile. Let $\dot{R}$ be the rate of erosion and $T_{\exp}$ the total exposure time of the rock. At depths $\gg \dot{R}T_{\exp}$ the effect of erosion is negligible while for depths $\ll \dot{R}T_{\exp}$ the effects are extremely important. Fortunately, as we will now show, the net effect of the erosion in the affected layer is simply to reduce the exponent in the power law by one unit.

Let X represent the distance of a test crystal below the *present* surface of a rock. In the case of a constant erosion rate $\dot{R}$, the distance below the surface at any time t in the past is simply $R(t) = X + \dot{R}t$. The rate of track production as a function of time is $\dot{\rho}(t) = KR(t)^{-\gamma} = K[X + \dot{R}t]^{-\gamma}$. Integrating this expression over the total exposure time of the rock gives $\rho(X) = [K/\dot{R}(1 - \gamma)][X^{-\gamma+1} - (X + \dot{R}T_{\exp})^{-\gamma+1}]$. When $X \ll \dot{R}T_{\exp}$, that is, when we are considering a depth where erosion is dominant, this reduces simply to $\rho(X) = [K/\dot{R}(1 - \gamma)][X^{-\gamma+1}]$.

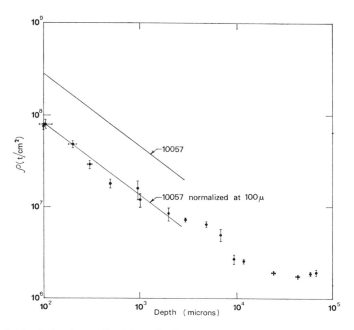

Fig. 8. Track density vs. depth in rock 10057 compared to 12063. The upper solid curve gives the direct experimental values while the lower solid curve is normalized to agree at 100 μ. The line given for 10057 represents the average of five independent depth profiles starting at high track density positions.

Table 2. Values of the slope α in the expression $\rho = CR^{-\alpha}$ for different samples (not corrected for galactic cosmic rays)

Sample	Range	Slope
10058	5 μ–30 μ	0.9 ± 0.1
(oriented surface crystal)		
10017	300 μ–2000 μ	1.1 ± 0.4
(grain mounts)		
10003†	100 μ–1000 μ	1.0 ± 0.2
12013	7.5–20 μ	1.0 ± 0.2
(polished section)		
10057 (1)	3 μ–300 μ	0.8 ± 0.1
10057 (2)	1 μ–300 μ	0.7 ± 0.1
10057 (3)	5 μ–260 μ	1.1 ± 0.2
10057 (4)	3 μ–3000 μ	0.8 ± 0.1
10057 (5)	2 μ–500 μ	0.8 ± 0.1
(different regions of		
single polished section)		
12063 (22)	100 μ–1000 μ	0.9 ± 0.1
(polished section)		
12022‡	10 μ–1000 μ	1.1 ± 0.1
Surveyor glass	20 μ–200 μ	2.4 ± 0.3
Kapoeta*	5 μ–40 μ	0.9 ± 0.2

* PELLAS et al. (1969; † PRICE and O'SULLIVAN (1970);
‡ BARBER et al. (1971).

This expression has several interesting consequences. First, in erosion equilibrium the *slope* of the track density vs. depth is independent of erosion rate. Second, the index describing the energy spectrum of the solar flare particles producing the tracks is given simply by adding one to the slope on a log–log plot of track density vs. depth. Finally, the *absolute value* of the track density at any given depth depends directly on the solar activity (the constant C is αk) and inversely on the erosion rate $\dot{R}$.

Consider now the application of these results to our data on lunar rocks. First, as shown in Table 2 the slopes of density vs. depth on log-log plots for all samples lie close to unity. As previously noted, these are distinctly different from the slope of ~ 2.5 found for the Surveyor glass. There are at least three possible explanations for this difference: (1) In the lunar rocks we are not looking at the true surface. Either material was removed from the rocks between the time they were on the moon and they were studied by us, or they were covered with a dust layer on the moon. The effect of such a layer is to underestimate the true slope. (2) Erosion of the rock has modified the density vs. depth curve. (3) The properties of solar flares during the period of exposure of the Surveyor glass are not characteristic of the long time average properties of solar flares.

Although Surveyor photographs show that lunar rock surfaces are usually clean, dust is observed in pockets and depressions (SHOEMAKER, 1970). It is also certain that some abrasion of the rocks occurs during their voyage from the moon to the laboratory. As far as we know the only way to decide on the importance of (1) is to make a large number of measurements in different rocks. If each time a high density region is chosen, the same gradient is found, then such regions are likely to be true surface crystals. As we show in Table 2 all rocks so far studied appear to give similar slopes when regions of maximum density are measured.

In our earlier Apollo 11 paper (CROZAZ *et al.*, 1970a) we pointed out the clear necessity for including erosion if solar flare particles continued to follow a simple power law spectrum down to energies of ~ 0.5 MeV/nuc. At that time, however, there existed the possibility of a sharp flattening of the energy spectrum at ~ 10 MeV/nuc. The Surveyor data now seem to make this latter possibility unlikely (though it is always important to keep in mind the fact that the Surveyor data represent a very small time sample) and make the presence of erosion essential.

Although there seems to be unanimity of opinion between track groups on the importance of erosion in determining the shape of the track density vs. depth curves in lunar rocks, the absolute magnitude of this erosion, as well as the physical processes involved, are not agreed upon (CROZAZ *et al.*, 1971; FLEISCHER *et al.*, 1971; and BARBER *et al.*, 1971b). This disagreement is apparently based partly on different input data, and partly on differences in interpretation. For these reasons we will treat this subject in some detail.

There are two general approaches to the problem of calculating erosion rates based on measurements of solar flare tracks. In the first approach, the rock data are matched, both in slope and absolute value, with computer calculations that use the Surveyor data as input. It is essential to the following discussion to realize that the final value of the inferred erosion rate depends inversely on the ratio of the sun's activity during the Surveyor period to its long-time activity averaged over the total exposure time of the rock. The inferred erosion rate also depends on the assumed sensitivity of the Surveyor glass relative to lunar materials.

In our earlier preprint, CROZAZ et al. (1971), we stated that the data in rock 12063 were consistent with an erosion rate of $\sim 10^{-7}$ cm/yr—provided that the activity during the Surveyor period was 2.5 times that of the long-term average. Two developments have occurred to modify this statement. First, private conversations with R. L. FLEISCHER and P. B. PRICE have convinced us that we assumed too low a sensitivity by a factor of three for the Surveyor glass. This would lower our inferred erosion rate. Second, however, additional data are now available indicating that the Surveyor period was not as active as we originally assumed. This has the compensating effect of raising the inferred erosion rate.

The best evidence on the relative activity of the sun during the Surveyor exposure comes from a comparison of short-lived and long-lived radioisotopes in lunar rocks. The results and analysis of FINKEL et al. (1971) show that the proton flux responsible for the production of Na^{22} (half-life 2.6 yrs) is $\sim 1.8 \pm 0.8$ times the average flux of protons responsible for the production of Al^{26} (half-life 7×10^5 yrs). Since the Surveyor exposure period is also 2.6 yrs, this gives a crude first estimate of the ratio of the activity during the Surveyor period to the long-term average.

However, two additional points must be kept in mind. First, it is estimated (ARNOLD, private communication) that $\sim 25\%$ of the Na^{22} activity was induced in a giant flare in 1956. Inclusion of this factor reduced the activity ratio to $\sim 1.4 \pm 0.6$. Second, and more important, as pointed out by RANCITELLI et al. (1971), the measured Al^{26} activity values (and hence inferred long-time average proton fluxes) can be lowered considerably by the inclusion of erosion. In particular, these authors estimate that the measured values can be reduced by a factor of 2.7 with an erosion rate of 10^{-7} cm/yr. (More recent calculations (RANCITELLI, private communication) have lowered this estimate to ~ 1.4 in the energy range of interest. This supports the view that the Surveyor exposure period may have been representative of the long-term solar average.) It is thus possible that the average solar activity during the Surveyor period was actually *less* by the factor (1.4/2.7) than the long-term value.

In view of this uncertainty we prefer to state our erosion rates based on the assumption that the activity during the Surveyor period was the *same* as the long-term average. In Fig. 7 we show that the near-surface data on rock 12063 can be well fitted with a long term solar flare spectrum αE^{-2} and an erosion rate of 10^{-7} cm/yr. The choice of the E^{-2} spectrum is discussed in the next section. Our choice of a (relatively) high erosion rate is based on a second, approximate approach that is independent of the detailed matching of absolute track densities. We have shown above that the net effect of erosion is to flatten the slope of the track density vs. depth curve in a region whose extent is $\sim \dot{R} T_{exp}$. We have further shown that T_{exp} for surface A of rock 12063 is $\lesssim 1.5 \times 10^6$ yrs. Since the slope in the region of 100 μ to 1 mm is similar to that observed in other rocks (see Table 2) we infer that 12063 is in erosion equilibrium to a depth of 1 mm and hence that $\dot{R} \geqslant 7 \times 10^{-8}$ cm/yr.

Although the slope of the track density vs. depth curve in 12063 is the same as in other rocks, the absolute value of the track density normalized at a depth of 100 μ is considerably less than in other rocks. For example, as can be seen in Fig. 8, rock

10057 has a track density ~ 3.5 as high. Track densities higher by factors of 2 to 3 than 12063 have also been reported for rock 10003 (PRICE and O'SULLIVAN, 1970) and rock 12022 (BARBER et al., 1971b). Fitting Surveyor data to these rocks as we have done above for rock 12063 would therefore give erosion rates of 3×10^{-8} to 5×10^{-8} cm/yr.

We wish now to compare our conclusions on erosion rates to those of other workers. FLEISCHER et al., 1971 show comparisons of calculated curves with rock data that lead them to conclude that erosion rates are $\leqslant 2 \times 10^{-8}$ cm/yr. It should first be pointed out that their curves are based on the assumption that the activity during the Surveyor period was ~ 2 times the long-term average. As they noted themselves, their results are compatible with an erosion rate of $\leqslant 4 \times 10^{-8}$ cm/yr when normalized to equal activity. In their comparison they show shaded regions of rock data without identifying the source of data. The most reliable results using polished sections as done here and by BARBER et al. (1971) lie along the lower edge of their curves, towards the higher erosion rates. Finally, they show no results for depths less than 100μ. Near-surface data given by CROZAZ et al. (1970) as well as by BARBER et al. (1971) would all lie close to their calculated curve of 10^{-7} cm/yr. We therefore conclude that our results are not incompatible with theirs and that erosion rates of 3×10^{-8} to 10^{-7} cm/yr are reasonable.

As discussed briefly in an earlier paper on the Surveyor glass (CROZAZ and WALKER, 1971), we do not agree with the extremely low erosion rates given by BARBER et al. (1971b).

It is also interesting to compare these results with erosion rates determined by independent methods. In our earlier Apollo 11 paper (CROZAZ et al., 1970) we showed that the erosion rate of lunar rocks must be $\leqslant 10^{-7}$ cm/yr. This number was based on the near equivalence of the proton spallation age and galactic iron age of rock 10057. A similar study of rock 10049 (COMSTOCK et al., 1971) gives $\leqslant 3 \times 10^{-8}$ cm/yr. The radioisotope measurements of FINKEL et al. (1971) and RANCITELLI et al. (1971) have also been interpreted as giving erosion rates of $\sim 2 \times 10^{-8}$ cm/yr and $> 10 \times 10^{-8}$ cm/yr respectively. Finally, the most detailed study of lunar impact pits to date (HÖRZ et al., 1970) has led to a "lower limit" of 2×10^{-8} to 4×10^{-8} cm/yr.

In comparing erosion rates, it is important to realize that there is a fundamental difference between the rates measured by fitting solar flare data and those described in the previous paragraph. All of the latter values represent a true mass wastage type of erosion. The solar flare rates on the other hand are determined by deliberately seeking out the regions of highest track density. Thus those regions that have been recently chipped away would be ignored. The net effect of this is that the solar flare determined rates represent lower *limits* for the mass wastage loss which could be considerably higher. The solar flare rates also set limits on any *atomic* or near atomic (e.g., solar wind fracturing) process. In our paper on the Surveyor glass we have in fact argued that the fact that the maximum track density observed in surface crystals from rocks at a depth of 10μ is $\sim$ only 2×10^9 t/cm^2 (compared to $\sim 10^6$ t/cm^2/yr from the Surveyor data) is a strong indication that such processes are important.

In summary, it is our present view that erosion is almost certainly important in determining the surface track profiles in lunar rocks. The rates of erosion may be variable from rock to rock, but lie in the range of $\sim 3 \times 10^{-8}$ to 10^{-7} cm/yr. The major uncertainty in these numbers lies in the uncertainty in the ratio of solar activity during the Surveyor period to the long-term average of the sun. The erosion measured by solar flare tracks is probably an atomic or near atomic erosion process. The mass wastage erosion rates for lunar rocks which are relevant for impact calculations and measurements of radioactivity profiles could be considerably higher. Our results are thus consistent with the most recent estimates of impact erosion.

SOLAR FLARE MAPS: CONSTANCY OF THE SUN

The simple analytic argument given previously showed that the net effect of erosion was to lower the spectral index of track density vs. range by one unit. In Table 2 it can be seen that the spectral index in many lunar (and meteoritic) samples is ~ 1, indicating an uneroded index of ~ 2. This is somewhat different than the index of 2.5 measured in the Surveyor glass and suggests that the long term energy spectrum of cosmic rays may be somewhat harder than that measured for the Surveyor exposure period.

The analytic treatment also showed that the absolute value of the track density at any given depth for a rock in erosion equilibrium depends on the ratio $K/\dot{R}$, where K is proportional to the average value of the solar flare flux and $\dot{R}$ is the erosion rate. It is extremely interesting to note that in those cases where careful maps have been made (12063, 12022, 10057), the maximum variation in track density at an equivalent depth along a maximum density profile is only 3.5. This suggests, but of course does not prove, that neither the erosion rate nor the average solar activity integrated over periods of 10^6 to 10^7 years have changed by more than this amount during the exposure periods of these rocks.

Now that the Surveyor data are available for comparison, it is extremely interesting to extend the solar flare mapping technique to a large number of different rocks, particularly those that have been irradiated at different times in the past, to investigate the possible variations in both erosion rates and solar activity.

Acknowledgments—G. C. is deeply indebted to Dr. E. PICCIOTTO for his support and interest. Drs. J. JEDWAB and R. RASMONT are gratefully acknowledged for allowing generous use of the SEM of the University of Brussels. It is a pleasure to acknowledge the assistance of P. SWAN who did much of the cosmic-ray mapping. The work was also greatly aided by M. HOPPE.

This work was supported by NASA Contract NAS 9-8165.

REFERENCES

ARNOLD J. R., HONDA M., and LAL D. (1961) Record of cosmic-ray intensity in the meteorites. *J. Geophys. Res.* **66,** 3519–3531.

BANDHARI N., BHAT S., LAL G., RAJAGOPALAN G., TAMHANE A. S., and VENKATAVARADAN V. S. (1971) Fossil track studies in lunar materials. II: The near-surface and post depositional exposure history of regolith components at the Apollo 12 site including the double core 25,28. Second Lunar Science Conference (unpublished proceedings).

BARBER D. J., HUTCHEON I., and PRICE P. B. (1971a) Extralunar dust in Apollo cores? *Science* **171,** 372–374.

BARBER D. J., HUTCHEON I., PRICE P. B., RAJAN R. S., and WENK H. R. (1971b) Exotic particle tracks and lunar history. Second Lunar Science Conference (unpublished proceedings).

BORG J., DRAN J. C., DURRIEU L., JOURET C., and MAURETTE M. (1970) High voltage electron microscope studies of fossil nuclear particle tracks in extraterrestrial matter. *Earth Planet. Sci. Let.* **8**, 379–386.

CROZAZ G., HAACK U., HAIR M., HOYT H., KARDOS J., MAURETTE M., MIYAJIMA M., SEITZ M., SUN S., WALKER R., WITTELS M., and WOOLUM D. (1970a) Solid state studies of the radiation history of lunar samples. *Science* **167**, 563–566.

CROZAZ G., HAACK U., HAIR M., MAURETTE M., WALKER R., and WOOLUM D. (1970) Nuclear track studies of ancient solar radiations and dynamic lunar surface processes. *Proc. Apollo 11 Lunar Sci. Conf., Geochim. Cosmochim. Acta* Suppl. 1, Vol. 3, pp. 2051–2080. Pergamon.

CROZAZ G., WALKER R., and WOOLUM D. (1971) Cosmic ray studies of "recent" dynamic processes on the surface of the moon. Second Lunar Science Conference (unpublished proceedings).

CROZAZ G. and WALKER R. (1971) Solar particle tracks in glass from the Surveyor III spacecraft. Second Lunar Science Conference (unpublished proceedings). Also, *Science* **171**, 1237–1239 (1971).

COMSTOCK G. M., EVWARAYE A. O., FLEISCHER R. L., and HART H. R., JR. (1971) The particle track record of the Ocean of Storms. Second Lunar Science Conference (unpublished proceedings).

FINKEL R. C., ARNOLD J. R., REEDY R. C., FRUCHTER J. S., LOOSLI H. H., EVANS J. C., SHEDLOVSKY J. P., IMAMURA M., and DELANY A. C. (1971) Depth variations of cosmogenic nuclides in a lunar surface rock. Second Lunar Science Conference (unpublished proceedings).

FLEISCHER R. L., PRICE P. B., WALKER R. M., and MAURETTE M. (1967) Origins of fossil charged particle tracks in meteorites. *J. Geophys. Res.* **72**, 331–353.

FLEISCHER R. L., HAINES E. L., HART H. R., JR., WOODS R. T., and COMSTOCK G. M. (1970) The particle track record of the Sea of Tranquillity. *Proc. Apollo 11 Lunar Sci. Conf., Geochim. Cosmochim. Acta* Suppl. 1, Vol. 3, pp. 2103–2120. Pergamon.

FLEISCHER R. L., HART H. R., JR., and COMSTOCK G. M. (1971) Very heavy solar cosmic rays: Energy spectrum and implications for lunar erosion. Second Lunar Science Conference (unpublished proceedings). Also *Science* **171**, 1240–1242 (1971).

HÖRZ F., HARTUNG J. B., and GAULT D. E. (1970) Micrometeorite craters on lunar rock surfaces. Lunar Science Institute contribution no. 9.

HOYT H. P., JR., KARDOS J. L., MIYAJIMA M., WALKER R. M., and ZIMMERMAN D. W. (1971) Thermoluminescence and stored energy measurements of Apollo 12 samples. Second Lunar Science Conference (unpublished proceedings).

LAL D., MACDOUGALL D., WILKENING L., and ARRENHIUS G. (1970) Mixing of the lunar regolith and cosmic ray spectra: Evidence from particle-track studies. *Proc. Apollo 11 Lunar Sci. Conf., Geochim. Cosmochim. Acta* Suppl. 1, Vol. 3, pp. 2295–2303. Pergamon.

LAL D. and RAJAN R. S. (1969) Observations on space irradiation of individual crystals of gas-rich meteorites. *Nature* **223**, 269–271.

MAURETTE M., THRO P., WALKER R., and WEBBINK R. (1969) Fossil tracks in meteorites and the chemical abundance and energy spectrum of extremely heavy cosmic rays. In *Meteorite Research* (editor P. Millman), pp. 286–315. D. Reidel.

PELLAS P., POUPEAU G., LORIN J. C., REEVES H., and AUDOUZE J. (1969) Primitive low-energy particle irradiation of meteoritic crystals. *Nature* **223**, 272–274.

PRICE P. B. and O'SULLIVAN D. (1970) Lunar erosion rate and solar flare paleontology. *Proc. Apollo 11 Lunar Sci. Conf., Geochim. Cosmochim. Acta* Suppl. 1, Vol. 3, pp. 2351–2359. Pergamon.

RANCITELLI L. A., PERKINS R. W., FELIX W. D., and WOGMAN N. A. (1971) Cosmogenic and primordial radionuclide measurements in Apollo 12 lunar samples by nondestructive analysis. Second Lunar Science Conference (unpublished proceedings).

SHOEMAKER E. Private communication.

Proceedings of the Second Lunar Science Conference, Vol. 3, pp. 2559–2568
The M.I.T. Press, 1971.

The particle track record of the Ocean of Storms

R. L. Fleischer, H. R. Hart, Jr., G. M. Comstock, and A. O. Evwaraye*

General Electric Research and Development Center, Schenectady, New York 12301

(Received 22 February 1971; accepted in revised form 29 March 1971)

Abstract—In Apollo 12 rocks the numbers of tracks from the solar and galactic iron group cosmic rays imply surface residence times that range from $< 10,000$ years to ~ 30 million years. The presence of steep track gradients at exposed surfaces shows that some rocks have been on the lunar surface in only one position, while others have been turned over and moved more than once. For example, rock 12017 was raised to within one meter of the surface, later thrown to the very surface, then flipped over and recently splattered with molten glass (just 9000 years ago). The abundance of nuclear interaction (spallation) tracks induced by the penetrating galactic protons provides residence times for different rocks in the top meter of soil of ~ 20 to 750 millions of years. The erosion of lunar rocks is estimated by comparing the cosmic ray track distributions in lunar rocks with the one found in an uneroded glass detector exposed in Surveyor III. Erosion at a rate of about one atomic layer per year is inferred. By inducing uranium-235 fission tracks we have measured widely ranging uranium concentrations: less than 10^{-3} parts per million in pyroxenes, ~ 1 ppm in glass, and up to 170 ppm in zircon. The fossil track abundance in the zircon gives no evidence for the presence of extinct radioactivity by plutonium-244 or by super-heavy nuclei.

Introduction

The abundant particle tracks found in most lunar samples constitute a highly detailed record of the diverse chronology of lunar samples. Solidification ages are recorded by uranium-238 fission tracks (Price and Walker, 1963a; Fleischer and Price, 1964a, b); times of exposure on the lunar surface (*surface residence* times) are given by tracks or iron group nuclei in the cosmic rays (Crozaz et al., 1970a; Fleischer et al., 1970a, later Price and O'Sullivan, 1970; Lal et al., 1970) and we shall see that nuclear interaction (spallation) tracks (Fleischer et al., 1970a, b) measure the total time spent near and at the lunar surface.

Procedures

The primary new technique employed in this work is the use of spallation tracks to measure the total time near the lunar surface. We have found that spallation tracks yield ages that agree with radiometrically measured cosmic ray exposure times. For other procedures, including etchants used, our previous work (Fleischer et al., 1970b) should be consulted.

Technique for measuring spallation ages

Previously (Fleischer et al., 1970b) we demonstrated that the short, nearly featureless tracks (Fleischer et al., 1967) produced by nuclear interactions of penetrating primary cosmic ray particles increase in number with the time of exposure in the top 1–2 meters of soil. At the same time we noted that the observed density of these tracks varied appreciably from grain to grain, even though there is little variation with position in rocks of the sizes that have been available from Apollo 11 and Apollo

---

* Present address: Department of Physics, Antioch College, Yellow Springs, Ohio.

12. The observed grain-to-grain variation in rocks is most likely a variation of etching efficiency that depends primarily (as judged by etching different portions of a fragmented crystal) on the crystallographic orientation of the etched surface and very likely on compositional variations, but only secondarily on the etching time, and very little (FLEISCHER *et al.*, 1967) on the orientation relative to the incident cosmic ray nuclei. A reproducible measurement is made by etching a number of grains of the minerals of interest (in this case, pyroxenes) and choosing the highest track densities present as representative of the highest etching efficiency.

As long as this procedure is used in obtaining both the natural track counts (ρ_{sp}) and the calibration counts (P) that measure production rates, the ratio ρ_{sp}/P will be reproducible on a given rock. Ideally the irradiation and the subsequent calibration would be done for the same crystals as the natural track counts.

The production rate P is measured for individual surface samples by bombarding annealed (track free after 17 hours at 820°C in platinum boats) samples with 3 GeV protons from the Princeton-Pennsylvania Accelerator in order to simulate the nuclear-active cosmic rays incident upon the moon. In reality these active particles include not only protons but primary helium and heavier nuclei and secondary high energy neutrons and pions. At the very surface the protons and alphas dominate. Although minor errors are introduced in calibrating with a single type of particle at a single energy, they are small relative to the variations inherent in the uncertainties of individual rock histories as to depths of burial and detailed geometries and compositions of shielding material. As judged by the etching of annealed samples following irradiation with fission fragments, the thermal treatment has not altered the registration properties of the pyroxenes.

RESULTS

Spallation ages and proton doses

In the preceding section we indicated how reproducible values are measured for the ratio of the spallation track density ρ_{sp} to the production rate P of recoil tracks caused by high energy protons. The ratio ρ_{sp}/P is the dose of protons needed to produce the observed spallation track density. In Table 1 measured values of ρ_{sp}/P are given for a group of lunar samples in the column headed Proton Exposure. By assuming a flux ϕ of $3 \times 10^7/\text{cm}^2$-yr of primary cosmic ray nucleons (BAZILEVSKAYA *et al.*, 1968) we calculate spallation Surface Ages given in the next column of Table 1 by $\rho_{sp}/P\phi$, the spallation ages if the entire proton exposure occurred on the lunar surface.

For samples such as those listed, which were all picked up at the surface, these ages should be to first approximation directly comparable with the radiometrically measured spallation ages—listed in the right-hand column of Table 1. In general the agreement is excellent, well within the scatter among different radiometric ages where more than one such age is available. In a higher approximation the agreement between the radiometric age and the track spallation surface age depends on the detailed burial history of the rock. Different track and radiometric ages for samples with complex burial histories will result if the cross-sections for the individual relevant spallation reactions vary differently with depth of burial (HONDA and ARNOLD, 1964). In addition, differences can arise if a sample is exposed at depth and then moved to the surface at a time that is recent with respect to the half life of the nuclide being used to measure the production rate radiometrically.

The true proton exposure ages are uncertain, however, because we do not know the samples' depths of burial during proton exposure. As accelerator experiments have shown (HONDA and ARNOLD, 1964), a cascade of nuclear-active, secondary

Table 1. Track spallation ages of lunar pyroxenes.

Sample Number	Production rate (P)* (tracks/10^9 protons)	Observed track density (ρ_{sp}) (cm^{-2})	Proton exposure (ρ_{sp}/P) (protons/cm^2)	Surface age (10^6 yr)	Minimum spallation age (10^6 yr)	Radiometric spallation ages (10^6 yr)
10017	1.7†	$2.1(\pm0.1) \times 10^7$	1.2×10^{16}	420	170	200–640[a,e,f,g,i,j]
10044	1.0	$8.2(\pm1.2) \times 10^6$	8.2×10^{15}	270	110	56–100[f]
10049	2.39	$1.49(\pm0.15) \times 10^6$	6.2×10^{14}	21	8.5	22.5–25[g,l]
12002	1.7†	$2.66(\pm0.25) \times 10^6$	1.6×10^{15}	55	20	50–145[b,d,h]
12017	1.45	$4.57(\pm0.32) \times 10^6$	3.2×10^{15}	105	40	—
12021	2.31	$5.1(\pm0.3) \times 10^7$	2.2×10^{16}	740	300	300[h]
12065	1.35	$6.81(\pm0.28) \times 10^6$	5.1×10^{15}	170	70	160–200[e,k]

* Absolute values uncertain to $\pm30\%$, but relative values are valid for 10049, 12002, 12017, 12021, and 12065.

† Average of other values.

a ALBEE et al. (1970); b ALEXANDER et al. (1971); c BLOCH et al. (1971); d D'AMICO et al. (1971); e EBERHARDT et al. (1970); f FIREMAN et al. (1970); g HINTENBERGER et al. (1970); h MARTI and LUGMAIR (1971); i MARTI et al. (1970); j O'KELLEY et al. (1970); k STOENNER et al. (1971); l From measurements by FUNKHOUSER et al. (1970).

protons and neutrons builds up with depth and then attenuates. From the data of FLEISCHER et al. (1967) we estimate that in order to produce optically visible, etched tracks from spallation recoil nuclei, reactions are required in which at least 5 nucleons are ejected from the struck nucleus. In such reactions the maximum flux is ~2.5 times the primary proton flux and occurs at a depth of about 20 to 25 cm of soil (or 10 to 12 cm of rock) (KOHMAN and BENDER, 1967). The column labeled Minimum Spallation Age in Table 1 gives minimum times corresponding to burial at a 25 cm soil-equivalent depth. It should also be evident that the observed track densities could have been produced by much longer exposures than we have listed, if the samples were located at greater depths where the high-energy particle flux is corresponding lower.

At present, track spallation ages are of relatively low precision. They do, however, give reasonable agreement with the radiometric ages, and they give track workers a new tool for assessing radiation exposure histories. So far, we report spallation results only for pyroxenes. However, by measuring spallation ages for another mineral with an identical track registration threshold, an improved internal check on the accuracy of such ages will be possible. By measuring ages using minerals of different thresholds and therefore different depth variations of the track production rate, data on depths of burial will be obtained. As we shall see, comparison of spallation track ages with surface residence ages calculated from the track densities from heavy cosmic rays also yields such data.

Tracks of heavy cosmic rays: rock 12017

We have shown previously (CROZAZ et al., 1970; FLEISCHER et al., 1970a) how the dominant cosmic ray tracks from the iron-group nuclei can be used to measure the surface residence times of rocks and rock fragments, and how from steep track density gradients near space-exposed surfaces, former orientations of rocks can be inferred. As an example, the results shown in Fig. 1 for rock 12017 allow us to derive solely from track data the complicated and varied history given in Table 2.

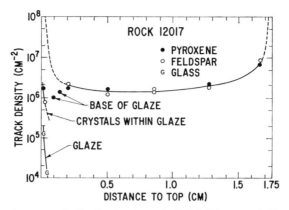

Fig. 1. Cosmic ray track distribution in rock 12017. The top of this rock was coated with glass of maximum thickness 0.15 cm. Tracks in crystals within the glass show it to be recently formed. Pyroxene and glass track densities are corrected for the measured etching efficiencies of 0.7 and 0.08.

Spallation tracks indicate the period over which the rock was exposed to galactic cosmic rays, and they are responsible for the first two entries in Table 2. The increases in cosmic-ray track density near both surfaces show that both sides have been exposed to space, and the slight asymmetry in the profile shows that the bottom received the longer exposure, roughly 1 million years, as compared to 7×10^5 years for the top. At the very top is a glass coating that apparently was splashed on after the rock was positioned with that side up. From (presumably annealed) crystals trapped within the coating its space exposure is inferred to be only ~ 9000 years, using the track production rate given by Fleischer *et al.* (1970b). The glass itself (microprobe analysis by wt. %: SiO_2, 46.4; FeO, 17.7; Al_2O_3, 10.5; MgO, 11.0; CaO, 9.30; TiO_2, 2.80) has the track retention characteristics given in Fig. 2. This glass is not highly retentive, allowing fading in 2 years at $400°K$ and 500 years at $350°K$. We estimate that with the thermal cycling that occurs on the moon, tracks in a glazed rock with estimated peak temperature $360°K$, would only be preserved over (very roughly) the last 500 years, as explained further in the caption to Fig. 2. We note parenthetically that in the glass there is a pronounced additional track fading that produces a decrease in track density toward the surface in the top 30 μ, a distance that corresponds to two or three optical depths for visible light. The cause of this effect has not been identified.

Table 2. Simplest track chronology for rock 12017.

Time (years before present)	Event
up to $\sim$ 105,000,000	Buried > 200 cm
$\sim$ 105,000,000	Moved to < 200 cm and >15 cm
$\sim$ 1,700,000	Moved to surface
$\sim$ 700,000	Flipped over
$\sim$ 9000	Splattered with hot glass
$\sim$ 500 to 0	Glass records solar flare particles

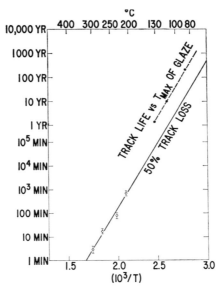

Fig. 2. Track retentivity in the glaze on rock 12017. Extrapolation of the data from lower temperatures predicts 50% track fading after one month continuously at 400°K. The dotted line indicates probable track life on the surface of the moon as a function of the maximum temperature reached by the glaze at lunar noon. Temperature vs. time data from the infra red measurements of SINTON (1962) were used.

In summary the low-energy cosmic rays (dominantly solar flare particles) have been recorded over different time intervals: the glaze over the last 40 to 50 solar cycles, the crystals within the glaze over the last ~ 800, and the bottom of the rock over a more ancient group of $\sim 500,000$ cycles. Track distributions in these three sites should allow the proposed (PRICE and O'SULLIVAN, 1970) "solar flare paleontology," comparing ancient solar spectra at different periods of time.

Surface residence times

Table 3 Summarizes cosmic-ray track information for Apollo 12 rocks and gives the most current data on 10049. This is an Apollo 11 rock of special interest because its surface time of 29 m.y. agrees with the 24 m.y. inferred from radioactivity measurements of spallation-produced nuclides (FUNKHOUSER et al., 1970; HINTENBERGER et al., 1970) and the 21 m.y. inferred here from spallation tracks. In short this sample spent all of its near surface time directly exposed to space and underwent very little erosion (which would have lowered the track density). The limit on erosion ($< 3 \times 10^{-8}$ cm/yr) is consistent with what we will infer shortly in this paper from our Surveyor III results. Table 3 shows a wide range of surface exposure times— from $\sim 10^4$ to 3×10^7 years—for samples some of which have been in a single surface position, some in at least two positions, and one in at least three.

The Surveyor III data has also made possible more reliable measurements of short surface residence times of small grains. Here the most abundant tracks are solar

Table 3. Minimum cosmic ray track densities and surface residence times for lunar samples.

Sample	Mineral	Track density (cm^{-2})	Depth in sample (cm)	Surface residence time (million years) (top/bottom)
10049	Pyroxene	1.55×10^7	0.90 cm	(29 total)
12002	Pyroxene	2.8×10^6	5.0	(24/0)
12017	Feldspar	1.51×10^6	0.45	(0.7/1.0)
12017	Feldspar in glaze	8×10^5	0.02	(0.009/0)
12021	Pyroxene	5.0×10^6	4.0	(13/13)*
12065	Pyroxene	2.2×10^6	6.4	(14/0)
12025 } (Soil) 12028 }	Pyroxene } Feldspar }	3×10^7†	—	(110 total)‡
12025, 4, 54–8.5, 9, 9	Pyroxene	5×10^7	0.002	0.01/0)§

* This result disagrees with that of PRICE (personal communication); a mix-up in sample position designation (either his sample or ours) is suspected.
† Average of 100 μ–400 μ diameter grains.
‡ Average time in top 60 cm of soil, calculated in same manner as in FLEISCHER *et al.* (1970b).
§ Using solar spectrum from Surveyor III glass (FLEISCHER *et al.*, 1971) after adjustment for solar cycle.

heavy cosmic rays whose flux previously was highly uncertain. Because of the presence in most lunar samples of an unknown amount of erosion, the track production rate vs depth was only a lower limit. However, with the recent data (adjusted to solar cycle average) for the uneroded Surveyor III glass and the assumption that it represents a typical solar cycle, the ages can be computed for small grains such as that sketched in Fig. 3. The steep track gradients mapped around roughly half of its perimeter reveal that at one time this grain was exposed directly to space, as a small grain resting on what is drawn as its right side. By matching the steepest gradient in this grain with calculations based on the Surveyor III flux of solar heavies (FLEISCHER *et al.*, 1971) a surface residence time of 10,000 years is obtained. This age would be altered somewhat if the solar fluxes are used that have been inferred by CROZAZ and WALKER (1971) or PRICE *et al.* (1971) from the same material. Hence, the absolute surface time for this sample is subject to possible revision, but is roughly correct and clearly much shorter than the 0.5 to 30 m.y. that typifies most larger moon rocks.

Burial and burial depths

By comparing track spallation ages with track surface residence times and with radiometric spallation ages, permissible burial depths can be inferred. Thus for rock 10049, where all three agree, the entire exposure must have been at the surface. For the other rocks listed, where the surface residence age is shorter than the spallation age, the samples must have been buried over most of their spallation exposure times.

Lunar erosion

In a separate experiment (FLEISCHER *et al.*, 1971) using tracks etched in a glass optical filter from the Surveyor III television camera we have measured the energy spectrum of the iron group solar cosmic ray particles over the energy range 1 to 100 Mev/nucleon, finding for the differential flux $1.8 \times 10^3 E^{-3}$ particles/m^2-sec-str-MeV/nucleon. We have also observed high energy fission of Pb, induced by galactic cosmic ray protons and alpha particles.

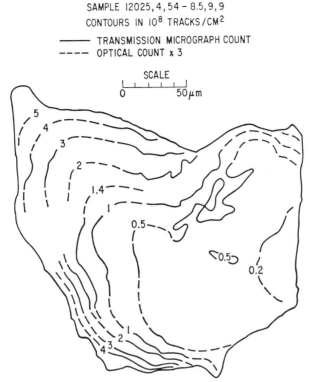

SAMPLE 12025, 4, 54 – 8.5, 9, 9

CONTOURS IN 10^8 TRACKS/CM^2

——— TRANSMISSION MICROGRAPH COUNT

– – – OPTICAL COUNT x 3

SCALE

0   50 μm

Fig. 3. Cosmic ray track density distribution in a grain found at 8.5 cm depth in core 12025. The gradients identify the surfaces exposed directly to space and to heavy cosmic ray nuclei from the sun. The surface exposure was for $\sim$ 10,000 years.

Using this energy spectrum and the track density profiles measured in lunar rocks we have obtained an estimate of the rate of fine scale erosion on the moon. Making the necessary corrections for the different properties of the glass and the lunar rocks, for the different solid angles involved, for the solar cycle variation, and assuming the present solar cycle to be typical of the last few million years, we find an erosion rate of 0 to 2×10^{-8} cm/yr to be consistent with the track profiles measured by four groups on rocks 10017 and 10003. The much lower track density profile of rock 10058 is consistent with the recent removal of a chip of appreciable thickness and does not contradict our remarkably low fine-scale erosion rate. Similar results have simultaneously been obtained by CROZAZ and WALKER (1971) and PRICE et al. (1971).

Uranium contents and fission track dating

Since fission track dating requires the presence of uranium (PRICE and WALKER, 1963b), the induced fission track measurements given in Table 4 are relevant. In the cases shown, uranium is too low to allow fission track dating of any of the samples except for zircon LZ where an upper limit can be given. Since the fossil track content

Table 4. Uranium content of lunar samples.

Mineral	Sample	Uranium* (wt fraction × 10⁹)	Notes
Augite	12017,17,6,3	0.4 ($\pm 60\%$)	Not including visible inclusions
Augite	12017,17,6,3	1.5 ($\pm 30\%$)	Including visible inclusions
Augite	12021,1,4,5	1.0 ($\pm 50\%$)	Not including visible inclusions
Augite	12021,1,4,5	4.5 ($\pm 20\%$)①	Including visible inclusions
Augite	12065,6,6	0.5 ($\pm 70\%$)	Including visible inclusions
Glass	12017,8,6	1,230 ($\pm 8\%$)	
Zircon†	LZ (from Apollo 11 fines)	167,000 ($\pm 12\%$)	
Zircon†	Z-2 (from Apollo 11 fines)	< 10,000	90% confidence

* Tracks observed on interior surfaces, except as noted.
† Tracks observed in Lexan adjacent to sample.
① Most uranium-rich inclusion scanned contained ∼ 3 × 10⁶ atoms of uranium.

was 1 to 3 × 10⁸/cm², the ages would be 1.3 to 3.3 × 10⁹ years if these were all fission tracks, and less if an appreciable fraction were of other origin. There is thus no evidence for an excess of fission tracks from presently extinct fission activity by Pu-244 or super-heavy elements.

CONCLUSIONS

Track spallation measurements and solar cosmic ray tracks in the Surveyor III glass add two new dimensions to lunar information available from track measurements. By comparing the spallation and surface residence ages inferred for individual rocks (using spallation and cosmic ray tracks), surface and near surface chronologies can be constructed. By comparing track gradients and abundances in the uneroded glass and in eroded rocks, erosion rates for individual rock surfaces can be determined. Not surprisingly, individual rock histories vary widely.

Acknowledgments—We are pleased to give thanks to H. ALLEN for proton irradiations at the Princeton-Pennsylvania Accelerator, J. FLOYD for neutron irradiation at Brookhaven National Laboratory, U. B. MARVIN of the Smithsonian Astrophysical Observatory for loan of two zircons, to N. NICKLE of the Jet Propulsion Laboratory for Surveyor III glass, and to M. F. CICCARELLI, M. D. McCONNELL, and E. STELLA for experimental assistance. This work was supported in part by NASA contract NAS 9-7898.

REFERENCES

ALBEE A. L., BURNETT D. S., CHODOS A. A., EUGSTER O. J., HUNEKE J. C., PAPANASTASSIOU D. A., PODOSEK F. A., RUSS PRICE G., II, SANZ H. G., TERA F., and WASSERBURG G. J. (1970) The Lunatic Asylum of the Charles Arms Laboratory of Geological Sciences, *Science,* **167,** 463–466.
ALEXANDER E. C., JR., DAVIS P. K., KAISER W. A., LEWIS R. S., and REYNOLDS J. H. (1971) Depth studies of rare gases in rock 12002. Second Lunar Science Conference (unpublished proceedings).
BAZILEVSKAYA G. A., CHARAKHCYAN A. N., CHARAKYCHYAN T. N., KVASHNIN A. N., PANKRATOV A. K., and STEPANYAN A. A. (1968) The energy spectrum of primary cosmic rays and the secondary radiation background in the vicinity of the earth. *Can. J. Physics,* **46,** S515–S517.

BLOCH M., FECHTIG H., FUNKHOUSER J., GENTNER W., JESSBERGER E., KIRSTEN T., MULLER O., NEUKUM G., SCHNEIDER E., STEINBRUNN F., and ZAHRINGER J. (1971) Location and variation of rare gases in Apollo 12 lunar samples. Second Lunar Science Conference (unpublished proceedings).

CROZAZ G., HAACK U., HAIR M., HOYT P., KARDOS J., MAURETTE M., MIYAJIMA M., SEITZ M., SUN S., WALKER R., WITTELS M., and WOLLUM D. (1970a) Radiation history of the moon. *Science* **167**, 563–566.

CROZAZ G., HAACK U., HAIR M., MAURETTE M., WALKER R., and WOLLUM D. (1970b) Nuclear track studies of ancient solar radiations and dynamic lunar surface processes. *Proc. Apollo 11 Lunar Sci. Conf., Geochim. Cosmochim. Acta* Suppl. 1, Vol. 3, pp. 2051–2080. Pergamon.

CROZAZ G. and WALKER R. M. (1971) Solar particle tracks in glass from the Surveyor III spacecraft. *Science* **171**.

D'AMICO J., DEFELICE J., FIREMAN E. L., JONES C., and SPANNAGEL G. (1971) Tritium and argon radioactivities and their depth variations in Apollo 12 samples. Second Lunar Science Conference (unpublished proceedings).

EBERHARDT P., GEISS J., GRAF H., GROEGLER N., KRAEHENBUEHL U., SCHWALLER H., SCHWARZMUELLER H., and STETTLER A. (1970) Trapped solar wind noble gases, radiation ages and K/Ar in lunar material. *Science* **167**, 558–560.

FIREMAN E. L., D'AMICO J., and DEFELICE J. (1970) Tritium and argon radioactivities in lunar material. *Science* **167**, 566–568.

FLEISCHER R. L., HAINES E. L., HANNEMAN R. E., HART H. R., JR., KASPER J. S., LIFSHIN E., WOODS R. T., and PRICE P. B. (1970a) Particle track, X-ray, and mass spectrometry studies of lunar material from the Sea of Tranquility. *Science* **167**, 568–571.

FLEISCHER R. L., HAINES E. L., HART H. R., JR., WOODS R. T., and COMSTOCK G. M. (1970b) The particle track record of the Sea of Tranquility. *Proc. Apollo 11 Lunar Sci. Conf., Geochim. Cosmochim. Acta* Suppl. 1, Vol. 3, pp. 2103–2120. Pergamon.

FLEISCHER R. L., HART H. R., JR., and COMSTOCK G. M. (1971) Very heavy solar cosmic rays: Energy spectrum and implications for lunar erosion. *Science* **171**, 1240–1242.

FLEISCHER R. L. and PRICE P. B. (1964a) Glass dating by fission fragment tracks. *J. Geophys. Res.* **69**, 331–339.

FLEISCHER R. L. and PRICE P. B. (1964a) Techniques for geological dating of minerals by chemical etching of fission fragment tracks. *Geochim. Cosmochim. Acta* **28**, 1705–1714.

FLEISCHER R. L., PRICE P. B., WALKER R. M., and MAURETTE M. (1967) Origins of fossil-charged particle tracks in meteorites. *J. Geophys. Res.* **72**, 331–353.

FUNKHOUSER J., SCHAEFFER O., BOGARD D., and ZAHRINGER J. (1970) Gas analysis of the lunar surface. *Science* **167**, 561–563.

HINTENBERGER H., WEBER H. W., VOSHAGE H., WANKE H., BEGEMANN F., VILSCEK E., and WLOTZKA F. (1970) Concentrations and isotopic compositions of rare gases, hydrogen, and nitrogen in lunar dust and rocks. *Science* **167**, 543–545.

HONDA M. and ARNOLD J. R. (1964) Effects of cosmic rays on meteorites. *Science* **143**, 203–212.

KOHMAN T. P. and BENDER M. L. (1967) Nuclide production by cosmic rays in meteorites and on the moon. In *High Energy Nuclear Reactions in Astrophysics* (editor B. S. P. Shen) pp. 169–245, Benjamin, Inc.

LAL D., MACDOUGALL D., WILKENING L., and ARRHENIUS G. (1970) Mixing of the lunar regolith and cosmic ray spectra: evidence from particle-track studies. *Proc. Apollo 11 Lunar Sci. Conf., Geochim. Cosmochim. Acta* Suppl. 1, Vol. 3, pp. 2295–2303. Pergamon.

MARTI K., LUGMAIR G. W., and UREY H. C. (1970) Solar wind gases, cosmic-ray effects and the irradiation history. *Science* **167**, 548–550.

MARTI K. and LUGMAIR G. W. (1971) $Kr^{81}-Kr$ and $K-Ar^{40}$ ages, cosmic-ray spallation products and neutron effects in Apollo 11 and Apollo 12 lunar samples. Second Lunar Science Conference (unpublished proceedings).

NAESER C. W. (1969) Etching fission tracks in zircons. *Science* **165**, 388–389.

O'KELLEY D. G., ELDRIDGE J. S., SCHONFELD E., and BELL P. R. (1970) Elemental compositions and ages of Apollo 11 lunar samples by nondestructive gamma-ray spectrometry. *Science* **167**, 580–582.

Price P. B., Hutcheon I., Cowsic R., and Barber D. J. (1971) Enhanced emission of iron nuclei in solar flares. Second Lunar Science Conference (unpublished proceedings).

Price P. B. and O'Sullivan D. (1970) Lunar erosion rate and solar flare paleontology. *Proc. Apollo 11 Lunar Sci. Conf., Geochim. Cosmochim. Acta* Suppl. 1, Vol. 3, pp. 2351–2359. Pergamon.

Price P. B. and Walker R. M. (1963a) Fossil tracks of charged particles in mica and the age of minerals. *J. Geophys. Res.* **68**, 4847–4862.

Price P. B. and Walker R. M. (1963b) A simple method of measuring low uranium concentrations in natural crystals. *Appl. Phys. Lett.* **2**, 23–25.

Sinton W. H. (1962) Temperatures on the lunar surface. In *Physics and Astronomy of the Moon* (editor Z. Kupal) pp. 407–428, Academic Press.

Stoenner R. W., Lyman W. J., and Davis R., Jr. (1971) Argon and tritium radioactivities in lunar rocks and in the sample return container. Second Lunar Science Conference (unpublished proceedings).

Proceedings of the Second Lunar Science Conference, Vol. 3, pp. 2569–2582
The M.I.T. Press, 1971.

The particle track record of lunar soil

G. M. Comstock, A. O. Evwaraye,* R. L. Fleischer,
and H. R. Hart, Jr.

General Electric Research and Development Center Schenectady, New York 12301

(Received 22 February 1971; accepted in revised form 29 March 1971)

Abstract—Measurements of primary cosmic-ray track densities and spallation-recoil track densities in the Apollo 12 deep-core sample are presented. Neither the primary nor the spallation track densities show any significant dependence on soil depth, while there is a great variation from grain to grain at a given depth. We conclude from this that the soil has been well stirred down to ~60 cm at the Apollo 12 site. This conclusion is quantitatively supported by computer models of the accumulation of cosmic ray tracks in lunar soil. A model in which thorough stirring is most frequent at shallow depths and less and less frequent at greater depths fits the observed track density distributions from both Apollo 11 and Apollo 12 core samples. Stirring ages of at least 1 to 2 billion years are required.

Introduction

Evidence from manned and unmanned lunar flights that the moon has a deep soil layer or regolith has led to great interest in the origin and mechanical history of this layer. This history has an intimate relationship with lunar surface features and the flux of solid matter in interplanetary space (Öpik, 1969, and others). Striking direct evidence for frequent mixing of this soil layer was first provided by fossil particle tracks left in Apollo 11 soil grains by primary heavy cosmic ray nuclei (Crozaz et al., 1970; Fleischer et al., 1970a, b). Further evidence has now been provided by primary cosmic ray and spallation-recoil track densities measured in the Apollo 12 deep core. These results are presented in the following section. We further report the results of Monte Carlo computer calculations which quantitatively support the hypothesis that the particle track distribution is due to frequent mixing of the lunar soil.

Apollo 12 Soil Samples

Cosmic ray tracks

We investigated 251 individual pyroxene, feldspar, and olivine soil grains of 50–500 μm diameter from 13 depths distributed through the 40 cm length of the connected core samples 12025 and 12028. These samples were etched and counted in the same manner as described for the Apollo 11 samples (Fleischer et al., 1970b).

Primary cosmic ray iron tracks could be counted or lower limits estimated in 73% of the grains sampled (Fig. 1). 61% were counted optically at 1350× magnification. This includes those marked as lower limits, most of which, in fact, probably fall below ~3 × 10⁸/cm². Transmission electron micrographs of track replicas were made at 7500× for 15% of the samples. Several of these were also counted optically, and the

---

* Present address: Department of Physics, Antioch College, Yellow Springs, Ohio.

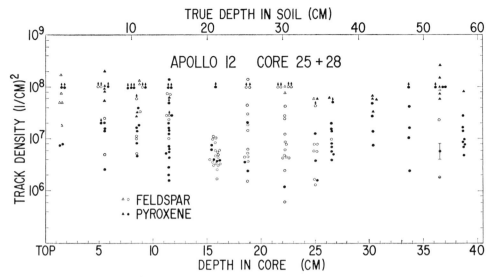

Fig. 1. Cosmic ray track distribution in core 12025 and 12028. The data points are observed optical microscope counts (circles) and adjusted electron micrograph counts (triangles).

micrograph counts were found to be 1 to 3 times greater than the optical counts, depending on track length distribution. The reasons for this have been discussed (Fleischer et al., 1970b), and each micrograph count has been reduced by an appropriate factor determined by the track length distribution before inclusion in Fig. 1. No other adjustments were made to the data in Fig. 1.

At least 20% of the samples that were counted contained track gradients (for example, Fig. 3 of Fleischer et al., 1971). In these cases, the counts were made from near the center of the grains; track densities near certain edges are often more than 10 times greater.

The statistical counting error for each sample is generally less than 10%. The error bars shown in Figs. 1, 2, and 3 represent the spread in counts obtained when larger samples were broken up and the fragments counted separately. The observed factor of 2 spread could arise from variations in etching efficiency and random orientation of the etched surface with respect to the cosmic ray flux.

The 40 cm length of the core is in fact sampling the top 60 cm of soil (Carrier et al., 1971). It is clear (Fig. 1) that there is a real spread in track densities of at least a factor of 100 at most depths and that there is no systematic decrease in track density with depth down to 60 cm. We did not obtain samples from the so-called gravel layer at soil depth 15–20 cm which has a low track density (Crozaz et al., 1971; Bhandari et al., 1971). Our samples at about 20 cm (Fig. 1) may be related to this layer.

The 27% of the grains sampled which are not included in Fig. 1 became rapidly overetched to a partially opaque condition which often, but not always, indicates a high track density. These may represent an additional high density population, so that at least 14% and possibly as much as 40% of the soil grains have $>10^8$ tracks/cm². BHANDARI et al. (1971) report that 45% of the soil samples have $>10^8$ tracks/cm². However, BARBER et al. (1971a) and CROZAZ et al. (1971) report a much higher fraction (80–90%) with $>10^8$ tracks/cm². Part of this discrepancy results from a lack of standardization: CROZAZ et al. (1970, 1971) are referring to deep pits viewed with the scanning electron microscope, which can reveal a density twice as great as the same region viewed optically. However, a real effect may exist, related to sample selection or track identification. Correlation with size is suggested by micron-size soil grains observed to have very high track densities (BORG et al., 1971; BARBER et al., 1971b). More detailed comparison of the various techniques used is necessary to clarify this situation.

Spallation tracks

In addition to primary cosmic ray tracks, the lunar soil grains also contain recoil tracks of heavy spallation products from high energy cosmic ray collisions with constituents of the minerals (FLEISCHER et al., 1970a). These tracks are much shorter than the cosmic ray iron tracks. They are clearly visible on micrographs of plastic replicas but tend to be obscured when regions of high cosmic-ray-track densities are etched sufficiently to reveal them optically.

We sought spallation tracks in 85 grains which had cosmic ray track densities $\leqslant 3 \times 10^7$ cm^{-2} (optically visible at $1000 \times$). The spallation track density is defined here to be the number of dots visible under reflected light minus the number of (cosmic ray) tracks with finite length under transmitted light.

Counts were obtained in this way from 80% of the grains sampled (Fig. 2). An additional 8% had no visible spallation tracks. Either these were not sufficiently etched, had a concentration $<10^6$ cm^{-2}, or were of an orientation with a low etching efficiency. Surface features could not be resolved under reflected light in 12% of the samples and this could be due to spallation track densities $\geqslant 5 \times 10^7$ cm^{-2}. We see no significant variation with depth and a spread at each depth of a factor of ~ 10.

There is no apparent correlation between cosmic ray and spallation track densities (Fig. 3). Those samples with cosmic ray gradients have been closer to the surface and so are displaced toward higher cosmic ray track densities; however, their spread in spallation counts is the same as samples without gradients. This gives us confidence that we are interpreting the spallation tracks properly and that samples with high cosmic-ray track density might tentatively be assumed to exhibit the same spallation distribution (cross-hatched region in Fig. 3).

Soil models

We now consider what these results can tell us about lunar soil history. The discussion is based on the depth dependence of the cosmic ray and spallation track production rates. The cosmic ray tracks are due almost entirely to primary iron nuclei. The flux of these particles is very high near the surface but falls off rapidly with depth due to a decreasing energy spectrum and an interaction mean path (in lunar soil) of 12 cm. Lighter cosmic ray nuclei are much more penetrating, however,

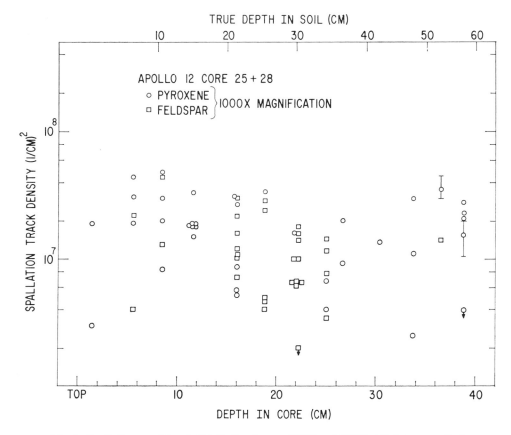

Fig. 2. Spallation-recoil track distribution in core 12025 and 12028. All data are optical microscope counts.

and the spallation track production rate shows a very different behavior. It first increases with depth due to the cosmic ray-induced production of secondaries which in turn induce spallation recoils. For the spallation products which are most likely to leave tracks, the production rate increases by a factor of ~2 from the surface down to 20–25 cm where it reaches a maximum (Kohman *et al.*, 1967).

By comparing these growth rates with the observed track distributions, we can learn about the mechanical history of the soil. We consider here three basic models of soil movement. First, a *quiet soil* model in which the individual soil grains have suffered no appreciable change of depth. In this case the track distributions should be proportional to the track production curves. The cosmic ray track density spread at any particular depth should be a factor of ~2 due to random orientation of the etched surface with respect to the attenuated cosmic ray flux. A similar spread for spallation tracks can be expected from variations in etching efficiency (Fleischer *et al.*, 1971). The quiet soil case is illustrated for cosmic ray iron tracks in Fig. 4,

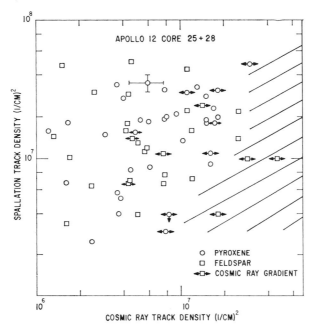

Fig. 3. Correlation between cosmic ray and spallation tracks. Spallation track densities
were not measured in the cross-hatched region.

where the curve marked "no mixing" refers to the quiet soil model after 1 billion
years.

We have also plotted in Fig. 4 the median track density observed at each depth
with limit bars that show the spread of the 70 % of the samples closest to the median.
The quiet soil model clearly does not satisfy the observed distribution. The spallation
track distribution (Fig. 2) also has too much variation at each depth and does not
show a depth dependence proportional to the production rate. A lack of correlation
between cosmic ray and spallation tracks is also expected if the grains have not spent
most of their time at the same depth.

Second, we consider a *burial model* in which an initially trackless soil experiences
a steady burial, with relatively insignificant mixing (e.g. BHANDARI *et al.*, 1971).
In this case, the track density expected at a given depth is (for uniform burial) the
integral of the production function from some initial depth to the observed depth.
The cosmic ray track density would increase rapidly in the first ~1 cm and approach
an essentially constant value below that depth, an observation which is consistent
with the observed median values (Fig. 4). We cannot state definitely what spread in
track density at a given depth would be expected from this model without specifying
the burial mechanism. The distribution of initial surface depths is critical; for example,
a wide spread could come from tracks formed while the samples were still part of
surface rocks and thus shielded to varying degrees.

The spallation track density, however, would continue to increase strongly throughout the depths sampled as a grain is buried. This is not what is observed (Fig. 2).

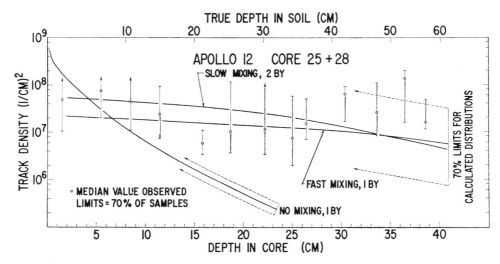

Fig. 4. Cosmic ray track distribution in core 12025 and 12028. The data points are derived from optical and (adjusted) electron microscope counts. The curves have been calculated for some of the models of soil history described in the text.

On the basis of the spallation tracks, therefore, we conclude that a burial model is not sufficient as long as the burial process involves layers that contain few spallation tracks when deposited. Preburial exposure of the grains (e.g. while in their parent rock) may weaken this conclusion.

Finally we consider *mixing models* which postulate that the major contribution to soil movement is frequent mixing by depth-dependent excavation and overlay of soil predominantly caused by small impact events. This model and a computer simulation are discussed in more detail in the following section, but the basic properties expected for cosmic ray iron tracks are indicated in Fig. 4 for two mixing rates. The "slow mixing" is derived from the maria crater distribution function (Shoemaker *et al.*, 1970); "fast mixing" uses a mixing rate ten times faster. The curves give calculated median values with the spread of 70% of the distribution as indicated, in good agreement with the data points. The spread in cosmic ray track densities in the mixing model is due to the distribution of residence times at depths ≲20 cm. The interpretation of layering in this model is discussed in the following section.

In this model, mixing of the top meter or so of soil would also produce an essentially depth-independent distribution of spallation tracks, as observed, with a spread at each depth reflecting the variation of production rate over the range of depths mixed, perhaps a factor of ~10.

The basic features of these models and the observations are summarized in Table 1.

LUNAR SOIL MIXING

Physical process

We now specify the mixing model favored by the evidence discussed in the last section.

Table 1. Lunar soil models.

Model	Track Source	Track density vs. depth above 60 cm	Spread at each depth	Cosmic ray-spallation correlation
Quiet soil	Cosmic ray Fe Spallation	Decreasing Peaks at $\approx$25 cm	$\sim \times2^*$ $\sim \times2^*$	Strong correlation
Frequent overlaying, sub-cm mixing only	Cosmic ray Fe Spallation	$\sim$Constant Increasing	$> \times100\,?\dagger$ $\sim \times2^*$	None
Frequent depth-dependent turnover	Cosmic ray Fe Spallation	$\sim$Constant $\sim$Constant	$\sim \times100\ddagger$ $\sim \times10\S$	None
Apollo 12 core observations	Cosmic ray Fe Spallation	$\sim$Constant $\sim$Constant	$\geqslant \times100$ $\geqslant \times10$	None

  * Due primarily to random orientation of etched surface.
  † Due to distribution of depths in overlay and broken rocks.
  ‡ Due to distribution of residence times at depths $\leq$20 cm.
  § Due to variation of production rate over range of mixed depths.

Theoretical analyses of surface activity generated by impact events have been made, for example, by ÖPIK (1969) and references therein. SHOEMAKER et al. (1970) have discussed the nature of the soil layer, or regolith, of Mare Tranquillitatus (landing site of Apollo 11). Experimental studies, such as GAULT et al. (1966) have been made of impact cratering mechanics. The physical processes indicated by such studies can be summarized as follows. Impact on the lunar surface by meteoroids causes soil and rock to be excavated from a crater and deposited on the surface in a blanket extending $\sim$2 crater diameters, with some material thrown much greater distances. Material thrown out of nearby craters will in turn overlay and bury this until excavation occurs again. After a given period of time the statistical nature of this activity will cause the soil layer to be completely turned over down to a certain median depth; during this time, the depth history of a particular soil grain will be quite complicated.

From the size-frequency distribution of craters observed at Mare Tranquillitatus, SHOEMAKER et al. (1970) have derived the time t required for turnover down to a median depth d ($\leqslant$10 cm) to be approximately $t = 0.0025Ad$ where A is the age of the mare surface. For $A = 4$ b.y. this gives t (m.y.) $= 10d$ (cm); since the cratering rate might have been greater in the past, we also consider here the effect of a model using t (m.y.) $= d$ (cm).

This near-surface stirring will be due primarily to events which produce craters $\leqslant$5 m in diameter. It is known that craters this small will disappear by erosion, overlay, and obliteration by larger craters in a lifetime small compared to the age of the maria. On this scale, the mare surface is said to be in equilibrium. In fact, many if not most of the craters which contribute to turning over the soil to a given depth d will disappear in the turnover time t, which is simply a restatement of the complicated excavation and burial history of a given soil grain over one mixing time.

Computational model and interpretation

We will be comparing this model to core sites where no large cratering event has occurred recently, so that the above discussion applies. It follows that for these sites we can assume that a large number of samples distributed randomly down to the mixing depth $d \leqslant 1$ m will again be distributed randomly after one mixing time. The present computations use this property to approximate the detailed depth history of a soil grain. We assume that periodic mixes down to three different mixing depths dc are characteristic of the near-surface stirring process in the following manner.

For the lower mixing rate described above we start with 1800 hypothetical samples distributed through the top 60 cm of soil. Every 100 m.y. (one mixing time for $dc = 10$ cm) those samples which have depths of 0–10 cm are assigned new depths at random between 0–10 cm. Similarly samples 0–25 cm deep and 0–60 cm deep are assigned new depths every 250 m.y. and 600 m.y., respectively. At each such "characteristic mix" we keep account of the accumulated tracks except for 1 % of the samples mixed which are reassigned 0 track density to simulate shock and impact-heat annealing (this is roughly the fraction of glass observed divided by the number of characteristic mixes over 4 b.y.). For the faster rate we mix depths of 0–10 cm, 0–30 cm, and 0–100 cm every 10, 30, and 100 m.y., respectively.

This model is designed to demonstrate the effects of periodically bringing material up from lower depths and of mixing the shallow depths more frequently. The characteristic mixes do not correspond to individual cratering events but simulate in one process the physically separate events of repeated excavation and burial by overlaying. The nature of these separate events is such that we would expect this computational model to underestimate the number of samples which spend time within a few hundred microns of the surface; i.e. to underestimate the high density tail of the track density distribution.

Next we consider the effect of mixing deeper than one meter, not included in the computations just described. Sample grains which remain below 1 m will acquire essentially no primary tracks. Nearby deep cratering events will lay down a blanket of such samples around a soil core site; if the blanket is deeper than the core sample, it will essentially restart the process of track accumulation. The annealing caused by deposits of lava and volcanic ash will have the same effect. We, therefore, start our computations with no tracks and interpret the derived cosmic ray age as the time since the last deep-cratering or track-annealing event in the vicinity of the core site. Since only a few such events have occurred at a given site, we expect this time to be long. Such an event might also lay down a shallower layer of trackless material. If subsequent mixing has not had time to obliterate such a layer, then soil grains below will record mixing history before the event.

Note that grains brought near the surface for the first time may physically be part of rocks (freshly broken from bedrock) which presumably are not as susceptible to burial and shallow mixing as the smaller soil grains. However, measured surface residence times (Crozaz et al., 1970; Fleischer et al., 1970b) and predicted destruction times (Shoemaker et al., 1970) indicate that such rocks are relatively quickly broken up and so should have little effect on the subsequent mixing history.

Track growth rate

As the mixing process described above evolves with time, the hypothetical samples accumulate tracks at a depth-dependent rate determined from the average primary cosmic ray spectrum. Primary particle tracks in lunar soil are due mostly to high-energy cosmic ray iron nuclei. The energy spectrum of these nuclei has been measured in interplanetary space by COMSTOCK et al. (1969) and extrapolated from balloon observations (FREIER and WADDINGTON, 1968). This solar-modulated energy spectrum (observed at the minimum of solar activity) has been corrected to the average level over the 11-year solar cycle using a model by WANG (1970). Observations in meteorites of primary particle tracks (FLEISCHER et al., 1967) and spallation products (HONDA and ARNOLD, 1964) support the assumption that this average spectrum has been roughly constant in the geologic past. From this energy spectrum, taking into account particle range and loss of iron nuclei by spallation in lunar material, distribution of path lengths in a semi-infinite soil, and etching characteristics of the soil minerals, we calculate the rate of track accumulation at a given depth. The track density per area is assumed to be observed in a plane parallel to the surface.

Results

Some of the results of these calculations are shown in Fig. 5. This figure shows the distribution of etched track densities after 2.8 b.y. for the samples in two depth intervals: the top 10.5 cm, corresponding to the Apollo 11 core depth, and the 30 to 40 cm interval, to compare with the Apollo 12 deep core. The scale is logarithmic, by factors of 2.

In this case the top 10 cm are mixed every 100 m.y., the top 25 cm every 250 m.y. and the top 60 cm every 600 m.y. For times earlier than about 2 b.y. the distributions at the two depths vary greatly from each other, with the top 10 cm initially acquiring tracks at a much higher rate. After ~2.5 b.y. the soil approaches uniformity down to a half meter, the distributions becoming very similar at different depths. At all later times each depth acquires tracks at the same average rate because of the frequent mixing.

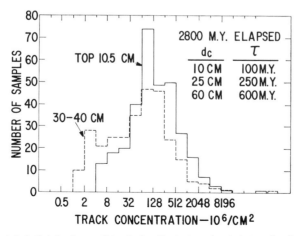

Fig. 5. Calculated distributions of track density in two depth intervals after 2.8 b.y. of soil mixing. The samples were periodically mixed down to three characteristic depths d_c with time periods τ.

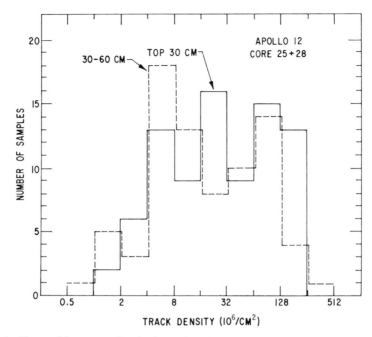

Fig. 6. Observed frequency distributions of track density in core 12025 and 12028 summed over two soil-depth intervals, 0–30 cm (solid line) and 30–60 cm (dashed line). The 0–30 cm distribution does not include the data from a possible special layer at ≈ 16 cm (Fig. 1). Lower limits have been smoothed over the next factor of 4.

The observed track density vs. frequency distribution summed over depth is shown in Fig. 6 for Apollo 12. The top half (0–30 cm) and bottom half (30–60 cm) of core 12025 + 12028 are summed separately to display any systematic change with depth. The samples at ≈ 16 cm (Fig. 1) were omitted since they may represent a layer that is not well mixed, as discussed later. Samples with lower limits (Fig. 1) have been smoothed over the next higher factor of 4; some of these may have still higher densities. Figure 7 shows the track density distribution observed in Apollo 11 core 10005. The track density shown in Figs. 6 and 7 should be increased by a factor of ~ 2 to correct for the random orientation of the etched crystal surface with respect to the cosmic ray flux. The distributions are similar to that shown in Fig. 5, both in absolute track density and in spread.

The time history of the calculated model is made clear in Fig. 8. Here we have plotted the medians of the calculated distributions as a function of time, as the system evolves, for two depth intervals. The arrows indicate the times and depths of the characteristic mixes used. The sharp breaks in the curves are due to the effects of the assumed mixes; in a real physical situation, the curves would have a much more irregular appearance. However, the envelopes of the model curves are meaningful and the approach to uniformity with depth at later times is clear. On the basis of the Apollo 11 results alone, in the top 10.5 cm, and for the mixing rate assumed, we could only conclude that the exposure age of the mixed soil is at least ~ 500 m.y.

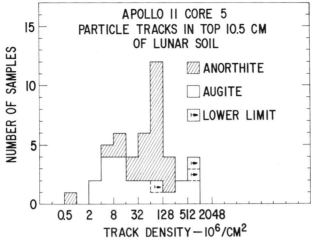

Fig. 7. Observed frequency distribution of track density for samples from throughout the top 10.5 cm of soil at Tranquillity Base. Each sample is a different soil grain. For comparison with Figs. 5, 8, and 9 these track densities should be increased by an average factor of 2 to correct for random orientation of the etched crystal surface.

On the basis of the deeper Apollo 12 core we can push this lower age limit up to about 2 b.y.

As indicated earlier, the average mixing rate is uncertain, and we also have investigated a faster mixing rate, shown in Fig. 9. Here the track densities approach uniformity much faster and the lower age limit is about 1 b.y. for the Apollo 12 site.

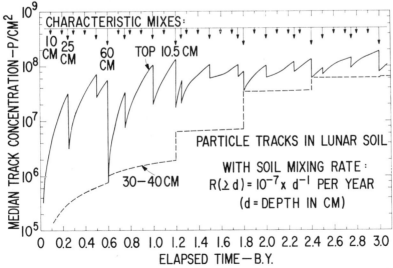

Fig. 8. Time history of the median value of the track density distributions for two depth intervals, 0–10 cm (solid line) and 30–40 cm (dashed line). Times and depths of the characteristic mixes used are also shown. A soil grain at depth d is mixed with rate $R(\geqslant d)$.

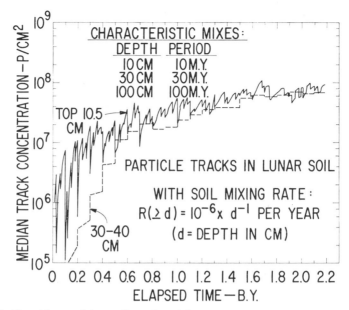

Fig. 9. Time history of the median value of the track density distributions for two depth intervals, 0–10 cm (solid line) and 30–40 cm (dashed line), for a fast mixing rate. Somewhat deeper characteristic mixing depths are used in this case.

When the results of these calculations are adjusted for average orientation and plotted vs. depth, we obtain the curves in Fig. 4 discussed in the preceding section.

Layers and surface residence

The mixing model discussed here does not contradict the existence of some layering. In this picture, the reported gravel layer at ≈13 cm and possibly the low-track-density samples at ≈16 cm are the bottom remnants of a fresh blanket brought up from ≈1 m, laid down over well-mixed soil and not yet completely obliterated. This blanket has since been covered by, and partially mixed with, other well-stirred soil; eventually it may be completely destroyed. The existence of such layers indicates the discrete nature of the excavation and overlay events which make up the mixing process and can be important in studying the details of this process.

Of equal importance is the occurrence of very high-track densities resulting from surface residence. We have indicated that this region is uncertain experimentally, and that the present computer model is probably underestimating the high density contribution. The role of high track densities needs to be studied further.

Conclusions

The distribution of cosmic ray and spallation-recoil tracks in the Apollo 12 deep core allow us to choose among three basic models for the lunar soil as summarized in Table 1. The evidence shows conclusively that the soil is not quiet and undisturbed.

The spallation tracks imply that simple burial by overlay or surface transport is not enough but rather that at least the top meter of soil has been subjected to mixing. Computer simulation of frequent soil turnover agrees well with the cosmic-ray track distribution and requires stirring times of at least 1 to 2 b.y. since the last cratering event large enough to lay down a deep blanket of material from well below 60 cm.

This mixing process involves discrete excavation and overlay events, and hence some temporary structure, due to shallow layers of previously deep material, might be expected to persist for a time. Detailed track studies of such structure and of deeper core samples would tell us more about the stirring mechanism and help to unravel the history of larger-scale ground movement such as the slumping of crater walls and rille formation.

Acknowledgments—We are pleased to acknowledge the able experimental assistance of E. F. KOCH, M. D. MCCONNELL, and E. STELLA. This work was supported in part by NASA under contract NAS 9-7898.

REFERENCES

ARRHENIUS G., ASUNMAA S. K., LIANG S., MACDOUGALL D., WILKENING L. (1971) Irradiation and impact in Apollo rock and soil samples. Second Lunar Science Conference (unpublished proceedings).

BARBER D. J., HUTCHEON I. D., PRICE P. B., RAJAN R. S., and WENK R. (1971a) Exotic particle tracks and lunar history. Second Lunar Science Conference (unpublished proceedings).

BARBER D. J., HUTCHEON I. and PRICE P. B. (1971b) Extralunar dust in Apollo cores? *Science* **171,** 372–374.

BHANDARI N., BHAT S., LAI D., RAJAGOPALAN G., TAMHANE A. S., and VENKATAVARADAN V. S. (1971) Fossil track studies in lunar material. II: The near-surface exposure history of regolith components. Second Lunar Science Conference (unpublished proceedings).

BORG J., DURRIEU L., DRAN J. C., JOURET C., and MAURETTE M. (1971) Irradiation, texture, and habit histories of the lunar dust grains. Second Lunar Science Conference (unpublished proceedings).

CARRIER W. D., III, JOHNSON S. W., WERNER R. A., and SCHMIDT R. (1971) Distortion in samples recovered with the Apollo core tubes. Second Lunar Science Conference (unpublished proceedings).

COMSTOCK G. M., FAN C. Y., and SIMPSON J. A. (1969) Energy spectra and abundances of the cosmic ray nuclei helium to iron from the OGO-1 satellite experiment. *Astrophys. J.* **155,** 609–617.

CROZAZ G., HAACK U., HAIR M., MAURETTE M., WALKER R., and WOOLUM D. (1970) Nuclear track studies of ancient solar radiations and dynamic lunar surface processes. *Proc. Apollo 11 Lunar Science Conf., Geochim. Cosmochim. Acta,* Suppl. 1, Vol. 3, pp. 2051–2080. Pergamon.

CROZAZ G., WALKER R., and WOOLUM D. (1971) Solar and galactic cosmic ray studies in fines and rocks. Second Lunar Science Conference (unpublished proceedings).

FLEISCHER R. L., PRICE P. B., WALKER R. M., MAURETTE M., and MORGAN G. (1967) Tracks of heavy primary cosmic rays in meteorites. *J. Geophys. Res.* **72,** 355–366.

FLEISCHER R. L., HAINES E. L., HANNEMAN R. E., HART H. R., JR., KASPER J. S., LIFSHIN E., WOODS R. T., and PRICE P. B. (1970a) Particle track, X-ray and mass spectrometry studies of lunar material from the Sea of Tranquillity. *Science* **167,** 568–571.

FLEISCHER R. L., HAINES E. L., HART H. R., JR., WOODS R. T., and COMSTOCK G. M. (1970b) The particle track record of the Sea of Tranquillity. *Proc. Apollo 11 Lunar Science Conf., Geochim. Cosmochim. Acta,* Suppl. 1, Vol. 3, pp. 2103–2120. Pergamon.

FLEISCHER R. L., HART H. R., JR., COMSTOCK G. M., and EVWARAYE A. O. (1971) Particle track record of the Ocean of Storms. Second Lunar Science Conference (unpublished proceedings).

FREIER P. S. and WADDINGTON C. J. (1968) Very heavy nuclei in the primary cosmic radiation. I. Observations of the energy spectrum. *Phys. Rev.* **175,** 1641–1648.

GAULT D. E., QUAIDE W. L., and OBERBECK V. R. (1968) Impacet cratering mechanics and structures. *Proc. Conf. on Shock Metamorphism of Natural Materials,* pp. 87–99. Mono Book, Baltimore.

Honda M. and Arnold J. R. (1964) Effects of cosmic rays on meteorites. *Science* **143**, 203–212.

Kohman T. D. and Bender M. L. (1967) Nuclide production by cosmic rays in meteorites and on the moon. In *High Energy Nuclear Reactions in Astrophysics* (Editor, B. S. P. Shen), pp. 169–245. Benjamin.

Lal D., MacDougall D., Wilkening L., and Arrhenius G. (1970). Mixing of the lunar regolith and cosmic ray spectra: Evidence from particle-track studies. *Proc. Apollo 11 Lunar Science Conf., Geochim. Cosmochim. Acta*, Suppl. 1, Vol. 3, pp. 2103–2120. Pergamon.

Öpik Ernst J. (1969) The moon's surface, in *Annual Review of Astronomy and Astrophysics* **7**, 473–526. Annual Reviews, Inc. Palo Alto, California.

Price P. B. and O'Sullivan D. (1970) Lunar erosion rate and solar flare paleontology. *Proc. Apollo 11 Lunar Science Conf., Geochim. Cosmochim. Acta*, Suppl. 1, Vol. 3, pp. 2351–2359. Pergamon.

Shoemaker, E. M., Hait M. H., Swann G. A., Schleicher D. L., Schaber G G., Sutton R. L., Dahlem D. H., Goddard E. N., and Waters A. C. (1970) Origin of the lunar regolith at Tranquillity Base. *Proc. Apollo 11 Lunar Science Conf., Geochim. Cosmochim. Acta*, Suppl. 1, Vol. 3, pp. 2399–2412. Pergamon.

Wang J. R. (1970) Dynamics of the 11-year modulation of galactic cosmic rays. *Astrophys. J.* **160**, 261–281.

Proceedings of the Second Lunar Science Conference, Vol. 3, pp. 2583–2598
The M.I.T. Press, 1971.

The exposure history of the Apollo 12 regolith

G. Arrhenius, S. Liang, D. Macdougall, and L. Wilkening

Scripps Institution of Oceanography, University of California, San Diego,
La Jolla, California 92037

and

N. Bhandari, S. Bhat, D. Lal, G. Rajagopalan,
A. S. Tamhane, and V. S. Venkatavaradan

Tata Institute of Fundamental Research, Homi Bhabha Road, Colaba, Bombay 5,
India

(*Received* 23 *February* 1971; *accepted in revised form* 30 *March* 1971)

Abstract—We report in this paper the results of fossil track studies in several Apollo 12 surface scoop samples and in samples from the double core 12025, 12028 to 60 cm (actual) depth. Detailed analysis of the frequency distribution of fossil tracks in crystals from the various samples has provided information on the dynamics and rates of mixing and deposition of the regolith during the last few hundred million years. Calculations are presented for estimating surface exposure ages based on fossil track production rates observed in five Apollo rocks and in the St. Severin meteorite. The multi-stage irradiation model considered indicates that there are two principal depths of irradiation important for track production in the regolith. Our model calculations also show that (1) the different double core strata have had a radiation, mixing, and deposition history closely similar to that of the surface over most of the Apollo 12 landing site; (2) most of the present surface regolith components have been excavated from depths greater than 10 cm some 10–50 m.y. ago, and (3) in general, vertical mixing is slower than deposition. Furthermore this deposition appears to occur in discrete episodes separated by long intervals of nondeposition. The time scale of this process is such that any grains (or a rock) lying on the top of the regolith remain there for periods of the order of a few million years before burial. The frequency distribution of very high-track density crystals, as well as the layering of the core, exclude mixing at a comparable or more rapid rate.

A comparison of the lunar regolith track distribution with that of gas-rich meteorites indicates significant differences between the conditions of irradiation in these two cases, and narrows the speculation on the environment of the meteorite parent body.

Introduction

The lunar regolith has a complex history of cratering over a wide dimensional range (Gault, 1970; Shoemaker *et al.*, 1970a and b; Hartmann, 1970; and others, and steady-state model calculations have been used to obtain relative chronological sequences for different craters. In this context, it is worth noting that one of the remarkable features about the regolith at the Apollo 12 site is the stratification observed in the double core (LSPET, 1970; Carrier *et al.*, 1971). If the current turnover models, based on a constant cratering rate, were valid, then it is difficult to understand the existence of numerous layers 2–4 cm thick unless the total time involved in the deposition of this entire sequence was of the order of 10 m.y. or less. It is therefore important to study the radiation history of the double core and the surface regolith components using as many methods as possible. Cosmogenic radio-active isotopes provide information about recent exposure ages for samples exposed to solar flares protons on the surface. The stable spallogenic nuclides, e.g., Ne^{21} (NYQUIST *et al.*, 1971; FUNKHOUSER *et al.*, 1971) and the isotopic changes in gadolinium produced by neutron capture (MARTI and LUGMAIR 1971; ALBEE *et al.*, 1971) are useful for estimating approximate exposure ages at depths up to a meter.

Our approach has been to study fossil cosmic ray tracks (FLEISCHER *et al.*, 1967a and b). The most important advantage associated with track studies is that there is a steep depth dependence in the track formation rate, so that, as we will discuss later, any tracks present in the regolith components were formed primarily during exposure of the grains at depths between 0 and 10 cm. Our conclusion with respect to mixing of the regolith are at variance with those of COMSTOCK *et al.* (1971) and CROZAZ *et al.* (1971).

METHODS

The mounting, grinding, polishing, and etching techniques used for fossil track observations are similar to those described earlier (LAL *et al.*, 1968); modifications and special procedures for reducing surface loss and for etching feldspar, pyroxene, and olivine are described in BHANDARI *et al.* (1971a and b). Individual crystals or grains were washed in alcohol and the adhering fine dust was separated. The coarser fraction was examined under a stereomicroscope and all large clear pyroxene and feldspar crystals were picked out and mounted in epoxy. Separate mounts were prepared for feldspars and pyroxenes, with 4 to 6 crystals per mount. From the remaining material, opaque grains were separated and 50 to 100 μ-size transparent crystals were mounted in groups of 30–50 grains per mount. Throughout this preparation precautions were taken to ensure that none of the grains or crystals was damaged. Except for possible damage in transit, the state of the grain surfaces investigated should therefore be identical to their condition in the lunar regolith at the time of sampling.

Track counts were made using high contrast optical microscopy and electron microscope replication techniques (MACDOUGALL *et al.*, 1971). The replica technique was employed for studying very high track densities, particularly for densities exceeding 5×10^8 cm^{-2}, which are difficult to count accurately using normal optical methods. Accurate replica counts were also used to calibrate a simple and quick approximation method for estimating high track densities by measuring loss from the surface of a crystal during controlled etching. This technique developed from our observations over several years of a large number of etched silicate crystals from the lunar regolith and gas-rich meteorites (LAL and RAJAN, 1969). We noticed that in track rich grains in gas-rich meteorites the very high track density borders incurred greater surface loss during etching than the central, lower track density regions. This differential surface loss arises because of a deeper penetration of etchant through the network of tracks in higher track density regions, and is approximately proportional to the track density. By measuring surface losses of high track density lunar grains, and measuring the track density of the same crystals by the replica technique, we have made a calibration which enables us to estimate track densities $> 10^8$ cm^{-2} simply by measuring surface loss during controlled etching.

Surface loss is measured microscopically by observing, before and after etching, the difference in the elevation of a crystal surface, relative to the epoxy in which it is potted. The epoxy surface is not attacked by the etchant, and provides a convenient reference level.

RESULTS

We have examined the fossil track record in 10 scoop samples and 23 double core samples. The positions of the samples analysed are indicated on the plan map of the Apollo 12 site (Fig. 1). Track density data are presented in Figs. 2 and 3 for the scoop and the double core samples respectively. In addition, track data for grains from the totebag (12060) and ALSRC (12057) are included in Fig. 2. These histograms show

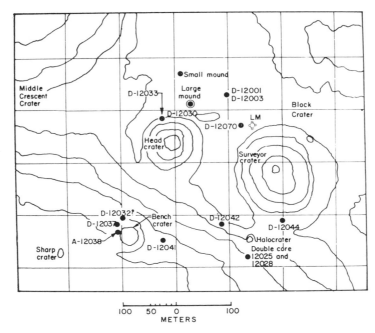

Fig. 1. Map of the Apollo 12 landing site showing sample locations. The symbols
D and *A* denote soil and rock samples, respectively.

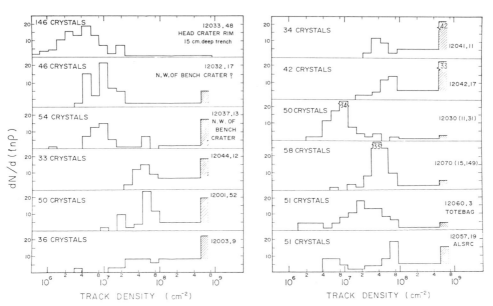

Fig. 2. Histograms showing track density frequency distributions in grains from the
surface scoop samples. For ease of comparison each distribution has been normalised to
100 crystals. The hatched bars represent crystals with densities greater than 5×10^8 cm^{-2}.

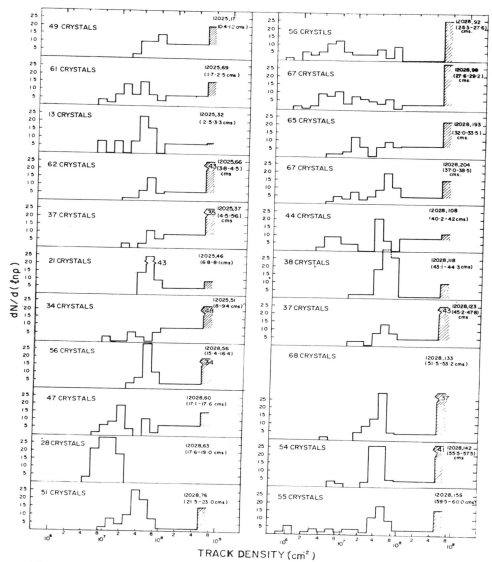

Fig. 3. Histograms showing track density frequency distributions in grains from the
double core 12025, 12028. Details as in Fig. 2.

track density versus number frequency, with the intervals in track density chosen so
that $\ln \rho_2 - \ln \rho_1 = \ln (2)^{1/2}$ or 0.346. This interval is comparable to or broader than
the errors of measurement.

We have analysed the track data for feldspar and pyroxene separately. In
Bhandari *et al.* 1971b, we discussed the differences in track densities between adjacent
feldspar and pyroxene grains in a section from the interior of a lunar rock, that is

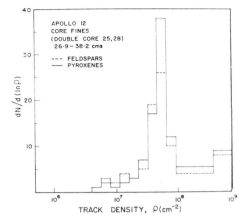

Fig. 4. Histogram showing the track density frequency distribution separately for feldspar and pyroxene crystals from the double core.

where both minerals are exposed to an identical flux and energy spectrum of cosmic radiation. We found that for low ρ values, the variable internal uranium concentrations in these crystals contributes to part of the differences. Furthermore, the track densities can differ up to 25% due to differences in the recording characteristics of the two mineral species. In the present discussion, however, we will use the combined track density distribution for both feldspars and pyroxenes. This approach is justified by the data in Fig. 4, where we have plotted feldspar and pyroxene results separately for grains from the same double core samples.

The data of Figs. 2 to 4 show that a wide range of track densities exists in the regolith grains. As an illustration we show microphotographs of low and high density crystals in Figs. 5a and 5b.

DISCUSSION

Scoop samples come from the top 3 or 15 cm of the regolith depending on the scoop used. The interval of sampling in the core is smaller, of the order of 1–2 cm. The tracks in the scoop and core samples reflect a several-stage irradiation history as described below:

(1) *Irradiation prior to the ultimate deposition*, which could be very complex, or could, in the simplest case, be negligible for deeply buried material which has always been shielded from cosmic rays. The shielding depth in the core (for track purposes) can conveniently be taken to be about 40 cm where a billion years' exposure would result in less than 10^6 tracks cm^{-2}. Crystals now exhibiting very high track densities may have been irradiated elsewhere on the surface and subsequently transported to the present site. A fraction of these crystals may have been irradiated as loose grains, bonded to their neighbors by the surface forces which give the lunar soil its characteristic cohesive structure (ARRHENIUS and ALFVÉN, 1971). Another fraction may derive its characteristic track distribution from irradiation while the grains were part of the surface layer of solid rocks, shedding their decomposition products into the soil as envisaged by CROZAZ *et al.* (1970).

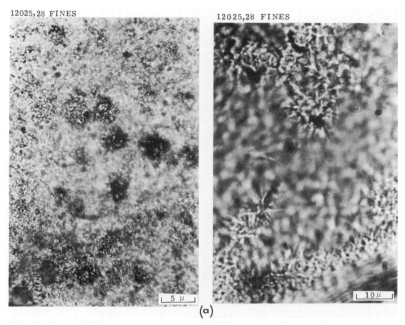

Fig. 5. Photomicrographs of fossil tracks in Apollo 11 and Apollo 12 fines: (a) tracks
feldspars with $\rho = (3–7) \times 10^7$ cm^{-2} (a pyroxene crystal in the lower left photograph
feldspar crystals with different densities

(2) *Postdepositional irradiation*, which in the simplest case would be an unshielded,
undisturbed surface exposure. A more complicated postdepositional irradiation
history would involve mixing to different depths by impact during irradiation. The
large number of parameters affecting the track record of regolith grains complicates
the interpretation in terms of regolith history, and in fact the Apollo 11 cores 10004
and 10005 looked hopeless in this respect (LAL *et al.*, 1970). However, the Apollo
12 data are easier to understand since the track density distributions in scoop samples
and core layers often seem to approach the simplest case of postdepositional irradiation, that is, an undisturbed surface exposure terminated by sudden blanketing. Below
we discuss model calculations for scoop and double core fines.

Calculation of surface irradiation ages

The relationship expected between ρ and $dN/d(\ln \rho)$, the number of grains in each
density interval, for the case of no predepositional irradiation history is

$$dN/d(\ln \rho) = \frac{K^{1/\alpha}}{B\alpha X_m} \rho^{-(1/\alpha)} \tag{1}$$

where X_m is the depth of the scoop; K, B, and α are dimensionless constants discussed
below.

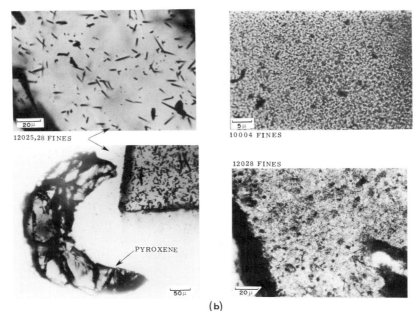

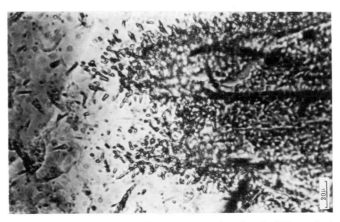

in feldspar crystals due to nuclei of $Z > 30$; (b) upper and lower left photographs show shows no tracks). Upper and lower right are feldspars with $\rho \gg 5 \times 10^8$ cm^{-2}; (c) two have apparently been melded together.

In deriving the above equation, we have assumed a depth (X) variation in the rate of formation of tracks, $\dot{\rho}(X)$, cm^{-2} m.y.$^{-1}$, in the regolith given by

$$\dot{\rho}(X) = K(A + BX)^{-\alpha} \qquad (2)$$

based on the spectral form for iron group nuclei as deduced from the study of fossil tracks in Apollo rocks and St. Severin meteorite (BHANDARI et al., 1971c) and theoretical calculations given by BHANDARI et al. (1971d). Relation (2) is valid for the moon for values of (BX) between 1.0 and 25 cm. Values of the other parameters are $A = 7.5$ cm, $K = 6.36 \times 10^{11}$, $\alpha = 6.15$. For a typical moon rock $B = 1$, and for the core $B = 0.53$. The latter value is the ratio of the bulk density of the soil (1.8, Scott *et al.*, 1970) to that of lunar rock (taken as 3.4). Relation (2) is also valid for BX between 0.1 and 1.0 cm with $K = 1.4 \times 10^6$, $A = 0$, and $\alpha = 0.7$.

Relation (1) holds for a scoop with cross-section varying linearly with depth. If, however, the cross-sectional area, C, of the scoop varies as

$$C(X) = C_0 \exp(-X/S) \tag{3}$$

where S is a constant, then the frequency distribution of track densities would be modified to

$$dN/d(\ln \rho) = \frac{K^{1/\alpha} \cdot e^{A/SB} \cdot e^{-(1/SB)(\rho/K)^{-1/\alpha}}}{B \cdot \alpha \cdot S \cdot (1 - e^{-X_m/S})} \cdot \rho^{-1/\alpha} \tag{4}$$

Quite clearly, information about surface and near surface irradiation lies in the analysis of grains with very high track densities ($\rho > 5 \times 10^8$ cm^{-2}). From relation (2), it is seen that $\dot{\rho}$(cm$^{-2} \cdot$ m.y.$^{-1}$) for values of X around 0.1 cm is about 10^7 and that ρ values of 10^9 cm^{-2} would represent a minimum of 10^8 years irradiation if the depth of irradiation exceeds 1 mm. From the observed track data in Apollo 12 rocks (Bhandari *et al.*, 1971c) we deduce that for 2×10^{-3} to 10^{-1} cm shielding, a good working relationship is

$$\dot{\rho}(2 \times 10^{-3} < X < 0.1 \text{ cm}) = 1.2 \times 10^6 \, X^{-0.75}, \tag{5}$$

where X is expressed in cm, and $\dot{\rho}$ in cm^{-2} m.y.$^{-1}$. Hence eqs. (2) and (5) cover the range of interest in the present work since all track density measurements we have made are for $X \geq 20 \, \mu$. Thus the very high track density grains seen in most samples have either been exposed on the surface (unshielded) for $\sim 10^7$ yr or at a mean depth of 0.1 to 1 cm for $\sim 10^9$ yr. Crystals exposed unshielded on the surface show well-defined gradients within grains of dimensions greater than 2×10^{-2} cm. Densities in the central regions of such crystals are usually $\leq 5 \times 10^7$ cm^{-2}.

The observations discussed above lead us to the conclusion that when high track density grains coexist with low track density grains at the same sample depth, the irradiation near the surface proceeded with only a partial vertical mixing, not ex-ceeding a few cm. The high density grains were exposed for periods of ~ 100 m.y. within 10^{-1} cm of the surface, a requirement which in itself puts a constraint on the rate of gardening.

Surface scoop samples

We now proceed to calculate the mean exposure ages of scoop samples, using relations (1) and (4), which are based on no gardening during irradiation. To reduce errors in the calculations due to neglect of gardening, we calculate the exposure ages based on the quartile track density values, $\rho_{0.25}$. The fraction of grains between $\rho_{\min}$

and $\rho_{0.25} = 0.25$. This exposure age, T(m.y.), is then given by the relation:

$$T = \rho_{0.25}/K(A + BX)^{-\alpha} \tag{6}$$

with $X = 0.75X_m$ in the case of a linear scoop, and

$$X = S \ln (1/[1.0 - 0.75(1 - e^{-X_m/S})])$$

in the case of the exponential cross-section model; X_m is the depth of the scoop in cm.

In the absence of gardening during irradiation, the observed distribution $N(\rho)$ is sufficient to determine both the value of X_m (which is not well known) and the exposure age T, but gardening complicates the situation. However, we find that the linear cross-section model or an exponential model with $S > 5$ cm accounts quite well for observed $N(\rho)$ distributions such as in Fig. 2. In Table 1 we list relevant details on the scoops as well as the calculated surface exposure ages, based on the simple model assuming no pre-deposition track record, and using the quartile track density value. If the *average* track density value is used in the calculations, the ages are 2 to 5 times greater. Reasons for using the quartile method are discussed below. Before discussing the implications of these calculations, we will discuss similar calculations for the double core.

Model for the core

Experimental data on the double core layers are given in Figs. 3 and 6, and in Table 2. If τ_{sn} is defined as the surface exposure time of the nth layer between deposition and later burial by the overlying layer $(n - 1)$, then τ_n, the integrated exposure time of the nth layer from deposition to the present is

$$\tau_n = \sum_{i=1}^{i=n} \tau_{si} \tag{7}$$

If the vertical depth X is taken as zero at the surface, and if the nth layer has a

Table 1. Observed average track densities in scoop* fines and calculated exposure ages.

Parent	Specific	No. crystals measured	Average track density (10^6 cm^{-2}) for crystals of $\rho < 10^8$ cm^{-2}	N_H/N (Fraction of crystals with $\rho > 10^8$ cm^{-2})‡	Quartile track densities (10^6 cm^{-2})	Surface irradiation age (m.y.)
12001	52	50	48	0.68	50	40
12003	9	36	38.5	0.78	45	36
12030	(11 + 31)	50	12	0.14	6	5
12032	17	46	13.5	0.17	5.8	5
12033†	48	146	5.6	0.03	2.5	80
12037	13	54	16	0.52	12	10
12041	11	34	46	0.82	100	80
12042	17	42	55	0.91	100	80
12044	12	33	48	0.79	80	65
12070	(15 + 149)	58	45	0.48	35	41

* The penetration depth of the scoop is assumed to be 3 cm in all cases except for 12070, where it is taken to be 15 cm.
† From a 15 cm deep trench. The age given in this case is the total in situ irradiation age.
‡ Also includes those crystals which have track gradients.

G. Arrhenius *et al.*

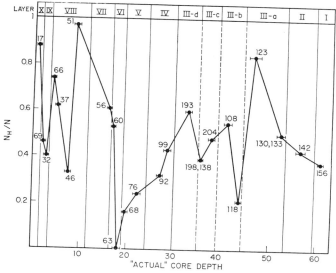

Fig. 6. The fraction (number of grains with $\rho > 10^8$ plus grains with gradients)/(all grains counted), or N_H/N, is plotted against location of sample in the double core. N_H can also be thought of as the number of grains irradiated without shielding.

thickness of $X_n - X_{n-1}$, then the expected track density in the nth layer is given by

$$\rho_n(\text{av}) = \tau_{sn}\dot{\rho}\,|0 \rightarrow (X_n - X_{n-1})| + \sum_{i=1}^{n-1} \tau_{si}\dot{\rho}\,|(X_{n-1} - X_{i-1}) \rightarrow (X_n - X_{i-1})| \quad (8)$$

where $\dot{\rho}\,|X_i \rightarrow X_j|$ refers to the average in-situ density production rate expected within a layer of thickness $X_j - X_i$:

$$\dot{\rho}\,|X_i \rightarrow X_j| = \frac{1}{X_j - X_i} \int_{X_i}^{X_j} \dot{\rho}(y)\,\mathrm{d}y \quad (9)$$

where $\dot{\rho}(y)$ is given by relation (2).

Depositional history of the core material

Using eq. 8 we have determined successive values of τ_{sn}. We have also modified eq. 8 for use with quartile density values and find that, as with the scoop samples, the quartile ages are systematically lower than ages calculated using average density values. We feel that the quartile method ages are more reliable since they are not affected by present uncertainties in the low energy cosmic ray flux nor are they affected if even 20 to 30% of the fines has had a predeposition irradiation history. Therefore only ages calculated using quartile densities are given in Table 2. What becomes apparent from Table 2 is that in no case is it obligatory to assume an appreciable predepositional irradiation, although in some cases such an irradiation cannot be excluded. In the following paragraphs we discuss our reasoning for such a conclusion.

In several cases we have studied more than one sample from within an individual layer as described in LSPET (1970). For layers IV and IX we have two samples in each case which show identical density distributions and the same average track density. These layers are apparently thoroughly mixed. In layers III-a, III-b, and VI the average track densities decrease with depth in a manner similar to, but less rapidly than, the track production function (see Fig. 6). Such a trend is also seen in the top three samples from layer VIII, although the deepest sample, 12028,51, does not fit the trend. It is possible that this latter sample represents a layer by itself which escaped recognition by the preliminary examination team. In any case, the decrease with depth in the average track density seen in the layers mentioned above would not be expected if gardening was (is) rapid or if the grains had acquired an appreciable fraction of their total track density before arriving at their present geographical position.

Four particular layers in the double core (IX, 1.6–3.1 cm; VI, 16–19.6 cm; IV, 24–29.6 cm; and I, 59–60 cm) have the following characteristics; (1) the lowest ρ values are $(1-5) \times 10^6$ cm^{-2}; (2) the quartile ρ value is $(5-20) \times 10^6$ cm^{-2}; (3) grains with very high track densities are few, less than 40% of the total (see Fig. 6); (4) several layers located within 10 cm above one of these four low track density layers contain a fraction of crystals with very high track densities. If it is assumed that for these four layers an appreciable fraction of the tracks were stored prior to deposition, then during the exposure of the immediately overlying layers, which have a high frequency of very high density grains, the crystals of the layer in question should have recorded many tracks and thus should have a higher track density than is observed

Table 2. Observed track densities in double core (25, 28) samples and model track exposures ages

Core	Specific	No. of crystals measured	Average track density (10^6 cm^{-2}) for crystals with $\rho < 10^8$ cm^{-2}	N_H/N^* (Fraction of crystals with $\rho > 10^8$ cm^{-2})	Quartile track density (10^6 cm^{-2})	Layer	Surface irradiation age*† (m.y.)
12025	17	49	56	0.88	50	X	25
12025	69	61	31	0.46	20⎱	IX	1
12025	32	13	33	0.40	32⎰		
12025	66	62	50	0.74	60⎱		
12025	37	37	53	0.62	70	VIII	45 ± 5
12025	46	21	53	0.33	40⎰		
12025	51	34	42	0.97	80	VIII(a ?)	35
12028	56	56	54	0.61	40	VII	40
12028	60	47	27	0.53	15⎱	VI	< 1
12028	63	28	11.5	0.0	7⎰		
12028	68	18	22	0.16	13⎱	V	10
12028	76	51	40	0.24	20⎰		
12028	92	56	21	0.32	5⎱	IV	5 ± 3
12028	99	67	15	0.43	6⎰		
12028	193	65	33	0.6	20	III-d	20
12028	(198 + 138)	18	43	0.39	9⎱	III-c	12 ± 10
12028	204	67	40	0.48	26⎰		
12028	108	44	24	0.55	11⎱	III-b	30 ± 20
12028	118	38	54	0.21	45⎰		
12028	123	37	49	0.84	44⎱	III-a	60 ± 20
12028	(130 + 133)	81	46	0.50	38⎰		
12028	142	54	43	0.43	35	II	50
12028	156	55	34	0.38	16	I	5

* Also includes those crystals which have track density gradients.
† The spread in calculated ages is given where it is large; in other cases errors are about ±10%.

and reflected in the calculated exposure ages given in Table 2. A check of this type is valid only for layers within a depth interval of about 5 to 10 cm, because the post-depositional bombardment rapidly becomes ineffective as a layer gets buried. This can be easily seen from relation (2) and in particular from the function $X \cdot \dot{\rho}(X)$, which signifies the unit time variation in production rate of tracks as a layer gets buried at a uniform rate. Between 0.5 and 5 cm depth this function varies only by a factor of 2, but beyond 10 cm it falls sharply with a power law of about 4. Thus, most of the tracks in layers now below 10 cm depth were formed when the layer was on or within 10 cm of the surface. One cannot exclude that a modest number of tracks were already present in some grains when they were deposited. In such cases one would overestimate surface exposure ages; the same consideration holds for the scoop samples. However, for layers presently buried deeper than 10 cm, it is impossible to produce the observed densities in situ even over the entire life of the moon, assuming present day fluxes.

From data in Table 2, we deduce that at the double core site the total deposition of the 60 cm (actual thickness) of regolith sampled occurred in 10 or more discrete steps at intervals of 5 to 50 million years. As discussed above, these intervals would be shorter if the grains had a predepositional irradiation history. The deduced surface exposure history for different layers in the double core is very similar to that deduced for the regolith surface scoops, a fact which should be considered as supporting the model calculations presented. Additional support is provided from observations on tracks in Apollo 12 (and Apollo 11) rocks. For five rocks, the surface exposure ages are deduced to be 2 to 5 m.y. (BHANDARI *et al.*, 1971c) based on fossil tracks and 100–500 m.y. (MARTI and LUGMAIR, 1971; BOGARD *et al.*, 1970) on the basis of spallogenic Ne^{21} concentrations. These "ages" show that the rocks analysed have typically been brought up from depths exceeding 10 cm two to five million years ago and that prior to this these rocks did not have a near-surface exposure history. Similar considerations are likely to apply to smaller regolith grains and we think that our model, which implicitly assumes no predepositional irradiation, is a satisfactory approximation and in accord with all observations.

Other models of regolith processes based on particle track evidence differ from ours in that they ascribe a much more important role to mixing and gardening than to layer-by-layer deposition (COMSTOCK *et al.*, 1971; CROZAZ *et al.*, 1971; CROZAZ *et al.*, 1970; FLEISCHER *et al.*, 1970). The most detailed treatments are the Monte Carlo mixing calculations by FLEISCHER *et al.* (1970) and COMSTOCK *et al.* (1971), which indicate that stirring of the regolith every few million years, with shallow stirrings more numerous than deep stirrings, would produce track density distributions similar to those observed. While we see no basic objection to such a model purely from a consideration of the Apollo 11 cores, the "easily recognizable stratigraphy" (LSPET, 1970) of both the Apollo 12 cores is clearly inconsistent with rapid stirring or gardening. Some of the layer boundaries have been described as "sharp contacts" (LSPET, 1970). Our model was conceived on the basis of the distinct core stratigraphy, and the consistency of the track data we have accumulated greatly reinforces the concept of layer-by-layer deposition with mixing rates that are small compared to the rate of deposition.

If our model calculations are a valid representation of exposure ages, then the calculated ages of several Apollo 12 scoop samples probably refer to crater-formation events. Observations of the astronauts indicate that some of the fresh craters, particularly Sharp and Bench craters at the Apollo 12 site, have ray structures; ejecta thrown out of craters are clearly visible in their vicinity (LSPET, 1970). The surface exposure ages for this material can therefore probably be related to the formation of the craters (SHOEMAKER et al., 1970b). Samples 12032 and 12037, both of which have small calculated exposure ages atypical of most of the regolith, were collected from the vicinity of Bench crater. Rock 12038, collected from the same area, has an exposure age of 1.5 m.y.

Implications for gas-rich meteorites

In view of the fact that we have now seen track density distributions in representative samples of the lunar regolith at two different locations (Apollo 11 and Apollo 12), it is interesting to consider the reasons for the striking differences between these distributions and those found in gas rich meteorites (LAL and RAJAN, 1969; PELLAS et al., 1969; WILKENING et al., 1971). In the lunar regolith, at depths up to 60 cm or so, adjacent grains from a given sample usually show different track densities, which normally range between approximately 10^7 and 5×10^8 for grains without track density gradients. The situation in the gas-rich meteorites, however, is quite different, and in general there are only two distinct track density groups in a given sample. The lower of the two values is observed in most of the grains and arises from in situ high-energy galactic cosmic ray irradiation. The other group of grains, which have higher densities, shows a steep gradient in track densities, decreasing from the borders, and although there may be some grain-to-grain variation in the density observed in the central regions of such crystals (due to varying size of the crystal) the track densities near the surface are always very similar (LAL and RAJAN, 1969; WILKENING et al., 1971). These observations are quite in contrast to what is seen in the case of lunar fines, where track densities at the borders of grains vary by large factors from one grain to another.

Another distinct difference lies in the distribution of near surface tracks around the grains. In the gas-rich meteorites the track-rich surface layer forms a remarkably uniform skin around the irradiated grains. In contrast, contours of equal track density in lunar grains do not follow crystal boundaries.

Two unique features observed in the gas-rich meteorites, (1) irradiation to a uniform dose, and (2) an isotropic irradiation of the track rich grains, require that before shielding by aggregation and burial the grains were either free to turn in a directional radiation or that they were suspended in an isotropic irradiation field. Another consequence of this observation concerns the mechanism postulated by CROZAZ et al. (1970), implying that the regolith grains with one sided irradiation, which are common among the surface exposed grains on the moon, owe this feature to exposure while they still were part of the surface of exposed rocks, later decomposed. If this mechanism is important, the absence or rarity of anisotropic irradiation in the meteorite grains implies that, in contrast to the moon, consolidated rocks did not serve as an important source of isolated grains in the environment where the parent bodies of the gas rich meteorites were formed.

There is also another aspect in which specifically the lunar breccias differ from the gas-rich meteorites. In the latter, the individual grains were brought together in such a gentle fashion that their heavily irradiated and hence fragile surface skins were preserved. In contrast, the processes on the moon which lead to the formation of breccias typically strongly modify or obliterate the track record by shock, fragmentation, partial melting, and heating.

The Kapoeta material also contains a considerable fraction of grains with features strongly suggestive of shock (Fredriksson and Keil, 1963), coexisting with the fragile track rich surface exposed crystals. It must then be concluded that both shock damage of some grains and irradiation of others predate the aggregation of the grains into their present configuration.

The differences between the lunar shock breccias and the gas rich meteorites indicate that the parent environments of the latter were characterized not only by lack of appreciable gravitation but also by freely turning, unshielded particles with very low relative velocities. The newtonian forces acting in such an environment have been discussed by ALFVÉN (1971) and TRULSEN (1971). Similar present day conditions, observed in the asteroidal jet streams, have been described by DANIELSSON (1971).

CONCLUSIONS

From the consistency of the track data for scoop, double core, and rock samples with the depositional model of the regolith considered here, we conclude that the core layers have been deposited at the Apollo 12 site at variable intervals of time over the last few hundred million years and have remained relatively undisturbed since. A probable source for the layers is material thrown out in crater forming events. Such a "throw out" model is different from most other regolith models proposed (e.g., COMSTOCK *et al.*, 1971; CROZAZ *et al.*, 1971; GAULT, 1970; SHOEMAKER, 1970a). These models consider gardening and mixing more important than layer-by-layer deposition. Inasmuch as our model indicates that mixing may be very slow, it resembles in part that of GOLD (1971). The clarification of the exposure history of the lunar regolith has made it possible to place realistic constraints also on the mode of accumulation of the grains which now constitute gas rich meteorites.

Acknowledgments—We wish to thank all those at NASA who had a hand in getting the samples to our laboratories, in particular the astronauts CHARLES CONRAD, RICHARD GORDON, and ALAN BEAN. Drs. D. H. ANDERSON, M. B. DUKE, and R. B. LAUGHON deserve special thanks for preparation and documentation of a large number of samples. We were assisted in the experimental work by Miss ASHA PADHYE, Miss NANDA PRABHU, Miss JOANNE GUY, and Mr. HOWARD SCHWARTZ. For skilled technical assistance we are grateful to Mr. P. B. BADEL and Mr. P. K. TALEKAR. This work was supported in part by NASA grants NGL 05-009-002 and NGL-05-009-154. We thank Dr. R. FRYXELL and G. HEIKEN for details of the double core stratigraphy.

REFERENCES

ALBEE A. L., BURNETT D. S., CHODOS A. A., HAINES E. L., HUNEKE J. C., PAPANASTASSIO D. A., PODOSEK F. A., RUSS G. P., TERA F., and WASSERBURG G. J. (1971) The irradiation history of lunar samples. Second Lunar Science Conference (unpublished proceedings).

ALFVÉN H. (1971) Apples in a spacecraft. *Science* (in press).

ARRHENIUS G. and ALFVÉN H. (1971) Asteroidal theories and observation. In *Physical Studies of Minor Planets* (editor T. Gehrels), Proc. 12th Coll. International Astronomical Union, Tucson, Arizona, March 8–10, 1971.

BHANDARI N., GOSWAMY J., KRISHNASWAMY S., LAL D., PRABHU N. and TAMHANE A. S. (1971a) Techniques for the study of fossil tracks in extraterrestrial and terrestrial samples, II. Study of fossil track densities in ordinary chondrites. *Geochem. J.* (Japan) (to be submitted).

BHANDARI N., BHAT S., LAL D., RAJAGOPALAN G., TAMHANE A. S., VENKATAVARADAN V. S., ARRHENIUS G., MACDOUGALL D., and WILKENING L. (1971b) *Proc. Indian Acad. Sci.* (to be submitted).

BHANDARI N., BHAT S., LAL D., RAJAGOPALAN G., TAMHANE A. S., and VENKATAVARADAN V. S. (1971c) Fossil track studies in lunar materials, I. High resolution time averaged (millions of years) data on chemical composition and energy spectrum of cosmic ray nuclei of $Z = 22$–28 at 1 A.U. Second Lunar Science Conference (unpublished proceedings).

BHANDARI N., LAL D., RAJAGOPALAN G., TAMHANE A. S., and VENKATAVARADAN V. S. (1971c) Formation rate of tracks due to cosmic ray "Iron" nuclei in meteorites and lunar rocks. *Moon* (to be submitted).

BOGARD D. D., FUNKHOUSER J. G., SCHAEFFER O. A., and ZÄHRINGER J. (1971) Noble gas abundances in lunar material—Cosmic ray spallation products and radiation ages from the Sea of Tranquility and the Ocean of Storms, *J. Geophys. Res.* **76**, 2757–2779.

CARRIER W. D., III, JOHNSON S. W., WARNER R. A., and SCHMIDT R. (1971) Disturbances in samples recovered with the Apollo 12 core tubes. Second Lunar Science Conference (unpublished proceedings).

COMSTOCK G. M., EVWARAYE A. O., FLEISCHER R. L., and HART H. R., JR. (1971) The particle track record of the Ocean of Storms. Second Lunar Science Conference (unpublished proceedings).

CROZAZ G., HAACK U., HAIR M., MAURETTE M., WALKER R., and WOOLUM D. (1970) Nuclear track studies of ancient solar radiations and dynamic lunar surface processes. *Proc. Apollo 11 Lunar Sci Conf.*, *Geochim. Cosmochim. Acta* Suppl. 1, Vol. 3, pp. 2051–2080. Pergamon.

CROZAZ G., WALKER R., and WOOLUM D. (1971) Cosmic ray studies of "recent" dynamic processes on the surface of the moon. Second Lunar Science Conference (unpublished proceedings).

DANIELSSON, L. (1971) Statistics of jet streams. In *Physical Studies of Minor Planets* (editor T. Gehrels). Proc. 12th Coll. International Astronomical Union, Tucson, Arizona, March 8–10, 1971.

FLEISCHER R. L., PRICE P. B., WALKER R. M., MAURETTE M., and MORGAN G. (1967a) Tracks of heavy primary cosmic rays in meteorites. *J. Geophys. Res.* **72**, 355–366.

FLEISCHER R. L., PRICE P. B., WALKER R. M., and MAURETTE M. (1967b) Origins of fossil charged-particle tracks in meteorites. *J. Geophys. Res.* **72**, 331–353.

FLEISCHER R. L., HAINES E. L., HART H. R., JR., WOODS R. T., and COMSTOCK G. M. (1970) The particle track record of the Sea Tranquillity. *Proc. Apollo 11 Lunar Sci. Conf.*, *Geochim. Cosmochim. Acta* Suppl. 1, Vol. 3, pp. 2103–2120. Pergamon.

FREDRIKSSON K. and KEIL K. (1963) The light-dark structure in the Pantar and Kapoeta stone meteorites. *Geochim. Cosmochim. Acta* **27**, 717–739.

FUNKHOUSER J., BOGARD D., and SCHAEFFER O. (1971) Noble gas analyses of core tube samples from Mare Tranquillitatus and Oceanus Procellarum. Second Lunar Science Conference (unpublished proceedings).

GAULT D. E. (1970) Saturation and equilibrium conditions for impact craters on the lunar surface: Criteria and implications. *Radio Science* **5**, 273–291.

GOLD T. (1971) Evolution of mare surface. Second Lunar Science Conference (unpublished proceedings).

HARTMANN W. K. (1970) Preliminary note on lunar cratering rates and absolute time scales. *Icarus* **12**, 131–133.

LAL D., MURALI A. V., RAJAN R. S., TAMHANE A. S., LORIN J. C., and PELLAS P. (1968) Techniques for proper revelation and viewing of etch-tracks in meteoritic and terrestrial minerals. *Earth Planet. Sci. Lett.* **5**, 111–119.

Lal D. and Rajan R. S. (1969) Observations relating to space irradiation of individual crystals of gas-rich meteorites. *Nature* **223**, 269–317.

Lal D., Macdougall D., Wilkening L., and Arrhenius G. (1970) Mixing of the lunar regolith and cosmic ray spectra: New evidence from fossil particle-track studies. *Proc. Apollo 11 Lunar Sci. Conf.*, *Geochim. Cosmochim. Acta* Suppl. 1, Vol. 3, pp. 2295–2303. Pergamon.

LSPET (Lunar Sample Preliminary Examination Team) (1970) Preliminary examination of the lunar samples from Apollo 12. *Science* **167**, 1325–1339.

Macdougall D., Lal D., Wilkening L., Bhat S., Arrhenius G., and Tamhane A. S. (1971) Techniques for the study of fossil tracks in extraterrestrial and terrestrial samples, I. Methods of high contrast and high resolution study. *Geochem. J.* (Japan) (submitted).

Marti K. and Lugmair G. W. (1971) Kr^{81}–Kr and K–Ar^{40} ages, cosmic ray spallation products and neutron effects in Apollo 11 and Apollo 12 lunar samples. Second Lunar Science Conference (unpublished proceedings).

Nyquist L. E. and Peppin R. O. (1971) Rare gases in Apollo 12 surface and subsurface fine materials. Second Lunar Science Conference (unpublished proceedings).

Pellas P., Poupeau G., Lorin J. C., Reeves H., and Adouze J. (1969) Primitive low-energy particle irradiation of meteoritic crystals. *Nature* **223**, 272–274.

Scott R. F., Carrier W. D., Costes N. C., and Mitchell J. K. (1970) Mechanical properties of the lunar regolith. NASA SP-235, 161–182.

Shoemaker E. M., Hait M. H., Swann G. A., Schleicher D. L., Dahlem D. H., Schaber G. G., and Sutton R. L. (1970a) Lunar regolith at Tranquility Base. *Science* **167**, 452–455.

Shoemaker E. M., Batson R. M., Bean A. L., Conrad C., Jr., Dahlem D. H., Goddard E. N., Hait M. H., Larson K. B., Schaber G. G., Schleischer D. L., Sutton R. L., Swann G. A., and Waters A. C. (1970b) Preliminary Apollo 12 Science Report. NASA SP-235.

Trulsen J. (1971) Collisional focusing of particles in space causing jet streams. In *Physical Studies of Minor Planets* (editor T. Gehrels). Proc. 12th Coll. International Astronomical Union, Tucson, Arizona, March 8–10, 1971.

Wilkening L., Lal D., and Reid A. M. (1971) The evolution of the Kapoeta Howardite based on fossil track studies. *Earth Planet. Sci. Lett.* **10**, 334–340.

Proceedings of the Second Lunar Science Conference, Vol. 3, pp. 2599–2609
The M.I.T. Press, 1971.

Spontaneous fission record of uranium and extinct transuranic elements in Apollo samples

N. Bhandari, S. Bhat, D. Lal, G. Rajagopalan,
A. S. Tamhane, and V. S. Venkatavaradan

Tata Institute of Fundamental Research
Homi Bhabha Road, Colaba, Bombay 5, India

(*Received* 24 *February* 1971; *accepted in revised form* 31 *March* 1971)

Abstract—Experimental evidence is given for the vestigial record of Pu^{244} and super-heavy transuranic elements in selected grains from lunar "fines" (Apollo 11, 12). It is based on the study of total recordable lengths of fossil tracks in *selected* silicate crystals (pyroxenes), where contributions to tracks due to cosmic ray nuclei are minimal. In these pyroxenes, tracks of about 13–25 μ length are found in considerable abundance over and above those expected due to cosmic ray nuclei and spontaneous fission of inherent U^{238}. No excess is found in the case of Apollo 11 or Apollo 12 rocks. We have determined that neutron-induced U^{235} fission as well as spontaneous fission of U^{238} in certain meteorites lead to tracks of mean recordable length, 14.3–15 μ in clinopyroxenes and ortho-pyroxenes, with a standard deviation of $<1.5 \mu$. Thus both U^{238} and Pu^{244} fissions lead to tracks of 14.3–15 μ range, and the longer tracks would not be inconsistent with the lengths for super-heavy elements as predicted theoretically. The implications of the data to relative track contributions due to spontaneous fission of U^{238}, Pu^{244} and super-heavy elements is discussed for the particular crystals analysed.

Introduction

We have studied fossil tracks in lunar pyroxenes with a view to evaluate the concentrations of the now extinct (?) super-heavy transuranic elements ($Z \geqslant 110$). It cannot be ruled out that today some of these elements, particularly from the theoretically predicted island of stability (see, for instance, Nilsson *et al.*, 1968; Nix, 1969) may exist in nonzero concentrations. Unsuccessful attempts to look for these elements in terrestrial minerals (Price *et al.*, 1968; Flerov and Perelygin, 1969) have been reported. The recent synthesis (Marinov *et al.*, 1971) of an element of $Z = 112$, eka-mercury, with half-life exceeding few months, lends credence to the hypothesised island of stability.

The origin of various fossil tracks in meteorites has been explicitly discussed by Fleischer *et al.* (1967a). In this paper, we will concern ourselves primarily with the fossil records of tracks due to spontaneous fission of uranium, plutonium, and heavier transuranic elements in lunar minerals. Tracks due to other sources, cosmic ray iron-group (and heavier) nuclei and spallation recoils, in fact, constitute the majority of tracks seen in most meteorites and lunar materials (Fleischer *et al.*, 1967b). These tracks, as far as we are concerned here, can obviously be considered as noise or background, and one has to know explicitly their frequency and length distributions. The problem of observing fossil tracks and identifying them with either cosmic ray nuclei or fission fragments is particularly beset with difficulties, if one examines the lengths of tracks which appear on a polished surface because one observes a chopped length distribution. If, for instance, fission fragments result in tracks of a unique recordable range R, the length distribution of surface tracks due to the fission fragments alone is expected to be flat between 0 and R, and an identification of a given track or even a statistical analysis of different nuclei recorded becomes impossible (LAL, 1969). The observations of surface tracks is mainly useful towards a measurement of the total track densities and, except in very favourable circumstances, this method is not suitable for studying fission fragment tracks, particularly when tracks due to cosmic ray nuclei are present at densities of (n) $\times$ 10^6 cm^{-2}. In the meteorite Toluca, FLEISCHER et al. (1968) showed that at a given location in the meteorite, diopside crystals had much higher track densities in contrast to that in other silicate minerals and the difference could be clearly attributed to a fission mechanism. Similarly, in St. Severin meteorite, while cosmic ray track densities in pyroxenes and feldspars decreased with depth by 3 orders of magnitude, the total surface track density remained nearly the same in whitlockite indicating appreciable fission contribution (CANTELAUBE et al., 1967). In examples such as cited above, the observations of surface tracks can be used for finding out if tracks are mainly fissiogenic—but here too, one cannot determine the magnitude of relative contributions due to U, Pu, and heavier elements.

A simple method has recently been developed for identifying the nuclei forming a given track, both of cosmic ray and fission origins. This method consists in revealing and studying the lengths of tracks over their entire etchable distance. The complete or the confined tracks are called TINTS or TINCLES, depending on whether the etching channel used to develop such a track is that provided by a surface track (a track-in-track event: TINT) or a canal/fissure/cleavage in the crystal (a track in the cleavage: TINCLE). For a discussion of the TINT method, reference is made to LAL (1969). Detailed studies of the extent of development of TINT/TINCLE track tips have been made. It is found that tracks up to 30 μ length, developed under controlled etching, do not increase in length on reetching by more than 0.5 to 1.0 μ. Furthermore, since track holes are plated with silver to increase optical contrast (see Fig. 1), the lengths can be measured with a high precision.

In order to obtain a high signal-to-noise ratio, we selected lunar crystals having as low a cosmic ray track density as possible; the contribution from fission are dependent only on the concentration of fissile elements in the crystals. In the case of lunar rocks, this criterion corresponds to the most interior samples from rocks of smallest radiation age. In the case of lunar fines, low-track density grains had to be selected from a rather large collection of etched grains, because track densities are often high. The track density of "fission" tracks, ρ_F, in these samples has been deduced to be small compared to ρ_{CR}, except in a few cases where uranium and other elements are present as trapped in inclusions.

A new technique recently employed (BHANDARI et al., 1971a) to find a fission-rich sample of fossil tracks consists of looking at exsolution and crystallographic flaw planes which form sites for uranium and other transuranic elements rejected from the crystal lattice. In a way, these regions can be considered as grain boundaries which are located within the crystals. By and large, most cleavages and flaw planes through which etchant penetrates to deeper reaches of crystals to reveal TINCLES, are merely

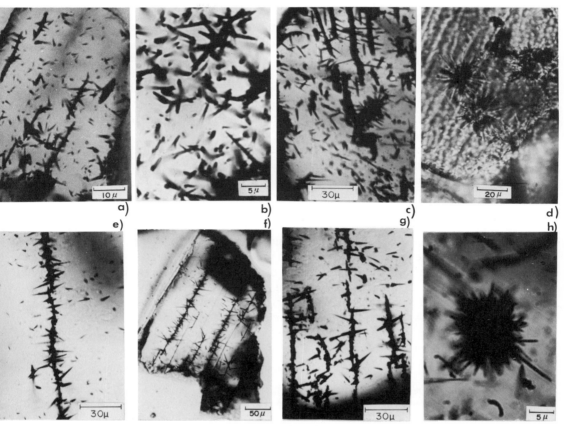

Fig. 1. Photomicrographs of fossil tracks (silver decorated) in meteorites and lunar samples: (a) in Apollo 12002 rock; (b), (c) and (d) in 12025, 12033, and 12028 fines, respectively; (e) and (g) in Crab Orchard meteorite; (f) and (h) in Moore County meteorite. Between different crystals from a given specimen, the cleavage number N_{cl} varies widely. Sunbursts, (d) and (h), as well as high N_{cl} TINCLES represent a predominantly fission origin.

weak points along cleavages or cracks. But now and then one clearly observes these rejection sites which manifest themselves as channels across which an unusually large number of tracks are etched, giving an appearance of a closely knit barbed wire. It can be easily seen that the number of TINCLES (of cosmic ray origin) expected to be revealed per 100 μ length along a line of cleavage, defined henceforth as *cleavage number*, $N_{cl} = 10^{-6} G \rho_s d$, where ρ_s is the "surface" density (cm^{-2}) of tracks due to cosmic ray nuclei and d, in microns, is the depth of focus for observing TINCLES along a cleavage plane. The value of G, the geometrical factor for etching tracks of different lengths, depends on the width of the cleavage plane; it is approximately equal to 1 since we are interested in TINCLES whose lengths (10–20 μ) are large compared to the cleavage gap ($\sim 1 \mu$). Typically d is of the order of 1 μ, and for $\rho_s =$ say, 5×10^6 cm^{-2}, N_{cl} is expected to be 5 (tracks/100 μ length). In Fig. 1 we show examples of TINCLES of high and low N_{cl} in lunar and meteoritic samples. Summarising, when N_{cl} values correspond to predicted values based on the above relation, the TINCLES and surface tracks represent registration of similar nuclei, but when this is not the case, i.e., $N_{cl} > 10^{-6} \rho_s d$, the excess TINCLES must be due to spontaneous fission of impurities rejected across the exsolution planes, which we will henceforth loosely term *cleavage plane* for convenience. Large variations in N_{cl} values within a given crystal and observations of neutron induced TINCLES support a primarily fissiogenic origin of natural TINCLES of high N_{cl}.

Clearly, for our work, it is preferable to study highest N_{cl} TINCLES, but here one faces the problem of identifying the continuity of a TINCLE across the cleavage plane. In practice we have measured TINCLES of up to 100 tracks per 100 μ length.

We have studied terrestrial, meteoritic and lunar pyroxenes in great detail for their track recording characteristics, and we also have studied their etching conditions towards a reliable measurement of the total recorded lengths of tracks. In this paper, we therefore confine ourselves to a discussion of results obtained for pyroxene crystals only. It should be noted here that a search for the presence of super-heavy elements by studying track lengths (in silicate crystals of iron meteorites) was attempted by PRICE and FLEISCHER (1969), who arrived at a negative answer. They studied the most probable etched track lengths by observing "incomplete" tracks on a fractured crystal surface. As discussed earlier, this method is not a sensitive one. Further, it is possible that super-heavy elements may have differentiated unfavourably in silicates present in iron meteorites.

EXPERIMENTAL

From a large number of grains processed for fossil tracks, a few suitable crystals from lunar fines were selected for track length measurements using the criteria of high N_{cl} numbers and low cosmic ray tracks. Large ($\geqslant 150 \mu$) transparent crystals selected from fines 10084, 12033, and 12028 (63, 99 and 193) were studied in this way. In case of lunar rocks, 12002, 12018, 12020 and 12038, the most interior samples were chosen.

The mounting, polishing, etching, and track decoration techniques are discussed elsewhere (LAL, 1969; LAL *et al.*, 1970). It has been shown that the tracks up to 30 μ, once etched, do not increase in length. In crystals of size exceeding 150–200 μ, all complete tracks of $\geqslant 3 \mu$ length were measured in a systematic scan, confining to tracks of dip angle $\geqslant 60°$ (w.r.t. vertical). In the case of TINTS, we considered all tracks whose tips nearer to the surface were within a depth of 10 μ from the surface. In the case of TINCLES, all tracks measured were chosen from cleavage points lying at depths of 10–15 μ from the surface and here also flat tracks were accepted to minimise bias in selection. There exists a slight experimental bias for shorter length; no corrections have however been made in the present analysis, the effect of which would be to increase the abundance of longer tracks, 15–25 μ length, by about 20% compared to those of 5–10 μ lengths.

RESULTS AND DISCUSSIONS

The results of measurements of track lengths are presented graphically in Fig. 2 for pyroxenes in Apollo rocks and fines. The given distributions refer to lengths of

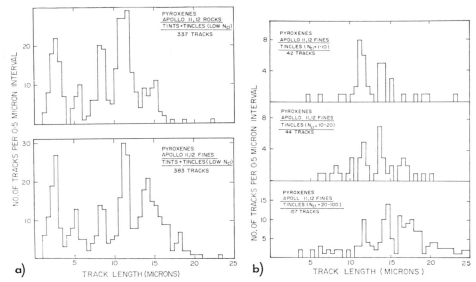

Fig. 2. (a) Length distribution of TINTS and TINCLES (low N_{cl}) in Apollo rocks and fines, (b) Length distribution of TINCLES in Apollo fines for three N_{cl} groups. TINTS and normal (low) N_{cl} TINCLES represent minimal fission contributions (due to U^{238} and Pu^{244} in the crystal lattice). High N_{cl} TINCLES, however, are predominantly due to fissions of exsolved impurities.

TINTS and TINCLES: 337 tracks in lunar rocks and 623 tracks in lunar fines. The following observations can be made from these histograms: (1) Both in rocks and in fines, there are well-defined peaks at 2.5, 5.5, 8.5, 11.5, and 15 μ. (2) The length distribution in rocks falls off to small values at 16 μ (Fig. 2a), and at 25 microns (Fig. 2b) in the case of lunar fines. (3) The relative abundance of tracks of length $> 15 \mu$ increases among TINCLES of higher N_{cl} in lunar fines.

We shall now briefly discuss the origin of these tracks as well as differences that exist between track data in lunar rocks and fines in relation to observations in meteorites. To distinguish the two prominent sources of fossil tracks, heavy cosmic ray nuclei (Z $\geqslant$ 22) and fission, we have analysed a number of meteorites in addition to lunar samples. Based on these studies, the following features seem to be well established about the total recordable lengths of tracks.

(1) It has been shown earlier (LAL et al., 1970; LAL, 1969) and has now been substantiated from studies in Apollo rocks (BHANDARI et al., 1971b) that cosmic ray nuclei, V, Cr, Mn, Fe, Co, and Ni, respectively, lead to tracks of mean etchable lengths about 2.5, 5.5, 8.5, 11.5, 15, and 18 μ. (2) From direct measurements of n-induced U^{235} fission fragment tracks in annealed pyroxenes from Angra Dos Reis, Moore County, and Kapoeta, the mean etchable lengths are deduced to lie between 14.3 and 15.0 μ, with a standard deviation of less than 1.5 μ (BHANDARI et al., 1971a). (3) In Angra Dos Reis, where cosmic ray tracks occur at ultra low densities, $<10^4$ cm^{-2}, more than 99% of natural tracks are of a fission origin. The length distribution is sharply peaked at 14.3 μ, which is mainly due to spontaneous fission of U^{238} and

Table 1. Calculated* total ranges of fission fragments (in μ) in some pyroxenes.

Mineral	U^{235} induced	U^{238} spontaneous	Pu^{244} spontaneous	Super-heavy (Sh)†			
				$Z = 114$		$Z = 126$	
				Sym.	Asym.	Sym.	Asym.
Pigeonite (specific gravity = 3.38) $Mg_{0.9} Fe_{0.96} Ca_{0.08} (Si_2O_6)$	18.0	18.2	18.3	20.9	23.6	23.3	25.2
Augite (specific gravity = 3.4) $Ca_{0.81} Mg_{0.75} Fe_{0.37} Na_{0.06} Al_{0.34} (Si_{1.66}O_6)$	17.5	17.7	17.8	20.2	22.9	22.6	24.5
Hypersthene (specific gravity = 3.65) $Mg_{0.6} Fe_{0.4} (SiO_3)$	16.5	16.7	16.7	19.1	21.6	21.2	23.0
Diopside (specific gravity = 3.27) $Ca\ Mg\ (Si_2O_6)$	17.7	17.9	18.0	20.5	23.2	22.9	24.8

* From BHANDARI et al. (1971a), based on Range-Energy Computer Code (RANGENER) of HENKE and Benton (1967). The fission energy release values are based on HYDE (1964), VIOLA and SEABORG (1966), and NIX (1969). Values of fission energy released: U^{235} (induced) = 167 MeV; U^{238} (spontaneous) = 170 MeV; Pu^{244} (spontaneous) = 172 MeV; $_{114}Sh^{298}$ (spontaneous) = 235 MeV; $_{126}Sh^{310}$ (spontaneous) = 293 MeV.

† Calculations for the symmetric and asymmetric fission modes are analogous to those discussed by RAO (1970).

Pu^{244} (BHANDARI et al., 1971a). In other meteorites, Kapoeta, Moore County, and Norton County, the fossil (natural) track lengths in a fission-rich zone, e.g., in sunburst events and TINCLES of high cleavage numbers (Fig. 1) peak at around 15 μ *but extend further to tracks of longer lengths*, 25 μ. These longer tracks have been interpreted (BHANDARI et al., 1971a) as due to the fission of super-heavy elements.

From (2) and (3), it therefore seems conclusively established that n-induced U^{235} and spontaneous U^{238} (or Pu^{244}) fission fragment tracks lead to a sharp peaked distribution in etchable lengths at around 15 μ. The total ranges of fission fragments have been calculated (BHANDARI et al., 1971a) for U^{235}, U^{238}, Pu^{244}, and super-heavy fission tracks (both fragments); numbers pertinent to our discussions are given in Table 1 for pyroxenes (of different compositions). The measured lengths of both fossil and induced tracks in case of U^{238}, Pu^{244}, U^{235} are found to be smaller by 3–4 μ from the calculated ranges indicating that about 2 μ are not recorded for each of the fragments towards the end of their range, corresponding to the point where primary ionisation drops below the critical value (FLEISCHER et al., 1967a, b; PRICE et al., (1968). This is in accord with the experimental data and track registration criterion discussed by PRICE et al. (1968). Following their theoretical treatment involving track registration above a certain primary ionisation value, we deduce that the mean recorded total lengths of tracks due to super-heavy elements in pyroxenes would lie between 17 and 20 μ, on the basis of certain assumed fission energy release values and simplified models for the distribution of energy between the fragments (Table 1). Based on the above discussion, it is possible to distinguish the contribution of various sources to tracks in lunar rocks and fines.

Cosmic ray contribution

The most prominent peak in Fig. 2 is at 11–12 μ and can be attributed to the recordable length for iron nuclei (the most abundant nuclei in iron group); smaller peaks are due to Mn, Cr, and V nuclei (at 8.5, 5.5, and 2.5 μ, respectively). Spallation recoil tracks (FLEISCHER *et al.*, 1967a) are also of about <2 μ in length, and they would overlap the vanadium region. All tracks (TINTS/TINCLES) of <13 μ are due to cosmic rays and do not include any fission contributions. As shown by the track length distribution of induced fission tracks, fission events give rise to tracks of $\geqslant 13$ μ where cosmic ray tracks are also present. Based on analyses of TINTS in pyroxenes from Patwar and other meteorites where cosmic ray tracks predominate ($\rho_s \sim 10^7$ cm^{-2}), we find that the relative ratio of tracks due to Fe (10–13 μ), Co (13–16 μ), and Ni–Cu (16–24 μ) is 1.0: $<$ 0.25 $<$ 0.18 (LAL, 1969).

U^{238} *and* Pu^{244} *contributions*

From the relative abundances of tracks due to cosmic ray nuclei, as given above, we determine excess tracks in the 13–25 μ region normalising to tracks in the 10–13 μ interval (iron nuclei). These excess tracks are attributed to fission, and we have estimated this fission contribution in Apollo rocks and fines as given in Table 2. It should be stressed here that the method of estimation of fission tracks is not dependant on the identification of the 10–13 μ tracks with iron nuclei; one is essentially subtracting off a cosmic ray background based on data from a predominantly cosmic ray track sample.

From data in Table 2, we note that the fission component is almost absent in Apollo rocks. However, in fines there exists a definite contribution (excess over cosmic ray level) which is higher for TINCLES of higher N_{cl} number. (Such a correlation with N_{cl} suggests a fissiogenic origin in view of the arguments given earlier.) The tracks in the 13–16 μ region can be due to fission of U^{238} and/or Pu^{244}; the recorded lengths for Pu^{244} fission are expected to be close to that for U^{238} (spont.) or n-induced U^{235} fission (Table 1). We have therefore calculated the expected U^{238} fission tracks/cm^2 (in 4.5 $\times$ 10^9 yr) on the basis of experimentally determined U^{238} concentration in the pyroxenes from Apollo fines. The average concentration of U^{238} in pyroxenes from Apollo 11, 12 fines is determined to be 8 ppb (weight) for crystals of $\geqslant 150$ μ, using the neutron activation method (FLEISCHER, 1968). The concentration of uranium varies by three orders of magnitude, being an approximate inverse function of crystal size (Fig. 3). However, the variations are small (well within a factor of 10) for the large size crystals studied, for which 8 ppb is an upper limit. Using this upper limit value, we estimate that the spontaneous fission of U^{238} over a storage time of 4.5 aeons falls short of the observed tracks in the 13–16 μ interval by a factor of 50. Thus $\rho(Pu^{244})/\rho(U^{238})$ in Apollo fines is $\geqslant 50$. The present work lends support to the earlier work of LAL *et al.* (1970), where presence of Pu^{244} in Apollo fines based on similar fossil track analysis was postulated. It should be pointed out here that since Apollo fines have a large amount of trapped solar wind krypton and xenon, the usual method of finding out Pu/U ratios has so far not been successful.

Super-heavy fission contribution

Neither of these two sources discussed above, cosmic rays and U^{238}, Pu^{244} fission can account for the longer tracks in the 16–25 μ region. However, in fines there exists

Table 2. Relative fission contributions to tracks in Apollo 11 and Apollo 12 samples.

Sample	Track Type Measured	N_{cl}	Fossil Track Density (cm^{-2})	Number of Tracks Measured		Total in 13–16 Microns Interval	Estimated† due to [U+Pu] fission (i.e., corrected for C.R. tracks)	Total in 16–22 μ Interval	Estimated* fission tracks from super-heavy nuclei	Relative Contributions To Tracks	
				Total	Due to Iron (10–13 μ)					$\dfrac{(U+Pu)}{Fe}$	$\dfrac{Superheavy}{U+Pu}$
Apollo 11, 12 Rocks	Tints and Tincles	5	(5–10) × 10^6	337	123	37	12	2	~0	0.12	~0
Apollo 11, 12 Fines	Tints and Tincles	5	(5–10) × 10^6	383	100	90	65	39	21	0.65	0.32
	Tincles	1–10	(5–10) × 10^6	42	22	59	34	23	5	0.34	0.15
	Tincles	10–20	(5–10) × 10^6	44	13	108	83	77	59	0.83	0.69
	Tincles	20–100	(5–10) × 10^6	154	25	136	111	270	252	1.1	2.3

Relative Track Abundances [Fe(10–13 μ) = 100 tracks]

* Corrected for contributions due to cosmic ray nuclei; see text.
† Estimated ratio of tracks due to Pu^{244} and U^{238} $\geq$ 50 in Apollo fines.

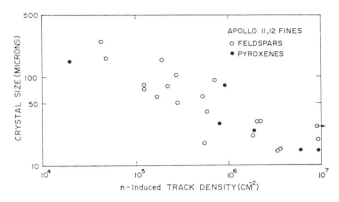

Fig. 3. Measured neutron-induced track densities in lunar fines irradiated to a dose of 1.4×10^{18} neutrons, as a function of crystal size.

a definite contribution which increases as N_{cl} increases. The excess tracks in the 16–24 μ length interval are appreciable in number (>100). Could these tracks have a non-spontaneous fission origin? It may be mentioned here that one often notices crystal defects which occur not infrequently in Moore County and Nakhla for instance, but these are easily distinguishable from tracks by various criteria, like angular distribution (they are generally parallel). Reference is made here to the detailed discussions of FLEISCHER et al. (1967a), who have considered possible origins of fossil tracks and given criteria for identifying them. We think that in the present work, many of the possibilities (howsoever remote) such as neutron-induced fission of known heavy elements (U, Th, Pb etc.) present in pyroxenes and monopoles (?) can be straightaway ruled out since our study is based on a direct experimental comparison on length distributions between lunar and meteoritic specimens exposed to varying dose of cosmic rays. It would be, for example, difficult to conceive of tracks of nonspontaneous fission origin as being present in Apollo fines, but not in rocks. Consider for example cosmic-ray alpha particle induced fission of heavy elements, such as thorium. Its contribution in crystals which have received identical cosmic ray doses, such as those from Apollo rocks and fines examined here, are expected to be similar. Furthermore, high energy particle induced fission is not expected to result in long tracks because experimentally one finds identical distribution of ranges in fission induced by protons, pions etc. over a wide range of energies (HYDE, 1964, p. 483). Thus all known sources of tracks seem inadequate to explain the longer ($>16 \mu$) tracks observed in lunar fines. It seems to us that they can be satisfactorily explained only as being due to fission of super-heavy nuclides ($Z \geqslant 110$). Absence of such tracks in lunar rocks which have younger ages (~ 3.5 b.y.) suggest that these nuclides must be extinct at the time of formation of these rocks. Also, within a group of crystals of $\rho \sim 7 \times 10^6$ cm^{-2} from Apollo fines, we have compared the length distributions for different N_{cl} values. Here the fact that excess tracks ($>13 \mu$ length) increase with N_{cl}, within the same crystal, and that the contribution goes up both in the 13–16 and 16–24 μ intervals, provides additional support to the hypothesis that these tracks are indeed of a spontaneous fission origin.

In Table 2 we have also given the estimated relative fossil track densities due to (U + Pu) and super-heavy elements. Similar ratios are observed (BHANDARI et al., 1971a) in the case of Moore County, Norton County, Steinbach, and Nakhla for the fission component. A good average value for the relative contributions to tracks due to U:Pu:super-heavy fissions is 0.03:1:1; the values scatter by factors of about 3–5.

The observed scatter in relative fission contributions clearly reflects effects due to chemical differentiations. Also, these ratios do not apply to the "total" lunar sample since we are analysing tracks in selected pyroxene crystals having low fossil track density; these shielded crystals may represent a component alien to the younger local mare surface. The existence of an older "magic" component has been suggested by several chemical and "dating" techniques (PAPANASTASSIOU and WASSERBURG, 1970; HUBBARD et al., 1971).

Thus, we believe that the experimental results presented above can be understood only if the super-heavy transuranic elements were present in lunar fines. From the present study, we do not know the charge(s) of the super-heavy element(s), but the measured fission ranges are not inconsistent with those expected for elements around $Z = 114$. Absence of super-heavy element tracks in lunar rocks, which are about 1 b.y. younger than the oldest dust grains examined suggest that the element should have a half-life short enough, $<10^8$ yr. to have become extinct by the time the rocks recrystallised.

Further useful criteria for an unambiguous identification of fission of super-heavy nuclei would be (1) relative frequency of ternary to binary fission (PRICE and FLEISCHER, 1969); (2) length distribution of individual fission fragments, measured from the cleavage plane; and (3) angular distribution. Work on these aspects of measurements is currently in progress. Finally, it must be pointed out here that prior to this work, several studies of fissiogenic gases in meteorites have made a case for the likely existence of super-heavy elements (ANDERS and HEYMANN, 1969; DAKOWSKI, 1969; RAO, 1970).

Acknowledgements—We are indebted to NASA for providing us with the lunar samples. Our special appreciation is due to Miss A. Padhye and Miss N. Prabhu for their skilful assistance in experimental work. We take this opportunity to heartily thank Drs. H. G. DE CARVALHO, ROY S. CLARKE, JR., WALTER DA SILVA CURVELLO, J. GEISS, KURT FREDRIKSSON, M. HONDA, CARLETON B. MOORE, P. PELLAS, and J. T. WASSON, who have generously provided us with valuable meteorite specimens for fossil track analyses. Finally we would like to thank Drs. D. S. BURNETT, R. L. FLEISCHER, P. PELLAS, P. B. PRICE, M. N. RAO, and R. M. WALKER for discussions.

REFERENCES

ANDERS E. and HEYMANN D. (1969) Elements 112 to 119: Were they present in meteorites? *Science* **164**, 821–823.

BHANDARI N., BHAT S., LAL D., RAJAGOPALAN G., TAMHANE A. S., and VENKATAVARADAN V. S. (1971a) Super-heavy elements in extraterrestrial samples. *Nature*, **230**, 219–224.

BHANDARI N., BHAT S., LAL D., RAJAGOPALAN G., TAMHANE A. S., and VENKATAVARADAN V. S. (1971b) High resolution time averaged (millions of years) data on chemical composition and energy spectrum of cosmic ray nuclei of $Z = 22$–28, at 1 A.U. based on fossil track studies in lunar materials. Second Lunar Science Conference (unpublished proceedings).

CANTELAUBE Y., MAURETTE M., and PELLAS P. (1967) Traces d'ions lourds dans les mineraux de la chondrite de Saint Severin in "Radioactive Dating and Methods of Low Level Counting," 215–229. International Atomic Energy Agency, Vienna.

DAKOWSKI M. (1969) The possibility of extinct superheavy elements occurring in meteorites, *Earth Planet. Sci. Lett.*, **6**, 152–154.

FLEISCHER R. L., PRICE P. B., and WALKER R. M. (1968) Identification of Pu²⁴⁴ fission tracks and the cooling of the parent body of the Toluca meteorite. *Geochim. Cosmochim. Acta* **32**, 21–31.

FLEISCHER R. L., PRICE P. B., WALKER R. M., and MAURETTE M. (1967a) Origins of fossil charged-particle tracks in meteorites. *J. Geophys. Res.* **72**, 331–353.

FLEISCHER R. L., PRICE P. B., WALKER R. M., MAURETTE M., and MORGAN G. (1967b) Tracks of heavy primary cosmic rays in meteorites, *J. Geophys. Res.* **72**, 355–366.

FLEISCHER R. L. (1968) Uranium distribution in stone meteorites by the fission track technique. *Geochim. Cosmochim. Acta* **32**, 989–998.

FLEROV G. N. and PERELYGIN V. P. (1969) On spontaneous fission of lead—search for very far transuranium elements. *Sov. Atom. Energy* **26** (No. 6), 603–605.

HENKE R. P. and BENTON E. V. (1967) A computer code for the computation of heavy-ion range energy relationship in any stopping material. Report USNRDL-TR-67-122.

HUBBARD N. J., MEYER C., JR., GAST P. W., and WIESMANN H. (1971) The composition and derivation of Apollo 12 soils, *Earth and Planet Sci. Lett.* **10**, 341–350.

HYDE E. K. (1964), *Fission Phenomena* in *Nuclear Properties of Heavy Elements*. Vol. III, p. 171, Prentice-Hall.

LAL D. (1969), Recent advances in the study of fossil tracks in meteorites due to heavy nuclei of the cosmic radiation. *Space Sci. Rev.* **9**, 623–650.

LAL D., MACDOUGALL D., WILKENING L., and ARRHENIUS G. (1970) Mixing of the lunar regolith and cosmic ray spectra, new evidence from fossil particle-track studies. *Proc. Apollo 11 lunar Sci. Conf., Geochim. Cosmochim. Acta* Suppl. 1, Vol. 3, pp. 2295–2303. Pergamon.

MARINOV A., BATTY C. J., KILVINGTON A. I., NEWTON G. W. A., ROBINSON V. J., and HEMINGWAY J. D. (1971) Evidence for the possible existence of a super-heavy element with atomic number 112. *Nature* **229**, 464–467.

NILSSON S. G., NIX J. R., SOBICZEWSKI A., SZYMANSKI Z., WYCECH S., GUSTAFSON C., and MOLLER P. (1968) On the spontaneous fission of nuclei with Z near 114 and N near 184. *Nuclear Physics*, **A115**, 545–562.

NIX J. R. (1969) Predicted properties of the fission of super-heavy nuclei. *Phys. Lett* **30B**, 1–4.

PAPANASTASSIOU D. A. and WASSERBERG G. J. (1970) Rb–Sr ages from the Ocean of Storms. *Earth and Planet. Sci. Lett.* **8**, 269–278.

PRICE P. B. and FLEISCHER R. L. (1969) Are fission tracks in meteorites from super-heavy elements? *Phys. Lett.* **30B**, 246–248.

PRICE P. B., FLEISCHER R. L., and MOAK C. D. (1968) Identification of very heavy cosmic ray tracks in meteorites. *Phys. Rev.*, **167**, 277–282.

PRICE P. B., FLEISCHER R. L., and WOODS R. T. (1970) Search for spontaneously fissioning elements in nature. *Phys. Rev.* (c), **1** (5), 1819–1821.

RAO M. N. (1970) On the existence of super-heavy elements near Z = 114 and N = 184 in meteorites. *Nucl. Phys.*, **A140**, 69–73.

TAMHANE A. S. (1971), Ph.D. thesis, in preparation.

VIOLA V. E. and SEABORG G. T. (1966) Nuclear systematics of heavy elements, I and II. *J. Inorg. Nucl. Chem.* **28**, 697.

Proceedings of the Second Lunar Science Conference, Vol. 3, pp. 2611–2619
The M.I.T. Press, 1971.

High resolution time averaged (millions of years) energy spectrum and chemical composition of iron-group cosmic ray nuclei at 1 A.U. based on fossil tracks in Apollo samples

N. Bhandari, S. Bhat, D. Lal, G. Rajagopalan,
A. S. Tamhane and V. S. Venkatavaradan

Tata Institute of Fundamental Research, Homi Bhabha Road, Colaba, Bombay 5, India

(Received 24 February 1971; accepted in revised form 31 March 1971)

Abstract—Four Apollo 12 rocks have been studied for fossil track records due to cosmic ray iron-group nuclei using a thick section technique. The technique allows a simultaneous measurement of fossil track densities in feldspar and pyroxene group of minerals along planes of different orientations in a rock, making it possible to deduce accurately the energy spectrum of cosmic rays in space.

Based on the track profiles at depths $\geqslant 10^{-1}$ cm, the form of energy spectrum of iron-group nuclei for the interval 0.06–0.50 BeV/n has been deduced. Absolute flux is based on normalisation to the mean contemporary cosmic ray and St. Severin fossil track based flux, at 0.50 BeV/n. At lower energies where erosion/attrition becomes important, the deduced spectral form is not unique. Even so, the spread can be fairly well restricted.

Fossil track exposure ages of rocks studied are discussed. It is shown that one can characterise two distinct exposure ages, (1) *sun-tan exposure time*, the period for which a rock has remained on the surface of the moon, and (2) *subdecimeter exposure time*, the period for which a given rock has been exposed as buried in the regolith up to depths of the order of 10 cm. The rocks studied are found to have sun-tan ages of the order of 2–5 m.y. and subdecimeter exposure ages of 0–40 m.y.

The erosion history of rocks is deduced to be a complex one warranting further detailed studies. However, it seems certain that there exist regions on the rock surface where discrete losses due to recent micrometeorite impacts or attrition have been small, $\leqslant 5\ \mu$. A continuous atomic erosion, e.g., due to solar wind bombardment, could be important, and a rate of the order of $(0.5-1.0) \times 10^{-8}$ cm/yr is allowed by the data. It is pointed out that for a given erosion rate (in space) the corresponding surface loss is much smaller considering the plausible model that the rock surface is perennially covered by a dust layer $n\mu$ thick: this dust layer gets eroded but it is continuously replaced and can physically be considered as an erosion shield. Using the TINT method, the 2–5 m.y. time averaged relative abundances of Cr:Mn:Fe have been deduced.

Introduction

We present in this paper results of extensive analysis of fossil track densities in silicate crystals from Apollo 12 rocks. Both spot samples and thick sections of 100–500 μ thickness have been etched and the number density (tracks/cm²) and total etchable lengths of tracks have been studied. The fossil track method has been previously applied for evaluating prehistoric composition and energy spectrum of iron-group nuclei in meteorites (Fleischer *et al.*, 1967; Cantelaube *et al.*, 1967; Lal *et al.*, 1969) and in lunar rocks (Crozaz *et al.*, 1970; Fleischer *et al.*, 1970; Lal *et al.*, 1970; Price and O'Sullivan, 1970). Our present work is analogous to that cited here except insofar as the employment of thick section technique which has permitted a very detailed examination of point by point variation of track densities in crystals, both feldspars and pyroxenes, along different planes of known orientation

2611

with respect to the surface of the rock. This technique has been particularly valuable for studying tracks due to primary iron-group nuclei of energies below 100 MeV/n.

The experimental techniques for mounting, polishing, etching, and decoration of silicate crystals for spot samples and thick sections are described in detail in BHANDARI et al. (1971a) and LAL et al. (1970).

RESULTS AND DISCUSSIONS

In Table 1, we give all relevant details of the four rocks studied. The location of samples is shown in Fig. 1 where the results of track counts in various samples are plotted. The orientation of β_{90} and β_0 planes as defined for meteorites (FLEISCHER et al., 1967) respectively correspond to (XZ, YZ) and XY planes, as shown in Fig. 1, for both through sections and surface chips. This XY or XZ assignment in the case of interior chips is arbitrary and only meant to denote orthogonal directions. XYZ denotes samples of unknown orientation.

At shallow depths track densities in pyroxenes (or feldspars) at given depth are higher for β_0 compared to β_{90} as expected (FLEISCHER et al., 1967; BHANDARI et al., 1971a). Track densities in feldspars are higher than in pyroxenes at all depths and this is due to somewhat larger recorded range in feldspar for a given charge. For track densities ρ (cm^{-2}) $> 5 \times 10^6$, the difference is 10–15%.

Using procedures discussed earlier (FLEISCHER et al., 1967; LAL, 1969; BHANDARI et al., 1971a), one can obtain from the observed track profiles, the energy spectrum of iron-group nuclei in space, for any specified orientation of rocks. In the case of lunar rocks, in contradistinction to meteorites where ablation losses are large, one does not a priori know the orientation of rocks during exposure and also the corresponding exposure time. Multiple exposure histories on the lunar surface or in near-surface

Table 1. Physical data on Apollo 12 rocks studied for fossil tracks.

Rock	Mass* (g)	Maximum l × b × h	Approximated Ellipsoid semiaxes† (a)	(b)	(c)	Type of sample	Specific number(s)	Depth interval (cm)	Remarks
			Dimensions (cm)				Sample details		
12002	1529	11 × 9 × 6	(5.4)	(3.75)	(3.15)	A through section	,61 ,62 ,63	0–4.75 4.7–6.05 5.6–6.3	" 'Top' " identified by counting spallogenic radioactivity.
12018	787	8 × 9 × 6	(4)	(3.25)	(3)	A through section	,20	Two orthogonal slices; 0–5.5 and 5.5–8.5	—
12020	312	8 × 6 × 6	(2.5)	(2)	(2.75)	(1) Exterior chip (2) Interior chip (3) Exterior chip	,18 ,20 ,19	0–0.57 1.4–2.5 4.5–5.5	—
12038	746	12.5 × 7.5 × 5.5	(3)	(2.4)	(3.6)	(1) Exterior chip (2) Interior chip (3) Exterior chip	,17 ,48 ,16	0–0.5 1.5–2.5 6.5–7.2	Documented, partly buried rock

* As given in NASA Lunar Sample Information Catalog-Report MSC–01512, 1970.
† Based on fossil track work with respect to orientation on the moon.

regions (as partly or fully buried in the regolith) are in fact indicated from the present and earlier studies (CROZAZ *et al.*, 1970; FLEISCHER *et al.*, 1970; LAL *et al.*, 1970). In cases where multiple breakups are indicated, the spallogenic He^3, Ne^{21} age suffice to ascertain the total cosmic ray exposure age. However, since (He^3, Ne^{21}) production and track formation have different *e*-fold lengths, the latter being much more depth sensitive, the two exposure ages refer to different shielding histories. The He^3, Ne^{21} refer to the integrated time of exposure between surface and $\leqslant$ (50–70) cm. We will

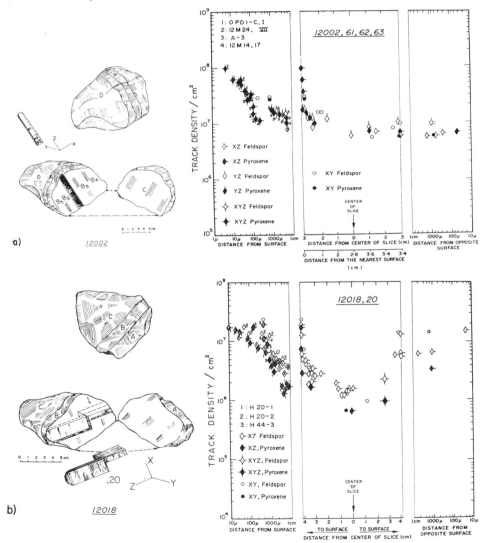

Fig. 1a, b, c, d. Measured fossil track densities (for tracks of length $> 1\ \mu$) based on thick section studies of four Apollo 12 rocks. The position of through slices and chips is shown in the schematics on L.H.S.

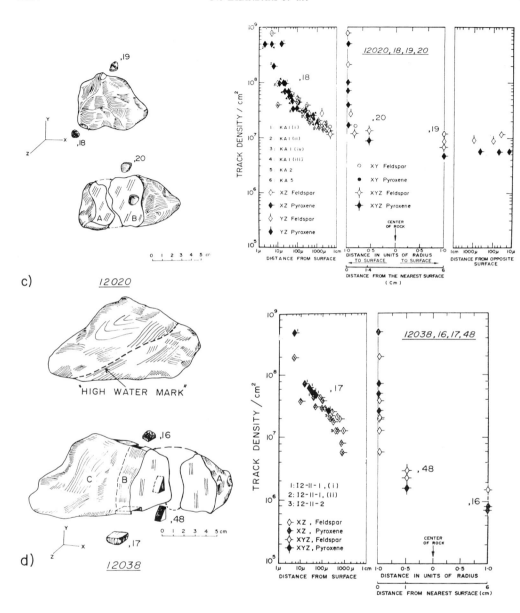

therefore denote in this paper, the spallogenic He³(Ne²¹) age as *submeter cosmic ray exposure age.*

In the case of fossil tracks, formation rate depends on the depth within rock, and the *e*-fold distance for depths < 10 cm in a large size rock varies between microns to centimeters. (In fact, a corollary to this statement is that one can ascertain the approximate depth (from surface) of a given specimen by studying gradients in the track densities.) Normally therefore if a rock has at all been "freely" exposed, it acquires more tracks at $\leqslant 1$ mm depth due to this exposure compared to those stored while it has been shielded. It therefore becomes meaningful to define what one may call the *sun-tan* exposure period—the period of sun or space-bathing, as completely unshielded.

It would be expected that the submeter exposure age (which is based on spallogenic He^3, Ne^{21}) equals or exceeds the sun-tan exposure age. If the difference in these ages is significant, it is possible that the rock has had a complex multiple irradiation history at different submeter levels in the regolith; in the simplest case one can consider a deep seated irradiation at > 1 m depth followed by outcropping and a sun-tan exposure.

The fossil track data are useful for estimating the *dwell time integral* of the rock when it was buried at shallow depths in the regolith. Considering the depth dependance in track formation rates (see FLEISCHER *et al.*, 1967; BHANDARI *et al.* 1971b, and also the observed track spectra in Fig. 1d), one observes that for depths > 10 cms, the track formation rates become increasingly unimportant compared to that around 1 cm depth. Therefore, in the event of a double exposure, a sun-tan exposure, and an irradiation when the rock is buried within the top 10 cm of the regolith, the resulting track profile in the depth interval of 0.1–5 cm would be flatter compared to the case when the rock is irradiated only once on the lunar surface (without any shielding). Thus, observations of the track data for the 0.1–5 cm depth interval are useful for evaluating the time integral for its single or multiple irradiation history within the top 10 cm of the regolith. This period of exposure will be designated as the *subdecimeter exposure age*.

In the framework of the above nomenclature and background, we now proceed to discuss the exposure history of rocks and the history of radiation received by the rocks.

In Fig. 2a and c, we show the expected track production rates for $X \leqslant 0.1$ cm, both for exponential rigidity and kinetic energy power law spectrum, of different hardness. Note that for depths $< 10^{-3}$ cm the track densities do not continue to rise asymptotically (Fig. 2a) because of the finite etchable range of the nuclei recorded (iron mostly). For the kinetic energy power law spectrum $J = dN/dE = $ const. $E^{-\gamma}$, the expected rate of formation of tracks, $P(X)$ varies with depth as $P(X) = $ const. $X^{-\alpha}$. The relation between α and γ is: $\alpha = 1.17\,\gamma^{0.605}$ for depths $5 \times 10^{-4} \leqslant X \leqslant 10^{-1}$ cms from surface, corresponding to approximate kinetic energy interval $0.4 \leqslant E \leqslant 60$ MeV/n.

For $\gamma = 3.0$, the calculated track profiles for different assumed attrition (ΔX) and erosion rates, ε (cm/m.y.) are shown in Fig. 2b. A comparison with the observed track profiles (Fig. 2d) indicates that attrition is not important ($< (5–10) \times 10^{-4}$ cm) in the case of rocks 12020, 12038, and 10017, but it is large for 12002 and 12018 through slices. The latter most probably represents surface losses during sawing. At least, this was found to be the case for 12002, where we found high track density crystals (with large gradients) in surface samples taken before cutting. For $X < 5 \times 10^{-3}$ cms, three rocks have nearly identical track profiles, and we have considered 12038 data for deriving energy spectrum at low energies.

At greater depths, $X > 10^{-1}$ cms, the track data for rocks 12018, 12038, and 10017 are consistent with a single energy spectrum, considering their geometrical shapes and

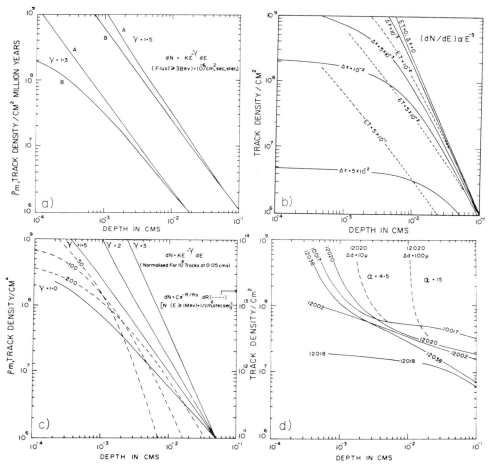

Fig. 2. Theoretically calculated rates of formation of tracks due to low-energy iron nuclei are in a semi-infinite body shown in Figs. (a) and (c) for power law and rigidity type spectra. In (a), curve B differs from curve A in as much as it includes the cut-off energy corresponding to the recordable range of iron. In (b), the effects on track spectrum for a postexposure discrete attrition of ΔX as well as continuous erosion ε (for different εT values) during irradiation, are shown. The experimental data for tracks in Apollo rocks is shown in (d); the dotted curve shows the shifted track spectra in rock 12020 if it is assumed that a dust layer of thickness Δd existed on the rock.

track based orientation on the lunar surface (see Table 2). In view of this we have assumed here that the subdecimeter exposure age of these rocks is small compared to the sun-tan ages.

The track-based energy spectrum for the energy range 5×10^{-4}–0.50 BeV/n, deduced from track profiles in 12038 for $X < 10^{-1}$ cm and in 12018, 12038, and 10017 for $X > 10^{-1}$ cm is shown in Fig. 3 for $\Delta X = 0$ and $\varepsilon T = 0$. The absolute flux is based on normalisation at 500 MeV to the mean contemporary cosmic ray (see VON

Table 2. Cosmic ray exposure ages of lunar rocks.

| Rock | Remarks on track profile | Approx. Orientation* | Sunny facet† | Fossil track based Exposure age (m.y.) | | Spallogenic submeter age (m.y.)‡ |
				Sun-tan age	Subdecimeter age	
12002	Large track gradients (L.T.G.) at one surface	$\eta = 0$	Surface of ,61	2.2	35	92
12018	Centrosymmetric (L.T.G.) at opposite surfaces	$\eta = \pi/2$	Both surfaces of ,20	1.7	0	150–190
12020	L.T.G. at one surface	$\eta = 0$	Surface of ,18	2.6	25	30–90
12038	L.T.G. at one surface	$\eta = 0$	Surface of ,17	1.3	0	90–190
10017	Asymmetric; L.T.G. at opposite surfaces	$\eta \leqslant 30°$	Surfaces of T4 and T3	4.2	0	350

* η = the angle between zenith and the normal to the rock surface (based on maximum track gradient direction).

† For all the rocks listed here, multiple orientation at the surface is not warranted by the track data. Observed solar proton produced excess Al^{26} and Na^{22} activities (FINKEL et al., 1971) are also consistent with this conclusion.

‡ BOGARD et al., 1971.

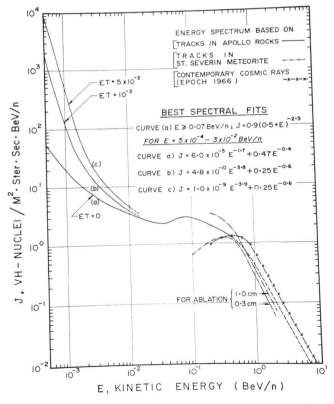

Fig. 3. The deduced differential flux of iron-group nuclei curve (a), based on Apollo 11 and Apollo 12 rocks has been normalised at 0.50 BeV/n to the mean St. Severin Fossil Track and EPOCH 1966 Cosmic Ray flux. The calculated low energy spectra for $\varepsilon T = 10^{-2}$ and 5×10^{-2} cm, with $\Delta X = 0$, is shown by curves (b) and (c), respectively.

Rosenvinge *et al.*, 1969) and St. Severin track based data of Lal *et al.* (1969); the latter has been multiplied by 0.67 to take into account a revision in the mean recorded length of iron-group nuclei in silicate crystals.

Based on the absolute fluxes (Fig. 3), we deduced the exposure ages of different rocks, considering track data for $X > 10^{-1}$ cm, where uncertainties due to erosion/ attrition, etc. are minimal. The calculated sun-tan and subdecimeter exposure ages are summarised in Table 2.

So far, we have tacitly assumed $\varepsilon T = 0$ (and also $\Delta X = 0$ which seems to be justified for 12038 as discussed earlier). For solar wind erosion, Wehner *et al.* (1963) have experimentally determined a value of 0.4 Å/yr for "stones" in near-earth orbits. To this must be added the contribution from micrometeorite erosion, which is unknown. Considering the mean exposure age of ~ 3 m.y., we have adopted 10^{-2} and 5×10^{-2} cm as representative values for εT. The corresponding deduced energy spectra are shown in Fig. 3, erosion affects the fluxes only at $E < 60$ MeV/n.

Chemical Composition of Iron Group Nuclei

We will now discuss our observations with respect to the relative abundances of different nuclei of charge 24, 25, and 26 which result in tracks of mean recordable lengths, 5.5, 8.5, and 11.5 μ, respectively (Lal, 1969).

From the measurements of lengths of tints and tincles (Bhandari *et al.*, 1971b) in Apollo rock and dust samples we have determined that the relative abundances of the iron, manganese and chromium nuclei recorded of primary energies < 350 and > 500 MeV/n are 1:0.60:0.42 and 1:0.62:0.72, respectively. The result is based on 250 tracks and is not corrected for contributions due to fragmentation ($< 15\%$).

The implications of the present data to the history of cosmic radiation with respect to the flux, energy spectra and chemical composition, will be discussed elsewhere.

Acknowledgements—It is our great pleasure to thank NASA for supplying us the lunar samples. This work would not have been possible but for the successful execution of the lunar field trip by the astronauts Charles Conrad, Richard Gordon, and Alan Bean. For the very careful preparation of a large number of samples and documentation thereof, we are very grateful to Daniel H. Anderson, Michael B. Duke, and to Robert B. Laughon for providing information on samples.

We are grateful to A. G. Padhye, N. R. Prabhu, P. B. Badle, P. S. Daudkhane, and P. K. Talekar for their technical assistance.

This section would not be complete without a special mention of the kind collaboration and assistance of J. R. Arnold, G. Arrhenius, L. Wilkening, and J. D. MacDougall. We would like to express our gratefulness to R. L. Fleischer for helpful comments and discussions.

References

Bhandari N., Bhat S., Lal D., MacDougall J. D., Tamhane A. S., Venkatavaradan V. S., Arrhenius G., and Wilkening L. (1971a) Fossil track record of cosmic ray nuclei in Apollo rocks. To be submitted to Proc. Ind. Acad. Sci.

Bhandari N., Bhat S., Lal D., Rajagopalan G., Tamhane A. S., and Venkatavaradan V. S. (1971b) Spontaneous fission record of uranium and extinct transuranic elements in Apollo samples. Second Lunar Science Conference (unpublished proceedings).

Bogard D. D., Funkhouser J. G., Schaeffer O. A., and Zahringer J. (1971) Noble gas abundances in lunar material, II. Cosmic ray spallation products and radiation ages from Mare Tranquillitatis and Oceanus Procellarum. *J. Geophys. Res.* (in press).

CANTELAUBE Y., MAURETTE M., and PELLAS P. (1967) Traces d'ions lourds dans les mineraux de la chondrite de Saint Severin. In *Radioactive Dating and Methods of Low Level Counting*, International Atomic Energy Agency, Vienna 215–229.

CROZAZ G., HAACK U., HAIR M., MAURETTE M., WALKER R., and WOOLUM D. (1970) Nuclear track studies of ancient solar radiations and dynamic lunar surface processes. *Proc. Apollo 11 Lunar Sci. Conf., Geochim. Cosmochim. Acta* Suppl. 1, Vol. 3, pp. 2051–2080. Pergamon.

CROZAZ G., WALKER R., and WOOLUM D. (1971) Cosmic ray studies of recent dynamic processes on the surface of the moon. Second Lunar Science Conference (unpublished proceedings).

FINKEL R. C., ARNOLD J. R., REEDY R. C., FRUCHTER J. S., LOOSLI H. H., EVANS J. C., SHEDLOVSKY J. P., and DELANY A. C. (1971) Depth variation of cosmogenic nuclides in a lunar surface rock. Second Lunar Science Conference (unpublished proceedings).

FLEISCHER R. L., PRICE P. B., WALKER R. M., and MAURETTE M. (1967) Origins of fossil charged-particle tracks in meteorites. *J. Geophys. Res.* **72**, 331–353.

FLEISCHER R. L., HAINES E. L., HART H. R., JR., WOODS R. T., and COMSTOCK G. M. (1970) The particle track record of the Sea of Tranquility. *Proc. Apollo 11 Lunar Sci. Conf. Geochim. Cosmochim. Acta* Suppl. 1, Vol. 3, pp. 2103–2120. Pergamon.

FLEISCHER R. L., HART H. R., JR., and COMSTOCK G. M. (1971) Very heavy solar cosmic rays: Energy spectrum and implications for lunar erosion. Second Lunar Science Conference (unpublished proceedings).

LAL, D. (1969) Recent advances in the study of fossil tracks in meteorites due to heavy nuclei of the cosmic radiation. *Space Sci. Rev.* **9**, 623–650.

LAL D., LORIN J. C., PELLAS P., RAJAN R. S., and TAMHANE A. S. (1969) On the energy spectrum of iron-group nuclei as deduced from fossil-track studies in meteoritic minerals. In *Meteorite Research*, R. Reidel, 275–285.

LAL D., MACDOUGALL D., WILKENING L., and ARRHENIUS G. (1970) Mixing of the lunar regolith and cosmic ray spectra, new evidence from fossil particle-track studies. *Proc. Apollo 11 Lunar Sci. Conf., Geochim. Cosmochim. Acta* Suppl. 1, Vol. 3, pp. 2295–2303. Pergamon.

PRICE P. B. and O'SULLIVAN D. (1970) Lunar erosion rate and solar flare paleontology. *Proc. Apollo 11 Lunar Sci. Conf., Geochim. Cosmochim. Acta* Suppl. 1, Vol. 3, 2351–2359. Pergamon.

TAMHANE A. S. (1971) Unpublished work.

VON ROSENVINGE T. T., WEBBER W. R., and ORMES J. F. (1969) A comparison of the energy spectra of cosmic ray helium and heavy nuclei. *Astrophys. Space Sci.* **5**, 342–359.

WEHNER G. K., KENKNIGHT C., and ROSENBERG D. L. (1963) Sputtering rates under solar-wind bombardment. *Planet. Space Sci.* **11**, 885–895.

Proceedings of the Second Lunar Science Conference, Vol. 3, pp. 2621–2627
The M.I.T. Press, 1971,

Ultra-heavy cosmic rays in the moon

P. B. PRICE, R. S. RAJAN, and E. K. SHIRK

Department of Physics, University of California, Berkeley, California 94720

(*Received* 22 *February* 1971; *accepted in revised form* 30 *March* 1971)

Abstract—Large (> 1 mm) pigeonite crystals in lunar rock 12021 contain tracks up to at least 1 mm long, some going through their entire thickness. We use these tracks to set upper limits of $\sim 3 \times 10^{-8}/\text{cm}^2$ year on the flux of multiply charged magnetic monopoles produced in high energy cosmic ray interactions in the moon and of $\sim 10^{-4}/\text{cm}^2$ year on the flux of super-heavy cosmic rays with $Z \geqslant 110$ stopping within moon crystals. We use the track length distributions to infer the existence of a finite flux of cosmic ray nuclei with $Z > 82$ over the last $\sim 10^7$ years. We conclude that the overall abundance pattern appears not to have changed drastically in 10^7 years. We use a 10 MeV/N Kr^{84} ion calibration to provide new information on the response of pigeonite crystals to monopoles and ultra-heavy nuclei. We briefly discuss reports by other workers of the existence of long-lived super-heavy elements.

INTRODUCTION

THE INITIAL motivation for this paper was to discuss the possibility that some of the extremely long tracks that we have discovered (BARBER *et al.*, 1971) in rock 12021 might have been made by magnetic monopoles or by super-heavy cosmic rays with $Z > 110$. Since the Apollo 12 conference we have been able to study the response of moon crystals to a beam of 10 MeV/N Kr^{84} ions. These results, which we describe below, show that the longest tracks need not have been produced by transuranic cosmic rays but could have been produced by nuclei with atomic number $Z \lesssim 92$.

In view of two very recent developments, we wish to emphasize that our Kr calibration does not rule out the possibility that our longest tracks were left by super-heavy cosmic rays, and we shall use them to obtain a limit on the flux of high energy nuclei with $Z > 92$ coming into the solar system. The developments to which we refer are the recently reported discovery (MARINOV *et al.*, 1971) of element 112, eka-mercury, with a half-life greater than a few months, which qualifies it for a position in the "island of stability," and the announcement at the Apollo 12 conference (BHAN-DARI *et al.*, 1971b) of evidence for the existence of spontaneous fission tracks from super-heavy elements that once existed in certain grains of the lunar soil.

The existence of cosmic rays with $Z > 30$ had previously been established from studies of fossil tracks in meteorites (FLEISCHER *et al.*, 1967) and confirmed by ob-servations in large area emulsion stacks (FOWLER *et al.*, 1967). Detailed studies in emulsions (FOWLER *et al.*, 1970) and plastic detectors (PRICE *et al.*, 1971; O'SULLIVAN *et al.*, 1971) have shown that the heaviest cosmic rays are greatly enriched in r-process elements relative to the interstellar medium. Balloon observations of cosmic rays with $Z > 92$ have been reported (FOWLER *et al.*, 1970; PRICE *et al.*, 1971), but the uncertainty in their atomic number is large. Because of their rapid depletion by nuclear collisions and ionization loss, the mere presence of a finite flux of extremely heavy ($Z > 82$) galactic cosmic ray nuclei at sub-relativistic energies (O'Sullivan *et al.*, 1971) suggests that they have traveled, on the average, for less than $\sim 10^5$ years. Since the shortest-lived nuclide whose decay products have been found (Reynolds, 1960) in solar system material, I^{129}, has a half-life of $\sim 2 \times 10^7$ years, the cosmic rays provide us with the possibility of detecting unstable elements with lifetimes between $\sim 10^5$ and $\sim 10^7$ years that cannot be detected in nature by other means. The observations of Bhandari *et al.* (1971a, b) suggest that super-heavy elements have been made in nature with a half-life $> 10^7$ years. At this point one is reminded of the papers by Dakowski (1969) and by Anders and Heymann (1969), who suggested that fission of a volatile super-heavy element with $112 \leq Z \leq 119$ (perhaps a longer-lived isotope of eka-mercury than Marinov *et al.* claim to have made) may have been responsible for the excess neutron-rich Xe isotopes in certain meteorites, whose presence correlates with mercury and other volatiles.

With this long preamble on what may be one of the most exciting developments in the history of nuclear and cosmochemistry, we present our own observations.

Observations

Using a solution consisting of $2HF : 1H_2SO_4 : 4H_2O$ we have etched tracks in thin sections from the top cm of rock 12021 containing pigeonite crystals with dimensions of at least 1 mm. Figure 1 gives the distribution of track lengths measured thus far. These are minimum lengths since at least one end of the track is lost at a crystal surface. The atomic numbers we assign to the cosmic rays that left these tracks are thus minimum values. Figure 2 shows examples (a) of a stopping heavy nucleus and (b) of one of our longest tracks, which passes entirely through a 1 mm crystal.

Two observations support the assertion that these long etch features are tracks and not crystal dislocations or linear inclusions: (1) The angular distribution of all long

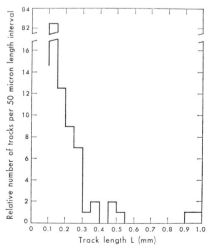

Fig. 1. Distribution of track lengths in rock 12021. Each length is a lower limit because the tracks extend beyond the edge of the crystal into the portion removed by grinding and polishing.

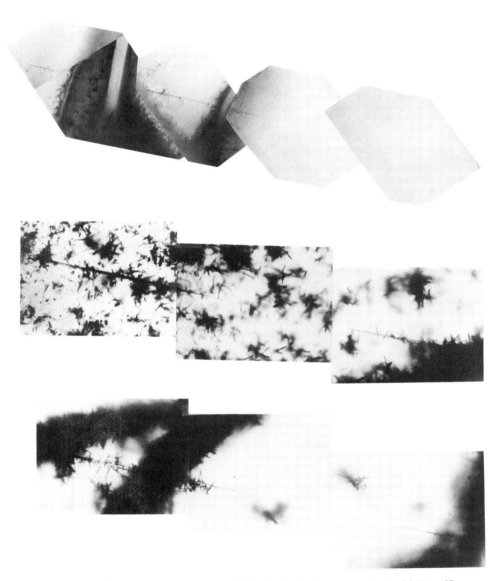

Fig. 2. Ultra-heavy cosmic rays in rock 12021. (a; top) A stopping nucleus with $Z \sim 65$. (b; middle and bottom) An energetic nucleus with $Z \sim 80$ passing through the entire thickness of a 1 mm wide pigeonite crystal. Six micrographs were lined up to make this photograph.

tracks, though peaked in a particular direction, is consistent with the expected distribution close to the surface of a rock and is not parallel to a low index crystallographic direction. (2) Annealing experiments show that Fe tracks disappear after one hour at 550°C, that most long tracks of high Z cosmic rays disappear after one hour at 630°C, and that fission fragment tracks disappear after one hour at 675°C. This behavior is consistent with the known fact that highly ionized regions anneal out at higher temperatures than do ionized region just above threshold (Perelygin et al., 1969).

Kr⁸⁴ Ion Calibration

In plastics (Price et al., 1968a) and in glasses (Fleischer et al., 1969) the etching rate along a track is a monotonic function of ionization rate and can be used to identify the energetic particle. In a crystalline solid a relationship between etching rate and ionization rate has not yet been established, partly because of the necessity to use beams of ions much heavier than have been available. However, assuming the entire portion of a particle's trajectory along which its ionization rate J exceeds a critical rate J_c is revealed by chemical etching, one can calculate the maximum etchable length L_{max} as a function of Z if one knows the way in which J depends on Z and β. This calculation is of course invalid if the mineral has ever been heated enough to relax the damaged regions along the latent tracks.

From irradiation of minerals with Fe, Br, and I ions of ~ 70 MeV, Price et al. (1968b) deduced a relation between L_{max} and Z in which L_{max} ranges from ~ 8 μm for Fe to ~ 2 mm for U. Using a track-in-track method Lal et al. (1969) and Lal (1969) have obtained track length histograms of stopping Fe-group cosmic rays and have assigned values of L_{max} for Cr, Mn, and Fe of 5 μm, 8 μm, and 11.5 μm, assuming the most abundant species to be Fe. From these combined results we predicted that 10 MeV/N Kr⁸⁴ ions have too low an ionization rate to leave etchable tracks. We find, however, that this is not the case. After an etching time such that tracks would have a diameter of ~ 1 μm, crystals of pigeonite, olivine and mica, as well as several glasses, record Kr⁸⁴ ions of 10.35 MeV/N from the Berkeley Heavy Ion Accelerator with 100% efficiency. The etching rate appears to be independent of beam orientation within crystals and to increase ionization rate. With further effort it may be possible to identify particles more reliably by the etch rate method than by the present method that relates L_{max} to Z.

Additional calibrations with beams of Fe and heavier nuclei are badly needed, because the Kr results appear inconsistent with the assignment by Lal et al. (1969) of 11.5 μm to the etchable length of Fe tracks. The Kr results suggest that Fe should record over ~ 15 μm. Until this conflict is resolved, we continue to use the relation of Price et al. (1968b) but with the threshold ionization rate lowered such that Fe tracks have an etchable length of 11.5 μm.

Abundances of Cosmic Rays with $Z > 40$

Table 1 summarizes our charge distributions and compares the present day cosmic ray abundances, measured in balloon-borne detectors (Fowler et al., 1970; O'Sullivan et al., 1971; Price et al., 1971), with the abundances averaged over the surface

Table 1. Abundance of cosmic rays with $Z > 40$.

Charge group relative to Fe	Abundance if 12021 was at surface for 10^7 years	Abundance if 12021 was at 35 g/cm² for 3×10^8 y and at surface for 5×10^6 y‡	Present-day abundance (emulsions and plastics)
$\dfrac{Z \geq 40}{Fe}$	$3.7 \times 10^{-5}(76)*$	$7 \times 10^{-5}(76)*$	$5.7 \times 10^{-5}(139)*$
$\dfrac{Z \geq 50}{Fe}$	$6.9 \times 10^{-6}(12)$	$2 \times 10^{-5}(12)$	$3.2 \times 10^{-5}(48)$
$\dfrac{Z \geq 70}{Fe}$	$1.3 \times 10^{-6}(5)$	$8 \times 10^{-6}(5)$	$1 \times 10^{-5}(33)$
$\dfrac{Z \geq 83}{Fe}$	$5 \times 10^{-7}(2)$	$5 \times 10^{-6}(2)$	$3 \times 10^{-6}(8)$
$\dfrac{Z \geq 92}{Fe}$	†	?	$1 \times 10^{-6}(2)$
$\dfrac{Z \geq 110}{Fe}$	†	?	$< 6 \times 10^{-7}(0)$

* Numbers of events are in parentheses.
† Relative abundances would equal 5×10^{-7} if the two 1 mm long tracks were made by cosmic rays with $Z > 92$ or > 110.
‡ All nuclei were assumed to have the same energy spectra in space. The greater energy loss rate and nuclear collision probability of the heavier nuclei were taken into account. The calculation is for 35 g/cm² burial beneath the surface of a semi-infinite body.

exposure time of rock 12021. Assuming all tracks were formed when the rock was on the very surface during 10^7 years gives the results in column 2. FLEISCHER *et al.* (1971) deduced a longer exposure time, $\sim 2.6 \times 10^7$ years, which would not seriously alter the results. Assuming half the tracks were formed at a depth of 35 g/cm² over a time of 3×10^8 years and half at the lunar surface gives the results in column 3. (MARTI and LUGMAIR (1971) have concluded from the abundances of Gd and Kr isotopes that rock 12021 was bombarded by cosmic rays for $\sim 3 > 10^8$ years while buried at ~ 35 g/cm². Our track density gradient at the surface of 12021 shows that it also was exposed at the surface for several million years).

DISCUSSION

Although the detailed abundances are still uncertain, this work establishes that there has been a significant flux of ultra-heavy cosmic rays, with Z up to and beyond 80, over times of $\sim 10^7$ to $\sim 10^8$ years. The overall abundance pattern over this period appears to be similar to that obtained in current balloon flights for the present-day flux, but we cannot claim that this similarity has been established until better calibrations in minerals are available.

If the two longest tracks were made by elements with $Z \gtrsim 110$, the flux of such super-heavy cosmic rays could be as high as $1/m^2$ year, which is $\sim 5 \times 10^{-7}$ times the Fe flux. This flux would be detectable in the present-day cosmic radiation after a long exposure in an orbiting laboratory or in a series of balloon exposures of plastics and emulsions of large area. In view of the work of MARINOV *et al.* (1971) and of BHANDARI *et al.* (1971a, b), further studies of long tracks in moon rocks and an increased effort in long, high altitude exposures would be extremely important.

The possibility that our charge distribution might be distorted because the heaviest cosmic rays were preferentially attenuated in a long subsurface exposure of rock 12021 was pointed out to us at the Apollo 12 conference by R. M. Walker. To avoid this possible distortion in the future, we have requested samples of 12004, which has large pigeonite crystals and whose short Kr^{81}/Kr^{83} exposure age (Bochsler et al., 1971) makes it more suitable than 12021 for our work.

The ionization rate of a hypothetical magnetic monopole is essentially constant and such a particle would leave a track with the same etching rate all along its length, which might even be as long as several meters. Our Kr results show that a track with a constant etch rate along its length was not necessarily produced by a monopole but could have been produced by a heavy nucleus. Assuming, however, that both of our 1 mm long tracks were made by monopoles, we arrive at an upper limit of $\sim 3 \times 10^{-8}/cm^2$ year on the flux of monopoles with magnetic charge $g \geq 3\hbar c/2e$. This flux is based on the total cosmic ray exposure age of $\sim 3 \times 10^8$ years obtained by Marti and Lugmair (1971). Previous limits on the flux of *primary* monopoles in the universe are far lower than this, $\sim 10^{-14}/cm^2$ year (Osborne, 1970; Fleischer et al., 1970). However, our limit is highly significant for *secondary* monopoles that might have been produced by high energy gamma rays resulting from cosmic ray interactions in the moon. The implications of this result will be discussed in detail in a separate publication, where we emphasize that the production of monopole pairs by high energy gamma rays should be more readily detectable in the moon than in the earth.

The fact that 10.35 MeV/N Kr ions leave etchable tracks in minerals may raise questions concerning the interpretation of the exciting observations of Bhandari et al. (1971a, b), who identify extinct super-heavy elements by the presence of tracks longer than fission tracks of U^{238} and Pu^{244} and longer than slowing Fe nuclei would be expected to leave. Without detracting from their important work, we simply emphasize that with any reasonable ionization equation we conclude from the Kr results that Fe ought to record over at least 15 μm.

Acknowledgments—We are grateful to A. Ghiorso for arranging for the 10 MeV/N irradiations. We are indebted to R. Cowsik, R. L. Fleischer, R. M. Walker, A. Ghiorso, and S. G. Thompson for useful discussions and to NASA, the Atomic Energy Commission, and the National Science Foundation for financial support.

REFERENCES

Anders E. and Heymann D. (1969). Elements 112 to 119: Were they present in meteorites? *Science* **164**, 821–822.

Barber D. J., Hutcheon I., Price P. B., Rajan R. S., and Wenk H. R. (1971). Exotic particle tracks and lunar history. Second Lunar Science Conference (unpublished proceedings).

Bhandari N., Bhat S. G., Lal D., Rajagopalan G., Tamhane A. S., and Venkatavaradan V. S. (1971a). Evidence for the existence of super-heavy elements in extraterrestrial samples, *Nature* **230**, 219–224.

Bhandari N., Bhat S. G., Lal D., Rajagopalan G., Tamhane A. S., and Venkatavaradan V. S. (1971b). The spontaneous fission record of uranium and extinct transuranci elements in Apollo 12 samples. Second Lunar Science Conference (unpublished proceedings).

Bochsler P., Eberhardt P., Geiss J., Graf N., Grogler N., Krahenbuhl V., Morgeli M., Schwaller H., and Stettler A. (1971). Potassium argon ages, exposure ages and radiation history of lunar rocks. Second Lunar Science Conference (unpublished proceedings).

DAKOWSKI M. (1969). The possibility of extinct super-heavy elements occurring in meteorites. *Earth Planet. Sci. Lett.* **6,** 152–154.

FLEISCHER R. L., PRICE P. B., WALKER R. M., MAURETTE M., and MORGAN G. (1967). Tracks in heavy primary cosmic rays in meteorites. *J. Geophys. Res.* **72,** 355–366.

FLEISCHER R. L., PRICE P. B., and WOODS R. T. (1969). Nuclear-particle-track identification in inorganic solids. *Phys. Rev.* **188,** 563–567.

FLEISCHER R. L., HART, H. R. JR., JACOBS I. S., PRICE P. B., SCHWARZ W. M., and WOODS R. J. (1970). Magnetic monopoles: Where are they and where aren't they? *J. of Appl. Phys.* **41,** 958–965.

FLEISCHER R. L., HART, H. R. JR., COMSTOCK G. M., and EVWARAYE (1971). The particle track record of the ocean of storms. Second Lunar Science Conference (unpublished proceedings).

FOWLER P. H., ADAMS R. A., COWEN V. G., and KIDD J. M. (1967). The charge spectrum of very heavy cosmic ray nuclei. *Proc. Roy. Soc.* **A301,** 39–45.

FOWLER P. H., CLAPHAM V. W., COWEN V. G., KIDD J. M., and MOSES R. T. (1970). The Charge spectrum of very heavy cosmic ray nuclei. *Proc. Roy. Soc.* **A318,** 1–43.

LAL D., RAJAN R. S., and TAMHANE A. S. (1969). On the study of the chemical composition of nuclei of $Z > 22$ in the cosmic radiation using meteoritic minerals as detectors. *Nature* **221,** 33–37.

LAL D. (1969). Recent advances in the study of fossil tracks in meteorites due to heavy nuclei of cosmic radiation. *Space Sci. Rev.* **9,** 623–650.

MARINOV A., BATTY C. J., KILVINGTON A. I., and HEMINGWAY J. D. (1971). Evidence for the possible existence of a superheavy element with atomic number 112. *Nature* **229,** 464–467.

MARTI K. and LUGMAIR G. W. (1971). Kr⁸¹–Kr and K–Ar⁴⁰ ages, cosmic ray spallation products and neutron effects in Apollo 11 and 12 lunar samples. Second Lunar Science Conference (unpublished proceedings).

OSBORNE W. Z. (1970). Limits on magnetic monopole fluxes in the primary cosmic radiation from inverse Compton scattering and muon poor extensive air showers. *Phys. Rev. Lett.* **24,** 1441–1445.

OSULLIVAN D., PRICE P. B., SHIRK E. K., FOWLER P. H., KIDD J. M., KOBETICH E. J., and THORNE R. (1971). High resolution measurements of slowing cosmic rays from Fe to U. *Phys. Rev. Lett.* **26,** 463–466.

PERELYGIN V. P., SHADIEVA N. H., TRETIAKOVA S. P., BOOS A. H., and BRANDT R. (1969). Ternary fission produced in Au, Bi, Th, and U with Ar ions. *Nucl. Phys.* **A127,** 577–585.

PRICE P. B., FLEISCHER R. L., PETERSON D. D., O'CELLAIGH C., O'SULLIVAN D., and THOMPSON A. (1968a). High resolution study of low-energy heavy cosmic rays with lexan track detectors. *Phys. Rev. Lett.* **21,** 630–633.

PRICE P. B., FLEISCHER R. L., and MOAK C. D. (1968b). On the identification of very heavy cosmic ray tracks in meteorites. *Phys Rev.* **167,** 277–282.

PRICE P. B., FOWLER P. H., KIDD J. M., KOBETICH E. J., FLEISCHER R. L., and NICHOLS G. E. (1971). Study of the charge spectrum of extremely heavy cosmic rays using combined plastic detectors and nuclear emulsions. *Phys. Rev. D.* **3,** 815–823.

REYNOLDS J. (1960). Determination of the ages of elements. *Phys. Rev. Lett.* **4,** 8–10.

Proceedings of the Second Lunar Science Conference, Vol. 3, pp. 2629–2638
The M.I.T. Press, 1971.

The lunar-surface orientation of some Apollo 12 rocks*

F. Hörz†

The Lunar Science Institute, Houston, Texas 77058

and

J. B. Hartung

NASA Manned Spacecraft Center, Houston, Texas 77058

(*Received* 16 *February* 1971; *accepted* 30 *March* 1971)

Abstract—Detailed studies of the distribution of microcraters on whole lunar rocks revealed that some rocks have large surface areas that are completely uncratered. It is demonstrated that the uncratered parts were buried in the lunar soil. Accordingly, the surface orientation of rocks 12017, 12021, 12038, and 12051 could be reconstructed.

Introduction

Investigations concerned with γ-ray spectroscopy, cosmic-ray and particle tracks, rare-gas analysis as well as magnetic and optical properties of lunar rocks require knowledge of the orientation of each specimen on the lunar surface. In most of these investigations, only information is required about which parts of the rock were exposed to the lunar environment and which ones were buried in the lunar soil. By using any of the listed techniques, one can reconstruct the surface orientation of the rocks, but doing so is usually a very time-consuming task. Therefore, it is desirable to know the surface orientation of each rock before the samples are cut and distributed to other investigators.

The principal technique used to obtain this information is the documented surface photography. Because of the limited time and adverse operational conditions, however, the astronauts are prevented from documenting individual specimens in an unambiguous way. We would like to present a new technique that supplements the lunar-surface photography. This technique is based on the distribution of glass-lined microcraters (Lspet, 1969, 1970; Hörz *et al.*, 1971a, 1971b).

Distribution and Significance of Cratered and Uncratered Surfaces

Detailed stereomicroscopic investigations of seven Apollo 12 rocks revealed marked differences in the distribution of craters on various surfaces of individual rocks (Hörz *et al.*, 1971a, 1971b). Most strikingly, rocks 12017, 12021, 12038, and 12051 exhibit relatively large surface areas that did *not display any microcraters* at all.

The demarcation line between cratered and uncratered surfaces is extremely sharp. If the uncratered areas were just freshly broken surfaces, the demarcation line would

---

* The Lunar Science Institute contribution No. 14.

† Present address: NASA Manned Spacecraft Center, Houston, Texas 77058.

be formed by sharp edges, corners, etc. The demarcation line in all four specimens, however, runs across relatively flat surfaces irrespective of rock geometry. The line can be traced over the whole circumference of the rock and clearly represents a "plane" separating cratered and uncratered surfaces. Such a spatial distribution indicates that uncratered areas are not confined to recently broken surfaces. The uncratered parts must have been shielded from the bombardment of projectiles responsible for the microcraters.

Various origins for the microcraters seem possible such as inflight collisions of "secondary" projectiles and collisions of "secondary" and/or "primary" projectiles when the rock was at rest on the lunar surface. For the first case, one would expect to find craters distributed over the entire surface of the rock because no basis exists to assume that the rocks will not tumble while being ejected in a large-scale impact event. However, the rocks display clearly uncratered surfaces of considerable size; that is, 20–50% of the overall surface area of the rocks. If one grants that the rocks do not tumble, then one certainly has to postulate that the rocks become oriented according to their aerodynamically most stable configuration (in a hypothetical, temporary "atmosphere" caused by impact-generated vaporization of silicates). Thus, the uncratered surfaces would result from "aerodynamic shielding." This effect, however, could not be observed in any of the four cases. The demarcation line is irrespective of rock geometry. Furthermore, the remarkably sharp contact between cratered and uncratered areas on relatively flat surfaces seems to be incompatible with inflight collisions.

Consequently, the areal distribution of cratered and uncratered surfaces was generated while the rock was resting on the lunar surface. Parts of the rock were buried and shielded from the bombardment of micrometeorites. Such an interpretation is supported by the presence of "ropy splashes" and "welded dust" (Hörz et al., 1971a, 1971b). Both features are interpreted as shock-melted materials (mixed with varying amounts of unshocked lunar dust) splashed against the rock as ejecta from microcratering events occurring in the lunar soil in the immediate vicinity of the rock. Though these features, in principle, could be observed on all cratered surfaces, they are remarkably concentrated close to the demarcation line. Never could these features be observed on uncratered surfaces.

On the basis of the geometrical position of the demarcation line with respect to the overall rock shape and the associated concentration of secondary, glassy ejecta, we conclude that the uncratered surfaces were buried in the lunar soil. The unburied parts were exposed to the lunar environment and bombarded by projectiles generating the microcraters. Though Hörz et al. (1971a, 1971b) advanced many arguments for an origin of the glass-lined pits as caused by primary micrometeorites, it is immaterial for the above conclusions whether the craters were actually formed by primary or secondary projectiles or any combination thereof. The only relevant point is that the uncratered parts were buried in the regolith and hence were shielded from the bombardment.

ORIENTATION OF ROCKS

Seven Apollo 12 rocks were examined. Rocks 12006, 12047, and 12073 displayed various crater densities on all surfaces; such a crater distribution indicates that the

Rock 12017

Fig. 1. Distribution of cratered and uncratered surfaces of rock 12017. Note sharp transition zone between cratered glassy surface (B) and completely uncratered glassy surface (A) (A, NASA-S-70-44099; B, NASA-S-70-45309).

Rock 12017

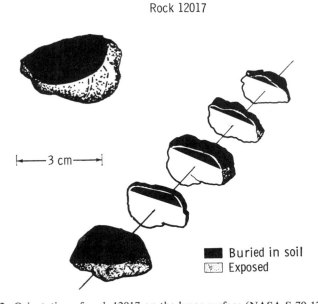

Fig. 2. Orientation of rock 12017 on the lunar surface (NASA-S-70-17450).

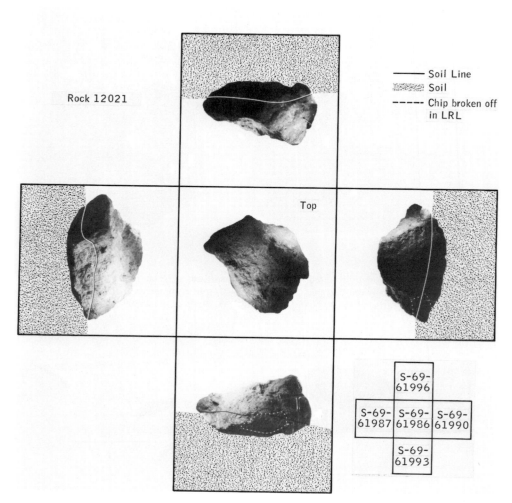

Fig. 3. Orthogonal views of rock 12021, indicating the position of the most recent soil line and the parts of the rocks buried in the lunar soil. (Central photograph shows the exposed upper surface; NASA photograph numbers are indicated in the inset.)

Rock 12021

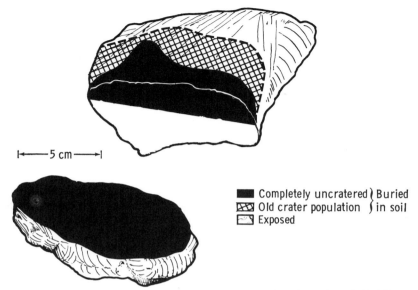

|←——5 cm——→|

■ Completely uncratered ⎫ Buried
⊠ Old crater population ⎬ in soil
▨ Exposed

Fig. 4. Orientation of rock 12021 on the lunar surface (NASA-S-70-44122).

Rock 12021

———————— Recent soil line ⎫ this
– – – – – – Old soil line   ⎬ work
–·—·—·—·— Soil line (Shoemaker et al., 1970)

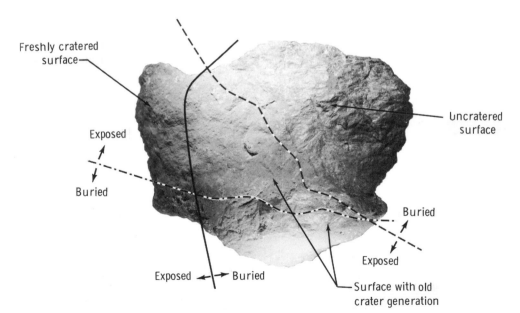

Freshly cratered
surface

Exposed

Buried

Exposed ←—·—→ Buried

Uncratered
surface

Buried

Exposed

Surface with old
crater generation

Fig. 5. "Bottom" surface of rock 12021, illustrating "recent" and "old" soil lines,
as well as soil line based on surface photography. Note completely uncratered surface
(NASA-S-70-61985).

rocks must have been moved on the lunar surface repeatedly. By no means an indication of the most recent orientation, the least cratered surface only represents a position held by the rock for a brief period at any possible stage during its "surface history." The surface orientation of rocks 12017, 12021, 12038, and 12051 are presented in the following sections. Unfortunately, the dust on the rocks, the subtlety of the features observed, and the limitations of the photography do not permit easy recognition of the described relationships. The orthogonal views are arranged in such a way that the lunar-exposed upper surface is always in the center. To facilitate comparison, the exploded sketches are based on the original drawings produced by the Lunar Receiving Laboratory (LRL) for the purpose of documenting the location of individual sample materials irrespective of actual surface orientation. The soil line indicated is accurate to ±1cm.

Rock 12017

On rock 12017 (Figs. 1 and 2), the demarcation line between cratered and uncratered surfaces is exceptionally well-developed on a large glass coating. The orientation

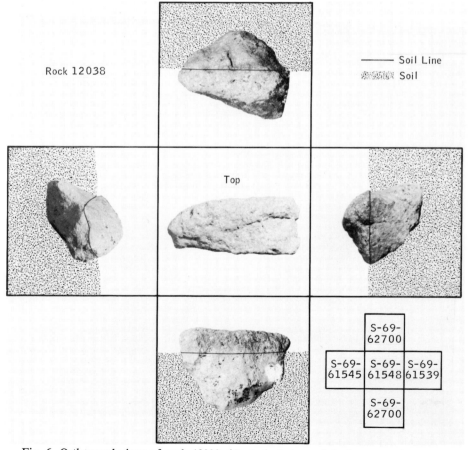

Fig. 6. Orthogonal views of rock 12038. (Central photograph is the exposed upper surface; NASA photograph numbers are indicated in the inset.)

shown is that *after* deposition of the glass. Before that event, the rock could have had a complex history.

Rock 12021

Rock 12021 (Figs. 3–5) has clearly experienced two different positions on the lunar surface. One area is completely uncratered and was therefore buried throughout the surface history of the rock. An adjacent surface displays an "old" generation of microcraters now partially covered with dust and clearly below the most recent soil line.

Rock 12038

Rock 12038 (Figs. 6 and 7) has a rather simple history. No evidence exists of possible previous positions, that is, of tumbling or multiple ejections. This rock is a prime example for studying the characteristics of the soil line and its relation to rock geometry and to the concentration of ropy splashes and welded dust around this line.

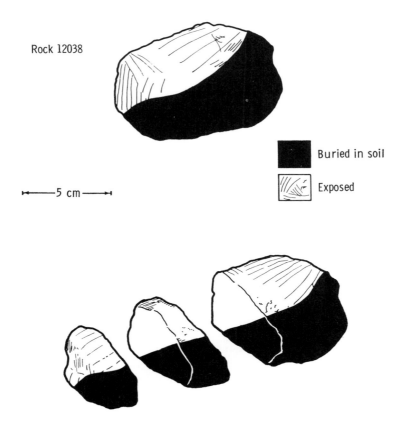

Rock 12038

Buried in soil

Exposed

├───5 cm───┤

Fig. 7. Orientation of rock 12038 on the lunar surface (NASA-S-70-17460).

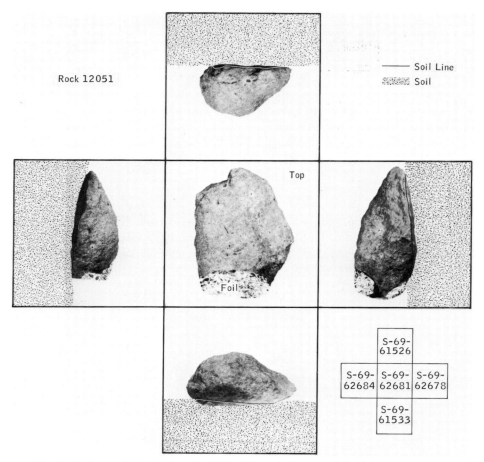

Fig. 8. Orthogonal views of rock 12051. (Central photograph is the exposed upper surface; NASA photograph numbers are indicated in the inset.)

Rock 12051

Rock 12051 (Figs. 8–10) is characterized by a large, extremely flat, uncratered surface. The soil line outlining this surface is especially prominent (Fig. 9).

DISCUSSION

On the basis of observational evidence, we conclude that the demarcation line between cratered and uncratered rock surfaces is synonymous with "soil line." The distribution of cratered and uncratered rock surfaces is incompatible with inflight collisions of secondary ejecta during large-scale cratering. The uncratered rock parts were buried in the lunar regolith.

Rock 12051

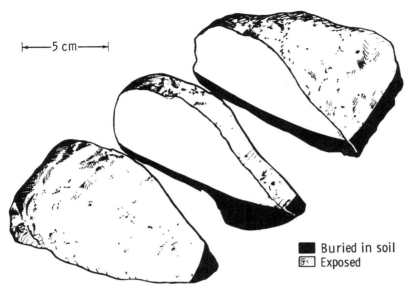

|—— 5 cm ——|

■ Buried in soil
▨ Exposed

Fig. 9. Orientation of rock 12051 on the lunar surface (NASA-S-70-44100).

—— Soil line this work
——— Soil line (Shoemaker et al., 1970)

Rock 12051

Buried ←— —→ Exposed

Exposed
↑
↓
Buried

Fig. 10. "Bottom" surface of rock 12051. Note relatively sharp boundary between cratered and uncratered areas. The soil line based on the surface photography is indicated (NASA-S-70-61519).

Our orientations of rocks 12021 and 12051 are in agreement with γ-ray spectroscopical investigations by SCHONFELD (private communication, 1970). No such data are available for rocks 12017 and 12038. The distribution of cosmic-ray tracks of rock 12021 are compatible with this orientation (PRICE, private communication, 1970). Thus, whenever comparisons between these three methods could be made, the agreement regarding which is a top and which is a bottom surface is good.

However, the orientations of rocks 12021 and 12051 (no data are available for rocks 12017 and 12038) presented here differ from those reported by the field geology team, who based their evaluations on lunar-surface photographs (SHOEMAKER *et al.*, 1970; SUTTON and SCHABER, 1971). The differences in interpretation are indicated in Figs. 5 and 10. For both rocks, the orientations presented by the field geology team would require the exposure of an uncratered surface to space. This requirement can only be satisfied by assuming that both rocks were rotated into such an orientation in extremely recent times. Based on presently available energy distributions of micrometeorites, however, the time required for an event causing rotation of a rock is about 100 times longer than the time required to produce an easily observable microcrater population (HÖRZ *et al.*, 1971b). Therefore, the probability of an event causing rotation of a rock without leaving time for the development of microcraters on an exposed surface is small. The photointerpretation would imply that *both* rocks tumbled in exceptionally recent time. On the basis of available sample comparisons, the frequency of such tumbling events postulated by the field geology team seems statistically highly improbable. The discrepancies are much more likely due to the methods applied, and the accumulation of more detailed data on other rocks will reveal the reliability of each technique.

Acknowledgments—We appreciate the stimulating discussions with E. SCHONFELD, R. L. SUTTON, P. H. PRICE, and R. W. WALKER, as well as the cooperation of curators D. H. ANDERSON and M. B. DUKE. These investigations were supported by the Lunar Science Institute, which is administered by the Universities Space Research Association in cooperation with the NASA Manned Spacecraft Center under NASA grant NSR 09-012-071. Official NASA photographs are used throughout this article.

REFERENCES

HÖRZ F., HARTUNG J. B., and GAULT D. E. (1971a) Micrometeorite craters and related features on lunar rock surfaces. *Earth Planet. Sci. Lett.*, **10**, 381386.

HÖRZ F., HARTUNG J. B., and GAULT D. E. (1971b) Micrometeorite craters on lunar rocks. *J. Geophys. Res.* in press.

LSPET (Lunar Sample Preliminary Examination Team) (1969) Preliminary examination of lunar samples from Apollo 11. *Science* **165**, 1211–1227.

LSPET (Lunar Sample Preliminary Examination Team) (1970) Preliminary examinaton of lunar samples from Apollo 12. *Science* **167**, 1325–1339.

SHOEMAKER E. M., BATSON R. M., BEAN A. L., CONRAD C., JR., DAHLEM D. H., GODDARD E. N., HAIT M. H., LARSON K. B., SCHABER G. G., SCHLEICHER D. L., SUTTON R. L., SWANN G. A., and WATERS A. C. (1970) Preliminary geologic investigation of the Apollo 12 landing site. *Apollo 12 Preliminary Science Report*, NASA SP-235, 113–156.

SUTTON R. L. and SCHABER G. G. (1971) Locations and orientations of rock samples from Apollo missions 11 and 12. Second Lunar Science Conference (unpublished proceedings).

Proceedings of the Second Lunar Science Conference, Vol. 3, pp. 2639–2652
The M.I.T. Press, 1971.

Meteorite impact craters, crater simulations, and the meteoroid flux in the early solar system

M. R. Bloch, H. Fechtig, W. Gentner, G. Neukum, and
E. Schneider

Max-Planck-Institut für Kernphysik, Heidelberg, Germany

(*Received* 22 *February* 1971; *accepted in revised form* 5 *April* 1971)

Abstract—Selected areas of lunar samples 12006, 12021, 12063, 10019, and 10046 were searched for impact craters of cosmic dust particles down to 1 μm diameter by means of a binocular microscope and an electron scanning microscope. The maximum crater number densities are 100 craters $\geqslant$ 0.1 mm diameter and 20 craters $\geqslant$ 0.3 mm diameter per cm^2 surface area.

The glass linings of several craters were analyzed chemically, using an electron microprobe, and compared with the composition in the direct neighborhood. We searched for projectile material different from lunar surface matter, especially for nickel. Generally, the chemical composition is similar to the composition of the surrounding material.

Experimental high velocity impact craters on quartz glass, boron silicate glass, and norite gave a crater to projectile diameter ratio of about 2 at 20 km/sec impact velocity.

The numbers of large craters in Mare Tranquillitatis, Oceanus Procellarum, and on selected Lunar Highland areas were compared with the meteoroid influx as known by direct observations. These comparisons lead to a time variable meteoroid flux of $\phi(m) \cdot e^{-Bt}$ with $\phi(m)$ = present flux, $B = 2.6$, and t = time in 10^9 years. From these measurements a crater/particle diameter ratio in the μm to km size range between 2 and 7 results. This is consistent with a meteoroid velocity of about 20 km/sec.

Mean crater lifetimes are calculated using the time variable flux model. For the 0.1 and 0.3 mm diameter craters these lifetimes are in the order of 10^6–10^7 years.

Introduction

Before the Apollo program, crater statistics on the lunar surface were only roughly known and meteoroid impact frequency values were only speculative. For the age of the lunar surface only relative estimates were available. Now, precise formation ages of different surface areas can be determined. Very little was known about the interaction of the interplanetary debris with the surface of the moon. With good photographs of the moon's surface and lunar samples to investigate in the laboratory, it is possible to count crater frequencies from the μ-size range to km-size range. Provided the appropriate laboratory simulations have been performed, these new measurements allow us to relate crater statistics to the flux of interplanetary matter in the past and today. Since the lunar surface has been exposed to interplanetary dust bombardment for a long time, we have the possibility to study and interpret the erosion process caused by cosmic dust bombardment.

Crater Statistics

The existence of mm- and μ-sized craters on lunar samples has been reported already by LSPET (1969, 1970), McKay *et al.* (1970), Carter and McGregor (1970),

NEUKUM et al. (1970), and HÖRZ et al. (1970). Selected areas of samples 12006, 12021, 12063, 10019, and 10046 have been scanned optically and by means of an electron scanning microscope (Stereoscan). Small impact craters down to about 1 μm diameter have been detected. Many craters found on samples 10019,13,1; 10046,19,3; and 12063,106a, top, are shown in BLOCH et al. (1971).

Figure 1 illustrates the types of small craters found on these lunar samples. Almost all craters on breccia 10019,13,1 show glass linings with numerous bubbles (Fig. 1, upper left), presumably produced during projectile impact by released gas. On the contrary, the glass linings of craters on crystalline rock 12063,106a, top, are mostly very smooth (Fig. 1, upper right). The lower positions of Fig. 1 show Stereoscan

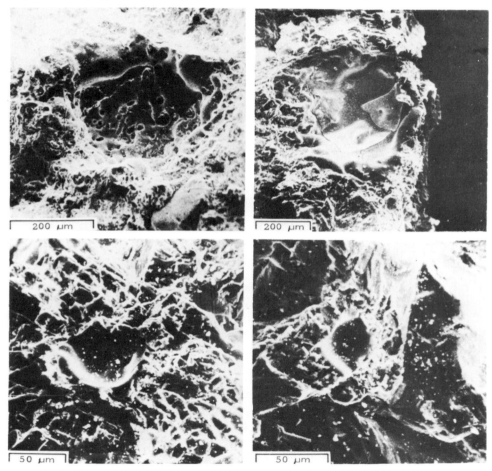

Fig. 1. Observed types of microcraters on lunar samples: *Upper left*: Bubbles in interior, typical for microcraters on investigated breccias 10019,13,1 and 10046,19,3. *Upper right*: Smooth glass-lined pit on crystalline sample 12063,106a, top. *Lower left*: Small crater on sample 12063,106a, top. *Lower right*: Micron-sized crater on crystalline material, sample 12063,106a, top.

photographs of craters on crystalline material 12063,106a, top, with irregularly formed spallation zones.

In Fig. 2, μ-sized craters are shown which are found on very smooth glasses of samples 12063,106a, top, and 10046,19,3. All these craters are produced by micron- and/or submicron-sized projectiles. Some show spallation features, some do not. CARTER (1971) has reported the existence of submicron-sized craters as small as 300 Å diameter on lunar glassy spherules.

In Table 1 crater statistics are given from these lunar samples. Selected areas of samples 12006, 12063, and 10019 show statistics which are in fairly good agreement with more extended statistics by HÖRZ et al. (1970, 1971). Samples 12021,16a and 10046,19,3, however, show only a few craters indicating shielding and/or a shorter time of exposure to meteoroid bombardment. Figure 3 is a diagram of the crater distribution from selected areas of sample 12006. We have found a maximum crater number density of 100 craters $\geqslant 0.1$ mm diameter and 20 craters $\geqslant 0.3$ mm diameter per cm² surface area in this sample, while the average number densities are as follows: (69 ± 12) craters $\geqslant 0.1$ mm pit diameter per cm² area, (17 ± 2) craters $\geqslant 0.3$ mm pit diameter per cm² area. The numbers of μ-sized craters (between 1 and 10 μm in diameter) on host craters on sample 12063,106a, top, range between 1 and 5 per host crater which totals up to 50 craters per mm² surface area in the micron-size range. Fresh appearing mm-size host craters show less μm-sized craters than older looking craters which would be consistent with shorter exposure times. Better statistics for μm-craters will be given later elsewhere since the work on μm-craters is not yet finished.

CHEMISTRY OF MICROCRATERS

Seven craters on sample 10019,13,1 have been analyzed chemically using an electron microprobe. The glass linings of the craters were scanned with the electron beam over a certain area at definite sites of the pit. The same was done in the surroundings of the crater just neighboring the sites inside. The X-ray intensities of the neighboring sites have been compared and averaged. The absorption of X-rays has been taken into account for the nonplanar geometry of the craters and the sample; the sample was inclined with respect to the electron beam in order to get a high take-off angle for the X-rays. Other geometry effects affecting the production of X-rays such as varying backscattering of electrons have been eliminated by normalizing the X-ray intensities to specimen currents. A more detailed discussion of the influence of geometry effects on X-ray measurements is given by NEUKUM (1969).

Overall analytical data show compositions of the crater interiors similar to the composition of the surroundings. This result agrees with the general conception that only part of the crater glass originates from the projectile. Therefore only small variations in the chemical compositions can be expected.

Detailed measurements were carried out for iron and nickel concentrations. Table 2 shows the percentage of iron and nickel in these craters compared to the respective concentrations in the surroundings. Only one of these craters shows both iron and nickel enhancements. Thus this crater was presumably produced by a particle with higher iron and nickel concentrations.

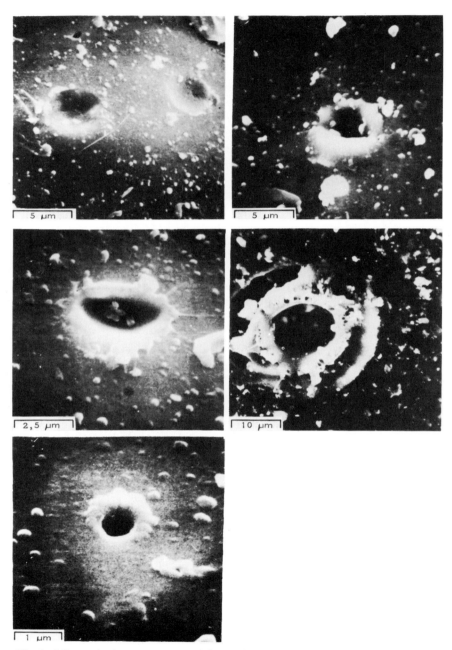

Fig. 2. Micron-sized craters on smooth lunar glasses (12063,106a, top and 10046,19,3).

Table 1. Crater statistics of selected areas from several lunar samples

Surface mm²	sample	10–50 μ	0.1 mm*	0.2	0.3	0.4	0.5	0.6	0.7	0.8	0.9	1.0	1.1	Observation technique
						pit diameter								
300	12006		39	75	50	30	12	8	3	1	—	—	1	Binocular
120	12021,16A		1	2	2	1	—	1	—	—	—	—	—	Binocular
90	12063,106		10	6	4	4	2	—	1	—	—	—	—	Binocular
40	12063,106a top	6	10	4	3	3	2	—	—	1	—	—	—	Stereoscan
70	12063,104 bottom		6	4	2	1	—	—	—	—	—	—	—	Binocular
17	10019,13,1		2	5	3	2	—	1	—	1	—	—	—	Stereoscan
13	10046,19,3		—	1	—	—	—	—	—	1	—	—	—	Stereoscan

* Counted were all craters with sizes within ±0.05 mm of the listed pit diameters.

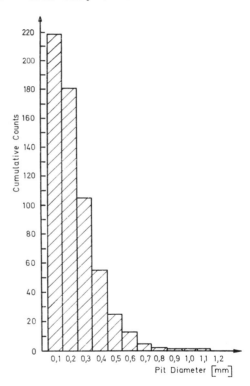

Sample 12006 (selected area)

Total Counts: 219

Scanned Area: 300 mm²

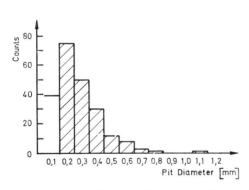

Fig. 3. Microcrater statistics of selected areas on sample 12006.

Table 2. Iron and nickel content of crater glasses and surroundings (error ±10% of the percentages tabulated)

Crater		1	2	3	4	5	6	7
Fe%	interior	9.0	8.5	12.5	12.0	8.5	8.5	7.0
	surroundings	11.0	10.5	10.5	10.0	10.0	9.5	11.5
	interior	< 0.2	< 0.2	1.8	< 0.2	< 0.2	< 0.2	< 0.2
Ni%	surroundings	0.6	< 0.2	< 0.2	< 0.8	< 0.2	< 0.2	< 0.2

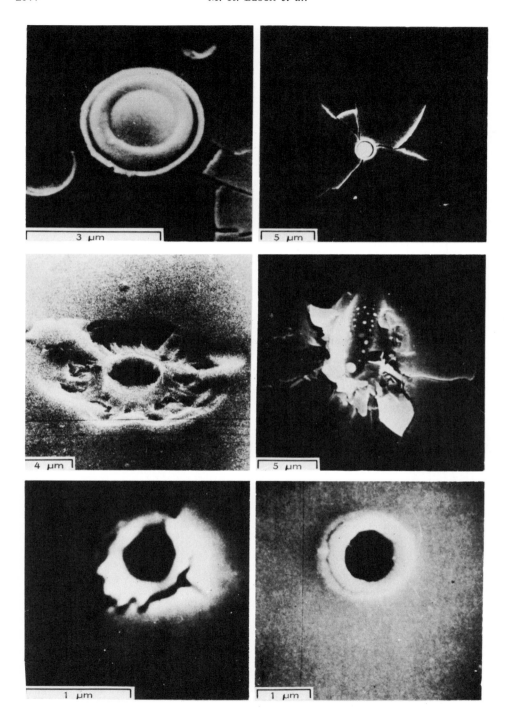

CRATER SIMULATIONS

Simulation experiments have been carried out using a 2 MV Van de Graaff accelerator in order to better interpret the morphology of microcraters on lunar samples and to obtain information about mass and velocity of the impacting particles. For this purpose quartz glass, boron silicate glass, and norite targets (norite has an equigranular structure of pyroxene and basic plagioclase with ore accessories and is very similar to lunar material) were bombarded by Fe- and Al-projectiles in the mass range between 10^{-14} and 10^{-9} g (0.1–10 μm diameter) at velocities between 0.5 and 30 km/sec.

Generally, one finds the following conditions: At low impact velocities (< 1 km/ sec), the target surface is plastically deformed, a flat hole remains and the projectile is generally reflected. At a velocity of about 1 km/sec, the projectile sticks more or less deformed in the impact hole. Usually in this case there is no spallation. Between about 1 and 4–5 km/sec the projectile is destroyed. Fragments of the projectile, sometimes partly melted during the impact process, can be detected in the pit; depending on the projectile diameter there may be slight spallation effects around the pit.

Hypervelocity impact craters (> 5 km/sec) show very smooth crater interiors indicating complete melting of the projectile and part of the target material. Sometimes there are small spherical features of melted material visible at the wall of the pit. Mostly the pits are hemispherically shaped and except for very small projectiles (0.1 μm–1 μm) there is well-developed spallation. Spallation, however, is not necessarily an indication for hypervelocity impact, but depends on mass and velocity of the projectile: for the same reason, the material of the spallation zone may have been ejected or not.

In the case of craters produced by obliquely impacting particles, there are often observed small spherules in the central pit which are formed during impact under the influence of surface tension from melted material. These spherules are similar to those found in lunar soil. Figure 4 shows Stereoscan photographs of some typical simulated impact features on glasses referring to the above described velocity ranges. Detailed specifications are given in the figure captions.

Measurements of the pit diameter, D, as a function of projectile mass, m, for three different constant velocities of Fe-projectiles on quartz glass targets showed the dependence

$$D \sim m^{1/3} \tag{1}$$

Fig. 4. Simulated microcraters on glass targets: *Upper left*: Flat hole of a Fe-projectile on a window glass target, velocity < 1 km/sec (left); crater with deformed particle sticking in the pit, no spallation zone, velocity ≈ 1 km/sec (right). *Upper right*: Fe-projectile on a quartz-glass target, particle diameter 1.2 μ, particle velocity 2.5 km/sec. *Center left*: Al-projectile on a quartz-glass target, particle diameter 4 μ, particle velocity 5 km/sec. *Center right*: Oblique impact of a Fe-projectile on a quartz-glass target, particle velocity 8 km/sec, impact angle 20°. *Lower left*: Fe-projectile on a quartz-glass target, particle diameter 0.15 μ, particle velocity 30 km/sec. *Lower right*: Fe-projectile on a quartz-glass target, particle diameter 1.1 μ, particle velocity 4.5 km/sec.

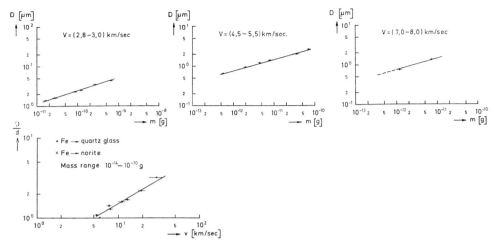

Fig. 5. Dependence of crater diameter on mass and velocity of projectile: *Upper diagrams*: Pit diameter D is plotted against projectile mass of Fe-projectiles on quartz-glass targets for three different velocities. *Lower diagram*: Pit diameter to projectile diameter ratio D/d as a function of impact velocity.

(Fig. 5, upper diagrams). In addition, D/d, the ratio of pit diameter to projectile diameter, is given as a function of the projectile velocity v for Fe-projectiles on quartz glass targets:

$$D/d \sim v^{2/3} \qquad (2)$$

(Fig. 5, lower diagram). According to RUDOLPH (1969) and EICHELBERGER and KINEKE (1967) one can assume

$$T \sim D$$

where T = depth of crater. (In the case of μm-craters on silicate targets approximately $T = D/2$.) Therefore, the crater volume V is proportional to D^3, and in connection with (1) and (2) it can easily be derived that V is proportional to the kinetic energy of the impacting particle. The same dependence has been found by RUDOLPH (1969) for Fe-projectiles on metal targets. Considering relationships (1) and (2), it can be stated that m and v are exactly separable parameters, at least for the micron-size range. By the well-defined dependences, one may obtain information about mass and velocity of the particles that impacted.

From these measurements the absolute value of D/d for Fe-projectiles on quartz glass is 2.2 at an impact velocity of 20 km/sec. This is unexpectedly low. RUDOLPH (1969) gives values of 10–12 for metal targets at the same velocity of Fe-projectiles and the same mass range. Apparently the D/d ratio shows a great difference between metal and silicate targets. But from other experiments with Fe- and Al-projectiles shot on several different kinds of glasses, we know that there is not too much variation of the D/d ratio for silicate targets. The two points of the diagram in Fig. 5 (D/d vs. v) referring to Fe-projectiles on norite targets also show that these values are not essentially different from the case of quartz glass targets.

METEOROID FLUXES IN THE EARLY SYSTEM

SHOEMAKER et al. (1970a, b) and GAULT (1970) have shown that the distribution of lunar craters follows two different characteristics: The number of larger meteoroid impacts ($>$ 140 m∅ in Mare Tranquillitatis and $>$ 50 m∅ in Oceanus Procellarum) is conserved in the distribution called "production curve." On the contrary, the smaller craters ($<$ 140 m∅ in Mare Tranquillitatis and $<$ 50 m∅ in Oceanus Procellarum) suffer from lunar erosion processes which are partially caused by meteoroid impacts, solar wind and cosmic-ray bombardment, and thermal gradient effects. The resulting crater distribution is called by SHOEMAKER et al. (1970a, b) the "steady state curve" and by GAULT (1970) the "equilibrium distribution." The following considerations refer to large craters of the "production curve" only.

In a previous paper (NEUKUM et al., 1970) we tried to compare the numbers of large craters with a time constant meteoroid flux. For the craters on the production curve in the Mare Tranquillitatis (MT) it is possible to explain this distribution if one assumes a crater (pit) to particle diameter ratio of about 20. GAULT (1970) gives a similar explanation using a D/d ratio $\leqslant$ 10 and an unrealistically high meteoroid flux. These ratios seem to us rather high for an assumed impact velocity of approximately 20 km/sec.

The greatest difficulty in the above mentioned models is explaining the far greater numbers of large craters in the lunar highlands relative to the Maria. SHOEMAKER (1966) published crater counts from Terrae close to Mare Tranquillitatis with a maximum crater number density about 10 times greater than the numbers of craters in Mare Tranquillitatis. The highest crater number densities were reported by HARTMANN (1966) from the Southern Lunar Highlands. These numbers are about 50 times higher than crater numbers in the Mare Tranquillitatis, which may be overestimated. Our rough counts from lunar maps of the highland in the neighborhood of the MT lead to about 10–15 times more craters in the km-size range, than in MT. This excess of craters must have been produced within less than 10^9 years, which is unexplainable with a time constant flux.

The most reliable data have been published by SHOEMAKER et al. (1970b) for the crater numbers on the production curves of MT and Oceanus Procellarum (OP). These are between 1.5 and 3.16 times higher in MT than in OP, with an average of 2.37. According to the age difference of MT and OP of only about 3.5×10^8 years (LUNATIC ASYLUM, 1970, 1971), it is again difficult to explain these numbers applying a time constant meteoroid influx. To interpret these data we propose a time variable flux model. Assuming that the number of meteoroids since the beginning of the solar system is reduced largely by impacts of these meteoroids on the planets and their moons, then it is obvious that this decrease is proportional to the number of meteoroids

$$- \frac{dN(m)}{dt} \sim N(m).$$

According to this definition of an exponential function one can possibly describe the flux of meteoroids in the solar system as

$$f(m, t) = e^{-Bt} \cdot \varphi(m),$$

with $\varphi(m)$ = differential flux today, m = mass, B = constant, and t = time, and accordingly the cumulative flux reads

$$F(m, t) = e^{-Bt} \cdot \int_{m_0}^{\infty} \varphi(m) \, dm = e^{-Bt} \cdot \phi(m_0),$$

with

$$\phi(m_0) = \int_{m_0}^{\infty} \varphi(m) \, dm = \text{cumulative flux today.}$$

The constant B can be determined by the crater number densities in different maria and lunar terrae. Since the ages for MT and OP are 3.65×10^9 years and 3.3×10^9 years, respectively (Lunatic Asylum, 1970, 1971),

$$\int_{-3.65}^{0} \int_{m_0}^{\infty} \varphi(m) \cdot e^{-Bt} \, dm \, dt = R_1 \int_{-3.3}^{0} \int_{m_0}^{\infty} \varphi(m) \cdot e^{-Bt} \, dm \, dt,$$

with $R_1 = 2.4$, the ratio of large craters in MT and OP, and T = time in 10^9 years, leading to $B = 2.5$.

For the Lunar Highlands with their model age of 4.6×10^9 years (Lunatic Asylum, 1970, 1971), and the crater numbers in MT.

$$\int_{-4.6}^{0} \int_{m_0}^{\infty} \varphi(m) \cdot e^{-Bt} \, dm \, dt = R_2 \int_{-3.65}^{0} \int_{m_0}^{\infty} \varphi(m) \cdot e^{-Bt} \, dm \, dt,$$

with $R_2 = 12$ leading to $B = 2.6$.

For $B = 2.6$ a half life of the meteoroid flux $T_{1/2} = 270$ million years results. Thus the time variable meteoroid flux is

$$f(m, t) = e^{-2.6t} \cdot \varphi(m)$$

or

$$F(m, t) = e^{-2.6t} \cdot \phi(m).$$

Here one should state that so far the special and still doubtful knowledge of the flux function $\varphi(m)$ or $\phi(m)$ is not used. Table 3 gives for the function $f(m, t)$ the time dependence of the meteoroid flux for the time of formation of the Lunar Terrae and the Lunar Maria MT and OP.

Similar results have been presented by Shoemaker and Hait (1971) during the Second Lunar Science Conference (unpublished proceedings), and Baldwin (1970) and Hartmann (1970). Shoemaker and Hait (1971) have calculated a half life of almost

Table 3. Time dependence of the time variable meteoroid flux function
$f(m, t)$.

Area	Age (t_0)	$f(m, t_0)$
Highland	$4.6 \times 10^9 \, a$	$1.6 \times 10^5 \times \varphi(m)$
MT	$3.65 \times 10^9 \, a$	$1.3 \times 10^4 \times \varphi(m)$
OP	$3.3 \times 10^9 \, a$	$5.3 \times 10^3 \times \varphi(m)$

half a billion years. This half life is still in agreement with our result since the errors in crater statistics and the ages of the maria lead to a variation of B between roughly 2.4 to 2.9.

COMPARISON OF CRATER FREQUENCIES AND FLUX OF PARTICLES

A direct comparison between craters of several kilometers in diameter on the moon and the particles impacted is not possible, because the dependence of crater diameter on particle mass and velocity are not experimentally known. Simulation experiments have been performed only in the μ- to cm-size range. From our own measurements we know best the dependence of projectile diameter to crater diameter at a certain velocity up to 30 km/sec. This knowledge is necessary because particle flux is a function of mass (or diameter) of particles.

In contrast to the good experimental knowledge of cratering in the μ-size range, we do not know the production curve for microcraters. But from the steady state distribution given by SHOEMAKER *et al.* (1970b) and the measurements by HÖRZ *et al.* (1970) and our own crater counting, we know that this distribution follows over many orders of magnitude a power function with the exponent 2.0. No great variation exists. Under the condition that the impact process does not rapidly and completely change in the case of particles of masses different by orders of magnitude, and that this change is accidentally compensated for by an opposite change in the flux of particles, the variations in the production curve cannot be too great. Otherwise, the steady state distribution should not have the same exponent throughout.

These considerations give us the basis to try to extrapolate the production curves found for large craters by SHOEMAKER *et al.* (1970b) to craters in the micron-size range. We assume a slope of -2.9 in the double logarithmic plot of the cumulative crater number densities vs. crater diameters for the production curve according to SHOEMAKER *et al.* (1970b).

The calculation from the extrapolated production curve to flux values is carried out with a D/d ratio of 2, with the assumption of a mean impact velocity of 20 km/sec and a mean density of the particles of 3 g/cm³. The direct comparison leads to a flux dependence of

$$\phi(m) = 10^{-14.6}\, m^{-1} \text{ per m}^2 \text{ sec}^1.$$

The flux function follows the direct meteoroid observations (WATSON, 1956; MILLMAN, 1970) and is in agreement with recent results in the dust mass range for the interplanetary space (ALEXANDER *et al.*, 1970; BERG and GERLOFF, 1970).

From the comparison we obtain directly the variation of crater (pit) diameter to projectile diameter with projectile masses or crater diameters. In the micron-size range D/d calculates to approximately 2 in agreement with our simulation results. In the meter range D/d comes out to 3 to 4 and for km sizes D/d is between 4 and 7. Contrary to some theoretical considerations these D/d values are somewhat low.

MEAN CRATER LIFETIMES

For craters in the equilibrium state ($\leqslant 140$ m∅ in Mare Tranquillitatis $\leqslant 50$ m∅ in Oceanus Procellarum), it is possible to compare the corresponding number densities with the number of particles impacted during the past. According to the equation

$$\phi(m)\int_{-t_0}^{0} e^{-Bt}\, dt = \frac{N}{F} \qquad \text{(crater number density)},$$

with $\dfrac{N}{F} \begin{cases} = 70/\text{cm}^2 \text{ for } D \geqslant 0.1 \text{ mm} \\ = 17/\text{cm}^2 \text{ for } D \geqslant 0.3 \text{ mm} \\ = 10^{4.9}/\text{km}^2 \text{ for } D \geqslant 1\text{m} \end{cases}$ (own measurements)

(SHOEMAKER *et al.*, 1970b),

$$\phi(m) = 10^{-14.6}\, m^{-1}/m^2 \text{ sec} \qquad \text{(present day flux)},$$

it is possible to calculate "mean crater lifetimes." Applying this formalism to 0.1 mm diameter craters, the lifetime calculates to 10^6 years and for 0.3 mm diameter craters to 10^7 years. For 1 m diameter craters the lifetime is 10^8–10^9 years.

These results can be compared with cosmic-ray exposure ages. Kirstein *et al.* (1971) has presented cosmic-ray exposure ages for Apollo 12 soil of 350 million years. With the average irradiation depth of about 80 cm this corresponds well with the mean crater lifetime of 10^8 to 10^9 years for 1 m diameter craters. FLEISCHER *et al.* (1971) and CROZAZ and WALKER (1971) have calculated a lunar erosion rate between 1 and 10 Å per year by comparison of the track distribution in lunar glass and in Surveyor III glass. These data are in good agreement with the crater lifetimes for small craters.

The main processes for lunar erosion are meteoroid bombardment, charged particles bombardment (solar wind protons and electrons, cosmic rays), and thermal gradient effects. At least for the uppermost layer of the lunar regolith a sputtering effect might be an important factor for erosion. Calculations using experimental data for the sputtering of silicates by argon (G. K. WEHNER *et al.* 1963) lead to an erosion rate of ≈ 3 Å per year for proton bombardment.

SUMMARY

The crater number densities on the lunar samples were found to be in agreement with the equilibrium distribution as defined by SHOEMAKER (1970b) and GAULT (1970). 69 ± 12 craters > 0.1 mm $\varnothing$ and 17 ± 2 craters > 0.3 mm $\varnothing$ per cm² surface area have been counted. The existence of micron- and submicron-sized craters was proved. Microcraters on lunar material could be simulated on various target materials down to crater diameters in the submicron-size range. Iron and aluminum projectiles in the mass range between 10^{-9} and 10^{-14} g have been accelerated up to 30 km/sec impact velocities. The appearance of spallation zones as a function of mass and impact velocity could be shown.

A direct comparison of the crater number densities in the Mare Tranquillitatis, the Oceanus Procellarum, and the Lunar Terrae with the meteoroid flux as known by direct observations could not be explained on the basis of a time constant meteoroid flux. On the contrary, only a time variable flux model can explain the crater frequencies satisfactorily. As a result there is an exponential decrease of the meteoroid flux from the formation of the moon and the maria until today. In the early time of the solar system the flux was 4 to 5 orders of magnitude higher than today.

Generally, we learned by the study of lunar samples in the laboratory that the moon suffers from an erosion process which is caused to a great deal by bombardment with interplanetary matter. The application of the time variable flux model to the crater distribution along the equilibrium curve leads to the calculation of mean crater lifetimes. The 0.1–0.3 mm size craters have a lifetime of 10^6 to 10^7 years, craters in the meter-size range have a lifetime of 10^8 to 10^9 years. These results are consistent with those obtained from studies of solar particles tracks in lunar and Surveyor III glasses and cosmic-ray exposure ages of lunar samples.

Acknowledgments—We thank the National Aeronautics and Space Administration for providing the lunar samples. We acknowledge the hospitality of the Lunar Science Institute, Houston during a short visit of H. FECHTIC. Special thanks to Dr. F. HÖRZ and Dr. J. HARTUNG for help and advice in crater counting.

REFERENCES

ALEXANDER W. M., ARTHUR C. W., and CORBIN J. D. (1970) Picrogram dust particle flux: 1967–1968 measurements in selenocentric, cislunar and interplanetary space. *Space Research* X, pp. 252–259. North-Holland.

BALDWIN R. B. (1970) Absolute ages of the lunar maria and large craters II. The viscosity of the moon's outer layers. *Icarus* **13**, 215–225.

BERG O. E. and GERLOFF U. (1970) Orbital elements of micrometeorites derived from Pioneer VIII measurements. *J. Geophys. Res.* **75**, 6932–6939.

BLOCH M. R., FECHTIG H., GENTNER W., NEUKUM G., SCHNEIDER E., and WIRTH H. (1971). Natural and simulated impact phenomena—A photo-documentation. Max-Planck-Institut für Kernphysik, Heidelberg, Germany (private printing).

CARTER J. L. and MACGREGOR J. D. (1970) Mineralogy, petrology and surface features of some Apollo 11 samples. *Proc. Apollo 11 Lunar Sci. Conf.*, *Geochim. Cosmochim. Acta* Suppl. 1., Vol. 1 pp. 247–265. Pergamon.

CARTER J. L. (1971) Chemistry, petrology, and morphology of some Apollo 12 materials. Second Lunar Science Conference (unpublished proceedings).

CROZAZ G. and WALKER R. M. (1971) Solar particle tracks in glass from the Surveyor III spacecraft. Second Lunar Science Conference (unpublished proceedings).

EICHELBERGER R. J. and KINEKE J. H. (1967) Hypervelocity impact. In *Kurzzeitphysik* (editors K. Vollrath and G. Thomer), Springer-Verlag Wien, New York. pp. 659–692.

FLEISCHER R. L., HART H. R., and COMSTOCK G. M. (1971) Very heavy solar cosmic rays: Energy spectrum and implications for lunar erosion. Second Lunar Science Conference (unpublished proceedings).

GAULT D. E. (1970) Saturation and equilibrium conditions for impact cratering on the lunar surface: Criteria and implications. *Radio Science* **5**, 272–291.

HARTMANN W. K. (1966) Martian cratering. *Commun. Lunar and Planet. Lab. Univ. of Arizona* **4**, part 4, no. 65, pp. 121–131.

HARTMANN W. K. (1970) Lunar cratering chronology. *Icarus* **13**, 299–301.

HÖRZ F., HARTUNG J. B., and GAULT D. E. (1970) Micrometeorite craters on lunar rock surfaces. Lunar Science Institute Contribution no. 09.

HÖRZ F., HARTUNG J. B., and GAULT D. E. (1971) Micrometeorite craters on lunar rock surfaces. Second Lunar Science Conference (unpublished proceedings).

KIRSTEN T., STEINBRUNN F., and ZÄHRINGER J. (1971) Location and variation of rare gases in Apollo 12 lunar samples. Second Lunar Science Conference (unpublished proceedings).

LSPET (Lunar Sample Preliminary Examination Team) (1969) Preliminary examination of lunar samples from Apollo 11. *Science* **165**, 1211–1227.

LSPET (Lunar Sample Preliminary Examination Team) (1970) Preliminary examination of lunar samples from Apollo 12. *Science* **167**, 1325–1339.

LUNATIC ASYLUM (1970) Ages, irradiation history, and chemical composition of lunar rocks from the Sea of Tranquillity. *Science* **167**, 463–466.

LUNATIC ASYLUM (1971) Rb–Sr ages, chemical abundance patterns and history of lunar rocks. Second Lunar Science Conference (unpublished proceedings).

McKAY D. S., GREENWOOD W. R., and MORRISON D. A. (1970) Origin of small lunar particles and breccia from the Apollo 11 site. *Proc. Apollo 11 Lunar Sci. Conf., Geochim. Cosmochim. Acta* Suppl. 1., Vol. 1, pp. 673–694. Pergamon.

MILLMAN P. M. (1970) Meteor showers and interplanetary dust. *Space Research* **10**, 260–265. North-Holland.

NEUKUM G. (1969) Investigation of projectile material in microcraters. NASA Technical Translation NASA-TT-F 12327.

NEUKUM G., MEHL A., FECHTIG H., and ZÄHRINGER J. (1970) Impact phenomena of micrometeorites on lunar surface materials *Earth Planet. Sci. Lett.* **8**, 31–35.

RUDOLPH V. (1969) Untersuchungen an Kratern von Mikroprojektilen im Geschwindigkeitsbereich von 0,5 bis 10 km/sec. *Z. Naturforsch.* **24a**, 326–331.

SHOEMAKER E. M. with contributions by ALDERMAN J. D., BORGESON W. T., CARR M. H., LUGN R. V., McCAULEY J. F., MILTON D. J., MOORE H. J., SCHMITT H. H., TRASK N. J., WILHELMS D. E., and WU S. S. C. (1966) *Progress in the Analysis of the Fine Structure and Geology of the Lunar Surface from the Ranger VIII and IX Photographs.* J.P.L. Technical Report no. 32-800, March 15, pp. 249–271.

SHOEMAKER E. M., HAIT M. H., SWANN G. A., SCHLEICHER D. L., DAHLEM D. H., SCHOBER G. G., and SUTTON R. L. (1970a) Lunar regolith at Tranquillity Base. *Science* **167**, 452–455.

SHOEMAKER E. M., BATSON R. M., BEAN A. L., CONRAD C., JR., DAHLEM D. H., GODDARD E. N., HART M. H., LARSON K. B., SCHOBER G. G., SCHLEICHER D. L., SUTTON R. L., SWANN G. A., and WATERS A. C. (1970b) *Preliminary Geologic Investigation of the Apollo 12 Landing Site, Part A: Geology of the Apollo 12 Landing Site.* Apollo 12 Preliminary Science Report, NASA SP-235.

SHOEMAKER E. M. and HAIT M. H. (1971) The bombardment of the lunar maria. Second Lunar Science Conference (unpublished proceedings).

WATSON F. G. (1956) *Between the Planets.* Harvard University Press.

WEHNER G. K., KNIGHT C. K., and ROSENBERG D. L. (1963) Sputtering rates under solar-wind bombardment. *Planet. Space Sci.* **11**, 885–895.

Proceedings of the Second Lunar Science Conference, Vol. 3, pp. 2653–2670
The M.I.T. Press, 1971.

Influence of target temperature on crater morphology and implications on the origin of craters on lunar glass spheres†

JAMES L. CARTER

University of Texas at Dallas, Geosciences Division, Division of Earth and
Planetary Sciences, P.O. Box 30365, Dallas, Texas 75230

and

DAVID S. McKAY

National Aeronautics and Space Administration, Manned Spacecraft Center,
Houston, Texas 77058

(*Received* 22 *February* 1971; *accepted in revised form* 7 *April* 1971)

Abstract—Studies of features produced by aluminum projectiles impacting silicate glass targets rich in calcium, sodium, and magnesium at 7 km/sec show that crater morphology varies as a function of target temperature. Pit depth below surface of the target, spall diameter, ratio of spall diameter to pit diameter, and volume of fractured material decrease with increasing target temperature from 500°C to 800°C. Pit diameter and the ratio of spall diameter to pit depth increase with temperature from 500°C to 800°C. Smooth, glass-lined pits with flowage features were formed at temperatures above 500°C. Hummocky craters were produced by trailing fragments of the sabot and sabot stopper at 700°C and 750°C. Glass spheres with surfaces showing a history of accumulation of metal vapors and attached angular glass fragments were produced at temperatures of 750°C and 800°C. The laboratory data suggest that the projectiles responsible for most of the craters on the lunar glass spheres were silicate fragments. These data favor the hypothesis that the silicate glass spheres and most of the craters on their surfaces are natural consequences of major meteoritic impacts rather than the craters resulting from a number of random isolated hypervelocity micrometeorite events. The impacting events recorded on the surfaces of the glass spheres thus occurred mainly within the debris cloud resulting from a large-scale meteorite impact.

INTRODUCTION

CARTER and MACGREGOR (1970a, b), FRONDEL *et al.* (1970a, b), McKAY *et al.* (1970a, b), and McKAY (1970) suggested from scanning electron microscopic examination of the surface of glass spheres from the lunar soil that some features (such as craters with central glassy depressions surrounded by splashed glassy rays) may have resulted from collision of the surface of the glass sphere with lunar-derived particles in a hot, impact-produced debris cloud. CARTER and MACGREGOR (1970a, b) suggested further that the physical state of both the projectiles and the targets varied from liquid to solid. TOLANSKY (1970) also suggested that the impacts occurred in a debris cloud and not while the particle rested on the lunar surface, or alternatively, that the sphere remained in orbit long enough for repeated cratering by micrometeorites to occur. If the cratering occurred in a hot impact-produced debris cloud, the majority of the projectiles must be lunar-derived materials and not primary micrometeorites.

---

† Contribution No. 173, Geosciences Division, The University of Texas at Dallas.

Table 1. Chemical composition of glass target in
wt.%.*

SiO_2	71.6
TiO_2	0.045
Al_2O_3	0.12
Fe_2O_3	0.30
MgO	3.08
CaO	10.8
Na_2O	13.1
K_2O	0.025
Total	99.07

* Atomic absorption analysist by K. V. Rodgers.

A survey project was initiated at NASA/MSC, Houston, employing a light gas (hydrogen) gun to examine the feasibility of creating smooth, glass-lined craters in glass targets at elevated temperatures. Because of limited availability of the gun facility, it was not possible to pursue a more systematic investigation of the effects of projectile mass, composition, and velocity, and the effects of target composition on crater morphology.

The study involved impacting 53 mm × 53 mm × 13 mm glass targets (composition listed in Table 1) with a 0.4 mm diameter (0.088 gm), type 2024 aluminum projectile at approximately 7 km/sec in vacuum (1×10^{-3} torr.). A single shot was made at a projectile velocity of 4.5 km/sec and a target temperature of 25°C. Other shots were made at 7 km/sec and target temperatures of 25°C, 500°C, 550°C, 600°C, 650°C, 700°C, 750°C, and 800°C. The target was heated with a nichrome wire furnace assemblage that was housed in a water-cooled stainless steel jacket. Stainless steel tubing was used as a collection tube for the ejecta, and a chromel-alumel type K thermocouple was used to monitor temperature. The aluminum projectile was launched in a Zelux sabot. The projectile velocity was monitored by high velocity shadow graph techniques.

CRATER MORPHOLOGY

Examination of the surface of the targets with a scanning electron microscope and a binocular light microscope revealed impacts produced by the aluminum projectile plus a wide variety of other materials that trailed the aluminum projectile. Craters produced by the aluminum projectile ranged from approximately 1.3 to 0.8 centimeters in diameter (Table 2). Craters resulting from other materials ranged in diameter from 600 μ to less than 1 μ.

Aluminum projectile craters

The crater produced by the impact of the aluminum projectile may be described broadly in terms of a pit surrounded by a remobilized glassy cylinder whose outer region grades into an intensely fractured zone. The next zone is characterized by less intense fracturing and a series of radiating fractures that terminate against sets of relatively low angle conical fractures extending to the surface of the target. The zone of the radiating fractures is contained within a zone of relatively high angle conical fractures that extend to the surface of the target (see HÖRZ *et al.*, 1971, for terminology used for impact-produced features on surfaces of the lunar rocks). Table 2 gives a general description of the features of the crater made by the aluminum projectile and craters made by other materials trailing the aluminum projectile together with measurements of the various parameters described previously. In addition, the depth of the pit below the target surface, the volume of fractured material, the ratio of the diameter of the spall zone to the diameter of the pit, and the ratio of the spall diameter to the pit depth are listed.

At 25°C and a projectile velocity of 4.5 km/sec only the inner sets of conical fractures spalled leaving a relatively deep hole surrounded by a shallow crater. The low angle conical fractures do not extend to the surface of the target (Fig. 1). As shown by Table 2, the morphology of the crater produced at 7 km/sec varies as a function of target temperature (i.e. target strength; MOORE *et al.*, 1965). At 25°C, and a projectile velocity of 7 km/sec, only the inner sets of conical fractures spalled leaving a relatively deep hole surrounded by a wide shallow crater that is itself surrounded by a much larger set of attached conical fractures (Fig. 2). The deep hole results from brittle fracture and subsequent target relaxation resulting in a central ejection of the crushed material. As target temperature is increased plastic deformation becomes more important.

The plot of spall diameter versus target temperature (Fig. 3) reveals an inverse correlation of spall diameter to target temperature from about 500°C to 800°C. The low value at 550°C results because the low angle conical fractures do not extend to the surface of the target. Fig. 4 shows a positive correlation of pit diameter to target temperature from 500°C to 800°C. The ratio of spall diameter to pit diameter shows an inverse correlation with temperature from 500°C to 800°C (Fig. 5). Unlike the diameter of the pit, which increases with temperature (Fig. 4), the outer diameter of the re-mobilized glassy cylinder remains essentially constant with temperature (Table 2). Also of interest is the relationship of pit depth below the surface of the target and target temperature (Fig. 6). There is an inverse correlation of depth of the pit to temperature of the target from 500°C to 800°C. These data suggest that at a target temperature of approximately 850°C the projectile will penetrate essentially only the diameter of the projectile (0.4 mm). A simple calculation reveals that at a target temperature of 850°C there would be a ratio of approximately 1 to 16 of volume of projectile to the solid volume given by pit depth and outside diameter of cylinder, and at 500°C it would be 1 to 47. Thus with complete mixing of projectile material and remobilized cylinder material the percentage of projectile material to cylinder material would increase from approximately 2% at a target temperature of 500°C to approximately 6% at a target temperature of 850°C. However there is probably non-random mixing of remobilized target material and projectile material, and a certain amount of both the target and projectile are vaporized (GAULT *et al.*, 1968). Also, the major portion of the mixing may occur near the point of maximum penetration resulting in areas considerably higher in projectile content than would result from homogeneous mixing.

The craters produced at a target temperature of 750°C and 800°C are two examples of pits showing nonrandom distribution of projectile materials (Figs. 7 to 12). Fig. 7

Table 2. Description and physical parameters of craters formed by

$T°C$	Velocity km/sec	Primary projectile spall diameter (mm)	Primary projectile Pit diameter (mm)	Ratio of spall diameter to pit diameter	Primary projectile Pit depth (mm)
25	4.5	7	0.6	11.7	0.92
25	7.10	10	0.6	17	1.18
500	7	12.5	0.5	24.5	1.18
550	7	11	0.7	15.6	1.02
600	7	12.4	0.6	17.1	1.05
650	7.34	11.2	0.9	12.5	0.84
700	7	10	1.0	10	0.70

the aluminum projectile and fragments of the sabot and sabot stopper

Ratio of Spall diameter to Pit depth	Fractured volume (mm³)	Description of the primary projectile crater	Material found in projectile crater
7.6	12	No evidence of plastic flow in pit. Central pit and walls of pit are highly fractured. Pit surrounded by conchoidally fractured area containing radial fractures that extend to a radius of 1.5 mm. Low-angle conical fractures do not extend to the surface	None
8.3	31	Central glassy cylinder spalled leaving a conical depression that is 2 mm wide at the top and 1.3 mm wide at the bottom. No clear evidence of plastic flow. Central part and walls of pit are highly fractured. Pit surrounded by conchoidally fractured area that extends out to a radius of 1.75 mm left by spallation of cylinder. Low-angle fractures mostly do not extend to the surface. Most low-angle spall fragments are attached.	None
10.4	53	Central glassy cylinder 1.3 mm in outside diameter is almost completely spalled. Pit shows some outward flowage of material. A large well-developed bowl-shaped spall area surrounds the pit and extends out to a radius of 3.25 mm to the low-angle fractures	Remobilized glassy material in bottom of pit
11.1	35	Central glassy cylinder 1.3 mm in outside diameter is about half spalled. Outward flowage of material in center of pit and along walls. Radial fractures poorly developed and extend to a radius of 2.25 mm. Radial fracture zone and glassy cylinder are intensely fractured. Low-angle fractures are attached. Low-angle conical fractures do not extend to the surface.	Remobilized glassy material and some glassy protuberances in pit. A few aluminum mounds are present in pit.
11.6	46.5	Central glassy cylinder 1.3 mm in outside diameter is mostly spalled. Central pit showing some outward flowage of material surrounded by brittle fracture envelope that extends out to 4.25 mm with well-developed radial fractures.	Glassy protuberances occur on the inner walls of the glassy cylinder and in the pit. Some aluminum mounds are present in pit.
12.8	24.5	Central glassy cylinder 1.3 mm in outside diameter shows moderate outward flowage of material. Radial fractures extend to a radius of 3.1 mm. Part of low-angle spall fragments attached.	Protuberance of remobilized glass in pit. Some aluminum mounds present.
14.3	20.5	Central glassy cylinder 1.3 mm in outside diameter shows considerable outward flowage of material. Well-developed radial fracture zone extends to a radius of 3 mm. Low-angle fractures spalled.	Glassy beads and protuberances occur on inner wall of cylinder and on pit. Aluminum mounds present on pit and inner wall of cylinder.

Table 2

$T^{\circ}C$	Velocity km/sec	Primary projectile spall diameter (mm)	Primary projectile Pit diameter (mm)	Ratio of spall diameter to Pit diameter	Primary projectile Pit depth (mm)
750	7.22	9.5	1.15	8.7	0.63
800	7	8.5	1.10	7.7	0.50

$T^{\circ}C$	Description of other craters	Material found in other craters
25	Craters range in diameter from 300 to 1200 μ. Entire set of conical fractures spalled, leaving a bowl-shaped pit with a central highly fractured area. Ratio of spall diameter to pit diameter varies from 3 to 4, increasing with crater diameter.	None
25	Craters range in diameter from 600 to 800 μ. Similar to aluminum projectile crater except ratio of spall diameter to pit diameter varies from 2.7 to 4.2.	None
500	Craters range in diameter from 100 to 500 μ. Similar to aluminum projectile crater except ratio of spall diameter to pit diameter varies from 6 to 18 increasing with increasing crater diameter. Low-angle conical fractures spalled.	Some contain iron mounds and dimples. Others are lined by sabot fragments showing considerable outward flowage of well developed glassy cylinder.
550	Craters range in diameter from 10 to 500 μ. Similar to aluminum projectile crater except ratio of spall diameter to pit diameter varies from 4 to 9. No clear relationship with size of crater. Well-developed glassy cylinder shows considerable outward flowage of material. Low angle conical fractures spalled. Some splashes of low viscosity material showing well-developed rayed features.	Most contain iron mounds.
600	Craters range in diameter from 50 to 300 μ. Similar to aluminum projectile crater except spall diameter to pit diameter varies from 5 to 10 increasing with increasing crater diameter. Well-developed glassy cylinder shows considerable outward flowage of material. Some splashes of low viscosity material showing well developed rayed features.	Iron mounds well developed in some craters.
650	Two craters formed by the two halves of the sabot are similar to the crater formed by the aluminum projectile. Two other craters are 1500 and 2500 m in diameter with spall diameter to pit diameter ratio of 8. The small craters have well developed glassy cylinders.	Iron mounds present in the two smaller craters.
700	Craters range in diameter from 10 to 600 μ. Similar to the crater formed by the aluminum projectile except spall diameter to pit diameter is approximately 7. Glassy cylinder is often missing.	Iron mounds present in some craters.

(*continued*)

Ratio of spall diameter to Pit depth	Fractured volume (mm³)	Description of the primary projectile crater	Material found in projectile crater
15.1	19	Central glassy cylinder 1.3 mm in outside diameter shows considerable outward flowage of material in bottom of pit and along cylinder wall. Well-developed radial fracture zone extends to a radius of 3.25 mm. Low-angle spall fragments mostly attached.	Protuberances and beads of glass on bottom of pit and on inner wall of cylinder. Numerous mounds of aluminum present on pit and inner wall of cylinder.
16.9	10	Central glassy cylinder 1.3 mm in outside diameter shows considerable outward flowage of material in-bottom of pit and along cylinder wall. Well-developed radial fracture zone extends to a radius of 3.5 mm. Low-angle spall fragment and cylinder spalled as single unit.	Protuberances and beads of glass on bottom of pit and on inner wall of cylinder. Considerable slumping of surface of cylinder. Numerous mounds of aluminum present on pit and inner wall of cylinder.

$T°C$	Description of other craters	Material found in other craters
750	Craters range in diameter from less than 1 to 600 μ. Similar to crater formed by the aluminum projectile. In one of the two craters formed by the sabot, the remobilized glassy cylinder is turned perpendicular to the target surface and attached to the pit. Spall diameter to pit diameter varies from 3 to 17, increasing with larger craters. Low-angle spall fragments sometimes attached.	Numerous attached glass beads and protuberances. Iron mounds present in some craters.
800	Craters range in diameter from 70 to 100 μ. Projectile is buried beneath surface of target. Attached conical fractures are upturned and their edges are rounded.	Glass beads are attached to pit.

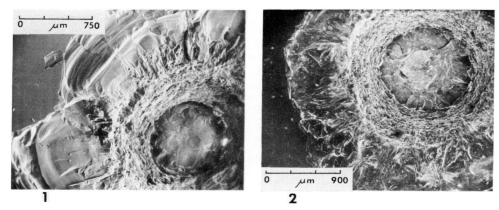

Fig. 1. Impact feature produced at a target temperature of 25°C and an aluminum projectile velocity of 4.5 km/sec. Fig. 2. Impact feature produced at a target temperature of 25°C and an aluminum projectile velocity of 7.1 km/sec.

shows the central region of the crater produced at a target temperature of 750°C. The photograph shows considerable flowage of material in the bottom of the pit. Individual beads of glass are present. The remobilized glass cylinder was attached to the bottom of the pit by the larger protuberances. Fig. 8 is a view of the glass cylinder that was removed from the crater for scanning electron microscopic examination. Considerable outward flowage of material is seen on the margin of the cylinder and trains of individual mounds and globules of the aluminum projectile are seen on the interior surface of the cylinder (Fig. 9). At 800°C a cylinder of remobilized glass showing considerable outward flowage of material was formed that displayed well-developed mounds and trains of aluminum, as well as attached beads of glass (Fig. 10). Fig. 11 is an enlarged view of the lower portion of the lip of the cylinder showing attached glass spheres and aluminum mounds. Fig. 12 is an enlarged view of the attached beads of glass in the central part of Fig. 11. The surface of the glass spheres show a history of accumulation and collision with aluminum vapor and angular fragments of glass. McKAY et al. (1970a, b), CARTER and MACGREGOR (1970a, b), and CARTER (1971a, b) have described surfaces of glass spheres from the lunar soil that show a complex history of accumulation of vapor deposition, and collision and attachment of particles similar to that shown in Figs. 11 and 12.

There is a positive correlation of the ratio of spall diameter to depth of pit below the surface of the target and temperature of the target (Fig. 13). The plot of the estimated volume of fractured material defined by the limits of the conical fractures versus temperature is shown by Fig. 14. The volume appears to increase from a target temperature of 25°C to 500°C and then decrease sharply to 800°C. The relationship from 500°C to 800°C is in contrast to the findings of MOORE et al. (1965) on the effect of target strength on spall diameter and excavated crater volume in basalt and metallic iron targets. Upon heating the glass target the mean deformation strength of the target should decrease, hence mass ejected from the crater should increase. The deformation strength of the glass target should be lowered because the confining pressure is lowered and the strain rate increased. Consequently, the compressive strength of a hot glass target may decrease faster than the tensile strength thus allowing hot glass to behave similarly to metal or water. However, as shown clearly by Figs. 3, 6, and 14, the spall diameter, pit depth, and fracture volume decrease with increasing target temperature from 500°C to 800°C. Since the amount of energy supplied by the projectile is relatively constant ($2.25 \pm 0.09 \times 10^7$ erg) at all target temperatures and

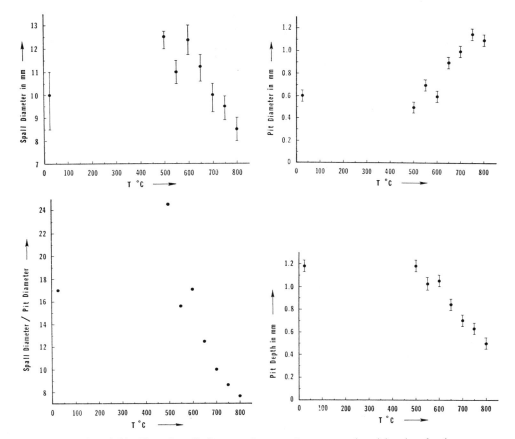

Fig. 3 (top left). Plot of spall diameter in mm of crater produced by the aluminum projectile and temperature of the glass target. There is an inverse correlation of spall diameter and target temperature. Fig. 4 (top right). Plot of pit diameter in mm produced by the aluminum projectile and temperature of the glass target. There is a positive correlation of pit diameter to target temperature. Fig. 5 (bottom left). Plot of the ratio of spall diameter to pit diameter and temperature of the glass target. There is an inverse correlation of this ratio to target temperature. Fig. 6 (bottom right). Plot of depth of pit in mm below the surface of the glass target produced by the aluminum projectile and temperature of the glass target. There is an inverse correlation of pit depth and target temperature.

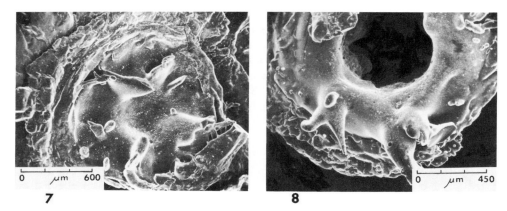

Fig. 7. Central portion of crater produced by the aluminum projectile at a target temperature of 750°C. Large protuberances are glass bridges that connected the remobilized glass cylinder to the pit. Small white dots are aluminum mounds. Fig. 8. Remobilized glass cylinder showing considerable outward movement of material produced by the aluminum projectile at a target temperature of 750°C. Individual mounds and trains of aluminum mounds are present on the interior surface (lower left) of the cylinder.

since the outer diameter of the remobilized glassy cylinder is essentially constant (Table 2), much of the energy is dissipated by plastic deformation, and, as shown by the decrease in cylinder wall thickness with increasing temperature (Fig. 4, Table 2), some of it occurs as outward flowage of the inner wall of the glass cylinder. Plastic deformation may also extend beyond the confines of the crater (TOLANSKY, 1971), and brittle fracture is present beneath the pit at all target temperatures.

Other craters

Some of the more interesting features observed on the surface of the targets were produced by fragments of the sabot and sabot stopper (see Table 2 for general description of these crater types). The velocity of these particles is unknown but as shown by GAULT *et al.* (1968) secondary particles produced when the projectile strikes a surface may have a wide range of velocities including velocities greater than the impacting projectile. The two halves of the sabot would travel at a lower velocity than the aluminum projectile (7 km/sec) because of gas drag. At 25°C, fragments of the sabot (aluminum projectile velocity was 4.5 km/sec) produced craters with the entire set of conical fractures spalled leaving a bowl-shaped pit with a central highly fractured area (Fig. 15). At 550°C some of the craters showed all the major sets of conical fractures attached with only the innermost small set spalled (Fig. 16). Fig. 17 shows the well-developed central glassy cylinder of a crater formed at 600°C by an iron fragment probably resulting from the sabot hitting the cast iron sabot stopper. The crater lies on the edge of the crater formed by the aluminum projectile. The inner surface and the bottom of the glassy cylinder are covered with spongy iron mounds. Dimples are also present in the bottom of the crater shown in Fig. 17.

At 700°C and higher, the edges of the conical fractures were often upturned. At 700°C, a crater resulting from the impact of an iron fragment displayed a central glassy hummocky depression covered with spongy iron mounds (Fig. 18). Fig. 19 is a view of the central glassy depression of another crater resulting from the impact of a fragment of iron. Target temperature was 750°C. Mounds and trains of spongy iron are present in the central glassy depression (Fig. 20). Iron concentrations beneath the surface are seen as faint light circular areas. The small deep central depression is filled with silica spherules (Figs. 20, 21). Vapor deposition of silica-rich material on the silica spherules is seen as white spots from 0.5 to less than 0.05 μ in diameter (Fig. 21). Since the target is a sodium-rich glass, this vapor-deposited material may be rich in sodium because it is the last material to be deposited from the vapor phase

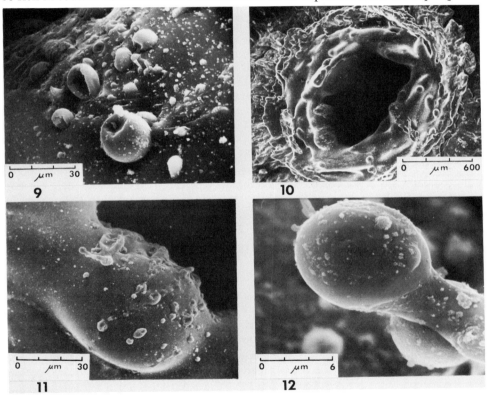

Fig. 9. Enlarged view of some aluminum mounds present on the interior surface of the cylinder shown in Fig. 8. Fig. 10. Remobilized glass cylinder showing considerable outward movement and slumping of material produced by the aluminum projectile at a target temperature of 800°C. Individual mounds and trains of aluminum mounds are present on the interior surface (center; lower left) of the cylinder. Fig. 11. Enlarged view of lower left portion of remobilized cylinder rim shown in Fig. 10. Mounds with indentations are droplets of aluminum. Glass spherules are attached to the glass stalk in the upper left portion of the photograph. Fig. 12. Enlarged view of glass spherules attached to the glass stalk in the upper left portion of Fig. 11. Numerous mounds of aluminum and embedded angular fragments of glass are seen on the glass spherules.

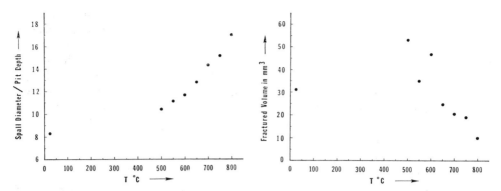

Fig. 13 (left). Plot of spall diameter to pit depth and temperature of the glass target. This ratio shows a positive correlation with temperature. Fig. 14 (right). Plot of estimated fracture volume in mm³ as given by the limits of the conical fractures and temperature of the glass target. There is an inverse correlation of fracture volume to target temperature from 500°C to 800°C.

and thus has the lowest vaporization point of the materials seen in the bottom of the depression.

A glass sphere attached to the side of this crater (Fig. 19) is shown enlarged in Fig. 22. The surface of the glass sphere shows a history of accumulation and collision with iron vapors, silicate liquids, and solid glass particles. Angular fragments of glass are embedded in the surface of the sphere and wormy objects of silica are present in the lower portion of the photograph. Spongy iron mounds are present mainly on the upper portion of the sphere which is the side of the glass sphere facing the central glassy depression. Also present on the central upper surface of the sphere are faint scrape marks and pits.

Two examples of craters produced on another glass target at a target temperature of 750°C by the two halves of the sabot are shown in Figs. 23 and 24. One crater displayed a doughnut-shaped glassy cylinder that has attached fragments of crushed glass around its periphery and attached beads of glass on its interior (Fig. 23). In the other example a portion of the remobilized glassy cylinder was turned perpendicular to the bottom of the crater and attached to the pit (Fig. 24). At 800°C the trailing fragmental sabot material was buried beneath the surface of the target leaving an attached conical fracture feature (Fig. 25). The edges of the attached upturned conical fractures were softened and rounded by the high temperature. Splashed rays were seen on only one crater 10 μ in diameter at 750°C. However, splashed rayed features of fragmental sabot material were sometimes seen (Fig. 26).

Discussion and Conclusions

This study has demonstrated the feasibility of producing smooth craters with flowage features in glass targets at moderate velocities when the target glass is at elevated temperatures. Additional experiments should be made to determine fully the effect on crater morphology of other parameters such as projectile mass, composition, velocity, and target composition. Data presented in this paper show that the target material surrounding the path of the projectile is crushed in a layered radiating pattern (Figs. 1, 2, 15–19, 23, 24) and a small amount of material immediately surrounding the projectile is fused (Figs. 7–11, 17–19, 23, 24). Target relaxation results in removal of this crushed material, layer by layer, in a cold brittle target. However, with increasing target temperature a larger portion of the area immediately surrounding the penetrating projectile is more readily mobilized and the cylindrical glass tube may remain attached to the target (Figs. 7–11, 17–19, 23, 24). This results

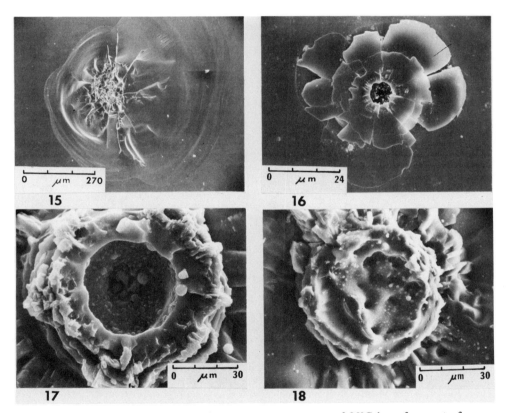

Fig. 15. Impact feature produced at a target temperature of 25°C by a fragment of the sabot. Aluminum projectile velocity was 4.5 km/sec. Fig. 16. Impact feature produced by an iron fragment at a target temperature of 550°C. Conical fractures are attached to glass-lines pit. Fig. 17. Enlarged view of the central glassy cylinder produced by an iron fragment at a target temperature of 600°C. The inner surface and bottom of the glassy cylinder are covered with spongy iron mounds. Dimples are also seen in the bottom of the glass-lined pit. Fig. 18. Enlarged view of the hummocky central glass-lined pit produced by an iron fragment at a target temperature of 700°C. The surface of the hummocky glass-lined pit is covered by iron mounds.

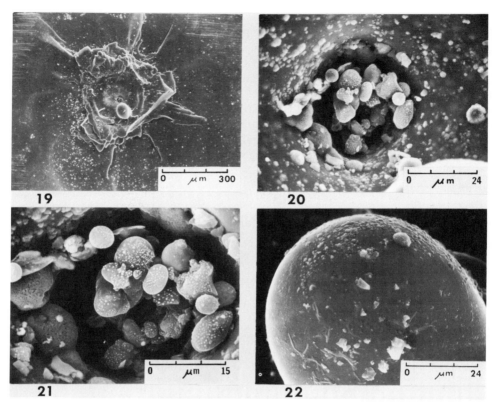

Fig. 19. Impact feature produced by an iron fragment at a target temperature of 750°C. A large glass sphere is attached to the lower right portion of the glass-lined pit. Fig. 20. Enlarged view of the central glass-lined pit shown in Fig. 18. Mounds and trains of spongy iron are seen on the surface of the central glassy depression. Iron concentrations beneath the surface are seen as faint light circular areas. The small deep central depression is filled with silica spherules. Vapor deposition of the silica-rich material as white spots from 0.5 to less than 0.05 μ in diameter is seen on the silica spherules. Fig. 21. Enlarged view of the small deep central depression shown in Fig. 19. Spongy iron mounds are attached to the wall of the depression. The bottom of the depression is filled with silica spherules. Vapor deposition of silica-rich material as white spots from 0.5 to less than 0.05 μ in diameter is seen on the silica spherules. Fig. 22. Enlarged view of the glass sphere attached to the lower right part of the crater shown by Fig. 19. The surface of the glass sphere shows a history of accumulation and collision with iron vapors, silicate liquids, and solid glass particles. Angular fragments of glass are embedded in the surface of the sphere and wormy objects of silica are present in the lower portion of the photograph. Spongy iron mounds are present mainly on the upper portion of the sphere. Faint scrape marks and pits are present on the central upper portion of the sphere.

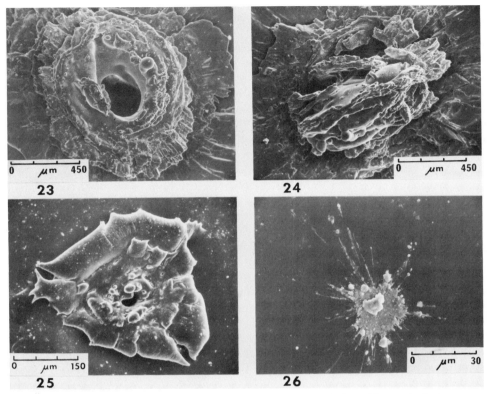

Fig. 23. Enlarged view of the Remobilized glass-lined cylinder with attached glass spherules produced by the sabot at a target temperature of 750°C. Fig. 24. Enlarged view of an impact feature produced by the sabot at a target temperature of 750°C. Remobilized glassy cylinder turned perpendicular to impact point and attached to pit. Fig. 25. Impact feature produced by a fragment of the sabot at a target temperature of 800°C. Projectile buried beneath surface of target. Edges of conical fractures are upturned and rounded. Fig. 26. Splash of low viscosity material at a target temperature of 550°C. Note well-developed splashed rays.

because of increased plastic deformation which prevents complete spallation of the cylindrical glass tube.

Similar features are described by CARTER and MACGREGOR (1970a, b), MCKAY *et al.* (1970a, b), MCKAY (1970), and CARTER (1971a, b) on the surfaces of glass spheres from the lunar soil. Experimental data show that smooth glass-lined pits are not produced by projectiles impacting at velocities less than about 10 km/sec (HÖRZ *et al.*, 1971, from the data of VEDDER). These experimental data apply only to cold targets, however, and it is clear from the data at 7 km/sec in heated targets that these results cannot be applied when the target is impacted at elevated temperatures. Temperature must clearly be considered as an important parameter along with velocity and projectile mass.

Since approximately 1 cm diameter craters with glass flowage features were produced at 7 km/sec at elevated target temperatures, it seems reasonable to infer from these data, the data of Bloch et al. (1971), and the data of Gault and Moore (1965) that craters less than 1 cm in diameter produced on heated targets should show plastic deformation at velocities less than 7 km/sec. Bloch et al.'s data show that small craters undergo plastic deformation at lower velocities than large craters. Thus a crater on the order of microns or tens of microns in diameter might be expected to show plastic flowage at impact velocities appreciably lower than 7 km/sec in heated targets. Experiments are underway to test this suggestion.

Glass-lined pits closely resembling the hummocky craters found by Carter and MacGregor (1970b, Fig. 16) on the lunar materials, were formed experimentally at 700°C and 750°C by fragments of the sabot and sabot stopper. Doughnut-shaped glassy fragments have been found in the lunar soil by one of us (JLC) and craters resembling Figs. 15 and 23 have been found on the lunar materials (Carter and MacGregor, 1970b, Figs. 12 and 15).

The apparent absence of splashed glassy rays surrounding the laboratory produced glass-lined pits suggests that in order to form central glassy depressions surrounded by splashed glassy rays, either the velocity of the projectile is greater than 7 km/sec or other parameters such as mass of the projectile, chemical composition of the glass target, chemical composition of the projectile, temperature (viscosity) of the projectile, geometry of the target, gravity, and presence of an atmosphere must be considered (McKay et al., 1971a).

Figs. 11, 12, 19–22 demonstrate the validity of the hypothesis that impact is a possible mechanism for the formation of the lunar glass spheres, the resulting vapor deposition and subsequent growth of iron mounds, and the collision and attachment of small particles to the surface of the lunar glass spheres. Also of possible importance to the interpretation of some iron bodies found inside lunar glass spheres (Agrell et al., 1970; Ramdohr and El Goresey, 1970) is the presence of iron bodies beneath the surface of the glassy central depressions formed as a result of mixing of molten globules of the iron projectile with the fused portion of the target (Fig. 20).

However, the absence of large concentrations of iron mounds arranged in patterns in the bottom of the glass-lined pits on the lunar materials suggests that iron particles were not the principal projectile material forming the small lunar craters (Carter, 1971a, b). On the other hand, the lack of major chemical differences between the glassy central depression, the splashed glassy rays, and the host glass sphere (Carter and MacGregor, 1970a, b; Bloch et al., 1971) suggests that the projectiles responsible for the craters on the lunar glass spheres were silicate fragments. These data and the presence of lunar composition glass in the pits of some craters on the "mini-moon" (Goldstein et al., 1970) suggest that lunar-derived silicate fragments may be mainly responsible for these craters.

Conclusive support for the interpretation of Carter and MacGregor (1970a, b) that the surface of the glass spheres displayed a wide variation in viscosity to the impacting projectiles is given by interferometric examination of glass spheres by Tolansky (1971). He showed a smooth-surfaced sphere (type II) that had been impacted by a small particle while the glass sphere was still slightly plastic and not yet frozen hard, resulting in distortion that extended to the center of the sphere.

These data together with the nonrandom distribution of craters (Carter, 1971a, b);

the almost complete absence of craters larger than about 10 μ in diameter on the type I silicate glass spheres (CARTER and MACGREGOR, 1970a, b; MCKAY et al., 1970a, b; MCKAY, 1970; CARTER, 1971a, b), the apparent absence of craters larger than about 1 μ in diameter on the glass-bonded agglutinates (MCKAY et al., 1971b; CARTER, 1971b), the almost complete absence of craters on the welded dust covered glassy objects (CARTER, 1971a, b), the variation in crater morphology that appears to be related to target strength (CARTER and MACGREGOR, 1970a, b) and the apparent restriction of crater formation to a distinct period of the glass sphere history (CARTER and MACGREGOR, 1970b; CARTER, 1971a, b), favor the hypothesis that the impact-produced smooth glass pits on the surface of glass spheres are natural consequences of major meteoritic impacts rather than a number of smaller isolated random hypervelocity events. Thus, most of the impacting events recorded on the surfaces of the glass spheres may have been formed within the debris cloud resulting from a meteorite impact. Consequently caution should be used when speculating on the abundance of micrometeorite projectiles that could be responsible for the impact-produced craters on glass spheres and on the exposure age of the glass spheres from crater counts.

This discussion applies only to the craters found on lunar glass spheres and related forms found in the soil. We have not considered the larger craters on the surfaces of rocks, and the results of our experiments would apply to the rocks only if it could be shown that the rocks were at elevated temperatures when impacted. We do not question the conclusions of HÖRZ et al. (1971) that these larger craters are mainly formed by extralunar primary micrometeorite projectiles.

In conclusion, this study has demonstrated the feasibility of making smooth, glass-lined craters with flowage features in glass targets at moderate velocities when the target is at elevated temperatures. These results indicate that smooth craters with glass flowage features on lunar glass spheres may be formed at moderate velocities providing that the glass sphere was at an elevated temperature at the time of impact. A crater on the order of microns or tens of microns in diameter might be expected to show plastic flowage at relatively low impact velocities on heated targets. This fact, when considered with independent evidence that lunar glass spheres were in many cases hot when impacted, and evidence of the nonrandom nature of impacts on small lunar particles is suggestive that some or even most craters on lunar glass spheres result from impacts of secondary particles in the debris cloud generated by a larger primary impact on the lunar surface.

Acknowledgments—Supported by NASA Contract NAS 9-10221 and NASA Grant NGL-44-004-001. We thank the Lunar Science Institute and NASA/MSC Houston for their support of the impact study. We acknowledge the help of BURTON G. COUR-PALAIS, who kindly provided and assisted with the light gas gun facility. We thank K. V. RODGERS, who performed the atomic absorption analysis of the glass target, and D. HEDER, E. PADOVANI, J. TONEY, and A. WHITE for technical assistance. Critical review by C. FRONDEL, C. HELSLEY, F. HÖRZ, and D. PRESNALL has contributed to the improvement of the manuscript.

REFERENCES

AGRELL S. O., SCOON J. H., MUIR I. D., LONG J. V. P., MCCONNELL J. D. C., and PECKETT A. (1970) Mineralogy and petrology of some lunar samples. *Science* **167**, 583–586.

Bloch M., Fechtig H., Funkhouser J., Gentner W., Jessberger E., Kirsten T., Muller D., Neukum G., Schneider E., Steinbrunn F., and Zähringer J. (1971) Meteorite impact craters, crater simulations, and the meteoritic flux in the early solar system. Second Lunar Science Conference (unpublished proceedings).

Carter J. L., MacGregor I. D. (1970a) Mineralogy, petrology, and surface features of lunar samples 10062,35; 10067,9; 10069,30; and 10085,16. *Science* **167**, 661–663.

Carter J. L. and MacGregor I. D. (1970b) Mineralogy, petrology, and surface features of some Apollo 11 samples. *Proc. Apollo 11 Lunar Sci. Conf., Geochim. Cosmochim. Acta* Suppl. 1, Vol. 1, pp. 247–275. Pergamon.

Carter J. L. (1971a) Chemistry and surface morphology of fragments from Apollo 12 soil and laboratory produced craters. Second Lunar Science Conference (unpublished proceedings).

Carter J. L. (1971b) Chemistry and surface morphology of fragments from Apollo 12 soil. Second Lunar Science Conference (unpublished proceedings).

Frondel C., Klein C., Jr., Ito J., and Drake J. C. (1970a) Mineralogy and composition of lunar fines and selected rocks. *Science* **167**, 681–683.

Frondel C., Klein C., Jr., Ito J., and Drake J. C. (1970b) Mineralogical and chemical studies of Apollo 11 lunar fines and selected rocks. *Proc. Apollo 11 Lunar Sci. Conf., Geochim. Cosmochim. Acta* Suppl. 1, Vol. 1, pp. 445–474. Pergamon.

Frondel C., Klein C., and Ito J. (1971) Mineralogical and chemical data on Apollo 12 lunar fines. Second Lunar Science Conference (unpublished proceedings).

Gault D. E. and Moore H. J. (1965) Scaling relationships for microscale to megascale impact craters. *Proc. 7th Hypervelocity Impact Symposium*, Tampa Fla., 1964, Vol. 1, pp. 341–351.

Gault D. E., Quaide W. L., and Oberbeck V. R. (1968) Impacting cratering mechanics and structures. In *Shock Metamorphism of Natural Material* (editors B. French and N. Short), pp. 87–99. Mono.

Goldstein J. I., Henderson E. P., and Yakowitz H. (1970) Investigation of lunar metal particles. *Proc. Apollo 11 Lunar Sci. Conf., Geochim. Cosmochim. Acta* Suppl. 1, Vol. 1, pp. 499–512. Pergamon.

Hörz F., Hartung J. B., and Gault D. E. (1971) Lunar microcraters. Second Lunar Science Conference (unpublished proceedings).

McKay D. S., Greenwood W. R., and Morrison D. A. (1970a) Morphology and related chemistry of small lunar particles from Tranquility Base. *Science* **167**, 654–656.

McKay D. S., Greenwood W. R., and Morrison D. A. (1970b) Origin of small lunar particles and breccia from the Apollo 11 site. *Proc. Apollo 11 Lunar Sci. Conf., Geochim. Cosmochim. Acta* Suppl. 1, Vol. 1, pp. 673–694. Pergamon.

McKay D. S. (1970) Microcraters in lunar samples. *Proc. Twenty-Eighth Annual Meeting Electron Microscopy Society of America* (editor C. J. Archeneaux), pp. 22–23. Claitors.

McKay D. S., Carter J. L., and Greenwood W. R. (1971a) Lunar metallic particle ("mini-moon"): An interpretation. *Science* **171**, 479–480.

McKay D. S., Morrison D. A., Lindsay J., and Ladle G. (1971b) Apollo 12 soil and breccia. Second Lunar Science Conference (unpublished proceedings).

Moore H. J., Gault D. E., and Heitowit E. D. (1965) Change of effective target strength with increasing size of hypervelocity impact craters. *Proc. 7th Symposium on Hypervelocity Impact*, Tampa, Fla., 1964, Vol. IV, pp. 35–45.

Ramdohr P., and El Goresey A. (1970) Opaque minerals of the lunar rocks and dust from Mare Tranquillitatis. *Science*, **167**, 615–618.

Tolansky S. (1970) Interferometric examination of small glassy spherules and related objects in a 5-gram lunar dust sample. *Science* **167**, 742–743.

Tolansky S. (1971) Interferometric examination of Apollo 12 lunar glass sperules. Second Lunar Science Conference (unpublished proceedings).

Proceedings of the Second Lunar Science Conference, Vol. 3, pp. 2671–2674
The M.I.T. Press, 1971.

Search for stable, fractionally charged particles (quarks) in lunar material

C. M. Stevens, J. P. Schiffer, and W. A. Chupka
Argonne National Laboratory, Argonne, Illinois 60439

(Received 19 February 1971; accepted in revised form 18 March 1971)

Abstract—Quarks are the postulated basic constituents of elementary particles. They would have charges of $\frac{1}{3}e$ or $\frac{2}{3}e$. Should free quarks indeed exist, they might well be stable. They could be produced by high-energy cosmic rays incident on the lunar surface. Because of very low chemical and geological activity on the moon, any such quarks should remain near the surface. Concentrations of $\leqslant 10^6$ quarks/gram appear possible on the basis of the experimental limit on direct measurements of quark flux on the Earth's surface. A search on 0.6 g of Apollo 11 and Apollo 12 fines has revealed no evidence for quarks, with a limiting concentration of 10^{-18}/nucleon ($= 5 \times 10^5$/gram).

INTRODUCTION

SINCE THE PROPOSAL by GELL-MANN (1964) of a truly elementary particle, which he named the quark, the quark-model of the known elementary particles has had considerable success (MURPURGO, 1970). Elementary-particle theorists think in terms of such a model but at the same time have some reservations about the existence of free quarks. Conclusive experimental proof is lacking. One of the intriguing properties of quarks is their fractional charge. Thus if they could be produced as free particles they would presumably remain fractionally charged forever. A positive quark would bind an electron, while a negative quark would be captured by a nucleus. In any case a fractional net charge would remain, and the Coulomb repulsion between positive quarks and positively charged nuclei would make recombination of quarks unlikely at anything less than stellar temperatures and pressures.

PAST QUARK SEARCHES

Accelerator searches

A number of experiments have made use of beams of protons with existing high-energy accelerators, the latest one at the 70-GeV Serpukhov accelerator (ANTIPOV et al., 1969). No evidence of fractional charge has been found to date.

Cosmic-ray experiments

A number of searches for quarks in cosmic rays have been carried out—generally with negative results. Two experiments reported possible observation of fractionally charged particles in cosmic radiation (CAIRNS et al., 1969; CHU et al., 1970), but those have raised considerable doubt (ADAIR and KASHA, 1969).

Chemical searches for stable quarks

A number of chemical experiments have been tried. Some depend on mass-spectrometric techniques, utilizing the permanent fractional charge on the quark or quarkic atom (Chupka *et al.*, 1966; Elbert *et al.*, 1970; Cook *et al.*, 1969; and others), and some on variations of the Millikan oil-drop experiment (Gallinaro and Morpurgo, 1969; Chupka *et al.*, 1966; and others). All of these results have been negative. Some searches have also been carried out in meteorites, which are probably too small to stop cosmic-ray quarks.

Advantages of the Lunar Surface as a Potential Medium for Quarks

If one postulates stable free quarks, it is difficult to escape the question as to why they do not occur in nature ab ovo in the same way as all known chemical elements do. Estimates of burnup in stellar interiors indicate that the abundance would have been reduced to anywhere between 10^{-9} and 10^{-18}. It is conceivable that a very peculiar element in such a concentration could have escaped chemical detection on earth.

The big uncertainty in chemical searches for quarks in terrestrial media is that the steady geological and chemical processes in the Earth's surface would easily remove quarks from the original surface strata and deposit them in unpredictable places. In the lunar surface such chemical uncertainties would be minimal. Once deposited by cosmic radiation, a quark would tend to remain near the surface. The depth to which quarks may penetrate by virtue of their initial energy will depend somewhat on their mass which, of course, is unknown. A reasonable estimate may put their range at a few kg/cm^2 or of the order of 10 m. Cairns *et al.* (1969) value of 5×10^{-6} quarks per m^2-sec-sr with an accumulation time of 10^9 years would yield a quark concentration of $10^5/cc$.

Preliminary Measurements

A preliminary search for fractionally charged particles was made in samples of Apollo 11 fines 10084 (0.07 and 0.30 g) and Apollo 12 fines 12070 (0.25 g) by heating samples in a platinum crucible to 1000°C and examining the negative ion spectrum of the effusing vapors in the mass range $A = 16-81$ with the Argonne 100-inch-radius mass spectrometer. The ion source and analyzer sections were believed to transmit nearly 100% of the ions effusing from the crucible orifice (diameter $= \frac{1}{18}$ inch). The collector slit was made $\frac{3}{8}$ inch wide to give a high detection efficiency while scanning and permit rapid scanning at low resolution (250). The ions were detected by an electron multiplier operated in both pulse-counting and current-integration modes. A 400-channel analyzer synchronized to the magnetic-field sweep was used for data collection.

In addition to the analyzes of untreated lunar fines, one sample was chemically treated to isolate the alkali fraction in 0.093 g of Apollo 11 10084 and analyzed. The alkali metals Li, Na, and K were isolated from the lunar sample by the following procedure. (1) The sample was decomposed by heating with a mixture of hydrofluoric and perchloric acids. Silicon was volatized as SiF_4. (2) Most of the transition metals were removed from the alkali and alkaline earth metal ions by liquid-liquid extraction using trioctyl methyl ammonium chloride. (3) The alkali metal ions were then

Table 1. Negative ions observed from direct heating of lunar soil.

Mass	Probable ion	Maximum intensity ions/sec	Temperature °C
16*	O	10^5	700–1000
17*	O, OH	10^3	700–1000
18*	O	10^3	700–1000
19*	F	$>10^4$	700–1000
23*	Na	10^2	900–1000
24*	?	10^4	700–1000
25*	?	10^3	1000
26*	CN, BO	10^4	700–1000
27*	CN, BO	10^4	700–1000
28*	CN	10^2	700–1000
29*	?	10^3	700–1000
29*	?	10^3	700–1000
30	?	10^2	
31	?	10^2	700–1000
32*	S, O_2	10^7	700–1000
33*	S	10^5	700–1000
34*	S	10^6	700–1000
35*	Cl	10^5	700–1000
36*	S	10^3	1000
37*	Cl	10^5	700–1000
39*	K	10^2	1000
40*	?	10^2	900
41*	K	10^1	900
42*	BO_2	10^4	700–1000
43*	BO_2	10^5	700–1000
44*	BO_2	10^3	700–1000
45*	BO_2	10^3	700–1000
46*	NO_2	10^5	800
47*	NO_2	10^3	800
48*	NO_2, O_3, SO	10^3	700–1000
49*	?	10^1	700
50*	?	10^2	700
51	?	10^2	700–1000
55*	?	10^2	1000
56	?	10^1	1000
58*	BO_3	10^4	700–1000
59*	BO_3	10^5	700–1000
60	BO_3	10^3	700–1000
61*	BO_3	10^3	700–1000
62*	NO_2	10^2	1000
63*	PO_2	10^3	800–1000
64*	S_2	10^5	800–1000
65*	S_2	10^3	800–1000
66*	S_2	10^4	800–1000
67	S_2	10^2	800–1000
68	S_2	10^2	1000
69	S_2	10^1	1000
74*	BS_2	10^2	900–1000
75*	BS_2	10^3	900–1000
76*	BS_2	10^2	900–1000
77*	BS_2	10^2	900–1000
78	?	10^1	900–1000
79*	Br, PO_3	10^2	700–1000
81*	Br	10^2	700–1000

* Observed in both untreated and alkali-enriched samples.

separated from the alkaline earth metal ions by means of anion exchange chromatography using dilute oxalic acid to complex the alkaline earths. In this procedure, chromium, titanium, and manganese remain in the alkaline earth fraction. (4) The final step involved the separation of the alkali metals, Li^+, Na^+, and K^+ from each other by use of cation exchange chromatography. The elutriant was dilute nitric acid and the order of elution was Li^+, Na^+, and K^+.

Atomic and molecular ion species were observed at 53 integral mass numbers with intensities from 10^{-18} to 10^{-12} amp, as seen in Table 1. All but fourteen of these peaks were identified as well known negative ions. For all other masses, we observed no peaks greater than 40 ions/sec, corresponding to a total of less than 10^5 charged particles emitted during a heating time of about one hour.

In several cases very weak peaks were observed at an odd mass number such as 30.5, but in every case these turned out to be "ghosts" or reflected ions resulting from a very intense peak nearby. The high sensitivity of the method is demonstrated by the detection of negative sodium and potassium ions.

Conclusions

Preliminary results in small samples indicate no evidence for quarks. A limit of perhaps 10^6/g can be placed for quarks with m/q nonintegral in the mass range 16–81; no useful limit can be set at present for integral values of m/q where normal ions appear in the mass spectrometer.

Acknowledgments—We are indebted to P. Horwitz of the Argonne Chemistry Division for carrying out the chemical separations. This work was performed under NASA contract T90267.

References

Adair R. K. and Kasha H. (1969) Analysis of some results of quark searches. *Phys. Rev. Lett.* **23**, 1355–1358.

Antipov Yu. M., Vishnevskii N. K., Ech F. A., Zaitsev A. M., Karpov I. I., Landsberg L. G., Lapshin V. G., Lebedev A. A., Morozov A. G., Prokoshkin Yu. D., Rodnov Yu. V., Rybakov V. G., Rykalin V. I., Senko V. A., Utochkin B. A., and Khromov V. P. (1969) Search for particles with fractional charge (quarks) in the 70 BeV accelerator at the high energy physics institute. *Yad. Fiz.* **10**, 346–353. English transl. *Soviet J. Nucl. Phys.* **10**, 199–203 (1970).

Cairns I., McCusker C. B. A., Peak L. S., and Woolcott R. L. S. (1969) Lightly ionizing particles in air-shower cores. *Phys. Rev.* **186**, 1394–1400.

Chu W. T., Kim Y. S., Beam W. J., and Kwak N. (1970) Evidence of a quark in a high-energy cosmic-ray bubble-chamber picture. *Phys. Rev. Lett.* **24**, 917–923.

Chupka W. A., Schiffer J. P., and Stevens C. M. (1966) Experimental search for stable, fractionally charged particles. *Phys. Rev. Lett.* **17**, 543–548.

Cook D. D., De Pasquali G., Frauenfelder H., Peacock R. N., Steinrisser F., and Wattenberg A. (1969) Search for fractionally charged particles. *Phys. Rev.* **188**, 2092–2097.

Elbert J. W., Erwin A. R., Herb R. G., Nielsen K. E., Petrilak M., Jr., and Weinberg A. (1970) A quark search in ordinary matter using simultaneous measurement of mass and charge. *Nuclear Physics* **B20**, 217–235.

Gallinaro G. and Murpurgo G. (1966) Preliminary results in the search for fractionally charged particles by the magnetic levitation electrometer. *Phys. Lett.* **23**, 609–613.

Gell-Mann M. (1964) A schematic model of baryons and mesons. *Phys. Lett.* **8**, 214–215.

Murpurgo G. (1970) A short guide to the quark model. In *Annual Review of Nuclear Science* (editors Emilio Segrè, J. Robb Grover, and H. Pierre Noyes), vol. 20, 105–146. Annual Reviews Inc., and references cited therein.

Proceedings of the Second Lunar Science Conference, Vol. 3, pp. 2675–2680
The M.I.T. Press, 1971.

Evolution of mare surface

T. GOLD
Center for Radiophysics and Space Research
Cornell University
Ithaca, New York 14850

(*Received* 23 *March* 1971; *accepted in revised form April* 26, 1971)

Abstract—Photographic and seismic evidence argue against bedrock lying closely beneath the lunar soil. Layers in the core sample and other features suggest that surface transportation of lunar fines is a continuing process. If this transportation is driven by electrostatic effects, then the difference between near and far sides of the moon can be understood as following from the electron bombardment in the magnetic tail of the earth.

No Evidence for General Bedrock at Shallow Depth

MANY INVESTIGATORS assume that the fine material found to cover most of the moon's surface represents in most areas a layer only a few meters thick, and is the pulverized material ground up from the bedrock below. On this basis one would have expected that strata of this underlying bedrock would frequently be exposed, especially on steep slopes. This does not appear to be the case. Photography of the lunar surface from orbit shows very many steep slopes, some in craters that look fresh and uneroded, others in regions where, despite erosion, the steep slope has been maintained; and yet it is very rare to see outcrops or strata. From fresh-looking craters one would judge that almost everywhere the material is very homogeneous to a depth of several kilometers. Craters are frequently very precisely circular and bowl-shaped. The cases where there are ledges or particularly densely strewn boulder fields around them may be a local exception. Alternatively, they may represent a compaction and induration of the lunar soil with depth, or they may demonstrate the processes of shock compaction in the case of sufficiently intense impact explosions. There is no clear evidence that anything other than lunar soil, differently compacted and cemented in different locations, fills all the mare basins.

OBERBECK and QUAIDE (1968) have argued that regional differences in crater morphology allow deductions to be made concerning the depth of regolith overlying bedrock in each region. Their investigations leave no doubt that significant regional differences exist, but the evidence that it is a solid rather than a more compacted subsurface is not convincing. Indeed in each region there are many fresh-looking craters that show no recognizable features of the kind that would be expected if the excavation had taken place partly in loose soil and partly in solid rock.

Seismic Evidence Favors Deep Deposit of Gradually Increasing Compaction

The seismic evidence cannot be understood except with an absence of widespread bedrock at shallow depth, and instead, the presence of a medium of much slower sonic

2675

velocity over an interval of several kilometers of depth. Lunar soil gradually compacted with depth would account very well for the whole range of seismic phenomena
seen (Gold and Soter, 1970).

Various attempts have been made to account for the seismic evidence in other
ways. These all involve multiple scattering in broken-up and acoustically very inhomogeneous material. The evidence makes clear, however, that the acoustic attenuation
in the lunar material is very low ($Q > 1000$) while acoustically very inhomogeneous
materials invariably lead to heavy acoustic attenuation. Boundaries between solids
and powders, or cracks in solids, or indeed any features where the wave velocity in
adjoining locations is very different, must be avoided if a model of low attenuation
is sought.

Embedding of Stones and Core Tube Layers Shows that Deposition of Soil Has Been Faster Process than Plowing Over by Meteorites

Stones on the lunar surface are in general clean, though partially embedded.
Material around them has been moved and filled in level to a rather sharp shoreline
around each stone. At the same time the stones have not been showered over by any
kind of spray. A surface transportation mechanism is required to account for this,
where the flow in general takes place within less than two centimeters of the surface
(Gold, 1970).

Plowing over by meteorite impact has been thought of as being the major surface
activity. It has been estimated, for example, that the ground has been plowed over a
hundred times to a depth of 40 cm in the lifetime of the mare surface (Shoemaker
et al., 1970). The core tube evidence makes clear that the actual material at that site
has not even been plowed over once to a depth of 40 cm. There is clear evidence that
the core has striations in height noticeable in optical properties, in chemical differences and in differences of the size distribution of the grains. This can only be understood by supposing that the surface has been added to at a rate that exceeds the
plowing by meteorites.

Whether the individual layers have been deposited ballistically as a result of a
nearby impact event, or whether they have been deposited by a surface transportation
process deriving its fresh material from an impact or other local event, is not argued
here. It is clear, however, that these layers, after being deposited by whatever means,
have not been mixed up again. The most striking evidence is that of Anders (1971),
who finds an increase in some trace elements in one layer by a factor of 10^5 over
their concentration above or below.

If one supposed the plowing rate by meteorite impact to be as fast as to dig
down to 40 cm depth once in 40 million years, one would require a deposition rate
faster than 1 cm per million years. Estimates for the meteorite frequency may, however, be greatly in error, and the mean rate in geologically recent epochs may have
been a great deal less. The average rate required to fill a mare basin in 4 billion years
would be 1 cm per 10,000 years, assuming an initial depth of 4 km in the basin after
the hydrostatic adjustment following the impact. It is, however, most likely that all
processes on the moon's surface, including surface transportation, took place at a

much higher rate, while the last stages of the accretion of the moon were still in progress. Higher bombardment rates, greater proximity to the earth, and perhaps some other factors, are all likely to have made for greater mobility of the surface material. The very clear evidence that cosmic ray surface exposure of the present surface material has been long (BARBER *et al.*, 1971; CROZAZ *et al.*, 1971; and FLEISCHER *et al.*, 1971) can be understood in those terms. The present rate of deposition of lunar soil on low ground may well be several orders of magnitude lower than the mean rate averaged over the age of the moon.

Mare surface in general is clearly not saturation bombarded since this would result in no statistically significant variations in crater density in different regions. It is, however, generally known that there are very large regional variations in crater density, and this again can be understood as a sign that the material is laid down in the mare regions and that the final appearance represents the equilibrium between this regionally different deposition process and the general impact cratering.

Rocks and Soil in the Same Locality Do Not Have the Same Origin

The compositional differences between soil and rocks in the same region make clear that the soil is not entirely local bedrock ground up with the rocks being pieces of that same bedrock. This would have been the expected situation if bedrock existed at a shallow depth and if the soil were merely the consequence of its local pulverization.

If instead the soil has suffered some surface creep over big distances, while the rocks are pieces thrown out from major craters and represent material originally at various depths, then such compositional differences can indeed be expected. The nuclear age dating of rocks and powder similarly require some differences in the origin of these two components (ALBEE *et al.*, 1970).

Discussion

The mare basins represent deep deposits 3 to 6 km in depth (judging from the seismic evidence and from the appearance of submerged craters), and those deposits may consist of similar material as is found on the surface, though somewhat compacted and cemented at the greater depths. There is crystalline rock below this, at least in some regions. Within this deep deposit rocks have been distributed by major impacts in which either they were generated, or previously existing crystalline rock at some depth was excavated. The origin of the crystalline rock may be early volcanism, or perhaps more likely it was produced by the impacts that made the mare basins. It is suggested that the deep dust layer has resulted mainly from a surface transport of material from high places on the moon to the low-lying mare basins (GOLD, 1955). Several lines of evidence, including the existence of a layered core, show that such a transport mechanism is currently operative. However, the present rate of transport cannot be representative of the entire period, and the process would have to have been much faster in an earlier epoch, to give accord with the surface exposure ages of rocks. During the entire accumulation process of the mare basin the surface would have looked much as it does now, with a sprinkling of rocks among the deposit of fine powder. What proportion of the dust is the result of pulverized lunar rocks and what proportion is directly accreted material in the last phase of the accumulation of the moon, cannot be established conclusively from present data. The high cosmic ray track densities observed in most of the fine material give a suggestion that infall has played a major part (BARBER *et al.*, 1971).

The required surface transportation and deposition mechanism is most likely electrostatic. Electrostatic agitation of a surface is readily produced in the laboratory by electron bombardment in the energy range of a few hundred electron volts. Higher or lower electron energies, or proton beams, have not led to any comparable effects, while electrons in the energy range of the secondary emission cross-over point are found to be remarkably effective in stirring up the surface and causing it to creep. The effect is due to the very large electric fields generated between adjacent particles when the potential of one is driven far positive through the emission of more than the primary number of electrons, while another, being to the other side of the cross-over point, is driven negative by the absorbed electrons. Large scale surface motion has

Fig. 1. Surface configuration of an insulating powder after being exposed to an electron beam of 2 kilovolt energy. The terrain has been changed totally several times over by electrostatic transportation processes so that no features have much to do with the original deposition of the powder. The actual surface area seen on the picture is
6 × 12 cm.

been demonstrated in the laboratory, both with a variety of powders and with actual lunar soil (Fig. 1).

A report of the laboratory studies of electrostatically produced surface creep will shortly be published elsewhere. A variety of complex effects have now been observed.

The difference between the front and the back of the moon

If indeed electron bombardment in the energy range of a few hundred volts is responsible for the major surface transportation process, then one would expect the front and the back of the moon to have had very different treatments. The moon is enveloped approximately four days each month in the magnetic tail of the earth, and it is there that electrons in this energy range occur as a consequence of the solar wind impingement on the magnetosphere. In the free solar wind stream electron energies are generally very low. The front of the moon is preferentially subjected to this electron bombardment, and any effects of it would thus be expected to dominate greatly on this side. The observed difference between the front and the back of the moon is indeed very great and demands an explanation. The difference is just of the kind that the mare ground predominates on the front side and is rare or absent on the back. If the back has any large low basins (which cannot be established with existing photography), they appear not to have been filled in. It is suggested therefore that the difference between the two hemispheres could be due to the surface transportation caused by magnetospheric tail electrons. A quantitative evaluation of the rates of transportation is difficult at the present time since it depends very sensitively on the surface electrical properties of the lunar soil, with regard to both the secondary emission coefficients and the conductivity; and measurements on material extraordinarily free from contamination and surface adsorption effects would be required.

Acknowledgments—My thanks are due to Mr. G. WILLIAMS for the laboratory demonstration of the electrostatic transportation effects. Work on lunar studies at Cornell University is supported by NASA Contract NASA NAS9–8018 and NASA Grant NGL–33–010–005.

REFERENCES

ALBEE A. L., BURNETT D. S., CHODOS A. A., EUGSTER O. J., HUNEKE J. C., PAPANASTASSIOU D. A., PODOSEK F. A., RUSS G. P. II, SANZ H. G., TERA F. and WASSERBURG G. I. (1970) Ages, irradiation history, and chemical composition of lunar rocks from the Sea of Tranquillity. *Science* **167**, 463–466.

ANDERS E., GANAPATHY R., KEAYS R. B., LAUL J. C., and MORGAN J. W. (1971) Volatile and siderophile elements in lunar rocks: Comparison with terrestrial and meteoritic basalts. Second Lunar Science Conference (unpublished proceedings).

BARBER D. J., HUTCHEON I., and PRICE P. B. (1971) Extralunar dust in Apollo cores? *Science* **171**, 372–374.

BARBER D. J., HUTCHEON I., PRICE P. B., RAJAN R. S., and WENK H. R. Exotic particle tracks and lunar history. Apollo 12 Lunar Science Conference (unpublished proceedings).

CROZAZ G., WALKER R., and WOOLUM D. (1971) Nuclear track studies of dynamic surface processes on the moon and the constancy of solar activity. Second Lunar Science Conference (unpublished proceedings).

FLEISCHER R. L., HART H. R., JR., COMSTOCK G. M., and EVWARAYE A. O. (1971) The particle track record of the Ocean of Storms. Second Lunar Science Conference (unpublished proceedings).

GOLD T. (1955) The lunar surface. *Min. Nat. Roy. Ast. Soc.* **115,** 585–604.

GOLD T. (1970) Apollo 11 and 12 close-up photography. *Icarus* **12,** 360–375.

GOLD T. and SOTER S. (1970) Apollo 12 seismic signal: Indication of a deep layer of powder. *Science* **169,** 1071–1075.

OBERBECK V. R. and QUAIDE W. L. (1968) Estimated thickness of a fragmental surface layer of Oceanus Procellarum. *J. Geophys. Res.* **72,** 4697–4704.

SHOEMAKER E. M., HAIT M. H., SWANN G. A., SCHLEICHER D. L., SCHABER G. G., GODDARD E. N., and WATERS A. C. (1970) Origin of the Lunar Regolith at Tranquillity Base. *Proc. Apollo 11 Lunar Sci. Conf., Geochim. Cosmochim. Acta* Suppl. 1, Vol. 3, pp. 2399–2412. Pergamon.

SURVEYOR III

Proceedings of the Second Lunar Science Conference, Vol. 3, pp. 2683–2697
The M.I.T. Press, 1971.

Surveyor III material analysis program

Neil L. Nickle

Jet Propulsion Laboratory, California Institute of Technology,
Pasadena, California 91103

(*Received* 22 *February* 1971; *accepted in revised form* 2 *April* 1971)

Abstract—The Surveyor III components returned from the moon by the Apollo 12 astronauts were released for scientific investigation by NASA on June 18, 1970. Prior to the release, an investigation plan was formulated to accommodate 40 investigators in 9 categories. The analysis plan was designed to obtain as much information as possible from each part, to maintain the integrity of each part for as long as possible, and to conform to the conditions imposed upon the program by NASA.

The analysis program was divided into first and second generation sets of investigations. The objective of the first set of investigations was to determine the effects of 31 months of lunar exposure using nondestructive analytical techniques on all the available components. Destructive tests, or those having some effect on the material, were permitted on approximately one-half of the returned parts. The remaining material is being held in abeyance for future, presently undefined tests.

INTRODUCTION

THE PROSPECT OF landing the Apollo 12 spacecraft in the vicinity of Surveyor III was first realized in July, 1969 when NASA's Apollo Lunar Exploration Office requested the Jet Propulsion Laboratory (JPL) to determine worthwhile activities that might be conducted at or around Surveyor III. Dr. Leonard Jaffe, Surveyor Project Scientist, contacted 63 scientists and engineers at JPL and other institutions and reviewed several hundred suggestions in an effort to provide a list of recommended tasks. An evaluation of the suggestions was based on (a) scientific value, (b) engineering value, (c) uniqueness of opportunity, (d) ease of accomplishment, and (e) hazards arising from the Surveyor III spacecraft. A majority of the recommended tasks were completed by the Apollo 12 crew, the balance being impractical due to time limitations, feasibility, and/or accessibility. In addition, the Manned Spacecraft Center (MSC) and Hughes Aircraft Company (HAC) submitted suggestions of their own which were incorporated.

Upon return of the Surveyor III material, an analysis and reporting plan was developed at MSC and presented at NASA Headquarters on December 18, 1969. The proposed engineering activities were approved and contracts negotiated between JPL and HAC for the testing and evaluation of the returned TV camera, and between MSC and HAC for the testing and evaluation of all other returned components. The science analysis portion was found to be incomplete for a proper evaluation of the effects of lunar exposure of this duration. Consequently, a comprehensive science and engineering testing plan was developed by JPL personnel with the intent of preserving the integrity of each part as long as possible. This necessitated the interspersement of science and engineering investigations in a sequential manner with each test having less effect on a particular part than the tests that followed.

This paper provides background information on return of the Surveyor III material and subsequent plans for analysis of the returned parts. Also discussed are the environmental conditions to which the returned material was subjected, exposure of the spacecraft and the returned parts to solar radiation, orientation of the spacecraft, results of investigations, and future plans for the remaining parts.

MATERIAL ANALYSIS PLAN

The material returned from the moon includes (1) the television camera with its associated optics, electronics, pieces of cabling, and support struts; (2) the scoop from the Surface Sampler/Soil Mechanics (SM/SS) device, which contained 6.54 grams of lunar soil; (3) a 19.7 cm section of unpainted, polished aluminum tube from the strut supporting the Radar Altitude Dopler Velocity Sensor (RADVS); and (4) a section of cabling and painted aluminum tube returned in an environmental sample container (SESC). Figure 1 shows the location of components removed from the spacecraft.

The analysis program was divided into a first and second generation set of investigations. The objective of the first set was to determine the effect of 31 months of lunar exposure using nondestructive analytical techniques on all available components. Destructive tests, or those having some effect on the material, were permitted on no

Fig. 1. Location of the parts removed from the Surveyor III spacecraft. The Apollo 12 Lunar Module is seen on the northwest rim of "Surveyor Crater."

more than one-half of the returned parts. The remaining material is being stored for future second generation investigations which are presently undefined.

Science investigators were individually invited to submit proposals that were brief but included a statement of objectives, the amount and type of material of interest, the type of tests to be performed, and the degree of alteration to the material. The proposals were reviewed for their scientific merit by a JPL Review Committee,* and recommendations were made regarding the type and amount of material to be allocated. Another group, the Surveyor Parts Steering Group† (SPSG), was authorized to allocate material to those investigators planning destructive tests not previously included on the approved analysis plan.

The analysis plan accommodated 40 investigators in 9 categories as follows: Microbe survival; Soil properties; Micrometeorite impacts; Radiation damage; Particle tracks; Naturally induced radioactivity; Solar wind rare gases; Surface changes and characteristics; Various engineering investigations.

ENVIRONMENTAL CONDITIONS TO WHICH THE RETURNED MATERIAL WAS SUBJECTED

The returned Surveyor III components were collected on November 19, 1969, 942 earth days after the spacecraft landed on Oceanus Procellarum. Since that time the material removed from the spacecraft has been subjected to a variety of environmental conditions in the course of necessary transportation and handling. For the most part those conditions were detrimental to the material's integrity, but they have been documented and will be discussed below.

As the parts were cut from the spacecraft and handled by Astronauts Bean and Conrad, they were placed in pockets in the Surveyor parts tote-bag. The bag was constructed from beta-cloth, a woven quartz-glass fabric coated with FEP Teflon. The parts remained in the tote-bag during transport from the Surveyor spacecraft to the Lunar Module (LM) Intrepid, from the LM to the Command and Service Module in lunar orbit, from the Command Module (CM) to the Mobile Quarantine Facility (MQF) on board the recovery ship, and from the MQF to the Crew Reception Area (CRA) in the LRL. This treatment is known to have at least caused abrasion of the exposed outer surfaces with partial removal of adhering lunar fines, and contamination of exposed surfaces with beta-cloth fibers.

While in quarantine in the CRA, the returned material was removed from the tote-bag, the camera and scoop were photographed on a table top, and all parts were individually heat sealed in two polyethylene bags. The bagged parts were placed in bonded storage where they remained until quarantine was lifted on January 7, 1970. All parts were then transferred to the astronaut debriefing room, where a temporary laboratory had been prepared.

The bagged parts were inspected and photographed. The camera was taken to the low-level radiation counting laboratory in the LRL, where it remained overnight.

---

* Membership in the JPL Review Committee comprises W. Carroll, L. Jaffe, D. Nash, and C. Snyder.
† Membership in the SPSG comprises N. Nickle (Chairman) and W. Carroll of JPL, B. Doe of U.S.G.S., Denver (formerly of NASA Hq.), D. Senich of NASA Hq., S. Jacobs of MSC, and G. Wasserburg of CIT. F. Fanale of JPL serves as an alternate in G. Wasserburg's absence.

Most parts were then unbagged and documentary photography taken of each surface. The camera and unpainted aluminum tube were unbagged on a laminar flow bench and supported by special jigs. The scoop was not opened on the laminar flow bench for fear of loosing lunar fines contained in and on the scoop. The SESC was not removed from its bag.

The camera's lower shroud and support collar were removed to allow for a biological assay. Thirty-three biological samples were collected from various sites, and less than 0.5 mg of lunar fines collected from the collar for a preliminary emission spectograph analysis (JOHNSON *et al.*, 1970). The camera was reassembled, wrapped in FEP Teflon, and packed in a foam-lined shipping container seven days later. The

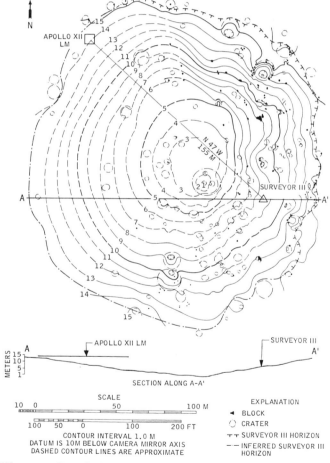

Fig. 2. "Surveyor Crater" showing the relative position of Surveyor III and Apollo 12 spacecrafts in plan and cross-section view. The LM was situated 155 m away from, north 47 deg west of, and at a ground level of 4.3 meters higher than the TV camera on Surveyor (JAFFE, 1971).

unpainted aluminum tube was sectioned into six pieces and individually packaged to protect the outer surface from further damage. The TV camera and unpainted aluminum tube received a cursory examination for micrometeoroid impacts.

All the material that was transferred to HAC was packaged for shipment and flown to Culver City, California on January 16, 1970. Hughes Aircraft Company provided a limited-access clean room (Class 100,000 as defined by HAC HP 10–220) for their multitude of engineering tests. The room contained two class-100 laminar flow benches which were used in all operations where a dust-free environment was desirable. All parts were placed in containers or covered with Teflon sheeting when not in actual use and stored in a floor vault for security.

Subsequently, parts called out in the Material Analysis Plan were transferred to JPL for distribution to engineering investigators outside of HAC, and to the science investigators in the United States and elsewhere. One condition imposed upon the investigators is the documentation of the treatment that each part received while in the investigator's possession. This information has been compiled at JPL and can be made available for specific parts upon request.

EXPOSURE OF THE SURVEYOR III SPACECRAFT TO SOLAR RADIATION

The Surveyor III spacecraft landed halfway down the eastern slope of a 200 meter diameter crater at 2.94 deg S latitude and 23.34 deg W longitude (STEINBACHER, 1967). The horizon visible to the spacecraft permitted an exposure to solar radiation between 7 deg and 178 deg to the local horizontal (Fig. 2). Therefore, the 171 deg of arc in the plane of the ecliptic is equivalent to 335 hours of exposure to solar radiation for each lunar day. Table 1 lists the values used in calculating the exposures found in Tables 2 and 3.

Material removed from the spacecraft and returned to earth remained on the lunar surface for 31.9 lunar days. None of the returned parts received the maximum 10,686 hours of exposure due to shadowing by the Planar Array Antenna, Solar Panel, Thermal Control Compartments, or other parts of the spacecraft.

In order to determine the actual exposure of specific parts to sunlight, six series of photographs were taken at JPL's Science and Engineering Testing Laboratory (SETL). A one-fifth scale model spacecraft was oriented to a collimated light source simulating the orientation of Surveyor with the sun (Fig. 3). Three cameras were set up to view different parts of the spacecraft and pictures were taken at the minimum illumination angle (2 deg), at each 10 deg interval through 170 deg, and at the maximum angle possible (178 deg). The data obtained from these photographs

Table 1. Data used to evaluate solar exposure of various parts of the Surveyor III spacecraft.

Synodical month	29.5 days
Angular velocity of sun from moon	0.51 deg/hr
Sunrise on the spacecraft	7 deg
Sunset on the spacecraft	178 deg
Maximum exposure to sunlight per lunation	171 deg, 335 hr.
Duration of stay on the lunar surface of returned parts	942 days, 31.9 lunations

Table 2. Exposure of selected parts of the TV camera to solar radiation.

Surface[1]	Exp/Lun[2] (hr)	Total Exp (hr)	Sun Angle @ 1st Exp	Angle Incid[3] @ 1st Exp	Sun Angle Last Exp	Angle Incid[3] Last Exp
			Lower Shroud			
A[4]	137	4383	7	45	77	10
B	26	814	122	2	135	11
	45	1440	155	25	178	39
C	167	5322	7	43	92	20
D[5]	20	626	125	8 (ave)	135	16 (ave)
	26	814	165	39 (ave)	178	64 (ave)
E	22	689	7	10	18	0
			Elevation Drive Housing[6]			
F	133	4256	7	80	75	30
G	12	376	172	5	178	6
H	33	1064	125	19	142	32
	14	438	171	64	178	73
I	120	3819	17	0	78	58
	35	1126	125	69	143	53
	14	438	171	26	178	19
J	20	626	7	11	17	0
			Vidicon Thermal Radiator			
Top[7]	145	4633	18	0	92	71
Bottom[8]	22	689	7	10	18	0
			Optical Filters			
Clear	141	4180[9]	18	0	90	69
Red	122	3554[9]	18	0	80	61
Green	108	3115[9]	30	76	85	64
Blue	3[9]	3[9]		Not Exposed		

[1] See Figs. 5, 6, and 7 for identification of surfaces.
[2] Assuming nonmoving surfaces, as opposed to the upper shroud, elevation drive mechanism, mirror, etc.
[3] The angle of incidence is measured from the plane of the surface.
[4] This surface was oriented within one deg of being tangent to the Apollo 12 LM (CARROLL, 1970).
[5] This surface has a radius of curvature of 6.95 cm. 95% of the area was continuously shaded by the mast and other supporting structures.
[6] This part rotated with the camera head assembly. The values listed in the Exp/Lun column are therefore too large for the first lunar day. It is estimated, therefore, that the values for the total exposure would lie within 10% of the values listed in the Exp/Lun column.
[7] Painted surface.
[8] Unpainted aluminum.
[9] It is difficult to assess the exposure during the first lunar day while the mission was in progress. The camera head assembly was oriented in all azimuth directions for various periods of time. The additional exposure experienced during this period, which can be obtained by the tedious reduction of the mission command tapes, must be added to these values. Of the 331 hours of daylight on the first lunar day, it is estimated that sunlight would have been incident upon one or more filters less than 1% of the time ($\leq$ 3 hr.) Therefore, 3 hours has been added to the Total Exp column for 31 lunations instead of 32.

Table 3. Exposure of the Soil Mechanics/Surface Sampler scoop to solar radiation.

Surface	Exp/Lun (hr)	Total Exp (hr)	Sun Angle @ 1st Exp	Angle Incid @ 1st Exp	Sun Angle Last Exp	Angle Incid Last Exp
Total	284	9078	7	variable	152	variable
	6	188	163	variable	166	variable

Fig. 3. A $\frac{1}{5}$ scale model spacecraft shown in the Surveyor III orientation as it existed at the termination of the mission. The axes of the mirrors represent the plane of the ecliptic. This configuration was used to evaluate the exposure of various parts of the spacecraft to solar radiation.

permitted an evaluation of the exposure of the TV camera and its parts, the Soil Mechanics/Surface Sampler (SM/SS) scoop, and the Radar Altitude Dopler Velocity Sensor (RADVS) strut to solar radiation.

TV camera

The Z-axis of the TV camera was tilted 23.5 deg from the local vertical in a direction N 43 W during the Surveyor III mission (STEINBACHER, 1967). The upper shroud of the camera and the normal to the plane of the TV mirror faced N 83 E. The pivot axis of the mirror is estimated to have been 1.5 m above the lunar surface. The surface of the lower shroud facing northeast was oriented parallel to the *x*-coordinate of the spacecraft (Fig. 4).

Various surfaces of the camera have been evaluated for their solar exposure and are tabulated in Table 2. Various external features of interest are defined on Figs. 5, 6, and 7.

The TV camera was equipped with four optical filters with the following specifications:

Filter Type	Manufacturer	Remarks
Clear	Bell & Howell	Dense Flint, $\rho = 3.60$ gm/cc, $n = 1.612$ (Fleischer, 1971), MgF$_2$ coating on both sides
Red	Corning	3–76
Green	Schott	OG–4 (Top; light yellow, bottom coated with inconel) (78.5% Ni, 14% Cr, 6.5% Fe)
	Chance	OGR–3 (Bottom; uncoated)
Blue	Schott	GG–15 (Top; bottom coated with inconel)
	Schott	BG–1 (Bottom; uncoated)

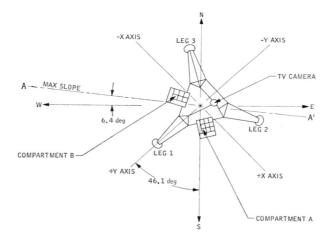

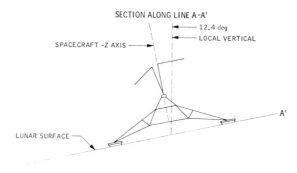

Fig. 4. Geometrical configuration of the Surveyor III spacecraft as it existed at the termination of the mission. The planar array antenna and solar panel shown in section view is not displayed in plan view.

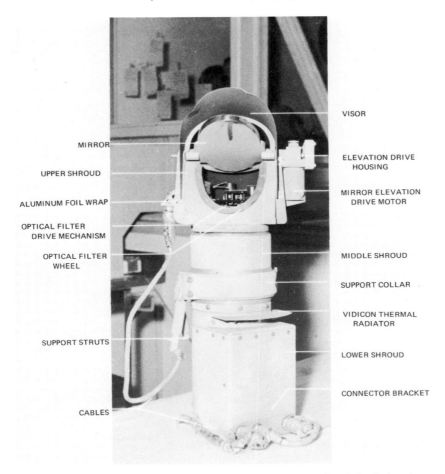

MIRROR

UPPER SHROUD

ALUMINUM FOIL WRAP

OPTICAL FILTER
DRIVE MECHANISM

OPTICAL FILTER
WHEEL

SUPPORT STRUTS

CABLES

VISOR

ELEVATION DRIVE
HOUSING

MIRROR ELEVATION
DRIVE MOTOR

MIDDLE SHROUD

SUPPORT COLLAR

VIDICON THERMAL
RADIATOR

LOWER SHROUD

CONNECTOR BRACKET

Fig. 5. Surveyor III TV camera as it was unbagged at the Lunar Receiving Laboratory.
Dents in the visor occurred during transport from the lunar surface.

The clear filter was situated over the lens of the camera at the termination of the mission; due to its relative position in the hood assembly, it received exposures of greater duration than the other filters. The exposure values listed in Table 2 do not hold for the entire surface of the filters since a portion of each filter was shaded at one time or another by either the front opening of the camera, the filter wheel drive mechanism, and/or the mirror. The values listed in Table 2 therefore represent the maximum exposure experienced by selected parts of each filter.

The upper portions of all filters were covered with varying amounts of lunar dust. Preliminary data from peels taken from the clear filter (ROBERTSON *et al.*, 1971) indicate the median grain size is 0.8 μ, with particles ranging in size from less than one to greater than 15 μ. 50% of all particles fall below one micron in size. Particle density averages 0.18 particles/μ^2; the surface area covered by particulate material entrapped in this peel, therefore, is 25.0%. This value is in agreement with the value

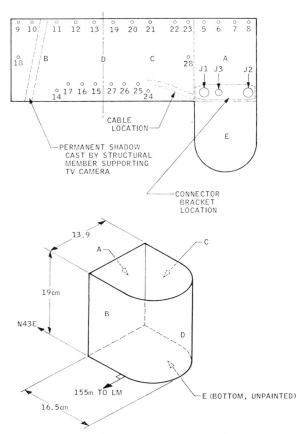

Fig. 6. Drawing of the TV camera's lower shroud identifying the various surfaces as used in Table 2 for the calculation of exposures to solar radiation. The numbered circles refer to the position of numbered screws and washers removed during disassembly.

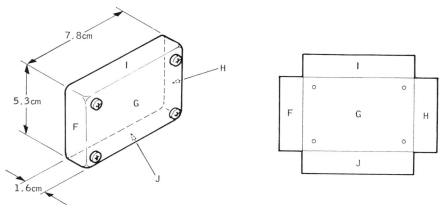

Fig. 7. Drawing of the TV camera's elevation drive housing identifying the various surfaces as used in Table 2 for the calculation of exposures to solar radiation.

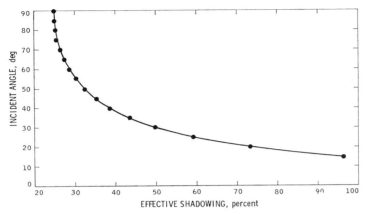

Fig. 8. Graph showing the effective shadowing created by particles adhering to the upper surface of the clear filter by an incident beam ranging from 0 to 90 degrees. Data have not been compiled for the other filters.

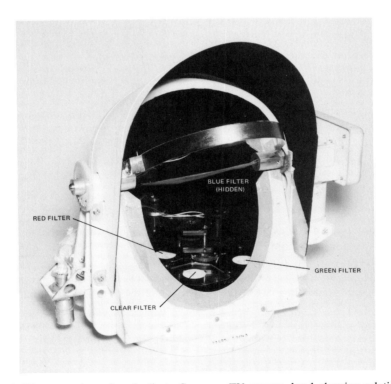

Fig. 9. Upper portion of a duplicate Surveyor TV camera head showing relative positions of the optical filters and the filter wheel drive assembly. The gummed labels and crosses show the central position from which visibility measurements were made (see text).

calculated by CARROLL (1970), and RENNILSON *et al.* (1971) from spectral transmission data taken from the clear filter. They determined that approximately 25% of
the surface area was covered with particles by comparing data taken before and after
cleaning the filter of its particulate material.

The effective shadowing of the upper surface of the filters by adhering particles is a
function of exposure geometry. If 25% of the surface of the clear filter is shadowed by
particles from an incident beam oriented 90 deg to the plane of the filter, then the
effective shadowing is going to be considerably greater at angles approaching that
plane.

Figure 8 is a plot showing the effective shadowing for incident angles ranging from
0 to 90 deg. The data was compiled by calculating areas of elliptical shadows cast by
hypothetical spheres from the particle characteristic data compiled by ROBERTSON
et al. (1971).

The visibility or solid angle of view that each filter had through the front opening
of the camera was determined to evaluate the results of particle track studies, alpha
radioactive fallout, and micrometeorite impact flux measurements. The visibility was
determined at the center of the filters, except for the blue filter which had no direct
view out of the camera. Figure 9 shows the relative position of the red (R), clear
(ND), and green (G) filters; their visibility was 0.49, 2.19, and 0.73 steradians
respectively. These values vary most with position on the red and green filters where
parts of the filter drive mechanism partially covered them.

Soil mechanics/surface sampler scoop

The arm of the scoop was left fully extended and at maximum elevation at the
termination of the Surveyor III mission. This geometry permitted considerably more
exposure to solar radiation than other returned parts. The relatively little shadowing
the scoop received came from the Planar Array Antenna, the Solar Panel, and, near
sundown, by the spacecraft itself. Exposure data are found in Table 3.

Unpainted aluminum tube

Determination of the exposure of the unpainted aluminum tube (RADVS strut)
to solar radiation was hampered by the absence of corroborative photography to
indicate precisely where along the 1-meter length of tubing the 19.7 cm section was
removed. If the section was obtained from the lower end of the tubing, then the
exposure value listed below is somewhat low. The lower end was not shadowed to the
same extent as higher portions due to the relative position of the adjacent thermal
control compartment-A. There is reason to believe that the section was removed from
approximately the center (CARROLL, 1970), however, and the exposure value reflects
that assumption.

A scratch was made by hand at the LRL along the length of the tube for orientation
(BLAIR and BUETTNER, 1971). The scratch was too light so a heavier scribe line was
subsequently made in the same area. This line represents the surface of maximum
exposure to solar radiation assuming the direction of maximum implantation of solar
wind rare gases is coincident with sunlight. Figure 10 shows a cross-section of the tube
and a trapped solar wind helium envelope along with the direction of sunrise, the

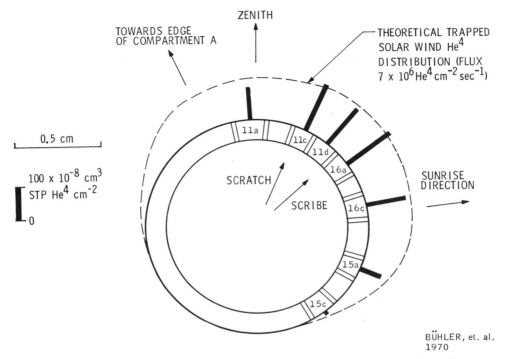

Fig. 10. Cross-section of an annulus of the polished aluminum tube showing the results of a preliminary examination designed to verify the rotational orientation of the tube on the lunar surface. The scribe line is coincident with the theoretical maxima and is situated 42 deg above the local horizontal, and facing an easterly direction.

zenith, and shadowing (BÜHLER *et al.*, 1970). The scribe line is believed to represent an orientation on the moon that is 42 deg above the local horizontal, and facing an easterly direction.

An attitude of N 15 E, plunge 18 S was determined for the axis of the tube by means of a Brunton Compass on a 1/5 scale model spacecraft (Fig. 3). Sunrise occurred on the tube at 7 deg and the tube became shadowed at 115 deg by compartment-A. A 72 deg sector therefore received no direct sunlight, while the remainder received amounts ranging from 0 to the maximum value at the scribe line and with incident angles ranging from grazing incidence to 75 deg (the maximum possible due to its spatial orientation) to the surface. The maximum exposure is 6,784 hours for areas in the vicinity of the scribe line.

ORIENTATION OF THE SURVEYOR III SPACECRAFT

Subsequent to the compilation of the solar exposure data just presented, the writer became aware of a change in orientation of the spacecraft since the termination of the Surveyor mission. Measurements made on Apollo 12 photographs of the Surveyor spacecraft indicate footpad 1 (downhill) was anchored while footpads 2 and 3 rotated 7.6 cm counter-clockwise about footpad 1 (Fig. 4). In addition, shock absorbers attached to legs 1 and 3 collapsed, causing the spacecraft to tilt an additional $2\frac{1}{2}$ deg (SCOTT *et al.*, 1971).

What effect this has on the exposure values depends on when the change in orientation occurred. If the change occurred early, the tabulated values will vary up or down depending on the spatial orientation of the surface in question. If the change occurred late in the 31 months residence time for the returned parts, then the values are considered good.

R. SCOTT informed the writer of the possibility that the change in orientation occurred late due to evidence seen in an Apollo 12 photograph (AS12–48–7124) of footpad 3. He contends that clumps of dirt emplaced on the white upper surface of the footpad during landing were moved and partially removed by an episodic event, such as a failure of the shock absorbers. The time at which the movement occurred can only be estimated from the comparison of the shielded and unshielded portions of footpad 3 and a knowledge of the mechanism and rate of the process which tans the painted surface.

An evaluation of the change in exposure to solar radiation due to later spacecraft movement will not be done except for specific requests, and then only if an estimate of the time of movement can be established. It is estimated that the change in total exposure will be negligible except for those surfaces that received little radiation at grazing angles of incidence.

RESULTS OF INVESTIGATIONS

Results of the engineering tests conducted on the returned parts have been reported by HAC in their final reports. These reports were distributed to all science investigators and all NASA centers, and a few remain available for distribution from JPL to interested parties.

Results of the science analyses are being published at the discretion of the individual investigators. In addition, a comprehensive Special Publication is being prepared by NASA that will include all engineering and science activities in the program to date. Early preliminary tests conducted in the LRL have been reported in NASA SP–235, Apollo 12 Preliminary Science Report.

FUTURE PLANS

The Surveyor III Material Analysis Program was separated into groups of engineering tests and science investigations. No more than half of the returned material was provided for these tests, the rationale being that later investigations could produce new or refined information from the remaining material which could lead to a better understanding of the lunar environment or to the effects the environment imposed upon the returned hardware.

The parts remaining from the first set of investigations and those not yet analyzed are being stored at JPL for distribution. While a definitive program for the analysis of this material has not been formed, it is likely that NASA will continue to accommodate qualified individuals with Surveyor III parts as long as an interest is maintained by the scientific community. Interested parties should submit brief proposals in writing to the writer at the Jet Propulsion Laboratory; they are reminded that no funds will be provided by NASA to perform these tests.

Acknowledgments—This paper represents the results of one phase of research carried out at the Jet Propulsion Laboratory, California Institute of Technology, under Contract NAS 7–100, sponsored by the National Aeronautics and Space Administration.

REFERENCES

BLAIR P. M. and BUETTNER D. G. (1971) Description and initial examination of returned parts. Surveyor III Parts and Materials/Evaluation of Lunar Effects, Hughes Aircraft Company, 3–21.

BÜHLER F., EBERHARDT P., GEISS J., and SCHWARZMÜLLER J. (1970) Trapped solar wind helium and neon in surveyor 3 material. Submitted to *Earth Planet. Sci. Lett.*

CARROLL W. F. (1970) Personal communication.

FLEISCHER R. L. (1971) Personal communication.

HAWTHORNE E. I. (1970) Test and evaluation of the surveyor III television camera returned from the moon by Apollo 12, Volumes I and II.

JAFFE L. D. (1971) Blowing of lunar soil by Apollo 12: Surveyor 3 evidence. Surveyor III Material Analysis Report, NASA Special Publication, in press.

JOHNSON P. H., BENSON R. E., COUR-PALAIS B. G., GIDDINS L. E., JR., JACOBS S., MARTIN J. R., MITCHELL F. J., and RICHARDSON K. A. (1970) Preliminary Results from Surveyor 3 Analysis. Apollo 12 Preliminary Science Report, NASA SP–235, 217–223.

RENNILSON J. J., HOLT H., and MOLL K. (1971) Change in the optical properties in Surveyor III's camera. Surveyor III Material Analysis Report, NASA Special Publication, in press.

ROBERTSON D. M., GAFFORD E. L., TENNY H., and STREBIN R. S., JR. (1971) Characterization of dust on clear optical filter from returned Surveyor III television camera. Battelle Memorial Institute Document BNWL–B–62.

SCOTT R. F., LU T.-D., and ZUKERMAN K. A. (1971) Movement of Surveyor 3 Spacecraft. *J. Geophys. Res.* In press.

STEINBACHER R. H. (1967) III orientation of camera with lunar surface and sun. Surveyor III Mission Report, Part III: Television Data, 15.

Proceedings of the Second Lunar Science Conference, Vol. 3, pp. 2699–2703
The M.I.T. Press, 1971.

Examination of returned Surveyor III camera visor for alpha radioactivity

THANASIS E. ECONOMOU and ANTHONY L. TURKEVICH

Enrico Fermi Institute, University of Chicago, Chicago, Illinois 60637

(*Received* 16 *February* 1971; *accepted in revised form* 24 *March* 1971)

Abstract—The Surveyor III camera visor brought back by the Apollo 12 astronauts has been examined for an alpha radioactive deposit formed by the decay of radon isotopes diffusing out of the lunar surface. An upper limit for the equilibrium amount of Po^{210} at Oceanus Procellarum has been set at less than 0.005 dis cm^{-2} sec^{-1}. This number is appreciably lower than the amount of Po^{210} observed by Surveyor V at Mare Tranquillitatis.

INTRODUCTION

ON APRIL 19, 1967, SURVEYOR III landed in the eastern part of Oceanus Procellarum at 23.34 deg W longitude and 2.99 deg S latitude (ACIC coordinate system). On November 20, 1970, the same site was revisited by Apollo 12 astronauts Alan Bean and Charles Conrad, who brought back the Surveyor III television camera together with 32 kg of moon rocks. Part of this camera, the visor, was made available to the University of Chicago to examine for the presence of a deposit of alpha radioactivity.

The possibility of such a radioactive deposit on the surface of the moon was suggested by several authors (e.g. KRANER *et al.*, 1966). Radon isotopes formed by the decay of uranium and thorium diffuse out of lunar material into space where they undergo further decay and some of their daughters end up on the lunar surface. In the thorium decay series the daughters have relatively short half-lifes, and all had decayed before the visor could be examined. On the other hand, the alpha emitting Po^{210} in the uranium decay series is held up by the 22-year half-life of its grandparent Pb^{210}. A measurement of the amount of Po^{210} (5.31 MeV) alpha activity on the visor, together with knowledge of the time spent on the moon, and on the earth before the measurement, provides a measure of the rate of radon decay product deposition on the lunar surface at Oceanus Procellarum.

The existence of such a deposit would be interesting in providing information on the emanating power of lunar material and on the amount of radon "atmosphere" on the moon. In addition, it may have an effect on the isotopic composition of the lead in lunar fines.

EXPERIMENTAL METHOD AND RESULTS

Measurements on the Surveyor III camera visor (Fig. 1) were started at the University of Chicago 0.64 years after it was taken off the Surveyor III spacecraft. The visor was placed in a vacuum chamber and examined for alpha radioactivity with the Alpha Scattering Instrument (TURKEVICH *et al.*, 1966). In order to increase the sensitivity, the proton system of the instrument was used. The active area of the proton detectors of this instrument is about ten times that of the alpha detectors and, in

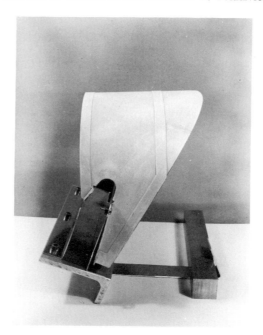

Fig. 1. The visor of Surveyor III television camera brought back by Apollo 12 astronauts.

addition, the examined visor could be placed closer to the proton than to the alpha detectors. The gold foils in front of proton detectors, which normally screen out the alpha particles in the Alpha Scattering Instrument were removed for these measurements and replaced by VYNS films. The visor was measured for a period of 9714 minutes using all 4 detectors and an additional period of 4475 minutes with less than the full complement of detectors in order to check on possible asymmetries in the deposit. The backgrounds in the instrument were negligible.

Figure 2a shows the experimental data obtained from the visor of the Surveyor III television camera. There are several unexpected results that characterize these data: (1) The continuous flat spectrum indicates that the source of alpha activity is not on the surface. (2) The intensity is too high—several orders of magnitude higher than was expected for a non-radiative material. (3) The presence of high energy alpha particles (higher than 6 MeV) indicates that the source is probably daughter products of Th^{228} or U^{234}. The surface of the visor, as most of the Surveyor parts, was covered with white paint for thermal control purposes. Because of the unavailability of a model television camera we were not able to measure directly the natural background from the visor. Figure 2b shows the results of measurements made on plates covered with the same paint and made at the same time as the visor itself. In these measurements, the plates were placed very close to each detector in a position where the absolute efficiency of detecting their activity could be calculated. This spectrum is very similar to that obtained from the visor. After comparing the absolute intensities, the conclusion was

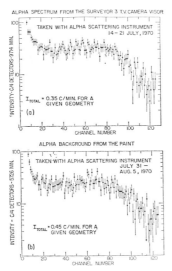

Fig. 2(a) (top). The alpha spectrum obtained from the Surveyor III television camera visor as measured by the Alpha Scattering Instrument. (b) (bottom). Background obtained from the paint used on visor.

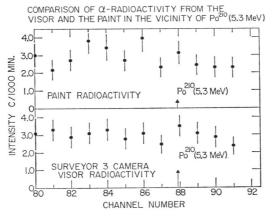

Fig. 3. Comparison of α-radioactivity from the visor and the paint in the vicinity of Po²¹⁰ (5.3 MeV).

reached that the gross activity on the visor returned from the moon was entirely due to the activity of the paint.

Although the presence of this alpha activity of the paint reduced the sensitivity of identifying an alpha radioactive deposit, the data can still be used to set upper limits for the Po^{210} radioactivity on the moon. Figure 3 shows a comparison of the alpha activity in the region of Po^{210} (5.3 MeV) for the paint and the visor. Using the gross alpha radioactivity as a measure of the relative efficiencies of detecting radiation from the visor and paint, the paint background could be subtracted from the visor data in the region of interest to get a net activity on the visor of $(0.7 \pm 3.5) \times 10^{-3}$ d min^{-1} cm^{-2}. The error quoted here and throughout this manuscript is the statistical one at the 1 σ level of confidence.

In order to estimate the amount of Po^{210} activity to be expected on a square centimeter of the lunar surface after a very long period, this net activity has to be corrected for the shadowing of the visor by assorted spacecraft parts (the "view factor" to space was 0.65), for the decay since removal from the moon (0.64 years) and for the fact that the visor was on the moon for only 2.58 years (keeping in mind the genetic relationship of the Rn^{222} decay chain). Application of these corrections leads to the measurement implying an activity, after infinite time, of

$$D = (0.88 \pm 4.43) \times 10^{-3} \text{ d sec}^{-1} \text{ cm}^{-2}$$

on the lunar surface at Oceanus Procellarum.

DISCUSSION

The limit on the Po^{210} alpha radioactivities obtained in this work is compared with the predictions of KRANER et al. (1966) and of subsequent attempts to measure this quantity in Table 1. The original prediction was an average for the whole moon, as was the limit set by YEH and VAN ALLEN (1969) from Explorer 35 observations. The only reported observation of the presence of the radioactivities is the work of TURKEVICH et al. (1970) from the data obtained by the Alpha Scattering Instrument on the Surveyor 5 mission to Mare Tranquillitatis.

Recently, LINDSTROM et al. (1971) determining the excess of Pb^{210} (over that in equilibrium with uranium) on the surface of the rock brought back by Apollo 11 astronauts, also from Mare Tranquillitatis, gave a limit which is 70 times lower than the value reported by TURKEVICH et al. (1970).

The limit set by the present work on such radioactivity in Oceanus Procellarum, a different site, but one at which the uranium content of the soil is actually appreciably higher than in Mare Tranquillitatis, is also lower than the reported value of TURKEVICH et al. (1970).

Although the results of LINDSTROM et al. (1971) and the present work appear to contradict the results reported by TURKEVICH et al., (1970) it must be remembered that the radioactive deposit should be confined to the topmost fraction of micron of the lunar surface. Any disturbance of the surface, such as the shaking off of a dust layer, or abrasion of the surface, would carry away the deposit also. Thus, although these two most recent attempts to detect the alpha radioactivities have failed, and therefore

Table 1. Equilibrium Po^{210} alpha radioactivity of lunar surface

	dis cm^{-2} sec^{-1}	
KRANER et al. (1966)	2.0*	whole moon
YEH and VAN ALLEN (1969)	< 0.16	whole moon
TURKEVICH et al. (1970)	0.03 ± 0.01	Mare Tranquillitatis
LINDSTROM et al. (1971)	< 0.0004	Mare Tranquillitatis
Present work (1971)	< 0.005	Oceanus Procellarum

* Predictions.

contradict the observation of TURKEVICH *et al.* (1970), there is some probability that they are not valid checks on the existence of the deposit. In both cases there is no assurance that the topmost layer was not removed. It may be that the Surveyor V mission, making an in situ measurement, was better able to detect this fragile deposit than were the examinations of samples brought back from the moon.

STOENNER *et al.* (1971), in examining the gas in the sealed sample return container (SRC) from the Apollo 12 mission, found evidence for about 5 d min^{-1} of Rn^{222} in the container atmosphere, although the confidence in their number is not large because of a high radon blank from the charcoal in their system. If it is assumed that 5 d min^{-1} of Rn^{222} represents roughly the equilibrium amount that would diffuse out of the 2.8 kg of lunar fines in the particular container, it would imply that 10^{-3} of the Rn^{222} diffuses out of lunar soil particles. Assuming that the same efficiency is applicable on the moon, the limit set for the Po^{210} deposit by the present work at Oceanus Procellarum means that radon diffuses out of less than about one meter of lunar soil on the moon. If the true equilibrium amount of Po^{210} is higher than found in the present work, the parent radon would be derived from greater depths.

Acknowledgements—This work was supported in part by research grant NGR-14-001-135 and contract NAS 9-7883 from the National Aeronautics and Space Administration. The authors thank Mr. NEIL NICKLE of the Jet Propulsion Laboratory for making available the Surveyor III camera visor and Mr. EDWIN BLUME for valuable help in the course of this work.

REFERENCES

KRANER H. W., SCHROEDER G. L., DAVIDSON G., and CARPERTER J. W. (1966) Radioactivity of the lunar surface. *Science* **152**, 1235–1237.

LINDSTROM R. M., EVANS J. C., JR., FINKEL R., and ARNOLD J. R. (1971) Radon emanation from the lunar surface. Private communication. *Earth Planet Sci. Lett.* (submitted).

STOENNER R. W., LYMAN W., and DAVIS R., JR. (1971) Radioactive rare gases in lunar rocks and in the lunar atmosphere. Second Lunar Science Conference (unpublished proceedings).

TURKEVICH A. L., KNOLLE K., EMMERT R. A., ANDERSON W. A., PATTERSON J. H., and FRANZGROTE E. J. (1966) Instrument for lunar surface chemical analysis. *Rev. Sci. Instrum.* **37**, 1681–1686.

TURKEVICH A. L., PATTERSON J. H., FRANZGROTE E. J., SOWINSKI K. P., and ECONOMOU T. E. (1970) Alpha radioactivity of the lunar surface at the landing sites of Surveyors 5, 6, and 7. *Science* **167**, 1722–1724.

YEH R. S. and VAN ALLEN J. A. (1969) Alpha particle emissivity of the moon: An observed upper limit. *Science* **166**, 370–372.

Proceedings of the Second Lunar Science Conference, Vol. 3, pp. 2705–2714
The M.I.T. Press, 1971.

Solar flares, the lunar surface, and gas-rich meteorites

D. J. Barber,* R. Cowsik†, I. D. Hutcheon, P. B. Price, and R. S. Rajan

Department of Physics, University of California, Berkeley, California 94720

(*Received* 22 *February* 1971; *accepted in revised form* 30 *March* 1971)

Abstract—An updated interplanetary energy spectrum of Fe nuclei from solar flares, based on tracks in the Surveyor III glass filter, is presented. From that spectrum and the track density profile in rock 12022, the average rock erosion rate over 10^7 years is estimated to be ~ 3 Å/year. This rate is $\sim$ three times higher than our initial estimate and is consistent with the rate inferred by Hörz *et al.*, 1971, from microcratering studies of 12022 and several other rocks. Track densities of $\sim 10^{10}$ to 10^{11}/cm² are quite common in the finest component of the soil at all depths down to at least 60 cm and have also been observed for the first time in interior grains of gas-rich meteorites. Erosion mechanisms and the origin of the lunar and meteoritic grains with high track densities are discussed.

Introduction

High track densities and steep track density gradients have been observed in interior grains of certain gas-rich meteorites (Lal and Rajan, 1969; Pellas *et al.*, 1969), in the top mm of lunar rocks and in crystals and glass from the lunar soil (Crozaz *et al.*, 1970; Fleischer *et al.*, 1970; Lal *et al.*, 1970; Price and O'Sullivan, 1970; Borg *et al.*, 1970; Barber *et al.*, 1971). The tracks were almost certainly produced by heavy nuclei ($Z \approx 26$) emitted in solar flares with a steeply falling energy spectrum. Heavy nuclei in the galactic cosmic rays have an energy spectrum that rises less steeply at low energies than does the solar particle spectrum but that penetrates much more deeply, down to several cm. The presence of tracks of solar origin in isolated interior grains were compacted into meteorites and that the peak shock pressure during compaction did not exceed ~ 100 kilobars, the value below which tracks made visible by chemical etching are not erased (Ahrens *et al.*, 1970). The presence of solar tracks in sub-surface lunar soil demands that those layers were once exposed at the surface.

If the rock surface was being eroded during its irradiation or was separated from the source of energetic particles by either solid or gaseous matter, the observed track density gradient would be lower than the predicted gradient (Price *et al.*, 1967). Until now the use of this concept to infer erosion rates and irradiation history has been impeded by ignorance of the average interplanetary energy spectrum of Fe-group nuclei ($Z \approx 26$).

Three recent developments make it profitable for us to re-examine lunar erosion, ancient solar flares, and the history of the lunar soil and gas-rich meteorites: (1) techniques for observing track densities up to $\sim 5 \times 10^{11}$/cm² with high voltage

---

* Present address: Department of Physics, Essex University, Colchester, U.K.

† On leave from Tata Institute of Fundamental Research, Bombay, India.

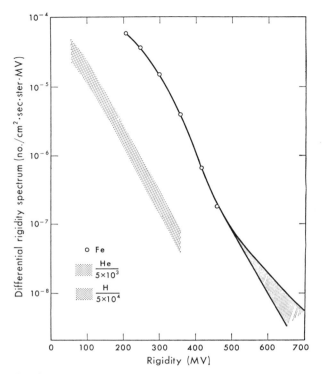

Fig. 1. The observed tracks in Surveyor glass are assumed to be solely due to Fe ions, and the rigidity spectrum is derived, using observed variation with depth. The hatched region at large rigidities represents uncertainty due to a background of fission tracks. The expected curve, with an indicated uncertainty factor of 2, was computed on the basis of satellite measurements of solar protons and alphas, using the proper photospheric abundance ratios. This predicted curve falls much below the observed curve, indicating enhanced emission of Fe-group nuclei.

electron microscopy; (2) direct measurement of the gradient of Fe tracks in glass from the Surveyor III camera after a 2.6 year exposure to solar flares during 1967–69; (3) calibration of the glass and of lunar minerals with beams of 10.3 MeV/N Ar^{40} and Kr^{84} ions.

ENERGY SPECTRUM OF INTERPLANETARY FE NUCLEI

Three groups have measured the track density from interplanetary Fe-group nuclei as a function of depth in portions of the flint glass lens filter that was exposed on the lunar surface from May 20, 1967 until the Apollo 12 astronauts brought the Surveyor 3 camera back to earth in November, 1969 (PRICE *et al.*, 1971; CROZAZ and WALKER, 1971; FLEISCHER *et al.*, 1971). We have critically compared the available data and in Fig. 1 we present a revised differential rigidity spectrum for that 2.6 year period that takes into account several factors not previously considered by all three groups: (1) Using beams of Ar^{40} and Kr^{84} ions, we have determined the dependence of cone angles of etched tracks in the glass on ionization rate. On the basis of this we estimate that the track of an Fe ion can be recognized between 2 and 35 microns of its residual range. (2) We have calculated an accurate range-energy relation for the composition of the flint glass. (3) FLEISCHER et al. (1971) have shown that there is a uniform background of cosmic-ray induced fission of lead atoms in the glass that distorts the highest energy portion of the spectrum. (4) Neil Nickle (private communication) has provided us with a recently measured size distribution of lunar dust particles on the glass surface that distorts the lowest energy portion of the spectrum. We have taken into account the degrading effect of these dust grains, which are discontinuously distributed and allow some of the lowest energy particles to reach the glass without energy loss. The remaining, main source of uncertainty in the new spectrum is the human efficiency for observing etched tracks with various cone angles inclined at various angles to the glass surface. Very close to the filter's surface we measured, using optical and scanning electron microscopy, a track density of $1.5 \times 10^6/cm^2$ ster. The values obtained by the two methods were selfconsistent and agree with the value by CROZAZ et al. (1971). Deeper in the glass we use the data of FLEISCHER et al. (1971), who have a high efficiency for the observation of tracks, since they scan a surface which is normal to the mean direction of the flux of the Fe ions.

In our recent paper (PRICE et al., 1971) we pointed out that the flux of Fe nuclei in Fig. 1 is far greater than expected on the assumption that the sun emits energetic particles in the ratio of their photospheric abundances. In drawing this conclusion we have used the published satellite data for solar alpha particles and protons of LANZEROTTI (1969–70) and the proton data of BOSTROM et al. (1968–70) and of HSIEH and SIMPSON (1971). In the energy interval where they overlap, the spectra of these three groups agree reasonably well for most flares, with the data of Hsieh and Simpson being low by as much as a factor 2 for the April 12, 1969 flare. The agreement with the solar proton spectra inferred by FINKEL et al. (1971) from radiochemical measurements on Co^{56} and Mn^{54} in rock 12002 is fair, tending to be low by up to a factor 2 (J. R. ARNOLD, private communication).

As an illustration of the enhancement of Fe flux, the broad curve in Fig. 1 shows the predicted Fe rigidity spectrum given by the product of the alpha particle spectrum of Lanzerotti and the solar abundance ratio $(Fe/He)_\odot \approx 2 \times 10^{-4}$ calculated by ROSS (1970) from the recent solar Fe abundance deduced by WOLNIK et al. (1970). We have attributed the difference between the observed and predicted flux to the preferential leakage of low energy Fe nuclei from the accelerating region because of their incomplete ionization and consequent high magnetic rigidity (PRICE et al., 1971). Our present analysis places this observation of a heavy ion enhancement on an even firmer basis than before. The enhancement is far greater than the maximum uncertainty in solar proton spectrum based on satellite and radiochemical data.

In the remainder of this paper we assume that the spectrum in Fig. 1 represents the flux level during the active half of an 11 year solar cycle and that the flux drops to zero during the inactive half, so that in 10^6 years the accumulated number of Fe tracks would be $(10^6 \times 3.2 \times 10^7 \text{ sec/y})/5.5$ times higher, assuming that the intensity of the present solar cycle is equal to the average intensity over millions of years.

Lunar Erosion Rate

The method of determining rock erosion rate depends on a comparison of track density gradients in a rock and in the Surveyor glass and is completely independent of the ratio of Fe ions to protons in solar flares.

The track density gradient in a lunar rock is most reliably obtained from measurements on an etched, polished section rather than on individual grains removed from various locations. Figure 2 shows the track density profile taken in a region of rock 12022 that contains no impact pits. This is a particularly valuable rock because of its simple history; in contrast to 10017 (Price and O'Sullivan, 1970; Crozaz *et al.*, 1970; Fleischer *et al.*, 1970) and 12063 (Crozaz *et al.*, 1971), rock 12022 was irradiated only from one direction and appears not to have received a sub-surface exposure. Our measurements, which combine transmission electron microscopy and optical microscopy, cover four orders of magnitude of depth and extend from track densities of $\sim 3 \times 10^6/\text{cm}^2$ up to $\sim 6 \times 10^9/\text{cm}^2$ at a depth of $\sim 3 \ \mu\text{m}$ below the surface.

From the profile deep inside the rock, due to galactic Fe-group nuclei, we infer a surface residence time of $\sim 1 \times 10^7$ years. In Fig. 2 we show the expected track

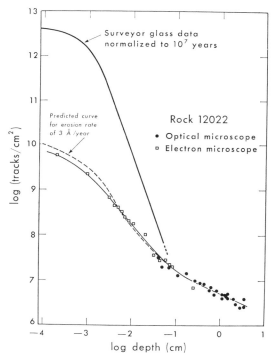

Fig. 2. The solar contribution to track densities during 10^7 years is estimated by assuming that the observed track densities in Surveyor glass represented the contribution over one-half of an 11-year solar cycle. On the basis of this and assuming a uniform erosion rate of 3 Å/year, the track density gradient predicted for a lunar rock is in good agreement with that measured for rock 12022.

densities at various depths deduced (making the aforementioned assumptions) from the measurements on the Surveyor glass filter. This dashed curve was derived assuming a mean erosion rate of 3 Å/year and is in excellent agreement with the measured track densities in rock 12022. The erosion rate of ~ 3 Å/year is reliable only to the extent that the spectrum in Fig. 1 truly represents the average solar flare spectrum over 10^7 years. From their radiochemical study of 12002, FINKEL et al. (1971) conclude that the proton intensity in the present solar cycle is representative of the average over $\sim 10^6$ years. Thus, ~ 3 Å/year appears to be a reasonable average value for recent lunar history, but of course may not be representative of erosional processes at an earlier epoch. For example, to build up a regolith ~ 7 meters thick at Oceanus Procellarum (SHOEMAKER and HAIT, 1971) in 3.3×10^9 years (ALBEE et al., 1971) would necessitate an *average* erosion rate of ~ 20 Å/year if the regolith were derived from comminution of local rock. We emphasize that these two rates are not incompatible if they apply to quite different epochs or if the regolith is built up dominantly by large meteorite collisions that would destroy rocks of the size brought back to earth.

At least three processes are responsible for the erosion of rock: (1) Sputtering of individual atoms by the solar wind (mainly hydrogen) may remove as much as 0.4 Å/year (WEHNER et al., 1963), depending on the average angle of inclination of the rock surface of the sun. (2) The flux of heavy nuclei emitted in solar flares is sufficiently great (Fig. 1) that, in the absence of other erosional processes, the outer 10 μm of rock would accumulate $\sim 10^{12}$ Fe-group tracks in a million years, as well as a considerably larger density of more lightly damaged regions produced by ions of abundant elements like C, O, Ne, Mg, Si, and S. At a dose of $\sim 10^{13}$/cm² Ar ions certain minerals develop extensive strains and fractures in regions where the ions have stopped (SEITZ et al., 1970), so that excessively radiation-damaged layers might flake off and contribute to the regolith. We estimate that the erosion rate by this mechanism might reach 0.1 or 0.2 Å/year for feldspars and certain other minerals that are especially susceptible to radiation damage. If however, sputtering removes as much as 0.4 Å/year, the density of solar flare tracks would be limited to $\sim 10^{11}$/cm² and flaking would probably not occur. (3) Micrometeorites certainly contribute to rock erosion. The magnitude of their contribution is uncertain because of uncertainty in the present-day flux of micrometeorites (known to within no better than $\pm 3\times$) and in the longtime constancy of the flux. In a very careful study of microcraters on lunar rocks, HÖRZ et al. (1971) arrive at an average erosion rate of ~ 1 to 2 Å/year and a surface lifetime of $\sim 10^7$ years before destruction by a large micrometeorite, subject to the above uncertainties. It is not clear from microscopic observations alone whether the crater distributions on Apollo 12 rocks have reached a steady state or not, but in order to account for the 3 Å/year inferred for rock 12022, microcratering must be a more important mechanism than sputtering, which cannot remove more than ~ 0.4 Å/year.

At the Apollo 12 conference we estimated the erosion rate of 12022 to be no more than 1 Å/year; FLEISCHER et al. (1971) quoted a rate of 0 to 2 Å/year in their analysis of track gradients reported by several groups at the Apollo 11 conference; and CROZAZ and WALKER (1971) quoted a value of ~ 10 Å/year, based on an apparent erosion equilibrium for the track gradient in 12063. The agreement between our revised value of ~ 3 Å/year for 12022 and the rate of 1 to 2 Å/year arrived at by Hörz *et al.* (1971) is sufficiently close that one can conclude that the micrometeorite flux over the last 10^7 years must have been fairly similar to the present-day rate, on which they based their calculations.

The effects of erosion undoubtedly depend to some extent on the size of the body being eroded and must be taken into account in attempting to understand the origin of the lunar fines and of the highly irradiated grains in gas-rich meteorites as well. Atomic sputtering should be essentially independent of the size of the body. Micrometeorite bombardments will not affect track gradients in submillimeter particles because none of these particles will survive a single collision. The only fine particles available for study are those that have avoided collision.

To summarize, track gradients in small particles are less steep than that predicted from the energy spectrum observed in the Surveyor glass. This discrepancy could arise from sputtering-type erosion or from coverage by a layer of matter but probably not from erosion by radiation-induced cracking. Micrometeorite bombardment erodes large particles and rocks but simply destroys small particles. This discussion is pertinent to the section that follows.

Highly Irradiated Grains in the Regolith and in Gas-Rich Meteorites

Any features common to both the moon and the meteorites contribute to our understanding of the origin of both. For example, the chemical composition and mineralogy of many of the lunar rocks are similar to those of eucrites (Wänke *et al.*, 1970). Ganapathy *et al.* (1970a, b) conclude from an analysis of enriched concentrations of certain trace elements that 2% of the lunar soil is of carbonaceous chondritic origin, presumably from accumulated infall. The enrichment is most pronounced in the small grain size fraction.

Using a 1 MeV electron microscope, Borg *et al.* (1970) and Dran *et al.* (1970) have found extremely high track densities ($>10^{11}$/cm²) in a large fraction of the finest grains in the lunar soil but have failed to find any tracks in grains taken from gas-rich meteorites. They have emphasized the differences in habit and texture features of lunar and meteoritic grains. Their inability to etch the tracks in the lunar grains and the predominance of high track densities in the *smallest* grains led them to suggest that solar suprathermal heavy ions, with damage rates below the threshold for etching, were responsible for the tracks. Suprathermal protons at high flux levels have been observed on several occasions by Frank (1970).

With a 650 kev electron microscope we have found extremely high densities of etchable tracks at all depths down to 60 cm in fines from Apollo cores (Barber *et al.*, 1971) as well as in thin sections of the Kapoeta and Fayetteville gas-rich meteorites. From both etching and dark field work we deduce that $\sim 20\%$ of fines < 5 μm dia have track densities $>10^{10}$/cm². In Fayetteville about 5–10$\%$ of the smaller grains have $\sim 10^{10}$ tracks/cm². Micrographs of unetched tracks in a particle of lunar soil and in the Fayetteville meteorite are presented in Fig. 3. Although the meteorite studies are still at a preliminary stage, our observations of track densities comparable to those in lunar grains and at least 20 times greater than had been originally reported in track-rich meteorite grains (LAL and RAJAN, 1969; PELLAS *et al.*, 1969) are highly significant because they remove one of the previous major distinctions (BORG *et al.*, 1970) between lunar grains and meteoritic grains and suggest the possibility of a similar origin. Track densities of $2 \times 10^{10}/\text{cm}^2$ or higher are present in the interior of crystals more than 10 μm diameter within sections of the Fayetteville meteorite which were thinned by ion-beam machining. Similar high track densities also exist in smaller ($\leqslant 0.5$ μm diameter) euhedral crystals. In previous optical microscope and scanning electron microscope studies of etched grains of gas-rich meteorites those grains with track densities exceeding $10^{10}/\text{cm}^2$ were never noticed because they completely dissolved in the standard etching process. The tracks in meteorites have been mainly studied by diffraction contrast imaging (i.e., without etching); the absence of the amorphous layer present on lunar grains makes etching less beneficial (BARBER *et al.*, 1971) in seeing tracks. And etching the meteorites is disadvantageous because it can cause grains which are barely held within the thinned and weakened fabric to drop out. We have established, however, that the tracks in most lunar and meteoritic minerals are etchable, under suitable conditions.

If radiation-induced flaking is a more important erosional process than is sputtering, we might be able to attribute the track-rich grains in the lunar soil to flaked-off surface layers of rocks. We believe, however, that there are difficulties with this simple picture. The surfaces of rocks show no evidence of extreme stress nor do they contain track densities as high as $10^{10}/\text{cm}^2$. The majority of the track-rich grains, both in the lunar soil and in the meteorites (Fig. 3), exhibit rather nice electron diffraction patterns that argue against extreme radiation damage. Admittedly, many of the fines have amorphous outer layers (thickness ~ 500 Å) attributed to accumulated damage by solar wind bombardment but the interiors are still crystalline, (Fig. 3(a)). These findings are in agreement with the work of DRAN *et al.* (1970) and BORG *et al.* (1970), who emphasized that the grains were not disordered. The observed euhedral habits of some of the meteoritic track-rich grains also argue against radiation stress-induced fracture. In the Kapoeta meteorite, however, electron microscopy reveals that many of the small grains contain minute cracks and microstructural features are severely distorted. The electron diffraction patterns correspondingly exhibit arcs and extended spots. So far we have failed to see tracks in the carbonaceous chondrites, Murray and Orgeuil, and other observations we have made suggest that tracks will not be found.

We have recently suggested that some of the highly irradiated lunar grains are fragments of infallen extra-lunar dust (BARBER *et al.*, 1971). It has previously been suggested that some of the gas-rich meteorites were assembled by sintering of circum-solar grains (LAL and RAJAN, 1969; PELLAS *et al.*, 1969). Continuing observations of ion-beam-thinned sections of gas-rich meteorites should provide severe constraints on their mode of origin. It would be especially useful to find a large grain with a high track density and a gradient that could be related to an erosional process.

We regard it as highly unlikely that suprathermal heavy ions were responsible for the observed high track densities. A suprathermal ion energy spectrum should continue to rise to a peak at low energy, so that the track length distribution should

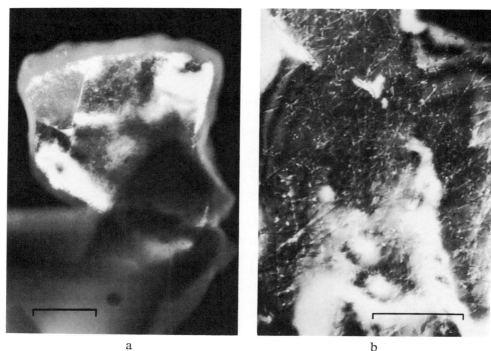

a   b

Fig. 3. (a) Dark field electron micrograph (650 KeV) of tracks of solar flare particles in
a grain from the Apollo 12 lunar fines. Bar mark 0.5 μm. (b) Dark field electron
micrograph (650 KeV) of fossil particle tracks in a section of the Fayetteville meteorite
which was thinned by sputter-etching with 5 KeV argon ions. Bar mark 0.5 μm.

be peaked at short lengths. Our electron microscope observations of both etched and
unetched tracks show that the track length distribution on lunar grains $\geqslant 10$ μm
thinned by sputter-etching is not peaked at short length and is typical of randomly
oriented tracks that penetrate the entire grain.

Acknowledgments—We are indebted to A. Ghiorso for the Ar and Kr ion irradiations, to N. Nickle
and D. Robertson for help with the Surveyor glass, to K. C. Hsieh, J. D. Sullivan, C. O. Bostrom,
L. J. Lanzerotti, J. R. Arnold, and R. C. Reedy for helpful discussions and data on solar protons
and alpha particles, and to NASA and the National Science Foundation for financial support.

References

Ahrens T. J., Fleischer R. L., Price P. B., and Woods R. T. (1970). Erasure of fission tracks in
glasses and silicates by shock waves, *Earth Planet. Sci. Lett.* **8,** 420–426.
Albee A. L., Burnett D. S., Chodos A. A., Haines E. L., Huneke J. C., Papanastassiou D. A.,
Podosek F. A., Russ G. P., Tera F., and Wasserburg G. J. (1971) Rb-Sr Ages, chemical abundance
patterns and history of lunar rocks. Second Lunar Science Conference (unpublished proceedings).
Barber D. J., Hutcheon I., and Price P. B. (1971) Extralunar dust in Apollo cores? *Science* **171,**
372–374.

BORG J., DRAN J. C., DURRIEU L., JOURET C., and MAURETTE M. (1970) High voltage electron microscope studies of fossil nuclear particle tracks in extra-terrestrial matter. *Earth Planet. Sci. Lett.* **8,** 379–386.

BOSTROM C. O., WILLIAMS D. J., and ARENS J. F. (1968–70) Solar proton monitoring. Unpublished data in ESSA Solar-Geophysical Data Bulletins.

CROZAZ G. and WALKER R. M. (1971) Solar particle tracks in glass from the Surveyor 3 spacecraft. *Science* **171,** 1237–1239.

CROZAZ G., WALKER R., and WOOLUM D. (1971) Cosmic ray studies of "recent" dynamic processes. Second Lunar Science Conference (unpublished proceedings).

CROZAZ G., HAACK U., HAIR M., MAURETTE M., WALKER R., and WOOLUM D. (1970) Nuclear track studies of ancient solar radiations and dynamic lunar processes. *Proc. Apollo 11 Lunar Sci. Conf., Geochim. Cosmochim. Acta* Suppl. 1, Vol. 3, pp. 2051–2070. Pergamon.

DRAN J. C., DURRIEU L., JOURET C., and MAURETTE M. (1970) Habit and texture studies of lunar and meteoritic materials with a 1 MeV electron microscope. *Earth Planet. Sci. Lett.* **9,** 391–400.

FINKEL R. C., ARNOLD J. R., REEDY R. C., FRUCHTER J. S., LOOSLI H. H., EVANS J. C., SHEDLOVSKY J. P., IMAMURA M., and DELANY A. C. (1971) Depth Variation of cosmogenic nuclides in a lunar surface rock. Second Lunar Science Conference (unpublished proceedings).

FLEISCHER R. L., HAINES E. L., HART H. R., WOODS R. T., and COMSTOCK G. M. (1970) The particle track record of the Sea of Tranquility. *Proc. Apollo 11 Lunar Sci. Conf., Geochim. Cosmochim. Acta* Suppl. 1, Vol. 3, pp. 2103–2120. Pergamon.

FLEISCHER R. L., HART H. R., and COMSTOCK G. M. (1971) Very heavy solar cosmic rays: Energy spectrum and implications for lunar erosion. *Science* **171,** 1240–1242.

FRANK L. A. (1970) On the presence of low energy protons ($5 \leq E \leq 50$ keV) in the interplanetary medium. *J. Geophys. Res.* **75,** 707–716.

GANAPATHY R., KEAYS R. R., LAUL J. C., and ANDERS E. (1970a) Trace elements in Apollo 11 lunar rocks: implications for meteorite influx and origin of moon. *Proc. Apollo Lunar Sci. Conf., Geochim. Cosmochim. Acta* Suppl. 1, Vol. 2, pp. 1117–1142. Pergamon.

GANAPATHY R., KEAYS R. R., and ANDERS E. (1970b) Apollo 12 Lunar samples: trace element analysis of a core and the uniformity of the regolith. *Science* **170,** 533–535.

HÖRZ F., HARTUNG J. B., and GAULT D. E. (1971) Lunar microcraters. Second Lunar Science Conference (unpublished proceedings).

HSIEH K. C. and SIMPSON J. A. Private communication of unpublished results.

LAL D., MacDOUGALL D., WILKENING L., and ARRHENIUS G. (1970) Mixing of the lunar regolith and cosmic ray spectra: Evidence from particle-track studies. *Proc. Apollo 11 Lunar Sci. Conf., Geochim. Cosmochim. Acta* Suppl. 1, Vol. 3, pp. 2295–2303. Pergamon.

LAL D. and RAJAN R. S. (1969) Observations relating to space irradiation of individual crystals of gas-rich meteorites. *Nature* **223,** 269–271.

LANZEROTTI L. J. (1969–70) Unpublished data in World Data Center Reports.

PELLAS P., POUPEAU G., LORIN J. C., REEVES H., and AUDOUZE J. (1969) Primitive low-energy particle irradiation of meteoritic crystals. *Nature* **223,** 272–274.

PRICE P. B. and O'SULLIVAN D. (1970) Lunar erosion rate and solar flare paleontology. *Proc. Apollo 11 Lunar Sci. Conf., Geochim. Cosmochim. Acta* Suppl. 1, Vol. 3, pp. 2351–2359. Pergamon.

PRICE P. B., HUTCHEON I., COWSIK R., and BARBER D. J. (1971) Enhanced emission of iron nuclei in solar flares. *Phys. Rev. Lett.* **26,** 916–919.

PRICE P. B., RAJAN R. S., and TAMHANE A. S. (1967) On the preatmospheric size and maximum space erosion rate of the Patwar stony-iron meteorite. *J. Geophys. Res.* **72,** 1377–1388.

ROSS J. E. (1970) Abundance of iron in the solar photosphere. *Nature* **225,** 610–611.

SEITZ M., WITTELS M. C., MAURETTE M., and WALKER R. M. (1970) Accelerator irradiations of minerals: implications for track formation mechanisms and for studies of lunar and meteoritic minerals. *Rad. Effects* **5,** 143–148.

SHOEMAKER E. M. and HAIT M. H. (1971) The bombardment of the lunar maria. Second Lunar Science Conference (unpublished proceedings).

Wänke H., Rieder R., Baddenhausen H., Spettel B., Teschke F., Quijano-Rico M., and Balacescu A. (1970) Major and trace elements in lunar material. *Proc. Apollo 11 Lunar Sci. Conf.*, *Geochim. Cosmochim. Acta*, Suppl. 1, Vol. 2, pp. 1719–1727. Pergamon.

Wehner G. K., Kenknight C., and Rosenberg D. L. (1963) Sputtering rates under solar-wind bombardment. *Planet. Space Sci.* **11**, pp. 885–895.

Wolnik S. J., Berthel R. O., and Wares G. W. (1970) Shock-tube measurements of absolute gf-values for FeI. *Astrophys. J.* **162,** 1037–1047.

Proceedings of the Second Lunar Science Conference, Vol. 3, pp. 2715–2719
The M.I.T. Press, 1971.

Microbiological sampling of returned Surveyor III electrical cabling

M. D. Knittel, M. S. Favero*, and R. H. Green

California Institute of Technology, Jet Propulsion Laboratory, Pasadena, California 91103

(*Received* 24 *February* 1971; *accepted in revised form* 31 *March* 1971)

Abstract—A piece of electrical cabling was retrieved from the Surveyor III spacecraft by the crew of Apollo 12 and subjected to microbiological analysis for surviving terrestrial microorganisms. The experiment was done in a sealed environmental chamber to protect against contamination. No viable microorganisms were found on the wiring bundle samples.

Introduction

The survival of microorganisms in space environments has been studied by simulation of planet atmospheres (Green *et al.*, 1970) or by exposing selected cultures during spacecraft flights (deSenes, 1969). Others have studied microbial survival in space by considering a single parameter such as vacuum (Morelli *et al.*, 1962).

The mission plan of Apollo 12 to land near the site of the Surveyor III spacecraft offered a unique opportunity for retrieval of selected parts for scientific and engineering studies. The microbiological examination of parts of the spacecraft could provide information concerning the survival of microorganisms in the environment of space. The part selected for microbiological examination was a piece of electrical wiring bundle running from the TV camera to another part of the spacecraft. It was selected because (1) previous information obtained during the planetary quarantine monitoring of Mariner-Mars 1969 had shown that there was a high level of bacterial contamination associated with wiring bundles (Christensen, 1969), (2) Surveyor III spacecraft had not been sterilized prior to launch; and (3) the cable could easily be removed and packaged against contamination.

It was realized that this was not an "ideal" experiment because the kinds and number of microorganisms initially present on the cable were not known and there was not available a "control" such as an identical Surveyor cable exposed to terrestrial environment for the length of time Surveyor III had been on the moon.

Microbiological Materials and Methods

Media

The bacteriological media used in this study were Eugon and thioglycollate broth (Difco, Detroit, Michigan). The thioglycollate broth was chosen for isolation of any anaerobic bacteria, and the Eugon broth for the growth of aerobic bacteria.

Equipment

The sampling of the wiring bundle was carried out in a glove box manufactured by Blickman Co. (Wehauken, New Jersey). Tools used to remove pieces of the cable were forceps, wire strippers, scissors, wire cutters, and a vise.

---

* Phoenix Field Station, U.S. Public Health Department, Phoenix, Arizona.

Sterilization of materials

The interior of the glove box and surfaces of containers entering the glove box were sterilized with 2% paracetic acid according to NASA standards (NASA, 1968). Media to be used in the assay were placed in glass screw cap tubes and sterilized by autoclaving at 121°C for 20 minutes. Tools and other hardware were placed in metal cans and sterilized by dry heat at 180°C for 4 hours.

Electrical Wiring Bundle Sampling Methods

The electrical wiring bundle was dissected into its component parts and each piece placed into one of the aforementioned bacteriological culture media. A culture enrichment method was chosen over a dilution and plating procedure because the numbers of surviving microorganisms (if any) were expected to be low and would be missed if a serial dilution and culturing technique were used. The sampling proceeded first by removing the outside nylon ties that held the wrappings, then by removing pieces of the exterior wraps until the bundle of wires was entirely exposed. The insulation was removed from each individual wire with wire strippers, and pieces of the exposed stranded wire removed with wire cutters. These procedures were carried out inside a stainless steel glove box to isolate the experiment from airborne bacterial contamination.

The sampling procedure included a sterilized wiring bundle that was sampled at random times during the dissection of the Surveyor III wiring bundle. This "control" was an internal procedural control to detect contamination that might occur during the course of the Surveyor III wiring bundle assay. These sterile control samples represented 10% of the total and were an internal standard for contamination monitoring.

Results

The objective of this experiment was to determine if terrestrial microorganisms that were present on Surveyor III when it was launched could survive 27 months of lunar exposure. If, during the actual sampling of the wires a contaminant were accidently introduced, it would be impossible to separate it from a lunar survivor. Therefore, in order to perfect the technique, it was necessary to perform several simulated assays with a piece of sterile wiring bundle before the lunar sample was assayed (Table 1). During these simulated assays, all of the procedures that were to be applied in sampling of the Surveyor III cable were used to determine if the sampling could be done without contamination. These procedures increased the confidence that the Surveyor III cable could be examined without contamination.

Prior to opening the Sealed Environmental Sampling Container (SESC), containing the Surveyor III cable and other parts, it was checked and was found to have leaked. Also, when the SESC was opened, it was found to contain a high concentration of oxygen. Previous sampling plans were thus modified to swab sample the outside and inside of each wrap to determine if airborne bacteria had passed through the leak and contaminated the surface wraps of the Surveyor III cable.

Table 1. Results of dissection of sterilized cable during three separate simulated sampling runs.

Sampling number	Number of samples		Number positive
	Eugon	Thioglycollate	
1	20	20	0
2	20	20	0
3	20	20	0

Table 2. Results of swab sampling of sterile wiring bundle surface wraps inoculated with *Bacillus subtilis* var. *niger*.

Test cable number	Wrapping area samples		Colonies *B. subtilis*
1	Exterior wrap		
		Outside	2.6×10^3
		Inside	7.7×10^3*
	Interior wrap		
		Outside	0
		Inside	0
	Interior wires		0
2	Exterior wrap		
		Outside	9×10^2
		Inside	0
	Interior wrap		
		Outside	0
		Inside	0
	Interior wires		0

* During the unwrapping manipulation, the wrap slipped from the forceps and curled back upon itself.

There was concern whether airborne bacterial contamination of the exterior wraps would penetrate to the interior of the wiring bundle. This was resolved by inoculating the surface wraps of a sterile wiring bundle with a spore suspension of *Bacillus subtilis* var. *niger* and performing a swab assay of the wraps surface on the bundle. This was done to learn whether bacterial contamination deposited on the surface would also contaminate the interior wires when manipulated during assay. Table 2 shows the results of this experiment and as can be seen if care is exercised during removal of the wraps, the contamination on the exterior surfaces remains on that surface.

The results from the inoculated control bundle gave confidence that if airborne bacteria did pass into the SESC through the leak, the wiring bundle wraps would protect the wires beneath it from contamination. During the assay of the Surveyor cable, the outside and inside of each wrap was swabbed and the swabs placed into either thioglycollate or Eugon broth. The results of this and the culture enrichment of the pieces of the Surveyor III wiring bundle are presented in Table 3. The swab samples of the surface wraps were all negative showing that either no airborne bacteria entered through the SESC leak or if they did they were not detected. The 69 samples of various parts of the wiring bundle which were placed in culture produced no

Table 3. Results of culture enrichment of Surveyor III wiring bundle

Number of samples	Sample description	Results
6	Nylon ties	0
7	Protective wrap	0
23	Insulation from wires	0
13	Wire	0
17	Wire and Insulation	0
3	Teflon sleeving	0
9	Wire or Insulation⎫ Sterile controls ⎭	0
8	Wrapping swabs	0

growth after 6 weeks of incubation at 25°C. The sterile control samples and environmental fallout plates were also all negative.

In order to determine if the procedure contained processes that would be inhibitory to the isolation of any survivors on the Surveyor cable, a piece of unsterilized control wiring bundle was sampled using the same procedures. This wiring bundle was prepared in the electronics shop at this laboratory according to the original specifications for the Surveyor spacecraft. The bundle was placed in a sterile screw-cap test tube and subjected to the same assay procedures as the Surveyor III wiring bundle. The bacteria isolated from this bundle do not represent the population that survived 27 months of terrestrial storage but rather it shows the types of bacteria that may have been on the Surveyor III wiring bundle during its assembly. A total of 30 positive samples were obtained from 40 samples taken. Twenty-one samples yielded gram-positive cocci, 6 spore-forming rods, 2 gram-positive nonspore-forming rods and 1 gram-negative rod. Random numbered sterile controls taken at the same time as the sampling were all negative. Agar plates exposed during various times of the sampling were also negative.

DISCUSSION

The results show that no viable microorganisms were recovered from that portion of Surveyor III cable that was sampled. Some factors that could have contributed to the sterility of the cable are prelaunch testing such as thermal vacuum testing, die away, the change in pressure during launch, and lunar vacuum and temperature.

The thermal and vacuum testing of the Mariner Mars '71 spacecraft has been found to reduce the number of viable microorganisms. The vacuum is in the order of 1×10^{-5} torr and temperatures range between -127 and $+127$°C (WANG, 1971). It has been found that an approximate 80% reduction in the number of spore-forming bacteria occurred as a result of this testing, and more than 90% reduction in the number of viable nonspore-forming bacteria had also occurred. The prelaunch thermal and vacuum testing of the Surveyor III spacecraft could have accounted for a major reduction in the bacterial contamination.

Recently, in our laboratory, it has been shown that when a surface is protected from a redeposition of microorganisms, such as within the layers of a thermal blanket, the initial population of microorganisms is reduced to near zero during 100 days of storage. The wiring bundle was most likely prepared and wrapped during assembly and not reopened again prior to launch; thus a redeposition of microorganisms could not have occurred during preflight testing. The initial population of microorganisms may have been high during assembly of the wiring bundle but because of die off this number may have been considerably reduced.

The remaining population of microorganisms on the wiring bundle would have been further reduced by the change in atmospheric pressure during the launch of Surveyor III. Research at this laboratory (KNITTEL, 1971) has shown that when dried bacteria are subjected to a rapid change in pressure from 760 torr to 1×10^{-5} torr within a 12-minute period, a loss of viability does occur and can cause a decrease in population of 10% with spore-forming bacteria and up to 50% with nonspore-forming bacteria.

The exposure of bacteria to high vacuum (10^{-10} torr) has shown that vacuum itself is not sterilizing even though a reduction in numbers of viable bacteria does occur. However, if during the vacuum exposure the cells or spores are also heated to 60°C and above, death of the bacteria is accelerated. For instance, *Bacillus subtilis* var. *niger* spores when exposed to 10^{-10} torr vacuum and heated to 60°C for 14 days lost 69 % of viability. *Stapholococcus epidermiditis* exposed to the same conditions lost 99 % of viability. During the 27 months of lunar exposure of Surveyor III, the spacecraft was exposed to vacuum and to temperature cycling which could reduce the numbers of any surviving bacteria even more.

The conclusion that is reached is that no microorganisms survived on the wiring bundle during its lunar exposure. This is not to say that a microorganism cannot survive exposure to lunar environment, but only that none were found on the returned piece. However, if a larger sample of wiring bundle had been examined for surviving microorganisms, the probability of finding a survivor would have been greater. A future controlled experiment should be considered to expose microorganisms to the lunar environment during an Apollo mission so that a more definitive answer can be obtained about survival of microorganisms in the lunar environment.

Acknowledgements—The authors wish to thank the crew of Apollo 12 for their successful flight and return of material used in this study. The authors would also like to thank G. M. RENNINGER, MARK ADAMS, J. H. STEVENS, and D. C. SCHNEIDER for their assistance in planning for this experiment. A special note of thanks is extended to Dr. D. M. TAYLOR for his helpful discussion during the course of this experiment.

REFERENCES

CHRISTENSEN M. R. (1969) Microbiological Monitoring of Mariner VI and VII Spacecraft. Environmental Requirements Section Informal Report No. 604–54.

GREEN R. H., TAYLOR D. M., GUSTAN E. A., FRASER S. J., and OLSON R. L. (1970) Survival of microorganisms in a simulated martian environment. *Space Life Science* **11**, 43–55.

KNITTEL M. D., GODREY J. F., HAGEN C. A., and TAYLOR D. M. (1971) Survival of spores and nonspore-forming bacteria during simulation of spacecraft launch pressure profile. American Society of Microbiology, *Proceedings*, p. 15.

MORELLI F. A., FEHLNER F. D., and STEMBRIDGE C. H. (1962) Effect of ultra-high vacuum on *Bacillus subtilis* var. *niger*. *Nature* **196**, 106–107.

DESENES F. J. (1969) Effects of radiation during spaceflight on microorganisms and plants on Biosatellite II and Gemini missions. *Life Science and Space Research* **7**, 62–66.

NASA (1968) Procedures for the Microbiological Examination of Space Hardware NHB 5340.1A. Washington, D.C.

WANG J. T. (1971) personal communication.

Proceedings of the Second Lunar Science Conference, Vol. 3, pp. 2721–2733
The M.I.T. Press, 1971.

Surveyor III: Bacterium isolated from lunar-retrieved TV camera

F. J. Mitchell*

Lunar Receiving Laboratory, Manned Spacecraft Center, Houston, Texas 77058

and

W. L. Ellis†

Brown and Root-Northrop, Manned Spacecraft Center, Houston, Texas 77058

(*Received* 9 *February* 1971; *accepted in revised form* 31 *March* 1971)

Abstract—Selected components of the unmanned Surveyor III spacecraft which had remained on the lunar surface for $2\frac{1}{2}$ years were collected and returned to earth by the crew of Apollo 12. A bacterium, *Streptococcus mitis*, was isolated from a sample of foam taken from the interior of the retrieved TV camera. The available data suggests that the bacterium was deposited in the camera prior to the Surveyor III spacecraft launch. The authors suggest that lyophilizing conditions existing during prelaunch vacuum testing and later on the lunar surface may have been instrumental in the apparent survival of this microorganism.

INTRODUCTION

On 20 April 1967, the unmanned Surveyor III spacecraft successfully landed near the eastern shore of Oceanus Procellarum on the lunar surface. On 20 November 1969, two Apollo 12 crew members walked from their lunar module to inspect and photograph the Surveyor III spacecraft. The entire TV camera and other selected components were then retrieved for return to earth (Fig. 1). Upon return to earth, the TV camera and lunar soil samples were placed in quarantine in the Lunar Receiving Laboratory (LRL) at the NASA Manned Spacecraft Center (MSC) at Houston, Texas. The quarantine was lifted on 7 January 1970, and inspection and disassembly of the retrieved TV camera began the next day.

Microbial analysis was the first of several studies of the retrieved TV camera and was performed immediately after the camera was opened. A serious constraint placed upon this analysis was the need to obtain samples without compromising any planned subsequent studies. As a consequence, not all desired microbial samples could be obtained. The emphasis of the microbial analysis was placed, therefore, upon isolating microorganisms which might be potentially pathogenic for man.

Decontamination measures taken before the Surveyor III launch did not eliminate the possibility that the spacecraft carried microorganisms to the moon. The following statement reflects the decontamination guidelines which were current at the time of the Surveyor spacecraft launches (Hall, 1966): "The precautions against the contamination of the moon, once strict have now been relaxed in view of our developing

---

* A United States Air Force Major currently on assignment to the Preventive Medicine Division at the NASA Manned Spacecraft Center, DC72, Houston, Texas 77058.

† Mailing address: Brown and Root-Northrop, P.O. Box 34416, Houston, Texas 77034.

Fig. 1. Surveyor III spacecraft on its lunar landing site with Astronaut Conrad inspecting the TV camera. The lunar module Intrepid appears in the background.

knowledge of the inhospitable environment for terrestrial life that exists on the lunar surface and the belief that landed contamination, if it survives, will remain localized. For these reasons, lunar landing spacecraft may have on board a low level of microbial life—they must be decontaminated, but not sterile."

The extensive experience gained from Apollo 11 and Apollo 12 indicated that extraterrestrial microorganisms would not be isolated from Surveyor III (OYAMA *et al.*, 1970, 1971; TAYLOR *et al.*, 1970, 1971). The recovery of terrestrial microorganisms originally present in the TV camera would be possible if these microorganisms had been able to survive in the lunar environment. However, verifying the origin of any isolate would be complicated by the possibility of postretrieval contamination.

It had not been anticipated at launch in 1967 that the TV camera would be returned to earth at some future time. Consequently, no prelaunch microbial analysis of the camera interior was performed, and therefore, no appropriate experimental control was available for comparison. However, substitute controls were available. Several identical backup Surveyor TV cameras had been held in bonded storage during the same time period that Surveyor III had remained on the lunar surface. One backup TV camera was used to refine techniques for disassembly and microbial sampling prior to performing any definitive procedures on the retrieved Surveyor III camera. A second backup TV camera, designated the type approval test camera (TAT–1), was disassembled after the Surveyor III camera and sampled identically to the retrieved Surveyor III TV camera.

DISASSEMBLY AND SAMPLING PROCEDURES

The retrieved TV camera was placed in a laminar-outflow hood equipped with high-efficiency particulate air filters (Fig. 2) in the LRL astronaut debriefing room, which has an air-conditioning system separate from the system used by the rest of the LRL. Every surface of the laminar flow hood which would be exposed to the camera was thoroughly washed twice with isopropyl alcohol prior to the camera being placed into the hood. A sterile cloth was placed on the floor of the hood to retain any lunar material which might accumulate as a result of the disassembly procedures. Only those personnel directly responsible for disassembling and sampling the TV camera were permitted in the room. They were clothed in laboratory attire, including surgical caps, face masks, and sterile gloves. Other participating personnel observed and coordinated activities from behind a viewing window.

Fig. 2. The retrieved Surveyor III TV camera, complete with shroud, original collar, and cables, as it appeared in the laminar-flow hood of the Lunar Receiving Laboratory. Sampling sites are numbered.

To prepare the TV camera for disassembly, the original collar of the camera was removed and replaced with a special tripod permitting easy manipulation of the camera in the laminar-flow hood. To remove the camera shroud, the outer aluminized and inner clear Teflon wrappings were removed from the cable connectors. The cable connectors were sampled and then washed with isopropyl alcohol. Retaining screws on the shroud were removed and the cable connectors pushed inside the shroud. The shroud was then removed from the bottom of the camera and the biological samples were immediately taken. The shroud fit very tightly on the camera, and although the camera was not hermetically sealed the interior of the camera was extremely clean. No evidence of lunar material was observed within the TV camera when the shroud was removed (Fig. 3). The only evidence that the camera had been launched to the moon and retrieved was a small number of particles (no larger than 1 mm³) which had accumulated in the bottom of the shroud. These particles were determined to be bits of ceramic insulation which had shaken loose during the flight to the moon or during the return flight (RIGLIN, CARROL, private communication, 1970).

Identical procedures were used for sampling the Surveyor III and the TAT–1 TV cameras. Three sterile calcium alginate swabs were arranged with the swab heads in tandem, moistened with sterile phosphate-buffered saline (0.0003 M $PO_4{}^{3-}$,

Fig. 3. Surveyor III camera interior with shroud and cables removed. Sampling sites are numbered.

Table 1. Microbial sampling sites of the Surveyor III and TAT–1 TV cameras. Sampling sites 1 and 2 are exterior camera samples pertaining to the Surveyor III TV camera only. The TAT–1 camera had no collar or cables; consequently, no sample of site 1 was taken. Site 2 included all three exterior metal cable connector surfaces.

| | Tube number | | |
Sampling site	TSB	THIO	YMB
1. Metal surface under front half of collar	1	11	21
2. Nylon ties, Teflon wrapping, cable connector suface	2	12	22
3. Surface area on support studs	3	13	23
4. Surface are on electronic conversion unit	4	14	24
5. Circuit board support-plate edges and screw studs	5	15	25
6. Surface area of all three cable connectors inside camera	6	16	26
7. Nylon ties and cable wrappings	7	17	27
8. Debris in bottom of shroud	8	18	28
9. Large area on inside of shroud	9	19	29
10. Top surface of exposed circuit boards	10	20	30
11. Foam samples from between circuit boards	31	32	33

0.147 M NaCl), and used to swab the maximum surface area of each site (Table 1). The swabs were then separated. One each was placed into 5 ml of trypticase soy broth (TSB) for aerobic analysis, 5 ml of thioglycollate broth (THIO) for anaerobic analysis, and 5 ml of yeast malt broth (YMB) containing 33 units/ml of penicillin G and 62 μg/ml of streptomycin for mycological analysis. In confined areas where this method of swabbing could not be used, three sequential samples were taken and placed in the appropriate media. The first such sample was always placed into TSB, the second into THIO, and the third into YMB.

Dry swabs, arranged as described previously, were employed at three sampling sites because of the nature of the material to be sampled or the requirements of prescribed followup studies. These samples included the bits of ceramic debris in the camera shroud base, the cable surfaces in the camera interior, and the top surface of the circuit boards. Samples numbered 31, 32, and 33 consisted of bits of polyurethane foam. This foam had been used as insulation between the two aluminum plates of the circuit boards. The space between the aluminum plates was approximately 4 mm. This thin layer of foam was accessible only where holes had been cut into the plates for the placement of electronic components. Only by using long, curved, needle-nosed forceps could one reach through the hole and into the space between the aluminum plates to obtain bits of the foam. The largest bit of foam that was extracted was approximately 1 mm³. Samples obtained with forceps or with dry swabs were cultured according to the same procedures and in the same media as prescribed for wet-swab samples.

The protocol established for the aerobic and anaerobic analyses (Fig. 4) maximized the possibility of detecting and quantitating low numbers of microorganisms in a sample while at the same time yielding valuable clues as to the source of any micro-organism detected. The protocol inherently contained a system of redundancy and cross-checks designed to identify suspected laboratory contamination. For example, growth on any BA plate from the 10^2 dilution tube without simultaneous growth in the original tube, the two dilution tubes, and on the BA plates from the 10^1 dilution

AEROBIC OR ANAEROBIC FLOW

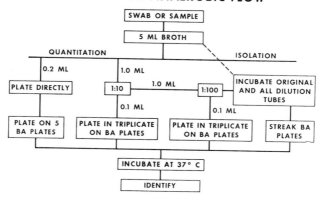

Fig. 4. Protocol established for the aerobic and anaerobic samples. This protocol was followed for both the retrieved Surveyor III TV camera and the backup TAT-1 TV camera.

tube and from the original tube containing the foam sample would be suspect. Growth on any BA plate without growth in the tube from which aliquots were taken to place on the BA plate would require extreme care in interpretation, and in this case one would probably suspect contaminated BA plates. Growth in either the 10^1 or the 10^2 dilution tube without growth in the original tube would require some logical reason why growth did not occur in the undiluted tube. The replicate BA plates provide the requirement for consistent results, within experimental error, on each plate and provide a check on techniques used in making dilutions. Streaking fresh BA plates with aliquots from each tube after 24 hours of incubation was intended to provide an opportunity for early isolation and separation in case the sample contained more than one microorganism with the result that one specie might overgrow another specie if the tube was allowed to incubate to full turbidity without examination. Again this operation provided another opportunity to cross-check with the results from the previous dilutions and plating. With growth on the 10^2 dilution BA streak plate one would also expect growth on the 10^1 dilution BA streak plate and on the BA streak plate from the original undiluted tube. Whatever the results, any observed growth would have to be consistent and logical in view of the redundancy and cross checks built into the protocol. Obvious cases of laboratory contamination could easily be identified and reported as such.

The swab or sample was placed into 5 ml of the selected broth and vortexed. One ml of this broth was spread in 0.2 ml aliquots onto five blood agar (BA) plates. Aliquots of 0.1 ml were taken from 10^1 and 10^2 dilutions of the broth containing the selected sample and spread on BA plates in replicates of three. The tubes containing the original samples, the two dilution tubes, and all plates were incubated for 24 hours. The TSB tubes and a set of BA plates containing 5% sheep (lamb) blood were incubated aerobically. The THIO tubes and a second set of BA plates were incubated anaerobically, using GasPak (BBL) systems in stainless steel jars. Aliquots from each tube were streaked onto fresh BA plates, and all plates and tubes were returned for incubation at 37°C for 30 days. Any observed growth on the plates was quantitated and identified. Growth in the incubated tubes was also identified. The YMB tubes were handled according to the established LRL procedures for mycological analysis.

SURVEYOR III CAMERA RESULTS

The only sample to produce visible microbial growth was sample number 32, a 1 mm^3 piece of foam incubated in undiluted thioglycollate broth. The initial growth was observed on the fourth day of incubation as a white "tail" of growth 2–3 mm in length, hanging from the piece of foam which was floating in the middle of the tube. No other growth was observed on that day. The next day this tube was turbid with growth and the 10^1 dilution tube exhibited approximately 100 foci of growth scattered predominantly at the top of the tube. No growth was observed in the 10^2 dilution tube or on any BA plate or BA streak plate for the remainder of the study.

In both tubes containing growth, only a single cellular morphology was observed, that of a Gram positive coccus in chains. Since the initially observed growth had required 4 days of incubation in THIO and since no growth was observed on the initial five anaerobic BA plates, these media were again inoculated with the isolate. In addition, TSB and aerobic BA plates were inoculated with the isolate. Growth was observed in both THIO and TSB within 24 hours. Growth was observed on the aerobic BA plates within 24 hours and on the anaerobic BA plates within 72 hours (first examination). As a precaution, 1 ml aliquots containing respectively 10^3, 10^4, 10^5, and 10^6 viable cells of the isolate were injected intraperitoneally (in replicates of five) into 5-week-old white male CD–1 mice, with no observed effect.

The isolate was identified, with confirmation from the U.S. Public Health Service Center for Disease Control in Atlanta, Georgia, as alpha hemolytic *Streptococcus mitis* (FACKLAM, private communication, 1970).

TAT–1 TV CAMERA RESULTS

The results from the backup TAT–1 camera, sampled identically as the retrieved Surveyor III camera, provide observations on microbial survival at ambient atmospheric pressure and room temperature for the same period of time that the Surveyor III camera rested on the lunar surface. The TAT–1 camera was held undisturbed in bonded storage in its original shipping container for this time period. Terrestrial microorganisms were isolated in very low numbers from one exterior and five interior locations. One bacterial isolation and five mycological isolations were made after long incubation periods varying from 6 to 27 days. All six isolations were made from accessible metallic and nonmetallic sampling sites.

From a Teflon-covered cable within the TAT–1 camera, a *Bacillus* species was isolated in THIO after 6 days of incubation. Growth appeared only in the tube containing the undiluted sample. An *Aureobasidium* species was isolated after a sampling of the TAT–1 exterior metal cable connectors was incubated in YMB for 14 days. The same species was also isolated after a sampling of the metal electronic conversion unit within the TAT–1 camera was incubated in THIO for 27 days. *Aspergillus pulvinus* was isolated from three sites in the interior of the camera. This isolate was detected after a sampling taken from the top surface of the nonmetallic circuit board was incubated in THIO for 12 days. A second isolation was made from a sampling of the metal cable connectors after 14 days of incubation in TSB. The third isolation was made from a sampling of the metal electronic conversion unit after 21 days of incubation in YMB.

In the five fungal isolations, growth appeared only in the tube containing the undiluted sample, indicating very low numbers of microorganisms originally present on the sampled surfaces. To illustrate, when three swabs were used in tandem to sample one of the selected sites, *Aspergillus pulvinus* was isolated from only one of the swabs; an *Aureobasidium* species was isolated from a second swab; and the third swab was negative.

Discussion

Every step in the retrieval of the Surveyor III TV camera was analyzed for possible contamination sources, including camera contact by the astronauts; ingassing in the lunar module and command module during the mission or at "splashdown"; and handling during quarantine, disassembly, and analysis at the LRL.

Contact by the astronauts during retrieval on the moon was not considered a probable source of contamination. Microorganisms were undoubtedly present on exterior surfaces of the astronauts' space suits during each lunar landing and selenological sample collection excursion. However, no viable terrestrial microorganism has ever been detected in the selenological samples collected by the astronauts (Oyama *et al.*, 1970, 1971; Taylor *et al.*, 1970, 1971).

After the TV camera was removed from the Surveyor III spacecraft, it was placed into a back pack carried by one of the astronauts. The pack was zipper-closed although there was no capability for sealing it. The pack was placed in storage first aboard the lunar module and then the command module, and finally was flown to the LRL by jet aircraft. At the LRL, the TV camera was removed from the pack and placed in a Teflon bag. The bag was heat sealed and then the camera and first bag were placed into a second Teflon bag which was also heat sealed. The double-bagged camera was then placed in bonded storage at room temperature until the lunar sample was released on 7 January 1970.

When the Apollo 12 lunar module landed on the moon, lunar dust was disturbed with such force that it traveled approximately 155 m with a reported velocity of at least 70 m/sec and "sandblasted" the Surveyor III spacecraft (Jaffe, 1971). Shadows in the exterior paint of the Surveyor III TV camera were clearly visible wherever a strut or other part had shielded the camera from this hail of lunar particles caused by the lunar module rocket exhaust.

While the TV camera was being disassembled it was observed that barely visible particles of lunar dust had accumulated underneath the camera collar. The presence of this fine dust in this protected area is a reflection of the minute size of some lunar particles and the "sandblasting" force which caused the penetration. It has already been noted that no such presence or accumulation of lunar particles was found in the interior of the camera protected by the shroud despite the "sandblasting." This suggests that the camera shroud may have provided a formidable barrier to ingassing carrying fine particles, perhaps even the size of a bacterium, from the environment into the camera interior.

The lunar material under the camera collar was sampled for viable microorganisms. None was recovered. Further, as the two layers of Teflon wrappings were removed from the exterior of the metal cable connector, a sampling was made of both layers of the Teflon wrappings as well as the metal surface of the cable connector. Again, no viable microorganisms were detected. This was a deliberate attempt to detect any microorganisms which might have been available in the external environment and which might have entered the camera interior during ingassing.

The Apollo 12 astronauts, spacecraft, and space suits were sampled prior to launch and after recovery. All three astronauts carried species of a number of genera of microorganisms, including *S. mitis* (FERGUSON, private communication, 1970). As a result, the cabin air of both the lunar and command modules undoubtedly contained a number of different bacteria as an aerosol load.

Assuming that microorganisms had entered the camera interior during ingassing, a representation of the entire microbial population available would be expected rather than a single species. This representative population of microorganisms would be expected to be randomly distributed in the camera. Therefore, if large surface areas of the camera interior were sampled, microbial contamination due to ingassing should be detected. Even if *S. mitis* was the only one of the population carried in by ingassing to survive, it should have been found randomly distributed over large surface areas instead of in the only relatively inaccessible location that was sampled.

On a unit area basis at least 10,000 times the area in which the isolate was detected was sampled, and this area represents large exposed surface areas of different types of materials throughout the camera interior. That *S. mitis* cells (alone from all the microorganisms available in the external environment) could enter the camera and find their way to the least accessible sampling site without being detected in 10,000 times that area of readily exposed surface area is difficult to envision. In the absence of any other microorganisms isolated and in view of the large sampling area it is considered improbable that ingassing at any point in the retrieval could be responsible for depositing *S. mitis* in the relatively inaccessible location where it was isolated.

Extreme precautions were taken at all times during the analysis to prevent any handling errors which might have caused contamination. Experimental controls of the implements and media used in the analysis did not initiate microbial growth.

To determine whether low numbers of organisms alone could cause the delay in initial growth, a dilution series of THIO containing the isolate was prepared. From each dilution tube, 0.1 ml was transferred to a THIO tube and to 5 aerobic BA plates. Visible growth appeared within 24 hours, even in the dilution tube initially containing less than 10 viable cells as determined by the colony count on the 5 BA plates. Furthermore, the presence of the foam sample did not account for the initial delay in growth, since growth was not delayed when the isolate was cultured in a dilution series of THIO containing foam sections the same size and composition as the original samples.

The fact that no growth was observed until the fourth day of incubation in liquid broth indicated that the isolated bacterium required an adaptation period. Growth delays are not uncommon in bacteria recovering from lyophilization (Sinsky and Silverman, 1970). No colonies were found on the first set of five anaerobic BA plates, indicating either that no viable cells were placed on the BA plates, or that the cells could not adapt and replicate on the solid agar surface as they had in the liquid broth media.

The "tail" of growth which streamed from the underside of the foam on the fourth day of incubation indicates a direct relationship between the organism and the foam sample and is an important observation. When a control dilution series of the broth containing the isolate was made with similarly sized foam sections, no such relationship (no "tail") was observed in any of the dilution tubes indicating no spontaneous association of the bacteria with the foam.

The initial delay in growth of the isolate, the direct association of the bacterium with the foam sample from which it was isolated, the relatively inaccessible location from which the isolate was obtained, and the absence of any other isolates in the large sampling area are, in our opinion, not consistent with the hypothesis that the Surveyor III TV camera was contaminated with the isolate during or after its retrieval.

It is inadequate to simply imply that the foam sample or the thioglycollate tube became contaminated and that this readily explains the growth in the original undiluted tube and the 10^1 dilution tube. That would not be examining all the data and it would require unsupported assumptions; for example, the assumption that somehow the contaminant came into intimate contact with and remained in association with the foam sample despite vortexing so that it eventually grew as a "tail" to the foam. It would have to assume that for some reason the S. mitis cells were damaged and growth was delayed 4 days, that of all the tubes in the experiment contamination occurred only in this particular tube despite the control data, or that contamination occurred in the sample taken from the most inaccessible of all the sampling sites. Still other such assumptions would be required for such a simple explanation. No one single observation is adequate. Every bit of data must be considered. In the opinion of the authors, the total data is consistent with the hypothesis that the isolated bacterium was in intimate association with and isolated from the piece of foam sample which was taken from the camera interior and processed in an asceptic manner under controlled conditions.

The isolated bacterium, S. mitis, is a spherical microorganism measuring from 0.5 to 1.0 μ in diameter and is a frequent, normal, benign inhabitant of the respiratory tract. Man constantly sheds microorganisms into the air, a large portion of which comes from the respiratory tract. Although normal talking drives out considerable numbers of organisms a good healthy sneeze may dispense as many as 20,000 aerosol droplets, which may vary in diameter from 10 μ to 2 mm and the larger of which may travel 15 feet before reaching the ground. These larger droplets settle rapidly, adhering to particles of dirt, and dry leaving organisms attached to the particles (Smith, Conant, and Overman, 1964).

A single aerosol droplet could contain large numbers of organisms. It has been estimated that saliva contains an average of 750 millions of organisms/ml (Rosebury, 1962). In addition saliva contains many organic constituents, the major portion of which is protein and the principal salivary protein of which is mucin. "It seems that mucin exerts much of its effect on the oral microbiota by physical localization of bacterial growth. Mucin probably protects bacteria primarily by a coating effect with the formation of a temporary artificial capsule about the cell; this has been demonstrated with such oral microorganisms as staphylococci, streptococci, and lactobacilli" (BURNET and SCHERP, 1968).

Other organic constituents of saliva are carbohydrates, including hexosamine, methyl pentose, galactose, mannose, deoxyribose, and glucose (BURNET and SCHERP, 1968). "The synthesis of intracellular glycogen in the presence of excess carbohydrate, and its rapid catabolism to lactate in the absence of exogenous carbohydrate, has been observed in *S. mitis*. The polysaccharide appears to function as the sole reserve of energy of this organism and may provide the cell with energy in a utilizable form. The conclusion seems to be justified that the possession of glycogen by *S. mitis* favors its survival during starvation" (VAN HOUTE and JANSEN, 1970). In addition, when drying bacteria the presence of glucose in the suspending fluid in concentrations of between 5 and 10% greatly increased the survival rate both immediately and after storage (FRY and GREAVES, 1951).

As noted in the Hughes Aircraft Company Report, dated 22 January 1971, *Surveyor III Parts and Materials/Evaluation of Lunar Effects Returned from the Moon by Apollo XII*, "There were opportunities for contaminants to deposit on the camera prior to launch." A number of these opportunities came while the shroud of the camera was removed for prelaunch inspections or repairs. In addition, the prelaunch thermal vacuum testing of the camera provided conditions conducive to lyophilization. The Surveyor III and TAT–1 TV cameras were subjected to a series of thermal vacuum tests following inspections and repairs. Information provided by personnel of the Hughes Aircraft Company, El Segundo, California, where the Surveyor III TV camera was tested before launch indicates that prior to launch the Surveyor III TV camera was exposed, under a 10^5 torr vacuum, at least 12 times to temperatures of $-29°C$ and at least three times each to temperatures of $-45°C$ and $-118°C$. Exposure at these temperatures was for at least one hour, and in many cases longer. The highest temperature attained during any testing cycle was 52°C. The last thermal vacuum test of the camera before it was placed on the spacecraft occurred late in January 1967 leaving approximately 90 days before launch. After the Surveyor III TV camera was attached to the Surveyor III spacecraft, it was again exposed to extreme temperature and vacuum conditions in the course of spacecraft thermal vacuum testing.

If the bacterium was deposited in the camera prior to launch one can only speculate as to how many of these lyophilizing cycles the bacterium experienced. In one report a paracolon bacillus culture was subjected to repeated lyophilization and reconstitution without allowing for further growth. Approximately the same percentage of cells survived each cycle of lyophilization and reconstitution (FRY and GREAVES, 1951). It is certain that, if deposited in the camera, the bacterium would have experienced at least one cycle when the TV camera was attached to the Surveyor III spacecraft and the spacecraft underwent its thermal vacuum testing. In addition, since the TV camera was not maintained under a continuous vacuum, ambient pressure returned to the camera for approximately 90 days while the spacecraft awaited its launch to the moon. The survival of the bacteria inside an aerosol droplet in the foam would depend, it would seem, on the amount of protective substances which might surround the bacteria and the effect the lyophilizing conditions had on the dried droplet. Considering the fact that tubercule bacilli can survive in dried sputum for at least 8 months (SMITH, CONANT, and OVERMAN, 1964) it would seem possible that if the bacteria were encapsulated in a protective coating and dried they might survive until they experienced the continuous vacuum of space after launch. "The haemolytic streptococcus group B is very resistant to drying, and one strain, which shows a survival rate of 100% even in serum water, was, in another experiment, not entirely killed 18 months after drying in distilled water. It seems impossible to kill this strain by drying" (FRY and GREAVES, 1951).

It has been reported that when bacteria and viruses are dry they require, like isolated enzymes, a higher temperature for irreversible damage (DAVIS et al., 1968). Engineering estimates at the MSC suggest the maximum temperature experienced inside the TV camera while on the lunar surface at 70°C (ERB, private communication, 1970). Perhaps in such a dried state and under the high continuous vacuum of space, survival of lyophilized bacteria is possible. It has been shown that several Streptococcus species have remained viable for at least 20 years after lyophilization under routine laboratory conditions (RHOADES, 1970). Finally, in dealing with large numbers of microorganisms, even the loss of 99.9+% of the original population can still leave considerable numbers of survivors. It is estimated that between 2 and 50 cells or clumps of cells (chains) of S. mitis were isolated from the foam sample.

It would be very desirable to be able to define the exact conditions under which the isolated bacterium may have been deposited on the foam, the amount of protection which may have been provided by its source in the respiratory tract, the tolerance of bacteria contained in an aerosol droplet to heat and high vacuum, and the initial concentration of bacteria. Although the literature contains many reports of experiments which at first appear to be applicable, they all seem to suffer from the same shortcomings: the test species were different, the vacuum or temperature was not high enough, and most common of all, the experiment did not last long enough.

The isolated bacterium was lyophilized upon its initial isolation and is available for further testing as time, money, and facilities are available. The bacterium will be submitted for addition to the American Type Culture Collection.

The available data indicates that Streptococcus mitis was isolated from the foam sample and suggest that the bacterium was deposited in the Surveyor III TV camera before spacecraft launch. It is suggested that the bacterium may have been provided some protection from its source in the respiratory tract, and that lyophilizing conditions to which the TV camera was subjected before launch and later on the lunar surface may have been instrumental in the apparent survival of this terrestrial microorganism.

Acknowledgments—Grateful acknowledgment is extended to M. D. KNITTEL of the Jet Propulsion Laboratory, Pasadena, California for supervising the selection of the sampling sites and assisting in the sampling of all cameras; R. G. RIGLIN of the Hughes Aircraft Company for his thorough and intimate knowledge of the Surveyor TV cameras and especially for his fortuitous choice and extraction of the foam sample containing the isolate; T. C. MOLINA, BROWN and ROOT-NORTHROP, for competently performing all laboratory microbiological analyses; E. I. HAWTHORNE, R. G. RIGLIN, and

P. M. BLAIR of the Hughes Aircraft Company, as well as W. F. CARROL of the Jet Propulsion Laboratory, for their expert disassembling of the retrieved TV camera.

Our sincere thanks are extended also for technical and administrative assistance to R. H. GREEN of the Jet Propulsion Laboratory; P. A. VOLZ of Eastern Michigan University; T. C. ALLISON of Pan American College; and the following personnel of the NASA Manned Spacecraft Center at Houston, Texas: A. D. CATTERSON; J. L. McQUEEN; J. K. FERGUSON; G. R. TAYLOR; B. C. WOOLEY; S. JACOBS; L. G. LEGER; and K. L. SUIT.

REFERENCES

BURNET G. W. and SCHERP H. W. (1968) The microbial flora of the oral cavity. In *Oral Microbiology and Infectious Disease*, Chap. 20, pp. 273–327, Williams & Wilkins.

DAVIS B. D., DULBECCO R., EISEN H. N., GINSBERG H. S., and WOOD W. B., JR. (1968) Sterilization and disinfection. In *Principles of Microbiology and Immunology*, Chap. 11, pp. 335–353, Harper & Row.

FRY R. J. and GREAVES R. I. N. (1951) The survival of bacteria during and after drying. *J. Hyg.* **49**, 220–246.

HALL L. B. (1966) NASA requirements for the sterilization of spacecraft. NASA SP-108, 25–36.

JAFFE L. D. (1971) Blowing of lunar soil by Apollo 12: Surveyor III evidence. *Science* **171**, 798–799.

OYAMA V. I., MEREK E. L., and SILVERMAN M. P. (1970) A search for viable organisms in a lunar sample. *Science* **167**, 773–775.

OYAMA V. I., MEREK E. L., SILVERMAN M. P., and BOYLEN C. (1971) Search for viable organisms in lunar samples: Further biological studies on Apollo 11 core, 12 bulk, and 12 core samples. Second Lunar Science Conference (unpublished proceedings).

RHOADES H. E. (1970) Effects of 20 years storage on lyophilized cultures of bacteria, molds, viruses, and yeasts. *Am. J. Vet. Res.* **31**, 1867–1870.

ROSEBURY T. (1962) The indigenous cocci. In *Microorganisms Indigenous to Man*, Chap. 2, pp. 9–47, The Blakiston Division, McGraw-Hill.

SINSKY T. J. and SILVERMAN G. J. (1970) Characterization of injury incurred by *Escherichia coli* upon freeze-drying. *J. Bact.* **101**, 429–437.

SMITH D. T., CONANT N. F., and OVERMAN J. R. (1964) Microbiological ecology and flora of the normal human body. In *Zinsser Microbiology*, Chap. 10, pp. 162–172, Meredith.

TAYLOR G. R., FERGUSON J. K., and TRUBY C. P. (1970) Methods used to monitor the microbial load of returned lunar material. *Appl. Micro.* **20**, 271–272.

TAYLOR G. R. (1971) Microbial assay of lunar samples. Second Lunar Science Conference (unpublished proceedings).

VAN HOUTE J. and JANSEN H. M. (1970) Role of glycogen in survival of *Streptococcus mitis. J. Bact.* **101**, 1083–1085.

Proceedings of the Second Lunar Science Conference, Vol. 3, pp. 2735–2742
The M.I.T. Press, 1971.

Discoloration and lunar dust contamination of Surveyor III surfaces

W. F. CARROLL

Jet Propulsion Laboratory, Pasadena, California 91103

and

P. M. BLAIR

Hughes Aircraft Corp., El Segundo, California

(Received 23 February 1971; accepted in revised form 22 March 1971)

Abstract—The discoloration of Surveyor III surfaces observed by the Apollo 12 Astronauts during their examination of the spacecraft on the moon and clearly evident on the returned hardware has been analyzed and shown to be due to expected radiation darkening and a heavier than expected layer of lunar fines. Lunar surface material disturbed by the Apollo 12 Lunar Module (LM) landing 155 m away reached the Surveyor and produced significant changes in the spacecraft surfaces.

INTRODUCTION

THE TAN COLOR of the Surveyor spacecraft which surprised Apollo 12 Astronauts CONRAD and BEAN (1969) was no surprise to engineers familiar with Surveyor III and with radiation darkening of coatings in space—until the returned parts were first examined in the Lunar Receiving Laboratory (LRL).

Radiation discoloration of the white paint used on Surveyor had been measured in simulation tests by BLAIR and BLAIR (1964) and ZERLAUT and GILLIGAN (1969) and verified by HAGEMEYER (1967) from temperature measurements on Surveyor I. Since the magnitude of discoloration is proportional to the amount of solar irradiation, patterns of discoloration related to solar illumination geometry were expected.

The abnormal landing of Surveyor III had resulted in veiling glare and substantial loss of contrast in the pictures taken during spacecraft operation, presumably due to dust on part of the mirror. The protruding upper portion of the mirror was significantly more affected than the lower, recessed part. A similar coating of lunar dust was expected on other surfaces of the camera as well, and with comparable variations in quantity.

No direct effects of the LM landing were expected since there was ". . . preflight consideration that the landing occur outside of a 500-foot radius of the target to minimize contamination of the Surveyor vehicle by descent engine exhaust and any attendant dust excitation" (Mission Evaluation Team, 1970). Furthermore, BEAN, et al. (1970) detected no directional pattern of dust contamination on Surveyor related to the LM.

When examined in the LRL, the camera exterior was found to be a dirty gray to tan color with varying shades and tones but without evidence of the expected patterns of radiation damage and dust accumulation.

All external surfaces of the camera were discolored or contaminated in varying degrees with considerable disturbance due to handling during retrieval and return.

Fig. 1. Returned Surveyor III television camera.

The discoloration was tentatively attributed to radiation effects, lunar dust and photolyzed organic outgassing products from adjacent spacecraft components.

The only noticeable discoloration pattern was a series of "shadows" which did not correspond to solar illumination or other identifiable spacecraft geometry except immediately adjacent camera hardware. These patterns were later shown to have originated from the direction of the Lunar Module (LM) landing site and are described by JAFFE (1971) and COUR-PALAIS et al. (1971).

When the support collar (Fig. 1) was removed from the camera, an offset image of the inspection hole was noted on the camera body and a quantity of dark particulate material was found inside the collar recess (Fig. 2).

ORGANIC CONTAMINATION

There are ample sources for organic contamination and the amounts present could affect some types of analysis and may be contributory to the condition of the camera optics. The contribution to the overall discoloration of the external surfaces is absent or negligible compared to dust and radiation effects.

RADIATION DAMAGE

There are painted surfaces which have large differences in total solar irradiation due to camera geometry, sun incidence and shadowing by the spacecraft antenna and solar panel. In the absence of the heavy dust coverage, the differences in radiation darkening would have been readily discernible during inspection.

During the course of the evaluation, spectral reflectance measurements have been made on representative portions of the camera external surface and a method devised to analytically separate effects of dust and radiation from the spectral data. Results of the analysis show the magnitude of the radiation damage proportional to the extent of solar exposure as reported by NICKLE (1971) and furthermore in reasonable quantitative agreement as a function of integrated solar illumination with simulated exposure results of ZERLAUT and GILLIGAN (1969).

Details of radiation effects, simulated exposure and the methods used to analyze the spectral data are beyond the scope of this paper.

POSSIBLE SOURCES OF LUNAR DUST

The major potential sources of lunar material on the camera exterior are (1) Surveyor landing; (2) Lunar transport (i.e., debris from meteoroid impacts); (3) Lunar Module approach and landing; and (4) Retrieval and return (including redistribution over the camera surfaces). The Surveyor III vernier descent engines remained on through the first two lunar touchdowns, causing the spacecraft to rebound from the surface each time. The engines were shut down by ground command

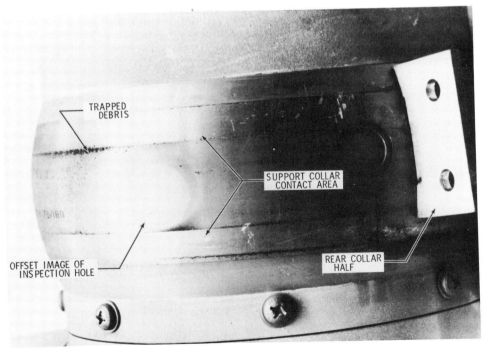

Fig. 2. Surveyor III camera body after removal of support front half. Rear collar half has been displaced upward and toward the right from its original position.

approximately 1 sec before the third touchdown and the spacecraft came to rest. The geometry and nature of the effect of the abnormal landing on the mirror, based on examination of pictures taken by Surveyor is described by Gault, *et al.* (1967).

The relatively heavy concentration of dust or pitting on the upper half and clean lower half of the mirror (inside camera housing) indicates that the source of the effect was directed upward from below the camera. A similar directional effect was expected on other camera surfaces.

Operable modes of lunar transport (i.e., debris from meteoroid impacts, electrostatic effects) would be expected to contribute to the dust coverage on the Surveyor camera. The magnitude of these effects over the 32 lunations cannot be adequately estimated but are probably minor compared to the results of the Surveyor and LM landings.

The subsequent approach and landing of the LM, described by the Mission Evaluation Team (1970) caused significant changes in the Surveyor camera surfaces, some of which are described by Jaffe (1971) and Cour-Palais *et al.* (1971).

The approaching LM passed outside the north rim of the Surveyor crater and landed approximately 155 m northwest of the Surveyor spacecraft. The first dust detected from photographic coverage occurred approximately 52 sec. before landing while the LM was at an altitude of about 100 feet; the crew first observed dust somewhat earlier, at an altitude of about 175 feet (Mission Evaluation Team, 1970).

The coordinates of the "sand-blast" effect which originated from the LM landing site during the final stages of landing were determined by Jaffe (1971). These coordinates, which have been independently verified during this work from other shadow patterns, indicate an origin elevated approximately 29° from camera "horizontal" in a plane normal to the flat portion of the NW side and parallel to the front (NE) surface of the lower shroud (see Fig. 1). Thus, the debris which produced the "sand-blast" effect shadows did not contact the front face.

Dust reported earlier by the astronauts, if it reached the Surveyor, would have contacted both the front face and the NW side.

Rennilson (1971) reports a heavier layer of dust on the mirror following return to Earth than was present during Surveyor operations in 1967. For dust disturbed by the LM to be deposited on the face of the camera mirror, it would have had to originate earlier and at a higher altitude than the first observed dust reported by the astronauts.

Sources of Observed Dust

Nearly all of the external surfaces of the camera are coated with an inorganic white paint. Thus, to study the causes of discoloration and the sources and degree of dust contamination, it was necessary to separate the effects and determine the quantitative contributions of each of the discoloration sources.

The porosity, surface roughness and friability of the paint precluded physical removal of the lunar material. Collodian and acetate peels removed only a portion of the lunar fines, along with part of the surface layer of the paint.

It was possible to examine particle size, shape and quantity of adhering particulate material (assumed to be mostly lunar fines) on unpainted surfaces in the Scanning

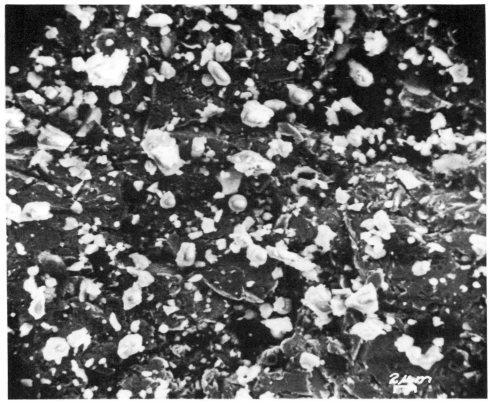

Fig. 3. Scanning electron microscope image of particulate (presumably lunar) debris on
Surveyor camera screw. Irregular background is forged surface of screw head.

Electron Microscope (SEM). The white features in Fig. 3 are particulate debris and
assumed to be mostly lunar material on the surface of the head of one of the screws
removed from the front of the lower shroud. The irregular background is the forged
surface of the screw head.

Similar direct examination of debris on the paint, silicate particles on a silicate
paint, was not possible. However, ANDERSON et al. (1971) have studied adhering
lunar material using nondispersive x-ray probe analysis, concentrating on elements
present in the lunar fines and absent in the paint.

Analysis of measured spectral reflectance of painted surfaces yields a relative
fraction of surface area covered by lunar material. Although this could be converted
to absolute amounts by making certain assumptions or by comparison to indirect
determinations by other investigators, it is meaningful for the present work only to
compare the amounts on various surfaces. The values in Table 1 are normalized to the
fractional area covered on the top of the visor (sample 906).

The samples measured on the N and NW side, facing the LM landing site (898,

907), exposed to the "sandblast" effect, indicate a substantially higher coverage by lunar material than the opposite side. Since the sandblasting produced a lighter color by removing material, the prior coverage was even higher. Although deposition of the heavy coating on the N and NW surfaces may have occurred during the Surveyor landing, such an explanation is inconsistent with the amount found on the NE (front) side.

There was nearly as much lunar material on the front (sample 893, facing NE) as on the side toward the LM landing site (NW). During the Surveyor landing, deposition on the front was unlikely; deposition without some shadowing and light dark contrast caused by protruding cable connectors would have been impossible. Deposition during final stages of LM landing (when detected by the astronauts) would likewise have produced contrast which were not evident.

Table 1. Comparison* of amount of lunar dust on various camera painted surfaces

Sample or Measurement	Location	Relative Quantity* of Lunar Dust
906	Top of visor	(1.0)*
907	Mirror hood—south† side (away LM)	0.5
908	Mirror hood—north side (toward LM)	1.0
898	Lower shroud, NW side (toward LM)	0.9
900	Lower shroud SE side (away from LM)	0.4
T-3	Lower shroud—SE side (small area adjacent to camera power cable)	≪0.1
893	Lower shroud—front (facing NE)	0.7
T-7	Lower shroud—rear (facing ∼ west)	1.1
T-8	Lower shroud—rear (facing ∼ south)	0.8

* Normalized to visor top (906).
† Lunar direction; for spacecraft orientation on the moon, see NICKLE (1971).

Long term deposition such as lunar surface debris disturbed by meteorite impact would be expected to produce uniformity on all sides, which was not observed. Redistribution during return to a degree necessary to obliterate directional effects from Surveyor landing or visibly detected LM dust would likewise have eliminated observed "sandblast" patterns.

Thus a major fraction of the lunar material on the NE (front) and NW sides must have arrived from a diffuse (multidirectional) source, disturbed by the approaching LM more or less uniformly over most of the last 1000 feet or more of its ground track.

There are camera surface areas with a covering of lunar dust which could not "see" the approaching LM. Therefore, the Surveyor landing and/or "lunar transport" must account for a portion of the dust contamination.

The lunar material on the returned polished tube section was concentrated on one side and heavier at one end than the other. The contaminant faced downward and slightly inboard (toward compartment A), and was therefore material disturbed by the vernier engines during the Surveyor landing.

The above discussion and resulting assignment of sources of lunar contaminant assumes essentially ballistic trajectories for particles disturbed by the LM.

Electrostatic or other effects which would alter the flight and deposition of particles may have been operable and, if so, would significantly affect the conclusions reached.

ADHESION

There are at least two distinct degrees of adhesion of lunar fines on returned Surveyor hardware, probably due to when the material was deposited and/or arrival conditions.

Most of the material on the bottom of the lower shroud had been disturbed by handling during retrieval and return (see Fig. 1) indicating poor adhesion, whereas the material on the polished tube was highly tenacious.

Acetate peels from the lower side of the vidicon radiator appeared to have removed all of the particulate material while multiple peels from the polished tube removed only part of the debris (Anderson *et. al.*, 1971).

Some of the material on the mirror was easily removed; when wiped by the astronauts on the moon, when accidently brushed during subsequent examination and by acetate peels. Following removal of material by acetate peels, the upper portion of the mirror remains partially diffuse and becomes less diffuse when rubbed. Until further analyses are complete, the presence of remaining highly tenacious lunar material can only be postulated. However, the fines removed by acetate peels are obviously less adherent than the contaminant on the polished tube section.

During examination of parts in the SEM, a number of particles were observed to move, probably due to electrostatic effects induced by the electron beam. Subsequent attempts to produce and observe similar movement were unsuccessful. The only particles which did move under these conditions were fiber segments which have been identified by ANDERSON *et al.* (1971) and COUR-PALAIS *et al.* (1971) as fragments from the backpack container.

Additional electrostatic adhesion measurements are planned for the future.

PARTICLE SIZE

Examination of external metallic surfaces in the SEM shows a wide variety of particle shape and sizes from submicron up to approximately 4 microns (see Fig. 3). Larger particles (assumed to be lunar material) are found in the slots of screws and in pores in the paint but not on any of the "external" metallic surfaces.

This size selection may be the result of initial deposition or adhesion or due to selective removal during return. If significant, a partial answer could be obtained by examination of fines in the camera compartment of the backpack. However, the backpack has not been made available for such investigation.

TRAPPED LM DEBRIS

The geometry of the offset image of the inspection hole observed during initial examination (see Figs. 1 and 2) has been shown to match the coordinates of the "sandblast" effect. Thus, the particulate material trapped inside the collar recess (see Fig. 2) is a small sample of the material disturbed by LM during the final phase of descent.

Based on the weight of trapped material determined by NICKLE (1971), and assuming that this is a representative sample, the final phase of the LM descent disturbed 10^6 gm/steradian with sufficient velocity to reach the Surveyor, 155 m away. If warranted, additional knowledge of the interaction of descent (or ascent) engine with the lunar surface could be obtained by further characterization (particle size distribution, density, etc.) of this sample and by refinement of estimates of amount lost due to rebound back out the hole.

Acknowledgments—This paper presented the results of one phase of research carried out at the Jet Propulsion Laboratory, California Institute of Technology, under Contract No. NAS 7-100, at Hughes Aircraft Co., Culver City, California under Contracts JPL 952792 and NAS 9-10492 and at NASA Manned Spacecraft Center, Houston, Texas, sponsored by the National Aeronautics and Space Administration.

The authors wish to acknowledge all of the people at NASA Headquarters, Manned Spacecraft Center, Jet Propulsion Laboratory and Hughes who made this program possible, especially Messrs. S. JACOBS and L. LEGER of MSC for their technical contribution and Messrs. J. DEVANEY and K. EVANS of JPL for SEM examination.

REFERENCES

ANDERSON D. L., CUNNINGHAM B. E., DAHMS R. G. and MORGAN R. G. (1971) X-Ray Probe, SEM and Optical Property Analysis of the Surface Features of the Surveyor III Materials. Second Lunar Science Conference (unpublished proceedings).

BEAN A. L., CONRAD C., JR., and GORDON R. F. (1970) Crew Observation, Apollo 12 Preliminary Science Report, NASA–SP–235.

BLAIR P. M. and BLAIR G. R. (1964) Summary Report on White Paint Development for Surveyor Spacecraft. Hughes Aircraft Company, TM800.

CONRAD C., JR. and BEAN A. L. (1969) Apollo 12 Mission Commentary.

COUR-PALAIS B. G., FLAHERTY R. E., HIGH R. W., KESSLER D. J., McKAY D. S., and ZOOK H. A. (1971) Results of Examination of the Returned Surveyor III Samples for Particulate Impacts. Second Lunar Science Conference (unpublished proceedings).

GAULT D., COLLINS R., GOLD T., GREEN J., KAYSER G. P., MASURSKY H., O'KEEFE J., PHINNEY R., and SHOEMAKER E. M. (1967) Lunar Theory and Processes in Surveyor III Mission Report, Part II Scientific Results. JPL Technical Report 32–1177.

HAGEMEYER W. A., JR., (1967) Surveyor White Paint Degradation. *J. Spacecraft Rockets* **4**, 828.

JAFFE L. D. (1971) Blowing of Lunar Soil by Apollo 12: Surveyor 3 Evidence. *Science* (in press).

Mission Evaluation Team (1970) Apollo 12 Mission Report, MSC–01855.

NICKLE N. (1971) Surveyor III Material Analysis Program, Second Lunar Science Conference (unpublished proceedings).

RENNILSON J. J., HOLT H., and MOLL K. (1971) Changes in the Optical Performance of Surveyor III's Camera (in preparation).

ZERLAUT G. A. and GULLIGAN J. E. (1969) Study of In-Situ Degradation of Thermal Control Surfaces, IITRI Report U6061.

Proceedings of the Second Lunar Science Conference, Vol. 3, pp. 2743–2751
The M.I.T. Press, 1971.

Examination of returned Surveyor III surface sampler

R. F. Scott and K. A. Zuckerman

Division of Engineering and Applied Science, California Institute of Technology
Pasadena, California 91109

(*Received* 27 *February* 1971; *accepted in revised form* 25 *March* 1971)

Abstract—The scoop and portions of the arms of the Surveyor III soil mechanics surface sampler were returned to earth by Apollo 12 Astronauts Conrad and Bean. A careful surface examination of the scoop has been made both visually and microscopically under a variety of lighting conditions. The blue of the original paint has been changed to a light tan color, but the change is not uniform on all surfaces. In places, the surface exhibits a blotchy appearance, probably due to differential thicknesses of a soil coating, which protected the surface from solar radiation. Although the surface is scratched and abraded, it is not known if this is due to preflight sandpapering or lunar surface operations. No micrometeorite pits were observed. Lunar soil adhered preferentially to (1) painted, (2) Teflon, and (3) metallic surfaces. The strength of the soil adhesion to the paint was in the order of 10^4 dynes/cm². Glassy spheres adhered more strongly to the paint than did other lunar granular material.

Introduction

THE SURVEYOR III spacecraft became operational at its landing site in Oceanus Procellarum on the moon 20 April 1967. It carried a Soil Mechanics Surface Sampler (SMSS) device for the purpose of performing mechanical tests of the lunar surface. The sampler was turned on the day after landing, and, after an initial calibration sequence, it was used to conduct the first controlled tests of the physical and mechanical properties of the lunar surface (SCOTT and ROBERSON, 1968). When the spacecraft was shut down for the lunar night on 3 May 1967, the surface sampler had been operated for $18\frac{1}{2}$ hours and had responded to a total of 1900 commands. During the surface operations, it was pushed into the lunar surface in 25 bearing and impact tests, and was dragged through the lunar granular material approximately 6 meters in the course of performing trenching tests. The spacecraft did not respond to commands sent on the second lunar day.

On 19 November 1969 Astronauts Conrad and Bean of the Apollo 12 mission landed close to Surveyor III and retrieved the scoop and part of three arms of the surface sampler along with other Surveyor III components. The returned portions of the SMSS were examined in detail at the Hughes Aircraft Company (HAC) facility in Culver City, California. This brief report presents some of the results of that examination.

Condition of Scoop before Examination

At the close of Surveyor III lunar surface operations a few grams of lunar soil remained inside the closed bucket of the sampler scoop; some soil also adhered to the scoop door mechanism and scoop exterior. After the scoop was retrieved by Conrad and Bean it was stored in the Apollo 12 lunar module and later in the command module under the atmospheric conditions prevailing inside these spacecraft. During quarantine in the NASA Lunar Receiving Laboratory (LRL) at Houston, Texas, the scoop was removed from its container bag at least once and exposed to the earth's atmosphere. It is not to be expected, therefore, that the lunar soil accompanying the scoop will exhibit the same properties as it possessed in the vacuum at the lunar surface. When the plastic bag containing the scoop was opened, it was found that the soil inside the bucket had spread over the exterior and interior surfaces of the scoop and the inside of the bag. The soil was observed adhering to various scoop surfaces to differing degrees, and some attempts were made to test the strength of this adhesion, although the mechanism of adhesion may be different from that existing on the moon.

EXAMINATION OF SCOOP SURFACE

The scoop and the attached arms were examined visually and microscopically up to about 100× magnification. In addition, the returned scoop and an essentially identical flight model, which had remained on earth, were photographed under white, ultraviolet, and infrared light conditions. The color photographs obtained in some of these studies are not reproduced here, but may be found in a more detailed report (SCOTT and ZUCKERMAN, 1971).

Figure. 1, an Apollo 12 photograph, shows the left side of the scoop (right and left as viewed by the Surveyor III camera) on the moon. In another picture, not shown, there appears to be a coating of soil on the bottom of the right side, near the scoop

Fig. 1. Apollo 12 view of left side of scoop on moon. NASA AS 12–48–7128.

Fig. 2. Laboratory photograph of right side of returned scoop. Width of field = 4.9 cm.

door. The left side (Fig. 1) shows a shading pattern in which the bottom of the scoop side is lighter. In the preliminary examination of the scoop at the LRL, the right side was heavily coated with soil in the approximate pattern shown in the photograph. This would appear to have been the original lunar soil picked up during surface sampler bearing tests. The shading pattern on the left side is also visible, particularly in the color photographs. The condition of the scoop as it appeared for the examination at HAC (Fig. 2) indicated that most of the soil adhering to the bottom of the right side had been rubbed off, but it can be seen that there is a lighter shading over the area to which it had adhered.

In appearance, a number of changes were manifest in the returned surface sampler. First of all, the blue paint which covers most of the surface appeared to have faded in color from the original light blue color to a whitish blue in the relatively protected or concealed areas of the arms and scoop. The original color of the paint is 5.0 PB 7/6 on the Munsell scale and the paint on the returned Surface Sampler was 10.0 B 8/2 on the cleaner (not soil-covered) areas and 10.0 B 7/2 on less clean parts. However, on the upper surfaces of the arms and on the upper and side surfaces of the scoop itself the color of the paint has been changed to a light tan. This tan is most pronounced on the upper surfaces and shades into a whitish blue on the underside of, for example, the arms. A microscopic examination of the paint surface at a magnification of 100× appears to indicate that the tan is a change in the painted surface rather than a light coating of particles covering the surface.

During transit from the moon and subsequent handling in the Lunar Receiving

Fig. 3. Left side of returned scoop tip. Width of field = 1.7 cm.

Laboratory and elsewhere, some of the paint around the edge of the scoop door may have been abraded and removed (Fig. 3) since some paint chips appeared in the associated soil. Some of the paint probably was also removed during operations on the lunar surface. It can be seen that the paint is covered with lunar soil particles, including a substantial proportion of small glassy spheres. The irregular bumpy texture of the painted surface is characteristic of the original painted coating. It is not clear if the change in color of the painted surface is due to a surface alteration of the paint or to a thin layer of fine particles on the surface. The color change is not everywhere uniform, and it seems to depend on the degree to which the surface was exposed to solar radiation. On the sides and top of the scoop, light tan blotchy patterns can be seen. In places, these patterns can be correlated with a protective covering of lunar soil apparent both in some of the original Surveyor III pictures and in the astronaut photographs (Fig. 1). The most pronounced color difference between the terrestrial and returned scoops is apparent in the ultraviolet pictures (Scott and Zuckerman, 1971). These observations are more consistent with a color change caused by solar radiation rather than a change resulting from a thin soil coating. It is not known why general gradational differences in the degree of the color change exist on apparently uniformly exposed sides of the scoop. It is possible that these are due to changes from place to place in the scoop paint thickness or composition, or may be due to the presence on the moon of differing thicknesses of dust coatings resulting from lunar surface operations. A further possibility is that the abrasion of the paint which took place before launch or during the lunar surface testing resulted in different sensitivities of the paint to the possible irradiation in different areas. It has also been shown (SCOTT *et al.*, 1971) that, at some time between the end of Surveyor III operations in May 1967 and the visit of the Apollo 12 astronauts in November 1969, two of the Surveyor spacecraft's shock absorbers had collapsed, probably moving the spacecraft slightly. Some soil could have been shaken from the left side of the scoop at this time. Alternatively, since the left side of the scoop was more exposed to the effects of the Apollo 12 descent engine, soil may have been removed from this side during final stages of the Apollo 12 landing.

A second item of interest concerning the painted surface is the crazing or cracking of the paint on the sides and base of the scoop door. Polygonal fracture patterns are apparent in Fig. 3. This portion of the scoop was made of a glass fiber-impregnated resin coated with the standard paint. The fracture pattern does not appear on the painted metallic surfaces of the rest of the scoop, and may therefore be related to the different thermal expansion characteristics of the paint, the resin, and the metal. It is also possible that radiation damage to the paint could have resulted in volume changes of the paint. In this case, the appearance of fracture patterns on the scoop door would be related to either the different thickness of the paint or different nature of bonding of the paint to that surface as compared with the other metallic surfaces. The chipping of the paint from the scoop door tips indicates that the bonding between the paint and the resin was weaker there than elsewhere. A careful study of the Surveyor III television pictures was inconclusive in regard to the presence of chipping or flaking at these points during the lunar surface operations in 1967. Observations during handling of the returned surface sampler indicate that the paint at the corners chips quite easily. Chips of paint were observed in the lunar soil which was collected from the inside and outside of the scoop.

The terrestrial flight model scoop has been employed in a variety of soil testing operations in a number of different soils on earth, and it was observed that the general effect of this soil contact has been to smooth down the irregularities in the painted surface without the development of scratches. Considerably less soil contact took place with the Surveyor III scoop, but it is evident, as shown in Fig. 4, that the surface has been scratched and abraded. A general smoothing of the surface of the paint is also evident in Fig. 4. It was initially thought that the scratches on the Surveyor III scoop apparent in Fig. 4 were due to the lunar surface operations, but it has since been learned that the painted surface of the scoop (and other portions of the space-craft) may have been lightly sandpapered (and in some instances repainted) prior to launch to remove defects in the paint. The direction and orientation of the scratches on the gear box (Fig. 4) and the right side of the scoop (Fig. 5) are consistent, however, with their production by lunar surface operations. This uncertainty may be resolved by a detailed comparison of scratch orientation on all scoop surfaces. The inside of the scoop, which was subjected to a great deal of lunar surface contact, was free from any signs of scratching or abrasion. Adhesion of the lunar soil to all surfaces of the returned scoop is readily apparent in a photograph of the Surveyor III scoop door mechanism (Fig. 6). Even the Teflon seal of the scoop door is relatively heavily coated with lunar soil particles (Fig. 7). The lunar soil scattered about the surface sampler appears to adhere preferentially to the different surfaces of the sampler.

Fig. 4. Screw head in gear box of returned scoop. Width of field = 1.7 cm.

Fig. 5. Scratches on right side of returned scoop. Width of field = 1.7 cm.

Fig. 6. Door of returned scoop. Width of field = 1.7 cm.

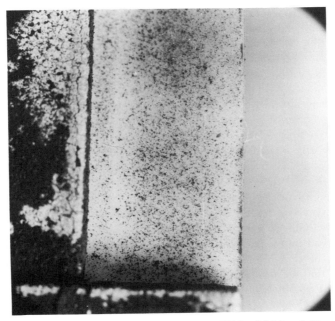

Fig. 7. Teflon on door of returned scoop. Width of field = 0.56 cm.

The most obvious observation is that the lunar material adheres more readily, in order, to (1) painted, (2) Teflon, and (3) metallic surfaces. It should be noticed, of course, that the metallic surfaces are not absolutely clear of lunar soil. It was not possible to tell in a superficial examination if there was selective adhesion of various components of the lunar soil.

Adhesion of the soil to the Teflon, and a slight color change of the Teflon itself was observed (Fig. 7). The Teflon appears slightly brown on its outer edges, shading to the original milky white appearance next to the metal part of the scoop door. It is apparent that this change took place rather quickly on the lunar surface since it is also visible in the Surveyor III pictures.

MEASUREMENT OF ADHESION OF LUNAR SOIL TO SURFACE OF RETURNED SCOOP

An attempt was made to measure the magnitude of the existing adhesion (whatever its nature) between the lunar soil and the various surfaces of the scoop by the following technique: A small vacuum-cleaning apparatus was built in order to remove the soil from the surface sampler surface. It consisted of a small pump, plastic hose, and two lucite chambers containing different sizes of filter papers. At the input end, a pen holder was used to retain a nozzle through which the soil was sucked. A number of different nozzle sizes was tested. In practice, the experiment and cleaning operation consisted of starting the vacuum pump and bringing the nozzle closer to the surface of interest while holding it at right angles to the surface. It was generally observed that at some particular distance from the surface a circular area underneath the nozzle tip would quite suddenly become clean leaving, in most cases, a very abrupt discontinuity between the clean surface and the adjacent soil-covered area. This result was interpreted to mean that the adhesion of the lunar soil to itself was somewhat greater than its adhesion to the scoop surface. Thus, when a critical surface shearing stress was reached due to the air flow over the surface, the soil detached itself from the surface and passed into the nozzle and thus into the collection chambers. By carefully measuring the distance of the nozzle from the surface of the scoop and the radius of the area which was made clean at the critical distance of approach, an estimate of the surface shearing stress required to remove the soil could be made. To make this estimation, the nozzle was calibrated by measuring the mass rate of flux of air into the nozzle at different distances of approach from various flat plates. From these tests it was estimated that the adhesive strength of the lunar soil to the painted surface was in the order of 10^4 dynes/cm². The adhesion of soil to the metallic surfaces of the sampler was somewhat less and was in the range of 2 to 3 × 10^3 dynes/cm².

It was observed that in an area of painted surface which had been cleaned off by this technique, the remaining particles consisted almost entirely of glassy spheres. It would appear that the adhesion of the spheres to the paint, at least, was considerably greater than that of other granular fragments, since one might expect that angular fragments would exhibit a greater degree of mechanical interlocking with a rough surface than spherical particles.

REFERENCES

SCOTT R. F. and ROBERSON F. I. (1968) Soil mechanics surface sampler: Lunar surface tests, results and analyses, *J. Geophys. Res.* **73,** 4045–4080.

SCOTT R. F. and ZUCKERMAN K. A. (1971) Examination of the Surveyor 3 surface sampler scoop returned by the Apollo 12 mission. Soil Mechanics Laboratory Report, Division of Engineering and Applied Science, California Institute of Technology.

SCOTT R. F., LU T.-D. and ZUCKERMAN K. A. (1971) Movement of Surveyor 3 spacecraft. *J. Geophys. Res.* (in press).

Proceedings of the Second Lunar Science Conference, Vol. 3, pp. 2753–2765
The M.I.T. Press, 1971.

X-ray probe, SEM, and optical property analysis of the surface features of Surveyor III materials

D. L. Anderson, B. E. Cunningham, R. G. Dahms, and R. G. Morgan

NASA Ames Research Center, Moffett Field, California 94035

(*Received* 28 *February* 1971; *accepted in revised form* 29 *March* 1971)

Abstract—The effects of the lunar environment on the surfaces of selected Surveyor III materials were determined. Several materials were examined: the thermal-control paint on the television camera, a polished aluminum tube, and unpainted stainless steel screws and aluminum washers. Scanning electron microscopy (SEM), X-ray probe, and spectral reflectance measurements were used to determine surface cratering and the change in surface spectral reflectance. It was determined that (1) none of the craters could be definitely identified as hypervelocity impact sites, (2) the solar absorptance (α_s) of the white paint degraded from a pre-flight value of 0.20 to post-flight values of 0.38 to 0.74, (3) the α_s of the aluminum changed from 0.15 to values from 0.26 to 0.75, and (4) the decrease in reflectance was due to a combination of the surface contamination (primarily lunar dust) and degradation of the paint from uv radiation.

Introduction

THE RECOVERY OF Surveyor III materials by the Apollo 12 astronauts has provided the scientific community with a unique opportunity to study the effects of the lunar environment on several types of engineering materials. As part of the overall NASA analysis program, a study of some of the painted and unpainted exterior surfaces was conducted at the Ames Research Center. Several types of surfaces were examined, namely, the white thermal-control paint on parts of the television camera (the elevation drive housing, a small 5.1 by 7.6 by 1.3 cm box, and the lower shroud), the polished surface of the unpainted aluminum RADVS (Radar Altimeter and Doppler Velocity Sensor) support tube (two 2.5 cm sections B and E as cut from a 19.7 cm length of the 1.3 cm diameter tube), and the unpainted surfaces of two stainless steel screws and two aluminum washers from the lower shroud. In this paper are presented results of this study as they pertain to surface cratering, to changes in surface spectral reflectance, and to a better definition of the lunar micrometeoroid and secondary particle environment (lunar ejecta). The study was conducted between July and December 1970. A more complete description of these parts and their handling prior to our investigation can be found in the Hughes Aircraft Company report (ANON., 1971).

Examination and Analysis

Examination of the camera parts and tube sections was carried out with three techniques: (1) optical and scanning electron microscopy (SEM), (2) energy dispersive X-ray probe analysis, and (3) spectral reflectance measurements. The microscopy techniques were used to examine the surface features of each part at magnifications up to 1,700× and 30,000× respectively. The normal practice of vapor-depositing a gold film over an insulating-paint surface for SEM was not permitted on the Surveyor parts due to constraints imposed for subsequent experiments to be conducted by other investigators. Therefore, some difficulty with charge buildup was encountered which limited the useful magnification for examination of the thermal-control paints up to $10,000\times$. The X-ray probe, an accessory to the SEM, was used to obtain the elemental composition of a surface. X-ray maps of a specimen were also obtained which showed the presence and spatial distribution of each element analyzed. The spectral reflectance of each part was measured in an integrating-sphere reflectometer with the sample located in the center of the sphere whenever possible. Spectral reflectance measurements were made at several locations on each part with special emphasis being given to the cleanest and dirtiest, or most contaminated, areas.

The relative positions of the parts examined in this study are shown in Figs. 1 and 2. Figure 1 shows the position of the parts on the Surveyor III spacecraft and the relative location of the Apollo Lunar Module landing site (Anon, 1967; Anon, 1970). Figure 2 shows the location of sections B and E on the 19.7 cm length of the RADVS support tube. The presentation of the results of the examination of these parts is divided into three categories: (1) discussion of the physical features of both the painted and unpainted surfaces, (2) discussion of the chemical composition of each type of surface before and after exposure to the lunar environment, and (3) discussion of the spectral reflectance of each type of surface.

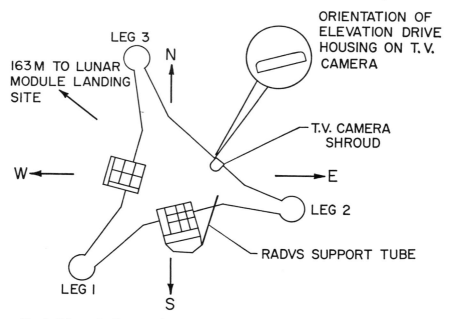

Fig. 1. Schematic diagram of orientation of Surveyor III spacecraft on lunar surface and with respect to Apollo 12 spacecraft, showing locations of television camera and Radar Altimeter and Doppler Velocity Sensor (RADVS) support tube examined in this study.

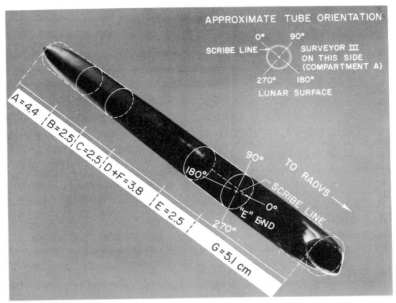

Fig. 2. Photograph of section of RADVS unpainted aluminum support tube as cut from Surveyor III spacecraft by Apollo 12 astronauts, showing locations of parts examined in this study, sections B and E.

In support of the first category, an ancillary laboratory program was conducted in the Ames Space Environment Simulator (CUNNINGHAM and EDDY, 1967) to produce hypervelocity impacts in materials similar to the Surveyor III parts described above. These materials were bombarded with micrometer-sized carbonyl-iron particles, electrostatically accelerated to velocities of up to 20 km/sec with a Van de Graaff accelerator, to obtain a better interpretation of possible impact features observed on the Surveyor III materials. It is believed that the carbonyl-iron microparticles, with a density of 7.8 g/cm^3 and at velocities in the 2–20 km/sec regime, can produce impact craters having shapes characteristic of impact craters produced at higher velocities by lower-density particles. Evidence supporting this viewpoint was presented by MORRISON (1970).

The two portions of the RADVS unpainted aluminum support tube, the E and B sections (Fig. 2) were examined for evidence of hypervelocity impact by micro-meteoroids or by secondary particles (lunar ejecta). The E section was examined by optical microscopy in two conditions: (1) in the "undisturbed" state as delivered to Ames via JPL from the Lunar Receiving Laboratory, and (2) after removal by previous investigators of most surface and imbedded contaminants with replicating film. In the undisturbed state, many micrometer-sized holes or cavities and foreign particles were observed. The particles were mainly concentrated on the side facing the spacecraft and the lunar surface; the holes were found with approximately the same distribution on all portions of the tube. Section E was again examined after removal of the contaminants; the surface that had been covered with the particles now appeared to have an eroded surface whereas the "clean" surface appeared to have the same features as were observed before contaminant removal. The B section was examined by both optical and scanning electron microscopy but only after removal of contaminants by prior investigators. In general, this section appeared, by optical microscopy, to have the same surface characteristics as observed on section E after contaminant removal.

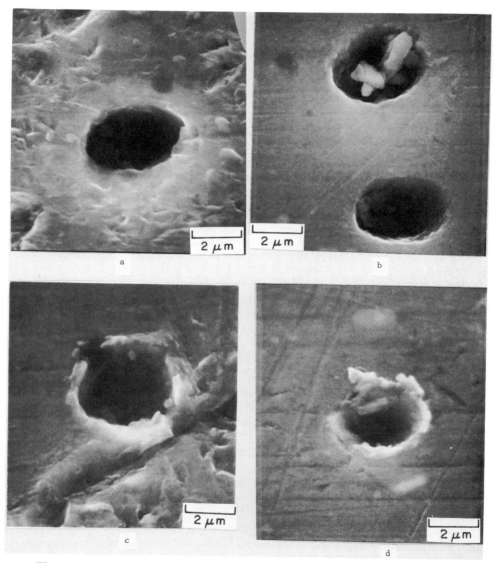

Fig. 3. Scanning electron micrographs of "dirty" and "clean" sides of RADVS aluminum support tube section B. (a) "unrimmed" hole on dirty side; (b) "unrimmed" holes on clean side; (c) "semi-rimmed" hole on clean side; (d) "rimmed" hole on clean side.

A total of about 200 mm^2 of the surface of section B was examined by SEM; typical micrographs of the "dirty" and "clean" sides of this section are shown in Fig. 3. Note that the micrograph of the "dirty" side (Fig. 3a) very clearly shows the eroded surface (after contaminant removal) caused by impacting rocket fumes and/or lunar ejecta. General erosion such as seen in this micrograph has not been produced in the previously mentioned ancillary laboratory program; furthermore, individual sites cannot be readily indentified for study to determine whether or not the erosion was caused by impact or by chemical action. Therefore, no attempt has been made as yet to characterize the cause of the erosion. It has been judged by the authors and by other investigators (CARROLL et al., 1971) that this "dirty" side faced the lunar surface and the Surveyor III spacecraft (Fig. 2). Since the tube was nearly horizontal on the spacecraft, no direct micrometeoroid impacts should be found on this side. The small hole shown in this micrograph (Fig. 3a) is representative of the numerous holes found over the entire surface of the tube by both optical and scanning electron microscopy (Figs. 3b, 3c, and 3d). It was therefore concluded that holes such as the ones shown in Fig. 3a may in some way be characteristic of the tube manufacturing or flight preparation processes.

The question of whether or not the holes observed on the "clean" side of the tube are hypervelocity impact sites was also considered. This was done by comparing SEM micrographs of section B (Fig. 3) with micrographs of aluminum surfaces which had been bombarded with 2–20 km/sec microparticles in the Ames SES and with similar holes which were found on the surface of a "control" specimen supplied by the Surveyor III manufacturer. Shown in Fig. 4 are micrographs of these laboratory-produced impact sites; these sites have well-defined features. (1) All have fully-developed, "flared-rim" craters; (2) oblique (non-perpendicular) entry produces "slanted," or elliptical, craters; (3) grazing entry produces long, elliptical "gouge" craters; and (4) crater walls and floors are smooth and, usually, free of residue except for small, spherical particles. A few of the holes found by SEM in section B have some of the above features to a limited degree. However, such "impact-like" holes were found all around the tube (as mentioned above in comments on features observed on the "dirty" side). Furthermore, not one of the holes examined resembled a "grazing" entry. Had micrometeoroid impacts occurred, a considerable portion of the impact sites would have been caused by grazing, or, at the least, oblique, entry. Therefore, it has been concluded that none of the holes examined by SEM could be definitely characterized as having been caused by hypervelocity impact of primary particles (micrometeoroids).

This conclusion is, of course, based on somewhat meager data. Although both sections B and E were examined rather thoroughly by optical microscopy at magnifications sufficiently high to easily resolve 100 μm diameter craters, none were found. On the other hand, only about 5% of the surface of section B was examined by SEM at magnifications sufficiently high to clearly resolve 1 μm diameter impact sites. However, it is warranted to use the results of this investigation to calculate an upper limit only for the lunar surface micrometeoroid environment, as was done by JAFFE (1970). This was done by calculating an area-time product, A_t, for the surface area examined and the time of exposure. It was assumed that micrometeoroids in the 1 μm diameter size range would be discrete, stony particles with a density of about 3.5 g/cm^3

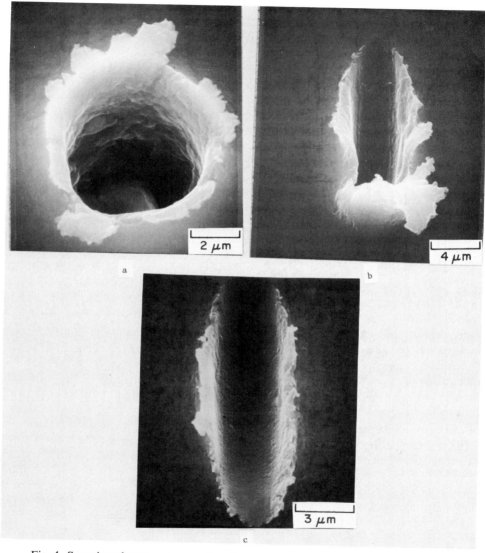

Fig. 4. Scanning electron micrographs of unpainted aluminum control tube (having same dimensions as sections cut from Surveyor III RADVS support tube), showing hypervelocity impact sites produced by carbonyl-iron microparticles at about 7 km/sec. (a) perpendicular impact; (b) oblique impact; (c) grazing impact.

(COSBY and LYLE, 1965). It was also assumed that impacts by such micrometeoroids would make 3 to 6 μm diameter craters. This is consistent with measurements obtained in the above SES experiments and by other investigators (RUDOLPH, 1967). For the 3.5 g/cm^3 density, the mass of a 1 μm particle would be about 1.8×10^{-12} g. For section B, the examined surface (~ 200 mm^2, 940-day exposure at a solid angle of about π steradians) has an A_t of about 1.6×10^4 m² sec, therefore the rate of impact was less than 6.2×10^{-5} particles/m² sec.

Two sets of unpainted screws and washers from the television camera lower shroud were examined by optical and scanning electron microscopy and X-ray probe analysis. The orientation of the screws on the spacecraft was such that one was facing and the other shielded from the landing site of the Lunar Module. The complete front surfaces and the sides of the screws as well as the exposed edges of the washers were examined for possible micrometeoroid impact sites. No surface features were found which were similar to the hypervelocity impacts observed in the ancillary experimental program (Fig. 4). It was concluded that none of the features observed could be definitely identified as hypervelocity impact sites.

The shapes of the material found on the surfaces of the screws were similar to those found on all other sections of the Surveyor III spacecraft studied. The only exceptions were rod-like materials, all 3–5 μm in diameter and up to 100 μm long, which were discovered on the elevation drive housing dust cover and on the screw that was shielded from the lunar module. Since similar rods, or fibers, were found in lunar material samples by investigators reporting at the Apollo 11 Lunar Science Conference (MORRISON et al., 1970; BARGHOORN et al., 1970; RAMDOHR and EL GORESY, 1970), it is appropriate to comment on the dispersive X-ray analysis of these fibers, conducted as part of this study. These rods were found by this analysis to be beta-cloth fibers from the Apollo 12 astronauts' gloves or back pack. The spectrum obtained in the X-ray analysis of the fiber was identical to the spectra of the rods found on the Surveyor III components.

Identification of possible impact sites on the painted surfaces was even more difficult than it was on the metallic surfaces. Very little was previously known about the physical characteristics of hypervelocity impacts in paints; therefore, paint samples prepared at Ames Research Center and standards prepared by the Surveyor III manufacturer and retained for control purposes were exposed to hypervelocity particle impacts in the previously mentioned ancillary test program. Figure 5a is a micrograph which shows a typical laboratory-produced hypervelocity impact in a paint sample; Fig. 5b is a micrograph of one of the standards. Note the lack of a flared crater rim or any apparent disturbance of the surrounding material. The holes due to the natural porosity of the paint are so similar to such impact holes that one often cannot be distinguished from the other. There are several holes on the standard that could easily be mistaken for impact sites.

Figure 5c is a micrograph of the elevation drive dust cover. All apparent impact sites were also examined at higher magnifications but none could be positively identified as being formed by a micrometeoroid. Because of this similarity between the pores and impact sites, a comparison was made between the number and size of holes on the unexposed paint standard and the elevation drive housing. Figure 6 shows the results of this statistical count. It can be seen that at least the majority of apparent impact sites found on this part could be due to natural paint porosity. The possibility that some of the apparent impacts could have been caused by dust from Apollo 12 was considered. The impacts from such dust would occur at a much lower velocity than would impacts from micrometeoroids (JAFFE, 1971) and the newly exposed paint

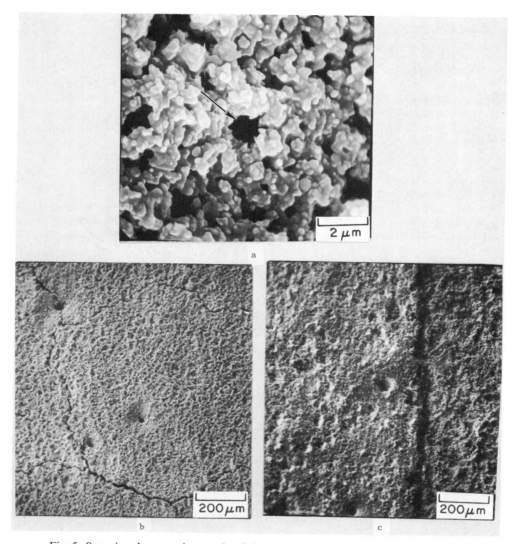

Fig. 5. Scanning electron micrographs of thermal control paints from Surveyor III television camera and of laboratory standard thermal control paint, showing hypervelocity impact sites and/or holes due to natural porosity of paint. (a) impact "crater" produced by carbonyl-iron microparticle at about 7 km/sec; (b) natural-porosity holes in paint standard furnished by Surveyor III manufacturer; (c) holes (or craters) in elevation-drive housing dust cover.

within the impact craters would differ from the "weathered" surface paint. Examination of small craters (30 to 100 μm in diameter) by optical microscopy showed many craters with very clean white walls—as would be expected if the crater was formed just before recovery with no time for additional contamination or degradation. In this

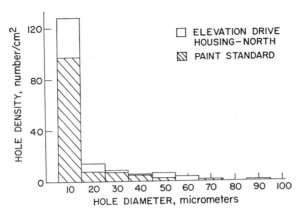

Fig. 6. Comparison of number and sizes of holes (or apparent impact sites) found in surfaces of thermal-control paint of elevation-drive housing dust cover of Surveyor III camera and laboratory standard furnished by Surveyor III manufacturer.

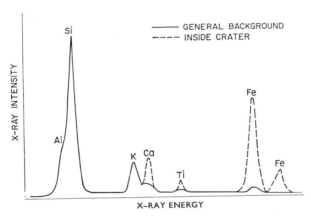

Fig. 7. Comparison of the X-ray spectra of the thermal-control paint of the Surveyor III television camera and the laboratory standard of this paint with the X-ray spectra of materials found with apparent impact craters.

size range, these white-walled craters account for approximately half of the observed difference in hole density between the standard and the exposed surface.

Identification of the chemical composition of residual debris in a hole was useful in determining the possible source of the impacting particle. The paint used on the Surveyor III television camera was a white inorganic paint composed of an aluminum silicate pigment and a potassium silicate binder. Figure 7 shows representative X-ray spectra typical of those obtained from several areas of the unexposed paint standard and the paint on the elevation drive housing. As expected, the aluminum, silicon, and potassium peaks found in the unexposed paint predominate these spectra. In addition, small amounts of calcium, titanium, and iron are evident. This is consistent with the

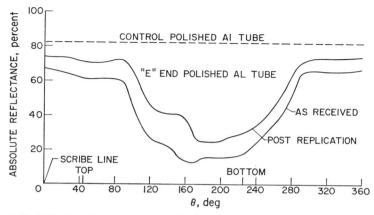

Fig. 8. Variation in absolute spectral reflectance around unpainted aluminum RADVS support tube compared with spectral reflectance of polished aluminum tube (from measurements, taken before and after removal of contaminants on section E).

composition of the dust layer found on all parts studied. Analysis of lunar soil by others (LSPET, 1970) show the presence of these elements. Several craters were found which contained residual material with greater relative amounts of the same three elements. No craters have yet been found which contain residual material of other chemical compositions. These results tend to indicate that those craters not accounted for by natural porosity are probably due to low velocity impacts of lunar material.

The spectral characteristics analysis of the two sections of the unpainted aluminum support tube was compared with a similar analysis of a section of polished aluminum tube made of the same alloy. This tube was polished by the Surveyor manufacturer using the same techniques as were used on the flight hardware. For this "control" specimen, the solar absorptance (α_s) was approximately 0.15. The post-flight values ranged from an α_s of 0.26 on the "clean" side to 0.75 on the "dirty" side with little variation along the axial length of each tube section. This variation in reflectance around tube section E is shown in Fig. 8 for a wavelength of 0.47 μm. The portion of the tube with the lowest reflection (greatest contamination) was oriented toward the lunar surface and slightly toward the spacecraft descent engine number 3. The reflectance was again measured after the surface was replicated as described earlier. As indicated in Fig. 8, this removal of loose material increased the reflectance on all sides of the tube. Figure 9 shows the total spectral distribution of reflectance around the tube. The contamination on the dirty side appears to be primarily of lunar origin or possibly from descent engine exhaust deposits. This contamination is not easily removed, however, since some traces of it remain even after repeated attempts by other investigators to remove it with ultrasonic cleaning (BUHLER *et al.*, 1971) or with normal replication processes for transmission microscopy experiments (BUVINGER, 1971).

As a basis for the spectral analysis of the painted surface, reflectance measurements were made on several test samples of this paint coated at the same time as the flight spacecraft, and the results indicated a pre-flight solar absorptance (α_s) of 0.20.

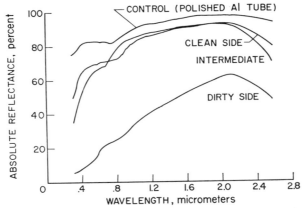

Fig. 9. Spectral distribution of reflectance (absolute reflectance as function of wavelength) on various portions of unpainted aluminum RADVS support tube (from measurements, taken before removal of contaminants, on section E).

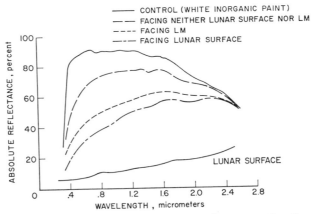

Fig. 10. Spectral distribution of reflectance (absolute reflectance as function of wavelength) for several areas of surface of thermal-control paint of elevation-drive housing dust cover of Surveyor III television camera.

The post-flight values depended upon the orientation of the surface relative to the sun, the lunar surface, and the landing site of the Apollo Lunar Module (LM). The final α_s values varied between 0.38 for a surface facing outer space to 0.74 for a surface facing directly toward the lunar surface. The spectral distribution of the reflectance is shown in Fig. 10. Note that the greatest change in reflectance occurred at the short wavelength end of the spectrum. The decrease in reflectance is due to a combination of the surface contamination and degradation of the paint from uv radiation.

Concluding Remarks

The results of this study indicate that the use of Surveyor materials as a means for definition of the lunar micrometeoroid environment on the lunar surface is very difficult. Optical microscopy provides only limited identification of microparticles and surface defects. Scanning electron microscopy provides excellent quantitative identification of surface characteristics. X-ray examination of residual material inside a hole can be a useful tool in determining the possible source of the impacting particle. Natural surface porosity and secondary impacts of lunar origin not associated with the normal lunar environment (landing of Surveyor and Apollo Lunar Module) account for most of the apparent impact craters studied. It was concluded that no sites were found which could definitely be characterized as being micrometeoroid impact craters. The results of the study, based on somewhat meager data, indicate a micro-meteoroid flux on the lunar surface for 1 μm diameter particles (mass of about 10^{-12} g) of less than about 6×10^{-5} particles/m² sec. None of the surfaces studied retained their initial optical properties; contamination appears to be the largest factor contributing to this change. It was found that the spectral reflectance and solar adsorptivity of both polished aluminum surfaces and thermal control paints were affected by the 940 day exposure to the lunar environment. In the case of the polished surfaces, the most significant effect was erosion by Surveyor III descent engines. The paints, however, were significantly damaged by solar radiation and surface contamination with a resulting change, for a surface facing outer space, in solar adsorptance from the pre-flight value of 0.20 to a post-flight value of 0.38.

Acknowledgments—The authors thank Mr. H. Y. LEM for performing the scanning electron micros-copy necessary for this investigation.

REFERENCES

ANONYMOUS (1967) Surveyor III mission report, Jet Propulsion Laboratory Technical Report 32–1177, Part 1.

ANONYMOUS (1970) *Test and Evaluation of the Surveyor III Television Camera Returned from the Moon by Apollo* 12. Hughes Aircraft Company Report SSD 00545, Vols. I and II.

ANONYMOUS (1971) Surveyor III parts and materials/evaluation of lunar effects. Hughes Aircraft Company Report P70–54.

BARGHOORN E. S., PHILLPOTT D., and TURNBILL C. (1970) Micropaleontological study of lunar material. *Science* **167**, 775.

BUHLER F., EBERHARDT P., GEISS J., and SCHWARZMULLER J. (1971) Trapped solar wind helium and neon in Surveyor III material. Second Lunar Science Conference (unpublished proceedings).

BUVINGER E. A. (1971) Replication electron microscopy on Surveyor III unpainted aluminum tubing. Second Lunar Science Conference (unpublished proceedings).

CARROLL W. F., BLAIR P. M., JACOBS S., and LEGER L. (1971) Discoloration and lunar dust contam-ination of Surveyor III surfaces. Second Lunar Science Conference (unpublished proceedings).

COSBY W. A. and LYLE R. G. (1965) *The Meteoroid Environment and Its Effects on Materials and Equipment*. NASA Special Publication, NASA SP–78.

CUNNINGHAM B. E. and EDDY R. E. (1967) Space environment simulator for materials studies at NASA Ames. *J. Spacecr. Rockets* **4**, 280–282.

JAFFE L. D. (1970) Lunar surface: Changes in 31 months and micrometeoroid flux. *Science* **167**, 1092–1094.

JAFFE L. D. (1971) Blowing of lunar soil by Apollo 12: Surveyor III evidence. Second Lunar Science Conference (unpublished proceedings).

LSPET (Lunar Sample Preliminary Examination Team) (1970) Preliminary examination of the lunar samples from Apollo 12. *Science* **167**, 1325–1339.

MORRISON G. H., GERARD J. T., KASHUBA A. T., GANGADHARAM E. V., ROTHENBERG A. M., POTTER N. M., and MILLER G. B. (1970) Multielement analysis of lunar soil and rocks. *Science* **167**, 505–507.

MORRISON R. H. (1970) Simulation of meteoroid-velocity impact by use of dense projectiles. *NASA Technical Note*, NASA TN D–5734.

RAMDOHR P. and EL GORESY A. (1970) Opaque minerals of the lunar rocks and dust from Mare Tranquillitatis. *Science* **167**, 615–618.

RUDOLPH V. (1967) Investigation of craters by microparticles at a velocity range of 0.5 to 10 km/sec. Ph.D. thesis, University of Heidelberg.

SCHOPF J. W. (1970) Micropaleontological studies of lunar samples. *Science* **167**, 779–780.

Proceedings of the Second Lunar Science Conference, Vol. 3, pp. 2767–2780
The M.I.T. Press, 1971.

Results of the Surveyor III sample impact examination conducted at the Manned Spacecraft Center

B. G. Cour-Palais, R. E. Flaherty, R. W. High, D. J. Kessler,
D. S. McKay, and H. A. Zook

NASA Manned Spacecraft Center, Houston, Texas 77058

(*Received* 19 *February* 1971; *accepted in revised form* 30 *March* 1971)

Abstract—The results of optical and scanning electron microscope (SEM) examinations of the Surveyor III TV camera shroud and sections of a polished aluminum tube cut from one of the radar antenna support struts for evidence of meteoroid and lunar ejecta impacts are presented. It is shown that the majority of the pits found on the shroud and tube sections were caused by low velocity impacts of particles that originated from the lunar surface. The lunar module descent is indicated as the cause of the numerous pits found on the shroud whereas the cause of those on the tube is more uncertain. Evidence is presented to show that the composition of particulate material in the craters is consistent with lunar soil. The pits on the polished tube are also shown to be associated with a brown contamination. In conclusion, the small number of probable meteoroid impacts is shown to be compatible with the latest estimates of meteoroid activity in the vicinity of the moon.

Introduction

The Apollo 12 mission plan included the return of a number of samples obtained from the Surveyor III spacecraft that had landed on the moon two and one-half years earlier. Among the samples retrieved were the TV camera and a 19.7 cm length of a polished aluminum tube. The tube was cut from one of the radar antenna support struts, but the exact position and orientation is unknown. The authors examined these components for evidence of meteoroid and lunar ejecta damage after the normal sample quarantine period had elapsed. This examination took place in the MSC Lunar Receiving Laboratory (LRL) between the 10th and 15th of January, 1970.

At the end of the examination period the polished aluminum tube was cut into six sections (Hughes Aircraft Company, 1971). Sections B and C were retained at MSC for detailed analyses by the authors and the remainder of the Surveyor samples returned to the Hughes facility in California. The sections were adjacent, each about 2.54 cm long, and cut from the less contaminated end of the tube. They were chosen because the earlier examination of the entire tube had indicated the presence of possible meteoroid craters.

This report describes the results of the examinations and the conclusions that were drawn by the authors. It is a part of the first generation Surveyor III Material Analysis Program prepared by NASA-JPL (Nickle, 1971).

Relevant Geometry

The Apollo 12 Lunar Module (LM) landed about 180 meters northwest of the Surveyor III spacecraft (NASA, 1970) and this closeness is dramatically shown in a

Fig. 1. A view of the Apollo 12 Lunar Module from the Surveyor III landing site. The arrow points to the radar altimeter support strut from which the polished tube section was cut.

photograph taken by the astronauts, Fig. 1. It can be seen that the LM landed on the rim of the Surveyor III Crater, and is sitting on the horizon relative to the spacecraft. The front, flat surface of the TV camera is approximately parallel to a line joining the Surveyor III and the LM which is confirmed by correlating some of the craters seen in the figure and earlier photographs taken by Surveyor III (NASA, 1967). This correlation puts the LM at a camera azimuth of approximately 90 deg. It is also apparent from the same source that the camera is leaning toward the LM, and that the horizon, in the direction of the LM, is at a camera elevation of 25 deg. A more detailed description of the relative positions of the various elements pertinent to this paper is available in a related report (NICKLE, 1971).

The radar antenna support strut from which the polished tube was cut is indicated by the arrow in Fig. 1. As mentioned previously, the position of the cuts and the orientation of the two ends is not known. However, the probable sequence of operations has been inferred as a result of detailed examinations of the severed ends and experiments duplicating the events on the lunar surface (Hughes Aircraft Company, 1971). These inferences are that (1) The cleaner end of the tube (section A) is uphill; (2) the apex at each end points away from the astronaut; and (3) the brownish contamination is on the side facing the interior of the spacecraft.

EXAMINATION OUTLINE

The preliminary examination took place in a temporary laboratory set up in the LRL and was accomplished in six days. During this time about 0.11 sq m of the TV camera shroud surface area of nearly 0.2 sq m was scanned at 25× magnification and every suspected impact crater was recorded. The remainder of the camera surface was scanned at lower magnifications to ensure that no significant meteoroid damage had been overlooked. In addition, the 19.7 cm long and 1.27 cm diameter polished aluminum tube was carefully scanned at a general level of 40× magnification. Local areas of interest were examined at much higher magnifications and typical surface effects and suspected impact craters were photographed for documentary purposes.

After the strut was sectioned and the samples returned to Hughes Aircraft, the two 2.54 cm long sections retained by MSC were examined in detail over a period of several months. They were both optically scanned at 100× magnification initially, and selected areas were later examined with a scanning electron microscope (SEM). The nondispersive X-ray detector was also used to analyze the material found in six fairly typical craters. Samples of the polished strut and the painted surface of the camera shroud supplied by Hughes Aircraft were also examined optically to determine surface backgrounds for comparative purposes.

RESULTS OF EXAMINATION

TV camera shroud

Although the time available only permitted a quick look for obvious impact craters, it is certainly true to say that there were no significantly damaging impacts on the camera shroud.

Typical surface effects and suspected impact craters are shown in Fig. 2. It is interesting to note that the paint surface differs around the periphery of the shroud. On the side facing toward the interior of the Surveyor spacecraft, the surface appears grainy as shown in the upper right-hand view. However, on the portions facing outward, the surface is cracked like a dry river-bed as may be seen in the upper left-hand view. Several holes and craters appear at the junction of cracks or along the cracks and these were not included in the tally of suspected impacts. In addition, there was evidence of a large number of shallow white craters covering the housing, with a definite concentration occurring around the periphery in a region directly in line with the LM. The cylindrical surface under the mirror head had 255 of these craters on the surface toward the LM, and only 2 on that facing away. The craters were obviously fresh because the original white color of the painted surfaces which had been discolored to a sandy brown, was displayed. Protruberances on the camera, such as screw heads and support struts, left dark shadows of unaffected paint on the camera pointing away from the LM.

Two of the craters that were identified as of possible meteoroid impact origin because of their hypervelocity appearance are shown in the lower views of Fig. 2. In all, there were five such craters ranging in size from 130 to 300 microns in diameter, although it is likely that not all of these were caused by meteoroids. This is especially true when it is considered that three of the suspected impacts occurred on the flat mirror gear-box housing, about 25 cm² in area.

The shroud surface also showed evidence of low-velocity impacts of irregular shape, as in the upper right hand view of Fig. 2, some with imbedded particulate material.

Polished aluminum tube

The polished aluminum tube obtained from the Surveyor III spacecraft was cut with a pair of long-handled shears with curved, overlapping blades. The cutting action

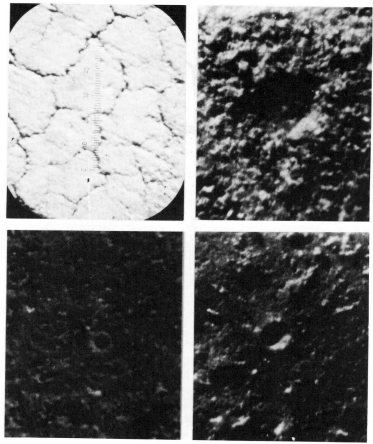

Fig. 2. Surface features found on the Surveyor III camera shroud. The upper views show the "dry river-bed" (left) and the "grainy" effect and a typical low-velocity indentation (right). The lower views show two of the possible meteoroid craters, 130 μ diameter (left) and 250 μ diameter (right).

partially flattened the ends of the tube as may be seen in Fig. 3. Also to be seen in this figure is an increase in contamination toward the left-hand end of the tube which appears brownish to the unaided eye. Under a microscope, it also appears brown and seems to be composed, at least partially, of crystals ranging in size up to a few microns. There is also a variation in the amount of the contamination one observes as the tube is rotated around the cylindrical axis. The 40× examination of the entire surface of the tube revealed only four craters larger than 25 μ diameter exhibiting possible characteristics of hypervelocity impacts at low magnification, i.e., a circular indentation surrounded by a smooth, raised lip. These craters were on the two sections of the tube obtained by MSC and subsequent detailed examination at higher magnifications discounted a meteoroid origin. Although the detailed examination of sections B and C at magnifications up to 600× revealed no obvious meteoroid impacts, a large number of other craters and pits were found. Figure 4 illustrates the number of craters 20 μ and larger in diameter that were observed in the field of view of an optical microscope at 100× magnification (corresponding to an area of about one square millimeter). Counts were taken as a function of angle around the tube from the scribe line which had been ruled along the tube prior to cutting and the histogram is an average of two trials on the "B" section of the tube. Very high pit densities (up to 40 per field of view) were obtained in two places but were obviously associated with polishing scratches and so are not included. The reduced count rate around 170 deg from the scribe line is not considered significant.

Also shown in Fig. 4 is a measure of the relative amounts of brown contamination on the B section as a function of angle around the tube. This curve was obtained by photographing the tube at each angular position as the tube was rotated and the lighting held constant. The contamination stood out in the photographs between the angles 100 and 280 deg and appeared to peak at about 190 deg. Outside of these angles, the B section was relatively clean. The relative heights of the ordinate of the contamination curve are not quantitatively significant. A high ordinate means that the photograph indicates "high" contamination relative to an angular position with a low ordinate. It is immediately evident that there is a close association between the pitting rate and density of the brown contamination.

In addition to the optical work, extensive analyses were carried out with the SEM as follows: (1) To look at craters found during the optical scan of the B and C tube sections in order to determine their origin. (2) To do a spot survey at high magnifications

Fig. 3. The polished aluminum tube section obtained from the Surveyor III showing the contamination darkening to the left. Sections B and C were cut 3.8 cm in from the right-hand end.

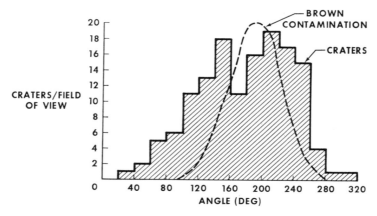

Fig. 4. Distribution of the contamination and low-velocity surface effects on sections B and C of the polished tube.

over the entire C section of the tube. (3) To determine by nondispersive X-ray analysis the composition of material in the craters and on the surface of the tube.

The results of the SEM work can be summarized as follows: (1) No craters showed evidence of hypervelocity impact origin. (It had not been possible by optical methods alone to determine whether or not some of the smaller craters had hypervelocity impact characteristics.) (2) All of the craters examined appeared to have a low-velocity impact origin, and many of them had material remaining in them. (3) The spot survey of section C confirmed the pitting density results of the optical scans but added little new information. (4) Analysis of the material in the craters strongly indicated that most of it was of lunar origin. (5) The brown contamination on the surface did not give any "peaks"; however, no elements with X-ray energies below about 1 KV are detected with the analyzer on this SEM. Hence, elements like oxygen, nitrogen, carbon, etc., would not have been seen in this analysis, if they were present.

Figure 5 is a composite optical and SEM photograph of 3 typical craters on the B section, located 280 deg from the scribe line. They were obviously not caused by hypervelocity impact as there is no smooth, raised lip entirely around the central indentation. However, it is clear from their shapes that material at a relatively low velocity, perhaps a few hundred meters per second, impacted the tube. The largest crater is approximately 30 μ in diameter and material is still imbedded in it. An X-ray pulse height analysis of this material showed it to be composed of silicon, calcium, and iron with significant traces of chromium and titanium. The upper left-hand view of Figure 6 illustrates a region of high pitting density at 220 degrees from the scribe on the B section. The crater in the center is about 8 microns in diameter, and the material in this crater has as its major components silicon, iron, calcium, and titanium. Titanium was also found in another crater on this tube. The white appearing material in the other three craters in Fig. 6 shows up dark brown under an optical microscope and the nondispersive X-ray analysis indicated it to be an iron–calcium silicate. Because only 6 typical craters were extensively analyzed by the SEM nondispersive X-ray analysis, the significant amounts of titanium found in three of them is quite indicative of a lunar origin.

From the mineralogical standpoint, at least three phases are present in the craters. One is a calcium–aluminum silicate which is undoubtedly plagioclase; the second is a calcium–iron–magnesium silicate with a trace of titanium, which is consistent with clinopyroxene; and the third contains calcium, iron, titanium, and silicon in varying amounts and may also contain aluminum and magnesium. This is probably glass and unresolvable mixtures of very fine fragments. The composition of the particulate material in the craters analyzed is completely consistent with lunar soil.

A crater that initially presented some excitement is the one shown in the lower right-hand view of Fig. 6 at 170 deg from the scribe. Its size is approximately 80 microns by 110 microns, one of the largest craters on the tube, and it contains "rods" 3.5 microns in diameter and 20 to 40 μ long. The SEM analysis subsequently revealed

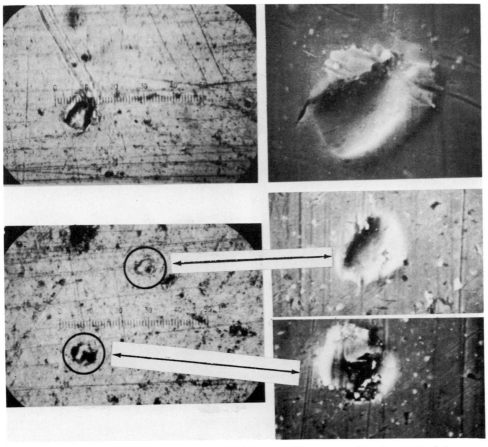

Fig. 5. Optical and SEM views of typical impact effects on sections B and C of the polished tube. The arrows in the lower views point to related craters.

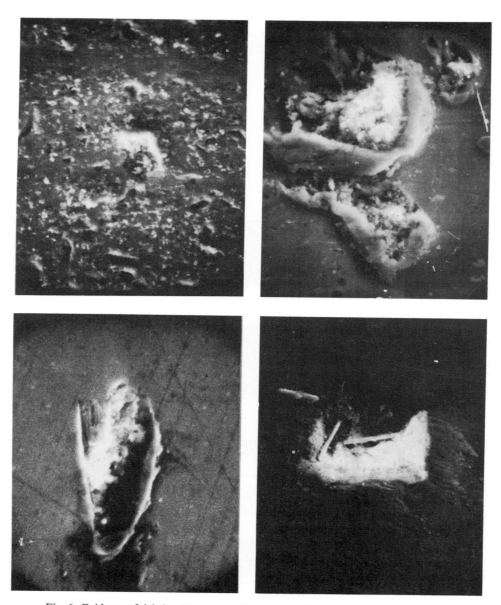

Fig. 6. Evidence of debris within some of the impact effects found on sections B and C of the polished tube. The rods in the lower right-hand view are glass fibers from the Surveyor parts tote-bag.

them to be identical in composition to the beta-cloth glass fibers in the astronauts' outer garments and in the backpack in which the Surveyor III parts were stowed. Experiments at MSC have shown that it is possible to break off a few fibers by jamming the end of a strand of beta-fiber into a crater of this size.

DISCUSSION

Meteoroid impacts

Since no meteoroid impacts were found on the tube for an exposure of 942 days, it is possible to set upper limits to the meteoroid flux at the moon. The detection threshold over the entire tube corresponds to craters about 50 μ in diameter. The highly contaminated region was sufficiently pitted and scarred as to make it impracticable to resolve features of smaller craters. On the nonpitted sides of sections B and C, the detection threshold corresponds to 25-μ and larger craters. The effective area of the nonpitted region is about half the area of these sections. If it is assumed that meteoroid impact craters are hemispherical in shape in the 2024-T3 tube aluminum alloy, then the threshold penetration depths are, respectively, 25 microns over the entire tube and 12.5 microns over one-half of each of the two 1-inch sections. A hypervelocity penetration equation for 2024 aluminum developed by Cour-Palais at MSC is given by

$$P = 0.34\, d^{1.06} \rho^{0.5} V^{0.67} \text{ cm}$$

where P = crater depth, cm; d = meteoroid diameter, cm; ρ = meteoroid mass density, gm/cm³; and V = meteoroid velocity, km/sec. The 50-μ diameter threshold corresponds to a meteoroid 14.6 μ in diameter and a mass of $10^{-8.79}$ gm. Similarly, the 25-μ diameter threshold corresponds to a meteoroid 7.6 μ in diameter and $10^{-9.64}$ gm mass. These masses were derived for a 20 km/sec impact velocity and a 1 g per cubic centimeter mass density.

The area of the entire tube is about 78.5 cm² and the area of the nonpitted regions of the B and C sections is 10.1 cm². Using a shielding factor of 0.5 due to the moon and another factor of 0.67 due to the fact that the Surveyor spacecraft cuts out about $\frac{1}{3}$ of the remaining solid angle that meteoroids could approach from, the effective area-times are 2.16 × 10⁵ m² sec for the entire tube and 0.28 × 10⁵ m² sec for the nonpitted regions of the B and C sections. The 95% upper confidence limits on the meteoroid flux for no impacts (RICKER, 1937) for these exposures are $10^{-4.75}$ impacts/ m² sec and $10^{-3.88}$ impacts/m² sec, respectively. To compare these upper limits of the moon with fluxes at Earth, one must allow for a gravitational flux increase factor at the Earth of 1.74. Hence, the corresponding upper limits at Earth would be $10^{-4.51}$ meteoroids/m² sec for masses larger than $10^{-8.79}$ gm and $10^{-3.64}$ meteoroids/m² sec for masses larger than $10^{-9.64}$ gm.

If the 5 craters found on the shroud are considered to be of meteoroid origin, then the flux allowing for lunar shielding and the spacecraft shielding one-quarter of the remaining solid angle, would be $10^{-5.83}$ impacts/m² second. Allowing for the gravitational attraction of the earth previously mentioned, this is a near-Earth flux of $10^{-5.59}$ meteoroids/m² second.

The mass associated with the smallest crater found in the shroud, 150 μ in diameter, is calculated to be $10^{-8.75}$ gm. This is based on the assumption that the crater diameter to meteoroid diameter ratio is 10 for the shroud surface. The ratio chosen was arrived at after consideration of laboratory hypervelocity impact data for targets varying over a wide range in ductility.

The 95% upper and lower confidence limits for the five probable impacts considered are 11.7 and 1.6, respectively (RICKER, 1937). Any uncertainty in the mass may be accounted for by adopting high and low values of $10^{-8.5}$ and $10^{-9.0}$ gm.

If we consider the largest crater on the shroud, 300 μ in diameter, then by similar reasoning it can be shown to be equivalent to a meteoroid mass of $10^{-7.85}$ gm and a near-Earth flux of $10^{-6.28}$ meteoroids/m² sec.

The meteoroid flux obtained from the polished tube and the TV camera shroud are shown plotted in relation to the near-Earth flux measurements and the current NASA meteoroid environment model (NASA, 1969) in Fig. 7. Also included on this plot are the flux estimates obtained from the photographic record of the Surveyor III foot-pad imprint (JAFFE, 1970) and the Gemini spacecraft windows (ZOOK *et al.*, 1970). In each instance where the data point has been obtained from the lack of impacts for the threshold detection size, the upper 95% confidence level is shown. Thus the Surveyor

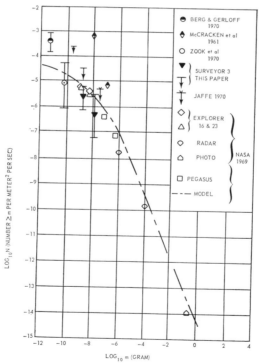

Fig. 7. Comparison of the meteoroid flux obtained from the examination of the Surveyor III camera shroud and polished tube section with other data obtained in deep space, near Earth, and from the lunar surface.

III pad imprint and the polished tube provide the upper limit to the meteoroid flux on the lunar surface. When considered in conjunction with the Surveyor III camera shroud and Gemini window impact data, it can be seen that there is good agreement with the statistically significant data obtained for penetration detectors in Earth orbit. The other data points shown are the results obtained by the acoustical detectors on Explorer VIII (McCracken *et al.*, 1961) and the impact ionization analyzer of Pioneer 8 and 9 (Berg and Gerloff, 1970). Whereas there is serious disagreement with the acoustical data of Explorer VIII, the Pioneer results are in agreement with the limits established by the Surveyor III cratering results. The Pioneer 8 and 9 instruments are much more sophisticated than the earlier Explorer VIII acoustical detectors, and the coincident detection devices and other means used to eliminate spurious signals lend credence to this data.

Lunar surface impacts

When the camera is viewed from the direction of the LM, shadowed areas not whitened by cratering are aligned directly behind protuberances such as bolts, screw heads, and other parts of the camera. These shadows are noticeable on the mirror hood, on the base of the camera where it was partly shielded by a plate, and near the screw heads on the mirror gear box as shown enlarged in Fig. 8. Note that the shadows are very well defined, and that numerous white surface craters are found outside of the shadowed region, indicating a "point" source. It is a simple matter to show that the origin of the particles responsible for the sand-blasting of the TV shroud is in the direction of the LM. Thus it is postulated that during the two and one-half years that the Surveyor III rested on the moon, the white surface of the camera became dis-colored and that dust accelerated by the LM as it landed sand-blasted the Surveyor spacecraft, removing much of the discoloration, except in areas that were shielded. The sharpness of the shadows created by the shielding and their direction indicates that the path of the lunar dust was only slightly curved by lunar gravity, suggesting that the dust was traveling in excess of 100 msec.

The close association between the brown contamination and the pits on section B of the polished tube and the fact that there is lunar material in the pits is evidence that this phenomena occurred while the Surveyor III spacecraft was on the moon. Three of the possibilities considered as an origin for contamination are as follows:

(1) Lunar secondary and tertiary ejecta stirred up by primary meteoroid impacts bombarded the exposed area of the tube causing the pitting, and the contamination is also composed of lunar material. This hypothesis is improbable. Studies on lunar secondary ejecta (Gault *et al.*, 1963), (Zook, 1967) and in-house meteoroid penetra-tion tests and analysis conducted by Cour-Palais at MSC have shown that there is unlikely to be a larger secondary flux capable of causing craters of a given size than the equivalent primary meteoroid flux. Also, the evidence from the shape and angles of the sheared ends of the tube suggest that the contaminated and pitted side of the tube points away from the astronauts doing the cutting. Fewer secondary ejecta particles would approach from this direction as it is partially shielded by the Surveyor space-craft. Finally, the SEM analysis of the contamination was unable to detect any

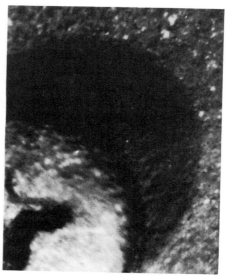

Fig. 8. Details of the shielding provided by the Surveyor III camera protuberances to the Lunar Module exhaust lunar dust storm. Note the distinction between lighting shadow and the shielding effect.

elements with atomic numbers greater than 11 (sodium). Hence, it is very unlikely that the brown contamination is composed of any component of the lunar soil.

(2) The pitting is due to lunar material blasted toward the Surveyor III spacecraft by the LM as it landed. This possibility cannot be ruled out since the TV camera shroud, as previously shown, was indeed blasted in this manner. For instance, it is likely that the astronauts held the shears at 45 deg to the horizontal to do the cutting, a natural position in their space suits. It has also been shown (Hughes Aircraft Company, 1971) that the cutters tend to pinch the tube more sharply away from the person doing the cutting than towards him. Examination of Fig. 2 shows the upper edge of each cut to be the most sharply pinched. If the uncontaminated end of the tube is to the right as in this figure and as suggested in the referenced Hughes report, the direction of maximum contamination points very nearly in the direction of the LM. Furthermore, experiments conducted by JPL (NICKLE, 1971; and Hughes Aircraft Company, 1971) have shown that parts of the tube are visible from the LM. Two problems arise with this hypothesis, however. One is that the pitting on the tube seems to be more intense than on the camera shroud and the other that the camera paint seems to have been brown before the LM landed and in a more or less uniform fashion. The effect of sand blasting by the LM was to expose the fresh white paint surface beneath. The pitted side of the tube on the other hand was darkened.

(3) The pitting is due to lunar material blasted toward the tube by the Surveyor's vernier thrusters and the contamination is due to incompletely burned propellant, unsymmetrical dimethyl hydrazine monohydrate combined with nitrogen tetroxide oxidizer with a little nitrous oxide added as a catalyst. This is also a possible source. If the tube shown in Fig. 2 is rotated through 180° such that the clean end of the tube is to the astronaut's lower left, the contaminated side of the tube points down to the lunar surface and in toward the Surveyor spacecraft. This orientation of the tube is not ruled out by geometry although it is not clear why the end closer to the lunar surface would be cleaner.

It would seem that the polished tube has material kicked up by either the LM descent stage or by the Surveyor III vernier thrusters. Also, the brown contamination could have come from either source as the propellants used are nearly identical. Of these two possible sources the Surveyor vernier thrusters are thought to be the more likely.

CONCLUSIONS

The several conclusions arising from the MSC examination of the Surveyor III TV camera housing and polished tube can be generalized as follows:

(1) The meteoroid flux impacting the lunar surface for masses within the range 10^{-7} to 10^{-10} grams is consistent with measurements made near Earth after allowing for gravitational focusing.

(2) Although it was not possible to separate the effect of meteoroid related lunar ejecta from the general background of low-velocity effects, it was shown that space-craft induced ejecta can have a considerable effect on the surface features of objects in the vicinity and as far as 180 meters away. The energy imparted to lunar surface dust was sufficient to cause pitting in aluminum and to crater and discolor the baked paint

surfaces of the camera shroud. It is suggested that the lower velocity limit for the dust accelerated by the LM is 100 meters/second.

REFERENCES

BERG O. E. and GERLOFF U. (1970) Orbital elements of micrometeorites derived from Pioneer 8 measurements. *J. Geophys. Res.* **75**, 6932–6939.

GAULT D. E., SHOEMAKER E. M., and MOORE H. J. (1963) Spray ejected from the lunar surface by meteoroid impact. NASA Technical Note No. D-1767.

HUGHES AIRCRAFT COMPANY (1971) Surveyor III parts and materials/Evaluation of lunar effects. Rep. No. P70-54.

JAFFE L. D. (1970) Lunar surface: Changes in 31 months and micrometeoroid flux. *Science* **170**, 1092–1094.

MCCRACKEN C. W., ALEXANDER W. M., and DUBIN M. (1961) Direct measurements of interplanetary dust particles in the vicinity of the Earth. NASA Technical Note No. D-1174.

NASA (1967) *Surveyor III: A Preliminary Report.* Office of Technology Utilization, NASA SP-146.

NASA (1969) *Meteoroid Environment Model—1969: Near-Earth to Lunar Surface.* Office of Advanced Research and Technology, NASA SP-8013.

NASA (1970) *Apollo 12: Preliminary Science Report.* Office of Technology Utilization, NASA SP-235.

NICKLE N. L. (1971) Surveyor III material analysis program. Second Lunar Science Conference (unpublished proceedings).

RICKER W. E. (1937) The concept of confidence of fiducial limits applied to the Poisson Frequency Distribution. *J. Amer. Stat. Assoc.* **32**, 349–356.

ZOOK H. A., FLAHERTY R. E. and KESSLER D. J. (1970) Meteoroid impacts on the Gemini windows. *Planet. Space Sci.* **18**, 953–964.

ZOOK H. A. (1967) The problem of secondary ejecta near the lunar surface. *Transactions of the 1967 National Symposium on Saturn V/Apollo and Beyond,* Vol. 1, paper EN-8. American Astronautical Society.

Proceedings of the Second Lunar Science Conference, Vol. 3, pp. 2781–2789
The M.I.T. Press, 1971.

Micrometeoroid flux from Surveyor glass surfaces

DONALD BROWNLEE, WILLIAM BUCHER, and PAUL HODGE

Department of Astronomy, University of Washington, Seattle, Washington 98105

(Received 22 February 1971; accepted in revised form 29 March 1971)

Abstract—The Surveyor III television camera's optical filters have been searched for micrometeoroid impact sites. Using optical techniques 30 square centimeters of glass surface were scanned microscopically with a small crater detection limit of 5 microns. Ten square millimeters of surface were examined with a scanning electron microscope with a detection limit of one micron. No hypervelocity craters were found and an upper limit is established for the flux of hypervelocity particles impacting the lunar surface. The implied flux is low but in general agreement with results of spacecraft Pioneer 8/9, Cosmos 163 and Pegasus. About thirty probable low velocity impact sites were found which are attributed to lunar ejecta.

INTRODUCTION

THE SURVEYOR III television camera was exposed to micrometeoroid bombardment for 2.6 years on the lunar surface before it was returned to the Earth by the Apollo 12 astronauts. The exposure time was nearly an order of magnitude longer than that of any other man-made object ever returned from space for analysis. Determination of the number of micrometeoroid impact craters on the camera provides a unique opportunity to make a very sensitive direct measurement of the flux of interplanetary dust particles impacting the lunar surface.

To make a meaningful flux measurement with this technique, adequate surfaces are required. For many types of surfaces, the crater resulting from the impact of a micron-sized hypervelocity particle is highly characteristic and readily distinguishable from pits, particles and other surface artifacts. Normally a surface is required that is smooth, that produces distinctive craters and that is relatively free of surface blemishes which might be confused with craters. Of the television camera surfaces the optical parts, the mirror and optical filters, are the most appropriate for the detection of small impact craters. This paper describes an investigation of the optical filters for micron-sized craters.

THE FILTERS

The camera contained four filters mounted in a rotatable filter wheel. The filter wheel was mounted directly below the mirror approximately in a horizontal plane (see Fig. 1). In Fig. 1 the paper disks with crosses indicate the centers of the red (R), green (G) and neutral density (ND) filters. The upward facing surfaces of the filters were exposed to impacts but only from a restricted portion of the sky because of partial shielding from the mirror and mirror hood. The red, green and neutral density filters were exposed to a segment of sky extending roughly from the lunar horizon to an elevation of 75°. The blue filter was completely shielded by internal components of the camera. The orientation of the camera relative to the local terrain is shown in Fig. 2.

The filters are made of various types of glass and are roughly 4.5 cm square and 0.3 cm thick. The quality of the surfaces is good and there are few scratches, pits or other crater-like artifacts. When the filters were returned from the moon they were covered with a substantial amount of particulate matter, most probably deposited during Surveyor's bouncing landing in 1967. Before the filters were released for analysis, one half of the top surface of each filter was cleaned of particulate matter at JPL using an acetate strippable film. After stripping, the cleaned portions of the filters were suitable for detection of craters as small as 0.5 μ.

That the filters are made of glass makes them highly suitable for crater searches because of the cratering properties of glass. When micron-sized particles impact glass at velocites in excess of roughly 2 km/sec, the shock wave produced by the impact produces stresses in the glass exceeding its tensile strength and extensive fracturing results. Typically the result of a micron-sized medium-density particle impacting glass at 2 to 20 km/sec is a hemispherical cup with fairly smooth walls surrounded by a region of fractured glass. Figures 3 and 4 show scanning electron microscope and optical photographs of typical craters produced in the laboratory by 6 km/sec iron spheres 1 μ in diameter. The morphology of the crater and surrounding spall zone is determined primarily by the velocity of impact, the angle of impact, and the physical properties of the particle (Vedder, 1971). The fracturing of glass around the crater provides an excellent characteristic facilitating crater detection. Using an optical microscope equipped with upper illumination, light scattered off the fractures surrounding craters enables efficient spotting at low magnification and distinction from particulate matter.

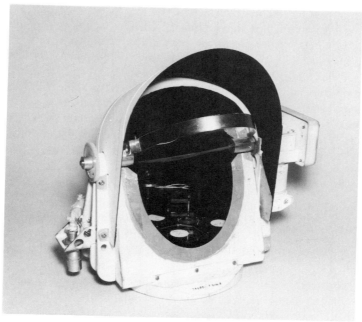

Fig. 1. Filter wheel and its position inside the camera.

Fig. 2. Position of the camera relative to the lunar terrain.

OPTICAL SCAN

The first study of the filters was an optical search for craters done with microscopes in a laminar flow class 100 cleanroom. The filters were mounted on $2'' \times 3''$ microscope slides to facilitate handling and to establish a coordinate reference. The scanning for detectable craters was done at $100\times$ using a Zeiss GFL microscope operating with upper dark field illumination. The stripped half of each filter was completely scanned at least once by three different microscopists. The object of the low power scan was to locate all detectable fractures in the glass. Normally glass fractures $10\,\mu$ and larger could be spotted because of their light scattering properties. When a suspected glass fracture was spotted it was then examined at higher magnifications. The dark field illumination control which alters the azimuthal angle of the illumination was varied and usually a glass fracture could be distinguished from other surface features by the manner in which light played around the fractures. On difficult features upper and lower bright field illumination was also utilized. By examination of craters produced in glass by Ames space simulation facility (CUNNINGHAM, *et al.*, 1966) it was determined that the scanning technique could reliably detect craters $5\,\mu$ in diameter and larger.

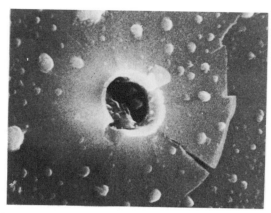

Fig. 3. SEM micrograph of a 3 μ crater in glass produced by a hypervelocity carbonyl iron sphere.

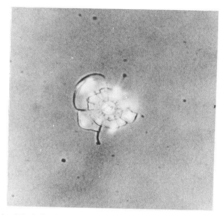

Fig. 4. Optical bright field micrograph of a crater similar to Fig. 3.

The optical scan located about ten glass fractures 10 μ and larger on the cleaned half of each filter. On a statistical basis alone few of these fractures could be considered the result of hypervelocity impact. The largest number of fractures was found on the blue filter, which, because it was shielded from impact, must be considered a control. An examination was also made of a set of back-up filters identical to the ones sent to the moon. The same density of fractures was found on these controls as on the flight surfaces.

In the hope that at least a few of the fractures were caused by craters, each one was extensively studied at high power with a Leitz Ortholux microscope and photographed with upper bright field illumination at 500×. To be identified as an impact site the fracture area was required to contain an area which at least resembled a cup-like crater (possibly greatly elongated) or the remnant of a crater partially spalled away. In at least 90% of the craters we have produced using microparticle accelerators

Fig. 5. SEM micrograph of a typical glass fracture detected in the low power optical scan.

the cup-like crater is easily identifiable and in all other cases they can be identified with difficulty. Of all the fractures on the Surveyor filters none contained an identifiable crater. Many of the fractures did, however, possess other features of hypervelocity impact. Many contained radial cracks and conchoidal fractures extending below the surface. These features, however, are not unique to hypervelocity impact and can be produced by simpler processes, for example, by pounding carborundum grains into a glass surface. It is therefore concluded that all fractures on the filters that were detected in the scanning process are defects in the glass produced by polishing procedures or other processing techniques. A typical glass fracture located in the optical scan is shown in Fig. 5 (a SEM photograph).

SCANNING ELECTRON MICROSCOPE (SEM) SCAN

After the optical study, the filters were broken up for more destructive analysis programs. Fortunately 50 mm² of the neutral density filter was saved for SEM work. Through the Planetology Branch of Ames Research Center we were given sufficient time on their SEM to study a large portion of this piece and also to study some artificially produced craters to determine the crater detection limit of the scanning technique. The Surveyor piece was scanned at 1000× and possible crater sites investigated at higher powers. Scanning of glass containing craters produced by a microparticle accelerator established that craters 1 μ in diameter and larger could be spotted reliably with the scanning technique. While scanning the filter many items were found that possibly were the result of low velocity impact of lunar ejecta. Fortunately the morphology transition from low velocity craters to hypervelocity occurs at approximately the lunar escape velocity so that impacts of extra-lunar particles can be distinguished from lunar ejecta (NEUKUM *et al.*, 1970). To be identified as a hypervelocity crater an object was required to have at least some of the following properties: cup-like depression, signs of melting or flow within the cup, lip structure, fracturing around the cup. No features were found that could be identified as hypervelocity impacts.

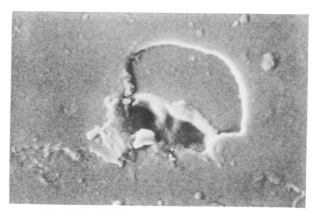

Fig. 6. Possible impact site of 1 μ ejecta particle.

About 30 potential low velocity impacts were found. Identification was based on apparent plastic flow and chipping of the 1500 Å MgF coating on the filter. Preliminary analysis indicates that most of these are low angle impacts whose shape suggests that they were produced by particles coming in the open area of the camera hood. A potential low velocity impact site showing flow and chipping is shown in Fig. 6.

FLUX

No craters were found in either the optical or the SEM studies, so only upper limits to the flux can be established. To calculate these limits the following assumptions are made:

1. Particle density = 2.5 gcm^{-3} (COSPAR standard)
2. The particle flux is isotropic

Strong evidence from Pioneer 8/9 (Berg 1970) and from zodiacal light doppler shifts (Reay, *et al.*, 1968) does not support this assumption but the error produced by anisotropy does not justify a more sophisticated treatment.

3. The ratio of crater diameter to projectile diameter is 1.6

This ratio was determined with the kind help of J. F. Vedder in connection with analysis of craters he has produced in soda lime glass with a 2 MeV electrostatic microparticle accelerator at the Ames Research Center. A study was made on craters in the 1 to 5 μ size that were produced by glass, aluminum and polystrene spheres of measured mass and velocity. For particles in the 5 to 10 km/sec velocity range the above crater/particle ratio was very representative. Calibrations using particles of this density are probably more realistic than those using conventional iron spheres.

4. That no hypervelocity impact craters, larger than the detection limits, exist on the scanned surfaces.

The assumption here is that natural craters are similar to ones produced artificially with microparticle accelerators. The predicted crater characteristics used here are based on craters produced by homogeneous spheres. VEDDER (1971) has shown that particle shape does affect crater morphology. Little is known about the effects of nonhomogeneity, unusual shape or low density.

The flux computation was done by taking the reciprocal of the time-area product (TAP) of the surfaces. This method assumes a 63% probability of having one impact (ALEXANDER, *et al.*, 1963). The time-area product was computed in the following manner:

$$TAP = TAK_\Omega K_\theta K_D$$

where T = exposure time

A = area examined

$K_\Omega = (2\pi)^{-1}$ (solid angle of sky seen by the filter)

$K_\theta = \text{Cos } \theta$, where θ is the average incidence angle of possible impact on the filter

K_D = fraction of the filter not covered by dust

For the three filters exposed to impact the following factors were used

Filter	A	K_Ω	K_θ	K_D
Neutral Density	9.1 cm²	0.35	0.60	0.5
Red	10.0 cm²	0.078	0.42	0.7
Green	9.8 cm²	0.116	0.56	0.7

K_Ω and K_θ were derived from data provided by Neil Nickle of JPL.

Flux for the optical scan

Summing the TAP's for the three filters yields an upper limit to the cumulative flux of 7.5×10^{-5} particles m^{-2} s^{-1} (2π steradian)$^{-1}$. The crater detection limit of 5 μ diameter implies that this limit is for particle masses 2×10^{-11} g and larger.

Flux for the SEM scan

A total of 10.86 mm² of surface was scanned with a detection limit of 1 μ. The computed flux limit is 1.1×10^{-2} (2π steradian)$^{-1}$ for masses 2×10^{-13} g and larger.

CONCLUSION

Because of flexibility in analysis, recoverable crater collection experiments are subject to fewer uncertainties in detection of impacts than are remote sensing experiments. Studies like this one and the S-10 and S-12 experiments flown on Gemini (HEMENWAY, 1968) provide a permanent record of impact events which can be analyzed under laboratory conditions to yield information on particle mass, density, shape, chemical composition and velocity. Crater collection experiments also record impacts of particles too small or of too low density to register on existing remote sensing experiments. Because of the low density sensitivity it is therefore reassuring that the derived optical upper limit to the flux is consistent with the models of both KERRIDGE (1970) and McDONNELL (1970), (see Fig. 7), which are primarily based on remote sensing measurements.

In Fig. 7 the "SEM" and "optical" points represent the upper limits derived in this paper. The dashed line is the 1963 average of satellite microphone data (ALEXANDER, *et al.*, 1963) and is included for historical comparison. The "footprint point" is a

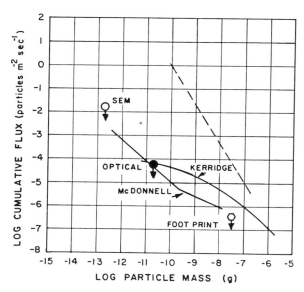

Fig. 7. Particle flux as a function of particle mass. The dashed line is the satellite micro-phone average published by ALEXANDER, et al., (1963) and is included for historical comparison.

flux limit derived from analysis of a Surveyor III footprint (JAFFE, 1970). The line marked "Kerridge" is an average of experimental data he selected as being reliable measurements of the flux at 1 A.U. (KERRIDGE, 1970). The line labeled "McDONNELL" is a model of the flux at the lunar surface based on controlled experiments (McDONNELL, 1970).

The SEM point is important because it represents a direct and accurate measure-ment in a mass range that has not been investigated by many other experiments. Pioneer 8/9 (BERG, et al., 1970) indicates a particle cutoff at about 10^{-11} g. A cutoff at this mass is of considerable interest because it corresponds to the dynamical radia-tion pressure cutoff for particles generated by short period comets (HARWIT, 1963). The discovery by NEUKUM, et al. (1970) of craters on lunar spherules produced by submicron particles contradicts this cutoff. It is hoped that further SEM work will yield increased sensitivity and provide additional information on this interesting submicron particle regime.

Acknowledgments—We would like to express appreciation to H. Y. LEM of the Ames Research Center for his skillful SEM analysis of the ND filter. This work was supported by NASA grant NGR 48-002-033.

REFERENCES

ALEXANDER W. M., McCRACKEN G. W., SECRETAN L., and BERG O. E. (1963) Review of direct measurements of interplanetary dust from satellites and probes. *Space Res.* **3**, 891–917.
BERG O. E. and GERLOFF U. (1970) More than two years of micrometeorite data from two Pioneer satellites. Paper presented to the XIII COSPAR meeting, Leningrad, May 1970.

CUNNINGHAM B. E. and EDDY R. E. (1966) Space environment simulator for studies of the effects of space environment on materials. In AIAA/IES/ASTM *Space Simulation Conference*, Amer. Inst. Aeronaut. Astronaut.

HARWIT M. (1963) Origins of the zodiacal dust cloud. *J. Geophys. Res.* **68**, 2171–2180.

HEMENWAY C. L., HALLGREN D. S., and KERRIDGE J. K., (1968) Results from Gemini S-10 and S-12 micrometeorite experiments. *Space Res.* **8**, 521–535.

JAFFE L. D. (1970) Lunar surface: changes in 31 months and micrometeoroid flux. *Science* **170**, 1092–1094.

KERRIDGE J. F. (1970) Micrometeorite environment at the earth's orbit. *Nature* **228**, 616–619.

McDONNELL J. A. M. (1970) Review of in situ measurements of cosmic dust particles in space. Paper presented to the XIII COSPAR meeting, Leningrad, May, 1970.

NEUKUM G., MEHL A., FECHTIG H., and ZÄHRINGER J. (1970) Impact phenomena of micrometeorites on lunar surface material. *Earth Planet. Sci. Lett.* **8**, 31–35.

REAY N. K. and RING J. (1968) Radial velocity measurement on the zodiacal light spectrum. *Nature* **219**, 710.

VEDDER J. F. (1971) Microcraters in glass and minerals. *Earth Planet. Sci. Lett.* (submitted).

Proceedings of the Second Lunar Science Conference, Vol. 3, pp. 2791–2795
The M.I.T. Press, 1971.

Replication electron microscopy on Surveyor III unpainted aluminum tubing

E. A. BUVINGER

Advanced Electronic Devices Branch
Air Force Avionics Laboratory
Air Force Systems Command
Wright-Patterson Air Force Base, Ohio 45433

(*Received* 22 *February* 1971; *accepted in revised form* 31 *March* 1971)

Abstract—The surface of the unpainted aluminum tubing (Sections C and E) from the RADVS strut of the Surveyor III Moonlander has been examined by replication electron microscopy. The lower (relative to the lunar surface) portions of the tubing exhibit erosion damage which probably occurred during the original landing maneuver. Many small (~ 1 μm) pits appear in the upper surface; at least one of these exhibits the characteristics of a hypervelocity impact. Little damage ascribed to solar-wind sputtering was found. Within the scope of this study, the greatest damage to the tubing is attributed to particle impact, and such damage is limited to a maximum depth of 2 μm.

INTRODUCTION

UNPAINTED, polished aluminum tubing from the RADVS strut of the Surveyor III spacecraft, after its return to Earth, was sectioned into six pieces identified as A, B, C, D & F, E, and G (NICKLE, 1971). End A is believed to have been adjacent to the body of the moonlander according to BÜHLER *et al.* (1971). Two sections, C and E (each ~ 1 in. long), have been examined in a transmission electron microscope utilizing replication techniques. Section C was received first; Section E was received about three months later, along with a piece of unused tubing for comparison purposes. The section of comparison tubing had been prepared recently in the same manner as the Surveyor tubing and was useful in determining positively that certain features could be ascribed to polishing-and-handling procedures.

The purpose of this investigation was to determine the type and degree of microscopic surface damage which the tubing incurred during its exposure to the lunar environment. Specifically, the surface was examined for evidence of ion bombardment (sputtering) and micrometeorite damage.

EXPERIMENTAL TECHNOLOGY

Upon receipt, the tubing sections were photographed for record and then washed with acetone to remove possible traces of residue from the soil-peel procedures used by previous investigators. For the replication process elvanol (polyvinyl alcohol) proved to be the most satisfactory material. The replication procedure used was as follows: A stripe of elvanol (15% solution) was dropped along the upper surface of the tubing, was dried, and was then stripped from the tubing. This stripe provided a "negative" replica of the tube surface. The replica was then shadowed for contrast by coating it with a heavy metal (in this case, platinum) at an oblique angle. Then 300 to 500 Å of carbon were deposited over the entire surface of the replica to provide support. This thin metal film was then cut into pieces $\sim \frac{1}{8}$ in. square, and the original underlying plastic replica was dissolved away. The squares were picked up on 200-mesh grids. These thin-film replica squares were then examined in the electron microscope. This procedure was repeated until the entire surface of the tubing had been replicated at least once.

The purpose of metal shadowing at an oblique angle is to cause the relatively higher portions of the plastic replica to shield the areas behind them from the metal. This provides contrast for transmission electron microscopy. The areas where the metal is thickest will scatter the most electrons, thus causing that area to appear darker on the phosphor viewing screen (or photographic print). Conversely, the areas which have been shielded from the metal will appear bright on photographic prints. It is important to remember that the replica being examined is a negative of the original tube surface. Thus, for example, microcraters in the tubing surface will appear to be rising above the tubing surface in photographs. The microcrater will cast a shadow whose length will be proportional to the depth of the original crater. The majority of the replicas were shadowed at 25 degrees. Some shadowing, however, was done at 10 degrees to enhance the fine details of the surface structure.

DISCUSSION

Both sections of tubing, when received, were contaminated on one side. One-half to two-thirds of the circumference of the tubing had a dull appearance to the naked eye, while the remainder of the surface appeared bright, as is expected with polished aluminum. The replicating plastic lifted much of the contaminant, which appeared globular under low-power optical microscopy. The areas where the first replicas had been lifted from the dull portion of the tubing were clearly discernible. Large pieces of contaminant which were imbedded in the replica could not be supported by the thin metal film when the plastic was dissolved and, therefore, they fell away. Smaller pieces adhered to the film and could be observed in the electron microscope. The material was generally opaque to 100-kV electrons. The particles range from a few hundred angstroms to more than 1 μm in size. The surface of the tubing in the same areas from which the contaminants were removed appears to be eroded extensively. Although this was true for both sections, it was particularly evident in Section E. The most likely explanation for this phenomenon would be sandblasting by lunar dust which could have occurred during the extended landing maneuver of the Surveyor III craft. The small particles would then be lunar debris which were either imbedded in the tubing or which adhered to it with the aid of plume contamination from the retro-rockets. The membrane-like material which covers most of the area shown in Fig. 1 (Section C) is assumed to be retro-rocket contamination. Some of this material was also present in the contaminant removed from Section E (Fig. 2), but here it was not so prevalent. Transmission electron diffraction was attempted on the contaminant particles, but meaningful results could not be obtained. The particle thickness exceeds that which is penetrable by 100-kV electrons. However, the contaminants are most likely a combination of organic material deposited during retro-fire and lunar debris. All of the contaminant material was quite stable in the electron beam. A comparison of the material removed from Sections C and E reveals that Section-C contaminants are not so massive as those from Section E and contain appreciably more of the membrane-like contamination and less solid particulate material.

The bright portions of the tubing in particular show extensive polishing-and-handling scratches which were apparent even under low-power optical microscopy. Upon examination in the electron microscope, some of these features were so extreme

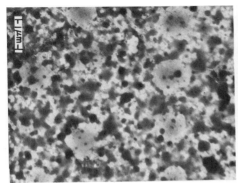

Fig. 1. Lower surface of section C.   Fig. 2. Lower surface of section E.

as to cause the thin-film replica to tear, destroying the area involved. In general, the surface appearance of the bright portions was similar for both sections (Fig. 3).

A number of microcraters were observed in the bright portions of the tubing. More were found from Section C than from Section E; however, since a more thorough investigation was conducted on Section C, this could account for the apparent difference.

One of the microcraters found was unique (Fig. 4). It exhibits a disturbed, raised area around its circumference and a relatively smooth central pit as defined by the shadow. The crater is ~ 1 μm across its smaller diameter and is ~ 0.7 μm deep. This microcrater resembles artificial impact craters in aluminum as described by WEIHRAUCH et al. (1968). Craters having similar features were found in lunar materials by DEVANEY and EVANS (1970).

In view of the many unknown factors involved (temperature at the time of formation, grain structure of the immediate area, size of the impinging particle, etc.), it is difficult to make absolute judgments concerning the cause of the microcrater. However, in the opinion of the author, it is probably the result of a hypervelocity impact.

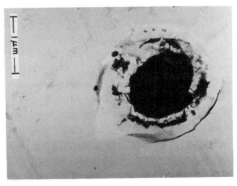

Fig. 3. Upper surface of section C.   Fig. 4. Pit in section C probably caused by
                                              hypervelocity impact.

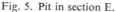

Fig. 5. Pit in section E.

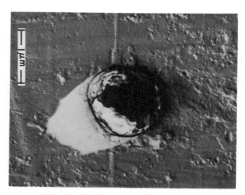

Fig. 6. Pit in section C.

Figure 5 shows a different type of microcrater. The area around the crater shows moderate disturbance but no splash lip. The crater itself has a shallow portion around its periphery and a deeper central core which has a relatively smooth bottom. The outer edge of the crater is ∼ 1.2 μm in diameter, and the maximum depth of the crater is ∼ 1.3 μm.

Another type of crater found is typified by Fig. 6. Again, the opaque central area is composed of particulate material. The outer diameter of this crater is ∼ 1.3 μm, and its depth is ∼ 0.6 μm. No definite area of disturbed material around the outside of the crater itself is apparent. The walls of the crater are relatively steep, and the bottom is somewhat rounded.

Apparently, several degrees of violence were involved in the formation of these craters. The size of the impinging particles could be responsible for some of the differences. It is not inconceivable that some of the more shallow craters could have been formed by small, hard—compared to aluminum tubing—particles being imbedded in the surface if a strong grip were used by the astronaut during the removal of the tubing from the spacecraft.

It has been previously conjectured by BENSON et al. (1970) that some of the pitting —which on the other Surveyor III components examined could not be attributed to polishing or to high-velocity impact—may have been caused by the blowing of dust and debris during the Apollo 12 LEM landing. For the case of the polished tubing, however, this is unlikely since available photographs show that the Surveyor craft itself was between the RADVS strut and the LEM; therefore, the strut would have been protected from such impacts.

All of the micropits found were from the bright areas of the tubing. The number found leads to the following approximations: less than 1 hyper-velocity impact/in.[2] and, for the other types of craters, less than 10/in.[2]

It is believed that craters up to about 5 μm in diameter could have been observed; features larger than this probably would tear out of the replica due to lack of support on the shadowed side. However, no crater was found which had a diameter greater than 2.5 μm. The maximum depth was 2 μm, and the majority had depths of less than 1 μm.

When considering possible damage due to solar-wind sputtering, it must be remembered that although a relatively smooth surface will develop a higher degree of surface roughness under sputtering conditions because of the slightly different sputtering rates of differently oriented crystallites, this tubing was mechanically buffed with rouge, and the resulting smeared surface was probably amorphous. This makes erosion-rate estimates very difficult. WEHNER and KENKNIGHT (1967), in an investigation of sputtering effects on the surface of the moon, calculated an erosion rate of ~ 0.25 Å/year due to full solar wind (H and He) striking a smooth, stony surface. This calculation was based upon solar-wind data from Mariner 2, Pioneer 6, and Explorer 18. For the purposes of the present study, one should consider the possibility of increased erosion due to such factors as increased probability of oblique ion incidence on the $\frac{1}{2}$-in.-diameter tubing, differences in the sputtering yield for various materials, the presence of heavy ions, etc. However, even if one assumed a factor of 100 increase over the original calculations, the total loss would be less than 65 Å. From the general appearance of the upper surface, it is believed that relatively little material was removed and that the actual loss was almost certainly no more than 65 Å. The underside of the tubing gives little assistance in these considerations since the erosion incurred during landing far exceeds that from sputtering by solar wind.

The greatest degree of damage incurred by the tubing during its 31-month stay on the lunar surface was the result of particle impact, and this damage is (within the scope of this study) limited to a maximum depth of 2 μm.

Acknowledgments—The helpful telephone discussions with NEIL L. NICKLE of the Jet Propulsion Laboratory at the California Institute of Technology are gratefully acknowledged. Information contained in this paper is the result of research performed within the Advanced Electronic Devices Branch of the Air Force Avionics Laboratory, Air Force Systems Command, Wright-Patterson Air Force Base, Ohio. Basic Research Project 4150 is the programming authority for this work.

REFERENCES

NICKLE N. L. (1971) Surveyor III Material Analysis Program. Second Lunar Science Conference (unpublished proceedings).

BÜHLER F., EBERHARDT P., GEISS J., and SCHWARZMÜLLER J. (1971) Trapped Solar Wind Helium and Neon in Surveyor 3 Material. Second Lunar Science Conference (unpublished proceedings).

WEIHRAUCH J. H., GERLOFF N., and FECHTIG H. (1968) Stereoscan Investigations of Metal Plates Exposed on LUSTER 1966, Gemini 9 and 12. In *COSPAR Space Research VIII* (editors A. P. Mitra, L. G. Jacchia and W. S. Newman), pp. 566–578. North-Holland.

DEVANEY J. R. and EVANS K. (1970) Characterization of Hyper-Velocity Particle Impact Craters on Apollo 11 Lunar Soil. In *Proceedings Electron Microscopy Society of America 28th Annual Meeting 1970* (editor C. J. Arceneaux), pp. 20–21. Claitor's Publishing Division.

BENSON R. E., COUR-PALAIS B. G., GIDDINGS L. E., JR., JACOBS S., JOHNSON P. H., MARTIN J. R., MITCHELL F. J., and RICHARDSON K. A. (1970) Preliminary Results from Surveyor 3 Analysis. NASA SP-235 (Apollo 12 Preliminary Science Report), pp. 217–223.

WEHNER G. K. and KENKNIGHT C. E. (1967) Investigation of Sputtering Effects on the Moon's Surface. NASA CR 88738.

Apollo 12 Lunar Sample Inventory

Sample No.	Mass in grams	Description	Sample No.	Mass in grams	Description
12001	2216.0	Fines	12033	450.0	Fines
12002	1529.5	Porphyritic basalt	12034	155.0	Breccia
12003	300.0	Fines and chips	12035	71.0	Ophitic basalt
12004	585.0	Porphyritic basalt	12036	75.0	Ophitic basalt
12005	482.0	Basalt	12037	145.0	Fines
12006	206.4	Ophitic basalt	12038	746.0	Ophitic basalt
12007	65.2	Basalt	12039	255.0	Ophitic basalt
12008	58.4	Porphyritic vitrophyre	12040	319.0	Ophitic basalt
12009	468.2	Porphyritic vitrophyre	12041	24.8	Fines
12010	360.0	Breccia	12042	255.0	Fines
12011	193.0	Porphyritic basalt	12043	60.0	Ophitic basalt
12012	176.2	Porphyritic basalt	12044	92.0	Fines
12013	82.3	Breccia	12045	63.0	Porphyritic basalt
12014	159.4	Porphyritic basalt	12046	166.0	Ophitic basalt
12015	191.2	Porphyritic vitrophyre	12047	193.0	Ophitic basalt
12016	2028.3	Basalt	12048	2.0	Fines
12017	53.0	Porphyritic basalt	12050	1.0	Fines
12018	787.0	Porphyritic basalt	12051	1660.0	Ophitic basalt
12019	462.4	Porphyritic basalt	12052	1866.0	Porphyritic basalt
12020	312.0	Porphyritic basalt	12053	879.0	Porphyritic basalt
12021	1876.6	Porphyritic basalt	12054	678.0	Basalt
12022	1864.3	Porphyritic basalt	12055	912.0	Porphyritic basalt
12023	269.3	Fines—LESC	12056	121.0	Ophitic basalt
12024	56.5	Fines—GASC	12057	650.0	Fines and chips
12025	56.1	Fines—double core tube, top	12060	20.7	Fines
			12061	9.5	Chips
12026	101.4	Fines—first single core tube	12062	738.7	Ophitic basalt
			12063	2426.0	Porphyritic basalt
12027	80.0	Fines—second single core tube	12064	1214.3	Ophitic basalt
			12065	2109.0	Porphyritic basalt
12028	189.6	Fines—double core tube, bottom	12070	1102.0	Fines
			12071	9.16	Chips
12029	6.5	Fines—in Surveyor scoop	12072	103.6	Basalt
			12073	407.65	Breccia
12030	75.0	Fines and chips	12075	232.5	Porphyritic basalt
12031	185.0	Ophitic basalt	12076	54.55	Porphyritic basalt
12032	310.5	Fines	12077	22.63	Basalt

Proceedings of the Second Lunar Science Conference, Vol. 3, pp. 2799–2800.
The M.I.T. Press, 1971.

Author Index

ABDEL-GAWAD M.   2125
ADAMS J. B.   2183
ADE P. A.   2203
ALEXANDER C. C.   2433
ALPERIN H.   2079
ANDERSON D. L.   2753
ANDERSON O. L.   2345
ARRHENIUS G.   2583

BALDRIDGE W. S.   2317
BARBER D. J.   2705
BASTIN J. A.   2203
BECKER K.   2009, 2057
BHANDARI H.
     2583, 2599, 2611
BHAT S.   2583, 2599, 2611
BILLINGSLEY F. C.   2301
BIRKEBAK R. C.   2197, 2311
BLAIR P. M.   2735
BLANDER M.   2125
BLOCH M. R.   2639
BORG J.   2027
BRITTON D.   2245
BROWNLEE D.   2781
BUCHER W.   2781
BUVINGER E. A.   2791

CAMPBELL M.   2173
CARRIER W. D.   1959
CARROLL W. F.   2735
CARTER J. L.   2653
CHAN S. I.   2515
CHATELAIN A.   2501
CHUNG D. H.   2381
CHUPKA W. A.   2671
COHRON G. T.   1973
COLBURN D. S.   2415
COLLETT L. S.   2367
COMSTOCK G. M.   2559, 2569
CONEL J. E.   2235
CONNELL G. L.   2083
COSTES N. C.   1973
COUR-PALAIS B. G.   2767
COWSIK R.   2705
CREMERS C. J.   2197, 2311
CROZAZ G.   2543
CUNNINGHAM B. E.   2753

DAHMNS R. G.   2753
DAWSON J. P.   2197

DOELL R. R.   2491
DOLLFUS A.   2285
DORETY N.   2477
DUNN J. R.   2461
DURRIEU L.   2027
DWORNIK E.   2433
DYAL P.   2391

ECONOMOU T. E.   2699
ELLIS W. L.   2721
ESTEP P. A.   2137
EVWARAYE A. O.   2559, 2569

FABEL G. W.   2213
FAILE S. P.   2069
FAVERO M. S.   2715
FECHTIG H.   2639
FENNER M. A.   2093
FISHER R. M.   2461
FLAHERTY R. E.   2767
FLEISCHER R. L.   2559, 2569
FREEMAN J. W.   2093
FULLER E. L.   2009
FULLER M. D.   2461

GAMMAGE R. B.   2009, 2057
GARLICK G. F. J.   2265, 2277
GEAKE J. E.
     2265, 2277, 2285
GENTNER W.   2639
GOETZ A. F. H.   2301
GOLD T.   2173, 2675
GÖRZ, H.   2021, 2213
GRANT R. W.   2125
GREEN R. H.   2715
GREENE C. H.   2049
GREENMAN N. N.   2223
GROMMÉ C. S.   2491
GROSS H. G.   2223
GROSSMAN J. J.   2153
GUPTA Y. P.   2083

HAMANO Y.   2323
HANEMAN D.   2529
HARGRAVES R. B.   2477
HART H. R.   2559, 2569
HARTUNG J. B.   2629
HEAD J. W.   2301
HELSLEY C. E.   2485

HEMINGWAY B. S.   2361
HENISCH H. K.   2213
HERZENBERG C. L.   2103
HEYWOOD H.   1989
HIGH R. W.   2767
HILLS H. K.   2093
HODGE P.   2781
HOLMES H. F.   2009
HÖRZ F.   2629
HOUSLEY R. M.   2125, 2337
HOUSTON W. N.   1953
HOYT H. P.   2245
HUTCHEON I. D.   2705

ISARD J. O.   2003

JOHNSON G. G.   2021
JOHNSON S. W.   1959
JOURET C.   2027

KANAMORI H.   2323
KARDOS J. L.   2245
KARR C.   2137
KATSUBE T. J.   2367
KAY, H. F.   2049
KESSLER D. J.   2767
KLINE D.   2501
KNITTEL M. D.   2715
KOLOPUS J. L.   2501
KOVACH J. J.   2137
KRAATZ P.   2083
KUMAZAWA M.   2345
KURTOSSY S.   2069

LAL D.   2583, 2599, 2611
LAMB W. E.   2277
LARSON E. E.   2451
LIANG S.   2583
LILLEY E. M.   2165

MACDOUGALL D.   2583
MANATT S. L.   2515
MARSTON A. C.   2203
MAURETTE M.   2027
MCCORD T. B.   2183, 2301
MCKAY D. S.   2653, 2767
MILLER D. J.   2529
MILLS A. A.   2265
MITCHELL F. J.   2721

MITCHELL J. K. 1953
MIYAJIMA M. 2245
MIZUTANI H. 2323
MOLER R. B. 2103
MORGAN R. G. 2753
MORRISON J. A. 2345
MOSS D. C. 1973
MUELLER G. 2041
MUIR A. H. 2125
MUKHERJEE N. R. 2153

NAGATA T. 2461
NASH D. B. 2235
NEUKUM G. 2639
NICKLE N. L. 2683
NORTON P. R. 2345

O'LEARY B. T. 2173

PANDYA S. J. 2203
PARKIN C. W. 2391
PEARCE G. W. 2451
PICKART S. J. 2079
PRICE P. B. 2621, 2705
PUPLETT E. 2203
PYE L. D. 2049

RAJAGOPALAN G.
2583, 2599, 2611
RAJAN R. S. 2621, 2705
RASE D. E. 2049

RILEY D. L. 2103
ROBIE R. A. 2361
ROY D. M. 2069
ROY R. 2021, 2069, 2213
RYAN J. A. 2153

SCHIFFER J. P. 2671
SCHMIDT R. 1959
SCHNEIDER E. 2639
SCHNEIDMILLER R. F. 2083
SCHOLZ C. 2345
SCHREIBER E. 2345
SCHUBERT G. 2415
SCHWARTZ K. 2415
SCHWERER F. C. 2461
SCOTT R. F. 2743
SENFTLE F. E. 2433
SHIRK E. K. 2621
SIMMONS G.
2317, 2327, 2381
SMITH B. F. 2415
SONETT C. P. 2415
STEIGMANN G. A. 2277
STEPHENS D. R. 2165
STEVENS C. M. 2671
STEVENS H. J. 2049
STRANGWAY D. W. 2451
SULLIVAN S. 2433

TAMHANE A. S.
2583, 2599, 2611

THORPE A. N. 2433
TITTMANN B. R. 2337
TITULAER C. 2285
TODD T. 2327
TSAY F. D. 2515
TURKEVICH A. L. 2699

VENKATAVARADAN V. S.
2583, 2599, 2611

WALKER G. 2265
WALKER R. M. 2245, 2543
WANG H. 2327
WARREN N. 2345
WEBER J. N. 2213
WEEKS R. A. 2501
WEGNER M. W. 2153
WEIDNER D. 2327
WERNER R. A. 1959
WESTPHAL W. B. 2381
WHITE E. W. 2021, 2213
WHITE W. B. 2213
WILKENING L. 2583
WOOLUM D. 2543

YOST E. 2301

ZIMMERMAN D. W. 2245
ZIMMERMAN J. 2245
ZOOK H. A. 2767
ZUCKERMAN K. A. 2743

SUBJECT INDEX

Prepared by

JEFFREY L. WARNER AND DALE FESSENDEN
NASA Manned Spacecraft Center
Houston, Texas 77058

This index was automatically extracted from keypunched titles, abstracts and introductions of the various papers contained in the Proceedings of the Second Lunar Science Conference by a new computer program. Details of the conventions and linguistic procedures involved in the automatic indexing are being set out in a NASA report now in preparation.

Some decisions had to be arrived at as to what would constitute the input text. Twenty papers were selected at random and analyzed as to what parts of each paper would index into useful phrases. The following conclusions were reached:

1. Titles and abstracts make good indexing text.

2. Introductions significantly add to the posted phrases in about half of the cases.

3. Conclusions and discussions, besides not adding significantly to the useful posted phrases, in some cases produced misleading phrases.

4. The body of the papers produced useful phrases, but the phrases were often over - specific.

Following these results, it was decided to keypunch titles, abstracts, and introductions. In addition, all elements, oxides, and isotopes listed in tables contained in the body of the papers were keypunched. This keypunched information served as input data to the indexing program.

Each article is identified by its first page. That is, all references are to the first page of a given paper and imply the complete paper.

It is essential when using this index to remember that if a concept does not appear in the title, abstract or introduction of a paper, that paper will not be indexed for that concept. For example, under the phrase "feldspar stoichiometry" there is only one posting. This does not mean that only one paper discusses feldspar stoichiometry; rather it signifies that only one paper included mention of feldspar stoichiometry in its title, abstract, or introduction. This constraint of the indexing applies to many other phrases (e.g., "achondrite" with only two postings).

ACKNOWLEDGMENTS — We wish to thank many individuals for their help and encouragement in completing this index. Fred Japp and Clara King, of the Computation and Analysis Division, are responsible for the excellent keypunching of the input text. Wally Stewart gave constant support to the programming effort. Conne Bender, Phyllis Richardson, Central Tatmon, of Lockheed Electronics Co., provided assistance in building the dictionary and analysis of the provisional indexings. And finally, we thank Paul Klingbiel, of the Defense Documentation Center, who taught us much about linguistics, and D. H. Anderson for assistance, discussions and encouragement.

abrasion 719
absorption 265, 2079, 2203, 2213, 2501, ... 2515
absorption band 2137, 2183, 2197, 2213, ... 2277
accretion 43, 449, 1021, 1037, 1083, 1139, ... 2027
accretion temperature 1021
accretionary lapilli 755
acetylene 1843
achondrite 617, 1281
acicular clinpyroxene rosette 141
acid derivative 973
acid etching 1843
acid hydrolysis 1397, 1843, 1879
acid leaching 1139
acid resistant lead 1547
acoustic velocity 2345
acoustic wave 2345
actinide 1571
actinide istope 1571
activator 2265
adhesion 2153
adsorbed gas 2529
age 17, 343, 393, 583, 737, 1159, 1337, 1471, 1487, 1493, 1503, 1521, 1547, 1591, 1607, 1729, 1825, 1959, 2559, ... 2569, 2611, 2639
age nucleosynthesis 1337
aggregate 775, 797, 817, 2317
albedo 2183, 2285
alkali feldspar 247
alkali glass 727
alkali metal 449, 1253
alkali metal chemistry 1159
alkali metal vapor 449
alkaline 973, 1101
alkane 1927
alpha ray 1209, 1571, 1577, 2699
alpha decay 1585
alpha radioactivity 2699
ALSEP solar wind 2093
ALSEP suprathermal ion detector 2093
alteration 1521
aluminum 1, 39, 59, 91, 43, 117, 135, 159, 193, 207, 219, 265, 285, 301, 343, 359, 377, 393, 413, 439, 449, 459, 481, 497, 507, 529, 559, 575, 583, 601, 617, 645, 679, 701, 727, 747, 755, 775, 797, 833, 855, 937, 973, 999, 1037, 1063, 1123, 1169, 1187, 1209, 1217, 1231, 1237, 1247, 1253, 1259, 1281, 1291, 1319, 1367, 1461, 1471, 1705, 1729, 1757, 2049, 2069, 2083, 2125, 2433, ... 2501, 2653, 2753, 2791
aluminum chromite 207

aluminum projectile 2653
aluminum tube 2753, 2767
aluminum-26, 559, 1747, 1773, 1791, 1797
Al/Si ratio 117
amino acid 1879
amino acid derivative 1913
amorphous coating 2027
amorphous radiation 2009
amorphous silicon 973
amphibole 59, 645
amplitudes 2079
ancient field 2451, 2461
ancient highland crust 583
ancient radiation 1651
ancient solar flare 2705
angular glass 2653
anhydrous moon 645
anisotropy 2327
annealed breccia 393
annealing 2069, 2277, 2501
anomalous density 2337
anorthosite 247, 319, 393, 439, 583, 679, 719, 737, 797, 817, 833, 893, 987, 999, ... 1139, 1503, 1651, 2183
anorthosite bedrock 679
anorthosite gabbro 285, 1037
anorthosite glass 937
anorthosite stratiform sheet 439
antiferromagnetism ilmenite 2461
antiferromagnetism ordering 2079
antimony 1169, 1253, 1277, 1281
antiperthite 247
apatite 247, 645, 1503
aqueous fluid 645
argon 775, 957, 1381, 1521, 1591, 1607, 1627, 1651, 1681, 1693, 1705, 1717, 1813, ... 1825, 2009, 2083
argon age 1159, 1521, 1591, 1627, 1693, ... 1825
argon radioactivity 1813, 1825
aromatic hydrocarbon 1879
asteroid 2285
astrochemistry 1367
astrology 1503
astronomy 2203
asymmetric resonance 2515
atmosphere pressure 617
atomic erosion 2611
atomic exchange 91
augite 59, 69, 247, 265, 301, 319, 359, ... 413, 559, 575, 617, 775, 1101, 1291
baddeleyite 159, 207, 319, 1063, 1503
barium 1, 43, 301, 319, 393, 439, 481, 507, 583, 679, 727, 747, 833, 973, 999,

1063, 1083, 1101, 1123, 1169, 1187, 1217, 1231, 1247, 1253, 1281, 1291, 1319, 1471, 1627, 1671, 1693

barium feldspar 43
basal plane twinning 817
basalt 39, 91, 43, 143, 151, 159, 167, 207, 219, 247, 301, 319, 343, 359, 393, 439, 449, 459, 469, 529, 559, 575, 583, 601, 617, 645, 679, 701, 727, 755, 833, 873, 893, 937, 973, 987, 999, 1021, 1037, 1169, 1187, 1209, 1247, 1261, 1307, 1319, 1331, 1337, 1343, 1351, 1381, 1397, 1407, 1421, 1493, 1547, 1953, 2057, 2137, 2165, 2183, 2311, 2323, 2327, 2337, 2381, 2415, 2451
basalt achondrite 459, 1123, 1169
basalt breccia 319
basalt cooling 645
basalt crystallization 583
basalt fines 973
basalt flow 999
basalt lava 583
basalt liquid 529
basalt magma 91
basalt magma fractionation 583
basalt mineralogy 43
basalt stratiform sheet 439
basesurge 1681
bedrock 665, 2675
bedrock debris 665
benzene 1879, 1913
beryllium 1451
beta ray 1337, 1585
binary silicate melt 957
biologic quarantine 1813, 1929, 1931
bismuth 1021, 1037, 1139, 1547, 1565
boron 1461
boron silicate glass 2639
breccia 17, 43, 143, 151, 177, 207, 285, 319, 359, 393, 439, 575, 583, 665, 679, 727, 755, 797, 817, 833, 873, 893, 909, 949, 1021, 1037, 1101, 1139, 1159, 1169, 1187, 1231, 1237, 1261, 1319, 1331, 1343, 1351, 1381, 1397, 1407, 1421, 1461, 1471, 1493, 1521, 1547, 1571, 1607, 1651, 1681, 1747, 1803, 1843, 1913, 1929, 2027, 2103, 2183, 2203, 2265, 2285, 2345, 2361, 2367, 2451, 2461, 2477
bremsstrahlung activation 1231, 1319
bromine 1, 1021, 1037, 1139, 1169, 1247, 1261, 1693, 2671
bubble pit 909
cadmium 1, 1021, 1037, 1139, 1231, 1319, 1891
calcium 1, 39, 47, 59, 91, 43, 117, 135, 143, 151, 159, 207, 219, 265, 285, 301, 359, 377, 393, 413, 439, 449, 459, 481, 507, 529, 559, 575, 583, 601, 617, 645,

679, 701, 719, 727, 747, 755, 775, 797, 833, 855, 937, 957, 973, 1063, 1123, 1169, 1187, 1209, 1217, 1231, 1247, 1253, 1259, 1291, 1301, 1319, 1367, 1471, 1705, 1773, 1813, 2049, 2069, 2083, 2125, 2433, 2653
calcium band 359
calcium clinopyroxene 91, 141
calcium iron silicate 47
calcium olivine 143
calcium pigeonitee 529, 775
calcium plagioclase 559, 775
calcium pyroxene 285, 343, 2183
carbide 1343, 1381, 1407, 1843
carbon 1343, 1351, 1381, 1397, 1407, 1421, 1843, 1879, 1901, 1913, 1927
carbon balance 1879
carbon chemistry 1397
carbon dioxide (see CO₂)
carbon hydrogen 1397
carbon mass balance 1901
carbon monoxide 1351, 1381, 1397, 1843, 1901, 2009
carbon oxide 1397, 1843, 1879, 1913
carbonaceous chondrite 1547, 1681
carbon isotopes 59, 1397, 1421, 1451, 1843, 1901, 1927
carbon isotopes alkane 1927
carbon/isotopes compound evolution 1931
carbon/isotopes hydrocarbon 1843
cation ordering 2137
Ca/Al ratio 1123
cell mass 449, 1367
centered cell 59
cerium 1, 727, 999, 1063, 1083, 1101, 1169, 1247, 1253, 1277, 1281, 1291, 1301, 1307, 1351, 1471, 1627, 1671
cesium 1, 1021, 1083, 1139, 1169, 1187, 1231, 1247, 1253, 1281, 1291, 1319
chalcopyrrhotite 219
chlorine 1, 285, 319, 583, 1169, 1187, 1247, 1261, 1281, 1547, 1773, 2083, 2671
chlorine apatite 319
chlorine chondrite 1037
chondrite 177, 529, 957, 1139, 1307, 1319, 1331, 1443, 1547
chondrule 957
chromian ulvospinel 193, 359
chromite 193, 207, 219, 301, 343, 923, 2083, 2485
chromite core 219, 359
chromite crystallization 193
chromite rim 2485
chromium 1, 39, 59, 91, 43, 143, 159, 193, 207, 219, 265, 285, 301, 359, 377, 393,

413, 439, 459, 481, 497, 529, 559, 575, 583, 601, 617, 645, 701, 719, 747, 775, 797, 833, 855, 973, 1063, 1083, 1123, 1169, 1187, 1217, 1231, 1247, 1253, 1259, 1277, 1281, 1291, 1301, 1319, 1367, 1451, 1471, 1651, 1705, 1773, 2049, 2083, 2153, 2611

chromium isotope 1451
chromium olivine 143
chromium spinel 343, 601, 617
chromium ulvospinel 207, 219, 377, 855
CIPW norm 459, 1259
cleavage 135, 2599
cleavage plane 2599
clinoenstatite 59
clinopyroxene 59, 91, 109, 141, 151, 193, 247, 265, 285, 301, 343, 559, 575, 775, 817, 833, 1291
clinopyroxene diffraction 109
CO$_2$ 973, 1101, 1343, 1351, 1381, 1397, 1407, 1717, 1773, 1843, 1879, 1891, 1901, 1913
cobalt 1, 193, 855
cobalt nickel-iron 207
coercivity 2477
cohenite 1351, 1843
comet 1037, 1865, 2543
cometary impact 1351
complexing anion 169
compressibility 2165
compressional velocity 2345
concentration gradient 343, 1757
cone 1973
contingency fines 1139
convection 2173, 2311
Copernicus ... 43, 665, 1139, 1421, 2285
Copernicus impact 665, 679
Copernicus ray 665
copper 1, 439, 583, 957, 973, 1021, 1169, 1187, 1217, 1247, 1253, 1277, 1281, 1471
core recovery 1959
core stratigraphy 665
core tube 665, 755, 1139, 1343, 1757, 1879, 1891, 1913, 1953, 1959, 1973, 2125, 2183, 2197
core tube fines 1959, 1973, 2057
core tube geometry 1959
core tube wall 1959
cosmic fines bombardment 2639
cosmic radiation 1757
cosmic ray 1461, 1503, 1591, 1607, 1671, 1797, 1803, 1825, 2559, 2569, 2599, 2611, 2621, 2629, 2671
cosmic ray age 1591
cosmic ray bombardment 1627, 1729
cosmic ray nuclei 2569, 2599
cosmic ray spallation .. 1421, 1591, 1671

cosmic ray track 1503, 2559
cosmochemistry 1037, 1139, 1337, 1367, 2621
cosmogenic gas 1651, 1693
cosmogenic isotope 1607, 1651, 1773
cosmogenic radioisotope 1729, 1747, 1757, 1791, 2583
cosmogenic xenon 1627
count crater 2639
crater ejecta 1343
crater morphology 873, 2653
crater rim 1139, 1187
crater rim fines 1139
crescent crater 1083
cristobalite 69, 117, 141, 247, 265, 301, 319, 359, 413, 645, 719, 775, 855, 1217
crustal norite 583
crystal accumulation 359
crystal cumulate 529
crystal diffraction 59
crystal fractionation 507
crystal habit 2009
crystal lattice 2599
crystal liquid 169
crystal structure 47, 117
crystal texture 1237
crystal x ray 59, 237
crystalline debris 701
crystalline homogenization 109
crystalline silicate cresecent 2041
crystallization 43, 143, 151, 159, 169, 193, 219, 285, 301, 343, 377, 413, 481, 507, 529, 559, 575, 601, 617, 645, 775, 923, 1063, 1351, 1417, 1503, 1671, 2213
crystallization age 583, 1627
crystallization cooling 301
crystallization metamorphism 43
crystallization temperature ... 91, 617
cumulate 343, 439, 529
cumlate texture 529
Curie 2433
Curie temperature 2491
curium 2153, 2599
debris 439, 701, 755, 1939
debris cloud 873, 2653
decay mode 1337
deformation 237, 285, 775, 1973, 2461
deformation band 177
deformation structure 833
deformation twin 775
demagnetization 2451, 2461
dendrite 177
dendritic texture 359
densification olivine glass 2069

density 167, 1209, 1381, 1521, 1547, 1591, 1717, 1953, 1973, 2027, 2041, 2049, 2153, 2165, 2197, 2311, 2317, 2337, 2345, 2367, 2415, 2529, 2543, 2599, 2611, 2639
density stratigraphy 439
depth structure 2337
Descartes 2301
descent engine exhaust 2735
deuterium 1407, 1421, 1843
deuterium exchange 1381
deuterium hydrocarbon 1343
devitrification 645, 949, 957
devitrification texture 949
devitrified glass 207, 575, 601, 949
devitrified glass spherule 949
devitrified rhyolite glass 949
diamagnetic vesicle 2415
diaplectic glass 893
dielectric conductivity 2367
differentiated crust 43
differentiated fines 1843
differentiated lava 617
differentiated ophitic basalt 469
differentiated porphyritic basalt 469
diffusion 775, 1607, 1651, 1681, 1813
diffusion hydrogen 1671
directional reflectance 2197
disequilibrium 855
dislocation 2153
dislocation substructure 69
disorder 47, 91, 247
disordered magnesium 109
disordered structure 135
diurnal heat wave 2245
divalent iron oxide 497
dolerite 377, 601
domain disorientation 237
domain structure 59, 69
Dreierketten 47
drive tube 17
drive tube core 665, 1929, 1939
DTA 59, 2245
dynamic pressure 2415
dysprosium 1, 999, 1083, 1101, 1169, 1187, 1247, 1253, 1281, 1291, 1301, 1307
ejecta 797, 1421, 2653, 2753, 2781
ejecta blanket 393, 2301
ejecta damage 2767
ejecta impact 2767
elastic surface wave 2337
elastic wave velocity 2323, 2337
electrical conductivity 2391, 2415, 2461
electromagnetic induction 2415
electromagnetic radiation 2083

electromagnetic transfer 2415
electron beam energy 2083
electron bombardment 2675
electron capture gas 1451
electron density 2529
electron diffraction 69, 117
electron emission 2083
electron excitation 2265
electron paramagnetic resonance 2277, 2501, 2529
electron petrology 69
electron scanning 937, 2639
electron spin resonance 2515
electron transmission 69
electrostatic adhesion 2153
emission 1217, 1247, 1253, 2057, 2083, 2265
emission band 247
emission wavelength 1865
endotherm 2245
energy balance 2197
equilibrium 91, 497, 617, 645, 855, 973, 1351, 1367, 1471, 1577, 2049, 2245, 2699
equilibrium atmosphere 1717
equilibrium crystallization 481
equilibrium residual glass 617
equilibrium temperature 91
erbium 1, 999, 1083, 1101, 1169, 1187, 1281, 1307, 1319
erosion 449, 1503, 1747, 1757, 1797, 2265, 2543, 2559, 2611, 2639, 2675, 2705
erosion damage 2791
erosion equilibrium 2543
eruption 617, 1351
eruption temperature 617
etching 159, 1651, 1843, 2057, 2153, 2599, 2611, 2705
ethylene 1843
eucrite 247, 439, 459, 617, 1021, 1123, 1281
euler angle 135
europium 1, 747, 999, 1063, 1083, 1101, 1123, 1169, 1187, 1253, 1277, 1281, 1291, 1301, 1307, 1319, 1351, 1627
europium anomaly (depletion) 999, 1083, 1101, 1169, 1187, 1307, 1319
evaporation 1729
excitation band 1865
exoelectron emission 2057
exotherm 2245
experimental petrology 601
exposure age 1187, 1607, 1651, 1671, 1747, 1797, 1803, 1813, 1825, 2543, 2583, 2611
exsolution 59, 69, 91, 109, 219, 359, 2599

exsolution lamella 109
exsolution plane 2599
extinct radioactivity 1627, 1693, 2559
extraterrestrial microorganisms .. 2721
fabric 893, 1953, 2027
fault 109, 775
feldspar 91, 117, 159, 193, 207, 247, 319,
   377, 393, 529, 583, 601, 817, 855, 873,
   1291, 1503, 2183, 2461, 2543, 2599, 2611
feldspar stoichiometry 413
ferromagnetic resonance 2501, 2529
ferromagnetism 2461, 2515
ferrous iron 117
fiber balance 937
field demagnetization 2477, 2491
fines accumulation 2735
fines adhesion 2743
fines benzene-methanol 1927
fines bombardment 1705
fines cloud 2027
fines coating 755, 2735, 2743
fines excitation 2735
fines exposure age 1651
fines lead 1547
fines location 17
fines mechanics 1973, 2743
fines oxygen isotope 1417
fission hypothesis 1021
fission track 151, 1503, 2599, 2621
fission track age 1503
fission track anomaly 1503
fission track map 1503
fission track uranium 151, 1301
fission xenon 1591, 1671, 1693
flow structure 343, 909
flow texture 393
fluid core 2485
fluid pressure 2165
fluorescence 1865, 1875, 2083
fluorine 1, 285, 319, 583, 645, 1169, 1187,
   1247, 1253, 1259, 1261, 2671, 2791
formation age 2639
fossil 2569, 2599
fossil track 1503, 1825, 2559, 2583, 2599,
   2611, 2621
fossil track density 2611
fossil xenon 1693
Fouriers Law 2311
Fra Mauro 393, 583, 701, 2301
fractional crystallization 343, 359, 529,
   617, 1063, 1217
fractionated basalt 583
fractionated crust 583
fractionated gas 1607

fractionated meteorite 1139
fractionation 43, 159, 207, 343, 377, 393,
   413, 439, 449, 481, 507, 583, 855, 923,
   987, 1021, 1063, 1101, 1253, 1381, 1503,
   1521, 1547, 1681, 1791
Funferketten 47
gabbro 47, 219, 247, 265, 285, 439, 719,
   817, 855, 873, 1421, 1451, 2125
gabbro anorthosite 135
gadolinium 1, 601, 999, 1083, 1101, 1169,
   1187, 1277, 1291, 1307, 1351, 1671, 1729,
   2583
gadolinium isotope 1591, 1671
galactic cosmic ray 1503, 1671, 1705,
   1747, 1757, 1773, 1791, 1813, 1825,
   2245, 2705
galactic iron 2559
galactic proton 1705, 1729, 1747, 2559
gallium 1, 1021, 1037, 1139, 1169, 1187,
   1217, 1247, 1253, 1277, 1281, 1471
gamma 1169, 2057, 2451, 2461, 2477,
   2485, 2491
gamma field 2391
gamma radiation 2057
gamma ray 1159, 1253, 1277, 1291, 1747,
   1757
gardening 665, 1773
garnet 601, 2083
gas (noble) 645, 755, 909, 1159, 1343,
   1351, 1367, 1381, 1397, 1451, 1591, 1607,
   1651, 1681, 1693, 1705, 1717, 1813, 1843,
   1879, 1901, 1913, 1931, 2057, 2153, 2529,
   2653
gas absorption 2529
gas conduction 2311
gas evolution 1351
gas exposure 2153
gas interaction 2009
gas krypton 1681
gas liquid 957, 1927
gas meteorite 2583, 2705
gas pressure 1381, 2003
gas retention age 1607
gaseous carbon 1343
gaseous conduction 2311
gaseous hydrocarbon 1843
gaseous oxygen 973
geochemistry 617, 1037, 1063, 1083, 1123,
   1159, 1169, 1565
geochemistry fractionation 1493
geography 1421
germanium 1, 1037, 1187, 1247, 1253,
   1277, 1291
glass 159, 207, 219, 247, 265, 285, 319,
   359, 393, 439, 459, 507, 575, 583, 601,
   665, 679, 701, 719, 727, 747, 755, 775,
   797, 817, 833, 855, 873, 893, 909, 949,

957, 1217, 1291, 1351, 1493, 1503, 1651, 2041, 2049, 2069, 2125, 2137, 2183, 2213, 2235, 2245, 2433, 2529, 2559, 2629, 2639, 2653, 2705, 2781
glass agglomerate 1989
glass aggregate 701, 797
glass clast 893
glass coating 1139
glass crater 449
glass crescent 2041
glass crust 957
glass crystallization exotherm 2245
glass depression 2653
glass forming melt 2049
glass lined crater 2653
glass matrix 285, 393
glass ray 2653
glass residue 159, 247, 645
glass shard 957
glass silicon 2433
glass spatter 2433
glass spherule 797, 817, 873, 937, 949, 957, 973, 2003, 2049, 2433, 2653, 2743
glass structure 2213
glazed aggregate 665, 755
globule 507, 909, 957, 2041
globule overgrowth 923
glow curve temperature 2245
gold 1, 1021, 1037, 1083, 1139, 1187, 1277, 1281, 1291, 1301, 1547
grain fines 2265
granite 43, 167, 413, 439, 727, 817, 1187, 2057
granite melt 507
granite vein 583
graphite 1343
gravitational anomaly 987
gravitational escape 1717
gravitational field 957
gravity 1973
hafnium 1, 39, 43, 893, 1063, 1083, 1187, 1231, 1253, 1277, 1281, 1291, 1301, 1319, 1627, 1901, 1927
hafnium demineralization 1901
halide 1261
halo crater 665
halo density 167
halogen 645, 1261, 1547
heat balance 987, 1159, 2197
heat capacity 2345
heat flow 2245, 2311
heat transfer 2197, 2203, 2311
helium 1, 1607, 1651, 1681, 1705, 1717, 1825

helium atmosphere 1343
helium diffusion 1705
helium exposure age 1747
helium ion radiation 167
highland 43, 135, 413, 583, 727, 737, 817, 987, 999, 1209, 2183, 2301, 2639
highland bedrock 583
highland regolith 679, 737
holmium 1, 1063, 1083, 1169, 1253, 1277, 1281, 1291, 1307
hydrocarbon 1343, 1381, 1843, 1879, 1891, 1901, 1913
hydrocarbon gas 1843, 1879
hydrogen 645, 1351, 1381, 1407, 1547, 1717, 1803, 1913, 1931, 2653
hydrogen diffusion 1803
hydrogen gas 1407, 1421
hydrogen isotope 645
hydrolysis 1343, 1843
hydrostatic pressure 1939, 2165
hydrostatic strain 2165
hydrous analog 645
hydroxyl 645
hypervelocity crater 2781
hypervelocity impact 719, 2753, 2791
hypervelocity micrometeorite 2653
igneous breccia 1521
igneous crystallization 43
igneous fractionation 319, 413
igneous glass 727
igneous texture 343
illumination angle 2197
ilmenite 69, 151, 169, 193, 207, 219, 237, 265, 301, 319, 377, 413, 507, 529, 601, 617, 679, 719, 775, 797, 817, 833, 855, 923, 1037, 1217, 1237, 1291, 1351, 1417, 1651, 1939, 2079, 2083, 2103, 2125, 2153, 2183, 2461
ilmenite antiferromagnetism structure 2079
ilmenite basalt 285, 797, 1421
ilmenite mare 2183
Imbrium Basin 27
Imbrium ejecta blanket 43
Imbrium impact 43
immiscibility 507
immiscible glass 957
immiscible liquid 507
immiscible melt 507
impact 43, 177, 449, 665, 679, 701, 755, 797, 833, 873, 909, 1139, 1331, 1351, 1521, 1681, 1717, 1747, 1939, 1959, 2003, 2069, 2093, 2125, 2165, 2327, 2477, 2543, 2653, 2743, 2767, 2781, 2791
impact breccia 1521

impact crater 797, 2781
impact debris cloud 893
impact energy 1717
impact explosion 2675
impact glass 285, 701, 2653
impact lithification 2477
impact melt 319, 529, 727, 1351
impact metamorphism 817
impact projectile 1681
impact shock 1547
impact velocity 2639
indium 1139, 1187, 1253, 1281, 1319
induction asymmetry 2415
induration 2675
infrared 2137, 2183, 2203
infrared electromagnetic absorption 2203
infrared electromagnetic radiation 2203
infrared emission band 247
inorganic carbon 1343, 1879
inorganic gas 1351
intergrowth 207
interplanetary debris 2639
interplanetary energy 2705
interplanetary fines bombardment 2639
interplanetary fines flux 2781
interplanetary flux 2639
interplanetary magnetic field 1717, 2415
intersertal basalt 469
interstitial glass 159, 207, 775
intracrystalline exchange 91
intrusion 59, 43, 135, 143
iodine 1693
ion beam 359, 2621
ion bombardment 2791
ion energy 2093
ion flux 2093
ion mass 1717
ionizing radiation 2501
iridium 1021, 1037, 1139, 1187, 1281, 1291, 1547, 2265
iron 1, 39, 47, 59, 91, 43, 109, 117, 135, 143, 151, 159, 193, 207, 219, 265, 285, 301, 343, 359, 377, 393, 413, 439, 449, 459, 481, 497, 507, 529, 559, 575, 583, 601, 617, 645, 679, 701, 719, 727, 747, 755, 775, 797, 833, 855, 937, 973, 999, 1063, 1123, 1139, 1169, 1187, 1209, 1217, 1231, 1237, 1247, 1259, 1277, 1281, 1291, 1301, 1319, 1367, 1443, 1471, 1705, 1773, 1797, 1813, 1931, 2049, 2069, 2083, 2103, 2125, 2137, 2153, 2183, 2433, 2451, 2461, 2485, 2515, 2529, 2611, 2653
iron crucibles 497
iron disorder 47

iron electron 2501
iron group nuclei 2559, 2611, 2705
iron ion 2057
iron mare basalt 393
iron metal (see iron-nickel)
iron meteorite 1331, 1337, 2125, 2599
iron meteorite silicate 2599
iron-nickel (see metal) 177, 207, 265, 301, 343, 719, 957, 1217, 2125
iron nuclei 2705
iron occupancy 91
iron oxide temperature 497
iron pigeonite 59
iron pyroxene 4113, 2079
iron reduction 2125
iron silicate 497
iron silicate liquid 497
iron spherule 2491
iron sulfide 1351
iron track gradient 2705
isochron 1471, 1521
isochron age 987, 1101, 1487, 1521
isochron (lead isotopes) 1493
isothermal decay 2277
isotope 359, 1337, 1397, 1407, 1421, 1443, 1461, 1565, 1571, 1577, 1585, 1591, 1607, 1643, 1651, 1681, 1705, 1717, 1747, 1773, 1803, 1813, 1825, 1843, 2583, 2621, 2699
isotope anomaly 1461
isotope fractionation 1417
isotope iron 2103
isotope mass fractionation 1717
kaolinite 973
kinetic 2069
kinetic energy 1585
KREEP 219, 393, 439, 755, 833, 999, 1101, 1123, 1139, 2103
KREEP breccia 393
KREEP glass 665, 727, 755, 817
KREEP glass 755
KREZP 583
krypton 1591, 1607, 1627, 1643, 1671, 1681, 1693, 1717, 1813, 2621
krypton age 1591
K/Ar age 775
K/Ba ratio 1123
K/Rb ratio ... 999, 1123, 1253, 1281, 1493
lamella 109, 923
lamella twinning 237, 797, 817
laminar flow 1301
lanthanum 1, 727, 999, 1063, 1083, 1123, 1169, 1187, 1217, 1247, 1253, 1277, 1281, 1291, 1301, 1307, 1319, 1471
lattice 237, 2265

lattice distortion 237
lava 91, 343, 449, 617, 1083, 1503, 2461,
.... 2477
lava temperature 507
leaching 1139, 1343, 1461, 1547
lead 1, 987, 1169, 1247, 1493, 1521, 1547,
  1565, 1585, 2183, 2543, 2599, 2639, 2699
lead age 727, 987
lead isotope evolution 151
lead isotopes 987, 1021, 1247, 1493, 1503,
.... 1521, 1547, 1565, 1577, 2699
lead lead isochron 1521
lead separation 1577
lead volatile transfer 1547
lineament 27
liquid core 2461
liquid helium 109, 2501
liquid jet 2003
liquid separation 507
liquidus 497, 559, 601
liquidus fractionation 775
liquidus temperature 169, 497, 2049
lithification 2477
lithium 1, 439, 727, 999, 1083, 1101, 1169,
  1187, 1217, 1247, 1261, 1277, 1291, 1351,
.... 1461
lithium isotope 1461
luminescence 247, 2213, 2223, 2235, 2265,
.... 2277
luminescence emission 2265
luminescence petrology 247
luminescence tridymite 247
Luna 16 1
Luny rock 43
lutecium 1, 999, 1063, 1083, 1101, 1169,
  1187, 1253, 1277, 1281, 1291, 1301,
.... 1307, 1319
mackinawite 219
magma 159, 285, 359, 439, 481, 507, 529,
  559, 617, 679, 1063, 1217, 1351, 1471,
.... 1521
magma crystallization 855
magma fractionation 481, 1421
magma gas 1381
magma liquid 413
magnesium 1, 39, 47, 59, 91, 43, 117, 143,
  159, 193, 207, 219, 265, 285, 301, 343,
  359, 377, 393, 413, 439, 459, 481, 497,
  507, 529, 559, 575, 583, 601, 617, 645,
  665, 679, 701, 719, 727, 747, 755, 775,
  797, 833, 855, 937, 957, 973, 999, 1037,
  1063, 1123, 1169, 1187, 1209, 1217, 1231,
  1237, 1247, 1253, 1259, 1261, 1291, 1319,
  1367, 1461, 1471, 1627, 1681, 1705, 1773,
  2049, 2069, 2083, 2125, 2235, 2433, 2653
magnesium cumulate 343

magnesium equilibrium 91
magnesium isotope 1461
magnesium pigeonite 59
magnesium residual liquid 343
magnetic evolution 1443
magnetic field 2391, 2415, 2451, 2461,
.... 2477, 2485, 2491
magnetic field magnitude 2491
magnetic monopole 2621
magnetic ordering 2079
magnetic resonance 2501, 2515
magnetism 2041, 2451, 2477, 2485, 2491
magnetometer 2391, 2415, 2451
manganese 1, 39, 47, 59, 117, 143, 159,
  193, 207, 219, 265, 285, 301, 359, 377,
  413, 439, 459, 481, 497, 507, 529, 559,
  575, 583, 601, 617, 645, 679, 701, 719,
  747, 775, 797, 833, 855, 973, 1063, 1123,
  1169, 1187, 1217, 1247, 1253, 1259, 1281,
  1291, 1301, 1319, 1367, 1471, 1757, 1773,
  1797, 2049, 2069, 2083, 2153, 2265, 2501,
.... 2515, 2611
manganese garnet 2083
manganese ion 2515
mantle 193, 583, 617, 1471
mare 1, 39, 91, 413, 617, 727, 737, 775,
  987, 999, 1209, 1231, 1521, 1547, 1681,
.... 1865, 2173, 2183, 2337, 2361
mare basalt 393, 413, 469, 583, 833, 999
mare basin 2675
mare crust 393
mare fines 1063
mare homogenization 413
mare regolith 999
mare surface evolution 2675
Mare Tranquillitatis 47, 91, 665, 727,
  1037, 1187, 1209, 1381, 2235, 2285,
.... 2639, 2699
mascon 617, 987
maskelynite 247
mass anomaly 2337
mass balance 617, 1901
mass fractional vaporization 1421
matrix clinopyroxene 141
matrix glass 833
matrix plagioclase 413
matrix texture 301
mechanical deformation 69
mechanical disintegration 719
mechanism interaction 1717
melanocratic olivine basalt 855
melt 151, 159, 169, 301, 343, 413, 449, 481,
  507, 529, 601, 719, 817, 909, 987, 1101,
.... 1351, 1493, 1503, 1681, 2041, 2049
melt temperature 2049

mercury 1169, 1261, 1381, 1717, 1927, 2599, 2621

mesostasis 159, 169, 439, 481

metal 47, 43, 177, 219, 301, 343, 617, 855, 973, 1187, 1331, 1451, 2103, 2125, 2743

metal cation 47

metal (copper) 219

metal (iron) (see iron-nickel) 207, 219, 377, 497, 601, 617, 855, 873, 1259, 2103, 2125, 2433, 2461, 2515

metal iron capsule 601

metal precipitation 301

metal spherule 957, 2433

metamorphism 43, 285, 1547

metastable 265, 377

metastable crystallization 265

meteorite 43, 141, 177, 247, 301, 319, 393, 459, 855, 1021, 1037, 1139, 1169, 1187, 1253, 1261, 1277, 1281, 1331, 1343, 1351, 1381, 1443, 1451, 1461, 1503, 1547, 1565, 1747, 1757, 1797, 1803, 1843, 1865, 1913, 2027, 2137, 2265, 2583, 2599, 2621, 2639, 2705, 2767

meteorite analog 459

meteorite bombardment 177, 1421, 1747

meteorite flux 2639

meteorite impact 449, 755, 873, 893, 1843, 1959, 2337, 2639, 2653, 2767

meteorite impact crater 2639, 2767

meteorite impact gardening 2173

meteorite lead 2599

meteorite metal 177, 1187

meteorite volatile 1351

methane 1343, 1397, 1843, 1879, 1891

mica 645

microbe 1931, 1939, 2715, 2721

microcrystalline zirconium oxide .. 169

microlite 301, 957

micrometeeorite 1421, 2653, 2753

micrometeorite bombardment 2781

micrometeorite damage 2791

micrometeorite flux 2781

micrometeorite impact 2223, 2611, 2781

micrometeorite impact crater 449, 2781

micrometeorite pit 2743

microorganisms 909, 1931, 1939, 2715, 2721

microstructural deformation 817

mineralogy 1, 39, 47, 167, 219, 265, 285, 301, 319, 343, 359, 377, 393, 413, 469, 481, 559, 583, 601, 679, 701, 719, 737, 747, 797, 855, 957, 1209, 1381, 1503, 1627, 1959, 2235, 2301

mineralogy heterogeneity 2301

mineralogy petrology 413

modal mineralogy .. 459, 601, 1217, 2103

modal olivine 285, 601

model age 999, 1101, 1247, 1471, 1487, 1493, 1547

model isochron 343, 1471

model lead isotopes 1547

molecular fossil 1927

molecular oxygen isotopes 1367

molecular structure 1879, 2137

molybdenum 1, 1083, 1169, 1247, 1301

monomict breccia 43

monopole 2621

morphology 529, 873, 949, 2041, 2653

Mossbauer 91, 109, 117, 377, 2103, 2125, 2461

Mossbauer absorption 91

mottled basalt 43

Na/K ratio 1253, 1281

neodymium 1, 999, 1063, 1083, 1101, 1169, 1187, 1247, 1253, 1291, 1307, 1351

neon 1, 27, 1381, 1607, 1627, 1643, 1651, 1681, 1705, 1717, 1747, 1825

neutron 1237, 1565, 1591, 1671, 1693, 1729, 1813, 2079, 2583, 2599

neutron absorption 1693

neutron bombardment 1337

neutron diffraction 2079

neutron emission 1585

neutron exposure 1671

neutron spin 2079

neutron xenon isotope 2621

nickel 1, 43, 177, 193, 207, 219, 265, 301, 319, 343, 359, 439, 481, 583, 601, 679, 719, 747, 797, 855, 957, 1037, 1083, 1169, 1187, 1217, 1247, 1253, 1291, 1471, 1773, 1939, 2049, 2083, 2125, 2515, 2639

nickel-iron (see iron-nickel)

niobium 39, 583, 1083, 1123, 1169, 1217, 1247, 1471

nitrogen 1343, 1351, 1381, 1717, 1913, 2009, 2153, 2245

nitrogen atmosphere 2367

noble gas 1351, 1381, 1591, 1717

noble gas isotope 1443 1717

norite 135, 285, 319, 439, 583, 679, 817, 2639

norite anorthosite crust 583

normative mineralogy 459

normative olivine 601

normative plagioclase 459

nuclear collision 2621

nuclear gamma resonance 2103

nuclear hyperfine coupling 2515

nuclear interaction 1607, 1825, 2559

nuclear magnetic resonance 2501

nuclear track 1773, 2543

nuclear track density 2543
nucleation 507, 559
nucleon 1729, 1773, 2671
ocean tholeiite basalt 2137
oceanic crust 413
octahedral aluminum 559
octahedral cation 59
olivine 91, 43, 143, 193, 207, 219, 247,
   265, 301, 319, 343, 359, 377, 481, 507,
   575, 601, 617, 645, 701, 719, 747, 775,
   855, 873, 957, 973, 1217, 1291, 1397,
   1417, 1651, 2041, 2057, 2079, 2083, 2103,
   ...... 2125, 2137, 2153, 2183, 2391, 2415
olivine accumulation 575, 583, 601
olivine augite basalt 601
olivine basalt 207, 601, 855, 923, 1291
olivine core 2415
olivine crystallization 1101, 1307
olivine dolerite 923, 2361
olivine gabbro 265, 665, 2485
olivine phenocryst 301, 617
olivine pigeonite basalt 797
olivine pyroxene cumulate 285
olivine vitrophyre 1417
opaque glass 957
opaque matrix 207
opaque oxide 207, 219
opaque pyroxene 973
ophitic basalt 69, 469, 1231, 1319
ophitic cristobalite basalt 159
ophitic olivine basalt 159
ophitic olivine diabase 69
optical absorption 2183, 2277
optical emission 1123
optical glass 2173
optical indicatrix 135
optical petrology 69, 2461
organic acid 973
organic outgassing 2735
organic solvent 1901
organism 1929, 1931
Orgueil 2027
orthoclase 393, 413, 645, 747, 755
orthopyroxene 319, 393, 529, 601, 755,
   .... 775, 797, 817, 2083
osmium 1261, 1331
overgrowth 207
oxidation .. 91, 143, 497, 1585, 2057, 2433
oxidation reduction 301
oxygen 47, 207, 1063, 1187, 1209, 1237,
   1367, 1407, 1421, 1879, 1913, 2009, 2083,
   ...... 2153, 2501, 2529, 2599, 2671
oxygen atmosphere 1343
oxygen depletion 1237
oxygen fugacity 193, 319, 1381

oxygen isotope fractionation 1417
oxygen isotopes 1351, 1421, 1417, 1717,
   .... 2057, 2671
oxygene isotopes residue 1421
oxygen potential 497
P symmetry 59
P wave 2323
palladium 1, 1021, 1037, 1187
paragenesis 301
paramagnetic manganese ion 2515
paramagnetic pyroxene 2461
paramagnetic resonance 2515
paramagnetism 2433, 2461, 2515
partial alteration 1521
partial fractionation 439
partial fusion 529
partial melt, 43, 319, 343, 481, 909, 1063,
   .... 1083, 1123, 1471
partial pressure 1367, 2345
partial thermoremanent magnetism
   .... 2485, 2491
patterned ground 27
penetrometer 1973
peridotite 319
peritectic 855
petrology 285, 319, 343, 359, 377, 469,
   481, 497, 507, 575, 601, 617, 665, 719,
   775, 797, 893, 1101, 1791, 1929, 2041,
   .... 2367, 2485
pH 973, 1939
phenocryst 91, 207, 247, 343, 507, 559,
   .... 617, 1101
phenocryst pyroxene 285
phosphate 43, 219, 413, 583, 1063, 1503
phosphate fusion 1101
phosphate glass 957
phosphorus 43, 265, 285, 319, 393, 439,
   459, 469, 507, 575, 583, 617, 679, 701,
   719, 747, 755, 797, 817, 973, 1123, 1259,
   .... 1471, 2049
phosphorus acid 1843
phosphorus chemistry 393
photoelectric photometry 2301
photons 2083, 2245
photosynthesis 973
picrite basalt 583
pigeonite 59, 69, 91, 109, 247, 319, 343,
   377, 413, 559, 601, 617, 775, 1101, 1291,
   2621
pigeonite augite exsolution 59
pigeonite augite solvus 59
pigeonite basalt 601
pigeonite core 247, 265, 285, 377
pigeonite phenocryst 91
pigeonite prophyry 141
pigeonite X-ray 59

pit 2543, 2653, 2767, 2781, 2791
plagioclase 69, 43, 117, 135, 141, 143, 151,
247, 265, 285, 301, 319, 343, 377, 413,
459, 481, 497, 507, 529, 559, 575, 583,
601, 617, 719, 737, 775, 797, 817, 833,
973, 1101, 1209, 1217, 1237, 1253, 1291,
1651, 2057, 2069, 2079, 2083, 2137, 2183,
.... 2235, 2265, 2277, 2461, 2501
plagioclase accumulation 1083
plagioclase clast 143
plagioclase crystal emission 2265
plagioclase fracturing 817
plagioclase glass 2213
plagioclase intergrowth 855
plagioclase rosette 141
plagioclase twin 135
planet 909, 1139, 1159, 1351, 1367, 1547
planet atmosphere 1717, 2715
planet evolution 1159
planet quarantine 2715
plasma 2415
plasma cloud 2093
plasma flow 2415
plastic deformation 775
plastic fission track 159
plastic flow 775
plastic silicate 873
platinum 1261, 1577, 1585
poikilitic olivine basalt 159
polarization 2285, 2381
polarization plane 1875
polycrystalline agglomerate 2203
polyhedra band 47
polymict breccia 43
polymorph 1651
polystrene 1891
porosity 2009, 2165
porous basalt 2165
porous texture 797
porphyrin 1865, 1875
porphyrin compound synthesis 1865
porphyritic basalt 469, 559
porphyritic diabase 69
porphyritic olivine basalt 413
porphyritic variolitic basalt 285
potassium 1, 39, 43, 117, 135, 159, 247,
265, 285, 301, 393, 413, 439, 459, 481,
507, 575, 583, 617, 645, 679, 701, 719,
727, 747, 755, 797, 817, 833, 855, 937,
973, 1159, 1217, 1231, 1253, 1259, 1471,
.... 1503, 1705, 2049, 2069, 2515, 2653
potassium feldspar (see orthoclase)
potassium granite 507
potassium halo 449
praseodymium 1, 1083, 1169, 1187, 1277

Precambrian basalt 507
precipitate 559
pressure 481, 559, 601, 617, 775, 1351,
2009, 2069, 2165, 2311, 2317, 2323, 2327,
.... 2337, 2345, 2671
pressure cotectic liquid 617
pressure density 2069
pressure thermal expansion 2317
pressure transfer 2311
pressure volume 2165
projectile mass 2653
projectile velocity 2653
promethium 1503, 1577, 1627, 2245, 2599
promethium fission 1503
promethium geochemistry 1503
promethium hypothesis 1627
promethium tube 2245
protohypersthene 617
proton bombardment 2265
proton excitation 2265
proton radiation 1451, 2285
provenance 285
pseudocubic lattice 39
pulse transmission 2327
pyrolysis 1343, 1351, 1397, 1407, 1879,
.... 1891
pyrolysis residue 1407
pyroxene 47, 59, 69, 91, 43, 109, 159,
207, 247, 265, 285, 301, 319, 343, 359,
377, 413, 481, 497, 507, 529, 583, 601,
617, 645, 719, 747, 775, 855, 957, 1101,
1217, 1291, 1503, 2079, 2083, 2125, 2137,
2153, 2183, 2461, 2477, 2559, 2599, 2611
pyroxene absorption band 2183
pyroxene basalt 359
pyroxene cleavage 343
pyroxene exsolution 69
pyroxene field 377
pyroxene mantle 529
pyroxene matrix 207
pyroxene phenocryst 285, 413, 559
pyroxene rim 265
pyroxferroite 47, 265, 319, 377, 413, 601
pyroxferroite mantle 343
pyroxferroite structure 47
quadrupole coupling 2501
quadrupole mass 1351
quartz 247, 775, 817
quartz glass 2639
quartz normative mineralogy 459
quartz tube 2529
radar 2753, 2767
radar reflection 2203
radiation 1169, 1253, 1443, 1565, 1591,
1643, 1671, 1747, 1773, 1803, 2027, 2041,

2057, 2223, 2245, 2277, 2311, 2501, 2583, 2705, 2735, 2753

radiation age 1607, 1705, 1729, 2599

radiation damage 1237, 2009, 2027, 2235, 2245, 2515, 2735

radiation energy 1643

radiation exposure 2485

radio halo 167

radio isotope 1607

radioactive argon 1813

radioactive decay 1461, 1577

radioactive equilibrium 1577

radioactive gas 1813

radioactive halo 167

radioactive isotope 1159, 1813

radioactive rare gas 1813

radioactive transformation 167

radioactivity 1159, 1237, 1337, 1461, 1577, 1773, 1791, 1797, 1813, 1825

radiogenic gas 1591, 1651, 1717

radioisotope 1159, 1337, 1443, 1565, 1729, 1747, 1757, 1773, 1791, 1797

radium 1585, 2699

radon atmosphere 2699

radon isotope decay 2699

Raman 2213

rare earth elements (see REE) 39, 43, 151, 393, 439, 469, 481, 701, 727, 747, 999, 1063, 1083, 1101, 1187, 1231, 1253, 1277, 1291, 1307, 1319, 1351, 1627

rare gas 1547, 1591, 1607, 1627, 1643, 1651, 1671, 1693, 1705, 1813, 2027, 2629

rare gas isotope .. 1607, 1643, 1693, 1705

rare gas nuclei 1607

rare gas radioactivity 1813

rare gas spallation 1671

Rb/Sr ratio .. 343, 1123, 1217, 1487, 1493

Rb/Sr fractionation 1493

Rb/Sr model age 1493

reciprocal lattice 237

recrystallization 207, 775, 817, 957, 2041

redox 301, 601, 1351

REE olivine 701

REE oxide 1101

REE whitlockite 583

reflectivity 2183, 2301

refractive index 39, 135, 747, 937, 973, 2069

regolith 1, 17, 27, 193, 343, 393, 439, 665, 679, 701, 737, 775, 797, 833, 999, 1021, 1037, 1139, 1301, 1319, 1337, 1343, 1351, 1397, 1471, 1487, 1547, 1671, 1681, 1705, 1757, 1791, 1797, 2057, 2125, 2203, 2245, 2361, 2543, 2569, 2583, 2611

regolith debris 665

regolith fines 701, 737

regolith outgassing 1681

regolith track 2583

remanent magnetic moment 2485

remanent magnetism 2451, 2461, 2477, 2485, 2491

residual carbon 1343

residual gas 2311

residual glass 617, 645

residual liquid 169, 343, 481, 507, 529, 583

residual magma 529

residual melt 855

residual nuclei 1729

residual temperature 2069

resonance absorption 2125

rhenium 1101, 1331

rhenium isotope 1337

rhodium 1

rhyolite 167, 583

rhyolite glass 583

rhyolite residue 583

rocket exhaust 1865

rocket impact 2327

rotational elongation 2041

rubidium 1, 43, 393, 439, 727, 999, 1021, 1101, 1123, 1139, 1169, 1187, 1217, 1231, 1247, 1253, 1281, 1319, 1351, 1471, 1487, 1493

rubidium-strontium 1493, 1503, 1521

rubidium-strontium age 1421, 1471, 1503

rubidium-strontium isotopes 1487

ruthenium 1, 583, 1261

rutile 219

S-wave velocity 2323, 2327

samarium 1, 747, 999, 1063, 1083, 1101, 1169, 1187, 1247, 1253, 1277, 1281, 1291, 1301, 1307, 1319, 1351

sanidine (see orthoclase)

saturation magnetism 2461

scandium 1, 439, 1063, 1083, 1169, 1187, 1217, 1231, 1247, 1253, 1277, 1281, 1291, 1301, 1319, 1757, 1773, 1939

scanning electron (microscope) 117, 755, 873, 909, 923, 2021, 2083, 2173, 2543, 2653, 2753, 2767, 2781

scoriaceous glass agglomerate 1989

sculptured glass 393

Sea of Plenty 1

secondary alteration 1651

secondary damage 2337

secondary ejecta 737

secondary electron energy 2083

secondary electron flux 2083

sedimentation 1989, 2173

seismic energy 2337

seismic ray 2327

selenium 1, 973, 1021, 1083, 1139, 1281

shear 1973

shear velocity (see S-wave velocity)

shock 207, 237, 343, 701, 775, 797, 817, 833, 855, 893, 957, 1503, 1681, 2069, 2103, 2337, 2461

shock compaction 2675

shock damage 247, 775

shock deformation 285, 775

shock impact 177

shock melt 833, 855, 2069

shock metamorphism 247, 833, 2069, 2461

shock pressure 177, 237, 2705

shock remanent magnetism 2461

shock welding 797

shocked breccia 797

shocked ilmenite 237

shocked ilmenite basalt 237

shocked plagioclase 893

sialic 727, 2223

Siebenerketten 47

silicate 39, 47, 169, 497, 645, 775, 855, 873, 923, 957, 1237, 2137, 2265, 2599, 2611, 2653

silicate chain 47

silicate glass 873

silicate glass spherule 873, 2653

silicate ion 957

silicate liquid immiscibility 507

silicate matrix 1351

silicate melt 301, 343, 507, 645, 957, 1351

silicate segregation 1021

silicate structure 957

silicon 1, 39, 43, 47, 59, 91, 117, 141, 159, 207, 219, 247, 265, 285, 301, 377, 393, 413, 439, 449, 481, 529, 559, 575, 583, 617, 679, 701, 719, 727, 747, 755, 775, 797, 855, 937, 957, 973, 1063, 1123, 1187, 1209, 1217, 1237, 1247, 1253, 1259, 1291, 1367, 1421, 1451, 1471, 1705, 1773, 2083, 2153, 2501

silicon capsule 497

silicon glass 719

silicon isotope 1421

silicon orthosilicate ion 957

silicon oxygen distance 47

silicon tetrahedra 47, 2137

silver 1, 617, 1021, 1139, 1253, 1281, 1301, 2599

sintered appendage 2041

sintered fines 755, 873

sintered glass 1927

SiO_2 polymorph 141

skeletal olivine phenocryst 601

slag 1291

sodium 39, 59, 43, 117, 135, 265, 285, 301, 393, 413, 439, 459, 481, 497, 507, 529, 559, 575, 583, 601, 617, 645, 679, 701, 719, 727, 755, 797, 833, 855, 937, 957, 973, 999, 1063, 1123, 1169, 1187, 1209, 1217, 1231, 1247, 1253, 1259, 1281, 1291, 1301, 1319, 1351, 1367, 1471, 1705, 1729, 1757, 2049, 2069, 2083, 2125, 2501, 2653, 2671

sodium ion 439, 559

sodium isotopes 1757

solar 987, 1083, 1367, 1443, 1451, 1461, 1503, 1547, 1591, 1651, 1705, 1729, 1747, 1757, 1773, 1791, 1813, 1843, 1913, 2027, 2285, 2391, 2543, 2559, 2621, 2639, 2705, 2735, 2753

solar cosmic interaction 1803

solar cosmic ray 1705

solar flare 1503, 1607, 1705, 1747, 1757, 1797, 1803, 1813, 1825, 2041, 2093, 2245, 2543, 2705

solar flare map 2543

solar flare proton 1705, 1747, 2583

solar flare radiation 1797, 2245

solar illumination 2735

solar lithium 1461

solar nebula 1021, 1503

solar proton 1421, 1747, 1797

solar proton bombardment 1757

solar proton flux 1705, 1773

solar radiation 1397, 1757, 2235, 2683, 2735, 2743

solar track 2705

solar wind 645, 1337, 1343, 1351, 1381, 1407, 1421, 1451, 1461, 1547, 1607, 1651, 1671, 1681, 1705, 1717, 1825, 1843, 1913, 1959, 2093, 2153, 2391, 2415, 2461, 2791

solar wind accretion 1717

solar wind bombardment 1651, 2611

solar wind gas 1591, 1607, 1651, 1681, 1717

solar wind helium 1381, 1651

solar wind hydrogen 1237, 1421

solar wind ion 1671, 1681

solar wind lithium 1461

solar wind proton 2285

solidus 601

solidus temperature 497

solvus 377, 855

sound velocity 2327, 2337

spallation 1407, 1451, 1461, 1591, 1607, 1643, 1671, 1729, 1757, 1797, 2559, 2569, 2599

spallation helium 1825

spallation isotope 1797, 2583

spallation lithium 1461
spallation neon 1643
spallation radioisotope 1747
spallation track density 2569
spherical glass 2003
spherical seed 2003
spherule 207, 701, 719, 747, 817, 937, 957, 1927, 2003, 2041, 2049, 2433, 2653
spherulite 949
spin lattice 2079
spin resonance 2277
spinel 193, 207, 219, 301, 359, 481, 497, 575 583, 601, 617, 679, 719, 855, 2485
static deformation 775
static pressure 2069
stoichiometry 1237
stony meteorite 1123
strain 237, 775
stratigraphy 413, 469, 665, 893, 1953, 1959, 2583
stress 1319
strontium 1, 43, 439, 583, 727, 747, 973, 999, 1083, 1101, 1123, 1169, 1187, 1217, 1247, 1253, 1281, 1291, 1471, 1487, 1627, 2083
strontium evolution 1471
strontium isotopes 601, 1101, 1471, 1487, 1493
subcalcic augite 319
subcalcic augite mantle 265
subcalcic augite rim 377
subcalcic clinpyroxenee 601
subhedral titanium chromite 413
subophitic basalt 159, 285
subophitic diabase 69
subophitic gabbro 285
subophitic matrix 343
subophitic texture 343
subsolidus 219, 601
subsolidus crystallization 559
subsolidus crystallization temperature 949
subsurface structure 2337
subsurface temperature 2203, 2337
suevite breccia 701
sulfide 193, 207, 219, 265, 301, 319, 343, 377, 413, 719, 855, 923, 1351, 2125, 2451, 2485
sulfide iron intergrowth 207
sulfide liquid 207
sulfur 207, 193, 1397, 1879, 2083
sulfur isotopes 1331, 1397, 2301, 2671
sun 1337, 1461, 1757, 2543, 2611
suprathermal ion detector 1717, 2093
suprathermal ion cloud 2093

surface artifact 2781
surface atom 2529
surface basalt 701
surface bedrock 343
surface coating 1421, 2235
surface cratering 2753
surface damage 2791
surface elastic wave 2337
surface electrical conductivity 2391
surface energy 2009
surface erosion 1773
surface exposure age 2515, 2543, 2583
surface fines 319, 665, 755, 1901, 1953, 1973, 2183, 2705
surface fractional crystallization 343
surface glass spherule 2003
surface hafnium demineralization 1901
surface ionization 1461
surface lineament 27
surface location 2125
surface magnetic field 2391
surface magnetometer 2391, 2415
surface morphology 873, 923, 937
surface oxidation 2057
surface oxygen 1421
surface radioactivity 1757
surface structure 2057
surface temperature 2203
surface tension 937
surface track density 2599
surface traverse 27
surface wave (velocity) 2337, 2381
Surveyor 1209, 1865, 1939, 2381, 2543, 2559, 2683, 2699, 2705, 2715, 2721, 2735, 2743, 2753, 2767, 2781, 2791
Surveyor Crater 285, 1139, 1187
Surveyor crater rim 285, 893, 1187
Surveyor fines 2735
Surveyor glass 2543, 2781
Surveyor recovery 2753
Surveyor vanadium 2381
tantalum 1063, 1169, 1187, 1277, 1281, 1291, 1301
tektite 909, 973, 1083, 2137
tektite glass 909
tellurium 1, 135, 1021, 1139, 1261
temperature gradient 2245
tension fracture 797
terbium 1, 1063, 1083, 1169, 1277, 1281, 1291, 1301, 1307
texture 43, 219, 265, 285, 301, 343, 393, 413, 439, 469, 575, 665, 797, 833, 855, 893, 949, 1217, 2027, 2125, 2323, 2337
thallium 1, 439, 1021, 1037, 1139, 1547, 1565, 2245

thallium age 2245
thermal annealing 2501
thermal balance (see heat balance)
thermal conductivity 2203, 2245, 2311,
.... 2345
thermal convection 2173
thermal cycling 449, 1671, 2337, 2485
thermal demagnetization 2485, 2491
thermal equilibrium 2245
thermal evolution 775
thermal expansion 2317
thermal flux 1729
thermal gradient 2245
thermal neutron (flux) 1187, 1585, 1591,
.... 1729
thermal synthesis 1879
thermodynamic equilibrium 1367
thermodynamic temperature 2203
thermoluminescence 2057, 2235, 2245,
.... 2265, 2277
thermophysical 2197
thermoremanent magnetism 2461, 2491
tholeiite 377, 999
thorium 393, 583, 1063, 1159, 1169, 1187,
    1231, 1247, 1253, 1261, 1277, 1281, 1291,
    1319, 1471, 1493, 1503, 1521, 1547, 1565,
    1571, 1577, 1757, 1791, 1797, 1813, 2245
thorium isotopes 1571, 1577, 1757, 1791,
.... 2767
thorium isotopes decay 1577, 2699
thorium-lead (ratio) 1493
thorium-lead isochron 1521
thorium-uranium 1503
thulium 1, 1083, 1169, 1277, 1291
tin 1, 1083
titaniferous basalt 987
titanium 1, 39, 59, 91, 43, 117, 143, 151,
    159, 193, 207, 219, 265, 285, 301, 319,
    359, 377, 393, 413, 439, 459, 469, 481,
    497, 507, 529, 559, 575, 583, 601, 617,
    645, 679, 701, 719, 727, 747, 755, 775,
    797, 833, 855, 937, 973, 1063, 1123,
    1169, 1187, 1209, 1231, 1237, 1247, 1253,
    1259, 1281, 1291, 1301, 1319, 1367, 1451,
    1461, 1471, 1503, 1651, 1705, 1757, 1773,
    1813, 2049, 2069, 2083, 2125, 2153, 2183,
    .... 2433, 2515, 2653
titanium chromite 219, 855
titanium isotope 1461
topography 1
torsional pendulum 2327
track 1443, 1503, 1757, 2027, 2153, 2543,
    2559, 2569, 2583, 2599, 2611, 2621, 2629,
    2705
track age 1503
track density 2543, 2569, 2583, 2599,
    .... 2611, 2705
track density gradient 2705
track gradient 2153, 2559
Tranquillitatis glass 727
Tranquillity Base 47, 247, 285, 1671,
    2361
tranquillityite 39, 319, 583
transition temperature 59, 2049
transmission electron (microscope) 109,
    .... 893, 2791
traverse 17, 1083
traverse map 17
tridymite 117, 141, 247, 265, 319, 413,
    .... 617, 645, 719, 833, 1217
tritium 1803, 1813, 1825
trivalent iron 91
trivalent REE 1083
tube geometry 1959
tungsten 1, 1083, 1169, 1187, 1281, 1291,
    2699
tungsten lineament 27
twin lamella 135, 855
twinning 237, 817
ultraviolet 2153
ultraviolet radiation 1939, 2235
ulvospinel 193, 207, 219, 265, 301, 617,
    .... 855, 2485
ulvospinel-chromite 265
ulvospinel core 2485
ulvopsinel rim 219, 301
undulatory extinction 855
unipolar induction field 2391
unmixing 377
uranium 39, 43, 135, 151, 159, 393, 439,
    469, 583, 797, 873, 999, 1083, 1159, 1169,
    1187, 1247, 1261, 1281, 1291, 1301, 1471,
    1493, 1503, 1521, 1547, 1565, 1571, 1577,
    1591, 1627, 1693, 1757, 1791, 1825, 2003,
    .... 2559, 2599, 2611
uranium basalt 151
uranium decay 2699
uranium fines 151
uranium fission (track) 2559
uranium fractionation 151
uranium geochemistry separation 1503
uranium isotopes 1571, 1577, 1585, 1757,
    2599
uranium isotopes fission 1693, 2599
uranium-lead age 1493
uranium-lead fractionation 1493
uranium-lead ratio 1521
uranium liquid 151
uranium-thorium-lead 1521
uranium-thorium ratio 1083, 1159, 1503
vacuum 109, 449, 507, 1351, 1407, 1547,
    1681, 1843, 1939, 2153, 2197, 2311, 2337,
    2415, 2491, 2501, 2529, 2653, 2715, 2721

vacuum pyrolysis 1343, 1351
vanadium 1, 219, 439, 601, 665, 855, 1063, 1083, 1169, 1217, 1231, 1247, 1253, 1281, 1291, 1319, 1443, 1471, 1757, 1773, ... 2083, 2381
vanadium isotope 1443
vapor 343, 449, 923, 1367, 2009, 2653
vapor condensation 449
vapor pressure 1351
vaporization 449, 1367, 1421
variolitic 285, 301, 343
variolitic matrix 343
variolitic texture 343
vein 439
velocity (wave) 2323, 2327, 2337, 2345, ... 2653, 2753
velocity—density 2345
velocity gradient 2337
velocity impact (high) 177, 909, 2767, ... 2781
velocity impact crater 2639
vesicle 141, 301, 469, 645, 923, 1351, 1381, ... 1681, 1973, 2003, 2311, 2367, 2529
vesicular basalt 2361
vesicular glass 817, 957
viable microorganisms 1939, 2715
viable organism 1931
vibration lattice mode 2203
viscosity 343, 413, 2049
viscous remanent magnetism 2461
visible absorption 2041
visible stratigraphy 1
visible wavelength 2301
vitrification 2183
volitile 439, 449, 469, 497, 617, 893, 973, 1021, 1037, 1083, 1101, 1139, 1159, 1169, 1253, 1261, 1351, 1381, 1451, 1493, 1547, ... 1891, 1913, 2621
volatile carbon 1407, 1901
volatile depletion 1169, 1547
volatile gas 1717
volatile metal 449, 1547
volatile retention 1547
volatile transfer 1493, 1547

volatile vapor 449
volume 39, 59, 247, 957, 1231, 1319, 1607, 1651, 1959, 1989, 2079, 2165, 2317, 2415, ... 2653
volume compressibility 2165
water 439, 645, 973, 1037, 1259, 1261, 1351, 1381, 1407, 1717, 2009, 2173, 2367, ... 2381, 2515, 2653
water absorption 1101
water sedimentation 2173
water vapor 1351, 2009, 2153, 2345
water vapor pressure 645
wave transmission 2323
weathering 17, 973
whitlockite 159, 219, 319, 393, 583, 679, ... 1503, 2599
wollastonite 47, 91
X-ray crystallography 301
X-ray emission 2213
X-ray precession 135, 237
X-ray structure 39
xenolith 583
xenon 1591, 1607, 1627, 1643, 1671, 1681, ... 1693, 1717
xenon age 1693
xenon exposure age 1627
ytterbium 1, 439, 999, 1063, 1083, 1101, 1123, 1169, 1187, 1217, 1247, 1253, 1277, ... 1281, 1291, 1301, 1307, 1319
yttrium-zirconium silicate 39
zinc 1, 583, 973, 1021, 1037, 1139, 1169, 1217, 1247, 1253, 1277, 1281, 1471, 1797
zircon 43, 319, 679, 727, 1063, 1503, 2559
zirconium 1, 39, 43, 151, 159, 169, 207, 219, 265, 319, 439, 481, 701, 719, 727, 747, 833, 1063, 1083, 1123, 1169, 1217, 1231, 1247, 1253, 1291, 1319, 1471, 1503, ... 1627, 2083
zirconium baddeleyite 583
zirconium fractionation 169
zirconium oxide 219, 169
zirconium silicate 39
zirconium-titanium silicate 1063
Zweierketten 47